Some Derivatives

$$\frac{d}{dz}\tan z = \sec^2 z \qquad \frac{d}{dz}\tanh z = \operatorname{sech}^2 z$$

$$\frac{d}{dz}\sinh z = \cosh z \qquad \frac{d}{dz}\cosh z = \sinh z$$

Some Integrals

$$\int \frac{dx}{1+x^2} = \arctan x \qquad \int \frac{dx}{1-x^2} = \operatorname{arctanh} x$$

$$\int \frac{dx}{\sqrt{1-x^2}} = \arcsin x \qquad \int \frac{dx}{\sqrt{1+x^2}} = \operatorname{arcsinh} x$$

$$\int \tan x \, dx = -\ln \cos x \qquad \int \tanh x \, dx = \ln \cosh x$$

$$\int \frac{dx}{x+x^2} = \ln\left(\frac{x}{1+x}\right) \qquad \int \frac{x\,dx}{1+x^2} = \frac{1}{2}\ln(1+x^2)$$

$$\int \frac{dx}{\sqrt{x^2-1}} = \operatorname{arccosh} x \qquad \int \frac{x\,dx}{\sqrt{1+x^2}} = \sqrt{1+x^2}$$

$$\int \frac{dx}{x\sqrt{x^2-1}} = \arccos(1/x) \qquad \int \frac{\sqrt{x}\,dx}{\sqrt{1-x}} = \arcsin(\sqrt{x}) - \sqrt{x(1-x)}$$

$$\int \frac{dx}{(1+x^2)^{3/2}} = \frac{x}{(1+x^2)^{1/2}} \qquad \int \ln(x)\,dx = x\ln(x) - x$$

$$\int_0^1 \frac{dx}{\sqrt{1-x^2}\sqrt{1-mx^2}} = K(m), \quad \text{complete elliptic integral of first kind}$$

Classical Mechanics

Classical Mechanics

John R. Taylor
UNIVERSITY OF COLORADO

University Science Books
Sausalito, California

University Science Books
www.uscibooks.com

PRODUCTION MANAGER   Christine Taylor
MANUSCRIPT EDITOR   Lee Young
DESIGNER   Melissa Ehn
ILLUSTRATORS   LineWorks
COMPOSITOR   Windfall Software, using ZzTEX
PRINTER AND BINDER   Edwards Brothers, Inc.

The circa 1918 front-cover photograph is reproduced
with permission from C P Cushing/Retrofile.com.

This book is printed on acid-free paper.

LIBRARY OF CONGRESS CATALOGING-IN-PUBLICATION DATA

Taylor, John R. (John Robert), 1939–
    Classical mechanics / John R. Taylor.
        p. cm.
    Includes bibliographical references
    ISBN-13: 978-1-891389-22-1 (acid-free paper)
    ISBN-10: 1-891389-22-X (acid-free paper)
    1. Mechanics.   I. Title.
    QC125.2.T39 2004
    531—dc22

                                                    2004054971

Printed in the United States of America
10   9   8   7   6   5   4

Contents

---

* Sections marked with an asterisk could be omitted on a first reading.

Preface

This book is intended for students of the physical sciences, especially physics, who have already studied some mechanics as part of an introductory physics course ("freshman physics" at a typical American university) and are now ready for a deeper look at the subject. The book grew out of the junior-level mechanics course which is offered by the Physics Department at Colorado and is taken mainly by physics majors, but also by some mathematicians, chemists, and engineers. Almost all of these students have taken a year of freshman physics, and so have at least a nodding acquaintance with Newton's laws, energy and momentum, simple harmonic motion, and so on. In this book I build on this nodding acquaintance to give a deeper understanding of these basic ideas, and then go on to develop more advanced topics, such as the Lagrangian and Hamiltonian formulations, the mechanics of noninertial frames, motion of rigid bodies, coupled oscillators, chaos theory, and a few more.

Mechanics is, of course, the study of how things move — how an electron moves down your TV tube, how a baseball flies through the air, how a comet moves round the sun. Classical mechanics is the form of mechanics developed by Galileo and Newton in the seventeenth century and reformulated by Lagrange and Hamilton in the eighteenth and nineteenth centuries. For more than two hundred years, it seemed that classical mechanics was the *only* form of mechanics, that it could explain the motion of all conceivable systems.

Then, in two great revolutions of the early twentieth century, it was shown that classical mechanics cannot account for the motion of objects traveling close to the speed of light, nor of subatomic particles moving inside atoms. The years from about 1900 to 1930 saw the development of relativistic mechanics primarily to describe fast-moving bodies and of quantum mechanics primarily to describe subatomic systems. Faced with this competition, one might expect classical mechanics to have lost much of its interest and importance. In fact, however, classical mechanics is now, at the start of the twenty-first century, just as important and glamorous as ever. This resilience is due to three facts: First, there are just as many interesting physical systems as ever that are best described in classical terms. To understand the orbits of space vehicles and of charged particles in modern accelerators, you have to understand classical mechanics. Second, recent developments in classical mechanics, mainly associated with the growth of chaos theory, have spawned whole new branches of physics and mathematics and have changed our understanding of the notion of causality. It is these new ideas that have attracted some of the best minds in physics back to the study of classical mechanics. Third, it is as true today as ever that a good understanding of classical mechanics is a prerequisite for the study of relativity and quantum mechanics.

Physicists tend to use the term "classical mechanics" rather loosely. Many use it for the mechanics of Newton, Lagrange, and Hamilton; for these people, "classical mechanics" excludes relativity and quantum mechanics. On the other hand, in some areas of physics, there is a tendency to include relativity as a part of "classical mechanics"; for people of this persuasion, "classical mechanics" means "non-quantum mechanics." Perhaps as a reflection of this second usage, some courses called "classical mechanics" include an introduction to relativity, and for the same reason, I have included one chapter on relativistic mechanics, which you can use or not, as you please.

An attractive feature of a course in classical mechanics is that it is a wonderful opportunity to learn to use many of the mathematical techniques needed in so many other branches of physics — vectors, vector calculus, differential equations, complex numbers, Taylor series, Fourier series, calculus of variations, and matrices. I have tried to give at least a minimal review or introduction for each of these topics (with references to further reading) and to teach their use in the usually quite simple context of classical mechanics. I hope you will come away from this book with an increased confidence that you can really use these important tools.

Inevitably, there is more material in the book than could possibly be covered in a one-semester course. I have tried to ease the pain of choosing what to omit. The book is divided into two parts: Part I contains eleven chapters of "essential" material that should be read pretty much in sequence, while Part II contains five "further topics" that are mutually independent and any of which can be read without reference to the others. This division is naturally not very clear cut, and how you use it depends on your preparation (or that of your students). In our one-semester course at the University of Colorado, I found I needed to work steadily through most of Part I, and I only covered Part II by having students choose one of its chapters to study as a term project. (An activity they seemed to enjoy.) Some of the professors who taught from a preliminary version of the book found their students sufficiently well prepared that they could relegate the first four or five chapters to a quick review, leaving more time to cover some of Part II. At schools where the mechanics course lasts two quarters, it proved possible to cover all of Part I and much of Part II as well.

Because the chapters of Part II are mutually independent, it is possible to cover some of them before you finish Part I. For example, Chapter 12 on chaos could be covered immediately after Chapter 5 on oscillations, and Chapter 13 on Hamiltonian mechanics could be read immediately after Chapter 7 on Lagrangian mechanics. A number of sections are marked with an asterisk to indicate that they can be omitted without loss of continuity. (This is not to say that this material is unimportant. I certainly hope you'll come back and read it later!)

As always in a physics text, it is crucial that you do lots of the exercises at the end of each chapter. I have included a large number of these to give both teacher and student plenty of choice. Some of them are simple applications of the ideas of the chapter and some are extensions of those ideas. I have listed the problems by section, so that as soon as you have read any given section you could (and probably should) try a few problems listed for that section. (Naturally, problems listed for a given section usually require knowledge of earlier sections. I promise only that you shouldn't need material from later sections.) I have tried to grade the problems to indicate their level of difficulty, ranging from one star ($\star$), meaning a straightforward exercise usually involving just one main concept, to three stars ($\star\star\star$), meaning a challenging problem that involves several concepts and will probably take considerable time and effort. This kind of classification is quite subjective, very approximate, and surprisingly difficult to make; I would welcome suggestions for any changes you think should be made.

Several of the problems require the use of computers to plot graphs, solve differential equations, and so on. None of these requires any specific software; some can be done with a relatively simple system such as MathCad or even just a spreadsheet like Excel; some require more sophisticated systems, such as Mathematica, Maple, or Matlab. (Incidentally, it is my experience that the course for which this book was written is a wonderful opportunity for the students to learn to use one of these fabulously useful systems.) Problems requiring the use of a computer are indicated thus: [Computer]. I have tended to grade them as $\star\star\star$ or at least $\star\star$ on the grounds that it takes a lot of time to set up the necessary code. Naturally, these problems will be easier for students who are experienced with the necessary software.

Each chapter ends with a summary called "Principal Definitions and Equations of Chapter *xx*." I hope these will be useful as a check on your understanding of the chapter as you finish reading it and as a reference later on, as you try to find that formula whose details you have forgotten.

There are many people I wish to thank for their help and suggestions. At the University of Colorado, these include Professors Larry Baggett, John Cary, Mike Dubson, Anatoli Levshin, Scott Parker, Steve Pollock, and Mike Ritzwoller. From other institutions, the following professors reviewed the manuscript or used a preliminary edition in their classes:

Meagan Aronson, U of Michigan
Dan Bloom, Kalamazoo College
Peter Blunden, U of Manitoba
Andrew Cleland, UC Santa Barbara
Gayle Cook, Cal Poly, San Luis Obispo
Joel Fajans, UC Berkeley
Richard Fell, Brandeis University
Gayanath Fernando, U of Connecticut
Jonathan Friedman, Amherst College
David Goldhaber-Gordon, Stanford
Thomas Griffy, U of Texas
Elisabeth Gwinn, UC Santa Barbara
Richard Hilt, Colorado College
George Horton, Rutgers
Lynn Knutson, U of Wisconsin

Jonathan Maps, U of Minnesota, Duluth
John Markert, U of Texas
Michael Moloney, Rose-Hulman Institute
Colin Morningstar, Carnegie Mellon
Declan Mulhall, Cal Poly, San Luis Obispo
Carl Mungan, US Naval Academy
Robert Pompi, SUNY Binghamton
Mark Semon, Bates College
James Shepard, U of Colorado
Richard Sonnenfeld, New Mexico Tech
Edward Stern, U of Washington
Michael Weinert, U of Wisconsin, Milwaukee
Alma Zook, Pomona College

I am most grateful to all of these and their students for their many helpful comments. I would particularly like to thank Carl Mungan for his amazing vigilance in catching typos, obscurites, and ambiguities, and Jonathan Friedman and his student, Ben Heidenreich, who saved me from a really embarassing mistake in Chapter 10. I am especially grateful to my two friends and colleagues, Mark Semon at Bates College and Dave Goodmanson at the Boeing Aircraft Company, both of whom reviewed the manuscript with the finest of combs and gave me literally hundreds of suggestions; likewise to Christopher Taylor of the University of Wisconsin for his patient help with Mathematica and the mysteries of LaTeX. Professor Manuel Fernando Ferreira da Silva at the Universidade da Beira Interior, Portugal, read the first printing with amazing thoroughness and gave me innumerable suggestions, many small but several crucial, and all most valued. Bruce Armbruster and Jane Ellis of University Science Books are an author's dream come true. My copy editor, Lee Young, is a rarity indeed, an expert in English usage *and* physics; he suggested many significant improvements. Finally and most of all, I want to thank my wife Debby. Being married to an author can be very trying, and she puts up with it most graciously. And, as an English teacher with the highest possible standards, she has taught me most of what I know about writing and editing. I am eternally grateful.

For all our efforts, there will surely be several errors in this book, and I would be most grateful if you could let me know of any that you find. Ancillary material, including an instructors' manual, and other notices will be posted at the University Science Books website, www.uscibooks.com.

John R. Taylor
Department of Physics
University of Colorado
Boulder, Colorado 80309, USA
John.Taylor@Colorado.edu

Essentials

Part I of this book contains material that almost everyone would consider essential knowledge for an undergraduate physics major. Part II contains optional further topics from which you can pick according to your tastes and available time. The distinction between "essential" and "optional" is, of course, arguable, and its impact on you, the reader, depends very much on your state of preparation. For example, if you are well prepared, you might decide that the first five chapters of Part I can be treated as a quick review, or even skipped entirely. As a practical matter, the distinction is this: The eleven chapters of Part I were designed to be read in sequence, and in writing each chapter, I assumed that you would be familiar with most of the ideas of the preceding chapters — either by reading them or because you had met them elsewhere. By contrast, I tried to make the chapters of Part II independent of one another, so that you could read any of them in any order, once you knew most of the material of Part I.

Newton's Laws of Motion

1.1 Classical Mechanics

Mechanics is the study of how things move: how planets move around the sun, how a skier moves down the slope, or how an electron moves around the nucleus of an atom. So far as we know, the Greeks were the first to think seriously about mechanics, more than two thousand years ago, and the Greeks' mechanics represents a tremendous step in the evolution of modern science. Nevertheless, the Greek ideas were, by modern standards, seriously flawed and need not concern us here. The development of the mechanics that we know today began with the work of Galileo (1564–1642) and Newton (1642–1727), and it is the formulation of Newton, with his three laws of motion, that will be our starting point in this book.

In the late eighteenth and early nineteenth centuries, two alternative formulations of mechanics were developed, named for their inventors, the French mathematician and astronomer Lagrange (1736–1813) and the Irish mathematician Hamilton (1805–1865). The Lagrangian and Hamiltonian formulations of mechanics are completely equivalent to that of Newton, but they provide dramatically simpler solutions to many complicated problems and are also the taking-off point for various modern developments. The term *classical mechanics* is somewhat vague, but it is generally understood to mean these three equivalent formulations of mechanics, and it is in this sense that the subject of this book is called classical mechanics.

Until the beginning of the twentieth century, it seemed that classical mechanics was the *only* kind of mechanics, correctly describing all possible kinds of motion. Then, in the twenty years from 1905 to 1925, it became clear that classical mechanics did not correctly describe the motion of objects moving at speeds close to the speed of light, nor that of the microscopic particles inside atoms and molecules. The result was the development of two completely new forms of mechanics: relativistic mechanics to describe very high-speed motions and quantum mechanics to describe the motion of microscopic particles. I have included an introduction to relativity in the "optional" Chapter 15. Quantum mechanics requires a whole separate book (or several books), and I have made no attempt to give even a brief introduction to quantum mechanics.   3

Although classical mechanics has been replaced by relativistic mechanics and by quantum mechanics in their respective domains, there is still a vast range of interesting and topical problems in which classical mechanics gives a complete and accurate description of the possible motions. In fact, particularly with the advent of chaos theory in the last few decades, research in classical mechanics has intensified and the subject has become one of the most fashionable areas in physics. The purpose of this book is to give a thorough grounding in the exciting field of classical mechanics. When appropriate, I shall discuss problems in the framework of the Newtonian formulation, but I shall also try to emphasize those situations where the newer formulations of Lagrange and Hamilton are preferable and to use them when this is the case. At the level of this book, the Lagrangian approach has many significant advantages over the Newtonian, and we shall be using the Lagrangian formulation repeatedly, starting in Chapter 7. By contrast, the advantages of the Hamiltonian formulation show themselves only at a more advanced level, and I shall postpone the introduction of Hamiltonian mechanics to Chapter 13 (though it can be read at any point after Chapter 7).

In writing the book, I took for granted that you have had an introduction to Newtonian mechanics of the sort included in a typical freshman course in "General Physics." This chapter contains a brief review of the ideas that I assume you have met before.

1.2 Space and Time

Newton's three laws of motion are formulated in terms of four crucial underlying concepts: the notions of space, time, mass, and force. This section reviews the first two of these, space and time. In addition to a brief description of the classical view of space and time, I give a quick review of the machinery of vectors, with which we label the points of space.

Space

Each point P of the three-dimensional space in which we live can be labeled by a position vector $\mathbf{r}$ which specifies the distance and direction of P from a chosen origin O as in Figure 1.1. There are many different ways to identify a vector, of which one of the most natural is to give its components (x, y, z) in the directions of three chosen perpendicular axes. One popular way to express this is to introduce three unit vectors, $\hat{\mathbf{x}}, \hat{\mathbf{y}}, \hat{\mathbf{z}}$, pointing along the three axes and to write

$$\mathbf{r} = x\hat{\mathbf{x}} + y\hat{\mathbf{y}} + z\hat{\mathbf{z}}. \tag{1.1}$$

In elementary work, it is probably wise to choose a single good notation, such as (1.1), and stick with it. In more advanced work, however, it is almost impossible to avoid using several different notations. Different authors have different preferences (another popular choice is to use $\mathbf{i}, \mathbf{j}, \mathbf{k}$ for what I am calling $\hat{\mathbf{x}}, \hat{\mathbf{y}}, \hat{\mathbf{z}}$) and you must get used to reading them all. Furthermore, almost every notation has its drawbacks, which can

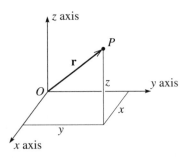

Figure 1.1   The point P is identified by its position vector $\mathbf{r}$, which gives the position of P relative to a chosen origin O. The vector $\mathbf{r}$ can be specified by its components (x, y, z) relative to chosen axes $Oxyz$.

make it unusable in some circumstances. Thus, while you may certainly choose your preferred scheme, you need to develop a tolerance for several different schemes.

It is sometimes convenient to be able to abbreviate (1.1) by writing simply

$$\mathbf{r} = (x, y, z). \tag{1.2}$$

This notation is obviously not quite consistent with (1.1), but it is usually completely unambiguous, asserting simply that $\mathbf{r}$ is the vector whose components are x, y, z. When the notation of (1.2) is the most convenient, I shall not hesitate to use it. For most vectors, we indicate the components by subscripts x, y, z. Thus the velocity vector $\mathbf{v}$ has components v_x, v_y, v_z and the acceleration $\mathbf{a}$ has components a_x, a_y, a_z.

As our equations become more complicated, it is sometimes inconvenient to write out all three terms in sums like (1.1); one would rather use the summation sign $\sum$ followed by a single term. The notation of (1.1) does not lend itself to this shorthand, and for this reason I shall sometimes relabel the three components x, y, z of $\mathbf{r}$ as r_1, r_2, r_3, and the three unit vectors $\hat{\mathbf{x}}, \hat{\mathbf{y}}, \hat{\mathbf{z}}$ as $\mathbf{e}_1, \mathbf{e}_2, \mathbf{e}_3$. That is, we define

$$r_1 = x, \qquad r_2 = y, \qquad r_3 = z,$$

and

$$\mathbf{e}_1 = \hat{\mathbf{x}}, \qquad \mathbf{e}_2 = \hat{\mathbf{y}}, \qquad \mathbf{e}_3 = \hat{\mathbf{z}}.$$

(The symbol $\mathbf{e}$ is commonly used for unit vectors, since $\mathbf{e}$ stands for the German "eins" or "one.") With these notations, (1.1) becomes

$$\mathbf{r} = r_1\mathbf{e}_1 + r_2\mathbf{e}_2 + r_3\mathbf{e}_3 = \sum_{i=1}^{3} r_i\mathbf{e}_i. \tag{1.3}$$

For a simple equation like this, the form (1.3) has no real advantage over (1.1), but with more complicated equations (1.3) is significantly more convenient, and I shall use this notation when appropriate.

Vector Operations

In our study of mechanics, we shall make repeated use of the various operations that can be performed with vectors. If $\mathbf{r}$ and $\mathbf{s}$ are vectors with components

$$\mathbf{r} = (r_1, r_2, r_3) \qquad \text{and} \qquad \mathbf{s} = (s_1, s_2, s_3),$$

then their **sum** (or resultant) $\mathbf{r} + \mathbf{s}$ is found by adding corresponding components, so that

$$\mathbf{r} + \mathbf{s} = (r_1 + s_1, \ r_2 + s_2, \ r_3 + s_3). \tag{1.4}$$

(You can convince yourself that this rule is equivalent to the familiar triangle and parallelogram rules for vector addition.) An important example of a vector sum is the resultant force on an object: When two forces $\mathbf{F}_a$ and $\mathbf{F}_b$ act on an object, the effect is the same as a single force, the resultant force, which is just the vector sum

$$\mathbf{F} = \mathbf{F}_a + \mathbf{F}_b$$

as given by the vector addition law (1.4).

If c is a scalar (that is, an ordinary number) and $\mathbf{r}$ is a vector, the *product* $c\mathbf{r}$ is given by

$$c\mathbf{r} = (cr_1, cr_2, cr_3). \tag{1.5}$$

This means that $c\mathbf{r}$ is a vector in the same direction[1] as $\mathbf{r}$ with magnitude equal to c times the magnitude of $\mathbf{r}$. For example, if an object of mass m (a scalar) has an acceleration $\mathbf{a}$ (a vector), Newton's second law asserts that the resultant force $\mathbf{F}$ on the object will always equal the product $m\mathbf{a}$ as given by (1.5).

There are two important kinds of product that can be formed from any pair of vectors. First, the **scalar product** (or **dot product**) of two vectors $\mathbf{r}$ and $\mathbf{s}$ is given by either of the equivalent formulas

$$\mathbf{r} \cdot \mathbf{s} = rs \cos\theta \tag{1.6}$$

$$= r_1 s_1 + r_2 s_2 + r_3 s_3 = \sum_{n=1}^{3} r_n s_n \tag{1.7}$$

where r and s denote the magnitudes of the vectors $\mathbf{r}$ and $\mathbf{s}$, and θ is the angle between them. (For a proof that these two definitions are the same, see Problem 1.7.) For example, if a force $\mathbf{F}$ acts on an object that moves through a small displacement $d\mathbf{r}$, the work done by the force is the scalar product $\mathbf{F} \cdot d\mathbf{r}$, as given by either (1.6) or (1.7). Another important use of the scalar product is to define the magnitude of a vector: The magnitude (or length) of any vector $\mathbf{r}$ is denoted by $|\mathbf{r}|$ or r and, by Pythagoras's theorem is equal to $\sqrt{r_1^2 + r_2^2 + r_3^2}$. By (1.7) this is the same as

$$r = |\mathbf{r}| = \sqrt{\mathbf{r} \cdot \mathbf{r}}. \tag{1.8}$$

The scalar product $\mathbf{r} \cdot \mathbf{r}$ is often abbreviated as $\mathbf{r}^2$.

---

[1] Although this is what people usually say, one should actually be careful: If c is negative, $c\mathbf{r}$ is in the *opposite* direction to $\mathbf{r}$.

The second kind of product of two vectors **r** and **s** is the **vector product** (or **cross product**), which is defined as the vector $\mathbf{p} = \mathbf{r} \times \mathbf{s}$ with components

$$\left.\begin{array}{l} p_x = r_y s_z - r_z s_y \\ p_y = r_z s_x - r_x s_z \\ p_z = r_x s_y - r_y s_x \end{array}\right\} \tag{1.9}$$

or, equivalently

$$\mathbf{r} \times \mathbf{s} = \det \begin{bmatrix} \hat{\mathbf{x}} & \hat{\mathbf{y}} & \hat{\mathbf{z}} \\ r_x & r_y & r_z \\ s_x & s_y & s_z \end{bmatrix},$$

where "det" stands for the determinant. Either of these definitions implies that $\mathbf{r} \times \mathbf{s}$ is a vector perpendicular to both **r** and **s**, with direction given by the familiar right-hand rule and magnitude $rs \sin\theta$ (Problem 1.15). The vector product plays an important role in the discussion of rotational motion. For example, the tendency of a force **F** (acting at a point **r**) to cause a body to rotate about the origin is given by the torque of **F** about O, defined as the vector product $\mathbf{\Gamma} = \mathbf{r} \times \mathbf{F}$.

Differentiation of Vectors

Many (maybe most) of the laws of physics involve vectors, and most of these involve *derivatives* of vectors. There are so many ways to differentiate a vector that there is a whole subject called vector calculus, much of which we shall be developing in the course of this book. For now, I shall mention just the simplest kind of vector derivative, the time derivative of a vector that depends on time. For example, the velocity $\mathbf{v}(t)$ of a particle is the time derivative of the particle's position $\mathbf{r}(t)$; that is, $\mathbf{v} = d\mathbf{r}/dt$. Similarly the acceleration is the time derivative of the velocity, $\mathbf{a} = d\mathbf{v}/dt$.

The definition of the derivative of a vector is closely analogous to that of a scalar. Recall that if $x(t)$ is a scalar function of t, then we define its derivative as

$$\frac{dx}{dt} = \lim_{\Delta t \to 0} \frac{\Delta x}{\Delta t}$$

where $\Delta x = x(t + \Delta t) - x(t)$ is the change in x as the time advances from t to $t + \Delta t$. In exactly the same way, if $\mathbf{r}(t)$ is any vector that depends on t, we define its derivative as

$$\frac{d\mathbf{r}}{dt} = \lim_{\Delta t \to 0} \frac{\Delta \mathbf{r}}{\Delta t} \tag{1.10}$$

where

$$\Delta \mathbf{r} = \mathbf{r}(t + \Delta t) - \mathbf{r}(t) \tag{1.11}$$

is the corresponding change in $\mathbf{r}(t)$. There are, of course, many delicate questions about the existence of this limit. Fortunately, none of these need concern us here: All of the vectors we shall encounter will be differentiable, and you can take for granted that the required limits exist. From the definition (1.10), one can prove that the derivative has all of the properties one would expect. For example, if $\mathbf{r}(t)$ and $\mathbf{s}(t)$

are two vectors that depend on t, then the derivative of their sum is just what you would expect:

$$\frac{d}{dt}(\mathbf{r} + \mathbf{s}) = \frac{d\mathbf{r}}{dt} + \frac{d\mathbf{s}}{dt}. \tag{1.12}$$

Similarly, if $\mathbf{r}(t)$ is a vector and $f(t)$ is a scalar, then the derivative of the product $f(t)\mathbf{r}(t)$ is given by the appropriate version of the product rule,

$$\frac{d}{dt}(f\mathbf{r}) = f\frac{d\mathbf{r}}{dt} + \frac{df}{dt}\mathbf{r}. \tag{1.13}$$

If you are the sort of person who enjoys proving these kinds of proposition, you might want to show that they follow from the definition (1.10). Fortunately, if you do not enjoy this kind of activity, you don't need to worry, and you can safely take these results for granted.

One more result that deserves mention concerns the components of the derivative of a vector. Suppose that $\mathbf{r}$, with components x, y, z, is the position of a moving particle, and suppose that we want to know the particle's velocity $\mathbf{v} = d\mathbf{r}/dt$. When we differentiate the sum

$$\mathbf{r} = x\hat{\mathbf{x}} + y\hat{\mathbf{y}} + z\hat{\mathbf{z}}, \tag{1.14}$$

the rule (1.12) gives us the sum of the three separate derivatives, and, by the product rule (1.13), each of these contains two terms. Thus, in principle, the derivative of (1.14) involves six terms in all. However, the unit vectors $\hat{\mathbf{x}}$, $\hat{\mathbf{y}}$, and $\hat{\mathbf{z}}$ do not depend on time, so their time derivatives are zero. Therefore, three of these six terms are zero, and we are left with just three terms:

$$\frac{d\mathbf{r}}{dt} = \frac{dx}{dt}\hat{\mathbf{x}} + \frac{dy}{dt}\hat{\mathbf{y}} + \frac{dz}{dt}\hat{\mathbf{z}}. \tag{1.15}$$

Comparing this with the standard expansion

$$\mathbf{v} = v_x\hat{\mathbf{x}} + v_y\hat{\mathbf{y}} + v_z\hat{\mathbf{z}}$$

we see that

$$v_x = \frac{dx}{dt}, \qquad v_y = \frac{dy}{dt}, \qquad \text{and} \qquad v_z = \frac{dz}{dt}. \tag{1.16}$$

In words, the rectangular components of $\mathbf{v}$ are just the derivatives of the corresponding components of $\mathbf{r}$. This is a result that we use all the time (usually without even thinking about it) in solving elementary mechanics problems. What makes it especially noteworthy is this: It is true only because the unit vectors $\hat{\mathbf{x}}$, $\hat{\mathbf{y}}$, and $\hat{\mathbf{z}}$ are constant, so that their derivatives are absent from (1.15). We shall find that in most coordinate systems, such as polar coordinates, the basic unit vectors are *not* constant, and the result corresponding to (1.16) is appreciably less transparent. In problems where we need to work in nonrectangular coordinates, it is considerably harder to write down velocities and accelerations in terms of the coordinates of $\mathbf{r}$, as we shall see.

Time

The classical view is that time is a single universal parameter t on which all observers agree. That is, if all observers are equipped with accurate clocks, all properly synchronized, then they will all agree as to the time at which any given event occurred. We know, of course, that this view is not exactly correct: According to the theory of relativity, two observers in relative motion do *not* agree on all times. Nevertheless, in the domain of classical mechanics, with all speeds much much less than the speed of light, the differences among the measured times are entirely negligible, and I shall adopt the classical assumption of a single universal time (except, of course, in Chapter 15 on relativity). Apart from the obvious ambiguity in the choice of the origin of time (the time that we choose to label $t = 0$), all observers agree on the times of all events.

Reference Frames

Almost every problem in classical mechanics involves a choice (explicit or implicit) of a *reference frame*, that is, a choice of spatial origin and axes to label positions as in Figure 1.1 and a choice of temporal origin to measure times. The difference between two frames may be quite minor. For instance, they may differ only in their choice of the origin of time — what one frame labels $t = 0$ the other may label $t' = t_o \neq 0$. Or the two frames may have the same origins of space and time, but have different orientations of the three spatial axes. By carefully choosing your reference frame, taking advantage of these different possibilities, you can sometimes simplify your work. For example, in problems involving blocks sliding down inclines, it often helps to choose one axis pointing down the slope.

A more important difference arises when two frames are in relative motion; that is, when one origin is moving relative to the other. In Section 1.4 we shall find that not all such frames are physically equivalent.[2] In certain special frames, called **inertial frames**, the basic laws hold true in their standard, simple form. (It is because one of these basic laws is Newton's first law, the law of inertia, that these frames are called inertial.) If a second frame is *accelerating or rotating* relative to an inertial frame, then this second frame is noninertial, and the basic laws — in particular, Newton's laws — do not hold in their standard form in this second frame. We shall find that the distinction between inertial and noninertial frames is central to our discussion of classical mechanics. It plays an even more explicit role in the theory of relativity.

1.3   Mass and Force

The concepts of mass and force are central to the formulation of classical mechanics. The proper definitions of these concepts have occupied many philosophers of science and are the subject of learned treatises. Fortunately we don't need to worry much about

---

[2] This statement is correct even in the theory of relativity.

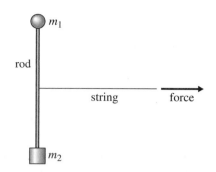

Figure 1.2   An inertial balance compares the masses m_1 and m_2
of two objects that are attached to the opposite ends of a rigid rod.
The masses are equal if and only if a force applied at the rod's
midpoint causes them to accelerate at the same rate, so that the
rod does not rotate.

these delicate questions here. Based on your introductory course in general physics,
you have a reasonably good idea what mass and force mean, and it is easy to describe
how these parameters are defined and measured in many realistic situations.

Mass

The mass of an object characterizes the object's inertia — its resistance to being
accelerated: A big boulder is hard to accelerate, and its mass is large. A little stone
is easy to accelerate, and its mass is small. To make these natural ideas quantitative
we have to define a unit of mass and then give a prescription for measuring the mass
of any object in terms of the chosen unit. The internationally agreed unit of mass is
the kilogram and is defined arbitrarily to be the mass of a chunk of platinum–iridium
stored at the International Bureau of Weights and Measures outside Paris. To measure
the mass of any other object, we need a means of comparing masses. In principle, this
can be done with an inertial balance as shown in Figure 1.2. The two objects to be
compared are fastened to the opposite ends of a light, rigid rod, which is then given
a sharp pull at its midpoint. If the masses are equal, they will accelerate equally and
the rod will move off without rotating; if the masses are unequal, the more massive
one will accelerate less, and the rod will rotate as it moves off.

   The beauty of the inertial balance is that it gives us a method of mass comparison
that is based directly on the notion of mass as resistance to being accelerated. In
practice, an inertial balance would be very awkward to use, and it is fortunate that
there are much easier ways to compare masses, of which the easiest is to weigh the
objects. As you certainly recall from your introductory physics course, an object's
mass is found to be exactly proportional to the object's weight[3] (the gravitational force
on the object) provided all measurements are made in the same location. Thus two

---

[3] This observation goes back to Galileo's famous experiments showing that all objects are
accelerated at the same rate by gravity. The first modern experiments were conducted by the
Hungarian physicist Eötvös (1848–1919), who showed that weight is proportional to mass to within

objects have the same mass if and only if they have the same weight (when weighed at the same place), and a simple, practical way to check whether two masses are equal is simply to weigh them and see if their weights are equal.

Armed with methods for comparing masses, we can easily set up a scheme to measure arbitrary masses. First, we can build a large number of standard kilograms, each one checked against the original 1-kg mass using either the inertial or gravitational balance. Next, we can build multiples and fractions of the kilogram, again checking them with our balance. (We check a 2-kg mass on one end of the balance against two 1-kg masses placed together on the other end; we check two half-kg masses by verifying that their masses are equal and that together they balance a 1-kg mass; and so on.) Finally, we can measure an unknown mass by putting it on one end of the balance and loading known masses on the other end until they balance to any desired precision.

Force

The informal notion of force as a push or pull is a surprisingly good starting point for our discussion of forces. We are certainly conscious of the forces that we exert ourselves. When I hold up a sack of cement, I am very aware that I am exerting an upward force on the sack; when I push a heavy crate across a rough floor, I am aware of the horizontal force that I have to exert in the direction of motion. Forces exerted by inanimate objects are a little harder to pin down, and we must, in fact, understand something of Newton's laws to identify such forces. If I let go of the sack of cement, it accelerates toward the ground; therefore, I conclude that there must be another force — the sack's weight, the gravitational force of the earth — pulling it downward. As I push the crate across the floor, I observe that it does not accelerate, and I conclude that there must be another force — friction — pushing the crate in the opposite direction. One of the most important skills for the student of elementary mechanics is to learn to examine an object's environment and identify all the forces on the object: What are the things touching the object and possibly exerting contact forces, such as friction or air pressure? And what are the nearby objects possibly exerting action-at-a-distance forces, such as the gravitational pull of the earth or the electrostatic force of some charged body?

If we accept that we know how to identify forces, it remains to decide how to measure them. As the unit of force we naturally adopt the newton (abbreviated N) defined as the magnitude of any single force that accelerates a standard kilogram mass with an acceleration of 1 m/s^2. Having agreed what we mean by one newton, we can proceed in several ways, all of which come to the same final conclusion, of course. The route that is probably preferred by most philosophers of science is to use Newton's second law to define the general force: A given force is 2 N if, by itself, it accelerates a standard kilogram with an acceleration of 2 m/s^2, and so

---

a few parts in 10^9. Experiments in the last few decades have narrowed this to around one part in 10^{12}.

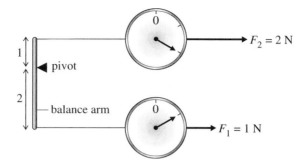

Figure 1.3   One of many possible ways to define forces of any magnitude. The lower spring balance has been calibrated to read 1 N. If the balance arm on the left is adjusted so that the lever arms above and below the pivot are in the ratio 1 : 2 and if the force F_1 is 1 N, then the force F_2 required to balance the arm is 2 N. This lets us calibrate the upper spring balance for 2 N. By readjusting the two lever arms, we can, in principle, calibrate the second spring balance to read any force.

on. This approach is not much like the way we usually measure forces in practice,[4] and for our present discussion a simpler procedure is to use some spring balances. Using our definition of the newton, we can calibrate a first spring balance to read 1 N. Then by matching a second spring balance against the first, using a balance arm as shown in Figure 1.3, we can define multiples and fractions of a newton. Once we have a fully calibrated spring balance we can, in principle, measure any unknown force, by matching it against the calibrated balance and reading off its value.

So far we have defined only the magnitude of a force. As you are certainly aware, forces are vectors, and we must also define their directions. This is easily done. If we apply a given force **F** (and no other forces) to any object at rest, the direction of **F** is defined as the direction of the resulting acceleration, that is, the direction in which the body moves off.

Now that we know, at least in principle, what we mean by positions, times, masses, and forces, we can proceed to discuss the cornerstone of our subject — Newton's three laws of motion.

---

[4] The approach also creates the confusing appearance that Newton's second law is just a consequence of the definition of force. This is not really true: Whatever definition we choose for force, a large part of the second law is experimental. One advantage of defining forces with spring balances is that it separates out the definition of force from the experimental basis of the second law. Of course, all commonly accepted definitions give the same final result for the value of any given force.

1.4   Newton's First and Second Laws; Inertial Frames

In this chapter, I am going to discuss Newton's laws as they apply to a **point mass**. A point mass, or **particle**, is a convenient fiction, an object with mass, but no size, that can move through space but has no internal degrees of freedom. It can have "translational" kinetic energy (energy of its motion through space) but no energy of rotation or of internal vibrations or deformations. Naturally, the laws of motion are simpler for point particles than for extended bodies, and this is the main reason that we start with the former. Later on, I shall build up the mechanics of extended bodies from our mechanics of point particles by considering the extended body as a collection of many separate particles.

Nevertheless, it is worth recognizing that there are many important problems where the objects of interest can be realistically approximated as point masses. Atomic and subatomic particles can often be considered to be point masses, and even macroscopic objects can frequently be approximated in this way. A stone thrown off the top of a cliff is, for almost all purposes, a point particle. Even a planet orbiting around the sun can usually be approximated in the same way. Thus the mechanics of point masses is more than just the starting point for the mechanics of extended bodies; it is a subject with wide application itself.

Newton's first two laws are well known and easily stated:

> ### Newton's First Law (the Law of Inertia)
> In the absence of forces, a particle moves with constant velocity **v**.

and

> ### Newton's Second Law
> For any particle of mass m, the net force **F** on the particle is always equal to the mass m times the particle's acceleration:
>
> $$\mathbf{F} = m\mathbf{a}. \qquad (1.17)$$

In this equation **F** denotes the vector sum of all the forces on the particle and **a** is the particle's acceleration,

$$\mathbf{a} = \frac{d\mathbf{v}}{dt} \equiv \dot{\mathbf{v}}$$

$$= \frac{d^2\mathbf{r}}{dt^2} \equiv \ddot{\mathbf{r}}.$$

Here **v** denotes the particle's velocity, and I have introduced the convenient notation of dots to denote differentiation with respect to t, as in $\mathbf{v} = \dot{\mathbf{r}}$ and $\mathbf{a} = \dot{\mathbf{v}} = \ddot{\mathbf{r}}$.

Both laws can be stated in various equivalent ways. For instance (the first law): In the absence of forces, a stationary particle remains stationary and a moving particle continues to move with unchanging speed in the same direction. This is, of course, exactly the same as saying that the velocity is always constant. Again, **v** is constant if and only if the acceleration **a** is zero, so an even more compact statement is this: In the absence of forces a particle has zero acceleration.

The second law can be rephrased in terms of the particle's **momentum**, defined as

$$\mathbf{p} = m\mathbf{v}. \tag{1.18}$$

In classical mechanics, we take for granted that the mass m of a particle never changes, so that

$$\dot{\mathbf{p}} = m\dot{\mathbf{v}} = m\mathbf{a}.$$

Thus the second law (1.17) can be rephrased to say that

$$\mathbf{F} = \dot{\mathbf{p}}. \tag{1.19}$$

In classical mechanics, the two forms (1.17) and (1.19) of the second law are completely equivalent.[5]

Differential Equations

When written in the form $m\ddot{\mathbf{r}} = \mathbf{F}$, Newton's second law is a **differential equation** for the particle's position $\mathbf{r}(t)$. That is, it is an equation for the unknown function $\mathbf{r}(t)$ that involves *derivatives* of the unknown function. Almost all the laws of physics are, or can be cast as, differential equations, and a huge proportion of a physicist's time is spent solving these equations. In particular, most of the problems in this book involve differential equations — either Newton's second law or its counterparts in the Lagrangian and Hamiltonian forms of mechanics. These vary widely in their difficulty. Some are so easy to solve that one scarcely notices them. For example, consider Newton's second law for a particle confined to move along the x axis and subject to a constant force F_o,

$$\ddot{x}(t) = \frac{F_o}{m}.$$

This is a second-order differential equation for $x(t)$ as a function of t. (Second-order because it involves derivatives of second order, but none of higher order.) To solve it

---

[5] In relativity, the two forms are *not* equivalent, as we'll see in Chapter 15. Which form is correct depends on the definitions we use for force, mass, and momentum in relativity. If we adopt the most popular definitions of these three quantities, then it is the form (1.19) that holds in relativity.

one has only to integrate it twice. The first integration gives the velocity

$$\dot{x}(t) = \int \ddot{x}(t)\, dt = v_\text{o} + \frac{F_\text{o}}{m} t$$

where the constant of integration is the particle's initial velocity, and a second integration gives the position

$$x(t) = \int \dot{x}(t)\, dt = x_\text{o} + v_\text{o} t + \frac{F_\text{o}}{2m} t^2$$

where the second constant of integration is the particle's initial position. Solving this differential equation was so easy that we certainly needed no knowledge of the theory of differential equations. On the other hand, we shall meet lots of differential equations that do require knowledge of this theory, and I shall present the necessary theory as we need it. Obviously, it will be an advantage if you have already studied some of the theory of differential equations, but you should have no difficulty picking it up as we go along. Indeed, many of us find that the best way to learn this kind of mathematical theory is in the context of its physical applications.

Inertial Frames

On the face of it, Newton's second law includes his first: If there are no forces on an object, then $\mathbf{F} = 0$ and the second law (1.17) implies that $\mathbf{a} = 0$, which is the first law. There is, however, an important subtlety, and the first law has an important role to play. Newton's laws cannot be true in all conceivable reference frames. To see this, consider just the first law and imagine a reference frame — we'll call it S — in which the first law is true. For example, if the frame S has its origin and axes fixed relative to the earth's surface, then, to an excellent approximation, the first law (the law of inertia) holds with respect to the frame S: A frictionless puck placed on a smooth horizontal surface is subject to zero force and, in accordance with the first law, it moves with constant velocity. Because the law of inertia holds, we call S an **inertial frame**. If we consider a second frame S' which is moving relative to S with constant velocity and is not rotating, then the same puck will also be observed to move with constant velocity relative to S'. That is, the frame S' is also inertial.

If, however, we consider a third frame S'' that is accelerating relative to S, then, as viewed from S'', the puck will be seen to be accelerating (in the opposite direction). Relative to the accelerating frame S'' the law of inertia does not hold, and we say that S'' is **noninertial**. I should emphasize that there is nothing mysterious about this result. Indeed it is a matter of experience. The frame S' could be a frame attached to a high-speed train traveling smoothly at constant speed along a straight track, and the frictionless puck, an ice cube placed on the floor of the train, as in Figure 1.4. As seen from the train (frame S'), the ice cube is at rest and remains at rest, in accord with the first law. As seen from the ground (frame S), the ice cube is moving with the same velocity as the train and continues to do so, again in obedience to the first law. But now consider conducting the same experiment on a second train (frame S'') that is accelerating forward. As this train accelerates forward, the ice cube is left behind, and, relative to S'', the ice cube accelerates backward, even though subject to no net

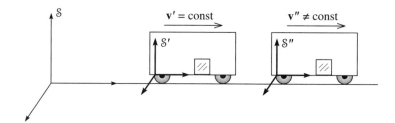

Figure 1.4   The frame $\mathcal{S}$ is fixed to the ground, while $\mathcal{S}'$ is fixed to a train traveling at constant velocity $\mathbf{v}'$ relative to $\mathcal{S}$. An ice cube placed on the floor of the train obeys Newton's first law as seen from both $\mathcal{S}$ and $\mathcal{S}'$. If the train to which $\mathcal{S}''$ is attached is accelerating forward, then, as seen in $\mathcal{S}''$, an ice cube placed on the floor will accelerate backward, and the first law does not hold in $\mathcal{S}''$.

force. Clearly the frame $\mathcal{S}''$ is noninertial, and neither of the first two laws can hold in $\mathcal{S}''$. A similar conclusion would hold if the frame $\mathcal{S}''$ had been attached to a rotating merry-go-round. A frictionless puck, subject to zero net force, would not move in a straight line as seen in $\mathcal{S}''$, and Newton's laws would not hold.

Evidently Newton's two laws hold only in the special, inertial (nonaccelerating and nonrotating) reference frames. Most philosophers of science take the view that the first law should be used to identify these inertial frames — a reference frame $\mathcal{S}$ is inertial if objects that are clearly subject to no forces are seen to move with constant velocity relative to $\mathcal{S}$.[6] Having identified the inertial frames by means of Newton's first law, we can then claim as an experimental fact that the second law holds in these same inertial frames.[7]

Since the laws of motion hold only in inertial frames, you might imagine that we would confine our attention exclusively to inertial frames, and, for a while, we shall do just that. Nevertheless, you should be aware that there are situations where it is necessary, or at least very convenient, to work in noninertial frames. The most important example of a noninertial frame is in fact the earth itself. To an excellent approximation, a reference frame fixed to the earth is inertial — a fortunate circumstance for students of physics! Nevertheless, the earth rotates on its axis once a day and circles around the sun once a year, and the sun orbits slowly around the center of the Milky Way galaxy. For all of these reasons, a reference frame fixed to the earth is not exactly inertial. Although these effects are very small, there are several phenomena — the tides and the trajectories of long-range projectiles are examples —

---

[6] There is some danger of going in a circle here: How do we know that the object is subject to no forces? We'd better not answer, "Because it's traveling at constant velocity"! Fortunately, we can argue that it is possible to identify all sources of force, such as people pushing and pulling or nearby massive bodies exerting gravitational forces. If there are no such things around, we can reasonably say that the object is free of forces.

[7] As I mentioned earlier, the extent to which the second law is an experimental statement depends on how we choose to define force. If we define force by means of the second law, then to some extent (though certainly not entirely) the law becomes a matter of definition. If we define forces by means of spring balances, then the second law is clearly an experimentally testable proposition.

that are most simply explained by taking into account the noninertial character of a frame fixed to the earth. In Chapter 9 we shall examine how the laws of motion must be modified for use in noninertial frames. For the moment, however, we shall confine our discussion to inertial frames.

Validity of the First Two Laws

Since the advent of relativity and quantum mechanics, we have known that Newton's laws are not universally valid. Nevertheless, there is an immense range of phenomena — the phenomena of classical physics — where the first two laws are for all practical purposes exact. Even as the speeds of interest approach c, the speed of light, and relativity becomes important, the first law remains exactly true. (In relativity, just as in classical mechanics, an inertial frame is *defined* as one where the first law holds.)[8] As we shall see in Chapter 15, the two forms of the second law, $\mathbf{F} = m\mathbf{a}$ and $\mathbf{F} = \dot{\mathbf{p}}$, are no longer equivalent in relativity, although with $\mathbf{F}$ and $\mathbf{p}$ suitably defined the second law in the form $\mathbf{F} = \dot{\mathbf{p}}$ is still valid. In any case, the important point is this: In the classical domain, we can and shall assume that the first two laws (the second in either form) are universally and precisely valid. You can, if you wish, regard this assumption as defining a model — the classical model — of the natural world. The model is logically consistent and is such a good representation of many phenomena that it is amply worthy of our study.

1.5   The Third Law and Conservation of Momentum

Newton's first two laws concern the response of a single object to applied forces. The third law addresses a quite different issue: Every force on an object inevitably involves a second object — the object *that exerts the force*. The nail is hit *by the hammer*, the cart is pulled *by the horse*, and so on. While this much is no doubt a matter of common sense, the third law goes considerably beyond our everyday experience. Newton realized that if an object 1 exerts a force on another object 2, then object 2 always exerts a force (the "reaction" force) back on object 1. This seems quite natural: If you push hard against a wall, it is fairly easy to convince yourself that the wall is exerting a force back on you, without which you would undoubtedly fall over. The aspect of the third law which certainly goes beyond our normal perceptions is this: According to the third law, the reaction force of object 2 on object 1 is always equal and opposite to the original force of 1 on 2. If we introduce the notation $\mathbf{F}_{21}$ to denote the force exerted on object 2 by object 1, Newton's third law can be stated very compactly:

---

[8] However, in relativity the relationship between different inertial frames — the so-called Lorentz transformation — is different from that of classical mechanics. See Section 15.6.

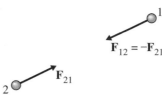

Figure 1.5   Newton's third law asserts that the reaction force exerted on object 1 by object 2 is equal and opposite to the force exerted on 2 by 1, that is, $\mathbf{F}_{12} = -\mathbf{F}_{21}$.

Newton's Third Law

If object 1 exerts a force $\mathbf{F}_{21}$ on object 2, then object 2 always exerts a reaction force $\mathbf{F}_{12}$ on object 1 given by

$$\mathbf{F}_{12} = -\mathbf{F}_{21}. \tag{1.20}$$

This statement is illustrated in Figure 1.5, which you could think of as showing the force of the earth on the moon and the reaction force of the moon on the earth (or a proton on an electron and the electron on the proton). Notice that this figure actually goes a little beyond the usual statement (1.20) of the third law: Not only have I shown the two forces as equal and opposite; I have also shown them acting along the line joining 1 and 2. Forces with this extra property are called **central forces**. (They act along the line of centers.) The third law does not actually require that the forces be central, but, as I shall discuss later, most of the forces we encounter (gravity, the electrostatic force between two charges, etc.) do have this property.

As Newton himself was well aware, the third law is intimately related to the law of conservation of momentum. Let us focus, at first, on just two objects as shown in Figure 1.6, which might show the earth and the moon or two skaters on the ice. In addition to the force of each object on the other, there may be "external" forces exerted by other bodies. The earth and moon both experience forces exerted by the sun, and both skaters could experience the external force of the wind. I have shown the net external forces on the two objects as $\mathbf{F}_1^{\text{ext}}$ and $\mathbf{F}_2^{\text{ext}}$. The total force on object 1 is then

$$(\text{net force on 1}) \equiv \mathbf{F}_1 = \mathbf{F}_{12} + \mathbf{F}_1^{\text{ext}}$$

and similarly

$$(\text{net force on 2}) \equiv \mathbf{F}_2 = \mathbf{F}_{21} + \mathbf{F}_2^{\text{ext}}.$$

We can compute the rates of change of the particles' momenta using Newton's second law:

$$\dot{\mathbf{p}}_1 = \mathbf{F}_1 = \mathbf{F}_{12} + \mathbf{F}_1^{\text{ext}} \tag{1.21}$$

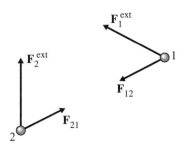

Figure 1.6   Two objects exert forces on each other and may also be subject to additional "external" forces from other objects not shown.

and

$$\dot{\mathbf{p}}_2 = \mathbf{F}_2 = \mathbf{F}_{21} + \mathbf{F}_2^{\text{ext}}. \tag{1.22}$$

If we now define the total momentum of our two objects as

$$\mathbf{P} = \mathbf{p}_1 + \mathbf{p}_2,$$

then the rate of change of the total momentum is just

$$\dot{\mathbf{P}} = \dot{\mathbf{p}}_1 + \dot{\mathbf{p}}_2.$$

To evaluate this, we have only to add Equations (1.21) and (1.22). When we do this, the two internal forces, $\mathbf{F}_{12}$ and $\mathbf{F}_{21}$, cancel out because of Newton's third law, and we are left with

$$\dot{\mathbf{P}} = \mathbf{F}_1^{\text{ext}} + \mathbf{F}_2^{\text{ext}} \equiv \mathbf{F}^{\text{ext}}, \tag{1.23}$$

where I have introduced the notation $\mathbf{F}^{\text{ext}}$ to denote the total external force on our two-particle system.

The result (1.23) is the first in a series of important results that let us construct a theory of many-particle systems from the basic laws for a single particle. It asserts that as far as the total momentum of a system is concerned, the internal forces have no effect. A special case of this result is that if there are no external forces ($\mathbf{F}^{\text{ext}} = 0$) then $\dot{\mathbf{P}} = 0$. Thus we have the important result:

$$\text{If} \quad \mathbf{F}^{\text{ext}} = 0, \quad \text{then} \quad \mathbf{P} = \text{const.} \tag{1.24}$$

In the absence of external forces, the total momentum of our two-particle system is constant — a result called the principle of conservation of momentum.

Multiparticle Systems

We have proved the conservation of momentum, Equation (1.24), for a system of two particles. The extension of the result to any number of particles is straightforward in principle, but I would like to go through it in detail, because it lets me introduce some

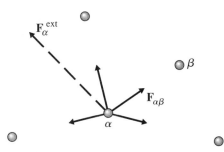

Figure 1.7   A five-particle system with particles labelled by α or $\beta = 1, 2, \cdots, 5$. The particle α is subject to four internal forces, shown by solid arrows and denoted $\mathbf{F}_{\alpha\beta}$ (the force on α by β). In addition particle α may be subject to a net external force, shown by the dashed arrow and denoted $\mathbf{F}_\alpha^{\text{ext}}$.

important notation and will give you some practice using the summation notation. Let us consider then a system of N particles. I shall label the typical particle with a Greek index α or β, either of which can take any of the values $1, 2, \cdots, N$. The mass of particle α is m_α and its momentum is $\mathbf{p}_\alpha$. The force on particle α is quite complicated: Each of the other $(N - 1)$ particles can exert a force which I shall call $\mathbf{F}_{\alpha\beta}$, the force on α by β, as illustrated in Figure 1.7. In addition there may be a net external force on particle α, which I shall call $\mathbf{F}_\alpha^{\text{ext}}$. Thus the net force on particle α is

$$\text{(net force on particle } \alpha) = \mathbf{F}_\alpha = \sum_{\beta \neq \alpha} \mathbf{F}_{\alpha\beta} + \mathbf{F}_\alpha^{\text{ext}}. \tag{1.25}$$

Here the sum runs over all values of β not equal to α. (Remember there is no force $\mathbf{F}_{\alpha\alpha}$ because particle α cannot exert a force on itself.) According to Newton's second law, this is the same as the rate of change of $\mathbf{p}_\alpha$:

$$\dot{\mathbf{p}}_\alpha = \sum_{\beta \neq \alpha} \mathbf{F}_{\alpha\beta} + \mathbf{F}_\alpha^{\text{ext}}. \tag{1.26}$$

This result holds for each $\alpha = 1, \cdots, N$.

Let us now consider the total momentum of our N-particle system,

$$\mathbf{P} = \sum_\alpha \mathbf{p}_\alpha$$

where, of course, this sum runs over all N particles, $\alpha = 1, 2, \cdots, N$. If we differentiate this equation with respect to time, we find

$$\dot{\mathbf{P}} = \sum_\alpha \dot{\mathbf{p}}_\alpha$$

or, substituting for $\dot{\mathbf{p}}_\alpha$ from (1.26),

$$\dot{\mathbf{P}} = \sum_\alpha \sum_{\beta \neq \alpha} \mathbf{F}_{\alpha\beta} + \sum_\alpha \mathbf{F}_\alpha^{\text{ext}}. \tag{1.27}$$

The double sum here contains $N(N-1)$ terms in all. Each term $\mathbf{F}_{\alpha\beta}$ in this sum can be paired with a second term $\mathbf{F}_{\beta\alpha}$ (that is, $\mathbf{F}_{12}$ paired with $\mathbf{F}_{21}$, and so on), so that

$$\sum_{\alpha}\sum_{\beta\neq\alpha}\mathbf{F}_{\alpha\beta}=\sum_{\alpha}\sum_{\beta>\alpha}(\mathbf{F}_{\alpha\beta}+\mathbf{F}_{\beta\alpha}). \tag{1.28}$$

The double sum on the right includes only values of α and β with $\alpha<\beta$ and has half as many terms as that on the left. But each term is the sum of two forces, $(\mathbf{F}_{\alpha\beta}+\mathbf{F}_{\beta\alpha})$, and, by the third law, each such sum is zero. Therefore the whole double sum in (1.28) is zero, and returning to (1.27) we conclude that

$$\dot{\mathbf{P}}=\sum_{\alpha}\mathbf{F}_{\alpha}^{\text{ext}}\equiv\mathbf{F}^{\text{ext}}. \tag{1.29}$$

The result (1.29) corresponds exactly to the two-particle result (1.23). Like the latter, it says that the internal forces have no effect on the evolution of the total momentum $\mathbf{P}$ — the rate of change of $\mathbf{P}$ is determined by the net *external* force on the system. In particular, if the net external force is zero, we have the

Principle of Conservation of Momentum

If the net external force $\mathbf{F}^{\text{ext}}$ on an N-particle system is zero, the system's total momentum $\mathbf{P}$ is constant.

As you are certainly aware, this is one of the most important results in classical physics and is, in fact, also true in relativity and quantum mechanics. If you are not very familiar with the sorts of manipulations of sums that we used, it would be a good idea to go over the argument leading from (1.25) to (1.29) for the case of three or four particles, writing out all the sums explicitly (Problems 1.28 or 1.29). You should also convince yourself that, conversely, if the principle of conservation of momentum is true for all multiparticle systems, then Newton's third law must be true (Problem 1.31). In other words, conservation of momentum and Newton's third law are equivalent to one another.

Validity of Newton's Third Law

Within the domain of classical physics, the third law, like the second, is valid with such accuracy that it can be taken to be exact. As speeds approach the speed of light, it is easy to see that the third law cannot hold: The point is that the law asserts that the action and reaction forces, $\mathbf{F}_{12}(t)$ and $\mathbf{F}_{21}(t)$, *measured at the same time t*, are equal and opposite. As you certainly know, once relativity becomes important the concept of a single universal time has to be abandoned — two events that are seen as simultaneous by one observer are, in general, *not* simultaneous as seen by a second observer. Thus, even if the equality $\mathbf{F}_{12}(t)=-\mathbf{F}_{21}(t)$ (with both times the same) were true for one observer, it would generally be false for another. Therefore, the third law cannot be valid once relativity becomes important.

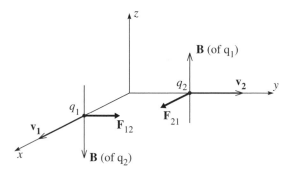

Figure 1.8   Each of the positive charges q_1 and q_2 produces
a magnetic field that exerts a force on the other charge. The
resulting magnetic forces $\mathbf{F}_{12}$ and $\mathbf{F}_{21}$ do not obey Newton's
third law.

Rather surprisingly, there is a simple example of a well-known force — the magnetic force between two moving charges — for which the third law is not exactly true,
even at slow speeds. To see this, consider the two positive charges of Figure 1.8, with
q_1 moving in the x direction and q_2 moving in the y direction, as shown. The exact
calculation of the magnetic field produced by each charge is complicated, but a simple
argument gives the correct directions of the two fields, and this is all we need. The
moving charge q_1 is equivalent to a current in the x direction. By the right-hand rule
for fields, this produces a magnetic field which points in the z direction in the vicinity
of q_2. By the right-hand rule for forces, this field produces a force $\mathbf{F}_{21}$ on q_2 that is
in the x direction. An exactly analogous argument (check it yourself) shows that the
force $\mathbf{F}_{12}$ on q_1 is in the y direction, as shown. Clearly these two forces do not obey
Newton's third law!

This conclusion is especially startling since we have just seen that Newton's third
law is equivalent to the conservation of momentum. Apparently the total momentum
$m_1\mathbf{v}_1 + m_2\mathbf{v}_2$ of the two charges in Figure 1.8 is not conserved. This conclusion, which
is correct, serves to remind us that the "mechanical" momentum $m\mathbf{v}$ of particles is not
the only kind of momentum. Electromagnetic fields can also carry momentum, and in
the situation of Figure 1.8 the mechanical momentum being lost by the two particles
is going to the electromagnetic momentum of the fields.

Fortunately, if both speeds in Figure 1.8 are much less than the speed of light
($v \ll c$), the loss of mechanical momentum and the concomitant failure of the third
law are completely negligible. To see this, note that in addition to the magnetic force
between q_1 and q_2 there is the electrostatic Coulomb force[9] kq_1q_2/r^2, which *does* obey
Newton's third law. It is a straightforward exercise (Problem 1.32) to show that the
magnetic force is of order v^2/c^2 times the Coulomb force. Thus only as v approaches
c — and classical mechanics must give way to relativity anyway — is the violation of

---

[9] Here k is the Coulomb force constant, often written as $k = 1/(4\pi\epsilon_0)$.

the third law by the magnetic force important.[10] We see that the unexpected situation of Figure 1.8 does not contradict our claim that in the classical domain Newton's third law is valid, and this is what we shall assume in our discussions of nonrelativistic mechanics.

1.6   Newton's Second Law in Cartesian Coordinates

Of Newton's three laws, the one that we actually use the most is the second, which is often described as the *equation of motion*. As we have seen, the first is theoretically important to define what we mean by inertial frames but is usually of no practical use beyond this. The third law is crucially important in sorting out the internal forces in a multiparticle system, but, once we know the forces involved, the second law is what we actually use to calculate the motion of the object or objects of interest. In particular, in many simple problems the forces are known or easily found, and, in this case, the second law is all we need for solving the problem.

As we have already noted, the second law,

$$\mathbf{F} = m\ddot{\mathbf{r}}, \tag{1.30}$$

is a second-order, differential equation[11] for the position vector $\mathbf{r}$ as a function of the time t. In the prototypical problem, the forces that comprise $\mathbf{F}$ are given, and our job is to solve the differential equation (1.30) for $\mathbf{r}(t)$. Sometimes we are told about $\mathbf{r}(t)$, and we have to use (1.30) to find some of the forces. In any case, the equation (1.30) is a *vector* differential equation. And the simplest way to solve such equations is almost always to resolve the vectors into their components relative to a chosen coordinate system.

Conceptually the simplest coordinate system is the Cartesian (or rectangular), with unit vectors $\hat{\mathbf{x}}$, $\hat{\mathbf{y}}$, and $\hat{\mathbf{z}}$, in terms of which the net force $\mathbf{F}$ can then be written as

$$\mathbf{F} = F_x\,\hat{\mathbf{x}} + F_y\,\hat{\mathbf{y}} + F_z\,\hat{\mathbf{z}} \tag{1.31}$$

and the position vector $\mathbf{r}$ as

$$\mathbf{r} = x\,\hat{\mathbf{x}} + y\,\hat{\mathbf{y}} + z\,\hat{\mathbf{z}}. \tag{1.32}$$

As we noted in Section 1.2, this expansion of $\mathbf{r}$ in terms of its Cartesian components is especially easy to differentiate because the unit vectors $\hat{\mathbf{x}}$, $\hat{\mathbf{y}}$, $\hat{\mathbf{z}}$ are constant. Thus we can differentiate (1.32) twice to get the simple result

$$\ddot{\mathbf{r}} = \ddot{x}\,\hat{\mathbf{x}} + \ddot{y}\,\hat{\mathbf{y}} + \ddot{z}\,\hat{\mathbf{z}}. \tag{1.33}$$

---

[10] The magnetic force between two steady currents is not necessarily small, even in the classical domain, but it can be shown that this force *does* obey the third law. See Problem 1.33.

[11] The force $\mathbf{F}$ can sometimes involve derivatives of $\mathbf{r}$. (For instance the magnetic force on a moving charge involves the velocity $\mathbf{v} = \dot{\mathbf{r}}$.) Very occasionally the force $\mathbf{F}$ involves a higher derivative of $\mathbf{r}$, of order $n > 2$, in which case the second law is an nth-order differential equation.

That is, the three Cartesian components of $\ddot{\mathbf{r}}$ are just the appropriate derivatives of the three coordinates x, y, z of $\mathbf{r}$, and the second law (1.30) becomes

$$F_x\,\hat{\mathbf{x}} + F_y\,\hat{\mathbf{y}} + F_z\,\hat{\mathbf{z}} = m\ddot{x}\,\hat{\mathbf{x}} + m\ddot{y}\,\hat{\mathbf{y}} + m\ddot{z}\,\hat{\mathbf{z}}. \qquad (1.34)$$

Resolving this equation into its three separate components, we see that F_x has to equal $m\ddot{x}$ and similarly for the y and z components. That is, in Cartesian coordinates, the single vector equation (1.30) is equivalent to the three separate equations:

$$\mathbf{F} = m\ddot{\mathbf{r}} \qquad \Longleftrightarrow \qquad \begin{cases} F_x = m\ddot{x} \\ F_y = m\ddot{y} \\ F_z = m\ddot{z}. \end{cases} \qquad (1.35)$$

This beautiful result, that, in Cartesian coordinates, Newton's second law in three dimensions is equivalent to three one-dimensional versions of the same law, is the basis of the solution of almost all simple mechanics problems in Cartesian coordinates. Here is an example to remind you of how such problems go.

EXAMPLE 1.1 A Block Sliding down an Incline

A block of mass m is observed accelerating from rest down an incline that has coefficient of friction μ and is at angle θ from the horizontal. How far will it travel in time t?

Our first task is to choose our frame of reference. Naturally, we choose our spatial origin at the block's starting position and the origin of time ($t = 0$) at the moment of release. As you no doubt remember from your introductory physics course, the best choice of axes is to have one axis (x say) point down the slope, one (y) normal to the slope, and the third (z) across it, as shown in Figure 1.9. This choice has two advantages: First, because the block slides straight down the slope, the motion is entirely in the x direction, and only x varies. (If we had chosen the x axis horizontal and the y axis vertical, then both x and y would vary.) Second, two of the three forces on the block are unknown (the normal force $\mathbf{N}$ and friction $\mathbf{f}$; the weight, $\mathbf{w} = m\mathbf{g}$, we treat as known), and with our choice of axes, each of the unknowns has only one nonzero component, since $\mathbf{N}$ is in the y direction and $\mathbf{f}$ is in the (negative) x direction.

We are now ready to apply Newton's second law. The result (1.35) means that we can analyse the three components separately, as follows:

There are no forces in the z direction, so $F_z = 0$. Since $F_z = m\ddot{z}$, it follows that $\ddot{z} = 0$, which implies that $\dot{z}$ (or v_z) is constant. Since the block starts from rest, this means that $\dot{z}$ is actually zero for all t. With $\dot{z} = 0$, it follows that z is constant, and, since it too starts from zero, we conclude that $z = 0$ for all t. As we would certainly have guessed, the motion remains in the xy plane.

Since the block does not jump off the incline, we know that there is no motion in the y direction. In particular, $\ddot{y} = 0$. Therefore, Newton's second law implies that the y component of the net force is zero; that is, $F_y = 0$. From Figure 1.9 we see that this implies that

$$F_y = N - mg\cos\theta = 0.$$

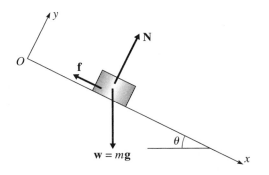

Figure 1.9   A block slides down a slope of incline θ. The three forces on the block are its weight, $\mathbf{w} = m\mathbf{g}$, the normal force of the incline, $\mathbf{N}$, and the frictional force $\mathbf{f}$, whose magnitude is $f = \mu N$. The z axis is not shown but points out of the page, that is, across the slope.

Thus the y component of the second law has told us that the unknown normal force is $N = mg \cos \theta$. Since $f = \mu N$, this tells us the frictional force, $f = \mu mg \cos \theta$, and all the forces are now known. All that remains is to use the remaining component (the x component) of the second law to solve for the actual motion.

The x component of the second law, $F_x = m\ddot{x}$, implies (see Figure 1.9) that

$$w_x - f = m\ddot{x}$$

or

$$mg \sin \theta - \mu mg \cos \theta = m\ddot{x}.$$

The m's cancel, and we find for the acceleration down the slope

$$\ddot{x} = g(\sin \theta - \mu \cos \theta). \tag{1.36}$$

Having found $\ddot{x}$, and found it to be constant, we have only to integrate it twice to find x as a function of t. First

$$\dot{x} = g(\sin \theta - \mu \cos \theta)t.$$

(Remember that $\dot{x} = 0$ initially, so the constant of integration is zero.) Finally,

$$x(t) = \tfrac{1}{2}g(\sin \theta - \mu \cos \theta)t^2$$

(again the constant of integration is zero) and our solution is complete.

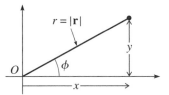

Figure 1.10   The definition of the polar coordinates r and ϕ.

1.7   Two-Dimensional Polar Coordinates

While Cartesian coordinates have the merit of simplicity, we are going to find that it is almost impossible to solve certain problems without the use of various non-Cartesian coordinate systems. To illustrate the complexities of non-Cartesian coordinates, let us consider the form of Newton's second law in a two-dimensional problem using polar coordinates. These coordinates are defined in Figure 1.10. Instead of using the two rectangular coordinates x, y, we label the position of a particle with its distance r from O and the angle ϕ measured up from the x axis. Given the rectangular coordinates x and y, you can calculate the polar coordinates r and ϕ, or vice versa, using the following relations. (Make sure you understand all four equations.[12])

$$\left. \begin{matrix} x = r\cos\phi \\ y = r\sin\phi \end{matrix} \right\} \longleftrightarrow \begin{cases} r = \sqrt{x^2 + y^2} \\ \phi = \arctan(y/x) \end{cases} \tag{1.37}$$

Just as with rectangular coordinates, it is convenient to introduce two unit vectors, which I shall denote by $\hat{\mathbf{r}}$ and $\hat{\boldsymbol{\phi}}$. To understand their definitions, notice that we can define the unit vector $\hat{\mathbf{x}}$ as the unit vector that points in the direction of increasing x when y is fixed, as shown in Figure 1.11(a). In the same way we shall define $\hat{\mathbf{r}}$ as the unit vector that points in the direction we move when r increases with ϕ fixed; likewise, $\hat{\boldsymbol{\phi}}$ is the unit vector that points in the direction we move when ϕ increases with r fixed. Figure 1.11 makes clear a most important difference between the unit vectors $\hat{\mathbf{x}}$ and $\hat{\mathbf{y}}$ of rectangular coordinates and our new unit vectors $\hat{\mathbf{r}}$ and $\hat{\boldsymbol{\phi}}$. The vectors $\hat{\mathbf{x}}$ and $\hat{\mathbf{y}}$ are the same at all points in the plane, whereas the new vectors $\hat{\mathbf{r}}$ and $\hat{\boldsymbol{\phi}}$ change their directions as the position vector $\mathbf{r}$ moves around. We shall see that this complicates the use of Newton's second law in polar coordinates.

Figure 1.11 suggests another way to write the unit vector $\hat{\mathbf{r}}$. Since $\hat{\mathbf{r}}$ is in the same direction as $\mathbf{r}$, but has magnitude 1, you can see that

$$\hat{\mathbf{r}} = \frac{\mathbf{r}}{|\mathbf{r}|}. \tag{1.38}$$

This result suggests a second role for the "hat" notation. For *any* vector $\mathbf{a}$, we can define $\hat{\mathbf{a}}$ as the unit vector in the direction of $\mathbf{a}$, namely $\hat{\mathbf{a}} = \mathbf{a}/|\mathbf{a}|$.

---

[12] There is a small subtlety concerning the equation for ϕ: You need to make sure ϕ lands in the proper quadrant, since the first and third quadrants give the same values for y/x (and likewise the second and fourth). See Problem 1.42.

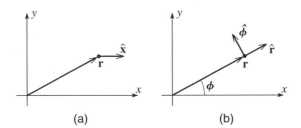

Figure 1.11 **(a)** The unit vector $\hat{\mathbf{x}}$ points in the direction of increasing x with y fixed. **(b)** The unit vector $\hat{\mathbf{r}}$ points in the direction of increasing r with ϕ fixed; $\hat{\boldsymbol{\phi}}$ points in the direction of increasing ϕ with r fixed. Unlike $\hat{\mathbf{x}}$, the vectors $\hat{\mathbf{r}}$ and $\hat{\boldsymbol{\phi}}$ change as the position vector $\mathbf{r}$ moves.

Since the two unit vectors $\hat{\mathbf{r}}$ and $\hat{\boldsymbol{\phi}}$ are perpendicular vectors in our two-dimensional space, any vector can be expanded in terms of them. For instance, the net force $\mathbf{F}$ on an object can be written

$$\mathbf{F} = F_r \hat{\mathbf{r}} + F_\phi \hat{\boldsymbol{\phi}}. \tag{1.39}$$

If, for example, the object in question is a stone that I am twirling in a circle on the end of a string (with my hand at the origin), then F_r would be the tension in the string and F_ϕ the force of air resistance retarding the stone in the tangential direction. The expansion of the position vector itself is especially simple in polar coordinates. From Figure 1.11(b) it is clear that

$$\mathbf{r} = r\hat{\mathbf{r}}. \tag{1.40}$$

We are now ready to ask about the form of Newton's second law, $\mathbf{F} = m\ddot{\mathbf{r}}$, in polar coordinates. In rectangular coordinates, we saw that the x component of $\ddot{\mathbf{r}}$ is just $\ddot{x}$, and this is what led to the very simple result (1.35). We must now find the components of $\ddot{\mathbf{r}}$ in polar coordinates; that is, we must differentiate (1.40) with respect to t. Although (1.40) is very simple, the vector $\hat{\mathbf{r}}$ changes as $\mathbf{r}$ moves. Thus when we differentiate (1.40), we shall pick up a term involving the derivative of $\hat{\mathbf{r}}$. Our first task is to find this derivative of $\hat{\mathbf{r}}$.

Figure 1.12(a) shows the position of the particle of interest at two successive times, t_1 and $t_2 = t_1 + \Delta t$. If the corresponding angles $\phi(t_1)$ and $\phi(t_2)$ are different, then the two unit vectors $\hat{\mathbf{r}}(t_1)$ and $\hat{\mathbf{r}}(t_2)$ point in different directions. The change in $\hat{\mathbf{r}}$ is shown in Figure 1.12(b), and (provided Δt is small) is approximately

$$\Delta\hat{\mathbf{r}} \approx \Delta\phi\,\hat{\boldsymbol{\phi}}$$

$$\approx \dot{\phi}\,\Delta t\,\hat{\boldsymbol{\phi}}. \tag{1.41}$$

(Notice that the direction of $\Delta\hat{\mathbf{r}}$ is perpendicular to $\hat{\mathbf{r}}$, namely the direction of $\hat{\boldsymbol{\phi}}$.) If we divide both sides by Δt and take the limit as $\Delta t \to 0$, then $\Delta\hat{\mathbf{r}}/\Delta t \to d\hat{\mathbf{r}}/dt$ and we find that

$$\frac{d\hat{\mathbf{r}}}{dt} = \dot{\phi}\,\hat{\boldsymbol{\phi}}. \tag{1.42}$$

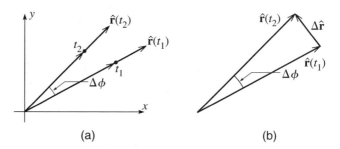

Figure 1.12   **(a)** The positions of a particle at two successive times, t_1 and t_2. Unless the particle is moving exactly radially, the corresponding unit vectors $\hat{\mathbf{r}}(t_1)$ and $\hat{\mathbf{r}}(t_2)$ point in different directions. **(b)** The change $\Delta\hat{\mathbf{r}}$ in $\hat{\mathbf{r}}$ is given by the triangle shown.

(For an alternative proof of this important result, see Problem 1.43.) Notice that $d\hat{\mathbf{r}}/dt$ is in the direction of $\hat{\boldsymbol{\phi}}$ and is proportional to the rate of change of the angle ϕ — both of which properties we would expect based on Figure 1.12.

Now that we know the derivative of $\hat{\mathbf{r}}$, we are ready to differentiate Equation (1.40). Using the product rule, we get two terms:

$$\dot{\mathbf{r}} = \dot{r}\hat{\mathbf{r}} + r\frac{d\hat{\mathbf{r}}}{dt},$$

and, substituting (1.42), we find for the velocity $\dot{\mathbf{r}}$, or $\mathbf{v}$,

$$\mathbf{v} \equiv \dot{\mathbf{r}} = \dot{r}\hat{\mathbf{r}} + r\dot{\phi}\,\hat{\boldsymbol{\phi}}. \tag{1.43}$$

From this we can read off the polar components of the velocity:

$$v_r = \dot{r} \quad \text{and} \quad v_\phi = r\dot{\phi} = r\omega \tag{1.44}$$

where in the second equation I have introduced the traditional notation ω for the angular velocity $\dot{\phi}$. While the results in (1.44) should be familiar from your introductory physics course, they are undeniably more complicated than the corresponding results in Cartesian coordinates ($v_x = \dot{x}$ and $v_y = \dot{y}$).

Before we can write down Newton's second law, we have to differentiate a second time to find the acceleration:

$$\mathbf{a} \equiv \ddot{\mathbf{r}} = \frac{d}{dt}\dot{\mathbf{r}} = \frac{d}{dt}(\dot{r}\hat{\mathbf{r}} + r\dot{\phi}\,\hat{\boldsymbol{\phi}}), \tag{1.45}$$

where the final expression comes from substituting (1.43) for $\dot{\mathbf{r}}$. To complete the differentiation in (1.45), we must calculate the derivative of $\hat{\boldsymbol{\phi}}$. This calculation is completely analogous to the argument leading to (1.42) and is illustrated in Figure 1.13. By inspecting this figure, you should be able to convince yourself that

$$\frac{d\hat{\boldsymbol{\phi}}}{dt} = -\dot{\phi}\hat{\mathbf{r}}. \tag{1.46}$$

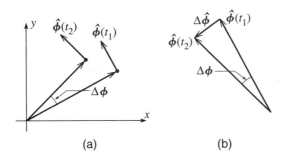

Figure 1.13   (a) The unit vector $\hat{\boldsymbol{\phi}}$ at two successive times t_1 and t_2. (b) The change $\Delta\hat{\boldsymbol{\phi}}$.

Returning to Equation (1.45), we can now carry out the differentiation to give the following five terms:

$$\mathbf{a} = \left(\ddot{r}\hat{\mathbf{r}} + \dot{r}\frac{d\hat{\mathbf{r}}}{dt}\right) + \left((\dot{r}\dot{\phi} + r\ddot{\phi})\hat{\boldsymbol{\phi}} + r\dot{\phi}\frac{d\hat{\boldsymbol{\phi}}}{dt}\right)$$

or, if we use (1.42) and (1.46) to replace the derivatives of the two unit vectors,

$$\mathbf{a} = \left(\ddot{r} - r\dot{\phi}^2\right)\hat{\mathbf{r}} + \left(r\ddot{\phi} + 2\dot{r}\dot{\phi}\right)\hat{\boldsymbol{\phi}}. \tag{1.47}$$

This horrible result is a little easier to understand if we consider the special case that r is constant, as is the case for a stone that I twirl on the end of a string of fixed length. With r constant, both derivatives of r are zero, and (1.47) has just two terms:

$$\mathbf{a} = -r\dot{\phi}^2\hat{\mathbf{r}} + r\ddot{\phi}\hat{\boldsymbol{\phi}}$$

or

$$\mathbf{a} = -r\omega^2\hat{\mathbf{r}} + r\alpha\hat{\boldsymbol{\phi}},$$

where $\omega = \dot{\phi}$ denotes the angular velocity and $\alpha = \ddot{\phi}$ is the angular acceleration. This is the familiar result from elementary physics that when a particle moves around a fixed circle, it has an inward "centripetal" acceleration $r\omega^2$ (or v^2/r) and a tangential acceleration, $r\alpha$. Nevertheless, when r is not constant, the acceleration includes all four of the terms in (1.47). The first term, $\ddot{r}$ in the radial direction is what you would probably expect when r varies, but the final term, $2\dot{r}\dot{\phi}$ in the ϕ direction, is harder to understand. It is called the Coriolis acceleration, and I shall discuss it in detail in Chapter 9.

Having calculated the acceleration as in (1.47), we can finally write down Newton's second law in terms of polar coordinates:

$$\mathbf{F} = m\mathbf{a} \qquad \Longleftrightarrow \qquad \begin{cases} F_r = m(\ddot{r} - r\dot{\phi}^2) \\ F_\phi = m(r\ddot{\phi} + 2\dot{r}\dot{\phi}). \end{cases} \tag{1.48}$$

These equations in polar coordinates are a far cry from the beautifully simple equations (1.35) for rectangular coordinates. In fact, one of the main reasons for taking the trouble to recast Newtonian mechanics in the Lagrangian formulation (Chapter 7) is that the latter is able to handle nonrectangular coordinates just as easily as rectangular.

You may justifiably be feeling that the second law in polar coordinates is so complicated that there could be no occasion to use it. In fact, however, there are many problems which are most easily solved using polar coordinates, and I conclude this section with an elementary example.

EXAMPLE 1.2   An Oscillating Skateboard

A "half-pipe" at a skateboard park consists of a concrete trough with a semicircular cross section of radius $R = 5$ m, as shown in Figure 1.14. I hold a frictionless skateboard on the side of the trough pointing down toward the bottom and release it. Discuss the subsequent motion using Newton's second law. In particular, if I release the board just a short way from the bottom, how long will it take to come back to the point of release?

Because the skateboard is constrained to move on a circular path, this problem is most easily solved using polar coordinates with origin O at the center of the pipe as shown. (At some point in the following calculation, try writing the second law in rectangular coordinates and observe what a tangle you get.) With this choice of polar coordinates, the coordinate r of the skateboard is constant, $r = R$, and the position of the skateboard is completely specified by the angle ϕ. With r constant, the second law (1.48) takes the relatively simple form

$$F_r = -mR\dot{\phi}^2 \tag{1.49}$$

and

$$F_\phi = mR\ddot{\phi}. \tag{1.50}$$

The two forces on the skateboard are its weight $\mathbf{w} = m\mathbf{g}$ and the normal force $\mathbf{N}$ of the wall, as shown in Figure 1.14. The components of the net force $\mathbf{F} = \mathbf{w} + \mathbf{N}$ are easily seen to be

$$F_r = mg\cos\phi - N \quad \text{and} \quad F_\phi = -mg\sin\phi.$$

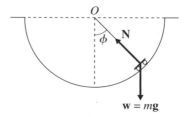

Figure 1.14   A skateboard in a semicircular trough of radius R. The board's position is specified by the angle ϕ measured up from the bottom. The two forces on the skateboard are its weight $\mathbf{w} = m\mathbf{g}$ and the normal force $\mathbf{N}$.

Substituting for F_r into (1.49) we get an equation involving N, ϕ, and $\dot{\phi}$. Fortunately, we are not really interested in N, and — even more fortunately — when we substitute for F_ϕ into (1.50), we get an equation that does not involve N at all:

$$-mg\sin\phi = mR\ddot{\phi}$$

or, canceling the m's and rearranging,

$$\ddot{\phi} = -\frac{g}{R}\sin\phi. \tag{1.51}$$

Equation (1.51) is the differential equation for $\phi(t)$ that determines the motion of the skateboard. Qualitatively, we can easily see the kind of motion that it implies. First, if $\phi = 0$, (1.51) says that $\ddot{\phi} = 0$. Therefore, if we place the board at rest ($\dot{\phi} = 0$) at the point $\phi = 0$, the board will never move (unless someone pushes it); that is, $\phi = 0$ is an equilibrium position, as you would certainly have guessed. Next, suppose that at some time, ϕ is not zero and, to be definite, suppose that $\phi > 0$; that is, the skateboard is on the right-hand side of the half-pipe. In this case, (1.51) implies that $\ddot{\phi} < 0$, so the acceleration is directed to the left. If the board is moving to the right it must slow down and eventually start moving to the left.[13] Once it is moving toward the left, it speeds up and returns to the bottom, where it moves over to the left. As soon as the board is on the left, the argument reverses ($\phi < 0$, so $\ddot{\phi} > 0$) and the board must eventually return to the bottom and move over to the right again. In other words, the differential equation (1.51) implies that the skateboard oscillates back and forth, from right to left and back to the right.

The equation of motion (1.51) cannot be solved in terms of elementary functions, such as polynomials, trigonometric functions, or logs and exponentials.[14] Thus, if we want more quantitative information about the motion, the simplest course is to use a computer to solve it numerically (see Problem 1.50). However, if the initial angle ϕ_o is *small*, we can use the small angle approximation

$$\sin\phi \approx \phi \tag{1.52}$$

and, within this approximation, (1.51) becomes

$$\ddot{\phi} = -\frac{g}{R}\phi \tag{1.53}$$

which *can* be solved using elementary functions. [By this stage, you have almost certainly recognized that our discussion of the skateboard problem closely parallels the analysis of the simple pendulum. In particular, the small-angle

---

[13] I am taking for granted that it doesn't reach the top and jump out of the trough. Since it was released from rest inside the trough, this is correct. Much the easiest way to prove this claim is to invoke conservation of energy, which we shan't be discussing for a while. Perhaps, for now, you could agree to accept it as a matter of common sense.

[14] Actually the solution of (1.51) is a Jacobi elliptic function. However, I shall take the point of view that for most of us the Jacobi function is not "elementary."

approximation (1.52) is what let you solve the simple pendulum in your introductory physics course. This parallel is, of course, no accident. Mathematically
the two problems are exactly equivalent.] If we define the parameter

$$\omega = \sqrt{\frac{g}{R}}, \tag{1.54}$$

then (1.53) becomes

$$\ddot{\phi} = -\omega^2 \phi. \tag{1.55}$$

This is the equation of motion for our skateboard in the small-angle approximation. I would like to discuss its solution in some detail to introduce several ideas
that we'll be using again and again in what follows. (If you've studied differential
equations before, just see the next three paragraphs as a quick review.)

We first observe that it is easy to find two solutions of the equation (1.55)
by inspection (that is, by inspired guessing). The function $\phi(t) = A \sin(\omega t)$ is
clearly a solution for any value of the constant A. [Differentiating $\sin(\omega t)$ brings
out a factor of ω and changes the sin to a cos; differentiating it again brings
out another ω and changes the cos back to $-\sin$. Thus the proposed solution
does satisfy $\ddot{\phi} = -\omega^2 \phi$.] Similarly, the function $\phi(t) = B \cos(\omega t)$ is another
solution for any constant B. Furthermore, as you can easily check, the sum of
these two solutions is itself a solution. Thus we have now found a whole family
of solutions:

$$\phi(t) = A \sin(\omega t) + B \cos(\omega t) \tag{1.56}$$

is a solution for any values of the two constants A and B.

I now want to argue that *every* solution of the equation of motion (1.55)
has the form (1.56). In other words, (1.56) is the *general solution* — we have
found *all* solutions, and we need seek no further. To get some idea of why
this is, note that the differential equation (1.55) is a statement about the second
derivative $\ddot{\phi}$ of the unknown ϕ. Now, if we had actually been told what $\ddot{\phi}$ is, then
we know from elementary calculus that we could find ϕ by two integrations,
and the result would contain two unknown constants — the two constants of
integration — that would have to be determined by looking (for example) at the
initial values of ϕ and $\dot{\phi}$. In other words, knowledge of $\ddot{\phi}$ would tell us that
ϕ itself is one of a family of functions containing precisely two undetermined
constants. Of course, the differential equation (1.55) does not actually tell us
$\ddot{\phi}$ — it is an equation for $\ddot{\phi}$ in terms of ϕ. Nevertheless, it is plausible that such
an equation would imply that ϕ is one of a family of functions that contain
precisely two undetermined constants. If you have studied differential equations,
you know that this is the case; if you have not, then I must ask you to accept it
as a plausible fact: For any given second-order differential equation [in a large
class of "reasonable" equations, including (1.55) and all of the equations we
shall encounter in this book], the solutions all belong to a family of functions

containing precisely two independent constants — like the constants A and B in (1.56). (More generally, the solutions of an nth-order equation contain precisely n independent constants.)

This theorem sheds a new light on our solution (1.56). We already knew that any function of the form (1.56) is a solution of the equation of motion. Our theorem now guarantees that *every* solution of the equation of motion is of this form. This same argument applies to all the second-order differential equations we shall encounter. If, by hook or by crook, we can find a solution like (1.56) involving two arbitrary constants, then we are guaranteed that we have found the general solution of our equation.

All that remains is to pin down the two constants A and B for our skateboard. To do so, we must look at the initial conditions. At $t = 0$, Equation (1.56) implies that $\phi = B$. Therefore B is just the initial value of ϕ, which we are calling ϕ_o, so $B = \phi_o$. At $t = 0$, Equation (1.56) implies that $\dot{\phi} = \omega A$. Since I released the board from rest, this means that $A = 0$, and our solution is

$$\phi(t) = \phi_o \cos(\omega t). \tag{1.57}$$

The first thing to note about this solution is that, as we anticipated on general grounds, $\phi(t)$ oscillates, moving from positive to negative and back to positive periodically and indefinitely. In particular, the board first returns to its initial position ϕ_o when $\omega t = 2\pi$. The time that this takes is called the period of the motion and is denoted τ. Thus our conclusion is that the period of the skateboard's oscillations is

$$\tau = \frac{2\pi}{\omega} = 2\pi \sqrt{\frac{R}{g}}. \tag{1.58}$$

We were given that $R = 5\,\text{m}$, and $g = 9.8\,\text{m/s}^2$. Substituting these numbers, we conclude that the skateboard returns to its starting point in a time $\tau = 4.5$ seconds.

Principal Definitions and Equations of Chapter 1

Dot and Cross Products

$$\mathbf{r} \cdot \mathbf{s} = rs \cos\theta = r_x s_x + r_y s_y + r_z s_z \qquad \text{[Eqs. (1.6) \& (1.7)]}$$

$$\mathbf{r} \times \mathbf{s} = (r_y s_z - r_z s_y, \, r_z s_x - r_x s_z, \, r_x s_y - r_y s_x) = \det \begin{bmatrix} \hat{\mathbf{x}} & \hat{\mathbf{y}} & \hat{\mathbf{z}} \\ r_x & r_y & r_z \\ s_x & s_y & s_z \end{bmatrix} \qquad \text{[Eq. (1.9)]}$$

Inertial Frames

An inertial frame is any reference frame in which Newton's first law holds, that is, a nonaccelerating, nonrotating frame.

Unit Vectors of a Coordinate System

If (ξ, η, ζ) are an orthogonal system of coordinates, then

$$\hat{\xi} = \text{ unit vector in direction of increasing } \xi \text{ with } \eta \text{ and } \zeta \text{ fixed}$$

and so on, and any vector $\mathbf{s}$ can be expanded as $\mathbf{s} = s_\xi \hat{\xi} + s_\eta \hat{\eta} + s_\zeta \hat{\zeta}$.

Newton's Second Law in Various Coordinate Systems

Vector Form	Cartesian (x, y, z)	2D Polar (r, ϕ)	Cylindrical Polar (ρ, ϕ, z)
$\mathbf{F} = m\ddot{\mathbf{r}}$	$\begin{cases} F_x = m\ddot{x} \\ F_y = m\ddot{y} \\ F_z = m\ddot{z} \end{cases}$	$\begin{cases} F_r = m(\ddot{r} - r\dot{\phi}^2) \\ F_\phi = m(r\ddot{\phi} + 2\dot{r}\dot{\phi}) \end{cases}$	$\begin{cases} F_r = m(\ddot{\rho} - \rho\dot{\phi}^2) \\ F_\phi = m(\rho\ddot{\phi} + 2\dot{\rho}\dot{\phi}) \\ F_z = m\ddot{z} \end{cases}$
	Eq. (1.35)	Eq. (1.48)	Problem 1.47 or 1.48

Problems for Chapter 1

The problems for each chapter are arranged according to section number. A problem listed for a given section requires an understanding of that section and earlier sections, but not of later sections. Within each section problems are listed in approximate order of difficulty. A single star (★) indicates straightforward problems involving just one main concept. Two stars (★★) identify problems that are slightly more challenging and usually involve more than one concept. Three stars (★★★) indicate problems that are distinctly more challenging, either because they are intrinsically difficult or involve lengthy calculations. Needless to say, these distinctions are hard to draw and are only approximate.

Problems that need the use of a computer are flagged thus: [Computer]. These are mostly classified as ★★★ on the grounds that it usually takes a long time to set up the necessary code — especially if you're just learning the language.

SECTION 1.2   Space and Time

1.1 ★ Given the two vectors $\mathbf{b} = \hat{\mathbf{x}} + \hat{\mathbf{y}}$ and $\mathbf{c} = \hat{\mathbf{x}} + \hat{\mathbf{z}}$ find $\mathbf{b} + \mathbf{c}$, $5\mathbf{b} + 2\mathbf{c}$, $\mathbf{b} \cdot \mathbf{c}$, and $\mathbf{b} \times \mathbf{c}$.

1.2 ★ Two vectors are given as $\mathbf{b} = (1, 2, 3)$ and $\mathbf{c} = (3, 2, 1)$. (Remember that these statements are just a compact way of giving you the components of the vectors.) Find $\mathbf{b} + \mathbf{c}$, $5\mathbf{b} - 2\mathbf{c}$, $\mathbf{b} \cdot \mathbf{c}$, and $\mathbf{b} \times \mathbf{c}$.

1.3 ★ By applying Pythagoras's theorem (the usual two-dimensional version) twice over, prove that the length r of a three-dimensional vector $\mathbf{r} = (x, y, z)$ satisfies $r^2 = x^2 + y^2 + z^2$.

1.4 ★ One of the many uses of the scalar product is to find the angle between two given vectors. Find the angle between the vectors $\mathbf{b} = (1, 2, 4)$ and $\mathbf{c} = (4, 2, 1)$ by evaluating their scalar product.

1.5 ★ Find the angle between a body diagonal of a cube and any one of its face diagonals. [*Hint:* Choose a cube with side 1 and with one corner at O and the opposite corner at the point $(1, 1, 1)$. Write down the vector that represents a body diagonal and another that represents a face diagonal, and then find the angle between them as in Problem 1.4.]

1.6 ★ By evaluating their dot product, find the values of the scalar s for which the two vectors $\mathbf{b} = \hat{\mathbf{x}} + s\hat{\mathbf{y}}$ and $\mathbf{c} = \hat{\mathbf{x}} - s\hat{\mathbf{y}}$ are orthogonal. (Remember that two vectors are orthogonal if and only if their dot product is zero.) Explain your answers with a sketch.

1.7 ★ Prove that the two definitions of the scalar product $\mathbf{r} \cdot \mathbf{s}$ as $rs\cos\theta$ (1.6) and $\sum r_i s_i$ (1.7) are equal. One way to do this is to choose your x axis along the direction of $\mathbf{r}$. [Strictly speaking you should first make sure that the definition (1.7) is independent of the choice of axes. If you like to worry about such niceties, see Problem 1.16.]

1.8 ★ (a) Use the definition (1.7) to prove that the scalar product is distributive, that is, $\mathbf{r} \cdot (\mathbf{u} + \mathbf{v}) = \mathbf{r} \cdot \mathbf{u} + \mathbf{r} \cdot \mathbf{v}$. **(b)** If $\mathbf{r}$ and $\mathbf{s}$ are vectors that depend on time, prove that the product rule for differentiating products applies to $\mathbf{r} \cdot \mathbf{s}$, that is, that

$$\frac{d}{dt}(\mathbf{r} \cdot \mathbf{s}) = \mathbf{r} \cdot \frac{d\mathbf{s}}{dt} + \frac{d\mathbf{r}}{dt} \cdot \mathbf{s}.$$

1.9 ★ In elementary trigonometry, you probably learned the law of cosines for a triangle of sides a, b, and c, that $c^2 = a^2 + b^2 - 2ab\cos\theta$, where θ is the angle between the sides a and b. Show that the law of cosines is an immediate consequence of the identity $(\mathbf{a} + \mathbf{b})^2 = a^2 + b^2 + 2\mathbf{a} \cdot \mathbf{b}$.

1.10 ★ A particle moves in a circle (center O and radius R) with constant angular velocity ω counterclockwise. The circle lies in the xy plane and the particle is on the x axis at time $t = 0$. Show that the particle's position is given by

$$\mathbf{r}(t) = \hat{\mathbf{x}}R\cos(\omega t) + \hat{\mathbf{y}}R\sin(\omega t).$$

Find the particle's velocity and acceleration. What are the magnitude and direction of the acceleration? Relate your results to well-known properties of uniform circular motion.

1.11 ★ The position of a moving particle is given as a function of time t to be

$$\mathbf{r}(t) = \hat{\mathbf{x}}b\cos(\omega t) + \hat{\mathbf{y}}c\sin(\omega t),$$

where b, c, and ω are constants. Describe the particle's orbit.

1.12 ★ The position of a moving particle is given as a function of time t to be

$$\mathbf{r}(t) = \hat{\mathbf{x}}b\cos(\omega t) + \hat{\mathbf{y}}c\sin(\omega t) + \hat{\mathbf{z}}v_0 t$$

where b, c, v_0 and ω are constants. Describe the particle's orbit.

1.13 ★ Let $\mathbf{u}$ be an arbitrary fixed unit vector and show that any vector $\mathbf{b}$ satisfies

$$b^2 = (\mathbf{u} \cdot \mathbf{b})^2 + (\mathbf{u} \times \mathbf{b})^2.$$

Explain this result in words, with the help of a picture.

1.14 ★ Prove that for any two vectors **a** and **b**,

$$|\mathbf{a} + \mathbf{b}| \leq (a + b).$$

[*Hint:* Work out $|\mathbf{a} + \mathbf{b}|^2$ and compare it with $(a + b)^2$.] Explain why this is called the triangle inequality.

1.15 ★ Show that the definition (1.9) of the cross product is equivalent to the elementary definition that **r** × **s** is perpendicular to both **r** and **s**, with magnitude $rs \sin\theta$ and direction given by the right-hand rule. [*Hint:* It is a fact (though quite hard to prove) that the definition (1.9) is independent of your choice of axes. Therefore you can choose axes so that **r** points along the x axis and **s** lies in the xy plane.]

1.16 ★★ (a) Defining the scalar product **r** · **s** by Equation (1.7), $\mathbf{r} \cdot \mathbf{s} = \sum r_i s_i$, show that Pythagoras's theorem implies that the magnitude of any vector **r** is $r = \sqrt{\mathbf{r} \cdot \mathbf{r}}$. **(b)** It is clear that the length of a vector does not depend on our choice of coordinate axes. Thus the result of part (a) guarantees that the scalar product **r** · **r**, as defined by (1.7), is the same for any choice of orthogonal axes. Use this to prove that **r** · **s**, as defined by (1.7), is the same for any choice of orthogonal axes. [*Hint:* Consider the length of the vector **r** + **s**.]

1.17 ★★ (a) Prove that the vector product **r** × **s** as defined by (1.9) is distributive; that is, that **r** × (**u** + **v**) = (**r** × **u**) + (**r** × **v**). **(b)** Prove the product rule

$$\frac{d}{dt}(\mathbf{r} \times \mathbf{s}) = \mathbf{r} \times \frac{d\mathbf{s}}{dt} + \frac{d\mathbf{r}}{dt} \times \mathbf{s}.$$

Be careful with the order of the factors.

1.18 ★★ The three vectors **a**, **b**, **c** are the three sides of the triangle ABC with angles α, β, γ as shown in Figure 1.15. **(a)** Prove that the area of the triangle is given by any one of these three expressions:

$$\text{area} = \tfrac{1}{2}|\mathbf{a} \times \mathbf{b}| = \tfrac{1}{2}|\mathbf{b} \times \mathbf{c}| = \tfrac{1}{2}|\mathbf{c} \times \mathbf{a}|.$$

(b) Use the equality of these three expressions to prove the so-called law of sines, that

$$\frac{a}{\sin\alpha} = \frac{b}{\sin\beta} = \frac{c}{\sin\gamma}.$$

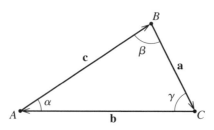

Figure 1.15   Triangle for Problem 1.18.

1.19 ★★ If $\mathbf{r}$, $\mathbf{v}$, $\mathbf{a}$ denote the position, velocity, and acceleration of a particle, prove that

$$\frac{d}{dt}[\mathbf{a} \cdot (\mathbf{v} \times \mathbf{r})] = \dot{\mathbf{a}} \cdot (\mathbf{v} \times \mathbf{r}).$$

1.20 ★★ The three vectors $\mathbf{A}$, $\mathbf{B}$, $\mathbf{C}$ point from the origin O to the three corners of a triangle. Use the result of Problem 1.18 to show that the area of the triangle is given by

$$(\text{area of triangle}) = \tfrac{1}{2}|(\mathbf{B} \times \mathbf{C}) + (\mathbf{C} \times \mathbf{A}) + (\mathbf{A} \times \mathbf{B})|.$$

1.21 ★★ A parallelepiped (a six-faced solid with opposite faces parallel) has one corner at the origin O and the three edges that emanate from O defined by vectors $\mathbf{a}$, $\mathbf{b}$, $\mathbf{c}$. Show that the volume of the parallelepiped is $|\mathbf{a} \cdot (\mathbf{b} \times \mathbf{c})|$.

1.22 ★★ The two vectors $\mathbf{a}$ and $\mathbf{b}$ lie in the xy plane and make angles α and β with the x axis. **(a)** By evaluating $\mathbf{a} \cdot \mathbf{b}$ in two ways [namely using (1.6) and (1.7)] prove the well-known trig identity

$$\cos(\alpha - \beta) = \cos\alpha \cos\beta + \sin\alpha \sin\beta.$$

(b) By similarly evaluating $\mathbf{a} \times \mathbf{b}$ prove that

$$\sin(\alpha - \beta) = \sin\alpha \cos\beta - \cos\alpha \sin\beta.$$

1.23 ★★ The unknown vector $\mathbf{v}$ satisfies $\mathbf{b} \cdot \mathbf{v} = \lambda$ and $\mathbf{b} \times \mathbf{v} = \mathbf{c}$, where λ, $\mathbf{b}$, and $\mathbf{c}$ are fixed and known. Find $\mathbf{v}$ in terms of λ, $\mathbf{b}$, and $\mathbf{c}$.

SECTION 1.4   Newton's First and Second Laws; Inertial Frames

1.24 ★ In case you haven't studied any differential equations before, I shall be introducing the necessary ideas as needed. Here is a simple exercise to get you started: Find the general solution of the first-order equation $df/dt = f$ for an unknown function $f(t)$. [There are several ways to do this. One is to rewrite the equation as $df/f = dt$ and then integrate both sides.] How many arbitrary constants does the general solution contain? [Your answer should illustrate the important general theorem that the solution to any nth-order differential equation (in a very large class of "reasonable" equations) contains n arbitrary constants.]

1.25 ★ Answer the same questions as in Problem 1.24, but for the differential equation $df/dt = -3f$.

1.26 ★★ The hallmark of an inertial reference frame is that any object which is subject to zero net force will travel in a straight line at constant speed. To illustrate this, consider the following: I am standing on a level floor at the origin of an inertial frame $\mathcal{S}$ and kick a frictionless puck due north across the floor. **(a)** Write down the x and y coordinates of the puck as functions of time as seen from my inertial frame. (Use x and y axes pointing east and north respectively.) Now consider two more observers, the first at rest in a frame $\mathcal{S}'$ that travels with constant velocity v due east relative to $\mathcal{S}$, the second at rest in a frame $\mathcal{S}''$ that travels with constant *acceleration* due east relative to $\mathcal{S}$. (All three frames coincide at the moment when I kick the puck, and $\mathcal{S}''$ is at rest relative to $\mathcal{S}$ at that same moment.) **(b)** Find the coordinates x', y' of the puck and describe the puck's path as seen from $\mathcal{S}'$. **(c)** Do the same for $\mathcal{S}''$. Which of the frames is inertial?

1.27 ★★ The hallmark of an inertial reference frame is that any object which is subject to zero net force will travel in a straight line at constant speed. To illustrate this, consider the following experiment: I am standing on the ground (which we shall take to be an inertial frame) beside a perfectly flat horizontal turntable, rotating with constant angular velocity ω. I lean over and shove a frictionless puck so that it slides across the turntable, straight through the center. The puck is subject to zero net force and, as seen from my inertial frame, travels in a straight line. Describe the puck's path as observed by someone sitting at rest on the turntable. This requires careful thought, but you should be able to get a qualitative picture. For a quantitative picture, it helps to use polar coordinates; see Problem 1.46.

SECTION 1.5   The Third Law and Conservation of Momentum

1.28 ⋆ Go over the steps from Equation (1.25) to (1.29) in the proof of conservation of momentum, but treat the case that $N = 3$ and write out all the summations explicitly to be sure you understand the various manipulations.

1.29 ⋆ Do the same tasks as in Problem 1.28 but for the case of four particles ($N = 4$).

1.30 ⋆ Conservation laws, such as conservation of momentum, often give a surprising amount of information about the possible outcome of an experiment. Here is perhaps the simplest example: Two objects of masses m_1 and m_2 are subject to no external forces. Object 1 is traveling with velocity $\mathbf{v}$ when it collides with the stationary object 2. The two objects stick together and move off with common velocity $\mathbf{v}'$. Use conservation of momentum to find $\mathbf{v}'$ in terms of $\mathbf{v}$, m_1, and m_2.

1.31 ⋆ In Section 1.5 we proved that Newton's third law implies the conservation of momentum. Prove the converse, that if the law of conservation of momentum applies to every possible group of particles, then the interparticle forces must obey the third law. [*Hint:* However many particles your system contains, you can focus your attention on just two of them. (Call them 1 and 2.) The law of conservation of momentum says that if there are no external forces on this pair of particles, then their total momentum must be constant. Use this to prove that $\mathbf{F}_{12} = -\mathbf{F}_{21}$.]

1.32 ⋆⋆ If you have some experience in electromagnetism, you could do the following problem concerning the curious situation illustrated in Figure 1.8. The electric and magnetic fields at a point $\mathbf{r}_1$ due to a charge q_2 at $\mathbf{r}_2$ moving with constant velocity $\mathbf{v}_2$ (with $v_2 \ll c$) are[15]

$$\mathbf{E}(\mathbf{r}_1) = \frac{1}{4\pi\epsilon_o} \frac{q_2}{s^2} \hat{\mathbf{s}} \quad \text{and} \quad \mathbf{B}(\mathbf{r}_1) = \frac{\mu_o}{4\pi} \frac{q_2}{s^2} \mathbf{v}_2 \times \hat{\mathbf{s}}$$

where $\mathbf{s} = \mathbf{r}_1 - \mathbf{r}_2$ is the vector pointing from $\mathbf{r}_2$ to $\mathbf{r}_1$. (The first of these you should recognize as Coulomb's law.) If $\mathbf{F}_{12}^{\text{el}}$ and $\mathbf{F}_{12}^{\text{mag}}$ denote the electric and magnetic forces on a charge q_1 at $\mathbf{r}_1$ with velocity $\mathbf{v}_1$, show that $F_{12}^{\text{mag}} \leq (v_1 v_2/c^2) F_{12}^{\text{el}}$. This shows that in the non-relativistic domain it is legitimate to ignore the magnetic force between two moving charges.

1.33 ⋆⋆⋆ If you have some experience in electromagnetism and with vector calculus, prove that the magnetic forces, $\mathbf{F}_{12}$ and $\mathbf{F}_{21}$, between two steady current loops obey Newton's third law. [*Hints:* Let the two currents be I_1 and I_2 and let typical points on the two loops be $\mathbf{r}_1$ and $\mathbf{r}_2$. If $d\mathbf{r}_1$ and $d\mathbf{r}_2$ are short segments of the loops, then according to the Biot–Savart law, the force on $d\mathbf{r}_1$ due to $d\mathbf{r}_2$ is

$$\frac{\mu_o}{4\pi} \frac{I_1 I_2}{s^2} d\mathbf{r}_1 \times (d\mathbf{r}_2 \times \hat{\mathbf{s}})$$

where $\mathbf{s} = \mathbf{r}_1 - \mathbf{r}_2$. The force $\mathbf{F}_{12}$ is found by integrating this around both loops. You will need to use the "$BAC - CAB$" rule to simplify the triple product.]

---

[15] See, for example, David J. Griffiths, *Introduction to Electrodynamics*, 3rd ed., Prentice Hall, (1999), p. 440.

1.34 ★★★ Prove that in the absence of external forces, the total *angular* momentum (defined as $\mathbf{L} = \sum_\alpha \mathbf{r}_\alpha \times \mathbf{p}_\alpha$) of an N-particle system is conserved. [*Hints*: You need to mimic the argument from (1.25) to (1.29). In this case you need more than Newton's third law: In addition you need to assume that the interparticle forces are *central*; that is, $\mathbf{F}_{\alpha\beta}$ acts along the line joining particles α and β. A full discussion of angular momentum is given in Chapter 3.]

SECTION 1.6   Newton's Second Law in Cartesian Coordinates

1.35 ★ A golf ball is hit from ground level with speed v_o in a direction that is due east and at an angle θ above the horizontal. Neglecting air resistance, use Newton's second law (1.35) to find the position as a function of time, using coordinates with x measured east, y north, and z vertically up. Find the time for the golf ball to return to the ground and how far it travels in that time.

1.36 ★ A plane, which is flying horizontally at a constant speed v_o and at a height h above the sea, must drop a bundle of supplies to a castaway on a small raft. **(a)** Write down Newton's second law for the bundle as it falls from the plane, assuming you can neglect air resistance. Solve your equations to give the bundle's position in flight as a function of time t. **(b)** How far before the raft (measured horizontally) must the pilot drop the bundle if it is to hit the raft? What is this distance if $v_o = 50$ m/s, $h = 100$ m, and $g \approx 10$ m/s^2? **(c)** Within what interval of time ($\pm\Delta t$) must the pilot drop the bundle if it is to land within ± 10 m of the raft?

1.37 ★ A student kicks a frictionless puck with initial speed v_o, so that it slides straight up a plane that is inclined at an angle θ above the horizontal. **(a)** Write down Newton's second law for the puck and solve to give its position as a function of time. **(b)** How long will the puck take to return to its starting point?

1.38 ★ You lay a rectangular board on the horizontal floor and then tilt the board about one edge until it slopes at angle θ with the horizontal. Choose your origin at one of the two corners that touch the floor, the x axis pointing along the bottom edge of the board, the y axis pointing up the slope, and the z axis normal to the board. You now kick a frictionless puck that is resting at O so that it slides across the board with initial velocity $(v_{ox}, v_{oy}, 0)$. Write down Newton's second law using the given coordinates and then find how long the puck takes to return to the floor level and how far it is from O when it does so.

1.39 ★★ A ball is thrown with initial speed v_o up an inclined plane. The plane is inclined at an angle ϕ above the horizontal, and the ball's initial velocity is at an angle θ above the plane. Choose axes with x measured up the slope, y normal to the slope, and z across it. Write down Newton's second law using these axes and find the ball's position as a function of time. Show that the ball lands a distance $R = 2v_o^2 \sin\theta \cos(\theta + \phi)/(g\cos^2\phi)$ from its launch point. Show that for given v_o and ϕ, the maximum possible range up the inclined plane is $R_{max} = v_o^2/[g(1 + \sin\phi)]$.

1.40 ★★★ A cannon shoots a ball at an angle θ above the horizontal ground. **(a)** Neglecting air resistance, use Newton's second law to find the ball's position as a function of time. (Use axes with x measured horizontally and y vertically.) **(b)** Let $r(t)$ denote the ball's distance from the cannon. What is the largest possible value of θ if $r(t)$ is to increase throughout the ball's flight? [*Hint:* Using your solution to part (a) you can write down r^2 as $x^2 + y^2$, and then find the condition that r^2 is always increasing.]

SECTION 1.7 Two-Dimensional Polar Coordinates

1.41 ★ An astronaut in gravity-free space is twirling a mass m on the end of a string of length R in a circle, with constant angular velocity ω. Write down Newton's second law (1.48) in polar coordinates and find the tension in the string.

1.42 ★ Prove that the transformations from rectangular to polar coordinates and vice versa are given by the four equations (1.37). Explain why the equation for ϕ is not quite complete and give a complete version.

1.43 ★ (a) Prove that the unit vector $\hat{\mathbf{r}}$ of two-dimensional polar coordinates is equal to

$$\hat{\mathbf{r}} = \hat{\mathbf{x}} \cos \phi + \hat{\mathbf{y}} \sin \phi \qquad (1.59)$$

and find a corresponding expression for $\hat{\boldsymbol{\phi}}$. **(b)** Assuming that ϕ depends on the time t, differentiate your answers in part (a) to give an alternative proof of the results (1.42) and (1.46) for the time derivatives $\hat{\mathbf{r}}$ and $\hat{\boldsymbol{\phi}}$.

1.44 ★ Verify by direct substitution that the function $\phi(t) = A \sin(\omega t) + B \cos(\omega t)$ of (1.56) is a solution of the second-order differential equation (1.55), $\ddot{\phi} = -\omega^2 \phi$. (Since this solution involves two arbitrary constants — the coefficients of the sine and cosine functions — it is in fact the general solution.)

1.45 ★★ Prove that if $\mathbf{v}(t)$ is any vector that depends on time (for example the velocity of a moving particle) but which has *constant magnitude*, then $\dot{\mathbf{v}}(t)$ is orthogonal to $\mathbf{v}(t)$. Prove the converse that if $\dot{\mathbf{v}}(t)$ is orthogonal to $\mathbf{v}(t)$, then $|\mathbf{v}(t)|$ is constant. [*Hint:* Consider the derivative of $\mathbf{v}^2$.] This is a very handy result. It explains why, in two-dimensional polars, $d\hat{\mathbf{r}}/dt$ has to be in the direction of $\hat{\boldsymbol{\phi}}$ and vice versa. It also shows that the speed of a charged particle in a magnetic field is constant, since the acceleration is perpendicular to the velocity.

1.46 ★★ Consider the experiment of Problem 1.27, in which a frictionless puck is slid straight across a rotating turntable through the center O. **(a)** Write down the polar coordinates r, ϕ of the puck as functions of time, as measured in the inertial frame $\mathcal{S}$ of an observer on the ground. (Assume that the puck was launched along the axis $\phi = 0$ at $t = 0$.) **(b)** Now write down the polar coordinates r', ϕ' of the puck as measured by an observer (frame $\mathcal{S}'$) at rest on the turntable. (Choose these coordinates so that ϕ and ϕ' coincide at $t = 0$.) Describe and sketch the path seen by this second observer. Is the frame $\mathcal{S}'$ inertial?

1.47 ★★ Let the position of a point P in three dimensions be given by the vector $\mathbf{r} = (x, y, z)$ in rectangular (or Cartesian) coordinates. The same position can be specified by **cylindrical polar coordinates**, ρ, ϕ, z, which are defined as follows: Let P' denote the projection of P onto the xy plane; that is, P' has Cartesian coordinates $(x, y, 0)$. Then ρ and ϕ are defined as the two-dimensional polar coordinates of P' in the xy plane, while z is the third Cartesian coordinate, unchanged. **(a)** Make a sketch to illustrate the three cylindrical coordinates. Give expressions for ρ, ϕ, z in terms of the Cartesian coordinates x, y, z. Explain in words what ρ is ("ρ is the distance of P from _____"). There are many variants in notation. For instance, some people use r instead of ρ. Explain why this use of r is unfortunate. **(b)** Describe the three unit vectors $\hat{\boldsymbol{\rho}}, \hat{\boldsymbol{\phi}}, \hat{\mathbf{z}}$ and write the expansion of the position vector $\mathbf{r}$ in terms of these unit vectors. **(c)** Differentiate your last answer twice to find the cylindrical components of the acceleration $\mathbf{a} = \ddot{\mathbf{r}}$ of the particle. To do this, you will need to know the time derivatives of $\hat{\boldsymbol{\rho}}$ and $\hat{\boldsymbol{\phi}}$. You could get these from the corresponding two-dimensional results (1.42) and (1.46), or you could derive them directly as in Problem 1.48.

1.48 ★★ Find expressions for the unit vectors $\hat{\rho}$, $\hat{\phi}$, and $\hat{z}$ of cylindrical polar coordinates (Problem 1.47) in terms of the Cartesian $\hat{x}$, $\hat{y}$, $\hat{z}$. Differentiate these expressions with respect to time to find $d\hat{\rho}/dt$, $d\hat{\phi}/dt$, and $d\hat{z}/dt$.

1.49 ★★ Imagine two concentric cylinders, centered on the vertical z axis, with radii $R \pm \epsilon$, where ϵ is very small. A small frictionless puck of thickness 2ϵ is inserted between the two cylinders, so that it can be considered a point mass that can move freely at a fixed distance from the vertical axis. If we use cylindrical polar coordinates (ρ, ϕ, z) for its position (Problem 1.47), then ρ is fixed at $\rho = R$, while ϕ and z can vary at will. Write down and solve Newton's second law for the general motion of the puck, including the effects of gravity. Describe the puck's motion.

1.50 ★★★ [Computer] The differential equation (1.51) for the skateboard of Example 1.2 cannot be solved in terms of elementary functions, but is easily solved numerically. **(a)** If you have access to software, such as Mathematica, Maple, or Matlab, that can solve differential equations numerically, solve the differential equation for the case that the board is released from $\phi_o = 20$ degrees, using the values $R = 5$ m and $g = 9.8$ m/s^2. Make a plot of ϕ against time for two or three periods. **(b)** On the same picture, plot the approximate solution (1.57) with the same $\phi_o = 20°$. Comment on your two graphs. Note: If you haven't used the numerical solver before, you will need to learn the necessary syntax. For example, in Mathematica you will need to learn the syntax for "NDSolve" and how to plot the solution that it provides. This takes a bit of time, but is something that is very well worth learning.

1.51 ★★★ [Computer] Repeat all of Problem 1.50 but using the initial value $\phi_o = \pi/2$.

CHAPTER 2

Projectiles and Charged Particles

In this chapter, I present two topics: the motion of projectiles subject to the forces of gravity and air resistance, and the motion of charged particles in uniform magnetic fields. Both problems lend themselves to solution using Newton's laws in Cartesian coordinates, and both allow us to review and introduce some important mathematics. Above all, both are problems of great practical interest.

2.1 Air Resistance

Most introductory physics courses spend some time studying the motion of projectiles, but they almost always ignore air resistance. In many problems this is an excellent approximation; in others, air resistance is obviously important, and we need to know how to account for it. More generally, whether or not air resistance is significant, we need some way to estimate how important it really is.

Let us begin by surveying some of the basic properties of the resistive force, or **drag**, **f** of the air, or other medium, through which an object is moving. (I shall generally speak of "air resistance" since air is the medium through which most projectiles move, but the same considerations apply to other gases and often to liquids as well.) The most obvious fact about air resistance, well known to anyone who rides a bicycle, is that it depends on the speed, v, of the object concerned. In addition, for many objects, the direction of the force due to motion through the air is opposite to the velocity **v**. For certain objects, such as a nonrotating sphere, this is exactly true, and for many it is a good approximation. You should, however, be aware that there are situations where it is certainly not true: The force of the air on an airplane wing has a large sideways component, called the **lift**, without which no airplanes could fly. Nevertheless, I shall assume that **f** and **v** point in opposite directions; that is, I shall consider only objects for which the sideways force is zero, or at least small enough

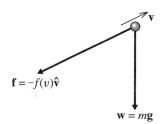

Figure 2.1   A projectile is subject to two forces, the force
of gravity, $\mathbf{w} = m\mathbf{g}$, and the drag force of air resistance,
$\mathbf{f} = -f(v)\hat{\mathbf{v}}$.

to be neglected. The situation is illustrated in Figure 2.1 and is summed up in the
equation

$$\mathbf{f} = -f(v)\hat{\mathbf{v}}, \tag{2.1}$$

where $\hat{\mathbf{v}} = \mathbf{v}/|\mathbf{v}|$ denotes the unit vector in the direction of $\mathbf{v}$, and $f(v)$ is the magnitude
of $\mathbf{f}$.

The function $f(v)$ that gives the magnitude of the air resistance varies with v in
a complicated way, especially as the object's speed approaches the speed of sound.
However, at lower speeds it is often a good approximation to write[1]

$$f(v) = bv + cv^2 = f_{\text{lin}} + f_{\text{quad}} \tag{2.2}$$

where f_{lin} and f_{quad} stand for the linear and quadratic terms respectively,

$$f_{\text{lin}} = bv \quad \text{and} \quad f_{\text{quad}} = cv^2. \tag{2.3}$$

The physical origins of these two terms are quite different: The linear term, f_{lin}, arises
from the viscous drag of the medium and is generally proportional to the viscosity of
the medium and the linear size of the projectile (Problem 2.2). The quadratic term,
f_{quad}, arises from the projectile's having to accelerate the mass of air with which it is
continually colliding; f_{quad} is proportional to the density of the medium and the cross-
sectional area of the projectile (Problem 2.4). In particular, for a spherical projectile
(a cannonball, a baseball, or a drop of rain), the coefficients b and c in (2.2) have the
form

$$b = \beta D \quad \text{and} \quad c = \gamma D^2 \tag{2.4}$$

where D denotes the diameter of the sphere and the coefficients β and γ depend
on the nature of the medium. For a spherical projectile in air at STP, they have the
approximate values

$$\beta = 1.6 \times 10^{-4} \text{ N·s/m}^2 \tag{2.5}$$

---

[1] Mathematically, Equation (2.2) is, in a sense, obvious. Any reasonable function is expected to
have a Taylor series expansion, $f = a + bv + cv^2 + \cdots$. For low enough v, the first three terms
should give a good approximation, and, since $f = 0$ when $v = 0$ the constant term, a, has to be zero.

and

$$\gamma = 0.25 \text{ N·s}^2/\text{m}^4. \tag{2.6}$$

(For calculation of these two constants, see Problems 2.2 and 2.4.) You need to remember that these values are valid only for a sphere moving through air at STP. Nevertheless, they give at least a rough idea of the importance of the drag force even for nonspherical bodies moving through different gases at any normal temperatures and pressures.

It often happens that we can neglect one of the terms in (2.2) compared to the other, and this simplifies the task of solving Newton's second law. To decide whether this does happen in a given problem, and which term to neglect, we need to compare the sizes of the two terms:

$$\frac{f_{\text{quad}}}{f_{\text{lin}}} = \frac{cv^2}{bv} = \frac{\gamma D}{\beta}v = \left(1.6 \times 10^3 \, \frac{\text{s}}{\text{m}^2}\right) Dv \tag{2.7}$$

if we use the values (2.5) and (2.6) for a sphere in air. In a given problem, we have only to substitute the values of D and v into this equation to find out if one of the terms can be neglected, as the following example illustrates.

EXAMPLE 2.1   A Baseball and Some Drops of Liquid

Assess the relative importance of the linear and quadratic drags on a baseball of diameter $D = 7$ cm, traveling at a modest $v = 5$ m/s. Do the same for a drop of rain ($D = 1$ mm and $v = 0.6$ m/s) and for a tiny droplet of oil used in the Millikan oildrop experiment ($D = 1.5 \, \mu$m and $v = 5 \times 10^{-5}$ m/s).

When we substitute the numbers for the baseball into (2.7) (remembering to convert the diameter to meters), we get

$$\frac{f_{\text{quad}}}{f_{\text{lin}}} \approx 600 \qquad [\text{baseball}]. \tag{2.8}$$

For this baseball, the linear term is clearly negligible and we need consider only the quadratic drag. If the ball is traveling faster, the ratio $f_{\text{quad}}/f_{\text{lin}}$ is even greater. At slower speeds the ratio is less dramatic, but even at 1 m/s the ratio is 100. In fact if v is small enough that the linear term is comparable to the quadratic, both terms are so small as to be negligible. Thus, for baseballs and similar objects, it is almost always safe to neglect f_{lin} and take the drag force to be

$$\mathbf{f} = -cv^2\hat{\mathbf{v}}. \tag{2.9}$$

For the raindrop, the numbers give

$$\frac{f_{\text{quad}}}{f_{\text{lin}}} \approx 1 \qquad [\text{raindrop}]. \tag{2.10}$$

Thus for this raindrop the two terms are comparable and neither can be neglected — which makes solving for the motion more difficult. If the drop were a lot larger or were traveling much faster, then the linear term would be negligible; and if the drop were much smaller or were traveling much slower, then the quadratic term would be negligible. But in general, with raindrops and similar objects, we are going to have to take both f_{lin} and f_{quad} into account.

For the oildrop in the Millikan experiment the numbers give

$$\frac{f_{quad}}{f_{lin}} \approx 10^{-7} \qquad \text{[Millikan oildrop].} \qquad (2.11)$$

In this case, the quadratic term is totally negligible, and we can take

$$\mathbf{f} = -bv\hat{\mathbf{v}} = -b\mathbf{v}, \qquad (2.12)$$

where the second, very compact form follows because, of course, $v\hat{\mathbf{v}} = \mathbf{v}$.

The moral of this example is clear: First, there are objects for which the drag force is dominantly linear, and the quadratic force can be neglected — notably, very small liquid drops in air, but also slightly larger objects in a very viscous fluid, such as a ball bearing moving through molasses. On the other hand, for most projectiles, such as golf balls, cannonballs, and even a human in free fall, the dominant drag force is quadratic, and we can neglect the linear term. This situation is a little unlucky because the linear problem is much easier to solve than the quadratic. In the following two sections, I shall discuss the linear case, precisely because it is the easier one. Nevertheless, it *does* have practical applications, and the mathematics used to solve it is widely used in many fields. In Section 2.4, I shall take up the harder but more usual case of quadratic drag.

To conclude this introductory section, I should mention the Reynolds number, an important parameter that features prominently in more advanced treatments of motion in fluids. As already mentioned, the linear drag f_{lin} can be related to the viscosity of the fluid through which our projectile is moving, and the quadratic term f_{quad} is similarly related to the inertia (and hence density) of the fluid. Thus one can relate the ratio f_{quad}/f_{lin} to the fundamental parameters η, the viscosity, and ϱ, the density, of the fluid (see Problem 2.3). The result is that the ratio f_{quad}/f_{lin} is of roughly the same order of magnitude as the dimensionless number $R = Dv\varrho/\eta$, called the **Reynolds number**. Thus a compact and general way to summarize the foregoing discussion is to say that the quadratic drag f_{quad} is dominant when the Reynolds number R is large, whereas the linear drag dominates when R is small.

2.2   Linear Air Resistance

Let us consider first a projectile for which the quadratic drag force is negligible, so that the force of air resistance is given by (2.12). We shall see directly that, because the drag force is linear in $\mathbf{v}$, the equations of motion are very simple to solve. The two forces on the projectile are the weight $\mathbf{w} = m\mathbf{g}$ and the drag force $\mathbf{f} = -b\mathbf{v}$, as shown in Figure 2.2. Thus the second law, $m\ddot{\mathbf{r}} = \mathbf{F}$, reads

$$m\ddot{\mathbf{r}} = m\mathbf{g} - b\mathbf{v}. \qquad (2.13)$$

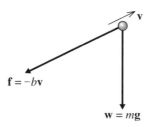

Figure 2.2   The two forces on a projectile for which the force of air resistance is linear in the velocity, $\mathbf{f} = -b\mathbf{v}$.

An interesting feature of this form is that, because neither of the forces depends on $\mathbf{r}$, the equation of motion does not involve $\mathbf{r}$ itself (only the first and second derivatives of $\mathbf{r}$). In fact, we can rewrite $\ddot{\mathbf{r}}$ as $\dot{\mathbf{v}}$, and (2.13) becomes

$$m\dot{\mathbf{v}} = m\mathbf{g} - b\mathbf{v}, \tag{2.14}$$

a first-order differential equation for $\mathbf{v}$. This simplification comes about because the forces depend only on $\mathbf{v}$ and not $\mathbf{r}$. It means we have to solve only a first-order differential equation for $\mathbf{v}$ and then integrate $\mathbf{v}$ to find $\mathbf{r}$.

Perhaps the most important simplifying feature of linear drag is that the equation of motion separates into components especially easily. For instance, with x measured to the right and y vertically downward, (2.14) resolves into

$$m\dot{v}_x = -bv_x \tag{2.15}$$

and

$$m\dot{v}_y = mg - bv_y. \tag{2.16}$$

That is, we have two separate equations, one for v_x and one for v_y; the equation for v_x does not involve v_y and vice versa. It is important to recognize that this happened only because the drag force was linear in $\mathbf{v}$. For instance, if the drag force were quadratic,

$$\mathbf{f} = -cv^2\hat{\mathbf{v}} = -cv\mathbf{v} = -c\sqrt{v_x^2 + v_y^2}\,\mathbf{v}, \tag{2.17}$$

then in (2.14) we would have to replace the term $-b\mathbf{v}$ with (2.17). In place of the two equations (2.15) and (2.16), we would have

$$\left.\begin{array}{l} m\dot{v}_x = -c\sqrt{v_x^2 + v_y^2}\,v_x \\[2mm] m\dot{v}_y = mg - c\sqrt{v_x^2 + v_y^2}\,v_y. \end{array}\right\} \tag{2.18}$$

Here, each equation involves *both* of the variables v_x and v_y. These two *coupled* differential equations are much harder to solve than the uncoupled equations of the linear case.

Because they are uncoupled, we can solve each equation for linear drag separately and then put the two solutions together. Further, each equation defines a problem that is interesting in its own right. Equation (2.15) is the equation of motion for an object

Figure 2.3   A cart moves on a horizontal frictionless track in a medium that produces a linear drag force.

(a cart with frictionless wheels, for instance) coasting horizontally in a medium that causes linear drag. Equation (2.16) describes an object (a tiny oil droplet for instance) that is falling vertically with linear air resistance. I shall solve these two separate problems in turn.

Horizontal Motion with Linear Drag

Consider an object such as the cart in Figure 2.3 coasting horizontally in a linearly resistive medium. I shall assume that at $t = 0$ the cart is at $x = 0$ with velocity $v_x = v_{xo}$. The only force on the cart is the drag $\mathbf{f} = -b\mathbf{v}$, thus the cart inevitably slows down. The rate of slowing is determined by (2.15), which has the general form

$$\dot{v}_x = -kv_x, \tag{2.19}$$

where k is my temporary abbreviation for $k = b/m$. This is a first-order differential equation for v_x, whose general solution must contain exactly one arbitrary constant. The equation states that the derivative of v_x is equal to $-k$ times v_x itself, and the only function with this property is the exponential function

$$v_x(t) = Ae^{-kt} \tag{2.20}$$

which satisfies (2.19) for any value of the constant A (Problems 1.24 and 1.25). Since this solution contains one arbitrary constant, it is the *general* solution of our first-order equation; that is, *any* solution must have this form. In our case, we know that $v_x(0) = v_{xo}$, so that $A = v_{xo}$, and we conclude that

$$v_x(t) = v_{xo}e^{-kt} = v_{xo}e^{-t/\tau}, \tag{2.21}$$

where I have introduced the convenient parameter

$$\tau = 1/k = m/b \qquad \text{[for linear drag]}. \tag{2.22}$$

We see that our cart slows down exponentially, as shown in Figure 2.4(a). The parameter τ has the dimensions of time (as you should check), and you can see from (2.21) that when $t = \tau$, the velocity is $1/e$ of its initial value; that is, τ is the "$1/e$" time for the exponentially decreasing velocity. As $t \to \infty$, the velocity approaches zero.

   To find the position as a function of time, we have only to integrate the velocity (2.21). Integrations of this kind can be done using the definite or indefinite integral. The definite integral has the advantage that it automatically takes care of the constant

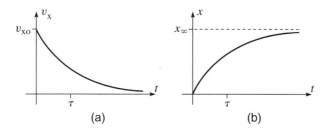

Figure 2.4   **(a)** The velocity v_x as a function of time, t, for a cart moving horizontally with a linear resistive force. As $t \to \infty$, v_x approaches zero exponentially. **(b)** The position x as a function of t for the same cart. As $t \to \infty$, $x \to x_\infty = v_{xo}\tau$.

of integration: Since $v_x = dx/dt$,

$$\int_0^t v_x(t')\,dt' = x(t) - x(0).$$

(Notice that I have named the "dummy" variable of integration t' to avoid confusion with the upper limit t.) Therefore

$$x(t) = x(0) + \int_0^t v_{xo}e^{-t'/\tau}\,dt'$$

$$= 0 + \left[-v_{xo}\tau e^{-t'/\tau}\right]_0^t$$

$$= x_\infty \left(1 - e^{-t/\tau}\right). \tag{2.23}$$

In the second line, I have used our assumption that $x = 0$ when $t = 0$. And in the last, I have introduced the parameter

$$x_\infty = v_{xo}\tau, \tag{2.24}$$

which is the limit of $x(t)$ as $t \to \infty$. We conclude that, as the cart slows down, its position approaches x_∞ asymptotically, as shown in Figure 2.4(b).

Vertical Motion with Linear Drag

Let us next consider a projectile that is subject to linear air resistance and is thrown vertically downward. The two forces on the projectile are gravity and air resistance, as shown in Figure 2.5. If we measure y vertically down, the only interesting component of the equation of motion is the y component, which reads

$$m\dot{v}_y = mg - bv_y. \tag{2.25}$$

With the velocity downward ($v_y > 0$), the retarding force is upward, while the force of gravity is downward. If v_y is small, the force of gravity is more important than the drag force, and the falling object accelerates in its downward motion. This will

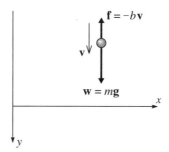

Figure 2.5   The forces on a projectile that is thrown vertically down, subject to linear air resistance.

continue until the drag force balances the weight. The speed at which this balance
occurs is easily found by setting (2.25) equal to zero, to give $v_y = mg/b$ or

$$v_y = v_{ter}$$

where I have defined the **terminal speed**

$$v_{ter} = \frac{mg}{b} \qquad \text{[for linear drag].} \qquad (2.26)$$

The terminal speed is the speed at which our projectile will eventually fall, if given
the time to do so. Since it depends on m and b, it is different for different bodies. For
example, if two objects have the same shape and size (b the same for both), the heavier
object (m larger) will have the higher terminal speed, just as you would expect. Since
v_{ter} is inversely proportional to the coefficient b of air resistance, we can view v_{ter} as
an inverse measure of the importance of air resistance — the larger the air resistance,
the smaller v_{ter}, again just as you would expect.

EXAMPLE 2.2   Terminal Speed of Small Liquid Drops

Find the terminal speed of a tiny oildrop in the Millikan oildrop experiment
(diameter $D = 1.5\,\mu$m and density $\varrho = 840\,\text{kg/m}^3$). Do the same for a small
drop of mist with diameter $D = 0.2$ mm.

From Example 2.1 we know that the linear drag is dominant for these objects,
so the terminal speed is given by (2.26). According to (2.4), $b = \beta D$ where
$\beta = 1.6 \times 10^{-4}$ (in SI units). The mass of the drop is $m = \varrho\,\pi\,D^3/6$. Thus (2.26)
becomes

$$v_{ter} = \frac{\varrho\,\pi\,D^2 g}{6\,\beta} \qquad \text{[for linear drag].} \qquad (2.27)$$

This interesting result shows that, for a given density, the terminal speed is proportional to D^2. This implies that, once air resistance has become important, a large sphere will fall faster than a small sphere of the same density.[2]
    Putting in the numbers, we find for the oildrop

$$v_{ter} = \frac{(840) \times \pi \times (1.5 \times 10^{-6})^2 \times (9.8)}{6 \times (1.6 \times 10^{-4})} = 6.1 \times 10^{-5}\,\text{m/s} \qquad \text{[oildrop]}.$$

In the Millikan oildrop experiment, the oildrops fall exceedingly slowly, so their speed can be measured by simply watching them through a microscope.
    Putting in the numbers for the drop of mist, we find similarly that

$$v_{ter} = 1.3\,\text{m/s} \qquad \text{[drop of mist]}. \qquad (2.28)$$

This speed is representative for a fine drizzle. For a larger raindrop, the terminal speed would be appreciably larger, but with a larger (and hence also faster) drop, the quadratic drag would need to be included in the calculation to get a reliable value for v_{ter}.

So far, we have discussed the terminal speed of a projectile (moving vertically), but we must now discuss how the projectile approaches that speed. This is determined by the equation of motion (2.25) which we can rewrite as

$$m\dot{v}_y = -b(v_y - v_{ter}). \qquad (2.29)$$

(Remember that $v_{ter} = mg/b$.) This differential equation can be solved in several ways. (For one alternative see Problem 2.9.) Perhaps the simplest is to note that it is almost the same as Equation (2.15) for the horizontal motion, except that on the right we now have $(v_y - v_{ter})$ instead of v_x. The solution for the horizontal case was the exponential function (2.20). The trick to solving our new vertical equation (2.29) is to introduce the new variable $u = (v_y - v_{ter})$, which satisfies $m\dot{u} = -bu$ (because v_{ter} is constant). Since this is *exactly* the same as Equation (2.15) for the horizontal motion, the solution for u is the same exponential, $u = Ae^{-t/\tau}$. [Remember that the constant k in (2.20) became $k = 1/\tau$.] Therefore,

$$v_y - v_{ter} = Ae^{-t/\tau}.$$

When $t = 0$, $v_y = v_{yo}$, so $A = v_{yo} - v_{ter}$ and our final solution for v_y as a function of t is

$$v_y(t) = v_{ter} + (v_{yo} - v_{ter})e^{-t/\tau} \qquad (2.30)$$

$$= v_{yo}e^{-t/\tau} + v_{ter}\left(1 - e^{-t/\tau}\right). \qquad (2.31)$$

---

[2] We are here assuming that the drag force is linear, but the same qualitative conclusion follows for a quadratic drag force. (Problem 2.24.)

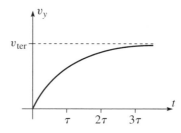

Figure 2.6   When an object is dropped in a medium with linear resistance, v_y approaches its terminal value v_ter as shown.

This second expression gives $v_y(t)$ as the sum of two terms: The first is equal to v_{yo} when $t = 0$, but fades away to zero as t increases; the second is equal to zero when $t = 0$, but approaches v_ter as $t \to \infty$. In particular, as $t \to \infty$,

$$v_y(t) \to v_\text{ter} \tag{2.32}$$

just as we anticipated.

Let us examine the result (2.31) in a little more detail for the case that $v_{yo} = 0$; that is, the projectile is dropped from rest. In this case (2.31) reads

$$v_y(t) = v_\text{ter}\left(1 - e^{-t/\tau}\right). \tag{2.33}$$

This result is plotted in Figure 2.6, where we see that v_y starts out from 0 and approaches the terminal speed, $v_y \to v_\text{ter}$, asymptotically as $t \to \infty$. The significance of the time τ for a falling body is easily read off from (2.33). When $t = \tau$, we see that

$$v_y = v_\text{ter}(1 - e^{-1}) = 0.63 v_\text{ter}.$$

That is, in a time τ, the object reaches 63% of the terminal speed. Similar calculations give the following results:

time t	percent of v_ter
0	0
τ	63%
2τ	86%
3τ	95%

Of course, the object's speed never actually reaches v_ter, but τ is a good measure of how fast the speed approaches v_ter. In particular, when $t = 3\tau$ the speed is 95% of v_ter, and for many purposes we can say that after a time 3τ the speed is essentially equal to v_ter.

EXAMPLE 2.3   Characteristic Time for Two Liquid Drops

Find the characteristic times, τ, for the oildrop and drop of mist in Example 2.2.

The characteristic time τ was defined in (2.22) as $\tau = m/b$, and v_{ter} was defined in (2.26) as $v_{\text{ter}} = mg/b$. Thus we have the useful relation

$$v_{\text{ter}} = g\tau. \tag{2.34}$$

Notice that this relation lets us interpret v_{ter} as the speed a falling object *would* acquire in a time τ, *if it had a constant acceleration equal to g*. Also note that, like v_{ter}, the time τ is an inverse indicator of the importance of air resistance: When the coefficient b of air resistance is small, both v_{ter} and τ are large; when b is large, both v_{ter} and τ are small.

For our present purposes, the importance of (2.34) is that, since we have already found the terminal velocities of the two drops, we can immediately find the values of τ. For the Millikan oildrop, we found that $v_{\text{ter}} = 6.1 \times 10^{-5}$ m/s, therefore

$$\tau = \frac{v_{\text{ter}}}{g} = \frac{6.1 \times 10^{-5}}{9.8} = 6.2 \times 10^{-6}\,\text{s} \qquad \text{[oildrop]}.$$

After falling for just 20 microseconds, this oildrop will have acquired 95% of its terminal speed. For almost every purpose, the oildrop *always* travels at its terminal speed.

For the drop of mist of Example 2.2, the terminal speed was $v_{\text{ter}} = 1.3$ m/s and so $\tau = v_{\text{ter}}/g \approx 0.13$ s. After about 0.4 s, the drop will have acquired 95% of its terminal speed.

Whether or not our falling object starts from rest, we can find its position y as a function of time by integrating the known form (2.30) of v_y,

$$v_y(t) = v_{\text{ter}} + (v_{yo} - v_{\text{ter}})e^{-t/\tau}.$$

Assuming that the projectile's initial position is $y = 0$, it immediately follows that

$$y(t) = \int_0^t v_y(t')\,dt'$$

$$= v_{\text{ter}}t + (v_{yo} - v_{\text{ter}})\tau\left(1 - e^{-t/\tau}\right). \tag{2.35}$$

This equation for $y(t)$ can now be combined with Equation (2.23) for $x(t)$ to give us the orbit of any projectile, moving both horizontally and vertically, in a linear medium.

2.3   Trajectory and Range in a Linear Medium

We saw at the begining of the last section that the equation of motion for a projectile moving in any direction resolves into two separate equations, one for the horizontal and one for the vertical motion [Equations (2.15) and (2.16)]. We have solved each of these separate equations in (2.23) and (2.35), and we can now put these solutions together to give the trajectory of an arbitrary projectile moving in any direction. In this discussion it is marginally more convenient to measure y vertically *upward*, in which case we must reverse the sign of v_{ter}. (Make sure you understand this point.) Thus the two equations of the orbit become

$$\left. \begin{aligned} x(t) &= v_{xo}\tau \left(1 - e^{-t/\tau}\right) \\ y(t) &= (v_{yo} + v_{ter})\tau \left(1 - e^{-t/\tau}\right) - v_{ter}t. \end{aligned} \right\} \tag{2.36}$$

You can eliminate t from these two equations by solving the first for t and then substituting into the second. (See Problem 2.17.) The result is the equation for the trajectory:

$$y = \frac{v_{yo} + v_{ter}}{v_{xo}}x + v_{ter}\tau \ln\left(1 - \frac{x}{v_{xo}\tau}\right). \tag{2.37}$$

This equation is probably too complicated to be especially illuminating, but I have plotted it as the solid curve in Figure 2.7, with the help of which you can understand some of the features of (2.37). For example, if you look at the second term on the right of (2.37), you will see that as $x \to v_{xo}\tau$ the argument of the log function approaches zero; therefore, the log term and hence y both approach $-\infty$. That is, the trajectory has a vertical asymptote at $x = v_{xo}\tau$, as you can see in the picture. I leave it as an exercise (Problem 2.19) for you to check that if air resistance is switched off (v_{ter} and τ both approach infinity), the trajectory defined by (2.37) does indeed approach the dashed trajectory corresponding to zero air resistance.

Horizontal Range

A standard (and quite interesting) problem in elementary physics courses is to show that the horizontal range R of a projectile (subject to no air resistance of course) is

$$R_{vac} = \frac{2v_{xo}v_{yo}}{g} \qquad \text{[no air resistance]} \tag{2.38}$$

where R_{vac} stands for the range in a vacuum. Let us see how this result is modified by air resistance.

The range R is the value of x when y as given by (2.37) is zero. Thus R is the solution of the equation

$$\frac{v_{yo} + v_{ter}}{v_{xo}}R + v_{ter}\tau \ln\left(1 - \frac{R}{v_{xo}\tau}\right) = 0. \tag{2.39}$$

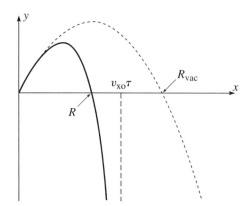

Figure 2.7   The trajectory of a projectile subject to a linear drag force (solid curve) and the corresponding trajectory in a vacuum (dashed curve). At first the two curves are very similar, but as t increases, air resistance slows the projectile and pulls its trajectory down, with a vertical asymptote at $x = v_{xo}\tau$. The horizontal range of the projectile is labeled R, and the corresponding range in vacuum R_{vac}.

This is a transcendental equation and cannot be solved analytically, that is, in terms of well known, elementary functions such as logs, or sines and cosines. For a given choice of parameters, it can be solved numerically with a computer (Problem 2.22), but this approach usually gives one little sense of how the solution depends on the parameters. Often a good alternative is to find some approximation that allows an *approximate* analytic solution. (Before the advent of computers, this was often the only way to find out what happens.) In the present case, it is often clear that the effects of air resistance should be *small*. This means that both v_{ter} and τ are large and the second term in the argument of the log function is small (since it has τ in its denominator). This suggests that we expand the log in a Taylor series (see Problem 2.18):

$$\ln(1 - \epsilon) = -\left(\epsilon + \tfrac{1}{2}\epsilon^2 + \tfrac{1}{3}\epsilon^3 + \cdots\right). \tag{2.40}$$

We can use this expansion for the log term in (2.39), and, provided τ is large enough, we can surely neglect the terms beyond ϵ^3. This gives the equation

$$\left[\frac{v_{yo} + v_{\text{ter}}}{v_{xo}}\right] R - v_{\text{ter}}\tau \left[\frac{R}{v_{xo}\tau} + \frac{1}{2}\left(\frac{R}{v_{xo}\tau}\right)^2 + \frac{1}{3}\left(\frac{R}{v_{xo}\tau}\right)^3\right] = 0. \tag{2.41}$$

This equation can be quickly tidied up. First, the second term in the first bracket cancels the first term in the second. Next, every term contains a factor of R. This implies that one solution is $R = 0$, which is correct — the height y is zero when $x = 0$. Nevertheless, this is not the solution we are interested in, and we can divide out the common factor of R. A little rearrangement (and replacement of v_{ter}/τ by g) lets us rewite the equation as

$$R = \frac{2v_{xo}v_{yo}}{g} - \frac{2}{3v_{xo}\tau}R^2. \tag{2.42}$$

This may seem a perverse way to write a quadratic equation for R, but it leads us quickly to the desired approximate solution. The point is that the second term on the right is very small. (In the numerator R is certainly no more than R_{vac} and we are assuming that τ in the denominator is very large.) Therefore, as a first approximation we get

$$R \approx \frac{2v_{xo}v_{yo}}{g} = R_{\text{vac}}. \tag{2.43}$$

This is just what we expected: For low air resistance, the range is close to R_{vac}. But with the help of (2.42) we can now get a second, better approximation. The last term of (2.42) is the required correction to R_{vac}; because it is already small, we would certainly be satisfied with an approximate value for this correction. Thus, in evaluating the last term of (2.42), we can replace R with the approximate value $R \approx R_{\text{vac}}$, and we find as our second approximation [remember that the first term in (2.42) is just R_{vac}]

$$R \approx R_{\text{vac}} - \frac{2}{3v_{xo}\tau}(R_{\text{vac}})^2$$

$$= R_{\text{vac}}\left(1 - \frac{4}{3}\frac{v_{yo}}{v_{\text{ter}}}\right). \tag{2.44}$$

(To get the second line, I replaced the second R_{vac} in the previous line by $2v_{xo}v_{yo}/g$ and τg by v_{ter}.) Notice that the correction for air resistance always makes R smaller than R_{vac}, as one would expect. Notice also that the correction depends only on the ratio v_{yo}/v_{ter}. More generally, it is easy to see (Problem 2.32) that the importance of air resistance is indicated by the ratio v/v_{ter} of the projectile's speed to the terminal speed. If $v/v_{\text{ter}} \ll 1$ throughout the flight, the effect of air resistance is very small; if v/v_{ter} is around 1 or more, air resistance is almost certainly important [and the approximation (2.44) is certainly no good].

EXAMPLE 2.4   Range of Small Metal Pellets

I flick a tiny metal pellet with diameter $D = 0.2$ mm and $\mathbf{v} = 1$ m/s at $45°$. Find its horizontal range assuming the pellet is gold (density $\varrho \approx 16\,\text{g/cm}^3$). What if it is aluminum (density $\varrho \approx 2.7\,\text{g/cm}^3$)?

In the absence of air resistance, both pellets would have the same range,

$$R_{\text{vac}} = \frac{2v_{xo}v_{yo}}{g} = 10.2\,\text{cm}.$$

For gold, Equation (2.27) gives (as you can check) $v_\text{ter} \approx 21$ m/s. Thus the correction term in (2.44) is

$$\frac{4}{3}\frac{v_\text{yo}}{v_\text{ter}} = \frac{4}{3} \times \frac{0.71}{21} \approx 0.05.$$

That is, air resistance reduces the range by 5% to about 9.7 cm. The density of aluminum is about 1/6 times that of gold. Therefore the terminal speed is one sixth as big, and the correction for aluminum is 6 times greater or about 30%, giving a range of about 7 cm. For the gold pellet the correction for air resistance is quite small and could perhaps be neglected; for the aluminum pellet, the correction is still small, but is certainly not negligible.

2.4   Quadratic Air Resistance

In the last two sections we have developed a rather complete theory of projectiles subject to a linear drag force, $\mathbf{f} = -b\mathbf{v}$. While we *can* find examples of projectiles for which the drag is linear (notably very small objects, such as the Millikan oildrop), for most of the more obvious examples of projectiles (baseballs, footballs, cannonballs, and the like) it is a far better approximation to say that the drag is pure quadratic, $\mathbf{f} = -cv^2\hat{\mathbf{v}}$. We must, therefore, develop a corresponding theory for a quadratic drag force. On the face of it, the two theories are not so very different. In either case we have to solve the differential equation

$$m\dot{\mathbf{v}} = m\mathbf{g} + \mathbf{f}, \tag{2.45}$$

and in both cases this is a first-order differential equation for the velocity $\mathbf{v}$, with $\mathbf{f}$ depending in a relatively simple way on $\mathbf{v}$. There is, however, an important difference. In the linear case ($\mathbf{f} = -b\mathbf{v}$), Equation (2.45) is a *linear* differential equation, inasmuch as the terms that involve $\mathbf{v}$ are all linear in $\mathbf{v}$ or its derivatives. In the quadratic case, Equation (2.45) is, of course, nonlinear. And it turns out that the mathematical theory of nonlinear differential equations is significantly more complicated than the linear theory. As a practical matter, we shall find that for the case of a general projectile, moving in both the x and y directions, Equation (2.45) cannot be solved in terms of elementary functions when the drag is quadratic. More generally, we shall see in Chapter 12 that for more complicated systems, nonlinearity can lead to the astonishing phenomenon of chaos, although this does not happen in the present case.

   In this section, I shall start with the same two special cases discussed in Section 2.2, a body that is constrained to move horizontally, such as a railroad car on a horizontal track, and a body that moves vertically, such as a stone dropped from a window (both now with quadratic drag forces). We shall find that in these two especially simple cases the differential equation (2.45) *can* be solved by elementary means, and the solutions introduce some important techniques and interesting results. I shall then discuss briefly the general case (motion in both the horizontal and vertical directions), which can be solved only numerically.

Horizontal Motion with Quadratic Drag

Let us consider a body moving horizontally (in the positive x direction), subject to a quadratic drag and no other forces. For example, you could imagine a cycle racer, who has crossed the finishing line and is coasting to a stop under the influence of air resistance. To the extent that the cycle is well lubricated and tires well inflated, we can ignore ordinary friction,[3] and, except at very low speeds, air resistance is purely quadratic. The x component of the equation of motion is therefore (I'll abbreviate v_x to v)

$$m\frac{dv}{dt} = -cv^2. \tag{2.46}$$

If we divide by v^2 and multiply by dt, we get an equation in which only the variable v appears on the left and only t on the right:[4]

$$m\frac{dv}{v^2} = -c\,dt. \tag{2.47}$$

This trick — of rearranging a differential equation so that only one variable appears on the left and only the other on the right — is called **separation of variables**. When it is possible, separation of variables is often the simplest way to solve a first-order differential equation, since the solution can be found by simple integration of both sides.

Integrating Equation (2.47) we find

$$m\int_{v_0}^{v}\frac{dv'}{v'^2} = -c\int_0^t dt'$$

where v_0 is the initial velocity at $t = 0$. Notice that I have written both sides as definite integrals, with the appropriate limits, so that I shan't have to worry about any constants of integration. I have also renamed the variables of integration as v' and t' to avoid

---

[3] As I shall discuss shortly, when the cyclist slows down to a stop, air resistance becomes smaller, and eventually friction becomes the dominant force. Nevertheless, at speeds around 10 mph or more, it is a fair approximation to ignore everything but the quadratic air resistance.

[4] In passing from (2.46) to (2.47), I have treated the derivative dv/dt as if it were the quotient of two separate numbers, dv and dt. As you are certainly aware this cavalier proceeding is not strictly correct. Nevertheless, it can be justified in two ways. First, in the theory of *differentials*, it is in fact true that dv and dt are defined as separate numbers (differentials), such that their quotient is the derivative dv/dt. Fortunately, it is quite unnecessary to know about this theory. As physicists we know that dv/dt is the limit of $\Delta v/\Delta t$, as both Δv and Δt become small, and I shall take the view that dv is just shorthand for Δv (and likewise dt for Δt), *with the understanding that it has been taken small enough that the quotient dv/dt is within my desired accuracy of the true derivative.* With this understanding, (2.47), with dv on one side and dt on the other, makes perfectly good sense.

confusion with the upper limits v and t. Both of these integrals are easily evaluated, and we find

$$m \left(\frac{1}{v_0} - \frac{1}{v} \right) = -ct \tag{2.48}$$

or, solving for v,

$$v(t) = \frac{v_0}{1 + cv_0 t/m} = \frac{v_0}{1 + t/\tau} \tag{2.49}$$

where I have introduced the abbreviation τ for the combination of constants

$$\tau = \frac{m}{cv_0} \qquad \text{[for quadratic drag].} \tag{2.50}$$

As you can easily check, τ is a time, with the significance that when $t = \tau$ the velocity is $v = v_0/2$. Notice that this parameter τ is different from the τ introduced in (2.22) for motion subject to linear air resistance; nevertheless, both parameters have the same general significance as indicators of the time for air resistance to slow the motion appreciably.

To find the bicycle's position x, we have only to integrate v to give (as you should check)

$$x(t) = x_0 + \int_0^t v(t')\, dt'$$

$$= v_0 \tau \ln\left(1 + t/\tau\right), \tag{2.51}$$

if we take the initial position x_0 to be zero. Figure 2.8 shows our results for v and x as functions of t. It is interesting to compare these graphs with the corresponding graphs of Figure 2.4 for a body coasting horizontally but subject to a linear resistance. Superficially, the two graphs for the velocity look similar. In particular, both go to zero as $t \to \infty$. But in the linear case v goes to zero *exponentially*, whereas in the quadratic case it does so only very slowly, like $1/t$. This difference in the behavior of v manifests itself quite dramatically in the behavior of x. In the linear case, we

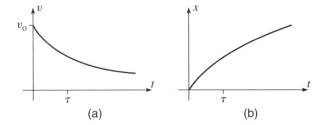

Figure 2.8   The motion of a body, such as a bicycle, coasting horizontally and subject to a quadratic air resistance. **(a)** The velocity is given by (2.49) and goes to zero like $1/t$ as $t \to \infty$. **(b)** The position is given by (2.51) and goes to infinity as $t \to \infty$.

saw that x approaches a finite limit as $t \to \infty$, but it is clear from (2.51) that in the quadratic case x increases without limit as $t \to \infty$.

The striking difference in the behavior of x for quadratic and linear drags is easy to understand qualitatively. In the quadratic case, the drag is proportional to v^2. Thus as v gets small, the drag gets *very* small — so small that it fails to bring the bicycle to rest at any finite value of x. This unexpected behavior serves to highlight that a drag force that is proportional to v^2 at *all* speeds is unrealistic. Although the linear drag and ordinary friction are very small, nevertheless as $v \to 0$ they must eventually become more important than the v^2 term and cannot be ignored. In particular, one or another of these two terms (friction in the case of a bicycle) ensures that no real body can coast on to infinity!

Vertical Motion with Quadratic Drag

The case that an object moves vertically with a quadratic drag force can be solved in much the same way as the horizontal case. Consider a baseball that is dropped from a window in a high tower. If we measure the coordinate y vertically down, the equation of motion is (I'll abbreviate v_y to v now)

$$m\dot{v} = mg - cv^2. \tag{2.52}$$

Before we solve this equation, let us consider the ball's terminal speed, the speed at which the two terms on the right of (2.52) just balance. Evidently this must satisfy $cv^2 = mg$, whose solution is

$$v_{\text{ter}} = \sqrt{\frac{mg}{c}}. \tag{2.53}$$

For any given object (given m, g, and c), this lets us calculate the terminal speed. For example, for a baseball it gives (as we shall see in a moment) $v_{\text{ter}} \approx 35$ m/s, or nearly 80 miles per hour.

We can tidy the equation of motion (2.52) a little by using (2.53) to replace c by mg/v_{ter}^2 and canceling the factors of m:

$$\dot{v} = g\left(1 - \frac{v^2}{v_{\text{ter}}^2}\right). \tag{2.54}$$

This can be solved by separation of variables, just as in the case of horizontal motion: First we can rewrite it as

$$\frac{dv}{1 - v^2/v_{\text{ter}}^2} = g\,dt. \tag{2.55}$$

This is the desired separated form (only v on the left and only t on the right) and we can simply integrate both sides.[5] Assuming the ball starts from rest, the limits of

---

[5] Notice that in fact any one-dimensional problem where the net force depends only on the velocity can be solved by separation of variables, since the equation $m\dot{v} = F(v)$ can always be

integration are 0 and v on the left and 0 and t on the right, and we find (as you should verify — Problem 2.35)

$$\frac{v_{\text{ter}}}{g} \operatorname{arctanh}\left(\frac{v}{v_{\text{ter}}}\right) = t \qquad (2.56)$$

where "arctanh" denotes the inverse hyperbolic tangent. This particular integral can be evaluated alternatively in terms of the natural log function (Problem 2.37). However, the hyperbolic functions, sinh, cosh, and tanh, and their inverses arcsinh, arccosh, and arctanh, come up so often in all branches of physics that you really should learn to use them. If you have not had much exposure to them, you might want to look at Problems 2.33 and 2.34, and study graphs of these functions.

Equation (2.56) can be solved for v to give

$$v = v_{\text{ter}} \tanh\left(\frac{gt}{v_{\text{ter}}}\right). \qquad (2.57)$$

To find the position y, we just integrate v to give

$$y = \frac{(v_{\text{ter}})^2}{g} \ln\left[\cosh\left(\frac{gt}{v_{\text{ter}}}\right)\right]. \qquad (2.58)$$

While both of these two formulas can be cleaned up a little (see Problem 2.35), they are already sufficient to work the following example.

EXAMPLE 2.5   A Baseball Dropped from a High Tower

Find the terminal speed of a baseball (mass $m = 0.15$ kg and diameter $D = 7$ cm). Make plots of its velocity and position for the first six seconds after it is dropped from a tall tower.

The terminal speed is given by (2.53), with the coefficient of air resistance c given by (2.4) as $c = \gamma D^2$ where $\gamma = 0.25$ N·s^2/m^4. Therefore

$$v_{\text{ter}} = \sqrt{\frac{mg}{\gamma D^2}} = \sqrt{\frac{(0.15\,\text{kg}) \times (9.8\,\text{m/s}^2)}{(0.25\,\text{N·s}^2/\text{m}^4) \times (0.07\,\text{m})^2}} = 35\,\text{m/s} \qquad (2.59)$$

or nearly 80 miles per hour. It is interesting to note that fast baseball pitchers can pitch a ball considerably *faster* than v_{ter}. Under these conditions, the drag force is actually *greater* than the ball's weight!

The plots of v and y can be made by hand, but are, of course, much easier with the help of computer software such as Mathcad or Mathematica that can make the plots for you. Whatever method we choose, the results are as shown in Figure 2.9, where the solid curves show the actual velocity and position while the dashed curves are the corresponding values in a vacuum. The actual velocity levels out,

---

written as $m\,dv/F(v) = dt$. Of course there is no assurance that this can be integrated analytically if $F(v)$ is too complicated, but it does guarantee a straightforward numerical solution at worst. See Problem 2.7.

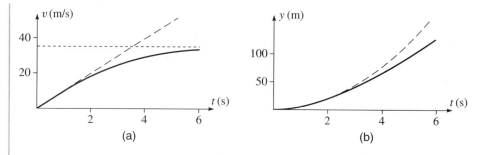

Figure 2.9   The motion of a baseball dropped from the top of a high tower (solid curves). The corresponding motion in a vacuum is shown with long dashes. (a) The actual velocity approaches the ball's terminal velocity $v_{ter} = 35$ m/s as $t \rightarrow \infty$. (b) The graph of position against time falls further and further behind the corresponding vacuum graph. When $t = 6$ s, the baseball has dropped about 130 meters; in a vacuum, it would have dropped about 180 meters.

approaching the terminal value $v_{ter} = 35$ m/s as $t \rightarrow \infty$, whereas the velocity in a vacuum would increase without limit. Initially, the position increases just as it would in a vacuum (that is, $y = \frac{1}{2}gt^2$), but falls behind as v increases and the air resistance becomes more important. Eventually, y approaches a straight line of the form $y = v_{ter}t + \text{const.}$ (See Problem 2.35.)

Quadratic Drag with Horizontal and Vertical Motion

The equation of motion for a projectile subject to quadratic drag,

$$m\ddot{\mathbf{r}} = m\mathbf{g} - cv^2\hat{\mathbf{v}}$$

$$= m\mathbf{g} - cv\mathbf{v}, \tag{2.60}$$

resolves into its horizontal and vertical components (with y measured vertically upward) to give

$$\left. \begin{array}{l} m\dot{v}_x = -c\sqrt{v_x^2 + v_y^2}\, v_x \\[2mm] m\dot{v}_y = -mg - c\sqrt{v_x^2 + v_y^2}\, v_y. \end{array} \right\} \tag{2.61}$$

These are two differential equations for the two unknown functions $v_x(t)$ and $v_y(t)$, but each equation involves *both* v_x and v_y. In particular, neither equation is the same as for an object that moves only in the x direction or only in the y direction. This means that we cannot solve these two equations by simply pasting together our two separate solutions for horizontal and vertical motion. Worse still, it turns out that the two equations (2.61) cannot be solved analytically at all. The only way to solve them is numerically, which we can only do for specified numerical initial conditions (that is, specified values of the initial position and velocity). This means that we cannot find the *general* solution; all we can do numerically is to find the particular solution corresponding to any chosen initial conditions. Before I discuss some general properties of the solutions of (2.61), let us work out one such numerical solution.

EXAMPLE 2.6   Trajectory of a Baseball

The baseball of Example 2.5 is now thrown with velocity 30 m/s (about 70 mi/h) at 50° above the horizontal from a high cliff. Find its trajectory for the first eight seconds of flight and compare with the corresponding trajectory in a vacuum. If the same baseball was thrown with the same initial velocity on horizontal ground how far would it travel before landing? That is, what is its horizontal range?

We have to solve the two coupled differential equations (2.61) with the initial conditions

$$v_{xo} = v_o \cos\theta = 19.3 \text{ m/s} \quad \text{and} \quad v_{yo} = v_o \sin\theta = 23.0 \text{ m/s}$$

and $x_o = y_o = 0$ (if we put the origin at the point from which the ball is thrown). This can be done with systems such as Mathematica, Matlab, or Maple, or with programming languages such as "C" or Fortran. Figure 2.10 shows the resulting trajectory, found using the function "NDSolve" in Mathematica.

Several features of Figure 2.10 deserve comment. Obviously the effect of air resistance is to lower the trajectory, as compared to the vacuum trajectory (shown dashed). For example, we see that in a vacuum the high point of the trajectory occurs at $t \approx 2.3$ s and is about 27 m above the starting point; with air resistance, the high point comes just before $t = 2.0$ s and is at about 21 m. In a vacuum, the ball would continue to move indefinitely in the x direction. The

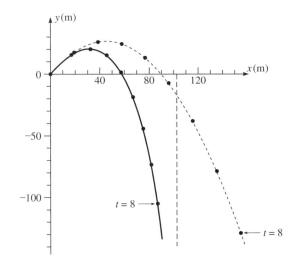

Figure 2.10   Trajectory of a baseball thrown off a cliff and subject to quadratic air resistance (solid curve). The initial velocity is 30 m/s at 50° above the horizontal; the terminal speed is 35 m/s. The dashed curve shows the corresponding trajectory in a vacuum. The dots show the ball's position at one-second intervals. Air resistance slows the horizontal motion, so that the ball approaches a vertical asymptote just beyond $x = 100$ meters.

effect of air resistance is to slow the horizontal motion so that x never moves to the right of a vertical asymptote near $x = 100$ m.

The horizontal range of the baseball is easily read off the figure as the value of x when y returns to zero. We see that $R \approx 59$ m, as opposed to the range in vacuum, $R_{vac} \approx 90$ m. The effect of air resistance is quite large in this example, as we might have anticipated: The ball was thrown with a speed only a little less than the terminal speed (30 vs 35 m/s), and this means that the force of air resistance is only a little less than that of gravity. This being the case, we should expect air resistance to change the trajectory appreciably.

This example illustrates several of the general features of projectile motion with a quadratic drag force. Although we cannot solve analytically the equations of motion (2.61) for this problem, we *can* use the equations to prove various general properties of the trajectory. For example, we noticed that the baseball reached a lower maximum height, and did so sooner, than it would have in a vacuum. It is easy to prove that this will always be the case: As long as the projectile is moving upward ($v_y > 0$), the force of air resistance has a *downward* y component. Thus the downward acceleration is greater than g (its value in vacuum). Therefore a graph of v_y against t slopes down from v_{yo} more quickly than it would in vacuum, as shown in Figure 2.11. This guarantees that v_y reaches zero sooner than it would in vacuum, and that the ball travels less distance (in the y direction) before reaching the high point. That is, the ball's high point occurs sooner, and is lower, than it would be in a vacuum.

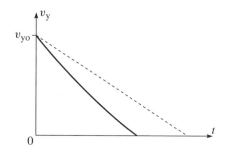

Figure 2.11   Graph of v_y against t for a projectile that is thrown upward ($v_{yo} > 0$) and is subject to a quadratic resistance (solid curve). The dashed line (slope $= -g$) is the corresponding graph when there is no air resistance. The projectile moves upward until it reaches its maximum height when $v_y = 0$. During this time, the drag force is downward and the downward acceleration is always greater than g. Therefore, the curve slopes more steeply than the dashed line, and the projectile reaches its high point sooner than it would in a vacuum. Since the area under the curve is less than that under the dashed line, the projectile's maximum height is less than it would be in a vacuum.

I claimed that the baseball of Example 2.6 approaches a vertical asymptote as $t \rightarrow \infty$, and we can now prove that this is always the case. First, it is easy to convince yourself that once the ball starts moving downward, it continues to accelerate downward, with v_y approaching $-v_{\text{ter}}$ as $t \rightarrow \infty$. At the same time v_x continues to decrease and approaches zero. Thus the square root in both of the equations (2.61) approaches v_{ter}. In particular, when t is large, the equation for v_x can be approximated by

$$\dot{v}_x \approx -\frac{cv_{\text{ter}}}{m} v_x = -kv_x$$

say. The solution of this equation is, of course, an exponential function, $v_x = Ae^{-kt}$, and we see that v_x approaches zero very rapidly (exponentially) as $t \rightarrow \infty$. This guarantees that x, which is the integral of v_x,

$$x(t) = \int_0^t v_x(t') \, dt',$$

approaches a finite limit as $t \rightarrow \infty$, and the trajectory has a finite vertical asymptote as claimed.

2.5   Motion of a Charge in a Uniform Magnetic Field

Another interesting application of Newton's laws, and (like projectile motion) an application that lets me introduce some important mathematical methods, is the motion of a charged particle in a magnetic field. I shall consider here a particle of charge q (which I shall usually take to be positive), moving in a uniform magnetic field **B** that points in the z direction as shown in Figure 2.12. The net force on the particle is just the magnetic force

$$\mathbf{F} = q\mathbf{v} \times \mathbf{B}, \tag{2.62}$$

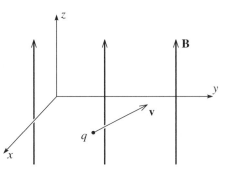

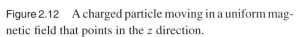

Figure 2.12   A charged particle moving in a uniform magnetic field that points in the z direction.

so the equation of motion can be written as

$$m\dot{\mathbf{v}} = q\mathbf{v} \times \mathbf{B}. \tag{2.63}$$

[As with projectiles, the force depends only on the velocity (not the position), so the second law reduces to a first-order differential equation for $\mathbf{v}$.]

As is so often the case, the simplest way to solve the equation of motion is to resolve it into components. The components of $\mathbf{v}$ and $\mathbf{B}$ are

$$\mathbf{v} = (v_x, v_y, v_z)$$

and

$$\mathbf{B} = (0, 0, B),$$

from which we can read off the components of $\mathbf{v} \times \mathbf{B}$:

$$\mathbf{v} \times \mathbf{B} = (v_y B, -v_x B, 0).$$

Thus the three components of (2.63) are

$$m\dot{v}_x = qBv_y \tag{2.64}$$

$$m\dot{v}_y = -qBv_x \tag{2.65}$$

$$m\dot{v}_z = 0. \tag{2.66}$$

The last of these says simply that v_z, the component of the particle's velocity in the direction of $\mathbf{B}$, is constant:

$$v_z = \text{const},$$

a result we could have anticipated since the magnetic force is always perpendicular to $\mathbf{B}$. Because v_z is constant, we shall focus most of our attention on v_x and v_y. In fact, we can even think of them as comprising a two-dimensional vector (v_x, v_y), which is just the projection of $\mathbf{v}$ onto the xy plane and can be called the *transverse velocity*,

$$(v_x, v_y) = \text{transverse velocity}.$$

To simplify the equations (2.64) and (2.65) for v_x and v_y, I shall define the parameter

$$\omega = \frac{qB}{m}, \tag{2.67}$$

which has the dimensions of inverse time and is called the **cyclotron frequency**. With this notation, Equations (2.64) and (2.65) become

$$\left.\begin{array}{l} \dot{v}_x = \omega v_y \\ \dot{v}_y = -\omega v_x. \end{array}\right\} \tag{2.68}$$

These two coupled differential equations can be solved in a host of different ways. I would like to describe one that makes use of complex numbers. Though perhaps

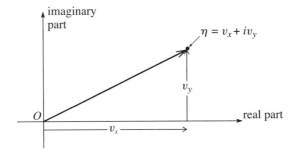

Figure 2.13   The complex number $\eta = v_x + i v_y$ is represented as a point in the complex plane. The arrow pointing from O to η is literally a picture of the transverse velocity vector (v_x, v_y).

not the easiest solution, this method has surprisingly wide application in many areas of physics. (For an alternative solution that avoids complex numbers, see Problem 2.54.)

The two variables v_x and v_y are, of course, real numbers. However, there is nothing to prevent us from defining a complex number

$$\eta = v_x + i v_y, \tag{2.69}$$

where i denotes the square root of -1 (called j by most engineers), $i = \sqrt{-1}$ (and η is the Greek letter eta). If we draw the complex number η in the complex plane, or Argand diagram, then its two components are v_x and v_y as shown in Figure 2.13; in other words, the representation of η in the complex plane is a picture of the two-dimensional transverse velocity (v_x, v_y).

The advantages of introducing the complex number η appear when we evaluate its derivative. Using (2.68), we find that

$$\dot{\eta} = \dot{v}_x + i \dot{v}_y = \omega v_y - i \omega v_x = -i \omega (v_x + i v_y)$$

or

$$\dot{\eta} = -i \omega \eta. \tag{2.70}$$

We see that the two coupled equations for v_x and v_y have become a single equation for the complex number η. Furthermore, it is an equation of the now familiar form $\dot{u} = ku$, whose solution we know to be the exponential $u = Ae^{kt}$. Thus we can immediately write down the solution for η:

$$\eta = Ae^{-i\omega t}. \tag{2.71}$$

Before we discuss the significance of this solution, I would like to review a few properties of complex exponentials in the next section. If you are very familiar with these ideas, by all means skip this material.

2.6   Complex Exponentials

While you are certainly familiar with the exponential function e^x for a real variable x, you may not be so at home with e^z when z is complex.[6] For the real case there are several possible definitions of e^x (for instance, as the function that is equal to its own derivative). The definition that extends most easily to the complex case is the Taylor series (see Problem 2.18)

$$e^z = 1 + z + \frac{z^2}{2!} + \frac{z^3}{3!} + \cdots . \tag{2.72}$$

For any value of z, real or complex, large or small, this series converges to give a well-defined value for e^z. By differentiating it, you can easily convince yourself that it has the expected property that it equals its own derivative. And one can show (not always so easily) that it has all the other familiar properties of the exponential function — for instance, that $e^z e^w = e^{(z+w)}$. (See Problems 2.50 and 2.51.) In particular, the function Ae^{kz} (with A and k any constants, real or complex) has the property that

$$\frac{d}{dz}\left(Ae^{kz} \right) = k \left(Ae^{kz} \right) . \tag{2.73}$$

Since it satisfies this same equation whatever the value of A, it is, in fact, the general solution of the first-order equation $df/dz = kf$. At the end of the last section, I introduced the complex number $\eta(t)$ and showed that it satisfied the equation $\dot\eta = -i\omega\eta$. We are now justified in saying that this guarantees that η must be the exponential function anticipated in (2.71).

We shall be particularly concerned with the exponential of a pure imaginary number, that is, $e^{i\theta}$ where θ is a real number. The Taylor series (2.72) for this function reads

$$e^{i\theta} = 1 + i\theta + \frac{(i\theta)^2}{2!} + \frac{(i\theta)^3}{3!} + \frac{(i\theta)^4}{4!} + \cdots . \tag{2.74}$$

Noting that $i^2 = -1$, $i^3 = -i$, and so on, you can see that all of the even powers in this series are real, while all of the odd powers are pure imaginary. Regrouping accordingly, we can rewrite (2.74) to read

$$e^{i\theta} = \left[1 - \frac{\theta^2}{2!} + \frac{\theta^4}{4!} + \cdots \right] + i \left[\theta - \frac{\theta^3}{3!} + \cdots \right] . \tag{2.75}$$

The series in the first brackets is the Taylor series for $\cos\theta$, and that in the second brackets is $\sin\theta$ (Problem 2.18). Thus we have proved the important relation:

$$e^{i\theta} = \cos\theta + i\sin\theta . \tag{2.76}$$

---

[6] For a review of some elementary properties of complex numbers, see Problems 2.45 to 2.49.

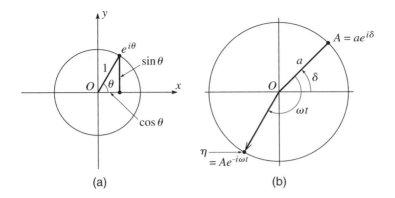

Figure 2.14   (a) Euler's formula, (2.76), implies that the complex number $e^{i\theta}$ lies on the unit circle (the circle of radius 1, centered on the origin O) with polar angle θ. (b) The complex constant $A = ae^{i\delta}$ lies on a circle of radius a with polar angle δ. The function $\eta(t) = Ae^{-i\omega t}$ lies on the same circle but with polar angle $(\delta - \omega t)$ and moves clockwise around the circle as t advances.

This result, known as **Euler's formula**, is illustrated in Figure 2.14(a). Note especially that the complex number $e^{i\theta}$ has polar angle θ and, since $\cos^2\theta + \sin^2\theta = 1$, the magnitude of $e^{i\theta}$ is 1; that is, $e^{i\theta}$ lies on the *unit circle*, the circle with radius 1 centered at O.

Our main concern is with a complex number of the form $\eta = Ae^{-i\omega t}$. The coefficient A is a fixed complex number, which can be expressed as $A = ae^{i\delta}$, where $a = |A|$ is the magnitude, and δ is the polar angle of A, as shown in Figure 2.14(b). (See Problem 2.45.) The number η can therefore be written as

$$\eta = Ae^{-i\omega t} = ae^{i\delta}e^{-i\omega t} = ae^{i(\delta-\omega t)}. \tag{2.77}$$

Thus η has the same magnitude as A (namely a), but has polar angle equal to $(\delta - \omega t)$, as shown in Figure 2.14(b). As a function of t, the number η moves clockwise around the circle of radius a with angular velocity ω.

It is important that you get a good feel for the role of the complex constant $A = ae^{i\delta}$ in (2.77): If A happened to equal 1, then η would be just $\eta = e^{-i\omega t}$, which lies on the unit circle, moving clockwise with angular velocity ω and starting from the real axis ($\eta = 1$) when $t = 0$. If $A = a$ is real but not equal to 1, then it simply magnifies the unit circle to a circle of radius a, around which η moves with the same angular speed and starting from the real axis, at $\eta = a$ when $t = 0$. Finally if $A = ae^{i\delta}$, then the effect of the angle δ is to rotate η through the fixed angle δ, so that η starts out at $t = 0$ with polar angle δ.

Armed with these mathematical results, we can now return to the charged particle in a magnetic field.

2.7   Solution for the Charge in a B Field

Mathematically, the solution for the velocity **v** of our charged particle in a B field is complete, and all that remains is to interpret it physically. We already know that v_z, the component along **B**, is constant. The components (v_x, v_y) transverse to **B** we have represented by the complex number $\eta = v_x + i v_y$, and we have seen that Newton's second law implies that η has the time dependence $\eta = A e^{-i\omega t}$, moving uniformly around the circle of Figure 2.14(b). Now, the arrow shown in that figure, pointing from O to η, is in fact a pictorial representation of the transverse velocity (v_x, v_y). Therefore this transverse velocity changes direction, turning clockwise, with constant angular velocity[7] $\omega = qB/m$ and with constant magnitude. Because v_z is constant, this suggests that the particle undergoes a spiralling, or helical, motion. To verify this, we have only to integrate **v** to find **r** as a function of t.

That v_z is constant implies that

$$z(t) = z_0 + v_{z0}t. \tag{2.78}$$

The motion of x and y is most easily found by introducing another complex number

$$\xi = x + iy$$

where ξ is the Greek letter xi. In the complex plane, ξ is a picture of the transverse position (x, y). Clearly, the derivative of ξ is η, that is, $\dot{\xi} = \eta$. Therefore,

$$\xi = \int \eta\, dt = \int A e^{-i\omega t}\, dt$$

$$= \frac{iA}{\omega} e^{-i\omega t} + \text{constant}. \tag{2.79}$$

If we rename the coefficient iA/ω as C and the constant of integration as $X + iY$, this implies that

$$x + iy = C e^{-i\omega t} + (X + iY).$$

By redefining our origin so that the z axis goes through the point (X, Y), we can eliminate the constant term on the right to give

$$x + iy = C e^{-i\omega t}, \tag{2.80}$$

and, by setting $t = 0$, we can identify the remaining constant C as

$$C = x_0 + iy_0.$$

This result is illustrated in Figure 2.15. We see there that the transverse position (x, y) moves clockwise round a circle with angular velocity $\omega = qB/m$. Meanwhile z as given by (2.78) increases steadily, so the particle actually describes a uniform helix whose axis is parallel to the magnetic field.

---

[7] I am assuming the charge q is positive; if q is negative, then $\omega = qB/m$ is negative, meaning that the transverse velocity rotates counterclockwise.

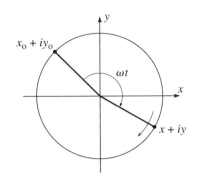

Figure 2.15   Motion of a charge in a uniform magnetic field in the z direction. The transverse position (x, y) moves around a circle as shown, while the coordinate z moves with constant velocity into or out of the page.

There are many examples of the helical motion of a charged particle along a magnetic field; for example, cosmic-ray particles (charged particles hitting the earth from space) can get caught by the earth's magnetic field and spiral north or south along the field lines. If the z component of the velocity happens to be zero, then the spiral reduces to a circle. In the cyclotron, a device for accelerating charged particles to high energies, the particles are trapped in circular orbits in this way. They are slowly accelerated by the judiciously timed application of an electric field. The angular frequency of the orbit is, of course, $\omega = qB/m$ (which is why this is called the cyclotron frequency). The radius of the orbit is

$$r = \frac{v}{\omega} = \frac{mv}{qB} = \frac{p}{qB}.\tag{2.81}$$

This radius increases as the particles accelerate, so that they eventually emerge at the outer edge of the circular magnets that produce the magnetic field.

The same method that we have used here for a charge in a magnetic field can also be used for a particle in magnetic *and electric* fields, but since this complication adds nothing to the method of solution, I shall leave you to try it for yourself in Problems 2.53 and 2.55.

Principal Definitions and Equations of Chapter 2

Linear and Quadratic Drags

Provided the speed v is well below that of sound, the magnitude of the drag force $\mathbf{f} = -f(v)\hat{\mathbf{v}}$ on an object moving through a fluid is usually well approximated as

$$f(v) = f_{\text{lin}} + f_{\text{quad}}$$

where

$$f_{\text{lin}} = bv = \beta Dv \quad \text{and} \quad f_{\text{quad}} = cv^2 = \gamma D^2 v^2. \quad \text{[Eqs. (2.2) to (2.6)]}$$

Here D denotes the linear size of the object. For a sphere, D is the diameter and, for a sphere in air at STP, $\beta = 1.6 \times 10^{-4}$ N·s/m^2 and $\gamma = 0.25$ N·s^2/m^4.

The Lorentz Force on a Charged Particle

$$\mathbf{F} = q(\mathbf{E} + \mathbf{v} \times \mathbf{B}). \quad \text{[Eq. (2.62) \& Problem 2.53]}$$

Problems for Chapter 2

Stars indicate the approximate level of difficulty, from easiest (⋆) to most difficult (⋆⋆⋆).

SECTION 2.1   **Air Resistance**

2.1 ⋆ When a baseball flies through the air, the ratio $f_{\text{quad}}/f_{\text{lin}}$ of the quadratic to the linear drag force is given by (2.7). Given that a baseball has diameter 7 cm, find the approximate speed v at which the two drag forces are equally important. For what approximate range of speeds is it safe to treat the drag force as purely quadratic? Under normal conditions is it a good approximation to ignore the linear term? Answer the same questions for a beach ball of diameter 70 cm.

2.2 ⋆ The origin of the linear drag force on a sphere in a fluid is the viscosity of the fluid. According to Stokes's law, the viscous drag on a sphere is

$$f_{\text{lin}} = 3\pi \eta D v \tag{2.82}$$

where η is the viscosity[8] of the fluid, D the sphere's diameter, and v its speed. Show that this expression reproduces the form (2.3) for f_{lin}, with b given by (2.4) as $b = \beta D$. Given that the viscosity of air at STP is $\eta = 1.7 \times 10^{-5}$ N·s/m^2, verify the value of β given in (2.5).

2.3 ⋆ **(a)** The quadratic and linear drag forces on a moving sphere in a fluid are given by (2.84) and (2.82) (Problems 2.4 and 2.2). Show that the ratio of these two kinds of drag force can be written as $f_{\text{quad}}/f_{\text{lin}} = R/48$,[9] where the dimensionless **Reynolds number** R is

$$R = \frac{D v \varrho}{\eta} \tag{2.83}$$

where D is the sphere's diameter, v its speed, and ϱ and η are the fluid's density and viscosity. Clearly the Reynolds number is a measure of the relative importance of the two kinds of drag.[10] When R is

---

[8] For the record, the viscosity η of a fluid is defined as follows: Imagine a wide channel along which fluid is flowing (x direction) such that the velocity v is zero at the bottom ($y = 0$) and increases toward the top ($y = h$), so that successive layers of fluid slide across one another with a velocity gradient dv/dy. The force F with which an area A of any one layer drags the fluid above it is proportional to A and to dv/dy, and η is defined as the constant of proportionality; that is, $F = \eta A \, dv/dy$.

[9] The numerical factor 48 is for a sphere. A similar result holds for other bodies, but the numerical factor is different for different shapes.

[10] The Reynolds number is usually defined by (2.83) for flow involving any object, with D defined as a typical linear dimension. One sometimes hears the claim that R is the ratio $f_{\text{quad}}/f_{\text{lin}}$. Since $f_{\text{quad}}/f_{\text{lin}} = R/48$ for a sphere, this claim would be better phrased as "R is roughly of the order of $f_{\text{quad}}/f_{\text{lin}}$."

very large, the quadratic drag is dominant and the linear can be neglected; vice versa when R is very small. **(b)** Find the Reynolds number for a steel ball bearing (diameter 2 mm) moving at 5 cm/s through glycerin (density 1.3 g/cm^3 and viscosity 12 N·s/m^2 at STP).

2.4 ★★ The origin of the quadratic drag force on any projectile in a fluid is the inertia of the fluid that the projectile sweeps up. **(a)** Assuming the projectile has a cross-sectional area A (normal to its velocity) and speed v, and that the density of the fluid is ϱ, show that the rate at which the projectile encounters fluid (mass/time) is $\varrho A v$. **(b)** Making the simplifying assumption that all of this fluid is accelerated to the speed v of the projectile, show that the net drag force on the projectile is $\varrho A v^2$. It is certainly not true that all the fluid that the projectile encounters is accelerated to the full speed v, but one might guess that the actual force would have the form

$$f_{\text{quad}} = \kappa \varrho A v^2 \tag{2.84}$$

where κ is a number less than 1, which would depend on the shape of the projectile, with κ small for a streamlined body, and larger for a body with a flat front end. This proves to be true, and for a sphere the factor κ is found to be $\kappa = 1/4$. **(c)** Show that (2.84) reproduces the form (2.3) for f_{quad}, with c given by (2.4) as $c = \gamma D^2$. Given that the density of air at STP is $\varrho = 1.29$ kg/m^3 and that $\kappa = 1/4$ for a sphere, verify the value of γ given in (2.6).

SECTION 2.2 **Linear Air Resistance**

2.5 ★ Suppose that a projectile which is subject to a linear resistive force is thrown vertically down with a speed v_{yo} which is *greater* than the terminal speed v_{ter}. Describe and explain how the velocity varies with time, and make a plot of v_y against t for the case that $v_{\text{yo}} = 2v_{\text{ter}}$.

2.6 ★ **(a)** Equation (2.33) gives the velocity of an object dropped from rest. At first, when v_y is small, air resistance should be unimportant and (2.33) should agree with the elementary result $v_y = gt$ for free fall in a vacuum. Prove that this is the case. [*Hint:* Remember the Taylor series for $e^x = 1 + x + x^2/2! + x^3/3! + \cdots$, for which the first two or three terms are certainly a good approximation when x is small.] **(b)** The position of the dropped object is given by (2.35) with $v_{\text{yo}} = 0$. Show similarly that this reduces to the familiar $y = \frac{1}{2}gt^2$ when t is small.

2.7 ★ There are certain simple one-dimensional problems where the equation of motion (Newton's second law) can always be solved, or at least reduced to the problem of doing an integral. One of these (which we have met a couple of times in this chapter) is the motion of a one-dimensional particle subject to a force that depends only on the velocity v, that is, $F = F(v)$. Write down Newton's second law and separate the variables by rewriting it as $m \, dv/F(v) = dt$. Now integrate both sides of this equation and show that

$$t = m \int_{v_o}^{v} \frac{dv'}{F(v')}.$$

Provided you can do the integral, this gives t as a function of v. You can then solve to give v as a function of t. Use this method to solve the special case that $F(v) = F_o$, a constant, and comment on your result. This method of *separation of variables* is used again in Problems 2.8 and 2.9.

2.8 ★ A mass m has velocity v_0 at time $t = 0$ and coasts along the x axis in a medium where the drag force is $F(v) = -cv^{3/2}$. Use the method of Problem 2.7 to find v in terms of the time t and the other given parameters. At what time (if any) will it come to rest?

2.9 ★ We solved the differential equation (2.29), $m\dot{v}_y = -b(v_y - v_{\text{ter}})$, for the velocity of an object falling through air, by inspection — a most respectable way of solving differential equations. Nevertheless, one would sometimes like a more systematic method, and here is one. Rewrite the equation in the "separated" form

$$\frac{m\,dv_y}{v_y - v_{\text{ter}}} = -b\,dt$$

and integrate both sides from time 0 to t to find v_y as a function of t. Compare with (2.30).

2.10 ★★ For a steel ball bearing (diameter 2 mm and density 7.8 g/cm^3) dropped in glycerin (density 1.3 g/cm^3 and viscosity 12 N·s/m^2 at STP), the dominant drag force is the linear drag given by (2.82) of Problem 2.2. **(a)** Find the characteristic time τ and the terminal speed v_{ter}. [In finding the latter, you should include the buoyant force of Archimedes. This just adds a third force on the right side of Equation (2.25).] How long after it is dropped from rest will the ball bearing have reached 95% of its terminal speed? **(b)** Use (2.82) and (2.84) (with $\kappa = 1/4$ since the ball bearing is a sphere) to compute the ratio $f_{\text{quad}}/f_{\text{lin}}$ at the terminal speed. Was it a good approximation to neglect f_{quad}?

2.11 ★★ Consider an object that is thrown vertically up with initial speed v_0 in a linear medium. **(a)** Measuring y *upward* from the point of release, write expressions for the object's velocity $v_y(t)$ and position $y(t)$. **(b)** Find the time for the object to reach its highest point and its position y_{max} at that point. **(c)** Show that as the drag coefficient approaches zero, your last answer reduces to the well-known result $y_{\text{max}} = \frac{1}{2}v_0^2/g$ for an object in the vacuum. [*Hint:* If the drag force is very small, the terminal speed is very big, so v_0/v_{ter} is very small. Use the Taylor series for the log function to approximate $\ln(1 + \delta)$ by $\delta - \frac{1}{2}\delta^2$. (For a little more on Taylor series see Problem 2.18.)]

2.12 ★★ Problem 2.7 is about a class of one-dimensional problems that can always be reduced to doing an integral. Here is another. Show that if the net force on a one-dimensional particle depends only on position, $F = F(x)$, then Newton's second law can be solved to find v as a function of x given by

$$v^2 = v_0^2 + \frac{2}{m}\int_{x_0}^x F(x')\,dx'. \tag{2.85}$$

[*Hint:* Use the chain rule to prove the following handy relation, which we could call the "$v\,dv/dx$ rule": If you regard v as a function of x, then

$$\dot{v} = v\frac{dv}{dx} = \frac{1}{2}\frac{dv^2}{dx}. \tag{2.86}$$

Use this to rewrite Newton's second law in the separated form $m\,d(v^2) = 2F(x)\,dx$ and then integrate from x_0 to x.] Comment on your result for the case that $F(x)$ is actually a constant. (You may recognise your solution as a statement about kinetic energy and work, both of which we shall be discussing in Chapter 4.)

2.13 ★★ Consider a mass m constrained to move on the x axis and subject to a net force $F = -kx$ where k is a positive constant. The mass is released from rest at $x = x_0$ at time $t = 0$. Use the result (2.85) in Problem 2.12 to find the mass's speed as a function of x; that is, $dx/dt = g(x)$ for some function $g(x)$. Separate this as $dx/g(x) = dt$ and integrate from time 0 to t to find x as a function of t. (You may recognize this as one way — not the easiest — to solve the simple harmonic oscillator.)

2.14 ★★★ Use the method of Problem 2.7 to solve the following: A mass m is constrained to move along the x axis subject to a force $F(v) = -F_0 e^{v/V}$, where F_0 and V are constants. **(a)** Find $v(t)$ if the initial velocity is $v_o > 0$ at time $t = 0$. **(b)** At what time does it come instantaneously to rest? **(c)** By integrating $v(t)$, you can find $x(t)$. Do this and find how far the mass travels before coming instantaneously to rest.

SECTION 2.3 Trajectory and Range in a Linear Medium

2.15 ★ Consider a projectile launched with velocity (v_{xo}, v_{yo}) from horizontal ground (with x measured horizontally and y vertically up). Assuming no air resistance, find how long the projectile is in the air and show that the distance it travels before landing (the horizontal range) is $2v_{xo}v_{yo}/g$.

2.16 ★ A golfer hits his ball with speed v_o at an angle θ above the horizontal ground. Assuming that the angle θ is fixed and that air resistance can be neglected, what is the minimum speed $v_o(\min)$ for which the ball will clear a wall of height h, a distance d away? Your solution should get into trouble if the angle θ is such that $\tan\theta < h/d$. Explain. What is $v_o(\min)$ if $\theta = 25°$, $d = 50$ m, and $h = 2$ m?

2.17 ★ The two equations (2.36) give a projectile's position (x, y) as a function of t. Eliminate t to give y as a function of x. Verify Equation (2.37).

2.18 ★ Taylor's theorem states that, for any reasonable function $f(x)$, the value of f at a point $(x + \delta)$ can be expressed as an infinite series involving f and its derivatives at the point x:

$$f(x + \delta) = f(x) + f'(x)\delta + \frac{1}{2!}f''(x)\delta^2 + \frac{1}{3!}f'''(x)\delta^3 + \cdots \tag{2.87}$$

where the primes denote successive derivatives of $f(x)$. (Depending on the function this series may converge for *any* increment δ or only for values of δ less than some nonzero "radius of convergence.") This theorem is enormously useful, especially for small values of δ, when the first one or two terms of the series are often an excellent approximation.[11] **(a)** Find the Taylor series for $\ln(1 + \delta)$. **(b)** Do the same for $\cos\delta$. **(c)** Likewise $\sin\delta$. **(d)** And e^δ.

2.19 ★ Consider the projectile of Section 2.3. **(a)** Assuming there is no air resistance, write down the position (x, y) as a function of t, and eliminate t to give the trajectory y as a function of x. **(b)** The correct trajectory, including a linear drag force, is given by (2.37). Show that this reduces to your answer for part (a) when air resistance is switched off (τ and $v_{ter} = g\tau$ both approach infinity). [*Hint:* Remember the Taylor series (2.40) for $\ln(1 - \epsilon)$.]

2.20 ★★ [Computer] Use suitable graph-plotting software to plot graphs of the trajectory (2.36) of a projectile thrown at 45°above the horizontal and subject to linear air resistance for four different values of the drag coefficient, ranging from a significant amount of drag down to no drag at all. Put all four trajectories on the same plot. [*Hint:* In the absence of any given numbers, you may as well choose convenient values. For example, why not take $v_{xo} = v_{yo} = 1$ and $g = 1$. (This amounts to choosing your units of length and time so that these parameters have the value 1.) With these choices, the strength of the drag is given by the one parameter $v_{ter} = \tau$, and you might choose to plot the trajectories for $v_{ter} = 0.3, 1, 3$, and ∞ (that is, no drag at all), and for times from $t = 0$ to 3. For the case that $v_{ter} = \infty$, you'll probably want to write out the trajectory separately.]

2.21 ★★★ A gun can fire shells in any direction with the same speed v_o. Ignoring air resistance and using cylindrical polar coordinates with the gun at the origin and z measured vertically up, show that

---

[11] For more details on Taylor's series see, for example, Mary Boas, *Mathematical Methods in the Physical Sciences* (Wiley, 1983), p. 22 or Donald McQuarrie, *Mathematical Methods for Scientists and Engineers* (University Science Books, 2003), p. 94.

the gun can hit any object inside the surface

$$z = \frac{v_0^2}{2g} - \frac{g}{2v_0^2}\rho^2.$$

Describe this surface and comment on its dimensions.

2.22 ★★★ [Computer] The equation (2.39) for the range of a projectile in a linear medium cannot be solved analytically in terms of elementary functions. If you put in numbers for the several parameters, then it *can* be solved numerically using any of several software packages such as Mathematica, Maple, and MatLab. To practice this, do the following: Consider a projectile launched at angle θ above the horizontal ground with initial speed v_0 in a linear medium. Choose units such that $v_0 = 1$ and $g = 1$. Suppose also that the terminal speed $v_{ter} = 1$. (With $v_0 = v_{ter}$, air resistance should be fairly important.) We know that in a vacuum, the maximum range occurs at $\theta = \pi/4 \approx 0.75$. **(a)** What is the maximum range in a vacuum? **(b)** Now solve (2.39) for the range in the given medium at the same angle $\theta = 0.75$. **(c)** Once you have your calculation working, repeat it for some selection of values of θ within which the maximum range probably lies. (You could try $\theta = 0.4, 0.5, \cdots, 0.8$.) **(d)** Based on these results, choose a smaller interval for θ where you're sure the maximum lies and repeat the process. Repeat it again if necessary until you know the maximum range and the corresponding angle to two significant figures. Compare with the vacuum values.

SECTION 2.4   Quadratic Air Resistance

2.23 ★ Find the terminal speeds in air of **(a)** a steel ball bearing of diameter 3 mm, **(b)** a 16-pound steel shot, and **(c)** a 200-pound parachutist in free fall in the fetal position. In all three cases, you can safely assume the drag force is purely quadratic. The density of steel is about 8 g/cm^3 and you can treat the parachutist as a sphere of density 1 g/cm^3.

2.24 ★ Consider a sphere (diameter D, density ϱ_{sph}) falling through air (density ϱ_{air}) and assume that the drag force is purely quadratic. **(a)** Use Equation (2.84) from Problem 2.4 (with $\kappa = 1/4$ for a sphere) to show that the terminal speed is

$$v_{ter} = \sqrt{\frac{8}{3}Dg\frac{\varrho_{sph}}{\varrho_{air}}}. \tag{2.88}$$

(b) Use this result to show that of two spheres of the same size, the denser one will eventually fall faster. **(c)** For two spheres of the same material, show that the larger will eventually fall faster.

2.25 ★ Consider the cyclist of Section 2.4, coasting to a halt under the influence of a quadratic drag force. Derive in detail the results (2.49) and (2.51) for her velocity and position, and verify that the constant $\tau = m/cv_0$ is indeed a time.

2.26 ★ A typical value for the coefficient of quadratic air resistance on a cyclist is around $c = 0.20$ N/(m/s)^2. Assuming that the total mass (cyclist plus cycle) is $m = 80$ kg and that at $t = 0$ the cyclist has an initial speed $v_0 = 20$ m/s (about 45 mi/h) and starts to coast to a stop under the influence of air resistance, find the characteristic time $\tau = m/cv_0$. How long will it take him to slow to 15 m/s? What about 10 m/s? And 5 m/s? (Below about 5 m/s, it is certainly not reasonable to ignore friction, so there is no point pursuing this calculation to lower speeds.)

2.27 ★ I kick a puck of mass m up an incline (angle of slope $= \theta$) with initial speed v_0. There is no friction between the puck and the incline, but there is air resistance with magnitude $f(v) = cv^2$. Write down and solve Newton's second law for the puck's velocity as a function of t on the upward journey. How long does the upward journey last?

2.28 ★ A mass m has speed v_0 at the origin and coasts along the x axis in a medium where the drag force is $F(v) = -cv^{3/2}$. Use the "$v\,dv/dx$ rule" (2.86) in Problem 2.12 to write the equation of motion in the separated form $m\,v\,dv/F(v) = dx$, and then integrate both sides to give x in terms of v (or vice versa). Show that it will eventually travel a distance $2m\sqrt{v_0}/c$.

2.29 ★ The terminal speed of a 70-kg skydiver in spread-eagle position is around 50 m/s (about 115 mi/h). Find his speed at times $t = 1, 5, 10, 20, 30$ seconds after he jumps from a stationary balloon. Compare with the corresponding speeds if there were no air resistance.

2.30 ★ Suppose we wish to approximate the skydiver of Problem 2.29 as a sphere (not a very promising approximation, but nevertheless the kind of approximation physicists sometimes like to make). Given the mass and terminal speed, what should we use for the diameter of the sphere? Does your answer seem reasonable?

2.31 ★★ A basketball has mass $m = 600$ g and diameter $D = 24$ cm. **(a)** What is its terminal speed? **(b)** If it is dropped from a 30-m tower, how long does it take to hit the ground and how fast is it going when it does so? Compare with the corresponding numbers in a vacuum.

2.32 ★★ Consider the following statement: If at all times during a projectile's flight its speed is much less than the terminal speed, the effects of air resistance are usually very small. **(a)** Without reference to the explicit equations for the magnitude of v_{ter}, explain clearly why this is so. **(b)** By examining the explicit formulas (2.26) and (2.53) explain why the statement above is even more useful for the case of quadratic drag than for the linear case. [*Hint:* Express the ratio f/mg of the drag to the weight in terms of the ratio v/v_{ter}.]

2.33 ★★ The hyperbolic functions $\cosh z$ and $\sinh z$ are defined as follows:

$$\cosh z = \frac{e^z + e^{-z}}{2} \quad \text{and} \quad \sinh z = \frac{e^z - e^{-z}}{2}$$

for any z, real or complex. **(a)** Sketch the behavior of both functions over a suitable range of real values of z. **(b)** Show that $\cosh z = \cos(iz)$. What is the corresponding relation for $\sinh z$? **(c)** What are the derivatives of $\cosh z$ and $\sinh z$? What about their integrals? **(d)** Show that $\cosh^2 z - \sinh^2 z = 1$. **(e)** Show that $\int dx/\sqrt{1+x^2} = \operatorname{arcsinh} x$. [*Hint:* One way to do this is to make the substitution $x = \sinh z$.]

2.34 ★★ The hyperbolic function $\tanh z$ is defined as $\tanh z = \sinh z / \cosh z$, with $\cosh z$ and $\sinh z$ defined as in Problem 2.33. **(a)** Prove that $\tanh z = -i \tan(iz)$. **(b)** What is the derivative of $\tanh z$? **(c)** Show that $\int dz \tanh z = \ln \cosh z$. **(d)** Prove that $1 - \tanh^2 z = \operatorname{sech}^2 z$, where $\operatorname{sech} z = 1/\cosh z$. **(e)** Show that $\int dx/(1 - x^2) = \operatorname{arctanh} x$.

2.35 ★★ **(a)** Fill in the details of the arguments leading from the equation of motion (2.52) to Equations (2.57) and (2.58) for the velocity and position of a dropped object subject to quadratic air resistance. Be sure to do the two integrals involved. (The results of Problem 2.34 will help.) **(b)** Tidy the two equations by introducing the parameter $\tau = v_{\text{ter}}/g$. Show that when $t = \tau$, v has reached 76% of its terminal value. What are the corresponding percentages when $t = 2\tau$ and 3τ? **(c)** Show that when $t \gg \tau$, the position is approximately $y \approx v_{\text{ter}}t + \text{const.}$ [*Hint:* The definition of $\cosh x$ (Problem 2.33) gives you a simple approximation when x is large.] **(d)** Show that for t small, Equation (2.58) for the position gives $y \approx \frac{1}{2}gt^2$. [Use the Taylor series for $\cosh x$ and for $\ln(1 + \delta)$.]

2.36 ★★ Consider the following quote from Galileo's *Dialogues Concerning Two New Sciences:*

> Aristotle says that "an iron ball of 100 pounds falling from a height of one hundred cubits reaches the ground before a one-pound ball has fallen a single cubit." I say that they arrive at the same time. You find, on making the experiment, that the larger outstrips the smaller by two finger-breadths, that is, when the larger has reached the ground, the other is short of it by two finger-breadths.

We know that the statement attributed to Aristotle is totally wrong, but just how close is Galileo's claim that the difference is just "two finger breadths"? **(a)** Given that the density of iron is about 8 g/cm³, find the terminal speeds of the two iron balls. **(b)** Given that a cubit is about 2 feet, use Equation (2.58) to find the time for the heavier ball to land and then the position of the lighter ball at that time. How far apart are they?

2.37 ★★ The result (2.57) for the velocity of a falling object was found by integrating Equation (2.55) and the quickest way to do this is to use the integral $\int du/(1 - u^2) = \operatorname{arctanh} u$. Here is another way to do it: Integrate (2.55) using the method of "partial fractions," writing

$$\frac{1}{1 - u^2} = \frac{1}{2}\left(\frac{1}{1 + u} + \frac{1}{1 - u}\right),$$

which lets you do the integral in terms of natural logs. Solve the resulting equation to give v as a function of t and show that your answer agrees with (2.57).

2.38 ★★ A projectile that is subject to quadratic air resistance is thrown vertically *up* with initial speed v_0. **(a)** Write down the equation of motion for the upward motion and solve it to give v as a function of t. **(b)** Show that the time to reach the top of the trajectory is

$$t_{\text{top}} = (v_{\text{ter}}/g)\arctan(v_0/v_{\text{ter}}).$$

(c) For the baseball of Example 2.5 (with $v_{\text{ter}} = 35$ m/s), find t_{top} for the cases that $v_0 = 1, 10, 20, 30,$ and 40 m/s, and compare with the corresponding values in a vacuum.

2.39 ★★ When a cyclist coasts to a stop, he is actually subject to two forces, the quadratic force of air resistance, $f = -cv^2$ (with c as given in Problem 2.26), and a constant frictional force f_{fr} of about 3 N. The former is dominant at high and medium speeds, the latter at low speed. (The frictional force is a combination of ordinary friction in the bearings and rolling friction of the tires on the road.) **(a)** Write down the equation of motion while the cyclist is coasting to a stop. Solve it by separating variables to give t as a function of v. **(b)** Using the numbers of Problem 2.26 (and the value $f_{\text{fr}} = 3$ N given above) find how long it takes the cyclist to slow from his initial 20 m/s to 15 m/s. How long to slow to 10 and 5 m/s? How long to come to a full stop? If you did Problem 2.26, compare with the answers you got there ignoring friction entirely.

2.40 ★★ Consider an object that is coasting horizontally (positive x direction) subject to a drag force $f = -bv - cv^2$. Write down Newton's second law for this object and solve for v by separating variables. Sketch the behavior of v as a function of t. Explain the time dependence for t large. (Which force term is dominant when t is large?)

2.41 ★★ A baseball is thrown vertically up with speed v_0 and is subject to a quadratic drag with magnitude $f(v) = cv^2$. Write down the equation of motion for the upward journey (measuring y vertically *up*) and show that it can be rewritten as $\dot{v} = -g[1 + (v/v_{\text{ter}})^2]$. Use the "$v\,dv/dx$ rule"

(2.86) to write $\dot{v}$ as $v\,dv/dy$, and then solve the equation of motion by separating variables (put all terms involving v on one side and all terms involving y on the other). Integrate both sides to give y in terms of v, and hence v as a function of y. Show that the baseball's maximum height is

$$y_{\text{max}} = \frac{v_{\text{ter}}^2}{2g} \ln \left(\frac{v_{\text{ter}}^2 + v_{\text{o}}^2}{v_{\text{ter}}^2} \right). \tag{2.89}$$

If $v_{\text{o}} = 20$ m/s (about 45 mph) and the baseball has the parameters given in Example 2.5 (page 61), what is y_{max}? Compare with the value in a vacuum.

2.42 ★★ Consider again the baseball of Problem 2.41 and write down the equation of motion for the downward journey. (Notice that with a quadratic drag the downward equation is different from the upward one, and has to be treated separately.) Find v as a function of y and, given that the downward journey starts at y_{max} as given in (2.89), show that the speed when the ball returns to the ground is $v_{\text{ter}}v_{\text{o}}/\sqrt{v_{\text{ter}}^2 + v_{\text{o}}^2}$. Discuss this result for the cases of very much and very little air resistance. What is the numerical value of this speed for the baseball of Problem 2.41? Compare with the value in a vacuum.

2.43 ★★★ [Computer] The basketball of Problem 2.31 is thrown from a height of 2 m with initial velocity $\mathbf{v}_{\text{o}} = 15$ m/s at 45° above the horizontal. **(a)** Use appropriate software to solve the equations of motion (2.61) for the ball's position (x, y) and plot the trajectory. Show the corresponding trajectory in the absence of air resistance. **(b)** Use your plot to find how far the ball travels in the horizontal direction before it hits the floor. Compare with the corresponding range in a vacuum.

2.44 ★★★ [Computer] To get an accurate trajectory for a projectile one must often take account of several complications. For example, if a projectile goes very high then we have to allow for the reduction in air resistance as atmospheric density decreases. To illustrate this, consider an iron cannonball (diameter 15 cm, density 7.8 g/cm^3) that is fired with initial velocity 300 m/s at 50 degrees above the horizontal. The drag force is approximately quadratic, but since the drag is proportional to the atmospheric density and the density falls off exponentially with height, the drag force is $f = c(y)v^2$ where $c(y) = \gamma D^2 \exp(-y/\lambda)$ with γ given by (2.6) and $\lambda \approx 10,000$ m. **(a)** Write down the equations of motion for the cannonball and use appropriate software to solve numerically for $x(t)$ and $y(t)$ for $0 \leq t \leq 35$ s. Plot the ball's trajectory and find its horizontal range. **(b)** Do the same calculation ignoring the variation of atmospheric density [that is, setting $c(y) = c(0)$], and yet again ignoring air resistance entirely. Plot all three trajectories for appropriate time intervals on the same graph. You will find that in this case air resistance makes a huge difference and that the variation of air resistance makes a small, but not negligible, difference.

SECTION 2.6   **Complex Exponentials**

2.45 ★ **(a)** Using Euler's relation (2.76), prove that any complex number $z = x + iy$ can be written in the form $z = re^{i\theta}$, where r and θ are real. Describe the significance of r and θ with reference to the complex plane. **(b)** Write $z = 3 + 4i$ in the form $z = re^{i\theta}$. **(c)** Write $z = 2e^{-i\pi/3}$ in the form $x + iy$.

2.46 ★ For any complex number $z = x + iy$, the **real** and **imaginary parts** are defined as the real numbers $\text{Re}(z) = x$ and $\text{Im}(z) = y$. The **modulus** or **absolute value** is $|z| = \sqrt{x^2 + y^2}$ and the **phase** or **angle** is the value of θ when z is expressed as $z = re^{i\theta}$. The **complex conjugate** is $z^* = x - iy$. (This last is the notation used by most physicists; most mathematicians use $\overline{z}$.) For each of the following complex numbers, find the real and imaginary parts, the modulus and phase, and the complex conjugate, and sketch z and z^* in the complex plane:

$$\textbf{(a)}\, z = 1 + i \qquad \textbf{(b)}\, z = 1 - i\sqrt{3}$$
$$\textbf{(c)}\, z = \sqrt{2}e^{-i\pi/4} \qquad \textbf{(d)}\, z = 5e^{i\omega t}.$$

In part (d), ω is a constant and t is the time.

2.47 ★ For each of the following two pairs of numbers compute $z + w$, $z - w$, zw, and z/w.

$$\textbf{(a)}\, z = 6 + 8i \;\text{ and }\; w = 3 - 4i \quad \textbf{(b)}\, z = 8e^{i\pi/3} \;\text{ and }\; w = 4e^{i\pi/6}.$$

Notice that for adding and subtracting complex numbers, the form $x + iy$ is more convenient, but for multiplying and especially dividing, the form $re^{i\theta}$ is more convenient. In part (a), a clever trick for finding z/w without converting to the form $re^{i\theta}$ is to multiply top and bottom by w^*; try this one both ways.

2.48 ★ Prove that $|z| = \sqrt{z^*z}$ for any complex number z.

2.49 ★ Consider the complex number $z = e^{i\theta} = \cos\theta + i\sin\theta$. **(a)** By evaluating z^2 two different ways, prove the trig identities $\cos 2\theta = \cos^2\theta - \sin^2\theta$ and $\sin 2\theta = 2\sin\theta\cos\theta$. **(b)** Use the same technique to find corresponding identities for $\cos 3\theta$ and $\sin 3\theta$.

2.50 ★ Use the series definition (2.72) of e^z to prove that[12] $de^z/dz = e^z$.

2.51 ★★ Use the series definition (2.72) of e^z to prove that $e^z e^w = e^{z+w}$. [*Hint:* If you write down the left side as a product of two series, you will have a huge sum of terms like $z^n w^m$. If you group together all the terms for which $n + m$ is the same (call it p) and use the binomial theorem, you will find you have the series for the right side.]

SECTION 2.7   Solution for the Charge in a B Field

2.52 ★ The transverse velocity of the particle in Sections 2.5 and 2.7 is contained in (2.77), since $\eta = v_x + iv_y$. By taking the real and imaginary parts, find expressions for v_x and v_y separately. Based on these expressions describe the time dependence of the transverse velocity.

2.53 ★ A charged particle of mass m and positive charge q moves in uniform electric and magnetic fields, $\mathbf{E}$ and $\mathbf{B}$, both pointing in the z direction. The net force on the particle is $\mathbf{F} = q(\mathbf{E} + \mathbf{v} \times \mathbf{B})$. Write down the equation of motion for the particle and resolve it into its three components. Solve the equations and describe the particle's motion.

2.54 ★★ In Section 2.5 we solved the equations of motion (2.68) for the transverse velocity of a charge in a magnetic field by the trick of using the complex number $\eta = v_x + iv_y$. As you might imagine, the equations can certainly be solved without this trick. Here is one way: **(a)** Differentiate the first of equations (2.68) with respect to t and use the second to give you a second-order differential equation for v_x. This is an equation you should recognize [if not, look at Equation (1.55)] and you can write down its general solution. Once you know v_x, (2.68) tells you v_y. **(b)** Show that the general solution you get here is the same as the general solution contained in (2.77), as disentangled in Problem 2.52.

---

[12] If you are the type who worries about mathematical niceties, you may be wondering if it is permissible to differentiate an infinite series. Fortunately, in the case of a power series (such as this), there is a theorem that guarantees the series can be differentiated for any z inside the "radius of convergence." Since the radius of convergence of the series for e^z is infinite, we can differentiate it for *any* value of z.

2.55 ★★★ A charged particle of mass m and positive charge q moves in uniform electric and magnetic fields, **E** pointing in the y direction and **B** in the z direction (an arrangement called "crossed E and B fields"). Suppose the particle is initially at the origin and is given a kick at time $t = 0$ along the x axis with $v_x = v_{xo}$ (positive or negative). **(a)** Write down the equation of motion for the particle and resolve it into its three components. Show that the motion remains in the plane $z = 0$. **(b)** Prove that there is a unique value of v_{xo}, called the drift speed v_{dr}, for which the particle moves undeflected through the fields. (This is the basis of velocity selectors, which select particles traveling at one chosen speed from a beam with many different speeds.) **(c)** Solve the equations of motion to give the particle's velocity as a function of t, for arbitrary values of v_{xo}. [*Hint:* The equations for (v_x, v_y) should look very like Equations (2.68) except for an offset of v_x by a constant. If you make a change of variables of the form $u_x = v_x - v_{dr}$ and $u_y = v_y$, the equations for (u_x, u_y) will have exactly the form (2.68), whose general solution you know.] **(d)** Integrate the velocity to find the position as a function of t and sketch the trajectory for various values of v_{xo}.

CHAPTER 3

Momentum and Angular Momentum

In this and the next chapter I shall describe the great conservation laws of momentum, angular momentum, and energy. These three laws are closely related to one another and are perhaps the most important of the small number of conservation laws that are regarded as cornerstones of all modern physics. Curiously, in classical mechanics the first two laws (momentum and angular momentum) are very different from the last (energy). It is a relatively easy matter to prove the first two from Newton's laws (indeed we already have proved conservation of momentum), whereas the proof of energy conservation is surprisingly subtle. I discuss momentum and angular momentum in this rather short chapter and energy in Chapter 4, which is appreciably longer.

3.1 Conservation of Momentum

In Chapter 1 we examined a system of N particles labeled $\alpha = 1, \cdots, N$. We found that as long as all the internal forces obey Newton's third law, the rate of change of the system's total linear momentum $\mathbf{P} = \mathbf{p}_1 + \cdots + \mathbf{p}_N = \sum \mathbf{p}_\alpha$ is determined entirely by the *external* forces on the system:

$$\dot{\mathbf{P}} = \mathbf{F}^{\text{ext}} \qquad (3.1)$$

where $\mathbf{F}^{\text{ext}}$ denotes the total external force on the system. Because of the third law, the internal forces all cancel out of the rate of change of the *total* momentum. In particular, if the system is isolated, so that the total external force is zero, we have the

> ### Principle of Conservation of Momentum
> If the net external force $\mathbf{F}^{\text{ext}}$ on an N-particle system is zero, the system's total mechanical momentum $\mathbf{P} = \sum m_\alpha \mathbf{v}_\alpha$ is constant.

If our system contains just one particle ($N = 1$), then all forces on the particle are external, and the conservation of momentum is reduced to the not very interesting statement that, in the absence of any forces, the momentum of a single particle is constant, which is just Newton's first law. However, if our system has two or more particles ($N \geq 2$), then momentum conservation is a nontrivial and often useful property, as the following simple and well-known example will remind you.

EXAMPLE 3.1   An Inelastic Collision of Two Bodies

Two bodies (two lumps of putty, for example, or two cars at an intersection) have masses m_1 and m_2 and velocities $\mathbf{v}_1$ and $\mathbf{v}_2$. The two bodies collide and lock together, so they move off as a single unit, as shown in Figure 3.1. (A collision in which the bodies lock together like this is said to be *perfectly inelastic*.) Assuming that any external forces are negligible during the brief moment of collision, find the velocity $\mathbf{v}$ just after the collision.

The initial total momentum, just before the collision, is

$$\mathbf{P}_{\text{in}} = m_1\mathbf{v}_1 + m_2\mathbf{v}_2,$$

and the final momentum, just after the collision, is

$$\mathbf{P}_{\text{fin}} = m_1\mathbf{v} + m_2\mathbf{v} = (m_1 + m_2)\mathbf{v}.$$

(Notice that this last equation illustrates the useful result that, once two bodies have locked together, we can find their momentum by considering them as a single body of mass $m_1 + m_2$.) By conservation of momentum these two momenta must be equal, $\mathbf{P}_{\text{fin}} = \mathbf{P}_{\text{in}}$, and we can easily solve to give the final velocity,

$$\mathbf{v} = \frac{m_1\mathbf{v}_1 + m_2\mathbf{v}_2}{m_1 + m_2}. \tag{3.2}$$

We see that the final velocity is just the weighted average of the original velocities $\mathbf{v}_1$ and $\mathbf{v}_2$, weighted by the corresponding masses m_1 and m_2.

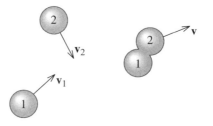

Figure 3.1   A perfectly inelastic collision between two lumps of putty.

An important special case is when one of the bodies is initially at rest, as when a speeding car rams a stationary car at a stop light. With $\mathbf{v}_2 = 0$, Equation (3.2) reduces to

$$\mathbf{v} = \frac{m_1}{m_1 + m_2}\mathbf{v}_1. \tag{3.3}$$

In this case the final velocity is always in the same direction as $\mathbf{v}_1$ but is reduced by the factor $m_1/(m_1 + m_2)$. The result (3.3) is used by police investigating car crashes, since it lets them find the unknown velocity $\mathbf{v}_1$ of a speeder who has rear-ended a stationary car, in terms of quantities that can be measured after the event. (The final velocity $\mathbf{v}$ can be found from the skid marks of the combined wreck.)

This sort of analysis of collisions, using conservation of momentum, is an important tool in solving many problems ranging from nuclear reactions, through car crashes, to collisions of galaxies.

3.2   Rockets

A beautiful example of the use of momentum conservation is the analysis of rocket propulsion. The basic problem that is solved by the rocket is this: With no external agent to push on or be pushed by, how does an object get itself moving? You can put yourself in the same difficulty by imagining yourself stranded on a perfectly frictionless frozen lake. The simplest way to get yourself to shore is to take off anything that is dispensible, such as a boot, and throw it as hard as possible away from the shore. By Newton's third law, when you push one way on the boot, the boot pushes in the opposite direction on you. Thus as you throw the boot, the reaction force of the boot on you will cause you to recoil in the opposite direction and then glide across the ice to shore. A rocket does essentially the same thing. Its motor is designed to hurl the spent fuel out of the back of the rocket, and by the third law, the fuel pushes the rocket forward.

To analyse a rocket's motion quantitatively we must examine the total momentum. Consider the rocket shown in Figure 3.2 with mass m, traveling in the positive x direction (so I can abbreviate v_x as just v) and ejecting spent fuel at the exhaust speed v_{ex} relative to the rocket. Since the rocket is ejecting mass, the rocket's mass m is steadily decreasing. At time t, the momentum is $P(t) = mv$. A short time later at[1] $t + dt$, the rocket's mass is $(m + dm)$, where dm is negative, and its momentum is $(m + dm)(v + dv)$. The fuel ejected in the time dt has mass $(-dm)$ and velocity

---

[1] Concerning the use of the small quantities like dt and dm, I recommend again the view that they are small *but nonzero* increments, with dt chosen sufficiently small that dm divided by dt is (within whatever we have chosen as our desired accuracy) equal to the derivative dm/dt. For more details, see the footnote immediately before Equation (2.47).

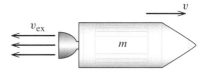

Figure 3.2   A rocket of mass m travels to the right with speed v and ejects spent fuel with exhaust speed v_{ex} relative to the rocket.

$v - v_{\text{ex}}$ relative to the ground. Thus the total momentum (rocket plus the fuel just ejected) at $t + dt$ is

$$P(t + dt) = (m + dm)(v + dv) - dm(v - v_{\text{ex}}) = mv + m\,dv + dm\,v_{\text{ex}}$$

where I have neglected the doubly small product $dm\,dv$. Therefore, the change in total momentum is

$$dP = P(t + dt) - P(t) = m\,dv + dm\,v_{\text{ex}}. \tag{3.4}$$

If there is a net external force F^{ext} (gravity, for instance), this change of momentum is $F^{\text{ext}}\,dt$. (See Problem 3.11.) Here I shall assume that there are no external forces, so that P is constant and $dP = 0$. Therefore

$$m\,dv = -dm\,v_{\text{ex}}. \tag{3.5}$$

Dividing both sides by dt, we can rewrite this as

$$m\dot{v} = -\dot{m}v_{\text{ex}} \tag{3.6}$$

where $-\dot{m}$ is the rate at which the rocket's engine is ejecting mass. This equation looks just like Newton's second law ($m\dot{v} = F$) for an ordinary particle, except that the product $-\dot{m}v_{\text{ex}}$ on the right plays the role of the force. For this reason this product is often called the **thrust**:

$$\text{thrust} = -\dot{m}v_{\text{ex}}. \tag{3.7}$$

(Since $\dot{m}$ is negative, this defines the thrust to be positive.)

Equation (3.5) can be solved by separation of variables. Dividing both sides by m gives

$$dv = -v_{\text{ex}}\frac{dm}{m}.$$

If the exhaust speed v_{ex} is constant, this equation can be integrated to give

$$v - v_{\text{o}} = v_{\text{ex}}\ln(m_{\text{o}}/m) \tag{3.8}$$

where v_{o} is the initial velocity and m_{o} is the initial mass of the rocket (including fuel and payload). This result puts a significant restriction on the maximum speed of the rocket. The ratio m_{o}/m is largest when all the fuel is burned and m is just the mass of rocket plus payload. Even if, for example, the original mass is 90% fuel, this ratio is only 10, and, since $\ln 10 = 2.3$, this says that the speed gained, $v - v_0$, cannot be more than 2.3 times v_{ex}. This means that rocket engineers try to make v_{ex} as big as possible and also design multistage rockets, which can jettison the heavy fuel tanks of the early stages to reduce the total mass for later stages.[2]

3.3   The Center of Mass

Several of the ideas of Section 3.1 can be rephrased in terms of the important notion of a system's center of mass. Let us consider a group of N particles, $\alpha = 1, \cdots, N$, with masses m_α and positions $\mathbf{r}_\alpha$ measured relative to an origin O. The **center of mass** (or **CM**) of this system is defined to be the position (relative to the same origin O)

$$\mathbf{R} = \frac{1}{M} \sum_{\alpha=1}^{N} m_\alpha \mathbf{r}_\alpha = \frac{m_1 \mathbf{r}_1 + \cdots + m_N \mathbf{r}_N}{M} \tag{3.9}$$

where M denotes the total mass of all of the particles, $M = \sum m_\alpha$. The first thing to note about this definition is that it is a vector equation. The CM position is a vector $\mathbf{R}$ with three components (X, Y, Z), and Equation (3.9) is equivalent to three equations giving these three components,

$$X = \frac{1}{M} \sum_{\alpha=1}^{N} m_\alpha x_\alpha, \qquad Y = \frac{1}{M} \sum_{\alpha=1}^{N} m_\alpha y_\alpha, \qquad Z = \frac{1}{M} \sum_{\alpha=1}^{N} m_\alpha z_\alpha.$$

Either way, the CM position $\mathbf{R}$ is a weighted average of the positions $\mathbf{r}_1, \cdots, \mathbf{r}_N$, in which each position $\mathbf{r}_\alpha$ is weighted by the corresponding mass m_α. (Equivalently, it is the sum of the $\mathbf{r}_\alpha$, each multiplied by the *fraction* of the total mass at $\mathbf{r}_\alpha$.)

To get a feeling for the CM, it may help to consider the case of just two particles $(N = 2)$. In this case, the definition (3.9) reads

$$\mathbf{R} = \frac{m_1 \mathbf{r}_1 + m_2 \mathbf{r}_2}{m_1 + m_2}. \tag{3.10}$$

It is easy to verify that the CM position has several familiar properties. For example, you can show (Problem 3.18) that the CM defined by (3.10) lies on the line joining the two particles, as shown in Figure 3.3. It is also easy to show that the distances of the CM from m_1 and m_2 are in the ratio m_2/m_1, so that the CM lies closer to the more massive particle. (In Figure 3.3 this ratio is 1/3.) In particular, if m_1 is much greater than m_2, the CM will be very close to $\mathbf{r}_1$. More generally, going back to Equation (3.9) for the CM of N particles, we see that if m_1 is much greater than any of the other masses (as is the case for the sun as compared to all the planets), then $m_1 \approx M$ while $m_\alpha \ll M$ for all other particles; this means that $\mathbf{R}$ is very close to $\mathbf{r}_1$. Thus, for example, the CM of the solar system is very close to the sun.

---

[2] Jettisoning the fuel tanks of stage 1 reduces the inital and final masses of stage 2 by the same amount. This *increases* the ratio m_0/m when we apply (3.8) to stage 2. See Problem 3.12.

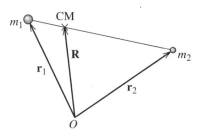

Figure 3.3   The CM of two particles lies at the position
$\mathbf{R} = (m_1\mathbf{r}_1 + m_2\mathbf{r}_2)/M$. You can prove that this lies on
the line joining m_1 to m_2, as shown, and that the distances
of the CM from m_1 and m_2 are in the ratio m_2/m_1.

We can now write the total momentum $\mathbf{P}$ of any N-particle system in terms of the
system's CM as follows:

$$\mathbf{P} = \sum_\alpha \mathbf{p}_\alpha = \sum_\alpha m_\alpha \dot{\mathbf{r}}_\alpha = M\dot{\mathbf{R}} \tag{3.11}$$

where the last equality is just the derivative of the definition (3.9) of $\mathbf{R}$ (multiplied by
M). This remarkable result says that the total momentum of the N particles is exactly
the same as that of a single particle of mass M and velocity equal to that of the CM.

We get an even more striking result when we differentiate (3.11). According to
(3.1), the derivative of $\mathbf{P}$ is just $\mathbf{F}^{\text{ext}}$. Therefore, (3.11) implies that

$$\mathbf{F}^{\text{ext}} = M\ddot{\mathbf{R}}. \tag{3.12}$$

That is, the center of mass $\mathbf{R}$ moves exactly as if it were a single particle of mass M,
subject to the net external force on the system. This result is the main reason why we
can often treat extended bodies, such as baseballs and planets, as if they were point
particles. Provided a body is small compared to the scale of its trajectory, its CM
position $\mathbf{R}$ is a good representative of its overall location, and (3.12) implies that $\mathbf{R}$
moves just like a point particle.

Given the importance of the CM, you need to feel comfortable calculating the CM
position for various systems. You may have had plenty of practice in introductory
physics or in a calculus course, but, in case you didn't, there are several exercises at
the end of this chapter. One important point to bear in mind is that when the mass
in a body is distributed continuously, the sum in the definition (3.9) goes over to an
integral

$$\mathbf{R} = \frac{1}{M} \int \mathbf{r}\, dm = \frac{1}{M} \int \varrho\, \mathbf{r}\, dV \tag{3.13}$$

where ϱ is the mass density of the body, dV denotes an element of volume, and the
integral runs over the whole body (that is, everywhere $\varrho \neq 0$). We shall be using similar
integrals to evaluate the moment of inertia tensor in Chapter 10. Meanwhile, here is
one example:

EXAMPLE 3.2   The CM of a Solid Cone

Find the CM position for the uniform solid cone shown in Figure 3.4.

It is perhaps obvious by symmetry that the CM lies on the axis of symmetry (the z axis), but this also follows immediately from the integral (3.13). For example, if you consider the x component of that integral, it is easy to see that the contribution from any point (x, y, z) is exactly cancelled by that from the point $(-x, y, z)$. That is, the integral for X is zero. Because the same argument applies to Y, the CM lies on the z axis. To find the height Z of the CM, we must evaluate the integral

$$Z = \frac{1}{M} \int \varrho \, z \, dV = \frac{\varrho}{M} \int z \, dx \, dy \, dz$$

where I could take the factor ϱ outside the integral since ϱ is constant throughout the cone (as long as we understand the integral is limited to the inside of the cone) and I have changed the volume element dV to $dx \, dy \, dz$. For any given z, the integral over x and y runs over a circle of radius $r = Rz/h$, giving a factor of $\pi r^2 = \pi R^2 z^2 / h^2$, so that

$$Z = \frac{\varrho \pi R^2}{M h^2} \int_0^h z^3 dz = \frac{\varrho \pi R^2}{M h^2} \frac{h^4}{4} = \frac{3}{4} h$$

where in the last step I replaced the mass M by ϱ times the volume or $M = \frac{1}{3} \varrho \pi R^2 h$. We conclude that the CM is on the axis of the cone at a distance $\frac{3}{4} h$ from the vertex (or $\frac{1}{4} h$ from the base).

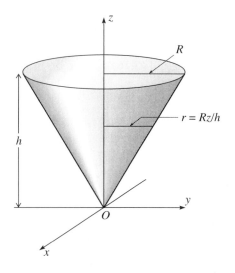

Figure 3.4   A solid cone, centered on the z axis, with vertex at the origin and uniform mass density ϱ. Its height is h and its base has radius R.

3.4   Angular Momentum for a Single Particle

In many ways the conservation of angular momentum parallels the conservation of ordinary (or "linear") momentum. Nevertheless, I would like to review the formalism in detail, first for a single particle and then for a multiparticle system. This will introduce several important ideas and some useful mathematics.

The **angular momentum** $\boldsymbol{\ell}$ of a single particle is defined as the vector

$$\boldsymbol{\ell} = \mathbf{r} \times \mathbf{p}. \tag{3.14}$$

Here $\mathbf{r} \times \mathbf{p}$ is the vector product of the particle's position vector $\mathbf{r}$, relative to the chosen origin O, and its momentum $\mathbf{p}$, as shown in Figure 3.5. Notice that because $\mathbf{r}$ depends on the choice of origin, the same is true of $\boldsymbol{\ell}$: The angular momentum $\boldsymbol{\ell}$ (unlike the linear momentum $\mathbf{p}$) depends on the choice of origin, and we should, strictly speaking, refer to $\boldsymbol{\ell}$ as the angular momentum *relative to O*.

The time rate of change of $\boldsymbol{\ell}$ is easily found:

$$\dot{\boldsymbol{\ell}} = \frac{d}{dt}(\mathbf{r} \times \mathbf{p}) = (\dot{\mathbf{r}} \times \mathbf{p}) + (\mathbf{r} \times \dot{\mathbf{p}}). \tag{3.15}$$

(You can easily check that the product rule can be used for differentiating vector products, as long as you are careful to keep the vectors in the right order. See Problem 1.17.) In the first term on the right, we can replace $\mathbf{p}$ by $m\dot{\mathbf{r}}$, and, because the cross product of any two parallel vectors is zero, the first term is zero. In the second term, we can replace $\dot{\mathbf{p}}$ by the net force $\mathbf{F}$ on the particle, and we get

$$\dot{\boldsymbol{\ell}} = \mathbf{r} \times \mathbf{F} \equiv \boldsymbol{\Gamma}. \tag{3.16}$$

Here $\boldsymbol{\Gamma}$ (Greek capital gamma) denotes the net torque about O on the particle, defined as $\mathbf{r} \times \mathbf{F}$. (Other popular symbols for torque are $\boldsymbol{\tau}$ and $\mathbf{N}$.) In words, (3.16) says that the rate of change of a particle's angular momentum about the origin O is equal to the net applied torque about O. Equation (3.16) is the rotational analog of the equation $\dot{\mathbf{p}} = \mathbf{F}$ for the linear momentum, and (3.16) is often described as the rotational form of Newton's second law.

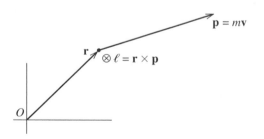

Figure 3.5   For any particle with position $\mathbf{r}$ relative to the origin O and momentum $\mathbf{p}$, the angular momentum about O is defined as the vector $\boldsymbol{\ell} = \mathbf{r} \times \mathbf{p}$. For the case shown, $\boldsymbol{\ell}$ points into the page.

Figure 3.6   A planet (mass m) is subject to the central force of the sun (mass M). If we choose the origin at the sun, then $\mathbf{r} \times \mathbf{F} = 0$, and the planet's angular momentum about O is constant.

In many one-particle problems one can choose the origin O so that the net torque $\boldsymbol{\Gamma}$ (about the chosen O) is zero. In this case, the particle's angular momentum about O is constant. Consider, for example, a single planet (or comet) orbiting the sun. The only force on the planet is the gravitational pull GmM/r^2 of the sun, as shown in Figure 3.6. A crucial property of the gravitational force is that it is **central**, that is, directed along the line joining the two centers. This means that $\mathbf{F}$ is parallel (actually, antiparallel) to the position vector $\mathbf{r}$ measured from the sun, and hence that $\mathbf{r} \times \mathbf{F} = 0$. Thus if we choose our origin at the sun, the planet's angular momentum about O is constant, a fact that greatly simplifies the analysis of planetary motion. For example, because $\mathbf{r} \times \mathbf{p}$ is constant, $\mathbf{r}$ and $\mathbf{p}$ must remain in a fixed plane; in other words, the planet's orbit is confined to a single plane containing the sun, and the problem is reduced to two dimensions, a result we shall exploit in Chapter 8.

Kepler's Second Law

One of the earliest triumphs for Newton's mechanics was that he was able to explain Kepler's second law as a simple consequence of conservation of angular momentum. Newton's laws of motion were published in 1687 in his famous book *Principia*. Nearly eighty years earlier, the German astronomer Johannes Kepler (1571□630) had published his three laws of planetary motion.[3] These laws are quite different from Newton's laws in that they are simply mathematical descriptions of the observed motion of the planets. For example, the first law states that the planets move around the sun in ellipses with the sun at one focus. Kepler's laws make no attempt to *explain* planetary motion in terms of more fundamental ideas; they are just summaries — brilliant summaries, requiring great insight, but nonetheless just summaries — of the observed motions of the planets. All three of Kepler's laws turn out to be consequences of Newton's laws of motion. I shall derive the first and third of the Kepler laws in Chapter 8. The second we are ready to discuss now.

---

[3] Kepler's first two laws appeared in his book *Astronomia Nova* in 1609 and the third in another book, *Harmonices Mundi*, published in 1619.

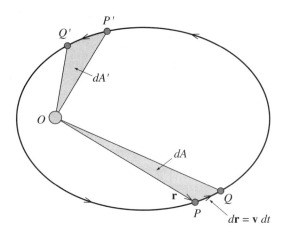

Figure 3.7   The orbit of a planet with the sun fixed at O. Kepler's second law asserts that if the two pairs of points P, Q and P', Q' are separated by equal time intervals, $dt = dt'$, then the two areas dA and dA' are equal.

Kepler's second law is generally stated something like this:

Kepler's Second Law

As each planet moves around the sun, a line drawn from the planet to the sun sweeps out equal areas in equal times.

This rather curious statement is illustrated in Figure 3.7, which shows the path of a planet or comet — the law applies to comets as well — orbiting about the sun at the origin O. (Throughout this discussion, I shall make the approximation that the sun is fixed; we shall see how to allow for the very small motion of the sun in Chapter 8.) The area "swept out" by the planet moving between any two points P and Q is just the area of the triangle OPQ. (Strictly speaking the "triangle" is the area between the two lines OP and OQ and the *arc* PQ. However, it is sufficient to consider pairs of points P and Q that are close together, in which case the difference between the arc PQ and the straight line PQ is negligible.) I shall denote the time elapsed between the planet's visiting P and Q by dt and the corresponding area of OPQ by dA. Kepler's second law asserts that if we choose any other pair of points P' and Q' separated by the same time interval ($dt' = dt$), then the area $OP'Q'$ will be the same as OPQ, or $dA' = dA$. Equivalently we can divide both sides of this equality by dt and assert that the *rate* at which the planet sweeps out area, dA/dt, is the same at all points on the orbit; that is, dA/dt is constant.

To prove this result, we note first that the line OP is just the position vector $\mathbf{r}$, and PQ is the displacement $d\mathbf{r} = \mathbf{v}\, dt$. Now, it is a well-known property of the vector product that if two sides of a triangle are given by vectors **a** and **b**, then the area of the triangle is $\frac{1}{2}|\mathbf{a} \times \mathbf{b}|$. (See Problem 3.24.) Thus the area of the triangle OPQ is

$$dA = \tfrac{1}{2}|\mathbf{r} \times \mathbf{v}\, dt|.$$

Replacing **v** by **p**/m and dividing both sides by dt, we find that

$$\frac{dA}{dt} = \frac{1}{2m}|\mathbf{r} \times \mathbf{p}| = \frac{1}{2m}\ell \tag{3.17}$$

where ℓ denotes the magnitude of the angular momentum $\boldsymbol{\ell} = \mathbf{r} \times \mathbf{p}$. Since the planet's angular momentum about the sun is conserved, this establishes that dA/dt is constant, which, as we have seen, is the content of Kepler's second law.

An alternative proof of the same result adds some additional insight: It is a straightforward exercise to show that (Problem 3.27)

$$\ell = mr^2\omega \tag{3.18}$$

where $\omega = \dot{\phi}$ is the planet's angular velocity around the sun. And it is an equally simple geometrical exercise to show that the rate of sweeping out area is

$$dA/dt = \tfrac{1}{2}r^2\omega. \tag{3.19}$$

Comparison of (3.18) and (3.19) shows that ℓ is constant if and only if dA/dt is constant. That is, conservation of angular momentum is exactly equivalent to Kepler's second law. In addition, we see that as the planet (or comet) approaches closer to the sun (r decreasing) its angular velocity ω necessarily increases. Specifically, ω is inversely proportional to r^2; for example, if the value of r at point P' is half that at P, then the angular velocity ω at P' is four times that at P.

It is interesting to note that our proof of Kepler's second law depended only on the fact that the gravitational force is central and hence that the planet's angular momentum about the sun is constant. Thus Kepler's second law is true for an object that moves under the influence of *any* central force. By contrast, we shall see in Chapter 8 that the first and third laws (in particular the first, which says that the orbits are ellipses with the sun at one focus) depend on the inverse-square nature of the gravitational force and are not true for other force laws.

3.5   Angular Momentum for Several Particles

Let us next discuss a system of N particles, $\alpha = 1, 2, \cdots, N$, each with its angular momentum $\boldsymbol{\ell}_\alpha = \mathbf{r}_\alpha \times \mathbf{p}_\alpha$ (with all of the $\mathbf{r}_\alpha$ measured from the same origin O, of course). We define the **total angular momentum L** as

$$\mathbf{L} = \sum_{\alpha=1}^{N} \boldsymbol{\ell}_\alpha = \sum_{\alpha=1}^{N} \mathbf{r}_\alpha \times \mathbf{p}_\alpha. \tag{3.20}$$

Differentiating with respect to t and using the result (3.16), we find that

$$\dot{\mathbf{L}} = \sum_\alpha \dot{\boldsymbol{\ell}}_\alpha = \sum_\alpha \mathbf{r}_\alpha \times \mathbf{F}_\alpha, \tag{3.21}$$

where, as usual, $\mathbf{F}_\alpha$ denotes the net force on particle α. This result shows that the rate of change of $\mathbf{L}$ is just the net torque on the whole system, an important result in its own right. However, my interest now is to separate the effects of the internal and external forces. As in Equation (1.25) we write $\mathbf{F}_\alpha$ as

$$\text{(net force on particle } \alpha) = \mathbf{F}_\alpha = \sum_{\beta \neq \alpha} \mathbf{F}_{\alpha\beta} + \mathbf{F}_\alpha^{\text{ext}} \tag{3.22}$$

where, as before, $\mathbf{F}_{\alpha\beta}$ denotes the force exerted on particle α by particle β, and $\mathbf{F}_\alpha^{\text{ext}}$ is the net force exerted on particle α by all agents outside our N-particle system. Substituting into (3.21), we find that

$$\dot{\mathbf{L}} = \sum_\alpha \sum_{\beta \neq \alpha} \mathbf{r}_\alpha \times \mathbf{F}_{\alpha\beta} + \sum_\alpha \mathbf{r}_\alpha \times \mathbf{F}_\alpha^{\text{ext}}. \tag{3.23}$$

Equation (3.23) corresponds to Equation (1.27) in our discussion of linear momentum back in Chapter 1, and we can rework it in much the same way as there, with one interesting additional twist. We can regroup the terms of the double sum, pairing each term $\alpha\beta$ with the corresponding term $\beta\alpha$, to give[4]

$$\sum_\alpha \sum_{\beta \neq \alpha} \mathbf{r}_\alpha \times \mathbf{F}_{\alpha\beta} = \sum_\alpha \sum_{\beta > \alpha} (\mathbf{r}_\alpha \times \mathbf{F}_{\alpha\beta} + \mathbf{r}_\beta \times \mathbf{F}_{\beta\alpha}). \tag{3.24}$$

If we assume that all the internal forces obey the third law ($\mathbf{F}_{\alpha\beta} = -\mathbf{F}_{\beta\alpha}$), then we can rewrite the sum on the right as

$$\sum_\alpha \sum_{\beta > \alpha} (\mathbf{r}_\alpha - \mathbf{r}_\beta) \times \mathbf{F}_{\alpha\beta}. \tag{3.25}$$

To understand this sum, we must examine the vector $(\mathbf{r}_\alpha - \mathbf{r}_\beta) = \mathbf{r}_{\alpha\beta}$, say. This is illustrated in Figure 3.8, where we see that $\mathbf{r}_{\alpha\beta}$ is the vector pointing toward particle α from particle β. If, in addition to satisfying the third law, the forces $\mathbf{F}_{\alpha\beta}$ are all *central*, then the two vectors $\mathbf{r}_{\alpha\beta}$ and $\mathbf{F}_{\alpha\beta}$ point along the same line, and their cross product is zero.

Returning to Equation (3.23), we conclude that, provided our various assumptions are valid, the double sum in (3.23) is zero. The remaining single sum is just the net external torque, and we conclude that

$$\dot{\mathbf{L}} = \boldsymbol{\Gamma}^{\text{ext}}. \tag{3.26}$$

In particular, if the net external torque is zero, we have the

---

[4] Be sure you understand what has happened here. For example, I have paired the term $\mathbf{r}_1 \times \mathbf{F}_{12}$ with the term $\mathbf{r}_2 \times \mathbf{F}_{21}$.

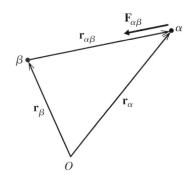

Figure 3.8   The vector $\mathbf{r}_{\alpha\beta} = (\mathbf{r}_\alpha - \mathbf{r}_\beta)$ points to particle α
from particle β. If the force $\mathbf{F}_{\alpha\beta}$ is central (points along the
line joining α and β), then $\mathbf{r}_{\alpha\beta}$ and $\mathbf{F}_{\alpha\beta}$ are collinear and their
cross product is zero.

Principle of Conservation of Angular Momentum

If the net external torque on an N-particle system is zero, the system's total
angular momentum $\mathbf{L} = \sum \mathbf{r}_\alpha \times \mathbf{p}_\alpha$ is constant.

The validity of this principle depends on our two assumptions that all internal forces
$\mathbf{F}_{\alpha\beta}$ are central and satisfy the third law. Since these assumptions are almost always
valid, the principle (as stated) is likewise. It is of the greatest utility in solving many
problems, as I shall illustrate shortly with a couple of simple examples.

The Moment of Inertia

Before discussing an example, it is worth noting that the calculation of angular
momenta does not always require one to go back to the basic definition (3.20). As you
probably recall from your introductory physics course, for a rigid body rotating about
a fixed axis (for example, a wheel rotating on its fixed axle), the rather complicated
sum (3.20) can be expressed in terms of the moment of inertia and the angular velocity
of rotation. Specifically, if we take the axis of rotation to be the z axis, then L_z, the z
component of angular momentum, is just $L_z = I\omega$, where I is the **moment of inertia**
of the body for the given axis, and ω is the angular velocity of rotation. We shall prove
and generalize this result in Chapter 10, or you can prove it yourself with the guidance
of Problem 3.30. For now, I shall ask you to carry it over from introductory physics.
In particular, as you may recall, the moments of inertia of various standard bodies
are known. For example, for a uniform disk (mass M, radius R) rotating about its
axis, $I = \frac{1}{2}MR^2$. For a uniform solid sphere rotating about a diameter, $I = \frac{2}{5}MR^2$. In
general, for any multiparticle system, $I = \sum m_\alpha \rho_\alpha^2$, where ρ_α is the distance of the
mass m_α from the axis of rotation.

EXAMPLE 3.3 Collision of a Lump of Putty with a Turntable

A uniform circular turntable (mass M, radius R, center O) is at rest in the xy plane and is mounted on a frictionless axle, which lies along the vertical z axis. I throw a lump of putty (mass m) with speed v toward the edge of the turntable, so it approaches along a line that passes within a distance[5] b of O, as shown in Figure 3.9. When the putty hits the turntable, it sticks to the edge, and the two rotate together with angular velocity ω. Find ω.

This problem is easily solved using conservation of angular momentum. Because the turntable is mounted on a frictionless axle, there is no torque on the table in the z direction. Therefore the z component of the external torque on the system is zero, and L_z is conserved. (This is true even if we include gravity, which acts in the z direction and contributes nothing to the torque in the z direction.) Before the collision, the turntable has zero angular momentum, while the putty has $\boldsymbol{\ell} = \mathbf{r} \times \mathbf{p}$, which points in the z direction. Thus the initial total angular momentum has z component

$$L_z^{\text{in}} = \ell_z = r(mv) \sin \theta = mvb.$$

After the collision, the putty and turntable rotate together about the z axis with total moment of inertia[6] $I = (m + M/2)R^2$, and the z component of the final angular momentum is $L_z^{\text{fin}} = I\omega$. Therefore, conservation of angular momentum in the form $L_z^{\text{in}} = L_z^{\text{fin}}$ tells us that

$$mvb = (m + M/2)R^2\omega,$$

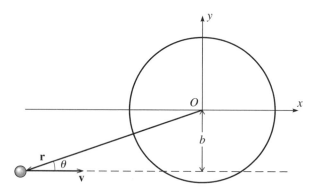

Figure 3.9   A lump of putty of mass m is thrown with velocity $\mathbf{v}$ at a stationary turntable. The putty's line of approach passes within the distance b of the table's center O.

---

[5] In collision theory — the theory of collisions, usually between atomic or subatomic particles — the distance b is called the *impact parameter*.

[6] This is mR^2 for the putty stuck at radius R plus $\frac{1}{2}MR^2$ for the uniform turntable.

or, solving for ω,

$$\omega = \frac{m}{(m + M/2)} \cdot \frac{vb}{R^2}. \tag{3.27}$$

This answer is not especially interesting. What *is* interesting is that we were able to find it with so comparatively little effort. This is typical of the conservation laws, that they can answer many questions so simply. The kind of analysis used here can be used in many situations (such as nuclear reactions) where an incident projectile is absorbed by a stationary target and its angular momentum is shared between the two bodies.

Angular Momentum about the CM

The conservation of angular momentum and the more general result (3.26), $\dot{\mathbf{L}} = \boldsymbol{\Gamma}^{\text{ext}}$, were derived on the assumption that all quantities were measured in an inertial frame, so that Newton's second law could be invoked. This required that both $\mathbf{L}$ and $\boldsymbol{\Gamma}^{\text{ext}}$ be measured about an origin O fixed in some inertial frame. Remarkably, the same two results also hold if $\mathbf{L}$ and $\boldsymbol{\Gamma}^{\text{ext}}$ are measured about the center of mass — even if the CM is being accelerated and so is not fixed in an inertial frame. That is,

$$\frac{d}{dt}\mathbf{L}(\text{about CM}) = \boldsymbol{\Gamma}^{\text{ext}}(\text{about CM}) \tag{3.28}$$

and hence, if $\boldsymbol{\Gamma}^{\text{ext}}(\text{about CM}) = 0$, then $\mathbf{L}(\text{about CM})$ is conserved. We shall prove this result in Chapter 10, or you can prove it yourself with the guidance of Problem 3.37. I mention it now, because it allows a very simple solution to various problems, as the following example illustrates.

EXAMPLE 3.4   A Sliding and Spinning Dumbbell

A dumbbell consisting of two equal masses m mounted on the ends of a rigid massless rod of length $2b$ is at rest on a frictionless horizontal table, lying on the x axis and centered on the origin, as shown in Figure 3.10. At time $t = 0$, the left mass is given a sharp tap, in the shape of a horizontal force $\mathbf{F}$ in the y direction, lasting for a short time Δt. Describe the subsequent motion.

There are actually two parts to this problem: We must find the initial motion immediately after the impulse, and then the subsequent, force-free motion. The initial motion is not hard to guess, but let us derive it using the tools of this chapter. The only external force is the force $\mathbf{F}$ acting in the y direction for the brief time Δt. Since $\dot{\mathbf{P}} = \mathbf{F}^{\text{ext}}$, the total momentum just after the impulse is $\mathbf{P} = \mathbf{F}\,\Delta t$. Since $\mathbf{P} = M\dot{\mathbf{R}}$ (with $M = 2m$), we conclude that the CM starts moving directly up the y axis with velocity

$$\mathbf{v}_{\text{cm}} = \dot{\mathbf{R}} = \mathbf{F}\,\Delta t/2m.$$

While the force $\mathbf{F}$ is acting, there is a torque $\Gamma^{\text{ext}} = Fb$ about the CM, and so, according to (3.28), the initial angular momentum (just after the impulse has

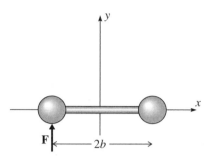

Figure 3.10   The left mass of the dumbbell is given a sharp tap in the y direction.

ceased) is $L = Fb\,\Delta t$. Since $L = I\omega$, with $I = 2mb^2$, we conclude that the dumbbell is spinning clockwise, with initial angular velocity

$$\omega = F\,\Delta t/2mb.$$

The clockwise rotation of the dumbbell means that the left mass is moving up relative to the CM with speed ωb, and its total initial velocity is

$$v_{\text{left}} = v_{\text{cm}} + \omega b = F\,\Delta t/m.$$

By the same token the right mass is moving down relative to the CM, and its total initial velocity is

$$v_{\text{right}} = v_{\text{cm}} - \omega b = 0.$$

That is, the right mass is initially stationary, while the left one carries all the momentum $F\,\Delta t$ of the system.

The subsequent motion is very straightforward. Once the impulse has ceased, there are no external forces or torques. Thus the CM continues to move straight up the y axis with constant speed, and the dumbbell continues to rotate with constant angular momentum about the CM and hence constant angular velocity.

Principal Definitions and Equations of Chapter 3

Equation of Motion for a Rocket

$$m\dot{v} = -\dot{m}v_{\text{ex}} + F^{\text{ext}}. \qquad \text{[Eqs. (3.6) \& (3.29)]}$$

The Center of Mass of Several Particles

$$\mathbf{R} = \frac{1}{M}\sum_{\alpha=1}^{N} m_\alpha \mathbf{r}_\alpha = \frac{m_1\mathbf{r}_1 + \cdots + m_N\mathbf{r}_N}{M} \qquad \text{[Eq. (3.9)]}$$

where M is the total mass of all particles, $M = \sum m_\alpha$.

Angular Momentum

For a single particle with position $\mathbf{r}$ (relative to an origin O) and momentum $\mathbf{p}$, the angular momentum about O is

$$\boldsymbol{\ell} = \mathbf{r} \times \mathbf{p}. \qquad\qquad \text{[Eq. (3.14)]}$$

For several particles, the total angular momentum is

$$\mathbf{L} = \sum_{\alpha=1}^{N} \boldsymbol{\ell}_\alpha = \sum_{\alpha=1}^{N} \mathbf{r}_\alpha \times \mathbf{p}_\alpha. \qquad\qquad \text{[Eq. (3.20)]}$$

Provided all the internal forces are central,

$$\dot{\mathbf{L}} = \boldsymbol{\Gamma}^{\text{ext}} \qquad\qquad \text{[Eq. (3.26)]}$$

where $\boldsymbol{\Gamma}^{\text{ext}}$ is the net external torque.

Problems for Chapter 3

Stars indicate the approximate level of difficulty, from easiest (★) to most difficult (★★★).

SECTION 3.1   Conservation of Momentum

3.1 ★ Consider a gun of mass M (when unloaded) that fires a shell of mass m with muzzle speed v. (That is, the shell's speed relative to the gun is v.) Assuming that the gun is completely free to recoil (no external forces on gun or shell), use conservation of momentum to show that the shell's speed relative to the ground is $v/(1 + m/M)$.

3.2 ★ A shell traveling with speed v_0 exactly horizontally and due north explodes into two equal-mass fragments. It is observed that just after the explosion one fragment is traveling vertically up with speed v_0. What is the velocity of the other fragment?

3.3 ★ A shell traveling with velocity $\mathbf{v}_0$ explodes into three pieces of equal masses. Just after the explosion, one piece has velocity $\mathbf{v}_1 = \mathbf{v}_0$ and the other two have velocities $\mathbf{v}_2$ and $\mathbf{v}_3$ that are equal in magnitude ($v_2 = v_3$) but mutually perpendicular. Find $\mathbf{v}_2$ and $\mathbf{v}_3$ and sketch the three velocities.

3.4 ★★ Two hobos, each of mass m_h, are standing at one end of a stationary railroad flatcar with frictionless wheels and mass m_{fc}. Either hobo can run to the other end of the flatcar and jump off with the same speed u (relative to the car). **(a)** Use conservation of momentum to find the speed of the recoiling car if the two men run and jump simultaneously. **(b)** What is it if the second man starts running only after the first has already jumped? Which procedure gives the greater speed to the car? [*Hint:* The speed u is the speed of either hobo, *relative to the car* just after he has jumped; it has the same value for either man and is the same in parts (a) and (b).]

3.5 ★★ Many applications of conservation of momentum involve conservation of energy as well, and we haven't yet begun our discussion of energy. Nevertheless, you know enough about energy from your introductory physics course to handle some problems of this type. Here is one elegant example: An *elastic collision* between two bodies is defined as a collision in which the total kinetic energy of the two bodies after the collision is the same as that before. (A familiar example is the collision between two billiard balls, which generally lose extremely little of their total kinetic energy.) Consider an elastic collision between two equal mass bodies, one of which is initially at rest. Let their velocities be $\mathbf{v}_1$ and $\mathbf{v}_2 = 0$ before the collision, and $\mathbf{v}'_1$ and $\mathbf{v}'_2$ after. Write down the vector equation representing conservation of momentum and the scalar equation which expresses that the collision is elastic. Use these to prove that the angle between $\mathbf{v}'_1$ and $\mathbf{v}'_2$ is 90°. This result was important in the history of atomic and nuclear physics: That two bodies emerged from a collision traveling on perpendicular paths was strongly suggestive that they had equal mass and had undergone an elastic collision.

SECTION 3.2   Rockets

3.6 ★ In the early stages of the Saturn V rocket's launch, mass was ejected at about 15,000 kg/s, with a speed $v_{ex} \approx 2500$ m/s relative to the rocket. What was the thrust on the rocket? Convert this to tons (1 ton $\approx$ 9000 newtons) and compare with the rocket's initial weight (about 3000 tons).

3.7 ★ The first couple of minutes of the launch of a space shuttle can be described very roughly as follows: The initial mass is 2×10^6 kg, the final mass (after 2 minutes) is about 1×10^6 kg, the average exhaust speed v_{ex} is about 3000 m/s, and the initial velocity is, of course, zero. If all this were taking place in outer space, with negligible gravity, what would be the shuttle's speed at the end of this stage? What is the thrust during the same period and how does it compare with the initial total weight of the shuttle (on earth)?

3.8 ★ A rocket (initial mass m_0) needs to use its engines to hover stationary, just above the ground. **(a)** If it can afford to burn no more than a mass λm_0 of its fuel, for how long can it hover? [*Hint:* Write down the condition that the thrust just balance the force of gravity. You can integrate the resulting equation by separating the variables t and m. Take v_{ex} to be constant.] **(b)** If $v_{ex} \approx 3000$ m/s and $\lambda \approx 10\%$, for how long could the rocket hover just above the earth's surface?

3.9 ★ From the data in Problem 3.7 you can find the space shuttle's initial mass and the rate of ejecting mass $-\dot{m}$ (which you may assume is constant). What is the minimum exhaust speed v_{ex} for which the shuttle would just begin to lift as soon as burn is fully underway? [*Hint:* The thrust must at least balance the shuttle's weight.]

3.10 ★ Consider a rocket (initial mass m_0) accelerating from rest in free space. At first, as it speeds up, its momentum p increases, but as its mass m decreases p eventually begins to decrease. For what value of m is p maximum?

3.11 ★★ **(a)** Consider a rocket traveling in a straight line subject to an external force F^{ext} acting along the same line. Show that the equation of motion is

$$m\dot{v} = -\dot{m}v_{ex} + F^{ext}. \tag{3.29}$$

[Review the derivation of Equation (3.6) but keep the external force term.] **(b)** Specialize to the case of a rocket taking off vertically (from rest) in a gravitational field g, so the equation of motion becomes

$$m\dot{v} = -\dot{m}v_{ex} - mg. \tag{3.30}$$

Assume that the rocket ejects mass at a constant rate, $\dot{m} = -k$ (where k is a positive constant), so that $m = m_0 - kt$. Solve equation (3.30) for v as a function of t, using separation of variables (that is, rewriting the equation so that all terms involving v are on the left and all terms involving t on the right). **(c)** Using the rough data from Problem 3.7, find the space shuttle's speed two minutes into flight, assuming (what is nearly true) that it travels vertically up during this period and that g doesn't change appreciably. Compare with the corresponding result if there were no gravity. **(d)** Describe what would happen to a rocket that was designed so that the first term on the right of Equation (3.30) was smaller than the initial value of the second.

3.12 ★★ To illustrate the use of a multistage rocket consider the following: **(a)** A certain rocket carries 60% of its initial mass as fuel. (That is, the mass of fuel is $0.6m_o$.) What is the rocket's final speed, accelerating from rest in free space, if it burns all its fuel in a single stage? Express your answer as a multiple of v_{ex}. **(b)** Suppose instead it burns the fuel in two stages as follows: In the first stage it burns a mass $0.3m_o$ of fuel. It then jettisons the first-stage fuel tank, which has a mass of $0.1m_o$, and then burns the remaining $0.3m_o$ of fuel. Find the final speed in this case, assuming the same value of v_{ex} throughout, and compare.

3.13 ★★ If you have not already done it, do Problem 3.11(b) and find the speed $v(t)$ of a rocket accelerating vertically from rest in a gravitational field g. Now integrate $v(t)$ and show that the rocket's height as a function of t is

$$ y(t) = v_{ex}t - \frac{1}{2}gt^2 - \frac{mv_{ex}}{k}\ln\left(\frac{m_o}{m}\right). $$

Using the numbers given in Problem 3.7, estimate the space shuttle's height after two minutes.

3.14 ★★ Consider a rocket subject to a linear resistive force, $\mathbf{f} = -b\mathbf{v}$, but no other external forces. Use Equation (3.29) in Problem 3.11 to show that if the rocket starts from rest and ejects mass at a constant rate $k = -\dot{m}$, then its speed is given by

$$ v = \frac{k}{b}v_{ex}\left[1 - \left(\frac{m}{m_o}\right)^{b/k}\right]. $$

SECTION 3.3   The Center of Mass

3.15 ★ Find the position of the center of mass of three particles lying in the xy plane at $\mathbf{r}_1 = (1, 1, 0)$, $\mathbf{r}_2 = (1, -1, 0)$, and $\mathbf{r}_3 = (0, 0, 0)$, if $m_1 = m_2$ and $m_3 = 10m_1$. Illustrate your answer with a sketch and comment.

3.16 ★ The masses of the earth and sun are $M_e \approx 6.0 \times 10^{24}$ and $M_s \approx 2.0 \times 10^{30}$ (both in kg) and their center-to-center distance is 1.5×10^8 km. Find the position of their CM and comment. (The radius of the sun is $R_s \approx 7.0 \times 10^5$ km.)

3.17 ★ The masses of the earth and moon are $M_e \approx 6.0 \times 10^{24}$ and $M_m \approx 7.4 \times 10^{22}$ (both in kg) and their center to center distance is 3.8×10^5 km. Find the position of their CM and comment. (The radius of the earth is $R_e \approx 6.4 \times 10^3$ km.)

3.18 ★★ **(a)** Prove that the CM of any two particles always lies on the line joining them, as illustrated in Figure 3.3. [Write down the vector that points from m_1 to the CM and show that it has the same direction as the vector from m_1 to m_2.] **(b)** Prove that the distances from the CM to m_1 and m_2 are in the ratio m_2/m_1. Explain why if m_1 is much greater than m_2, the CM lies very close to the position of m_1.

3.19 ★★ **(a)** We know that the path of a projectile thrown from the ground is a parabola (if we ignore air resistance). In the light of the result (3.12), what would be the subsequent path of the CM of the pieces if the projectile exploded in midair? **(b)** A shell is fired from level ground so as to hit a target 100 m away. Unluckily the shell explodes prematurely and breaks into two equal pieces. The two pieces land at the same time, and one lands 100 m beyond the target. Where does the other piece land? **(c)** Is the same result true if they land at different times (with one piece still landing 100 m beyond the target)?

3.20 ★★ Consider a system comprising two extended bodies, which have masses M_1 and M_2 and centers of mass at $\mathbf{R}_1$ and $\mathbf{R}_2$. Prove that the CM of the whole system is at

$$\mathbf{R} = \frac{M_1 \mathbf{R}_1 + M_2 \mathbf{R}_2}{M_1 + M_2}.$$

This beautiful result means that in finding the CM of a complicated system, you can treat its component parts just like point masses positioned at their separate centers of mass — even when the component parts are themselves extended bodies.

3.21 ★★ A uniform thin sheet of metal is cut in the shape of a semicircle of radius R and lies in the xy plane with its center at the origin and diameter lying along the x axis. Find the position of the CM using polar coordinates. [In this case the sum (3.9) that defines the CM position becomes a two-dimensional integral of the form $\int \mathbf{r}\sigma \, dA$ where σ denotes the surface mass density (mass/area) of the sheet and dA is the element of area $dA = r \, dr \, d\phi$.]

3.22 ★★ Use spherical polar coordinates r, θ, ϕ to find the CM of a uniform solid hemisphere of radius R, whose flat face lies in the xy plane with its center at the origin. Before you do this, you will need to convince yourself that the element of volume in spherical polars is $dV = r^2 dr \sin\theta \, d\theta \, d\phi$. (Spherical polar coordinates are defined in Section 4.8. If you are not already familiar with these coordinates, you should probably not try this problem yet.)

3.23 ★★★ [Computer] A grenade is thrown with initial velocity $\mathbf{v}_0$ from the origin at the top of a high cliff, subject to negligible air resistance. **(a)** Using a suitable plotting program, plot the orbit, with the following parameters: $\mathbf{v}_0 = (4, 4)$, $g = 1$, and $0 \le t \le 4$ (and with x measured horizontally and y vertically up). Add to your plot suitable marks (dots or crosses, for example) to show the positions of the grenade at $t = 1, 2, 3, 4$. **(b)** At $t = 4$, when the grenade's velocity is $\mathbf{v}$, it explodes into two equal pieces, one of which moves off with velocity $\mathbf{v} + \Delta\mathbf{v}$. What is the velocity of the other piece? **(c)** Assuming that $\Delta\mathbf{v} = (1, 3)$, add to your original plot the paths of the two pieces for $4 \le t \le 9$. Insert marks to show their positions at $t = 5, 6, 7, 8, 9$. Find some way to show clearly that the CM of the two pieces continues to follow the original parabolic path.

SECTION 3.4   Angular Momentum for a Single Particle

3.24 ★ If the vectors $\mathbf{a}$ and $\mathbf{b}$ form two of the sides of a triangle, prove that $\frac{1}{2}|\mathbf{a} \times \mathbf{b}|$ is equal to the area of the triangle.

3.25 ★ A particle of mass m is moving on a frictionless horizontal table and is attached to a massless string, whose other end passes through a hole in the table, where I am holding it. Initially the particle is moving in a circle of radius r_0 with angular velocity ω_0, but I now pull the string down through the hole until a length r remains between the hole and the particle. What is the particle's angular velocity now?

3.26 ★ A particle moves under the influence of a central force directed toward a fixed origin O. **(a)** Explain why the particle's angular momentum about O is constant. **(b)** Give in detail the argument that the particle's orbit must lie in a single plane containing O.

3.27 ★★ Consider a planet orbiting the fixed sun. Take the plane of the planet's orbit to be the xy plane, with the sun at the origin, and label the planet's position by polar coordinates (r, ϕ). **(a)** Show that the planet's angular momentum has magnitude $\ell = mr^2\omega$, where $\omega = \dot{\phi}$ is the planet's angular velocity about the sun. **(b)** Show that the rate at which the planet "sweeps out area" (as in Kepler's second law) is $dA/dt = \frac{1}{2}r^2\omega$, and hence that $dA/dt = \ell/2m$. Deduce Kepler's second law.

SECTION 3.5 Angular Momentum for Several Particles

3.28 ★ For a system of just three particles, go through in detail the argument leading from (3.20) to (3.26), $\dot{\mathbf{L}} = \mathbf{\Gamma}^{\text{ext}}$, writing out all the summations explicitly.

3.29 ★ A uniform spherical asteroid of radius R_0 is spinning with angular velocity ω_0. As the aeons go by, it picks up more matter until its radius is R. Assuming that its density remains the same and that the additional matter was originally at rest relative to the asteroid (anyway on average), find the asteroid's new angular velocity. (You know from elementary physics that the moment of inertia is $\frac{2}{5}MR^2$.) What is the final angular velocity if the radius doubles?

3.30 ★★ Consider a rigid body rotating with angular velocity ω about a fixed axis. (You could think of a door rotating about the axis defined by its hinges.) Take the axis of rotation to be the z axis and use cylindrical polar coordinates $\rho_\alpha, \phi_\alpha, z_\alpha$ to specify the positions of the particles $\alpha = 1, \cdots, N$ that make up the body. **(a)** Show that the velocity of the particle α is $\rho_\alpha\omega$ in the ϕ direction. **(b)** Hence show that the z component of the angular momentum ℓ_α of particle α is $m_\alpha\rho_\alpha^2\omega$. **(c)** Show that the z component L_z of the total angular momentum can be written as $L_z = I\omega$ where I is the moment of inertia (for the axis in question),

$$I = \sum_{\alpha=1}^{N} m_\alpha \rho_\alpha^2. \qquad (3.31)$$

3.31 ★★ Find the moment of inertia of a uniform disc of mass M and radius R rotating about its axis, by replacing the sum (3.31) by the appropriate integral and doing the integral in polar coordinates.

3.32 ★★ Show that the moment of inertia of a uniform solid sphere rotating about a diameter is $\frac{2}{5}MR^2$. The sum (3.31) must be replaced by an integral, which is easiest in spherical polar coordinates, with the axis of rotation taken to be the z axis. The element of volume is $dV = r^2 dr \sin\theta\, d\theta\, d\phi$. (Spherical polar coordinates are defined in Section 4.8. If you are not already familiar with these coordinates, you should probably not try this problem yet.)

3.33 ★★ Starting from the sum (3.31) and replacing it by the appropriate integral, find the moment of inertia of a uniform thin square of side $2b$, rotating about an axis perpendicular to the square and passing through its center.

3.34 ★★ A juggler is juggling a uniform rod one end of which is coated in tar and burning. He is holding the rod by the opposite end and throws it up so that, at the moment of release, it is horizontal, its CM is traveling vertically up at speed v_0 and it is rotating with angular velocity ω_0. To catch it, he wants to arrange that when it returns to his hand it will have made an integer number of complete rotations. What should v_0 be, if the rod is to have made exactly n rotations when it returns to his hand?

3.35 ★★ Consider a uniform solid disk of mass M and radius R, rolling without slipping down an incline which is at angle γ to the horizontal. The instantaneous point of contact between the disk and the incline is called P. **(a)** Draw a free-body diagram, showing all forces on the disk. **(b)** Find the linear acceleration $\dot{v}$ of the disk by applying the result $\dot{\mathbf{L}} = \mathbf{\Gamma}^{\text{ext}}$ for rotation about P. (Remember that $L = I\omega$ and the moment of inertia for rotation about a point on the circumference is $\frac{3}{2}MR^2$. The condition that the disk not slip is that $v = R\omega$ and hence $\dot{v} = R\dot{\omega}$.) **(c)** Derive the same result by applying $\dot{\mathbf{L}} = \mathbf{\Gamma}^{\text{ext}}$

to the rotation about the CM. (In this case you will find there is an extra unknown, the force of friction. You can eliminate this by applying Newton's second law to the motion of the CM. The moment of inertia for rotation about the CM is $\frac{1}{2}MR^2$.)

3.36 ★★ Repeat the calculations of Example 3.4 (page 97) for the case that the force $\mathbf{F}$ acts in a "northeasterly" direction at angle γ from the x axis. What are the velocities of the two masses just after the impulse has been applied? Check your answers for the cases that $\gamma = 0$ and $\gamma = 90°$.

3.37 ★★★ A system consists of N masses m_α at positions $\mathbf{r}_\alpha$ relative to a fixed origin O. Let $\mathbf{r}'_\alpha$ denote the position of m_α relative to the CM; that is, $\mathbf{r}'_\alpha = \mathbf{r}_\alpha - \mathbf{R}$. **(a)** Make a sketch to illustrate this last equation. **(b)** Prove the useful relation that $\sum m_\alpha \mathbf{r}'_\alpha = 0$. Can you explain why this relation is nearly obvious? **(c)** Use this relation to prove the result (3.28) that the rate of change of the angular momentum *about the CM* is equal to the total external torque about the CM. (This result is surprising since the CM may be accelerating, so that it is not necessarily a fixed point in any inertial frame.)

Energy

This chapter takes up the conservation of energy. You will see that the analysis of energy conservation is surprisingly more complicated than the corresponding discussions of linear and angular momenta in Chapter 3. The main reason for the difference is this: In almost all problems of classical mechanics there is only one kind of linear momentum ($\mathbf{p} = m\mathbf{v}$ for each particle), and one kind of angular momentum ($\boldsymbol{\ell} = \mathbf{r} \times \mathbf{p}$ for each particle). By contrast, energy comes in many different and important forms: kinetic, several kinds of potential, thermal, and more. It is the processes that transform energy from one kind to another that complicate the use of energy conservation. We shall see that conservation of energy is a quite subtle business, even for a system consisting of just a single particle.

One manifestation of the relative difficulty of the discussion of energy is that we shall need some new tools from vector calculus, namely, the concepts of the gradient and the curl. I shall introduce these important ideas as we need them.

4.1 Kinetic Energy and Work

As I have said, there are many different kinds of energy. Perhaps the most basic is **kinetic energy** (or KE), which for a single particle of mass m traveling with speed v is defined to be

$$T = \tfrac{1}{2}mv^2. \tag{4.1}$$

Let us imagine the particle moving through space and examine the change in its kinetic energy as it moves between two neighboring points $\mathbf{r}_1$ and $\mathbf{r}_1 + d\mathbf{r}$ on its path as shown in Figure 4.1. The time derivative of T is easily evaluated if we note that $v^2 = \mathbf{v} \cdot \mathbf{v}$, so that

$$\frac{dT}{dt} = \tfrac{1}{2}m\frac{d}{dt}(\mathbf{v} \cdot \mathbf{v}) = \tfrac{1}{2}m(\dot{\mathbf{v}} \cdot \mathbf{v} + \mathbf{v} \cdot \dot{\mathbf{v}}) = m\dot{\mathbf{v}} \cdot \mathbf{v}. \tag{4.2}$$

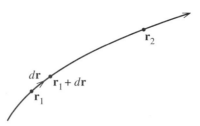

Figure 4.1   Three points on the path of a particle:
$\mathbf{r}_1, \mathbf{r}_1 + d\mathbf{r}$ (with $d\mathbf{r}$ infinitesimal) and $\mathbf{r}_2$.

By the second law, the factor $m\dot{\mathbf{v}}$ is equal to the net force $\mathbf{F}$ on the particle, so that

$$\frac{dT}{dt} = \mathbf{F} \cdot \mathbf{v}. \tag{4.3}$$

If we multiply both sides by dt, then since $\mathbf{v}\, dt$ is the displacement $d\mathbf{r}$ we find

$$dT = \mathbf{F} \cdot d\mathbf{r}. \tag{4.4}$$

The expression on the right, $\mathbf{F} \cdot d\mathbf{r}$, is defined to be the **work done by the force $\mathbf{F}$** in the displacement $d\mathbf{r}$. Thus we have proved the **Work–KE theorem**, that the change in the particle's kinetic energy between two neighboring points on its path is equal to the work done by the net force as it moves between the two points.[1]

So far we have proved the Work–KE theorem only for an infinitesimal displacement $d\mathbf{r}$, but it generalizes easily to larger displacements. Consider the two points shown as $\mathbf{r}_1$ and $\mathbf{r}_2$ in Figure 4.1. We can divide the path between these points 1 and 2 into a large number of very small segments, to each of which we can apply the infinitesimal result (4.4). Adding all of these results, we find that the total change in T going from 1 to 2 is the sum $\sum \mathbf{F} \cdot d\mathbf{r}$ of all the infinitesimal works done in all the infinitesimal displacements between points 1 and 2:

$$\Delta T \equiv T_2 - T_1 = \sum \mathbf{F} \cdot d\mathbf{r}. \tag{4.5}$$

In the limit that all the displacements $d\mathbf{r}$ go to zero, this sum becomes an integral:

$$\sum \mathbf{F} \cdot d\mathbf{r} \to \int_1^2 \mathbf{F} \cdot d\mathbf{r}. \tag{4.6}$$

---

[1] Two points that can be puzzling at first: The work $\mathbf{F} \cdot d\mathbf{r}$ can be *negative*, if for example $\mathbf{F}$ and $d\mathbf{r}$ point in opposite directions. While the notion of a force doing negative work conflicts with our everyday notion of work, it is perfectly consistent with the physicist's definition: A force in the opposite direction to the displacement *reduces* the KE, so, by the work–KE theorem, the corresponding work has to be negative. Second, if $\mathbf{F}$ and $d\mathbf{r}$ are perpendicular, then the work $\mathbf{F} \cdot d\mathbf{r}$ is zero. Again this conflicts with our everyday sense of work, but is consistent with the physicist's usage: A force that is perpendicular to the displacement does not change the KE.

This integral, called a **line integral**,[2] is a generalization of the integral $\int f(x)\,dx$ over a single variable x, and its definition as the limit of the sum of many small pieces is closely analogous. If you feel any doubt about the symbol $\int_1^2 \mathbf{F} \cdot d\mathbf{r}$ on the right of (4.6), think of it as being just the sum on the left (with all the displacements infinitesimally small). In evaluating a line integral, it is usually possible to convert it into an ordinary integral over a single variable, as the following examples show. Notice that, as the name implies, the line integral depends (in general) on the path that the particle followed from point 1 to point 2. For any force $\mathbf{F}$, the line integral on the right of (4.6) is called the **work done by the force $\mathbf{F}$** moving between points 1 and 2 along the path concerned.

EXAMPLE 4.1   Three Line Integrals

Evaluate the line integral for the work done by the two-dimensional force $\mathbf{F} = (y, 2x)$ going from the origin O to the point $P = (1, 1)$ along each of the three paths shown in Figure 4.2. Path a goes from O to $Q = (1, 0)$ along the x axis and then from Q straight up to P, path b goes straight from O to P along the line $y = x$, and path c goes round a quarter circle centered on Q.

The integral along path a is easily evaluated in two parts, if we note that on OQ the displacements have the form $d\mathbf{r} = (dx, 0)$, while on QP they are $d\mathbf{r} = (0, dy)$. Thus

$$W_a = \int_a \mathbf{F} \cdot d\mathbf{r} = \int_O^Q \mathbf{F} \cdot d\mathbf{r} + \int_Q^P \mathbf{F} \cdot d\mathbf{r} = \int_0^1 F_x(x, 0)\,dx + \int_0^1 F_y(1, y)\,dy$$

$$= 0 + 2 \int_0^1 dy = 2.$$

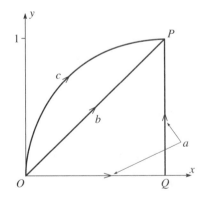

Figure 4.2   Three different paths, a, b, and c, from the origin to the point $P = (1, 1)$.

---

[2] Not an especially happy name for those of us who think of a line as something straight. However, there are curved lines as well as straight lines, and in general a line integral can involve a curved line, such as the path shown in Figure 4.1.

On the path b, $x = y$, so that $dx = dy$, and

$$W_b = \int_b \mathbf{F} \cdot d\mathbf{r} = \int_b (F_x \, dx + F_y \, dy) = \int_0^1 (x + 2x) dx = 1.5.$$

Path c is conveniently expressed parametrically as

$$\mathbf{r} = (x, y) = (1 - \cos\theta, \sin\theta)$$

where θ is the angle between OQ and the line from Q to the point (x, y), with $0 \leq \theta \leq \pi/2$. Thus on path c

$$d\mathbf{r} = (dx, dy) = (\sin\theta, \cos\theta) \, d\theta$$

and

$$W_c = \int_c \mathbf{F} \cdot d\mathbf{r} = \int_c (F_x \, dx + F_y \, dy)$$

$$= \int_0^{\pi/2} \left[\sin^2\theta + 2(1 - \cos\theta) \cos\theta \right] d\theta = 2 - \pi/4 = 1.21.$$

Some more examples can be found in Problems 4.2 and 4.3 and, if you have never studied line integrals, you may want to try some of these.

With the notation of the line integral, we can rewrite the result (4.5) as

$$\Delta T \equiv T_2 - T_1 = \int_1^2 \mathbf{F} \cdot d\mathbf{r} \equiv W(1 \rightarrow 2) \tag{4.7}$$

where I have introduced the notation $W(1 \rightarrow 2)$ for the work done by $\mathbf{F}$ moving from point 1 to point 2. The result is the **Work–KE theorem** for arbitrary displacements, large or small: The change in a particle's KE as it moves between points 1 and 2 is the work done by the net force.

It is important to remember that the work that appears on the right of (4.7) is the work done by the *net force* $\mathbf{F}$ on the particle. In general, $\mathbf{F}$ is the vector sum of various separate forces

$$\mathbf{F} = \mathbf{F}_1 + \cdots + \mathbf{F}_n \equiv \sum_{i=1}^n \mathbf{F}_i.$$

(For example, the net force on a projectile is the sum of two forces, the weight and air resistance.) It is a most convenient fact that to evaluate the work done by the net force $\mathbf{F}$, we can simply add up the works done by the separate forces $\mathbf{F}_1, \cdots, \mathbf{F}_n$. This claim is easily proved as follows:

$$W(1 \rightarrow 2) = \int_1^2 \mathbf{F} \cdot d\mathbf{r} = \int_1^2 \sum_i \mathbf{F}_i \cdot d\mathbf{r}$$

$$= \sum_i \int_1^2 \mathbf{F}_i \cdot d\mathbf{r} = \sum_i W_i(1 \rightarrow 2). \tag{4.8}$$

The crucial step, from the first line to the second, is justified because the integral of a sum of n terms is the same as the sum of the n individual integrals. The Work–KE theorem can therefore be rewritten as

$$T_2 - T_1 = \sum_{i=1}^{n} W_i(1 \to 2). \tag{4.9}$$

In practice, one almost always uses the theorem in this way: Calculate the work W_i done by each of the n separate forces on the particle and then set ΔT equal to the sum of all the W_i.

If the net force on a particle is zero, then the Work–KE theorem tells us that the particle's kinetic energy is constant. This simply says that the speed v is constant, which, though true, is not very interesting, since it already follows from Newton's first law.

4.2 Potential Energy and Conservative Forces

The next step in the development of the energy formalism is to introduce the notion of potential energy (or PE) corresponding to the forces on an object. As you probably recall, not every force lends itself to the definition of a corresponding potential energy. Those special forces that do have a corresponding potential energy (with the required properties) are called *conservative forces*, and we must discuss the properties that distinguish conservative from nonconservative forces. Specifically, we shall find that there are two conditions that a force must satisfy to be considered conservative.

To simplify our discussion, let us assume at first that there is only one force acting on the object of interest — the gravitational force on a planet by its sun, or the electric force $q\mathbf{E}$ on a charge in an electric field (with no other forces present). The force $\mathbf{F}$ may depend on many different variables: It may depend on the object's position $\mathbf{r}$. (The farther the planet is from the sun, the weaker the gravitational pull.) It may depend on the object's velocity, as is the case with air resistance; and it may depend on the time t, as would be the case for a charge in a time-varying electric field. Finally, if the force is exerted by humans, it will depend on a host of imponderables — how tired they are feeling, how conveniently they are situated to push, and so on.

The first condition for a force $\mathbf{F}$ to be conservative is that $\mathbf{F}$ depends only on the position $\mathbf{r}$ of the object on which it acts; it must not depend on the velocity, the time, or any variables other than $\mathbf{r}$. This sounds, and is, quite restrictive, but there are plenty of forces that have this property: The gravitational force of the sun on a planet (position $\mathbf{r}$ relative to the sun) can be written as

$$\mathbf{F}(\mathbf{r}) = -\frac{GmM}{r^2}\hat{\mathbf{r}}$$

which evidently depends only on the variable $\mathbf{r}$. (The parameters m and M — and, of course, the gravitational constant G — are constant for a given planet and given sun.) Similarly, the electrostatic force $\mathbf{F}(\mathbf{r}) = q\mathbf{E}(\mathbf{r})$ on a charge q by a static electric field $\mathbf{E}(\mathbf{r})$ has this property. Forces that do not satisfy this condition include the force of air

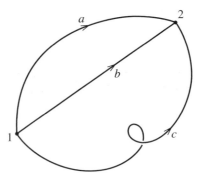

Figure 4.3   Three different paths, a, b, and c, joining the same two points 1 and 2.

resistance (which depends on the velocity), friction (which depends on the direction of motion), the magnetic force (which depends on the velocity), and the force of a time-varying electric field $\mathbf{E}(\mathbf{r}, t)$ (which obviously depends on time).

The second condition that a force must satisfy to be called conservative concerns the work done by the force as the object on which it acts moves between two points $\mathbf{r}_1$ and $\mathbf{r}_2$ (or just 1 and 2 for short),

$$W(1 \rightarrow 2) = \int_1^2 \mathbf{F} \cdot d\mathbf{r}. \tag{4.10}$$

Figure 4.3 shows two points, 1 and 2, and three different paths connecting them. It is entirely possible that the work done between points 1 and 2, as defined by the integral (4.10), has different values depending on which of the three paths, a, b, or c, the particle happens to follow. For example, consider the force of sliding friction as I push a heavy crate across the floor. This force has a constant magnitude, F_{fric} say, and is always opposite to the direction of motion. Thus the work done by friction as the crate moves from 1 to 2 is given by (4.10) to be

$$W_{\text{fric}}(1 \rightarrow 2) = -F_{\text{fric}}L,$$

where L denotes the length of the path followed. The three paths of Figure 4.3 have different lengths, and $W_{\text{fric}}(1 \rightarrow 2)$ will have a different value for each of the three paths.

On the other hand, there are forces with the property that the work $W(1 \rightarrow 2)$ is the same for *any* path connecting the same two points 1 and 2. An example of a force with this property is the gravitational force, $\mathbf{F}_{\text{grav}} = m\mathbf{g}$, of the earth on an object close to the earth's surface. It is easy to show (Problem 4.5) that, because $\mathbf{g}$ is a constant vector pointing vertically down, the work done in this case is

$$W_{\text{grav}}(1 \rightarrow 2) = -mgh, \tag{4.11}$$

where h is just the vertical height gained between points 1 and 2. This work is the same for *any* two paths between the given points 1 and 2. This property, the path independence of the work it does, is the second condition that a force must satisfy to be considered conservative, and we are now ready to state the two conditions:

Conditions for a Force to be Conservative

A force $\mathbf{F}$ acting on a particle is **conservative** if and only if it satisfies two conditions:

  (i) $\mathbf{F}$ depends only on the particle's position $\mathbf{r}$ (and not on the velocity $\mathbf{v}$, or the time t, or any other variable); that is, $\mathbf{F} = \mathbf{F}(\mathbf{r})$.
  (ii) For any two points 1 and 2, the work $W(1 \to 2)$ done by $\mathbf{F}$ is the same for all paths between 1 and 2.

The reason for the name "conservative" and for the importance of the concept is this: If all forces on an object are conservative, we can define a quantity called the potential energy (or just PE), denoted $U(\mathbf{r})$, a function only of position, with the property that the total mechanical energy

$$E = \text{KE} + \text{PE} = T + U(\mathbf{r}) \tag{4.12}$$

is constant; that is, E is *conserved*.

To define the potential energy $U(\mathbf{r})$ corresponding to a given conservative force, we first choose a reference point $\mathbf{r}_{\text{o}}$ at which U is defined to be zero. (For example, in the case of gravity near the earth's surface, we often define U to be zero at ground level.) We then define $U(\mathbf{r})$, the **potential energy** at an arbitrary point $\mathbf{r}$, to be[3]

$$U(\mathbf{r}) = -W(\mathbf{r}_{\text{o}} \to \mathbf{r}) \equiv - \int_{\mathbf{r}_{\text{o}}}^{\mathbf{r}} \mathbf{F}(\mathbf{r}') \cdot d\mathbf{r}'. \tag{4.13}$$

In words, $U(\mathbf{r})$ is minus the work done by $\mathbf{F}$ if the particle moves from the reference point $\mathbf{r}_{\text{o}}$ to the point of interest $\mathbf{r}$, as in Figure 4.4. (We shall see the reason for the minus sign shortly.) Notice that the definition (4.13) only makes sense because of the property (ii) of conservative forces. If the work integral in (4.13) were different for different paths, then (4.13) would not define a unique function[4] $U(\mathbf{r})$.

---

[3] Notice that I have called the variable of integration $\mathbf{r}'$ to avoid confusion with the upper limit $\mathbf{r}$.

[4] The definition (4.13) also depends on property (i) of conservative forces, but in a slightly subtler way. If $\mathbf{F}$ depended on another variable besides $\mathbf{r}$ (for instance, t or $\mathbf{v}$), then the right side of (4.13) would depend on when or how the particle moved from $\mathbf{r}_{\text{o}}$ to $\mathbf{r}$, and again there would be no uniquely defined $U(\mathbf{r})$.

Figure 4.4   The potential energy $U(\mathbf{r})$ at any point $\mathbf{r}$ is defined as minus the work done by $\mathbf{F}$ if the particle moves from the reference point $\mathbf{r}_\text{o}$ to $\mathbf{r}$. This gives a well-defined function $U(\mathbf{r})$ only if this work is independent of the path followed — that is, the force is conservative.

EXAMPLE 4.2   Potential Energy of a Charge in a Uniform Electric Field

A charge q is placed in a uniform electric field pointing in the x direction with strength E_o, so that the force on q is $\mathbf{F} = q\mathbf{E} = qE_\text{o}\hat{\mathbf{x}}$. Show that this force is conservative and find the corresponding potential energy.

The work done by $\mathbf{F}$ going between any two points 1 and 2 along any path is

$$W(1 \to 2) = \int_1^2 \mathbf{F} \cdot d\mathbf{r} = qE_\text{o} \int_1^2 \hat{\mathbf{x}} \cdot d\mathbf{r} = qE_\text{o} \int_1^2 dx = qE_\text{o}(x_2 - x_1). \quad (4.14)$$

This depends only on the two end points 1 and 2. (In fact it depends only on their x coordinates x_1 and x_2.) Certainly, it is independent of the path, and the force is conservative. To define the corresponding potential energy $U(\mathbf{r})$, we must first pick a reference point $\mathbf{r}_\text{o}$ at which U will be zero. A natural choice is the origin, $\mathbf{r}_\text{o} = 0$, in which case the potential energy is $U(\mathbf{r}) = -W(0 \to \mathbf{r})$ or, according to (4.14),

$$U(\mathbf{r}) = -qE_\text{o}x.$$

We can now derive a crucial expression for the work done by $\mathbf{F}$ in terms of the potential energy $U(\mathbf{r})$. Let $\mathbf{r}_1$ and $\mathbf{r}_2$ be any two points as in Figure 4.5. If $\mathbf{r}_\text{o}$ is the reference point at which U is zero, then it is clear from Figure 4.5 that

$$W(\mathbf{r}_\text{o} \to \mathbf{r}_2) = W(\mathbf{r}_\text{o} \to \mathbf{r}_1) + W(\mathbf{r}_1 \to \mathbf{r}_2)$$

and hence

$$W(\mathbf{r}_1 \to \mathbf{r}_2) = W(\mathbf{r}_\text{o} \to \mathbf{r}_2) - W(\mathbf{r}_\text{o} \to \mathbf{r}_1). \quad (4.15)$$

Each of the two terms on the right is (minus) the potential energy at the corresponding point. Thus we have proved that the work on the left is just the difference of these two potential energies:

$$W(\mathbf{r}_1 \to \mathbf{r}_2) = -[U(\mathbf{r}_2) - U(\mathbf{r}_1)] = -\Delta U. \quad (4.16)$$

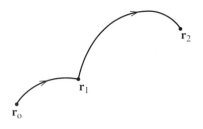

Figure 4.5   The work $W(\mathbf{r}_1 \to \mathbf{r}_2)$ going from $\mathbf{r}_1$ to $\mathbf{r}_2$ is the same as $W(\mathbf{r}_0 \to \mathbf{r}_2)$ minus $W(\mathbf{r}_0 \to \mathbf{r}_1)$. This result is independent of what path we use for either limb of the journey, provided the force concerned is conservative.

The usefulness of this result emerges when we combine it with the Work–KE theorem (4.7):

$$\Delta T = W(\mathbf{r}_1 \to \mathbf{r}_2). \tag{4.17}$$

Comparing this with (4.16), we see that

$$\Delta T = -\Delta U \tag{4.18}$$

or, moving the right side across to the left,[5]

$$\Delta(T + U) = 0. \tag{4.19}$$

That is, the **mechanical energy**

$$E = T + U \tag{4.20}$$

does not change as the particle moves from $\mathbf{r}_1$ to $\mathbf{r}_2$. Since the points $\mathbf{r}_1$ and $\mathbf{r}_2$ were *any* two points on the particle's trajectory, we have the important conclusion: If the force on a particle is conservative, then the particle's mechanical energy never changes; that is, the particle's energy is conserved, which explains the use of the adjective "conservative."

Several Forces

So far we have established the conservation of energy for a particle subject to a single conservative force. If the particle is subject to several forces, all of them conservative, our result generalizes easily. For instance, imagine a mass suspended from the ceiling by a spring. This mass is subject to two forces, the forces of gravity ($\mathbf{F}_{\text{grav}}$) and the spring ($\mathbf{F}_{\text{spr}}$). The force of gravity is certainly conservative (as I've already argued), and, provided the spring obeys Hooke's law, $\mathbf{F}_{\text{spr}}$ is likewise (see Problem 4.42). We

---

[5] We now see the reason for the minus sign in the definition of U. It gives the minus sign on the right of (4.18), which in turn gives the desired plus sign on the left of (4.19).

can define separate potential energies for each force, U_{grav} for $\mathbf{F}_{\text{grav}}$ and U_{spr} for $\mathbf{F}_{\text{spr}}$, each with the crucial property (4.16) that the change in U gives (minus) the work done by the corresponding force. According to the Work–KE theorem, the change in the mass's kinetic energy is

$$\Delta T = W_{\text{grav}} + W_{\text{spr}}$$

$$= -\left(\Delta U_{\text{grav}} + \Delta U_{\text{spr}}\right), \tag{4.21}$$

where the second line follows from the properties of the two separate potential energies. Rearranging this equation, we see that $\Delta(T + U_{\text{grav}} + U_{\text{spr}}) = 0$. That is, the total mechanical energy, defined as $E = T + U_{\text{grav}} + U_{\text{spr}}$, is conserved.

The argument just given extends immediately to the case of n forces on a particle, so long as they are all conservative. If for each force $\mathbf{F}_i$ we define a corresponding potential energy U_i, then we have the

Principle of Conservation of Energy for One Particle

If all of the n forces $\mathbf{F}_i$ $(i = 1, \cdots, n)$ acting on a particle are conservative, each with its corresponding potential energy $U_i(\mathbf{r})$, the **total mechanical energy**, defined as

$$E \equiv T + U \equiv T + U_1(\mathbf{r}) + \cdots + U_n(\mathbf{r}), \tag{4.22}$$

is constant in time.

Nonconservative Forces

If some of the forces on our particle are nonconservative, then we cannot define corresponding potential energies; nor can we define a conserved mechanical energy. Nevertheless, we can define potential energies for all of the forces that *are* conservative, and then recast the Work–KE theorem in a form that shows how the nonconservative forces change the particle's mechanical energy. First, we divide the net force on the particle into two parts, the conservative part $\mathbf{F}_{\text{cons}}$ and the nonconservative part $\mathbf{F}_{\text{nc}}$. For $\mathbf{F}_{\text{cons}}$ we can define a potential energy, which we'll call just U. By the Work–KE theorem, the change in kinetic energy between any two times is

$$\Delta T = W = W_{\text{cons}} + W_{\text{nc}}. \tag{4.23}$$

The first term on the right is just $-\Delta U$ and can be moved to the left side to give $\Delta(T + U) = W_{\text{nc}}$. If we define the mechanical energy as $E = T + U$, then we see that

$$\Delta E \equiv \Delta(T + U) = W_{\text{nc}}. \tag{4.24}$$

Mechanical energy is no longer conserved, but we have the next best thing. The mechanical energy changes to precisely the extent that the nonconservative forces do work on our particle. In many problems the only nonconservative force is the force of sliding friction, which usually does negative work. (The frictional force $\mathbf{f}$ is in the direction opposite to the motion, so the work $\mathbf{f} \cdot d\mathbf{r}$ is negative.) In this case W_{nc} is negative and (4.24) tells us that the object loses mechanical energy in the amount "stolen" by friction. All of these ideas are illustrated by the following simple example.

EXAMPLE 4.3   Block Sliding Down an Incline

Consider again the block of Example 1.1 and find its speed v when it reaches the bottom of the slope, a distance d from its starting point.

The setup and the forces on the block are shown in Figure 4.6. The three forces on the block are its weight, $\mathbf{w} = m\mathbf{g}$, the normal force of the incline, $\mathbf{N}$, and the frictional force $\mathbf{f}$, whose magnitude we found in Example 1.1 to be $f = \mu mg \cos\theta$. The weight $m\mathbf{g}$ is conservative, and the corresponding potential energy is (as you certainly recall from introductory physics, but see Problem 4.5)

$$U = mgy$$

where y is the block's vertical height above the bottom of the slope (if we choose the zero of PE at the bottom). The normal force does no work, since it is perpendicular to the direction of motion, so will not contribute to the energy balance. The frictional force does work $W_{\text{fric}} = -fd = -\mu mgd \cos\theta$. The change in kinetic energy is $\Delta T = T_{\text{f}} - T_{\text{i}} = \frac{1}{2}mv^2$ and the change in potential energy is $\Delta U = U_{\text{f}} - U_{\text{i}} = -mgh = -mgd \sin\theta$. Thus (4.24) reads

$$\Delta T + \Delta U = W_{\text{fric}}$$

or

$$\tfrac{1}{2}mv^2 - mgd \sin\theta = -\mu mgd \cos\theta.$$

Solving for v we find

$$v = \sqrt{2gd(\sin\theta - \mu \cos\theta)}.$$

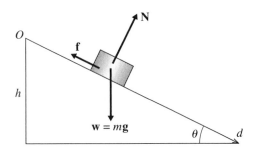

Figure 4.6   A block on an incline of angle θ. The length of the slope is d, and the height is $h = d \sin\theta$.

As usual, you should check that this answer agrees with common sense. For example, does it give the expected answer when $\theta = 90°$? What about $\theta = 0$? (The case $\theta = 0$ is a bit subtler.)

4.3   Force as the Gradient of Potential Energy

We have seen that the potential energy $U(\mathbf{r})$ corresponding to a force $\mathbf{F}(\mathbf{r})$ can be expressed as an integral of $\mathbf{F}(\mathbf{r})$ as in (4.13). This suggests that we should be able to write $\mathbf{F}(\mathbf{r})$ as some kind of derivative of $U(\mathbf{r})$. This suggestion proves correct, though to implement it we shall need some mathematics that you may not have met before. Specifically, since $\mathbf{F}(\mathbf{r})$ is a vector [while $U(\mathbf{r})$ is a scalar] we shall be involved in some *vector calculus.*

Let us consider a particle acted on by a conservative force $\mathbf{F}(\mathbf{r})$, with corresponding potential energy $U(\mathbf{r})$, and examine the work done by $\mathbf{F}(\mathbf{r})$ in a small displacement from $\mathbf{r}$ to $\mathbf{r} + d\mathbf{r}$. We can evaluate this work in two ways. On the one hand, it is, by definition,

$$W(\mathbf{r} \rightarrow \mathbf{r}+d\mathbf{r}) = \mathbf{F}(\mathbf{r}) \cdot d\mathbf{r}$$

$$= F_x\, dx + F_y\, dy + F_z\, dz, \tag{4.25}$$

for any small displacement $d\mathbf{r}$ with components (dx, dy, dz).

On the other hand, we have seen that the work $W(\mathbf{r} \rightarrow \mathbf{r}+d\mathbf{r})$ is the same as (minus) the change in PE in the displacement:

$$W(\mathbf{r} \rightarrow \mathbf{r}+d\mathbf{r}) = -dU = -[U(\mathbf{r} + d\mathbf{r}) - U(\mathbf{r})]$$

$$= -[U(x + dx, y + dy, z + dz) - U(x, y, z)]. \tag{4.26}$$

In the second line, I have replaced the position vector $\mathbf{r}$ by its components to emphasize that U is really a function of the three variables (x, y, z). Now, for functions of one variable, a difference like that in (4.26) can be expressed in terms of the derivative:

$$df = f(x + dx) - f(x) = \frac{df}{dx}dx. \tag{4.27}$$

This is really no more than the definition of the derivative.[6] For a function of three variables, such as $U(x, y, z)$, the corresponding result is

$$dU = U(x + dx, y + dy, z + dz) - U(x, y, z)$$

$$= \frac{\partial U}{\partial x}dx + \frac{\partial U}{\partial y}dy + \frac{\partial U}{\partial z}dz \tag{4.28}$$

where the three derivatives are the *partial derivatives* with respect to the three independent variables (x, y, z). [For example, $\partial U/\partial x$ is the rate of change of U as x

---

[6] Strictly speaking, this equation is exact only in the limit that $dx \rightarrow 0$. As usual, I take the view that dx is small enough (though nonzero) that the two sides are equal within our chosen accuracy target.

changes, with y and z fixed, and is found by differentiating $U(x, y, z)$ with respect to x treating y and z as constants. See Problems 4.10 and 4.11 for some examples.] Substituting (4.28) into (4.26), we find that the work done in the small displacement from $\mathbf{r}$ to $\mathbf{r} + d\mathbf{r}$ is

$$W(\mathbf{r} \rightarrow \mathbf{r}+d\mathbf{r}) = -\left[\frac{\partial U}{\partial x}dx + \frac{\partial U}{\partial y}dy + \frac{\partial U}{\partial z}dz\right]. \qquad (4.29)$$

The two expressions (4.25) and (4.29) are both valid for *any* small displacement $d\mathbf{r}$. In particular, we can choose $d\mathbf{r}$ to point in the x direction, in which case $dy = dz = 0$ and the last two terms in both (4.25) and (4.29) are zero. Equating the remaining terms, we see that $F_x = -\partial U/\partial x$. By choosing $d\mathbf{r}$ to point in the y or z directions, we get corresponding results for F_y and F_z, and we conclude that

$$F_x = -\frac{\partial U}{\partial x}, \qquad F_y = -\frac{\partial U}{\partial y}, \qquad F_z = -\frac{\partial U}{\partial z}. \qquad (4.30)$$

That is, $\mathbf{F}$ is the vector whose three components are minus the three partial derivatives of U with respect to x, y, and z. A slightly more compact way to write this result is this:

$$\mathbf{F} = -\hat{\mathbf{x}}\frac{\partial U}{\partial x} - \hat{\mathbf{y}}\frac{\partial U}{\partial y} - \hat{\mathbf{z}}\frac{\partial U}{\partial z}. \qquad (4.31)$$

Relationships like (4.31) between a vector ($\mathbf{F}$) and a scalar (U) come up over and over again in physics. For example, the electric field $\mathbf{E}$ is related to the electrostatic potential V in exactly the same way. More generally, given any scalar $f(\mathbf{r})$, the vector whose three components are the partial derivatives of $f(\mathbf{r})$ is called the **gradient** of f, denoted ∇f:

$$\nabla f = \hat{\mathbf{x}}\frac{\partial f}{\partial x} + \hat{\mathbf{y}}\frac{\partial f}{\partial y} + \hat{\mathbf{z}}\frac{\partial f}{\partial z}. \qquad (4.32)$$

The symbol ∇f is pronounced "grad f." The symbol ∇ by itself is called "grad," or "del," or "nabla." With this notation, (4.31) is abbreviated to

$$\mathbf{F} = -\nabla U. \qquad (4.33)$$

This important relation gives us the force $\mathbf{F}$ in terms of derivatives of U, just as the definition (4.13) gave U as an integral of $\mathbf{F}$. When a force $\mathbf{F}$ can be expressed in the form (4.33), we say that $\mathbf{F}$ is **derivable from a potential energy**. Thus, we have shown that any conservative force is derivable from a potential energy.[7]

---

[7] I am following standard terminology here. Notice that we have defined "conservative" so that a conservative force conserves energy *and* is derivable from a potential energy. This is occasionally confusing, since there are forces (such as the magnetic force on a charge or the normal force on a sliding object) that do no work and hence conserve energy, but are not "conservative" in the sense defined here, since they are not derivable from a potential energy. This unfortunate confusion seldom causes trouble, but you may want to register it somewhere in the back of your mind.

EXAMPLE 4.4   Finding F from U

The potential energy of a certain particle is $U = Axy^2 + B \sin Cz$, where A, B and C are constants. What is the corresponding force?

To find **F** we have only to evaluate the three partial derivatives in (4.31). In doing this, you must remember that $\partial U / \partial x$ is found by differentiating with respect to x, treating y and z as constant, and so on. Thus $\partial U / \partial x = Ay^2$, and so on, and the final result is

$$\mathbf{F} = -(\hat{\mathbf{x}} \, Ay^2 + \hat{\mathbf{y}} \, 2Axy + \hat{\mathbf{z}} \, BC \cos Cz).$$

It is sometimes convenient to remove the f from (4.32) and to write

$$\nabla = \hat{\mathbf{x}} \frac{\partial}{\partial x} + \hat{\mathbf{y}} \frac{\partial}{\partial y} + \hat{\mathbf{z}} \frac{\partial}{\partial z}. \tag{4.34}$$

In this view, ∇ is a vector differential operator that can be applied to any scalar f and produces the vector given in (4.32).

A very useful application of the gradient is given by (4.28), whose right-hand side you will recognize as $\nabla U \cdot d\mathbf{r}$. Thus, if we replace U by an arbitrary scalar f, we see that the change in f resulting from a small displacement $d\mathbf{r}$ is just

$$df = \nabla f \cdot d\mathbf{r}. \tag{4.35}$$

This useful relation is the three-dimensional analog of Equation (4.27) for a function of one variable. It shows the sense in which the gradient is the three-dimensional equivalent of the ordinary derivative in one dimension.

If you have never met the ∇ notation before, it will take a little getting used to. Meanwhile, you can just think of (4.33) as a convenient shorthand for the three equations (4.30). For practice using the gradient, you could look at Problems 4.12 through 4.19.

4.4   The Second Condition that F be Conservative

We have seen that one of the two conditions that a force **F** be conservative is that the work $\int_1^2 \mathbf{F} \cdot d\mathbf{r}$ which it does moving between any two points 1 and 2 must be independent of the path followed. You are certainly to be excused if you don't see how we could test whether a given force has this property. Checking the value of the integral for every pair of points and every path joining those points is indeed a formidable prospect! Fortunately, we never need to do this. There is a simple test, which can be quickly applied to any force that is given in analytic form. This test involves another of the basic concepts of vector calculus, this time the so-called *curl* of a vector.

It can be shown (though I shall not do so here[8]) that a force $\mathbf{F}$ has the desired property, that the work it does is independent of path, if and only if

$$\nabla \times \mathbf{F} = 0 \qquad (4.36)$$

everywhere. The quantity $\nabla \times \mathbf{F}$ is called the **curl** of $\mathbf{F}$, or just "curl $\mathbf{F}$," or "del cross $\mathbf{F}$." It is defined by taking the cross product of ∇ and $\mathbf{F}$ just as if the components of ∇, namely $(\partial/\partial x, \partial/\partial y, \partial/\partial z)$, were ordinary numbers. To see what this means, consider first the cross product of two ordinary vectors $\mathbf{A}$ and $\mathbf{B}$. In the table below, I have listed the components of $\mathbf{A}$, $\mathbf{B}$, and $\mathbf{A} \times \mathbf{B}$:

vector	x component	y component	z component
$\mathbf{A}$	A_x	A_y	A_z
$\mathbf{B}$	B_x	B_y	B_z
$\mathbf{A} \times \mathbf{B}$	$A_y B_z - A_z B_y$	$A_z B_x - A_x B_z$	$A_x B_y - A_y B_x$

$$(4.37)$$

The components of $\nabla \times \mathbf{F}$ are found in exactly the same way, except that the entries in the first row are differential operators. Thus,

vector	x component	y component	z component
∇	$\partial/\partial x$	$\partial/\partial y$	$\partial/\partial z$
$\mathbf{F}$	F_x	F_y	F_z
$\nabla \times \mathbf{F}$	$\frac{\partial}{\partial y} F_z - \frac{\partial}{\partial z} F_y$	$\frac{\partial}{\partial z} F_x - \frac{\partial}{\partial x} F_z$	$\frac{\partial}{\partial x} F_y - \frac{\partial}{\partial y} F_x$

$$(4.38)$$

No one would claim that (4.36) is *obviously* equivalent to the condition that $\int_1^2 \mathbf{F} \cdot d\mathbf{r}$ is path-independent, but it is, and it provides an easily applied test for the path-independence property, as the following example shows.

EXAMPLE 4.5   Is the Coulomb Force Conservative?

Consider the force $\mathbf{F}$ on a charge q due to a fixed charge Q at the origin. Show that it is conservative and find the corresponding potential energy U. Check that $-\nabla U = \mathbf{F}$.

The force in question is the Coulomb force, as shown in Figure 4.7(a),

$$\mathbf{F} = \frac{kqQ}{r^2}\hat{\mathbf{r}} = \frac{\gamma}{r^3}\mathbf{r} \qquad (4.39)$$

where k denotes the Coulomb force constant, often written as $1/(4\pi\epsilon_o)$, and γ is just an abbreviation for the constant kqQ. From the last expression we can read off the components of $\mathbf{F}$, and using (4.38) we can calculate the components of $\nabla \times \mathbf{F}$. For example, the x component is

$$(\nabla \times \mathbf{F})_x = \frac{\partial}{\partial y}F_z - \frac{\partial}{\partial z}F_y = \frac{\partial}{\partial y}\left(\frac{\gamma z}{r^3}\right) - \frac{\partial}{\partial z}\left(\frac{\gamma y}{r^3}\right). \qquad (4.40)$$

---

[8] The condition (4.36) follows from a result called Stokes's theorem. If you would like to explore this a little, see Problem 4.25. For more details, see any text on vector calculus or mathematical methods. I particularly like *Mathematical Methods in the Physical Sciences* by Mary Boas (Wiley, 1983), p. 260.

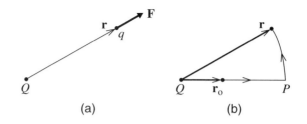

Figure 4.7   **(a)** The Coulomb force $\mathbf{F} = \gamma\hat{\mathbf{r}}/r^2$ of the fixed charge Q on the charge q. **(b)** The work done by $\mathbf{F}$ as q moves from $\mathbf{r}_0$ to $\mathbf{r}$ can be evaluated following a path that goes radially outward to P and then around a circle to $\mathbf{r}$.

The two derivatives here are easily evaluated: First, since $\partial z/\partial y = \partial y/\partial z = 0$, we can rewrite (4.40) as

$$(\mathbf{\nabla} \times \mathbf{F})_x = \gamma z \left(\frac{\partial}{\partial y} r^{-3} \right) - \gamma y \left(\frac{\partial}{\partial z} r^{-3} \right). \tag{4.41}$$

Next recall that

$$r = (x^2 + y^2 + z^2)^{1/2},$$

so that, for example,

$$\frac{\partial r}{\partial y} = \frac{y}{r}. \tag{4.42}$$

(Check this one using the chain rule.) We can now evaluate the two remaining derivatives in (4.41) to give (remember the chain rule again)

$$(\mathbf{\nabla} \times \mathbf{F})_x = \gamma z \left(\frac{-3}{r^4} \cdot \frac{y}{r} \right) - \gamma y \left(\frac{-3}{r^4} \cdot \frac{z}{r} \right) = 0.$$

The other two components work in exactly the same way (check it, if you don't believe me), and we conclude that $\mathbf{\nabla} \times \mathbf{F} = 0$. According to the result (4.36), this guarantees that $\mathbf{F}$ satisfies the second condition to be conservative. Since it certainly satisfies the first condition (it depends only on the variable $\mathbf{r}$), we have proved that $\mathbf{F}$ is conservative. (The proof that $\mathbf{\nabla} \times \mathbf{F} = 0$ is considerably quicker in spherical polar coordinates. See Problem 4.22.)

The potential energy is defined by the work integral (4.13),

$$U(\mathbf{r}) = - \int_{\mathbf{r}_0}^{\mathbf{r}} \mathbf{F}(\mathbf{r}') \cdot d\mathbf{r}' \tag{4.43}$$

where $\mathbf{r}_0$ is the (as yet unspecified) reference point where $U(\mathbf{r}_0) = 0$. Fortunately, we know that this integral is independent of path, so we can choose whatever path is most convenient. One possibility is shown in Figure 4.7(b), where I have chosen a path that goes radially outward to the point labeled P and then around a circle (centered on Q) to $\mathbf{r}$. On the first segment, $\mathbf{F}(\mathbf{r}')$ and

$d\mathbf{r}'$ are collinear, and $\mathbf{F}(\mathbf{r}') \cdot d\mathbf{r}' = (\gamma/r'^2)dr'$. On the second, $\mathbf{F}(\mathbf{r}')$ and $d\mathbf{r}'$ are perpendicular, so no work is done along this segment, and the total work is just that of the first segment,

$$U(\mathbf{r}) = -\int_{r_o}^{r} \frac{\gamma}{r'^2}\, dr' = \frac{\gamma}{r} - \frac{\gamma}{r_o}. \qquad (4.44)$$

Finally, it is usual in this problem to choose the reference point $\mathbf{r}_o$ at infinity, so that the second term here is zero. With this choice (and replacing γ by kqQ) we arrive at the well-known formula for the potential energy of the charge q due to Q,

$$U(\mathbf{r}) = U(r) = \frac{kqQ}{r}. \qquad (4.45)$$

Notice that the answer depends only on the magnitude r of the position vector $\mathbf{r}$ and not on the direction.

To check ∇U let us evaluate the x component:

$$(\nabla U)_x = \frac{\partial}{\partial x}\left(\frac{kqQ}{r}\right) = -\frac{kqQ}{r^2} \cdot \frac{\partial r}{\partial x} \qquad (4.46)$$

where the last expression follows from the chain rule. The derivative $\partial r/\partial x$ is x/r [compare Equation (4.42)], so

$$(\nabla U)_x = -kqQ\frac{x}{r^3} = -F_x,$$

as given by (4.39). The other two components work in exactly the same way, and we have shown that

$$\nabla U = -\mathbf{F} \qquad (4.47)$$

as required.

4.5 Time-Dependent Potential Energy

We sometimes have occasion to study a force $\mathbf{F}(\mathbf{r},t)$ that satisfies the second condition to be conservative ($\nabla \times \mathbf{F} = 0$), but, because it is *time-dependent*, does not satisfy the first condition. In this case, we can still define a potential energy $U(\mathbf{r}, t)$ with the property that $\mathbf{F} = -\nabla U$, but it is no longer the case that total mechanical energy, $E = T + U$, is conserved. Before I justify these claims, let me give an example of this situation. Figure 4.8 shows a small charge q in the vicinity of a charged conducting sphere (for example, a Van de Graaff generator) with a charge $Q(t)$ that is slowly leaking away through the moist air to ground. Because $Q(t)$ changes with time, the force that it exerts on the small charge q is explicitly time-dependent. Nevertheless, the spatial dependence of the force is the same as for the time-independent Coulomb force of Example 4.5 (page 119). Exactly the same analysis as in that example shows that $\nabla \times \mathbf{F} = 0$.

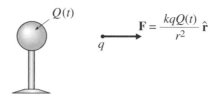

Figure 4.8   The charge $Q(t)$ on the conducting sphere is
slowly leaking away, so the force on the small charge q
varies with time, even if its position $\mathbf{r}$ is constant.

Let me now justify the claims made above. First, since $\nabla \times \mathbf{F}(\mathbf{r}, t) = 0$, the same
mathematical theorem quoted in connection with Equation (4.36) guarantees that the
work integral $\int_1^2 \mathbf{F}(\mathbf{r}, t) \cdot d\mathbf{r}$ (evaluated at any one time t) is path independent. This
means we can define a function $U(\mathbf{r}, t)$ by an integral exactly analogous to (4.13),

$$U(\mathbf{r}, t) = -\int_{\mathbf{r}_o}^{\mathbf{r}} \mathbf{F}(\mathbf{r}', t) \cdot d\mathbf{r}', \tag{4.48}$$

and, for the same reasons as before, $\mathbf{F}(\mathbf{r}, t) = -\nabla U(\mathbf{r}, t)$. (See Problem 4.27.) In this
case, we can say the force $\mathbf{F}$ is derivable from the time-dependent potential energy
$U(\mathbf{r}, t)$.

So far everything has gone through just as before, but now the story changes. We
can define the mechanical energy as $E = T + U$, but it is no longer true that E is
conserved. If you review carefully the argument leading to Equation (4.19), you may
be able to see what goes wrong, but we can in any case show directly that $E = T + U$
changes as the particle moves on its path. As before, consider any two neighboring
points on the particle's path at times t and $t + dt$. Exactly as in (4.4), the change in
kinetic energy is

$$dT = \frac{dT}{dt} dt = (m\dot{\mathbf{v}} \cdot \mathbf{v}) dt = \mathbf{F} \cdot d\mathbf{r}. \tag{4.49}$$

Meanwhile, $U(\mathbf{r}, t) = U(x, y, z, t)$ is a function of four variables (x, y, z, t) and

$$dU = \frac{\partial U}{\partial x} dx + \frac{\partial U}{\partial y} dy + \frac{\partial U}{\partial z} dz + \frac{\partial U}{\partial t} dt. \tag{4.50}$$

You will recognize the first three terms on the right as $\nabla U \cdot d\mathbf{r} = -\mathbf{F} \cdot d\mathbf{r}$. Thus

$$dU = -\mathbf{F} \cdot d\mathbf{r} + \frac{\partial U}{\partial t} dt. \tag{4.51}$$

When we add this to Equation (4.49) the first two terms cancel, and we are left with

$$d(T + U) = \frac{\partial U}{\partial t} dt. \tag{4.52}$$

Clearly it is only when U is independent of t (that is, $\partial U / \partial t = 0$) that the mechanical
energy $E = T + U$ is conserved.

Returning to the example of Figure 4.8, we can understand this conclusion and see what has happened to conservation of energy. Imagine that I hold the charge q stationary at the position of Figure 4.8, while the charge on the sphere leaks away. Under these conditions, the KE of q doesn't change, but the potential energy $kq\,Q(t)/r$ slowly diminishes to zero. Clearly $T + U$ is not constant. However, while mechanical energy is not conserved, *total* energy *is* conserved: The loss of mechanical energy is exactly balanced by the gain of thermal energy as the discharge current heats up the surrounding air. This example suggests, what is true, that the potential energy depends explicitly on time in precisely those situations where mechanical energy gets transformed to some other form of energy or to mechanical energy of other bodies external to the system of interest.

4.6   Energy for Linear One-Dimensional Systems

So far we have discussed the energy of a particle that is free to move in all three dimensions. Many interesting problems involve an object that is constrained to move in just one dimension, and the analysis of such problems is remarkably simpler than the general case. Oddly enough, there is some ambiguity in what a physicist means by a "one-dimensional system." Many introductory physics texts start out discussing the motion of a one-dimensional system, by which they mean an object (a railroad car, for instance) that is confined to move on a perfectly straight, or linear, track. In discussing such linear systems, we naturally take the x axis to coincide with the track, and the position of the object is then specified by the single coordinate x. In this section I shall focus on linear one-dimensional systems. However, there are much more complicated systems, such as a roller coaster on its curving track, that are also one-dimensional, inasmuch as their position can be specified by a single parameter (such as the distance of the roller coaster along its track). As I shall discuss in the next section, energy conservation for such curvilinear one-dimensional systems is just as straightforward as for a perfectly straight track.

To begin, let us consider an object constrained to move along a perfectly straight track, which we take to be the x axis. The only component of any force $\mathbf{F}$ that can do work is the x component, and we can simply ignore the other two components. Therefore the work done by $\mathbf{F}$ is the one-dimensional integral

$$W(x_1 \rightarrow x_2) = \int_{x_1}^{x_2} F_x(x)\,dx. \qquad (4.53)$$

If the force is to be conservative, F_x must satisfy the two usual conditions: (i) It must depend only on the position x [as I have already implied in writing the integral (4.53)]. (ii) The work (4.53) must be independent of path. The remarkable feature of one-dimensional systems is that the first condition already guarantees the second, so the latter is superfluous. To understand this property, you have only to recognize that in one dimension there is only a small choice of paths connecting any two points. Consider, for example, the two points A and B shown in Figure 4.9. The obvious path between points A and B is the path that goes from A directly to B (let's call this path

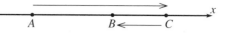

Figure 4.9   The path called $ABCB$ goes from A past B and on to C, then back to B.

"AB"). Another possibility, shown in the figure, is to go from A past B to C and then back to B (let's call this one "$ABCB$"). The work done along this path can be broken up as follows:

$$W(ABCB) = W(AB) + W(BC) + W(CB).$$

Now, provided the force depends only on the position x [condition (i)] each increment of work going from B to C is exactly equal (but of opposite sign) to the corresponding contribution going from C to B. That is, the last two terms on the right cancel, and we conclude that

$$W(ABCB) = W(AB),$$

as required. One can of course concoct a path from A to B that doubles back and forth many times, but a little thought should convince you that any such path can be broken into a number of segments some of which together traverse the direct path AB exactly once, and all the rest of which cancel in pairs. Thus the work done on *any* path between A and B is the same as that on the direct path AB, and we have proved that in one dimension the first condition for a force to be conservative guarantees the second.

Graphs of the Potential Energy

A second useful feature of one-dimensional systems is that with only one independent variable (x) we can plot the potential energy $U(x)$, and, as we shall see, this makes it easy to visualize the behavior of the system. Assuming all forces on the object are conservative, we define the potential energy as

$$U(x) = -\int_{x_0}^{x} F_x(x')\, dx' \tag{4.54}$$

where F_x is the x component of the net force on the particle. For example, for a mass on the end of a spring obeying Hooke's law, the force is $F_x = -kx$, and, if we choose the reference point $x_0 = 0$, Equation (4.54) gives the celebrated result

$$U = \tfrac{1}{2}kx^2$$

for any spring obeying Hooke's law.

Corresponding to the three-dimensional result $\mathbf{F} = -\nabla U$, we have the simpler result in one dimension

$$F_x = -\frac{dU}{dx}. \tag{4.55}$$

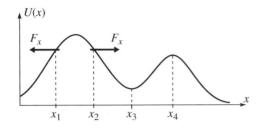

Figure 4.10   The graph of potential energy $U(x)$ against x for any one-dimensional system can be thought of as a picture of a roller coaster track. The force $F_x = -dU/dx$ tends to push the object "downhill" as at x_1 and x_2. At the points x_3 and x_4, where $U(x)$ is minimum or maximum, $dU/dx = 0$ and the force is zero; such points are therefore points of equilibrium.

If we plot the potential energy against x as in Figure 4.10, we can easily see qualitatively how the object has to behave. The direction of the net force is given by (4.55) as "downhill" on the graph of $U(x)$ — to the left at x_1 and to the right at x_2. It follows that the object always accelerates in the "downhill" direction — a property that reminds one of the motion of a roller coaster, which also always accelerates downhill. This analogy is not an accident: For a roller coaster, $U(x)$ is mgh (where h is the height above ground) and the graph of $U(x)$ against x has the same shape as a graph of h against x, which is just a *picture* of the track. For any one-dimensional system, we can always *think* about the graph of $U(x)$ as a picture of a roller coaster, and common sense will generally tell us the kind of motion that is possible at different places, as I now describe.

At points, such as x_3 and x_4, where $dU/dx = 0$ and $U(x)$ is minimum or maximum, the net force is zero, and the object can remain in equilibrium. That is, the condition $dU/dx = 0$ characterizes points of equilibrium. At x_3, where $d^2U/dx^2 > 0$ and $U(x)$ is minimum, a small displacement from equilibrium causes a force which pushes the object back to equilibrium (back to the left on the right of x_3, back to the right on the left of x_3). In other words, equilibrium points where $d^2U/dx^2 > 0$ and $U(x)$ is minimum are points of *stable* equilibrium. At equilibrium points like x_4 where $d^2U/dx^2 < 0$ and $U(x)$ is maximum, a small displacement leads to a force *away* from equilibrium, and the equilibrium is *unstable*.

If the object is *moving* then its kinetic energy is positive and its total energy is necessarily greater than $U(x)$. For example, suppose the object is moving somewhere near the equilibrium point $x = b$ in Figure 4.11. Its total energy has to be greater than $U(b)$ and could, for example, equal the value shown as E in that figure. If the object happens to be on the right of b and moving toward the right, its PE will increase and its KE must therefore decrease until the object reaches the **turning point** labeled c, where $U(c) = E$ and the KE is zero. At $x = c$ the object stops and, with the force back to the left, it accelerates back toward $x = b$. It cannot now stop until once again the KE is zero, and this occurs at the turning point a, where $U(a) = E$ and the object accelerates back to the right. Since the whole

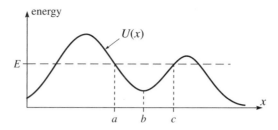

Figure 4.11   If an object starts out near $x = b$ with the energy E shown, it is trapped in the valley or "well" between the two hills and oscillates between the turning points at $x = a$ and c where $U(x) = E$ and the kinetic energy is zero.

cycle now repeats itself, we see that if the object starts out between two hills and its energy is lower than the crest of both hills, then the object is trapped in the valley or "well" and oscillates indefinitely between the two turning points where $U(x) = E$.

Suppose the object again starts out between the two hills but with energy higher than the crest of the right hill though still lower than the left. In this case, it will escape to the right since $E > U(x)$ everywhere on the right, and it can never stop once it is moving in that direction. Finally, if the energy is higher than both hills, the object can escape in either direction.

These considerations play an important role in many fields. An example from molecular physics is illustrated in Figure 4.12, which shows the potential energy of a typical diatomic molecule, such as HCl, as a function of the distance between the two atoms. This potential energy function governs the radial motion of the hydrogen atom (in the case of HCl) as it vibrates in and out from the much heavier chlorine atom. The zero of energy has been chosen where the two atoms are far apart (at infinity) and at rest. Notice that the independent variable is the interatomic distance r which, by its definition, is always positive, $0 \leq r < \infty$. As $r \to 0$, the potential energy gets very large, indicating that the two atoms repel one another when very close together (because of the Coulomb repulsion of the nuclei). If the energy is positive ($E > 0$) the H atom can escape to infinity, since there is no "hill" to trap it; the H atom can come in from infinity, but it will stop at the turning point $r = a$ and (in the absence of any mechanism to take up some of its energy) it will move away to infinity again. On the other hand, if $E < 0$, the H atom is trapped and will oscillate in and out between the two turning points shown at $r = b$ and $r = d$. The equilibrium separation of the molecule is at the point shown as $r = c$. It is the states with $E < 0$ that correspond to what we normally regard as the HCl molecule. To form such a molecule, two separate atoms (with $E > 0$) must come together to a separation somewhere near $r = c$, and some process, such as emission of light, must remove enough energy to leave the two atoms trapped with $E < 0$.

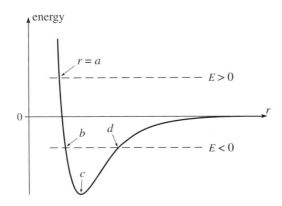

Figure 4.12 The potential energy for a typical diatomic molecule such as HCl, plotted as a function of the distance r between the two atoms. If $E > 0$, the two atoms cannot approach closer than the turning point $r = a$, but they can move apart to infinity. If $E < 0$, they are trapped between the turning points at b and d and form a bound molecule. The equilibrium separation is $r = c$.

Complete Solution of the Motion

A third remarkable feature of one-dimensional conservative systems is that we can — at least in principle — use the conservation of energy to obtain a complete solution of the motion, that is, to find the position x as a function of time t. Since $E = T + U(x)$ is conserved, with $U(x)$ a known function (in the context of a given problem) and E determined by the initial conditions, we can solve for $T = \frac{1}{2}m\dot{x}^2 = E - U(x)$ and hence for the velocity $\dot{x}$ as a function of x:

$$\dot{x}(x) = \pm\sqrt{\frac{2}{m}}\sqrt{E - U(x)}. \tag{4.56}$$

(Notice that there is an ambiguity in the sign since energy considerations cannot determine the *direction* of the velocity. For this reason, the method described here usually does not work in a truly three-dimensional problem. In one dimension, you can almost always decide the sign of $\dot{x}$ by inspection, though you must remember to do so.)

Knowing the velocity as a function of x, we can now find x as a function of t, using separation of variables, as follows: We first rewrite the definition $\dot{x} = dx/dt$ as

$$dt = \frac{dx}{\dot{x}}.$$

[Since $\dot{x} = \dot{x}(x)$, this separates the variables t and x.] Next, we can integrate between any initial and final points to give

$$t_\mathrm{f} - t_\mathrm{i} = \int_{x_\mathrm{i}}^{x_\mathrm{f}} \frac{dx}{\dot{x}}. \tag{4.57}$$

This gives the time for travel between any initial and final positions of interest. If we substitute for $\dot{x}$ from (4.56) (and assume, to be definite, that $\dot{x}$ is positive) then the time to go from the initial x_0 at time 0 to an arbitrary x at time t is

$$t = \int_{x_o}^{x} \frac{dx'}{\dot{x}(x')} = \sqrt{\frac{m}{2}} \int_{x_o}^{x} \frac{dx'}{\sqrt{E - U(x')}}. \tag{4.58}$$

(As usual, I've renamed the variable of integration as x' to avoid confusion with the upper limit x.) The integral (4.58) depends on the particular form of $U(x)$ in the problem at hand. Assuming we can do the integral [and we can at least do it numerically for any given $U(x)$], it gives us t as a function of x. Finally we can solve to give x as a function of t, and our solution is complete, as the following simple example illustrates.

EXAMPLE 4.6   Free Fall

I drop a stone from the top of a tower at time $t = 0$. Use conservation of energy to find the stone's position x (measured down from the top of the tower, where $x = 0$) as a function of t. Neglect air resistance.

The only force on the stone is gravity, which is, of course, conservative. The corresponding potential energy is

$$U(x) = -mgx.$$

(Remember x is measured downward.) Since the stone is at rest when $x = 0$, the total energy is $E = 0$, and according to (4.56) the velocity is

$$\dot{x}(x) = \sqrt{\frac{2}{m}} \sqrt{E - U(x)} = \sqrt{2gx}$$

(a result that is well known from elementary kinematics). Thus

$$t = \int_{0}^{x} \frac{dx'}{\dot{x}(x')} = \int_{0}^{x} \frac{dx'}{\sqrt{2gx'}} = \sqrt{\frac{2x}{g}}.$$

As anticipated, this gives t as a function of x, and we can solve to give the familiar result

$$x = \tfrac{1}{2}gt^2.$$

This simple example, involving the gravitational potential energy $U(x) = -mgx$, can be solved many different (and some simpler) ways, but the energy method used here can be used for *any* potential energy function $U(x)$. In some cases, the integral (4.58) can be evaluated in terms of elementary functions, and we obtain an analytic solution of the problem; for example, if $U(x) = \tfrac{1}{2}kx^2$ (as for a mass on the end of a spring), the integral turns out to be an inverse sine function, which implies that x oscillates sinusoidally with time, as we should expect (see Problem 4.28). For some potential energies, the integral cannot be done in terms of elementary functions, but can nonetheless be related to functions that are tabulated (see Problem 4.38). For some problems, the only way to do the integral (4.58) is to do it numerically.

4.7 Curvilinear One-Dimensional Systems

So far the only one-dimensional system I have discussed is an object constrained to move along a linear path, with position specified by the coordinate x. There are other, more general, systems that can equally be said to be one-dimensional, inasmuch as their position is specified by a single number. An example of such a one-dimensional system is a bead threaded on a curved rigid wire as illustrated in Figure 4.13. (Another is a roller coaster confined to a curved track.) The position of the bead can be specified by a single parameter, which we can choose as the distance s, measured along the wire, from a chosen origin O. With this choice of coordinate, the discussion of the curved one-dimensional track parallels closely that of the straight track, as I now show.

The coordinate s of our bead corresponds, of course, to x for a cart on a straight track. The speed of the bead is easily seen to be $\dot{s}$, and the kinetic energy is therefore just

$$T = \tfrac{1}{2}m\dot{s}^2$$

as compared to the familiar $\tfrac{1}{2}m\dot{x}^2$ for the straight track. The force is a little more complicated. As our bead moves on the curved wire the net normal force is not zero; on the contrary, the normal force is what constrains the bead to follow its assigned curving path. (For this reason, the normal force is called the *force of constraint*.) On the other hand, the normal force does no work, and it is the *tangential* component F_{tang} of the net force that is our chief concern. In particular, it is fairly easy to show (Problem 4.32) that

$$F_{\text{tang}} = m\ddot{s}$$

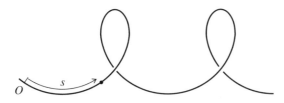

Figure 4.13 An object constrained to move on a curved track can be considered to be a one-dimensional system, with the position specified by the distance s (measured along the track) of the object from an origin O. The system shown is a bead threaded on a stiff wire, bent into a double loop-the-loop.

(just as $F_x = m\ddot{x}$ on a straight track). Further, if all the forces on the bead that have a tangential component are conservative, we can define a corresponding potential energy $U(s)$ such that $F_{\text{tang}} = -dU/ds$, and the total mechanical energy $E = T + U(s)$ is constant. The whole discussion of Section 4.6 can now be applied to the bead on a curved wire (or any other object constrained to move on a one-dimensional path). In particular, those points where $U(s)$ is a minimum are points of stable equilibrium, and those where $U(s)$ is maximum are points of unstable equilibrium.

There are many systems that appear to be much more complicated than the bead on a wire, but are nonetheless one-dimensional and can be treated in much the same way. Here is an example.

EXAMPLE 4.7   Stability of a Cube Balanced on a Cylinder

A hard rubber cylinder of radius r is held fixed with its axis horizontal, and a wooden cube of mass m and side $2b$ is balanced on top of the cylinder, with its center vertically above the cylinder's axis and four of its sides parallel to the axis. The cube cannot slip on the rubber of the cylinder, but it can of course rock from side to side, as shown in Figure 4.14. By examining the cube's potential energy, find out if the equilibrium with the cube centered above the cylinder is stable or unstable.

Let us first note that the system is one-dimensional, since its position as it rocks from side to side can be specified by a single coordinate, for instance the angle θ through which it has turned. (We could also specify it by the distance s of the cube's center from equilibrium, but the angle is a little more convenient. Either way the system's position is specified by a single coordinate, and our problem is definitely one-dimensional.) The constraining forces are the normal

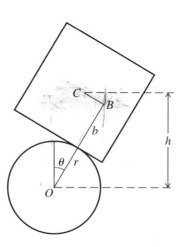

Figure 4.14   A cube, of side $2b$ and center C, is placed on a fixed horizontal cylinder of radius r and center O. It is originally put so that C is centered above O, but it can roll from side to side without slipping.

and frictional forces of the cylinder on the cube; that is, these two forces constrain the cube to move only as shown in Figure 4.14. Since neither of these does any work we need not consider them explicitly. The only other force on the cube is gravity, and we know from elementary physics that this is conservative and that the gravitational potential energy is the same as for a point mass at the center of the cube; that is, $U = mgh$, where h is the height of C above the origin, as shown in Figure 4.14. (See Problem 4.6.) The length of the line shown as OB is just $r + b$, while the length BC is the distance the cube has rolled around the cylinder, namely $r\theta$. Therefore $h = (r + b)\cos\theta + r\theta\sin\theta$ and the potential energy is

$$U(\theta) = mgh = mg[(r + b)\cos\theta + r\theta\sin\theta].\qquad(4.59)$$

To find the equilibrium position (or positions) we must find the points where $dU/d\theta$ vanishes. (Strictly speaking I haven't proved this very plausible claim yet for this kind of constrained system; I'll discuss it shortly.) The derivative is easily seen to be (check this for yourself)

$$\frac{dU}{d\theta} = mg[r\theta\cos\theta - b\sin\theta].$$

This vanishes at $\theta = 0$, confirming the obvious — that $\theta = 0$ is a point of equilibrium. To decide whether this equilibrium is stable, we have only to differentiate again and find the value of $d^2U/d\theta^2$ at the equilibrium position. This gives (as you should check)

$$\frac{d^2U}{d\theta^2} = mg(r - b)\qquad(4.60)$$

(at $\theta = 0$). If the cube is smaller than the cylinder (that is, $b < r$), this second derivative is positive, which means that $U(\theta)$ has a minimum at $\theta = 0$ and the equilibrium is stable; if the cube is balanced on the cylinder, it will remain there indefinitely. On the other hand, if the cube is larger than the cylinder ($b > r$), the second derivative (4.60) is negative, the equilibrium is unstable, and the smallest disturbance will cause the cube to roll and fall off the cylinder.

Further Generalizations

There are many other, more complicated systems that are still legitimately described as one dimensional. Such systems may comprise several bodies, but the bodies are joined by struts or strings in such a way that just one parameter is needed to describe the system's position. An example of such a system is the Atwood machine shown in Figure 4.15, which consists of two masses, m_1 and m_2, suspended from opposite ends of a massless, inextensible string that passes over a frictionless pulley. (To simplify the discussion, I shall assume the pulley is massless, although it is easy to allow for a mass of the pulley.) The two masses can move up and down, but the forces of the pulley on the string and the string on the masses constrain matters so that the mass m_2 can move up only to the extent that m_1 moves down by exactly the same distance.

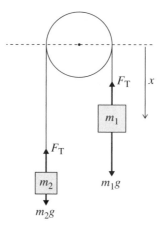

Figure 4.15   An Atwood machine consisting of two masses, m_1 and m_2, suspended by a massless inextensible string that passes over a massless, frictionless pulley. Because the string's length is fixed, the position of the whole system is specified by the distance x of m_1 below any convenient fixed level. The forces on the two masses are their weights $m_1 g$ and $m_2 g$, and the tension forces F_T (which are equal since the pulley and string are massless).

Thus the position of the whole system can be specified by a single parameter, for example the height x of m_1 below the pulley's center as shown, and the system is again one-dimensional.[9]

Let us consider the energies of the masses m_1 and m_2. The forces acting on them are gravity and the tension in the string. Since gravity is conservative, we can introduce potential energies U_1 and U_2 for the gravitational forces, and our previous considerations imply that in any displacement of the system,

$$\Delta T_1 + \Delta U_1 = W_1^{\text{ten}} \tag{4.61}$$

and

$$\Delta T_2 + \Delta U_2 = W_2^{\text{ten}} \tag{4.62}$$

where the terms W^{ten} denote the work done by the tension on m_1 and m_2. Now, in the absence of friction, the tension is the same all along the string. Thus, although the tension certainly does work on the two individual masses, the work done on m_1 is equal and opposite to that done on m_2, when m_1 moves down and m_2 moves an equal distance up (or vice versa). That is,

$$W_1^{\text{ten}} = -W_2^{\text{ten}}. \tag{4.63}$$

---

[9] You may object, correctly, that the masses can also move sideways. If this worries you, we can thread each mass over a vertical frictionless rod, but these rods are actually unnecessary: As long as we refrain from pushing the masses sideways, each will remain in a vertical line of its own accord.

Thus, if we add the two energy equations (4.61) and (4.62), the terms involving the tension in the string cancel and we are left with

$$\Delta(T_1 + U_1 + T_2 + U_2) = 0.$$

That is, the total mechanical energy

$$E = T_1 + U_1 + T_2 + U_2 \tag{4.64}$$

is conserved. The beauty of this result is that all reference to the constraining forces of the string and pulley has disappeared.

It turns out that many systems which contain several particles that are constrained in some way (by strings, struts, or a track on which they must move, etc.) can be treated in this same way: The constraining forces are crucially important in determining how the system moves, but they do no work on the system as a whole. Thus in considering the total energy of the system, we can simply ignore the constraining forces. In particular, if all other forces are conservative (as with our example of the Atwood machine), we can define a potential energy U_α for each particle α, and the total energy

$$E = \sum_{\alpha=1}^{N}(T_\alpha + U_\alpha)$$

is constant. If the system is also one-dimensional (position specified by just one parameter, as with the Atwood machine), then all of the considerations of Section 4.6 apply.

A careful discussion of constrained systems is far easier in the Lagrangian formulation of mechanics than in the Newtonian. Thus I shall postpone any further discussion to Chapter 7. In particular, the proof that a stable equilibrium normally corresponds to a minimum of the potential energy (for a large class of constrained systems) is sketched in Problem 7.47.

4.8   Central Forces

A three-dimensional situation that has some of the simplicity of one-dimensional problems is a particle that is subject to a central force, that is, a force that is everywhere directed toward or away from a fixed "force center." If we take the force center to be the origin, a central force has the form

$$\mathbf{F}(\mathbf{r}) = f(\mathbf{r})\hat{\mathbf{r}} \tag{4.65}$$

where the function $f(\mathbf{r})$ gives the magnitude of the force (and is positive if the force is outward and negative if it is inward). An example of a central force is the Coulomb force on a charge q due to a second charge Q at the origin; this has the familiar form

$$\mathbf{F}(\mathbf{r}) = \frac{kqQ}{r^2}\hat{\mathbf{r}}, \tag{4.66}$$

which is obviously an example of (4.65), with the magnitude function given by $f(\mathbf{r}) = kq\,Q/r^2$. The Coulomb force has two additional properties not shared by all central forces: First, as we have proved, it is conservative. Second, it is **spherically symmetric** or **rotationally invariant**; that is, the magnitude function $f(\mathbf{r})$ in (4.65) is independent of the direction of $\mathbf{r}$ and, hence, has the same value at all points at the same distance from the origin. A compact way to express this second property of spherical symmetry is to observe that the magnitude function $f(\mathbf{r})$ depends only on the magnitude of the vector $\mathbf{r}$ and not its direction, so can be written as

$$f(\mathbf{r}) = f(r). \qquad (4.67)$$

A remarkable feature of central forces is that the two properties just mentioned always go together: A central force that is conservative is automatically spherically symmetric, and, conversely, a central force that is spherically symmetric is automatically conservative. These two results can be proved in several ways, but the most direct proofs involve the use of spherical polar coordinates. Therefore, before offering any proofs, I shall briefly review the definition of these coordinates.

Spherical Polar Coordinates

The position of any point P is, of course, identified by the vector $\mathbf{r}$ pointing from the origin O to P. The vector $\mathbf{r}$ can be specified by its Cartesian coordinates (x, y, z), but in problems involving spherical symmetry it is almost always more convenient to specify $\mathbf{r}$ by its spherical polar coordinates (r, θ, ϕ), as defined in Figure 4.16. The first coordinate r is just the distance of P from the origin; that is, $r = |\mathbf{r}|$, as usual. The angle θ is the angle between $\mathbf{r}$ and the z axis. The angle ϕ, often called the **azimuth**, is the angle from the x axis to the projection of $\mathbf{r}$ on the xy plane, as shown.[10] It is a simple exercise (Problem 4.40) to relate the Cartesian coordinates (x, y, z) to the polar coordinates (r, θ, ϕ) and vice versa. For example, by inspecting Figure 4.16 you should be able to convince yourself that

$$x = r\sin\theta\cos\phi, \qquad y = r\sin\theta\sin\phi, \qquad \text{and} \qquad z = r\cos\theta. \qquad (4.68)$$

A beautiful use of spherical coordinates, which may help you to visualize them, is to specify positions on the surface of the earth. If we choose the origin at the center of the earth, then all points on the surface have the same value of r, namely the radius of the earth.[11] Thus positions on the surface can be specified by giving just the two angles (θ, ϕ). If we choose our z axis to coincide with the north polar axis, then it is easy to see from Figure 4.16 that θ gives the *latitude* of the point P, measured down from the north pole. (Since latitude is traditionally measured up from the equator, our angle θ is often called the *colatitude*.) Similarly, ϕ is the *longitude* measured east from the meridian of the x axis.

---

[10] You should be aware that, while the definitions given here are those always used by physicists, most mathematics texts reverse the roles of θ and ϕ.

[11] Actually the earth isn't perfectly spherical, so r isn't quite constant, but this doesn't change the conclusion that any position on the surface can be specified by giving θ and ϕ.

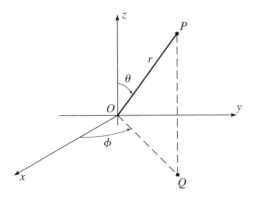

Figure 4.16   The spherical polar coordinates (r, θ, ϕ) of a
point P are defined so that r is the distance of P from the
origin, θ is the angle between the line OP and the z axis,
and ϕ is the angle of the line OQ from the x axis, where
Q is the projection of P onto the xy plane.

The statement that a function $f(\mathbf{r})$ is spherically symmetric is simply the statement
that, with $\mathbf{r}$ expressed in spherical polars, f is independent of θ and ϕ. This is what
we mean when we write $f(\mathbf{r}) = f(r)$, and the test for spherical symmetry is simply
that the two partial derivatives $\partial f/\partial\theta$ and $\partial f/\partial\phi$ are both zero everywhere.

The unit vectors $\hat{\mathbf{r}}, \hat{\boldsymbol{\theta}}$, and $\hat{\boldsymbol{\phi}}$ are defined in the usual way: First, $\hat{\mathbf{r}}$ is the unit vector
pointing in the direction of movement if r increases with θ and ϕ fixed. Thus, as shown
in Figure 4.17, the vector $\hat{\mathbf{r}}$ points radially outward, and is just the unit vector in the
direction of $\mathbf{r}$ as usual. (On the surface of the earth, $\hat{\mathbf{r}}$ points upward, in the direction
of the local vertical.) Similarly, $\hat{\boldsymbol{\theta}}$ points in the direction of increasing θ with r and ϕ
fixed, that is, southward along a line of longitude. Finally, $\hat{\boldsymbol{\phi}}$ points in the direction of
increasing ϕ with r and θ fixed, that is, eastward along a circle of latitude.

Since the three unit vectors $\hat{\mathbf{r}}, \hat{\boldsymbol{\theta}}$, and $\hat{\boldsymbol{\phi}}$ are mutually perpendicular, we can evaluate
dot products in spherical polars in just the same way as in Cartesians. Thus, if

$$\mathbf{a} = a_r\hat{\mathbf{r}} + a_\theta\hat{\boldsymbol{\theta}} + a_\phi\hat{\boldsymbol{\phi}}$$

and

$$\mathbf{b} = b_r\hat{\mathbf{r}} + b_\theta\hat{\boldsymbol{\theta}} + b_\phi\hat{\boldsymbol{\phi}}$$

then (make sure you see this)

$$\mathbf{a} \cdot \mathbf{b} = a_r b_r + a_\theta b_\theta + a_\phi b_\phi. \tag{4.69}$$

Like the unit vectors of two-dimensional polar coordinates, the unit vectors $\hat{\mathbf{r}}, \hat{\boldsymbol{\theta}}$,
and $\hat{\boldsymbol{\phi}}$ vary with position, and, as was the case in two dimensions, this variability
complicates many calculations involving differentiation, as we shall now see.

The Gradient in Spherical Polar Coordinates

In Cartesian coordinates, we have seen that the components of ∇f are precisely the partial derivatives of f with respect to x, y, and z,

$$\nabla f = \hat{\mathbf{x}}\frac{\partial f}{\partial x} + \hat{\mathbf{y}}\frac{\partial f}{\partial y} + \hat{\mathbf{z}}\frac{\partial f}{\partial z}. \tag{4.70}$$

The corresponding expression for ∇f in polar coordinates is not so straightforward. To find it, recall from (4.35) that, in a small displacement $d\mathbf{r}$, the change in any function $f(\mathbf{r})$ is

$$df = \nabla f \cdot d\mathbf{r}. \tag{4.71}$$

To evaluate the small vector $d\mathbf{r}$ in polar coordinates, we must examine carefully what happens to the point $\mathbf{r}$ when we change r, θ, and ϕ: A small change dr in r moves the point a distance dr radially out, in the direction of $\hat{\mathbf{r}}$. As you can see from Figure 4.17, a small change $d\theta$ in θ moves the point around a circle of longitude (radius r) through a distance $r\,d\theta$ in the direction of $\hat{\boldsymbol{\theta}}$. (Note well the factor of r — the distance is not just $d\theta$.) Similarly, a small change $d\phi$ in ϕ moves the point around a circle of latitude (radius $r\sin\theta$) through a distance $r\sin\theta\,d\phi$. Putting all this together, we see that

$$d\mathbf{r} = dr\,\hat{\mathbf{r}} + r\,d\theta\,\hat{\boldsymbol{\theta}} + r\sin\theta\,d\phi\,\hat{\boldsymbol{\phi}}.$$

Knowing the components of $d\mathbf{r}$, we can now evaluate the dot product in (4.71) in terms of the unknown components of ∇f,

$$df = (\nabla f)_r\,dr + (\nabla f)_\theta\,r\,d\theta + (\nabla f)_\phi\,r\sin\theta\,d\phi. \tag{4.72}$$

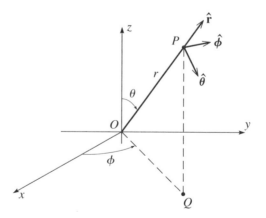

Figure 4.17   The three unit vectors of spherical polar co-ordinates at the point P. The vector $\hat{\mathbf{r}}$ points radially out, $\hat{\boldsymbol{\theta}}$ points "south" along a line of longitude, and $\hat{\boldsymbol{\phi}}$ points "east" around a circle of latitude.

Meanwhile, since f is a function of the three variables r, θ, ϕ, the change in f is, of course,

$$df = \frac{\partial f}{\partial r}dr + \frac{\partial f}{\partial \theta}d\theta + \frac{\partial f}{\partial \phi}d\phi. \tag{4.73}$$

Comparing (4.72) and (4.73), we conclude that the components of ∇f in spherical polars are

$$(\nabla f)_r = \frac{\partial f}{\partial r}, \qquad (\nabla f)_\theta = \frac{1}{r}\frac{\partial f}{\partial \theta}, \qquad \text{and} \qquad (\nabla f)_\phi = \frac{1}{r \sin\theta}\frac{\partial f}{\partial \phi} \tag{4.74}$$

or, a little more compactly,

$$\nabla f = \hat{\mathbf{r}}\frac{\partial f}{\partial r} + \hat{\boldsymbol{\theta}}\frac{1}{r}\frac{\partial f}{\partial \theta} + \hat{\boldsymbol{\phi}}\frac{1}{r \sin\theta}\frac{\partial f}{\partial \phi}. \tag{4.75}$$

Similar considerations apply to the curl and other operators of vector calculus, all of which are markedly more complicated in spherical polar coordinates (and all other non-Cartesian coordinates) than in Cartesian coordinates. Since the formulas for these operators are very hard to remember, I have listed the more important ones inside the back cover. Proofs can be found in any textbook of vector calculus.[12] Armed with these ideas, let us return to central forces.

Conservative and Spherically Symmetric, Central Forces

I claimed earlier that a central force is conservative if and only if it is spherically symmetric. This claim can be proved several different ways. The quickest proofs (though not necessarily the most insightful) use spherical polar coordinates. Let us assume first that the central force $\mathbf{F}(\mathbf{r})$ is conservative and try to prove that it must be spherically symmetric. Since it is conservative, it can be expressed in the form $-\nabla U$, which according to (4.75), has the form

$$\mathbf{F}(\mathbf{r}) = -\nabla U = -\hat{\mathbf{r}}\frac{\partial U}{\partial r} - \hat{\boldsymbol{\theta}}\frac{1}{r}\frac{\partial U}{\partial \theta} - \hat{\boldsymbol{\phi}}\frac{1}{r \sin\theta}\frac{\partial U}{\partial \phi}. \tag{4.76}$$

Since $\mathbf{F}(\mathbf{r})$ is central, only its radial component can be nonzero, and the last two terms in (4.76) must be zero. This requires that $\partial U/\partial \theta = \partial U/\partial \phi = 0$; that is, $U(\mathbf{r})$ is spherically symmetric, and (4.76) reduces to

$$\mathbf{F}(\mathbf{r}) = -\hat{\mathbf{r}}\frac{\partial U}{\partial r}.$$

Since U is spherically symmetric (depends only on r), the same is true of $\partial U/\partial r$, and we see that the central force $\mathbf{F}(\mathbf{r})$ is indeed spherically symmetric. I shall leave the proof of the converse result, that a central force which is spherically symmetric is necessarily conservative, to the problems at the end of this chapter. (See Problems 4.43

---

[12] See, for example, Mary L. Boas, *Mathematical Methods in the Physical Sciences*, John Wiley, 1983, p. 431.

and 4.44, but the simplest proof mimics almost exactly the analysis of the Coulomb force in Example 4.5.)

The importance of these results is this: First, because a force $\mathbf{F}(\mathbf{r})$ that is central and spherically symmetric has a magnitude that depends only on r, it is nearly as simple as a one-dimensional force. Second, although $\mathbf{F}(\mathbf{r})$ is certainly not actually a one-dimensional force (its *direction* still depends on θ and ϕ), we shall see in Chapter 8 that any problem involving this kind of force is mathematically equivalent to a certain related one-dimensional problem.

4.9   Energy of Interaction of Two Particles

Almost all of our discussion of energy has focused on the energy of a single particle (or any larger object that can be approximated as a particle). It is now time to extend the discussion to systems of several particles, and I shall naturally start with just two particles. In this section, I shall suppose that the two particles interact via forces $\mathbf{F}_{12}$ (on particle 1 by particle 2) and $\mathbf{F}_{21}$ (on particle 2 by particle 1), but that there are no other, external, forces. In general, the force $\mathbf{F}_{12}$ could depend on the positions of both particles, so can be written as

$$\mathbf{F}_{12} = \mathbf{F}_{12}(\mathbf{r}_1, \mathbf{r}_2),$$

and by Newton's third law

$$\mathbf{F}_{12} = -\mathbf{F}_{21}.$$

As an example of such a two-particle system we could consider an isolated binary star, in which case the only two forces are the gravitational attraction of each star for the other. If we denote the vector pointing to star 1 from star 2 by $\mathbf{r}$, as in Figure 4.18, the force $\mathbf{F}_{12}$ is just the familiar

$$\mathbf{F}_{12} = -\frac{Gm_1m_2}{r^2}\hat{\mathbf{r}} = -\frac{Gm_1m_2}{r^3}\mathbf{r}.$$

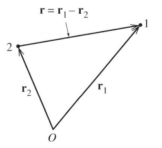

Figure 4.18   The vector $\mathbf{r}$ pointing to particle 1 from particle 2 is just $\mathbf{r} = (\mathbf{r}_1 - \mathbf{r}_2)$.

The vector $\mathbf{r}$ can be written in terms of the two positions $\mathbf{r}_1$ and $\mathbf{r}_2$. In fact, as can be seen in Figure 4.18,

$$\mathbf{r} = \mathbf{r}_1 - \mathbf{r}_2.$$

Thus the force $\mathbf{F}_{12}$, expressed as a function of $\mathbf{r}_1$ and $\mathbf{r}_2$, is

$$\mathbf{F}_{12} = -\frac{Gm_1m_2}{|\mathbf{r}_1 - \mathbf{r}_2|^3}(\mathbf{r}_1 - \mathbf{r}_2). \qquad (4.77)$$

A striking property of the force (4.77) is that it depends on the two positions $\mathbf{r}_1$ and $\mathbf{r}_2$ only through the particular combination $\mathbf{r}_1 - \mathbf{r}_2$. This property is not an accident, and is in fact true of any isolated two-particle system. The reason is that any isolated system must be **translationally invariant**: If we bodily translate the system to a new position, without changing the relative positions of the particles, the interparticle forces should remain the same. This is illustrated in Figure 4.19, which shows a pair of points $\mathbf{r}_1$ and $\mathbf{r}_2$ and a second pair of points $\mathbf{s}_1$ and $\mathbf{s}_2$, with $\mathbf{s}_1 - \mathbf{s}_2 = \mathbf{r}_1 - \mathbf{r}_2$. Since the two points $\mathbf{r}_1$ and $\mathbf{r}_2$ could be simultaneously translated to $\mathbf{s}_1$ and $\mathbf{s}_2$, the force $\mathbf{F}_{12}(\mathbf{r}_1, \mathbf{r}_2)$ must be the same as $\mathbf{F}_{12}(\mathbf{s}_1, \mathbf{s}_2)$ for *any* points satisfying $\mathbf{r}_1 - \mathbf{r}_2 = \mathbf{s}_1 - \mathbf{s}_2$. In other words, $\mathbf{F}_{12}(\mathbf{r}_1, \mathbf{r}_2)$ depends only on $\mathbf{r}_1 - \mathbf{r}_2$, as claimed, and we can write

$$\mathbf{F}_{12} = \mathbf{F}_{12}(\mathbf{r}_1 - \mathbf{r}_2). \qquad (4.78)$$

The result (4.78) greatly simplifies our discussion. We can learn almost everything about the force $\mathbf{F}_{12}$ by fixing $\mathbf{r}_2$ at any convenient point. In particular, let us temporarily fix $\mathbf{r}_2$ at the origin, in which case (4.78) reduces to just $\mathbf{F}_{12}(\mathbf{r}_1)$. (This maneuver amounts to translating both particles until particle 2 is at the origin, and we know that the force is unaffected by any such translation.) With $\mathbf{r}_2$ fixed, our discussion of the force on a single particle now applies. For example, if the force $\mathbf{F}_{12}$ on particle 1 is to be conservative, then it must satisfy

$$\nabla_1 \times \mathbf{F}_{12} = 0 \qquad (4.79)$$

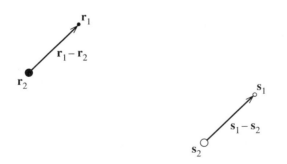

Figure 4.19   If $\mathbf{r}_1 - \mathbf{r}_2 = \mathbf{s}_1 - \mathbf{s}_2$, then two particles at $\mathbf{r}_1$ and $\mathbf{r}_2$ could be bodily translated to $\mathbf{s}_1$ and $\mathbf{s}_2$ without affecting their relative positions. This means that the force between the particles at $\mathbf{r}_1$ and $\mathbf{r}_2$ must be the same as that at $\mathbf{s}_1$ and $\mathbf{s}_2$.

where $\mathbf{V}_1$ is the differential operator

$$\mathbf{V}_1 = \hat{\mathbf{x}}\frac{\partial}{\partial x_1} + \hat{\mathbf{y}}\frac{\partial}{\partial y_1} + \hat{\mathbf{z}}\frac{\partial}{\partial z_1}$$

with respect to the coordinates (x_1, y_1, z_1) of particle 1. If (4.79) is satisfied, we can define a potential energy $U(\mathbf{r}_1)$ such that the force on particle 1 is

$$\mathbf{F}_{12} = -\mathbf{V}_1 U(\mathbf{r}_1).$$

This gives the force $\mathbf{F}_{12}$ for the case that particle 2 is at the origin. To find it for particle 2 anywhere else we have only to translate back to an arbitrary position by replacing $\mathbf{r}_1$ with $\mathbf{r}_1 - \mathbf{r}_2$ to give

$$\mathbf{F}_{12} = -\mathbf{V}_1 U(\mathbf{r}_1 - \mathbf{r}_2). \tag{4.80}$$

Notice that I don't have to change the operator $\mathbf{V}_1$, since an operator like $\partial/\partial x_1$ is unchanged by addition of a constant to x_1.

To find the reaction force $\mathbf{F}_{21}$ on particle 2, we have only to invoke Newton's third law, which says that $\mathbf{F}_{21} = -\mathbf{F}_{12}$. That is, we have only to change the sign of (4.80). We can re-express this by noticing that

$$\mathbf{V}_1 U(\mathbf{r}_1 - \mathbf{r}_2) = -\mathbf{V}_2 U(\mathbf{r}_1 - \mathbf{r}_2), \tag{4.81}$$

where $\mathbf{V}_2$ denotes the gradient with respect to the coordinates of particle 2. (To prove this, invoke the chain rule. See Problem 4.50.) So, instead of changing the sign of (4.80) to find $\mathbf{F}_{21}$, we can simply replace $\mathbf{V}_1$ by $\mathbf{V}_2$ to give

$$\mathbf{F}_{21} = -\mathbf{V}_2 U(\mathbf{r}_1 - \mathbf{r}_2). \tag{4.82}$$

Equations (4.80) and (4.82) are a beautiful result that generalizes to multiparticle systems. To emphasize what they say, let me rewrite them as

$$\left.\begin{array}{l} \text{(Force on particle 1)} = -\mathbf{V}_1 U \\ \text{(Force on particle 2)} = -\mathbf{V}_2 U. \end{array}\right\} \tag{4.83}$$

There is a *single* potential energy function U, from which we can derive *both* forces. To find the force on particle 1, we just take the gradient of U with respect to the coordinates of particle 1; to find the force on particle 2, we take the gradient with respect to the coordinates of particle 2.

Before generalizing this result to multiparticle systems, let us consider the conservation of energy for our two-particle system. Figure 4.20 shows the orbits of the two particles. During a short time interval dt, particle 1 moves through $d\mathbf{r}_1$ and particle 2 through $d\mathbf{r}_2$, and work is done on both particles by the corresponding forces. By the work–KE theorem

$$dT_1 = \text{(work on 1)} = d\mathbf{r}_1 \cdot \mathbf{F}_{12}$$

and similarly

$$dT_2 = \text{(work on 2)} = d\mathbf{r}_2 \cdot \mathbf{F}_{21}.$$

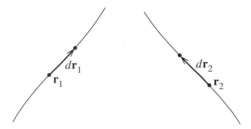

Figure 4.20   Motion of two interacting particles. During a
short time interval dt, particle 1 moves from $\mathbf{r}_1$ to $\mathbf{r}_1 + d\mathbf{r}_1$
and particle 2 from $\mathbf{r}_2$ to $\mathbf{r}_2 + d\mathbf{r}_2$.

Adding these, we find for the change in the *total* kinetic energy $T = T_1 + T_2$,

$$dT = dT_1 + dT_2 = (\text{work on 1}) + (\text{work on 2})$$

$$= W_{\text{tot}} \tag{4.84}$$

where

$$W_{\text{tot}} = d\mathbf{r}_1 \cdot \mathbf{F}_{12} + d\mathbf{r}_2 \cdot \mathbf{F}_{21}$$

denotes the total work done on both particles. Replacing $\mathbf{F}_{21}$ by $-\mathbf{F}_{12}$ and then
replacing $\mathbf{F}_{12}$ with (4.80), we can rewrite W_{tot} as

$$W_{\text{tot}} = (d\mathbf{r}_1 - d\mathbf{r}_2) \cdot \mathbf{F}_{12} = d(\mathbf{r}_1 - \mathbf{r}_2) \cdot [-\nabla_1 U(\mathbf{r}_1 - \mathbf{r}_2)]. \tag{4.85}$$

If we rename $(\mathbf{r}_1 - \mathbf{r}_2)$ as $\mathbf{r}$, then the right side of this equation can be seen to be just
(minus) the change in the potential energy, and we find that[13]

$$W_{\text{tot}} = -d\mathbf{r} \cdot \nabla U(\mathbf{r}) = -dU \tag{4.86}$$

where the last step follows from the property (4.35) of the gradient operator. It is
worth pausing to appreciate this important result. The total work W_{tot} is the sum of
two terms, the work done by $\mathbf{F}_{12}$ as particle 1 moves through $d\mathbf{r}_1$ *plus* the work done
by $\mathbf{F}_{21}$ as particle 2 moves through $d\mathbf{r}_2$. According to (4.86), the potential energy U
takes both of these terms into account and W_{tot} is simply $-dU$.

Returning to the total kinetic energy, we now see that according to (4.84) the change
dT is just $-dU$. Moving the term dU to the other side, we conclude that

$$d(T + U) = 0.$$

That is, the total energy,

$$E = T + U = T_1 + T_2 + U, \tag{4.87}$$

---

[13] If you invoke the chain rule for differentiation, you can see that it makes no difference whether
we write $\nabla_1 U(\mathbf{r})$ or $\nabla U(\mathbf{r})$.

of our two-particle system is conserved. Note well that the total energy of our two particles contains *two* kinetic terms (of course), but only *one* potential term, since U accounts for the work done by both of the forces $\mathbf{F}_{12}$ and $\mathbf{F}_{21}$.

Elastic Collisions

Elastic collisions give a simple application of these ideas. An elastic collision is a collision between two particles (or bodies that can be treated as particles) that interact via a conservative force that goes to zero as their separation $\mathbf{r}_1 - \mathbf{r}_2$ increases. Since the force goes to zero as $|\mathbf{r}_1 - \mathbf{r}_2| \to \infty$, the potential energy $U(\mathbf{r}_1 - \mathbf{r}_2)$ approaches a constant, which we may as well take to be zero. For example, the two particles could be an electron and a proton, or they could be two billiard balls. That the force between two billiard balls is conservative is not obvious, but it is a fact that billiard balls are manufactured so that they behave like almost perfect (that is, conservative) springs when they are forced together. It is certainly easy to think of other objects (such as lumps of putty), for which the interobject force is nonconservative, and the collisions of such objects are not elastic.

In a collision, the two particles start out far apart, approach one another, and then move apart again. Because the forces are conservative, the total energy is conserved; that is, $T + U = \text{constant}$ (where, of course, $T = T_1 + T_2$). But when the particles are far apart, U is zero. Thus if we use the subscripts "in" and "fin" to label the situations well before and well after the particles come together, then conservation of energy implies that

$$T_{\text{in}} = T_{\text{fin}} . \tag{4.88}$$

In other words, an elastic collision can be characterized as a collision in which two particles come together and re-emerge with their total kinetic energy unchanged. However, it is important to remember that there is no principle of conservation of kinetic energy. On the contrary, while the particles are close together their PE is nonzero and their KE certainly is changing. It is only when they are well separated that the PE is negligible and conservation of energy leads to the result (4.88).

The foregoing discussion may suggest that elastic collisions should be a very common occurence. All that is needed is two particles whose interaction is conservative. In practice, elastic collisions are not as widespread as this seems to imply. The trouble comes from the requirement that it be two *particles* that enter and leave the collision. For example, if we fire one billiard ball at a second with sufficient energy, the two balls may shatter. Similarly, if we fire an electron with sufficient energy at an atom, the atom may fall apart or, at least, change the internal motion of its constituents. Even in the collision of two genuine particles, such as an electron and a proton, relativity tells us that, with sufficient energy, new particles can be created. Clearly, at high enough energy, the assumption that the two objects entering a collision can be approximated as indivisible particles eventually breaks down, and we cannot assume that collisions will be elastic, even if all the underlying forces are conservative. Nevertheless, at reasonably low energies there are many situations where collisions are perfectly elastic:

At sufficiently low energy, collisions of an electron with an atom always are, and to a good approximation, the same is true of billiard balls.

Elastic collisions provide several simple illustrations of the uses of conservation of energy and momentum, of which the following is one.

EXAMPLE 4.8   An Equal-Mass, Elastic Collision

Consider an elastic collision between two particles of equal mass, $m_1 = m_2 = m$ (for example, two electrons, or two billiard balls), as shown in Figure 4.21. Prove that if particle 2 is initially at rest then the angle between the two outgoing velocities is $\theta = 90°$.

Conservation of momentum implies that $m\mathbf{v}_1 = m\mathbf{v}'_1 + m\mathbf{v}'_2$ or

$$\mathbf{v}_1 = \mathbf{v}'_1 + \mathbf{v}'_2. \tag{4.89}$$

That the collision is elastic implies that $\frac{1}{2}m\mathbf{v}_1^2 = \frac{1}{2}m\mathbf{v}'^2_1 + \frac{1}{2}m\mathbf{v}'^2_2$ or

$$\mathbf{v}_1^2 = \mathbf{v}'^2_1 + \mathbf{v}'^2_2.$$

Squaring (4.89), we find that

$$\mathbf{v}_1^2 = \mathbf{v}'^2_1 + 2\mathbf{v}'_1 \cdot \mathbf{v}'_2 + \mathbf{v}'^2_2,$$

and comparing the last two equations we see that

$$\mathbf{v}'_1 \cdot \mathbf{v}'_2 = 0;$$

that is, $\mathbf{v}'_1$ and $\mathbf{v}'_2$ are perpendicular (unless one of them is zero, in which case the angle between them is undefined). This result was useful in atomic and nuclear physics; when an unknown projectile hit a stationary target particle, the fact that the two emerged traveling at $90°$ was taken as evidence that the collision was elastic and the two particles had equal masses.

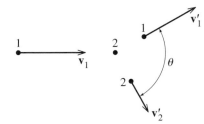

Figure 4.21   Elastic collision between two equal-mass particles. Particle 1 enters with velocity $\mathbf{v}_1$ and collides with the stationary particle 2. The angle between the two final velocities $\mathbf{v}'_1$ and $\mathbf{v}'_2$ is θ.

4.10   The Energy of a Multiparticle System

We can extend our discussion of two particles to N particles fairly easily. The main complication is notational: The large number of $\sum$ signs can make it hard to see clearly what is going on. For this reason, I shall start by considering the case of four particles ($N = 4$) and write out all of the various sums explicitly.

Four Particles

Let us consider, then, four particles, as shown in Figure 4.22. The particles can interact with each other (for example, they could be charged, so that each particle experiences the Coulomb force from the three others), and they may be subject to external forces, such as gravity or the Coulomb force of nearby charged bodies. In defining the energy of this system, the easy part is the kinetic energy T, which is, of course, the sum of four terms,

$$T = T_1 + T_2 + T_3 + T_4, \tag{4.90}$$

one term $T_\alpha = \frac{1}{2}m_\alpha v_\alpha^2$ for each particle.

To define the potential energy, we must examine the forces on the particles. First, there are the internal forces of the four particles interacting with each other. For each pair of particles there is an action–reaction pair of forces; for example, particles 3 and 4 produce the forces $\mathbf{F}_{34}$ and $\mathbf{F}_{43}$ shown in Figure 4.22. I shall take for granted that each of these interparticle forces $\mathbf{F}_{\alpha\beta}$ is unaffected by the presence of the other particles and any external bodies. For example, $\mathbf{F}_{34}$ is just the same as if particles 1

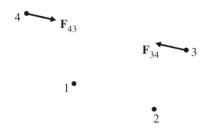

Figure 4.22   A system of four particles $\alpha = 1, 2, 3, 4$. For each pair of particles, $\alpha\beta$, there is an action–reaction pair of forces, $\mathbf{F}_{\alpha\beta}$ and $\mathbf{F}_{\beta\alpha}$, such as the pair $\mathbf{F}_{34}$ and $\mathbf{F}_{43}$ shown. In addition, each particle α may be subject to an external net force $\mathbf{F}_\alpha^{\text{ext}}$. The four particles could be charged dust motes floating in the air, with the forces $\mathbf{F}_{\alpha\beta}$ being electrostatic and $\mathbf{F}_\alpha^{\text{ext}}$ being gravity plus buoyancy of the air.

and 2 and all external bodies were removed.[14] Thus, we can treat the two forces $\mathbf{F}_{34}$ and $\mathbf{F}_{43}$ exactly as in Section 4.9. Provided the forces are conservative, we can define a potential energy

$$U_{34} = U_{34}(\mathbf{r}_3 - \mathbf{r}_4) \tag{4.91}$$

and the corresponding forces are the appropriate gradients as in (4.83)

$$\mathbf{F}_{34} = -\boldsymbol{\nabla}_3 U_{34} \quad \text{and} \quad \mathbf{F}_{43} = -\boldsymbol{\nabla}_4 U_{34}. \tag{4.92}$$

There are in all six distinct pairs of particles, 12, 13, 14, 23, 24, 34, and for each pair we can define a corresponding potential energy $U_{12}, \cdots, U_{34}$ from which the corresponding forces are obtained in the same way.

Each of the external forces $\mathbf{F}_{\alpha}^{\text{ext}}$ depends only on the corresponding position $\mathbf{r}_\alpha$. (The force $\mathbf{F}_1^{\text{ext}}$, for instance, depends on the position $\mathbf{r}_1$, but not on $\mathbf{r}_2$, $\mathbf{r}_3$, $\mathbf{r}_4$.) Therefore, we can handle $\mathbf{F}_{\alpha}^{\text{ext}}$ exactly as we did the force on a single particle. In particular, if $\mathbf{F}_{\alpha}^{\text{ext}}$ is conservative, we can introduce a potential energy $U_{\alpha}^{\text{ext}}(\mathbf{r}_\alpha)$ and the corresponding force is given by

$$\mathbf{F}_{\alpha}^{\text{ext}} = -\boldsymbol{\nabla}_\alpha U_{\alpha}^{\text{ext}}(\mathbf{r}_\alpha) \tag{4.93}$$

where, of course, $\boldsymbol{\nabla}_\alpha$ denotes differentiation with respect to the coordinates of particle α.

We can now put all the potential energies together and define the total potential energy as the sum

$$U = U^{\text{int}} + U^{\text{ext}} = (U_{12} + U_{13} + U_{14} + U_{23} + U_{24} + U_{34})$$
$$+ (U_1^{\text{ext}} + U_2^{\text{ext}} + U_3^{\text{ext}} + U_4^{\text{ext}}). \tag{4.94}$$

In this definition, U^{int} is the sum over all six pairs of particles of the pairwise potential energies, $U_{12}, \cdots, U_{34}$, and U^{ext} is the sum of the four potential energies, $U_1^{\text{ext}}, \cdots, U_4^{\text{ext}}$ arising from the external forces.

It is a fairly straightforward matter to show (see Problem 4.51 for more details) that the force on particle α is just (minus) the gradient of U with respect to the coordinates $(x_\alpha, y_\alpha, z_\alpha)$. Consider, for instance, the gradient $-\boldsymbol{\nabla}_1 U$. When $-\boldsymbol{\nabla}_1$ acts on the first line of (4.94), its action on the first three terms, $U_{12} + U_{13} + U_{14}$ gives precisely the three internal forces, $\mathbf{F}_{12} + \mathbf{F}_{13} + \mathbf{F}_{14}$. Acting on the last three terms, $U_{23} + U_{24} + U_{34}$, it produces zero, since none of these depend on $\mathbf{r}_1$. When $-\boldsymbol{\nabla}_1$ acts on the second line of (4.94), its action on the first term, U_1^{ext}, produces the external

---

[14] This is quite a subtle point. I am, of course, not denying that the extra particles exert extra forces on particle 3. I claim only that the force of particle 4 on particle 3 is independent of the presence or absence of particles 1 and 2 and any external bodies. One could imagine a world where this claim was false (the presence of particle 1 could somehow change the force of 4 on 3), but experiment seems to confirm that in our world my claim is true.

force $\mathbf{F}_1^{\text{ext}}$. Acting on the last three terms it produces zero, since none of them depend on $\mathbf{r}_1$. Accordingly,

$$-\nabla_1 U = \mathbf{F}_{12} + \mathbf{F}_{13} + \mathbf{F}_{14} + \mathbf{F}_1^{\text{ext}}$$

$$= \text{(net force on particle 1).} \tag{4.95}$$

In exactly the same way, we can prove that in general

$$-\nabla_\alpha U = \text{(net force on particle } \alpha) \tag{4.96}$$

as expected.

The second crucial property of our definition of potential energy U is that (provided all the forces concerned are conservative, so we can define U), the total energy, defined as $E = T + U$, is conserved. We prove this in the now familiar way (for more details, see Problem 4.52): Apply the work–KE theorem to each of the four particles and add the results to show that, in any short time interval, $dT = W_{\text{tot}}$ where W_{tot} denotes the total work done by all forces on all particles. Next show that $W_{\text{tot}} = -dU$, and conclude that $dT = -dU$, and hence

$$dE = dT + dU = 0.$$

That is, energy is conserved.

N Particles

The extension of these ideas to an arbitrary number of particles is now quite straight-forward, and I shall just write down the principal formulas. For N particles, labeled $\alpha = 1, \cdots, N$, the total kinetic energy is just the sum of the N separate kinetic energies

$$T = \sum_\alpha T_\alpha = \sum_\alpha \tfrac{1}{2} m_\alpha v_\alpha^2.$$

Assuming that all forces are conservative, for each pair of particles, $\alpha\beta$, we introduce the potential energy $U_{\alpha\beta}$ that describes their interaction, and for each particle α we introduce the potential energy U_α^{ext} that describes the net external force on that particle. The total potential energy is then

$$U = U^{\text{int}} + U^{\text{ext}} = \sum_\alpha \sum_{\beta > \alpha} U_{\alpha\beta} + \sum_\alpha U_\alpha^{\text{ext}}. \tag{4.97}$$

(Here the condition $\beta > \alpha$ in the double sum makes sure we don't double count the internal interactions $U_{\alpha\beta}$. For instance, we include U_{12} but not U_{21}.)

With the potential energy U defined in this way, the net force on any particle α is given by $-\nabla_\alpha U$, as in Equation (4.96), and total energy $E = T + U$ is conserved. Finally, if any forces are nonconservative, we can define U as the potential energy pertaining to the conservative forces and then show that, in this case, $dE = W_{\text{nc}}$ where W_{nc} is the work done by the nonconservative forces.

Rigid Bodies

While the formalism of the last two sections is fairly general and complicated, you can perhaps take some comfort that most applications of the formalism are much simpler than the formalism itself. As one simple example, consider a rigid body, such as a golf ball or a meteorite, made up of N atoms. The number N is typically very large, but the energy formalism just developed usually turns out to be very simple. As you probably recall from elementary physics, the total kinetic energy of the N particles rigidly bound together is just the kinetic energy of the center-of-mass motion plus the kinetic energy of rotation. (I'll be proving this in Chapter 10, but I hope you'll accept it for now.) The potential energy of the internal, interatomic forces as given by (4.97) is

$$U^{\text{int}} = \sum_{\alpha} \sum_{\beta > \alpha} U_{\alpha\beta}(\mathbf{r}_\alpha - \mathbf{r}_\beta). \tag{4.98}$$

If the interatomic forces are central (as is usually the case), then, as we saw in Section 4.8, the potential energy $U_{\alpha\beta}$ actually depends on just the magnitude of $\mathbf{r}_\alpha - \mathbf{r}_\beta$ (not its direction). Thus we can rewrite (4.98) as

$$U^{\text{int}} = \sum_{\alpha} \sum_{\beta > \alpha} U_{\alpha\beta}(|\mathbf{r}_\alpha - \mathbf{r}_\beta|). \tag{4.99}$$

Now, as a rigid body moves, the positions $\mathbf{r}_\alpha$ of its constituent atoms can, of course, move, but the distance $|\mathbf{r}_\alpha - \mathbf{r}_\beta|$ between any two atoms cannot change. (This is, in fact, the definition of a rigid body.) Therefore, if the body concerned is truly rigid, none of the terms in (4.99) can change. That is, the potential energy U^{int} of the internal forces is a constant and can, therefore, be ignored. Thus, in applying energy considerations to a rigid body we can entirely ignore U^{int} and have to worry only about the energy U^{ext} corresponding to the external forces. Since this latter energy is often a very simple function (see the following example), energy considerations as applied to a rigid body are usually very straightforward.

EXAMPLE 4.9   A Cylinder Rolling down an Incline

A uniform rigid cylinder of radius R rolls without slipping down a sloping track as shown in Figure 4.23. Use energy conservation to find its speed v when it reaches a vertical height h below its point of release.

In accordance with the preceding discussion we can ignore the internal forces that hold the cylinder together. The external forces on the cylinder are the normal and frictional forces of the track and gravity. The first two do no work, and gravity is conservative. As you certainly recall from introductory physics, the gravitational potential energy of an extended body is the same as if all the mass were concentrated at the center of mass. (See Problem 4.6.) Therefore,

$$U^{\text{ext}} = MgY,$$

where Y is the height of the cylinder's CM measured up from any convenient reference level. The kinetic energy of the cylinder is $T = \frac{1}{2}Mv^2 + \frac{1}{2}I\omega^2$, where

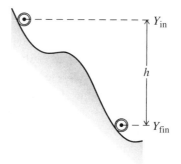

Figure 4.23   A uniform cylinder starts from rest and rolls without slipping down a slope through a total vertical drop $h = Y_{in} - Y_{fin}$ (with the CM coordinate Y measured vertically up).

I is its moment of inertia, $I = \frac{1}{2}MR^2$, and ω is its angular velocity of rolling, $\omega = v/R$. Thus the final kinetic energy is

$$T = \tfrac{3}{4}Mv^2$$

and the initial KE is zero. Therefore, conservation of energy in the form $\Delta T = -\Delta U^{ext}$ implies that

$$\tfrac{3}{4}Mv^2 = -Mg(Y_{fin} - Y_{in}) = Mgh$$

and hence that the final speed is

$$v = \sqrt{\frac{4gh}{3}}.$$

Principal Definitions and Equations of Chapter 4

Work–KE Theorem

The change in KE of a particle as it moves from point 1 to point 2 is

$$\Delta T \equiv T_2 - T_1 = \int_1^2 \mathbf{F} \cdot d\mathbf{r} \equiv W(1 \to 2) \qquad \text{[Eq. (4.7)]}$$

where $T = \frac{1}{2}mv^2$ and $W(1 \to 2)$ is the work which is done by the total force $\mathbf{F}$ on the particle and is defined by the preceding integral.

Conservative Forces and Potential Energy

A force $\mathbf{F}$ on a particle is **conservative** if (i) it depends only on the particle's position, $\mathbf{F} = \mathbf{F}(\mathbf{r})$, and (ii) for any two points 1 and 2, the work $W(1 \to 2)$ done by $\mathbf{F}$ is the same for all paths joining 1 and 2 (or equivalently, $\nabla \times \mathbf{F} = 0$). [Sections 4.2 & 4.4]

If $\mathbf{F}$ is conservative, we can define a corresponding **potential energy** so that

$$U(\mathbf{r}) = -W(\mathbf{r}_0 \to \mathbf{r}) \equiv - \int_{\mathbf{r}_0}^{\mathbf{r}} \mathbf{F}(\mathbf{r}') \cdot d\mathbf{r}' \qquad \text{[Eq. (4.13)]}$$

and

$$\mathbf{F} = -\nabla U. \qquad \text{[Eq. (4.33)]}$$

If all the forces on a particle are conservative with corresponding potential energies $U_1, \cdots, U_n$, then the **total mechanical energy**

$$E = T + U_1 + \cdots + U_n \qquad \text{[Eq. (4.22)]}$$

is constant. More generally if there are also nonconservative forces, $\Delta E = W_{\text{nc}}$, the work done by the nonconservative forces.

Central Forces

A force $\mathbf{F}(\mathbf{r})$ is **central** if it is everywhere directed toward or away from a "force center." If we take the latter to be the origin,

$$\mathbf{F}(\mathbf{r}) = f(\mathbf{r})\hat{\mathbf{r}}. \qquad \text{[Eq. (4.65)]}$$

A central force is spherically symmetric $[f(\mathbf{r}) = f(r)]$ if and only if it is conservative.
[Sec. (4.8)]

Energy of a Multiparticle System

If all forces (internal and external) on a multiparticle system are conservative, the total potential energy,

$$U = U^{\text{int}} + U^{\text{ext}} = \sum_{\alpha} \sum_{\beta > \alpha} U_{\alpha\beta} + \sum_{\alpha} U_{\alpha}^{\text{ext}} \qquad \text{[Eq. (4.97)]}$$

satisfies

$$(\text{net force on particle } \alpha) = -\nabla_{\alpha} U \qquad \text{[Eq. (4.96)]}$$

and

$$T + U = \text{constant}. \qquad \text{[Problem 4.52]}$$

Problems for Chapter 4

Stars indicate the approximate level of difficulty, from easiest (⋆) to most difficult (⋆⋆⋆).

SECTION 4.1　Kinetic Energy and Work

4.1 ⋆ By writing $\mathbf{a} \cdot \mathbf{b}$ in terms of components prove that the product rule for differentiation applies to the dot product of two vectors; that is,

$$\frac{d}{dt}(\mathbf{a} \cdot \mathbf{b}) = \frac{d\mathbf{a}}{dt} \cdot \mathbf{b} + \mathbf{a} \cdot \frac{d\mathbf{b}}{dt}.$$

4.2 ⋆⋆ Evaluate the work done

$$W = \int_O^P \mathbf{F} \cdot d\mathbf{r} = \int_O^P (F_x\, dx + F_y\, dy) \tag{4.100}$$

by the two-dimensional force $\mathbf{F} = (x^2, 2xy)$ along the three paths joining the origin to the point $P = (1, 1)$ as shown in Figure 4.24(a) and defined as follows: **(a)** This path goes along the x axis to $Q = (1, 0)$ and then straight up to P. (Divide the integral into two pieces, $\int_O^P = \int_O^Q + \int_Q^P$.) **(b)** On this path $y = x^2$, and you can replace the term dy in (4.100) by $dy = 2x\, dx$ and convert the whole integral into an integral over x. **(c)** This path is given parametrically as $x = t^3$, $y = t^2$. In this case rewrite x, y, dx, and dy in (4.100) in terms of t and dt, and convert the integral into an integral over t.

4.3 ⋆⋆ Do the same as in Problem 4.2, but for the force $\mathbf{F} = (-y, x)$ and for the three paths joining P and Q shown in Figure 4.24(b) and defined as follows: **(a)** This path goes straight from $P = (1, 0)$ to the origin and then straight to $Q = (0, 1)$. **(b)** This is a straight line from P to Q. (Write y as a function of x and rewrite the integral as an integral over x.) **(c)** This is a quarter-circle centered on the origin. (Write x and y in polar coordinates and rewrite the integral as an integral over ϕ.)

4.4 ⋆⋆ A particle of mass m is moving on a frictionless horizontal table and is attached to a massless string, whose other end passes through a hole in the table, where I am holding it. Initially the particle is moving in a circle of radius r_o with angular velocity ω_o, but I now pull the string down through the hole until a length r remains between the hole and the particle. **(a)** What is the particle's angular velocity now? **(b)** Assuming that I pull the string so slowly that we can approximate the particle's path by a

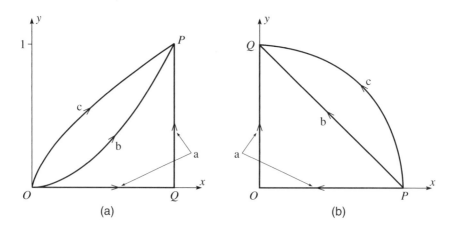

Figure 4.24　**(a)** Problem 4.2. **(b)** Problem 4.3

circle of slowly shrinking radius, calculate the work I did pulling the string. **(c)** Compare your answer to part (b) with the particle's gain in kinetic energy.

SECTION 4.2 Potential Energy and Conservative Forces

4.5 ★ (a) Consider a mass m in a uniform gravitational field $\mathbf{g}$, so that the force on m is $m\mathbf{g}$, where $\mathbf{g}$ is a constant vector pointing vertically down. If the mass moves by an arbitrary path from point 1 to point 2, show that the work done by gravity is $W_{\text{grav}}(1 \to 2) = -mgh$ where h is the vertical height gained between points 1 and 2. Use this result to prove that the force of gravity is conservative (at least in a region small enough so that $\mathbf{g}$ can be considered constant). **(b)** Show that, if we choose axes with y measured vertically up, the gravitational potential energy is $U = mgy$ (if we choose $U = 0$ at the origin).

4.6 ★ For a system of N particles subject to a uniform gravitational field $\mathbf{g}$ acting vertically down, prove that the total gravitational potential energy is the same as if all the mass were concentrated at the center of mass of the system; that is,

$$U = \sum_{\alpha} U_{\alpha} = MgY$$

where $M = \sum m_{\alpha}$ is the total mass and $\mathbf{R} = (X, Y, Z)$ is the position of the CM, with the y coordinate measured vertically up. [*Hint:* We know from Problem 4.5 that $U_{\alpha} = m_{\alpha} g y_{\alpha}$.]

4.7 ★ Near to the point where I am standing on the surface of Planet X, the gravitational force on a mass m is vertically down but has magnitude $m\gamma y^2$ where γ is a constant and y is the mass's height above the horizontal ground. **(a)** Find the work done by gravity on a mass m moving from $\mathbf{r}_1$ to $\mathbf{r}_2$, and use your answer to show that gravity on Planet X, although most unusual, is still conservative. Find the corresponding potential energy. **(b)** Still on the same planet, I thread a bead on a curved, frictionless, rigid wire, which extends from ground level to a height h above the ground. Show clearly in a picture the forces on the bead when it is somewhere on the wire. (Just name the forces so it's clear what they are; don't worry about their magnitude.) Which of the forces are conservative and which are not? **(c)** If I release the bead from rest at a height h, how fast will it be going when it reaches the ground?

4.8 ★★ Consider a small frictionless puck perched at the top of a fixed sphere of radius R. If the puck is given a tiny nudge so that it begins to slide down, through what vertical height will it descend before it leaves the surface of the sphere? [*Hint:* Use conservation of energy to find the puck's speed as a function of its height, then use Newton's second law to find the normal force of the sphere on the puck. At what value of this normal force does the puck leave the sphere?]

4.9 ★★ (a) The force exerted by a one-dimensional spring, fixed at one end, is $F = -kx$, where x is the displacement of the other end from its equilibrium position. Assuming that this force is conservative (which it is) show that the corresponding potential energy is $U = \frac{1}{2}kx^2$, if we choose U to be zero at the equilibrium position. **(b)** Suppose that this spring is hung vertically from the ceiling with a mass m suspended from the other end and constrained to move in the vertical direction only. Find the extension x_o of the new equilibrium position with the suspended mass. Show that the total potential energy (spring plus gravity) has the same form $\frac{1}{2}ky^2$ if we use the coordinate y equal to the displacement measured from the new equilibrium position at $x = x_o$ (and redefine our reference point so that $U = 0$ at $y = 0$).

SECTION 4.3   Force as the Gradient of Potential Energy

4.10 ⋆ Find the partial derivatives with respect to x, y, and z of the following functions: **(a)** $f(x, y, z) = ax^2 + bxy + cy^2$, **(b)** $g(x, y, z) = \sin(axyz^2)$, **(c)** $h(x, y, z) = ae^{xy/z^2}$, where a, b, and c are constants. Remember that to evaluate $\partial f/\partial x$ you differentiate with respect to x treating y and z as constants.

4.11 ⋆ Find the partial derivatives with respect to x, y, and z of the following functions: **(a)** $f(x, y, z) = ay^2 + 2byz + cz^2$, **(b)** $g(x, y, z) = \cos(axy^2z^3)$, **(c)** $h(x, y, z) = ar$, where a, b, and c are constants and $r = \sqrt{x^2 + y^2 + z^2}$. Remember that to evaluate $\partial f/\partial x$ you differentiate with respect to x treating y and z as constants.

4.12 ⋆ Calculate the gradient ∇f of the following functions, $f(x, y, z)$: **(a)** $f = x^2 + z^3$. **(b)** $f = ky$, where k is a constant. **(c)** $f = r \equiv \sqrt{x^2 + y^2 + z^2}$. [*Hint:* Use the chain rule.] **(d)** $f = 1/r$.

4.13 ⋆ Calculate the gradient ∇f of the following functions, $f(x, y, z)$: **(a)** $f = \ln(r)$, **(b)** $f = r^n$, **(c)** $f = g(r)$, where $r = \sqrt{x^2 + y^2 + z^2}$ and $g(r)$ is some unspecified function of r. [*Hint:* Use the chain rule.]

4.14 ⋆ Prove that if $f(\mathbf{r})$ and $g(\mathbf{r})$ are any two scalar functions of $\mathbf{r}$, then

$$\nabla(fg) = f\nabla g + g\nabla f$$

4.15 ⋆ For $f(\mathbf{r}) = x^2 + 2y^2 + 3z^2$, use the approximation (4.35) to estimate the change in f if we move from the point $\mathbf{r} = (1, 1, 1)$ to $(1.01, 1.03, 1.05)$. Compare with the exact result.

4.16 ⋆ If a particle's potential energy is $U(\mathbf{r}) = k(x^2 + y^2 + z^2)$, where k is a constant, what is the force on the particle?

4.17 ⋆ A charge q in a uniform electric field $\mathbf{E}_o$ experiences a constant force $\mathbf{F} = q\mathbf{E}_o$. **(a)** Show that this force is conservative and verify that the potential energy of the charge at position $\mathbf{r}$ is $U(\mathbf{r}) = -q\mathbf{E}_o \cdot \mathbf{r}$. **(b)** By doing the necessary derivatives, check that $\mathbf{F} = -\nabla U$.

4.18 ⋆⋆ Use the property (4.35) of the gradient to prove the following important results: **(a)** The vector ∇f at any point $\mathbf{r}$ is perpendicular to the surface of constant f through $\mathbf{r}$. (Choose a small displacement $d\mathbf{r}$ that lies in a surface of constant f. What is df for such a displacement?) **(b)** The direction of ∇f at any point $\mathbf{r}$ is the direction in which f increases fastest as we move away from $\mathbf{r}$. (Choose a small displacement $d\mathbf{r} = \epsilon\mathbf{u}$, where $\mathbf{u}$ is a unit vector and ϵ is fixed and small. Find the direction of $\mathbf{u}$ for which the corresponding df is maximum, bearing in mind that $\mathbf{a} \cdot \mathbf{b} = ab \cos\theta$.)

4.19 ⋆⋆ **(a)** Describe the surfaces defined by the equation $f = $ const, where $f = x^2 + 4y^2$. **(b)** Using the results of Problem 4.18, find a unit normal to the surface $f = 5$ at the point $(1, 1, 1)$. In what direction should one move from this point to maximize the rate of change of f?

SECTION 4.4   The Second Condition that F be Conservative

4.20 ⋆ Find the curl, $\nabla \times \mathbf{F}$, for the following forces: **(a)** $\mathbf{F} = k\mathbf{r}$; **(b)** $\mathbf{F} = (Ax, By^2, Cz^3)$; **(c)** $\mathbf{F} = (Ay^2, Bx, Cz)$, where A, B, C and k are constants.

4.21 ⋆ Verify that the gravitational force $-GMm\hat{\mathbf{r}}/r^2$ on a point mass m at $\mathbf{r}$, due to a fixed point mass M at the origin, is conservative and calculate the corresponding potential energy.

4.22 ⋆ The proof in Example 4.5 (page 119) that the Coulomb force is conservative is considerably simplified if we evaluate $\nabla \times \mathbf{F}$ using spherical polar coordinates. Unfortunately, the expression for $\nabla \times \mathbf{F}$ in spherical polar coordinates is quite messy and hard to derive. However, the answer is given inside the back cover, and the proof can be found in any book on vector calculus or mathematical methods.[15] Taking the expression inside the back cover on faith, prove that the Coulomb force $\mathbf{F} = \gamma \hat{\mathbf{r}}/r^2$ is conservative.

4.23 ★★ Which of the following forces is conservative? **(a)** $\mathbf{F} = k(x, 2y, 3z)$ where k is a constant. **(b)** $\mathbf{F} = k(y, x, 0)$. **(c)** $\mathbf{F} = k(-y, x, 0)$. For those which are conservative, find the corresponding potential energy U, and verify by direct differentiation that $\mathbf{F} = -\nabla U$.

4.24 ★★★ An infinitely long, uniform rod of mass μ per unit length is situated on the z axis. **(a)** Calculate the gravitational force $\mathbf{F}$ on a point mass m at a distance ρ from the z axis. (The gravitational force between two point masses is given in Problem 4.21.) **(b)** Rewrite $\mathbf{F}$ in terms of the rectangular coordinates (x, y, z) of the point and verify that $\nabla \times \mathbf{F} = 0$. **(c)** Show that $\nabla \times \mathbf{F} = 0$ using the expression for $\nabla \times \mathbf{F}$ in cylindrical polar coordinates given inside the back cover. **(d)** Find the corresponding potential energy U.

4.25 ★★★ The proof that the condition $\nabla \times \mathbf{F} = 0$ guarantees the path independence of the work $\int_1^2 \mathbf{F} \cdot d\mathbf{r}$ done by $\mathbf{F}$ is unfortunately too lengthy to be included here. However, the following three exercises capture the main points:[16] **(a)** Show that the path independence of $\int_1^2 \mathbf{F} \cdot d\mathbf{r}$ is equivalent to the statement that the integral $\oint_\Gamma \mathbf{F} \cdot d\mathbf{r}$ around any *closed* path Γ is zero. (By tradition, the symbol $\oint$ is used for integrals around a closed path — a path that starts and stops at the same point.) [*Hint:* For any two points 1 and 2 and any two paths from 1 to 2, consider the work done by $\mathbf{F}$ going from 1 to 2 along the first path and then back to 1 along the second in the reverse direction.] **(b)** Stokes's theorem asserts that $\oint_\Gamma \mathbf{F} \cdot d\mathbf{r} = \int (\nabla \times \mathbf{F}) \cdot \hat{\mathbf{n}} \, dA$, where the integral on the right is a surface integral over a surface for which the path Γ is the boundary, and $\hat{\mathbf{n}}$ and dA are a unit normal to the surface and an element of area. Show that Stokes's theorem implies that if $\nabla \times \mathbf{F} = 0$ everywhere, then $\oint_\Gamma \mathbf{F} \cdot d\mathbf{r} = 0$. **(c)** While the general proof of Stokes's theorem is beyond our scope here, the following special case is quite easy to prove (and is an important step toward the general proof): Let Γ denote a rectangular closed path lying in a plane perpendicular to the z direction and bounded by the lines $x = B$, $x = B + b$, $y = C$ and $y = C + c$. For this simple path (traced counterclockwise as seen from above), prove Stokes's theorem that

$$\oint_\Gamma \mathbf{F} \cdot d\mathbf{r} = \int (\nabla \times \mathbf{F}) \cdot \hat{\mathbf{n}} \, dA$$

where $\hat{\mathbf{n}} = \hat{\mathbf{z}}$ and the integral on the right runs over the flat, rectangular area inside Γ. [*Hint:* The integral on the left contains four terms, two of which are integrals over x and two over y. If you pair them in this way, you can combine each pair into a single integral with an integrand of the form $F_x(x, C + c, z) - F_x(x, C, z)$ (or a similar term with the roles of x and y exchanged). You can rewrite this integrand as an integral over y of $\partial F_x(x, y, z)/\partial y$ (and similarly with the other term), and you're home.]

SECTION 4.5 **Time-Dependent Potential Energy**

4.26 ★ A mass m is in a uniform gravitational field, which exerts the usual force $F = mg$ vertically down, but with g varying with time, $g = g(t)$. Choosing axes with y measured vertically up and defining $U = mgy$ as usual, show that $\mathbf{F} = -\nabla U$ as usual, but, by differentiating $E = \frac{1}{2}mv^2 + U$ with respect to t, show that E is not conserved.

---

[15] See, for example, *Mathematical Methods in the Physical Sciences* by Mary Boas (Wiley, 1983), p. 435.

[16] For a complete discussion see, for example, *Mathematical Methods*, Boas, Ch. 6, Sections 8–11.

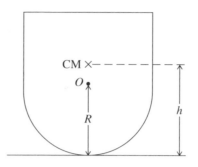

Figure 4.25   Problem 4.30

4.27 ★★ Suppose that the force $F(r, t)$ depends on the time t but still satisfies $\nabla \times F = 0$. It is a mathematical fact (related to Stokes's theorem as discussed in Problem 4.25) that the work integral $\int_1^2 F(r, t) \cdot dr$ (evaluated at any one time t) is independent of the path taken between the points 1 and 2. Use this to show that the time-dependent PE defined by (4.48), for any fixed time t, has the claimed property that $F(r, t) = -\nabla U(r, t)$. Can you see what goes wrong with the argument leading to Equation (4.19), that is, conservation of energy?

SECTION 4.6 Energy for Linear One-Dimensional Systems

4.28 ★★ Consider a mass m on the end of a spring of force constant k and constrained to move along the horizontal x axis. If we place the origin at the spring's equilibrium position, the potential energy is $\frac{1}{2}kx^2$. At time $t = 0$ the mass is sitting at the origin and is given a sudden kick to the right so that it moves out to a maximum displacement at $x_{\max} = A$ and then continues to oscillate about the origin. **(a)** Write down the equation for conservation of energy and solve it to give the mass's velocity $\dot{x}$ in terms of the position x and the total energy E. **(b)** Show that $E = \frac{1}{2}kA^2$, and use this to eliminate E from your expression for $\dot{x}$. Use the result (4.58), $t = \int dx'/\dot{x}(x')$, to find the time for the mass to move from the origin out to a position x. **(c)** Solve the result of part (b) to give x as a function of t and show that the mass executes simple harmonic motion with period $2\pi\sqrt{m/k}$.

4.29 ★★ [Computer] A mass m confined to the x axis has potential energy $U = kx^4$ with $k > 0$. **(a)** Sketch this potential energy and qualitatively describe the motion if the mass is initially stationary at $x = 0$ and is given a sharp kick to the right at $t = 0$. **(b)** Use (4.58) to find the time for the mass to reach its maximum displacement $x_{\max} = A$. Give your answer as an integral over x in terms of m, A, and k. Hence find the period τ of oscillations of amplitude A as an integral. **(c)** By making a suitable change of variables in the integral, show that the period τ is inversely proportional to the amplitude A. **(d)** The integral of part (b) cannot be evaluated in terms of elementary functions, but it can be done numerically. Find the period for the case that $m = k = A = 1$.

SECTION 4.7 Curvilinear One-Dimensional Systems

4.30 ★ Figure 4.25 shows a child's toy, which has the shape of a cylinder mounted on top of a hemisphere. The radius of the hemisphere is R and the CM of the whole toy is at a height h above the floor. **(a)** Write down the gravitational potential energy when the toy is tipped to an angle θ from the vertical. [You need to find the height of the CM as a function of θ. It helps to think first about the height of the hemisphere's center O as the toy tilts.] **(b)** For what values of R and h is the equilibrium at $\theta = 0$ stable?

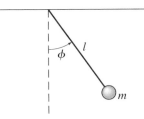

Figure 4.26   Problem 4.34

4.31 ★ (a) Write down the total energy E of the two masses in the Atwood machine of Figure 4.15 in terms of the coordinate x and $\dot{x}$. **(b)** Show (what is true for any conservative one-dimensional system) that you can obtain the equation of motion for the coordinate x by differentiating the equation $E = $ const. Check that the equation of motion is the same as you would obtain by applying Newton's second law to each mass and eliminating the unknown tension from the two resulting equations.

4.32 ★★ Consider the bead of Figure 4.13 threaded on a curved rigid wire. The bead's position is specified by its distance s, measured along the wire from the origin. **(a)** Prove that the bead's speed v is just $v = \dot{s}$. (Write $\mathbf{v}$ in terms of its components, dx/dt, etc., and find its magnitude using Pythagoras's theorem.) **(b)** Prove that $m\ddot{s} = F_{\text{tang}}$, the tangential component of the net force on the bead. (One way to do this is to take the time derivative of the equation $v^2 = \mathbf{v} \cdot \mathbf{v}$. The left side should lead you to $\ddot{s}$ and the right to F_{tang}.) **(c)** One force on the bead is the normal force $\mathbf{N}$ of the wire (which constrains the bead to stay on the wire). If we assume that all other forces (gravity, etc.) are conservative, then their resultant can be derived from a potential energy U. Prove that $F_{\text{tang}} = -dU/ds$. This shows that one-dimensional systems of this type can be treated just like linear systems, with x replaced by s and F_x by F_{tang}.

4.33 ★★ [Computer] **(a)** Verify the expression (4.59) for the potential energy of the cube balanced on a cylinder in Example 4.7 (page 130). **(b)** Make plots of $U(\theta)$ for $b = 0.9r$ and $b = 1.1r$. (You may as well choose units such that r, m, and g are all equal to 1.) **(c)** Use your plots to confirm the findings of Example 4.7 concerning the stability of the equilibrium at $\theta = 0$. Are there any other equilibrium points and are they stable?

4.34 ★★ An interesting one-dimensional system is the simple pendulum, consisting of a point mass m, fixed to the end of a massless rod (length l), whose other end is pivoted from the ceiling to let it swing freely in a vertical plane, as shown in Figure 4.26. The pendulum's position can be specified by its angle ϕ from the equilibrium position. (It could equally be specified by its distance s from equilibrium — indeed $s = l\phi$ — but the angle is a little more convenient.) **(a)** Prove that the pendulum's potential energy (measured from the equilibrium level) is

$$U(\phi) = mgl(1 - \cos\phi). \tag{4.101}$$

Write down the total energy E as a function of ϕ and $\dot{\phi}$. **(b)** Show that by differentiating your expression for E with respect to t you can get the equation of motion for ϕ and that the equation of motion is just the familiar $\Gamma = I\alpha$ (where Γ is the torque, I is the moment of inertia, and α is the angular acceleration $\ddot{\phi}$). **(c)** Assuming that the angle ϕ remains small throughout the motion, solve for $\phi(t)$ and show that the motion is periodic with period

$$\tau_\text{o} = 2\pi\sqrt{l/g}. \tag{4.102}$$

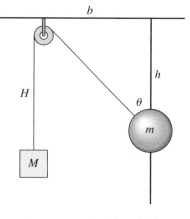

Figure 4.27   Problem 4.36

(The subscript "o" is to emphasize that this is the period for small oscillations.)

4.35 ★★ Consider the Atwood machine of Figure 4.15, but suppose that the pulley has radius R and moment of inertia I. **(a)** Write down the total energy of the two masses and the pulley in terms of the coordinate x and $\dot{x}$. (Remember that the kinetic energy of a spinning wheel is $\frac{1}{2}I\omega^2$.) **(b)** Show (what is true for any conservative one-dimensional system) that you can obtain the equation of motion for the coordinate x by differentiating the equation $E = $ const. Check that the equation of motion is the same as you would obtain by applying Newton's second law separately to the two masses and the pulley, and then eliminating the two unknown tensions from the three resulting equations.

4.36 ★★ A metal ball (mass m) with a hole through it is threaded on a frictionless vertical rod. A massless string (length l) attached to the ball runs over a massless, frictionless pulley and supports a block of mass M, as shown in Figure 4.27. The positions of the two masses can be specified by the one angle θ. **(a)** Write down the potential energy $U(\theta)$. (The PE is given easily in terms of the heights shown as h and H. Eliminate these two variables in favor of θ and the constants b and l. Assume that the pulley and ball have negligible size.) **(b)** By differentiating $U(\theta)$ find whether the system has an equilibrium position, and for what values of m and M equilibrium can occur. Discuss the stability of any equilibrium positions.

4.37 ★★★ [Computer] Figure 4.28 shows a massless wheel of radius R, mounted on a frictionless, horizontal axle. A point mass M is glued to the edge of the wheel, and a mass m hangs from a string wrapped around the perimeter of the wheel. **(a)** Write down the total PE of the two masses as a function of the angle ϕ. **(b)** Use this to find the values of m and M for which there are any positions of equilibrium. Describe the equilibrium positions, discuss their stability, and explain your answers in terms of torques. **(c)** Plot $U(\phi)$ for the cases that $m = 0.7M$ and $m = 0.8M$, and use your graphs to describe the behavior of the system if I release it from rest at $\phi = 0$. **(d)** Find the critical value of m/M on one side of which the system oscillates and on the other side of which it does not (if released from rest at $\phi = 0$).

4.38 ★★★ [Computer] Consider the simple pendulum of Problem 4.34. You can get an expression for the pendulum's period (good for large oscillations as well as small) using the method discussed in connection with (4.57), as follows: **(a)** Using (4.101) for the PE, find $\dot{\phi}$ as a function of ϕ. Next use (4.57), in the form $t = \int d\phi/\dot{\phi}$, to write the time for the pendulum to travel from $\phi = 0$ to its maximum value (the amplitude) Φ. Because this time is a quarter of the period τ, you can now write down the period. Show that

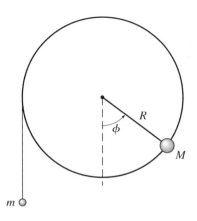

Figure 4.28   Problem 4.37

$$\tau = \tau_0 \frac{1}{\pi} \int_0^\Phi \frac{d\phi}{\sqrt{\sin^2(\Phi/2) - \sin^2(\phi/2)}} = \tau_0 \frac{2}{\pi} \int_0^1 \frac{du}{\sqrt{1-u^2}\sqrt{1-A^2u^2}}, \qquad (4.103)$$

where τ_0 is the period (4.102) (Problem 4.34) for small oscillations and $A = \sin(\Phi/2)$. [To get the first expression you will need to use the trig identity for $1 - \cos\phi$ in terms of $\sin^2(\phi/2)$. To get the second you need to make the substitution $\sin(\phi/2) = Au$.] These integrals cannot be evaluated in terms of elementary functions. However, the second integral is a standard integral called the *complete elliptic integral of the first kind*, sometimes denoted $K(A^2)$, whose values are tabulated[17] and are known to computer software such as Mathematica [which calls it EllipticK(A^2)]. **(b)** If you have access to computer software that knows this function, make a plot of τ/τ_0 for amplitudes $0 \le \Phi \le 3$ rad. Comment. What becomes of τ as the amplitude of oscillation approaches π? Explain.

4.39 ★★★ (a) If you have not already done so, do Problem 4.38(a). **(b)** If the amplitude Φ is small then so is $A = \sin(\Phi/2)$. If the amplitude is very small, we can simply ignore the last square root in (4.103). Show that this gives the familiar result for the small-amplitude period, $\tau = \tau_0 = 2\pi\sqrt{l/g}$. **(c)** If the amplitude is small but not very small, we can improve on the approximation of part (b). Use the binomial expansion to give the approximation $1/\sqrt{1-A^2u^2} \approx 1 + \frac{1}{2}A^2u^2$ and show that, in this approximation, (4.103) gives

$$\tau = \tau_0[1 + \tfrac{1}{4}\sin^2(\Phi/2)].$$

What percentage correction does the second term represent for an amplitude of 45°? (The exact answer for $\Phi = 45°$ is $1.040\,\tau_0$ to four significant figures.)

SECTION 4.8   Central Forces

4.40 ★ (a) Verify the three equations (4.68) that give x, y, z in terms of the spherical polar coordinates r, θ, ϕ. **(b)** Find expressions for r, θ, ϕ in terms of x, y, z.

---

[17] See, for example, M.Abramowitz and I.Stegun, *Handbook of Mathematical Functions*, Dover, New York, 1965. Be warned that different authors use different notations. In particular, some authors call the exact same integral $K(A)$.

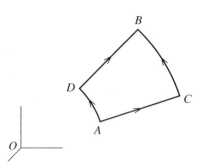

Figure 4.29   Problem 4.44

4.41 ★ A mass m moves in a circular orbit (centered on the origin) in the field of an attractive central force with potential energy $U = kr^n$. Prove the **virial theorem** that $T = nU/2$.

4.42 ★ In one dimension, it is obvious that a force obeying Hooke's law is conservative (since $F = -kx$ depends only on the position x, and this is sufficient to guarantee that F is conservative in one dimension). Consider instead a spring that obeys Hooke's law and has one end fixed at the origin, but whose other end is free to move in all three dimensions. (The spring could be fastened to a point in the ceiling and be supporting a bouncing mass m at its other end, for instance.) Write down the force $\mathbf{F}(\mathbf{r})$ exerted by the spring in terms of its length r and its equilibrium length r_o. Prove that this force is conservative. [*Hints:* Is the force central? Assume that the spring does not bend.]

4.43 ★★ In Section 4.8, I claimed that a force $\mathbf{F}(\mathbf{r})$ that is central and spherically symmetric is automatically conservative. Here are two ways to prove it: **(a)** Since $\mathbf{F}(\mathbf{r})$ is central and spherically symmetric, it must have the form $\mathbf{F}(\mathbf{r}) = f(r)\hat{\mathbf{r}}$. Using Cartesian coordinates, show that this implies that $\nabla \times \mathbf{F} = 0$. **(b)** Even quicker, using the expression given inside the back cover for $\nabla \times \mathbf{F}$ in spherical polars, show that $\nabla \times \mathbf{F} = 0$.

4.44 ★★ Problem 4.43 suggests two proofs that a central, spherically symmetric force is automatically conservative, but neither proof makes really clear *why* this is so. Here is a proof that is less complete but more insightful: Consider any two points A and B and two different paths ACB and ADB connecting them as shown in Figure 4.29. Path ACB goes radially out from A until it reaches the radius r_B of B, and then around a sphere (center O) to B. Path ADB goes around a sphere of radius r_A until it reaches the line OB, and then radially out to B. Explain clearly why the work done by a central, spherically symmetric force $\mathbf{F}$ is the same along both paths. (This doesn't prove that the work is the same along *any* two paths from A to B. If you want you can complete the proof by showing that any path can be approximated by a series of paths moving radially in or out and paths of constant r.)

4.45 ★★ In Section 4.8, I proved that a force $\mathbf{F}(\mathbf{r}) = f(\mathbf{r})\hat{\mathbf{r}}$ that is central and conservative is automatically spherically symmetric. Here is an alternative proof: Consider the two paths ACB and ADB of Figure 4.29, but with $r_B = r_A + dr$ where dr is infinitesimal. Write down the work done by $\mathbf{F}(\mathbf{r})$ going around both paths, and use the fact that they must be equal to prove that the magnitude function $f(\mathbf{r})$ must be the same at points A and D; that is, $f(\mathbf{r}) = f(r)$ and the force is spherically symmetric.

SECTION 4.9   Energy of Interaction of Two Particles

4.46 ★ Consider an elastic collision of two particles as in Example 4.8 (page 143), but with unequal masses, $m_1 \neq m_2$. Show that the angle θ between the two outgoing velocities satisfies $\theta < \pi/2$ if $m_1 > m_2$, but $\theta > \pi/2$ if $m_1 < m_2$.

4.47 ★ Consider a head-on elastic collision between two particles. (Since the collision is head-on, the motion is confined to a single straight line and is therefore one-dimensional.) Prove that the relative velocity after the collision is equal and opposite to that before. That is, $v_1 - v_2 = -(v'_1 - v'_2)$, where v_1 and v_2 are the initial velocities and v'_1 and v'_2 the corresponding final velocities.

4.48 ★ A particle of mass m_1 and speed v_1 collides with a second particle of mass m_2 at rest. If the collision is perfectly inelastic (the two particles lock together and move off as one) what fraction of the kinetic energy is lost in the collision? Comment on your answer for the cases that $m_1 \ll m_2$ and that $m_2 \ll m_1$.

4.49 ★★ Both the Coulomb and gravitational forces lead to potential energies of the form $U = \gamma/|\mathbf{r}_1 - \mathbf{r}_2|$, where γ denotes kq_1q_2 in the case of the Coulomb force and $-Gm_1m_2$ for gravity, and $\mathbf{r}_1$ and $\mathbf{r}_2$ are the positions of the two particles. Show in detail that $-\nabla_1 U$ is the force on particle 1 and $-\nabla_2 U$ that on particle 2.

4.50 ★★ The formalism of the potential energy of two particles depends on the claim in (4.81) that

$$\nabla_1 U(\mathbf{r}_1 - \mathbf{r}_2) = -\nabla_2 U(\mathbf{r}_1 - \mathbf{r}_2).$$

Prove this. (Use the chain rule for differentiation. The proof in three dimensions is notationally awkward, so prove the one-dimensional result that

$$\frac{\partial}{\partial x_1} f(x_1 - x_2) = -\frac{\partial}{\partial x_2} f(x_1 - x_2)$$

and then convince yourself that it extends to three dimensions.)

SECTION 4.10   The Energy of a Multiparticle System

4.51 ★★ Write out the arguments of all the potential energies of the four-particle system in (4.94). For instance $U = U(\mathbf{r}_1, \mathbf{r}_2, \cdots, \mathbf{r}_4)$, whereas $U_{34} = U_{34}(\mathbf{r}_3 - \mathbf{r}_4)$. Show in detail that the net force on particle 3 (for instance) is given by $-\nabla_3 U$. [You know that the separate forces, internal and external, are given by (4.92) and (4.93).]

4.52 ★★ Consider the four-particle system of Section 4.10. **(a)** Write down the work–KE theorem for each of the four particles separately and, by adding these four equations, show that the change in the total KE in a short time interval dt is $dT = W_{tot}$ where W_{tot} is the total work done on all particles by all forces. [This shouldn't take more than two or three lines.] **(b)** Next show that $W_{tot} = -dU$ where dU is the change in total PE during the same time interval. Deduce that the total mechanical energy $E = T + U$ is conserved.

4.53 ★★ **(a)** Consider an electron (charge $-e$ and mass m) in a circular orbit of radius r around a fixed proton (charge $+e$). Remembering that the inward Coulomb force ke^2/r^2 is what gives the electron its centripetal acceleration, prove that the electron's KE is equal to $-\frac{1}{2}$ times its PE; that is, $T = -\frac{1}{2}U$ and hence $E = \frac{1}{2}U$. (This result is a consequence of the so-called *virial theorem*. See Problem 4.41.) Now consider the following inelastic collision of an electron with a hydrogen atom: Electron number 1 is in a circular orbit of radius r around a fixed proton. (This is the hydrogen atom.) Electron 2 approaches from afar with kinetic energy T_2. When the second electron hits the atom, the first electron is knocked free, and the second is captured in a circular orbit of radius r'. **(b)** Write down an expression for the total energy of the three-particle system in general. (Your answer should contain five terms, three PEs but only two KEs, since the proton is considered fixed.) **(c)** Identify the values of all five terms and the total energy E long before the collision occurs, and again long after it is all over. What is the KE of the outgoing electron 1 once it is far away? Give your answers in terms of the variables T_2, r, and r'.

CHAPTER 5

Oscillations

Almost any system that is displaced from a position of stable equilibrium exhibits *oscillations*. If the displacement is *small*, the oscillations are almost always of the type called simple harmonic. Oscillations, and particularly simple harmonic oscillations, are therefore extremely widespread. They are also extremely useful. For example, all good clocks depend on an oscillator to regulate their time keeping: The first reliable clocks used a pendulum; the first accurate watches (historically crucial in navigation) used an oscillating balance wheel; modern watches use the oscillations of a quartz crystal; and today's most accurate clocks, such as the atomic clock at the National Institute for Standards and Technology in Boulder, Colorado, use the oscillations of an atom. In this chapter, we shall explore the physics and mathematics of oscillations. I shall begin with simple harmonic oscillations and then go on to damped oscillations (oscillations that die out because of resistive forces) and driven oscillations (oscillations that are maintained by an outside driving force, as in all clocks). The last three sections of this chapter describe the use of Fourier series in finding the motion of an oscillator driven by an arbitrary periodic driving force.

5.1 Hooke's Law

As you are certainly aware, a mass on the end of a spring that obeys Hooke's law executes oscillations of the type that we call simple harmonic. Before we review the proof of this claim, let us first ask why Hooke's law is so important and appears so frequently. Hooke's law asserts that the force exerted by a spring has the form (for now we'll restrict ourselves to a spring confined to the x axis)

$$F_x(x) = -kx \tag{5.1}$$

where x is the displacement of the spring from its equilibrium length and k is a positive number called the force constant. That k is positive means that the equilibrium at $x = 0$ is stable: When $x = 0$ there is no force, when $x > 0$ (displacement to the right) the force is negative (back to the left), and when $x < 0$ (displacement to the left) the force is positive (back to the right); either way, the force is a *restoring* force, and the equilibrium is stable. (If k were negative, the force would be away from the origin, and the equilibrium would be unstable, in which case we do not expect to see oscillations.) An exactly equivalent way to state Hooke's law is that the potential energy is

$$U(x) = \tfrac{1}{2}kx^2.$$

Consider now an arbitrary conservative one-dimensional system which is specified by a coordinate x and has potential energy $U(x)$. Suppose that the system has a stable equilibrium position $x = x_o$, which we may as well take to be the origin ($x_o = 0$). Now consider the behavior of $U(x)$ in the vicinity of the equilibrium position. Since any reasonable function can be expanded in a Taylor series, we can safely write

$$U(x) = U(0) + U'(0)x + \tfrac{1}{2}U''(0)x^2 + \cdots. \tag{5.2}$$

As long as x remains small, the first three terms in this series should be a good approximation. The first term is a constant, and, since we can always subtract a constant from $U(x)$ without affecting any physics, we may as well redefine $U(0)$ to be zero. Because $x = 0$ is an equilibrium point, $U'(0) = 0$ and the second term in the series (5.2) is automatically zero. Because the equilibrium is stable, $U''(0)$ is positive. Renaming $U''(0)$ as k, we conclude that for small displacements it is always a good approximation to take[1]

$$U(x) = \tfrac{1}{2}kx^2. \tag{5.3}$$

That is, for sufficiently small displacements from stable equilibrium, Hooke's law is always valid. Notice that if $U''(0)$ were negative, then k would also be negative, and the equilibrium would be unstable — a case we're not interested in just now. Hooke's law in the form (5.3) crops up in many situations, although it is certainly not necessary that the coordinate be the rectangular coordinate x, as the following example illustrates.

EXAMPLE 5.1   The Cube Balanced on a Cylinder

Consider again the cube of Example 4.7 (page 130) and show that for small angles θ the potential energy takes the Hooke's-law form $U(\theta) = \tfrac{1}{2}k\theta^2$.

We saw in that example that

$$U(\theta) = mg[(r + b)\cos\theta + r\theta\sin\theta].$$

If θ is small we can make the approximations $\cos\theta \approx 1 - \theta^2/2$ and $\sin\theta \approx \theta$, so that

$$U(\theta) \approx mg[(r + b)(1 - \tfrac{1}{2}\theta^2) + r\theta^2] = mg(r + b) + \tfrac{1}{2}mg(r - b)\theta^2,$$

---

[1] The only exception is if $U''(0)$ happens to be zero, but I shall not worry about this exceptional case here.

which, apart from the uninteresting constant, has the form $\frac{1}{2}k\theta^2$ with "spring constant" $k = mg(r - b)$. Notice that the equilibrium is stable (k positive) only when $r > b$, a condition we had already found in Example 4.7.

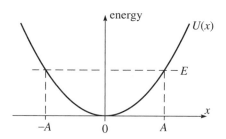

Figure 5.1   A mass m with potential energy $U(x) = \frac{1}{2}kx^2$ and total energy E oscillates between the two turning points at $x = \pm A$, where $U(x) = E$ and the kinetic energy is zero.

As discussed in Section 4.6, the general features of the motion of any one-dimensional system can be understood from a graph of $U(x)$ against x. For the Hooke's-law potential energy (5.3), this graph is a parabola, as shown in Figure 5.1. If a mass m has potential energy of this form and has any total energy $E > 0$, it is trapped and oscillates between the two turning points where $U(x) = E$, so that the kinetic energy is zero and the mass is instantaneously at rest. Because $U(x)$ is symmetric about $x = 0$, the two turning points are equidistant on opposite sides of the origin and are traditionally denoted $x = \pm A$, where A is called the amplitude of the oscillations.

5.2   Simple Harmonic Motion

We are now ready to examine the equation of motion (that is, Newton's second law) for a mass m that is displaced from a position of stable equilibrium. To be definite, let us consider a cart on a frictionless track attached to a fixed spring as sketched in Figure 5.2. We have seen that we can approximate the potential energy by (5.3) or, equivalently, the force by $F_x(x) = -kx$. Thus the equation of motion is $m\ddot{x} = F_x = -kx$ or

$$\ddot{x} = -\frac{k}{m}x = -\omega^2 x \qquad (5.4)$$

where I have introduced the constant

$$\omega = \sqrt{\frac{k}{m}},$$

which we shall see is the angular frequency with which the cart will oscillate. Although we have arrived at Equation (5.4) in the context of a cart on a spring moving along

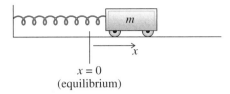

Figure 5.2   A cart of mass m oscillating on the end of a spring.

the x axis, we shall see eventually that it applies to many different oscillating systems in many different coordinate systems. For example, we have already seen in Equation (1.55) that the angle ϕ that gives the position of a pendulum (or a skateboard in a half-pipe) is governed by the same equation, $\ddot{\phi} = -\omega^2\phi$, at least for small values of ϕ. In this section I am going to review the properties of the solutions of (5.4). Unfortunately, there are many different ways to write the same solution, all of which have their advantages, and you should be comfortable with them all.

The Exponential Solutions

Equation (5.4) is a second-order, linear, homogeneous differential equation[2] and so has two independent solutions. These two independent solutions can be chosen in several different ways, but perhaps the most convenient is this:

$$x(t) = e^{i\omega t} \qquad \text{and} \qquad x(t) = e^{-i\omega t}.$$

As you can easily check, both of these functions do satisfy (5.4). Further, any constant multiple of either solution is also a solution, and likewise any sum of such multiples. Thus the function

$$x(t) = C_1 e^{i\omega t} + C_2 e^{-i\omega t} \tag{5.5}$$

is also a solution for any two constants C_1 and C_2. (That any linear combination of solutions like this is itself a solution is called the **superposition principle** and plays a crucial role in many branches of physics.) Since this solution (5.5) contains two arbitrary constants, it is the general solution of our second-order equation (5.4).[3] Therefore, *any* solution can be expressed in the form (5.5) by suitable choice of the coefficients C_1 and C_2.

---

[2] Linear because it contains no higher powers of x or its derivatives than the first power, and homogeneous because *every* term is a first power (that is, there is no term independent of x and its derivatives).

[3] Recall the result, discussed below Equation (1.56), that the general solution of a second-order differential equation contains precisely two arbitrary constants.

The Sine and Cosine Solutions

The exponential functions in (5.5) are so convenient to handle that (5.5) is often the best form of the solution. Nevertheless, this form does have one disadvantage. We know, of course, that $x(t)$ is real, whereas the two exponentials in (5.5) are complex. This means the coefficients C_1 and C_2 must be chosen carefully to ensure that $x(t)$ itself is real. I shall return to this point shortly, but first I shall rewrite (5.5) in another useful way. From Euler's formula (2.76) we know that the two exponentials in (5.5) can be written as

$$e^{\pm i\omega t} = \cos(\omega t) \pm i \sin(\omega t).$$

Substituting into (5.5) and regrouping we find that

$$x(t) = (C_1 + C_2) \cos(\omega t) + i(C_1 - C_2) \sin(\omega t)$$
$$= B_1 \cos(\omega t) + B_2 \sin(\omega t) \tag{5.6}$$

where B_1 and B_2 are simply new names for the coefficients in the previous line,

$$B_1 = C_1 + C_2 \qquad \text{and} \qquad B_2 = i(C_1 - C_2). \tag{5.7}$$

The form (5.6) can be taken as the definition of **simple harmonic motion** (or **SHM**): Any motion that is a combination of a sine and cosine of this form is called simple harmonic. Because the functions $\cos(\omega t)$ and $\sin(\omega t)$ are real, the requirement that $x(t)$ be real means simply that the coefficients B_1 and B_2 must be real.

We can easily identify the coefficients B_1 and B_2 in terms of the initial conditions of the problem. Clearly at $t = 0$, (5.6) implies that $x(0) = B_1$. That is, B_1 is just the initial position $x(0) = x_o$. Similarly, by differentiating (5.6), we can identify ωB_2 as the initial velocity v_o.

If I start the oscillations by pulling the cart aside to $x = x_o$ and releasing it from rest ($v_o = 0$), then $B_2 = 0$ in (5.6) and only the cosine term survives, so that

$$x(t) = x_o \cos(\omega t). \tag{5.8}$$

If I launch the cart from the origin ($x_o = 0$) by giving it a kick at $t = 0$, only the sine term survives, and

$$x(t) = \frac{v_o}{\omega} \sin(\omega t).$$

These two simple cases are illustrated in Figure 5.3. Notice that both solutions, like the general solution (5.6), are periodic because both the sine and cosine are. Since the argument of both sine and cosine is ωt, the function $x(t)$ repeats itself after the time τ for which $\omega \tau = 2\pi$. That is, the period is

$$\tau = \frac{2\pi}{\omega} = 2\pi \sqrt{\frac{m}{k}}. \tag{5.9}$$

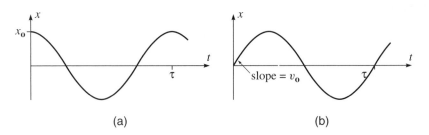

Figure 5.3   (a) Oscillations in which the cart is released from x_o at $t = 0$ follow a cosine curve. (b) If the cart is kicked from the origin at $t = 0$, the oscillations follow a sine curve with initial slope v_o. In either case the period of the oscillations is $\tau = 2\pi/\omega = 2\pi\sqrt{m/k}$ and is the same whatever the values of x_o or v_o.

The Phase-Shifted Cosine Solution

The general solution (5.6) is harder to visualize than the two special cases of Figure 5.3, and it can usefully be rewritten as follows: First, we define yet another constant

$$A = \sqrt{B_1^{\,2} + B_2^{\,2}}. \tag{5.10}$$

Notice that A is the hypotenuse to a right triangle whose other two sides are B_1 and B_2. I have indicated this in Figure 5.4, where I have also defined δ as the lower angle of that triangle. We can now rewrite (5.6) as

$$x(t) = A\left[\frac{B_1}{A}\cos(\omega t) + \frac{B_2}{A}\sin(\omega t)\right]$$

$$= A[\cos\delta\cos(\omega t) + \sin\delta\sin(\omega t)]$$

$$= A\cos(\omega t - \delta). \tag{5.11}$$

From this form it is clear that the cart is oscillating with amplitude A, but instead of being a simple cosine as in (5.8), it is a cosine which is shifted in phase: When $t = 0$ the argument of the cosine is $-\delta$, and the oscillations lag behind the simple cosine by the *phase shift* δ. We have derived the result (5.11) from Newton's second law, but, as so often happens, one can derive the same result in more than one way. In particular,

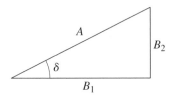

Figure 5.4   The constants A and δ are defined in terms of B_1 and B_2 as shown.

(5.11) can also be derived using the energy approach discussed in Section 4.6. (See Problem 4.28.)

Solution as the Real Part of a Complex Exponential

There is still another useful way to write our solution, in terms of the complex exponentials of (5.5). The coefficients C_1 and C_2 there are related to the coefficients B_1 and B_2 of the sine–cosine form by Equation (5.7), which we can solve to give

$$C_1 = \tfrac{1}{2}(B_1 - i B_2) \qquad \text{and} \qquad C_2 = \tfrac{1}{2}(B_1 + i B_2). \qquad (5.12)$$

Since B_1 and B_2 are real, this shows that both C_1 and C_2 are generally complex and that C_2 is the complex conjugate of C_1,

$$C_2 = C_1^*.$$

(Recall that for any complex number $z = x + iy$, the complex conjugate z^* is defined as[4] $z^* = x - iy$.) Thus the solution (5.5) can be written as

$$x(t) = C_1 e^{i\omega t} + C_1^* e^{-i\omega t} \qquad (5.13)$$

where the whole second term on the right is just the complex conjugate of the first. (See Problem 5.35 if this isn't clear to you.) Now, for any complex number $z = x + iy$,

$$z + z^* = (x + iy) + (x - iy) = 2x = 2\,\mathrm{Re}\,z$$

where $\mathrm{Re}\,z$ denotes the real part of z (namely x). Thus (5.13) can be written as

$$x(t) = 2\,\mathrm{Re}\,C_1 e^{i\omega t}.$$

If we define a final constant $C = 2C_1$, we see from Equation (5.12) and Figure 5.4 that

$$C = B_1 - i B_2 = A e^{-i\delta} \qquad (5.14)$$

and

$$x(t) = \mathrm{Re}\,C e^{i\omega t} = \mathrm{Re}\,A e^{i(\omega t - \delta)}.$$

This beautiful result is illustrated in Figure 5.5. The complex number $Ae^{i(\omega t-\delta)}$ moves counterclockwise with angular velocity ω around a circle of radius A. Its real part [namely $x(t)$] is the projection of the complex number onto the real axis. While the complex number goes around the circle, this projection oscillates back and forth on the x axis, with angular frequency ω and amplitude A. Specifically, $x(t) = A \cos(\omega t - \delta)$, in agreement with (5.11).

---

[4] While most physicists use the notation z^*, mathematicians almost always use $\bar{z}$ for the complex conjugate of z.

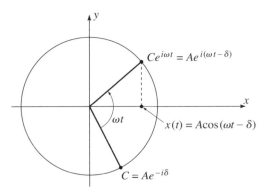

Figure 5.5   The position $x(t)$ of the oscillating cart is the real part of the complex number $Ae^{i(\omega t-\delta)}$. As the latter moves around the circle of radius A, the former oscillates back and forth on the x axis with amplitude A.

EXAMPLE 5.2   A Bottle in a Bucket

A bottle is floating upright in a large bucket of water as shown in Figure 5.6. In equilibrium it is submerged to a depth d_o below the surface of the water. Show that if it is pushed down to a depth d and released, it will execute harmonic motion, and find the frequency of its oscillations. If $d_o = 20$ cm, what is the period of the oscillations?

The two forces on the bottle are its weight mg downward and the upward buoyant force $\varrho g A d$, where ϱ is the density of water and A is the cross-sectional area of the bottle. (Remember that Archimedes' principle says that the buoyant force is ϱg times the volume submerged, which is just Ad.) The equilibrium depth d_o is determined by the condition

$$mg = \varrho g A d_o. \tag{5.15}$$

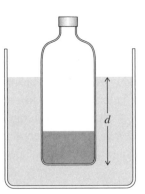

Figure 5.6   The bottle shown has been loaded with sand so that it floats upright in a bucket of water. Its equilibrium depth is $d = d_o$.

Suppose now the bottle is at a depth $d = d_o + x$. (This defines x as the distance *from equilibrium*, always the best coordinate to use.) Newton's second law now reads

$$m\ddot{x} = mg - \varrho g A(d_o + x).$$

By (5.15), the first and second terms on the right cancel, and we're left with $\ddot{x} = -\varrho g A x/m$. But again by (5.15), $\varrho g A/m = g/d_o$, so the equation of motion becomes

$$\ddot{x} = -\frac{g}{d_o}x,$$

which is exactly the equation for simple harmonic motion. We conclude that the bottle moves up and down in SHM with angular frequency $\omega = \sqrt{g/d_o}$. A remarkable feature of this result is that the frequency of oscillations does not involve m, ϱ, or A explicitly; also, the frequency is the same as that of a simple pendulum of length $l = d_o$. If $d_o = 20$ cm, then the period is

$$\tau = \frac{2\pi}{\omega} = 2\pi\sqrt{\frac{d_o}{g}} = 2\pi\sqrt{\frac{0.20 \text{ m}}{9.8 \text{ m/s}^2}} = 0.9 \text{ sec.}$$

Try this experiment yourself! But be aware that the details of the flow of water around the bottle complicate the situation considerably. The calculation here is a very simplified version of the truth.

Energy Considerations

To conclude this section on simple harmonic motion, let us consider briefly the energy of the oscillator (the cart on a spring or whatever it is) as it oscillates back and forth. Since $x(t) = A \cos(\omega t - \delta)$, the potential energy is just

$$U = \tfrac{1}{2}kx^2 = \tfrac{1}{2}kA^2 \cos^2(\omega t - \delta).$$

Differentiating $x(t)$ to give the velocity, we find for the kinetic energy

$$T = \tfrac{1}{2}m\dot{x}^2 = \tfrac{1}{2}m\omega^2 A^2 \sin^2(\omega t - \delta)$$

$$= \tfrac{1}{2}kA^2 \sin^2(\omega t - \delta)$$

where the second line results from replacing ω^2 by k/m. We see that both U and T oscillate between 0 and $\tfrac{1}{2}kA^2$, with their oscillations perfectly out of step — when U is maximum T is zero and vice versa. In particular, since $\sin^2\theta + \cos^2\theta = 1$, the total energy is constant,

$$E = T + U = \tfrac{1}{2}kA^2, \tag{5.16}$$

as it has to be for any conservative force.

5.3 Two-Dimensional Oscillators

In two or three dimensions, the possibilities for oscillation are considerably richer than in one dimension. The simplest possibility is the so-called **isotropic harmonic oscillator**, for which the restoring force is proportional to the displacement from equilibrium, with the same constant of proportionality in all directions:

$$\mathbf{F} = -k\mathbf{r}. \tag{5.17}$$

That is, $F_x = -kx$, $F_y = -ky$ (and $F_z = -kz$ in three dimensions), all with the same constant k. This force is a central force directed toward the equilibrium position, which we may as well take to be the origin, as sketched in Figure 5.7(a). Figure 5.7(b) shows an arrangement of four identical springs that would produce a force of this form; it is easy to see that if the mass at the center is moved away from its equilibrium position it will experience a net inward force, and it is not too hard to show (Problem 5.19) that this inward force has the form (5.17) for small displacements $\mathbf{r}$.[5] Another example of a two-dimensional isotropic oscillator is (at least approximately) a ball bearing rolling near the bottom of a large spherical bowl. Two important three-dimensional examples are an atom vibrating near its equilibrium in a symmetric crystal, and a proton (or neutron) as it moves inside a nucleus.

Let us consider a particle that is subject to this type of force and suppose, for simplicity, that it is confined to two dimensions. The equation of motion, $\ddot{\mathbf{r}} = \mathbf{F}/m$, separates into two independent equations:

$$\left.\begin{aligned} \ddot{x} &= -\omega^2 x \\ \ddot{y} &= -\omega^2 y \end{aligned}\right\} \tag{5.18}$$

where I have introduced the familiar angular frequency $\omega = \sqrt{k/m}$ (which is the same in both x and y equations because the same is true of the force constants). Each of

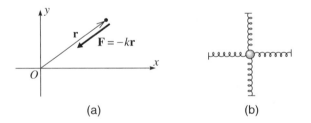

Figure 5.7   **(a)** A restoring force that is proportional to **r** defines the isotropic harmonic oscillator. **(b)** The mass at the center of this arrangement of springs would experience a net force of the form $\mathbf{F} = -k\mathbf{r}$ as it moves in the plane of the four springs.

---

[5] It is perhaps worth pointing out that one does *not* get a force of the form (5.17) by simply attaching a mass to a spring whose other end is anchored to the origin.

these two equations has exactly the form of the one-dimensional equation discussed in the last section, and the solutions are [as in (5.11)]

$$\left.\begin{array}{l} x(t) = A_x \cos(\omega t - \delta_x) \\ y(t) = A_y \cos(\omega t - \delta_y) \end{array}\right\} \tag{5.19}$$

where the four constants A_x, A_y, δ_x, and δ_y are determined by the initial conditions of the problem. By redefining the origin of time, we can dispose of the phase shift δ_x, but, in general, we cannot also dispose of the corresponding phase in the y solution. Thus the simplest form for the general solution is

$$\left.\begin{array}{l} x(t) = A_x \cos(\omega t) \\ y(t) = A_y \cos(\omega t - \delta) \end{array}\right\} \tag{5.20}$$

where $\delta = \delta_y - \delta_x$ is the *relative* phase of the y and x oscillations. (See Problem 5.15.)

The behavior of the solution (5.20) depends on the values of the three constants A_x, A_y, and δ. If either A_x or A_y is zero, the particle executes simple harmonic motion along one of the axes. (The ball bearing in the bowl rolls back and forth through the origin, moving in the x direction only or the y direction only.) If neither A_x nor A_y is zero, the motion depends critically on the relative phase δ. If $\delta = 0$, then $x(t)$ and $y(t)$ rise and fall in step, and the point (x, y) moves back and forth on the slanting line that joins (A_x, A_y) to $(-A_x, -A_y)$, as shown in Figure 5.8(a). If $\delta = \pi/2$, then x and y oscillate out of step, with x at an extreme when y is zero, and vice versa; the point (x, y) describes an ellipse with semimajor and semiminor axes A_x and A_y, as in Figure 5.8(b). For other values of δ, the point (x, y) moves around a slanting ellipse, as shown for the case $\delta = \pi/4$ in Figure 5.8(c). (For a proof that the path really is an ellipse, see Problem 8.11.)

In the **anisotropic oscillator**, the components of the restoring force are proportional to the components of the displacement, but with different constants of proportionality:

$$F_x = -k_x x, \qquad F_y = -k_y y, \qquad \text{and} \qquad F_z = -k_z z. \tag{5.21}$$

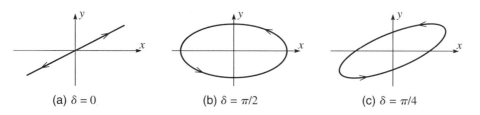

(a) $\delta = 0$ (b) $\delta = \pi/2$ (c) $\delta = \pi/4$

Figure 5.8   Motion of a two-dimensional isotropic oscillator as given by (5.20). **(a)** If $\delta = 0$, then x and y execute simple harmonic motion in step, and the point (x, y) moves back and forth along a slanting line as shown. **(b)** If $\delta = \pi/2$, then (x, y) moves around an ellipse with axes along the x and y axes. **(c)** In general (for example, $\delta = \pi/4$), the point (x, y) moves around a slanted ellipse as shown.

An example of such a force is the force felt by an atom displaced from its equilibrium position in a crystal of low symmetry, where it experiences different force constants along the different axes. For simplicity, I shall again consider a particle in two dimensions, for which Newton's second law separates into two separate equations just as in (5.18):

$$\left.\begin{aligned} \ddot{x} &= -\omega_x^2 x \\ \ddot{y} &= -\omega_y^2 y. \end{aligned}\right\} \tag{5.22}$$

The only difference between this and (5.18) is that there are now different frequencies for the different axes, $\omega_x = \sqrt{k_x/m}$ and so on. The solution of these two equations is just like (5.20):

$$\left.\begin{aligned} x(t) &= A_x \cos(\omega_x t) \\ y(t) &= A_y \cos(\omega_y t - \delta). \end{aligned}\right\} \tag{5.23}$$

Because of the two different frequencies, there is a much richer variety of possible motions. If ω_x/ω_y is a rational number, it is fairly easy to see (Problem 5.17) that the motion is periodic, and the resulting path is called a Lissajous figure (after the French physicist Jules Lissajous, 1822–1880). For example, Figure 5.9(a) shows an orbit of a particle for which $\omega_x/\omega_y = 2$ and the x motion repeats itself twice as often as the y motion. In the case shown, the result is a figure eight. If ω_x/ω_y is irrational, the motion is more complicated and never repeats itself. This case is illustrated, with $\omega_x/\omega_y = \sqrt{2}$, in Figure 5.9(b). This kind of motion is called **quasiperiodic**: The motion of the separate coordinates x and y is periodic, but because the two periods are incompatible, the motion of $\mathbf{r} = (x, y)$ is not.

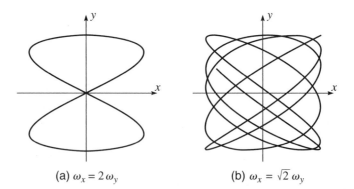

(a) $\omega_x = 2\omega_y$ (b) $\omega_x = \sqrt{2}\,\omega_y$

Figure 5.9   (a) One possible path for an anisotropic oscillator with $\omega_x = 2$ and $\omega_y = 1$. You can see that x goes back and forth twice in the time that y does so once, and the motion then repeats itself exactly. (b) A path for the case $\omega_x = \sqrt{2}$ and $\omega_y = 1$ from $t = 0$ to $t = 24$. In this case the path never repeats itself, although, if we wait long enough, it will come arbitrarily close to any point in the rectangle bounded by $x = \pm A_x$ and $y = \pm A_y$.

5.4   Damped Oscillations

I shall now return to the one-dimensional oscillator, and take up the possibility that there are resistive forces that will damp the oscillations. There are several possibilities for the resistive force. Ordinary sliding friction is approximately constant in magnitude, but always directed opposite to the velocity. The resistance offered by a fluid, such as air or water, depends on the velocity in a complicated way. However, as we saw in Chapter 2, it is sometimes a reasonable approximation to assume that the resistive force is proportional to v or (under different circumstances) to v^2. Here I shall assume that the resistive force is proportional to v; specifically, $\mathbf{f} = -b\mathbf{v}$. One of my main reasons is that this case leads to an especially simple equation to solve, and the equation is itself a very important equation that appears in several other contexts and is therefore well worth studying.[6]

Consider, then, an object in one dimension, such as a cart attached to a spring, that is subject to a Hooke's law force, $-kx$, and a resistive force, $-b\dot{x}$. The net force on the object is $-b\dot{x} - kx$, and Newton's second law reads (if I move the two force terms over to the left side)

$$m\ddot{x} + b\dot{x} + kx = 0. \tag{5.24}$$

One of the beautiful things about physics is the way the same mathematical equation can arise in totally different physical contexts, so that our understanding of the equation in one situation carries over immediately to the other. Before we set about solving Equation (5.24), I would like to show how the same equation appears in the study of LRC circuits. An LRC circuit is a circuit containing an inductor (inductance L), a capacitor (capacitance C), and a resistor (resistance R), as sketched in Figure 5.10. I have chosen the positive direction for the current to be counterclockwise, and the charge $q(t)$ to be the charge on the left-hand plate of the capacitor [with $-q(t)$ on the right], so that $I(t) = \dot{q}(t)$. If we follow around the circuit in the positive direction, the electric potential drops by $L\dot{I} = L\ddot{q}$ across the inductor, by $RI = R\dot{q}$ across the resistor, and by q/C across the capacitor. Applying Kirchhoff's second rule for circuits, we conclude that

$$L\ddot{q} + R\dot{q} + \frac{1}{C}q = 0. \tag{5.25}$$

This has exactly the form of Equation (5.24) for the damped oscillator, and anything that we learn about the equation for the oscillator will be immediately applicable to the LRC circuit. Notice that the inductance L of the electric circuit plays the role of the mass of the oscillator, the resistance term $R\dot{q}$ corresponds to the resistive force, and $1/C$ to the spring constant k.

---

[6] You should be aware, however, that although the case I am considering — that the resistive force $\mathbf{f}$ is linear in $\mathbf{v}$ — is very important, it is nevertheless a *very* special case. I shall describe some of the startling complications that can occur when $\mathbf{f}$ is not linear in $\mathbf{v}$ in Chapter 12 on nonlinear mechanics and chaos.

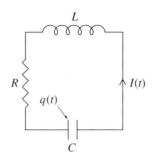

Figure 5.10   An LRC circuit.

Let us now return to mechanics and the differential equation (5.24). To solve this equation it is convenient to divide by m and then introduce two other constants. I shall rename the constant b/m as 2β,

$$\frac{b}{m} = 2\beta. \qquad (5.26)$$

This parameter β, which can be called the **damping constant**, is simply a convenient way to characterize the strength of the damping force — as with b, large β corresponds to a large damping force and conversely. I shall rename the constant k/m as ω_o^2, that is,

$$\omega_o = \sqrt{\frac{k}{m}}. \qquad (5.27)$$

Notice that ω_o is precisely what I was calling ω in the previous two sections. I have added the subscript because, once we admit resistive forces, various other frequencies become important. From now on, I shall use the notation ω_o to denote the system's **natural frequency**, the *frequency at which it would oscillate if there were no resistive force present*, as given by (5.27). With these notations, the equation of motion (5.24) for the damped oscillator becomes

$$\ddot{x} + 2\beta\dot{x} + \omega_o^2 x = 0. \qquad (5.28)$$

Notice that both of the parameters β and ω_o have the dimensions of inverse time, that is, frequency.

Equation (5.28) is another second-order, linear, homogeneous equation [the last was (5.4)]. Therefore, if by any means we can spot two independent[7] solutions, $x_1(t)$ and $x_2(t)$ say, then any solution must have the form $C_1 x_1(t) + C_2 x_2(t)$. What this

_____

[7] It is about time I gave you a definition of "independent." In general this is a little complicated, but for two functions it is easy: Two functions are independent if neither is a constant multiple of the other. Thus the two functions $\sin(x)$ and $\cos(x)$ are independent; likewise the two functions x and x^2; but the two functions x and $3x$ are not.

means is that we are free to play a game of inspired guessing to find ourselves two independent solutions; if by hook or by crook we can spot two solutions, then we have the general solution.

In particular, there is nothing to stop us *trying* to find a solution of the form

$$x(t) = e^{rt} \tag{5.29}$$

for which

$$\dot{x} = re^{rt}$$

and

$$\ddot{x} = r^2 e^{rt}.$$

Substituting into (5.28) we see that our guess (5.29) satisfies (5.28) if and only if

$$r^2 + 2\beta r + \omega_o^2 = 0 \tag{5.30}$$

[an equation sometimes called the **auxiliary equation** for the differential equation (5.28)]. The solutions of this equation are, of course, $r = -\beta \pm \sqrt{\beta^2 - \omega_o^2}$. Thus if we define the two constants

$$r_1 = -\beta + \sqrt{\beta^2 - \omega_o^2}$$
$$r_2 = -\beta - \sqrt{\beta^2 - \omega_o^2} \tag{5.31}$$

then the two functions $e^{r_1 t}$ and $e^{r_2 t}$ *are* two independent solutions of (5.28) and the general solution is

$$x(t) = C_1 e^{r_1 t} + C_2 e^{r_2 t} \tag{5.32}$$

$$= e^{-\beta t} \left(C_1 e^{\sqrt{\beta^2 - \omega_o^2}\, t} + C_2 e^{-\sqrt{\beta^2 - \omega_o^2}\, t} \right). \tag{5.33}$$

This solution is rather too messy to be especially illuminating, but, by examining various ranges of the damping constant β, we can begin to see what (5.33) entails.

Undamped Oscillation

If there is no damping then the damping constant β is zero, the square root in the exponents of (5.33) is just $i\omega_o$, and our solution reduces to

$$x(t) = C_1 e^{i\omega_o t} + C_2 e^{-i\omega_o t}, \tag{5.34}$$

the familiar solution for the undamped harmonic oscillator.

Weak Damping

Suppose next that the damping constant β is small. Specifically, suppose that

$$\beta < \omega_o, \tag{5.35}$$

a condition sometimes called **underdamping**. In this case, the square root in the exponents of (5.33) is again imaginary, and we can write

$$\sqrt{\beta^2 - \omega_o^2} = i\sqrt{\omega_o^2 - \beta^2} = i\omega_1,$$

where

$$\omega_1 = \sqrt{\omega_o^2 - \beta^2}. \tag{5.36}$$

The parameter ω_1 is a frequency, which is less than the natural frequency ω_o. In the important case of very weak damping ($\beta \ll \omega_o$), ω_1 is very close to ω_o. With this notation, the solution (5.33) becomes

$$x(t) = e^{-\beta t}\left(C_1 e^{i\omega_1 t} + C_2 e^{-i\omega_1 t}\right). \tag{5.37}$$

This solution is the product of two factors: The first, $e^{-\beta t}$, is a decaying exponential, which steadily decreases toward zero. The second factor has exactly the form (5.34) of undamped oscillations, except that the natural frequency ω_o is replaced by the somewhat lower frequency ω_1. We can rewrite the second factor, as in Equation (5.11), in the form $A\cos(\omega_1 t - \delta)$ and our solution becomes

$$x(t) = Ae^{-\beta t}\cos(\omega_1 t - \delta). \tag{5.38}$$

This solution clearly describes simple harmonic motion of frequency ω_1 with an exponentially decreasing amplitude $Ae^{-\beta t}$, as shown in Figure 5.11. The result (5.38) suggests another interpretation of the damping constant β. Since β has the dimensions of inverse time, $1/\beta$ is a time, and we now see that it is the time in which the amplitude function $Ae^{-\beta t}$ falls to $1/e$ of its initial value. Thus, at least for underdamped oscillations, β can be seen as the decay parameter, a measure of the rate at which the motion dies out,

$$\tau = 2\pi/\omega$$

(decay parameter) $= \beta$ [underdamped motion].

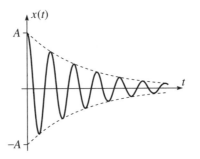

Figure 5.11   Underdamped oscillations can be thought of as simple harmonic oscillations with an exponentially decreasing amplitude $Ae^{-\beta t}$. The dashed curves are the envelopes, $\pm Ae^{-\beta t}$.

The larger β the more rapidly the oscillations die out, at least for the case $\beta < \omega_\text{o}$ that we are discussing here.

Strong Damping

Suppose instead that the damping constant β is large. Specifically suppose that

$$\beta > \omega_\text{o}, \tag{5.39}$$

a condition sometimes called **overdamping**. In this case, the square root in the exponents of (5.33) is real and our solution is

$$x(t) = C_1 e^{-\left(\beta - \sqrt{\beta^2 - \omega_\text{o}^2}\right)t} + C_2 e^{-\left(\beta + \sqrt{\beta^2 - \omega_\text{o}^2}\right)t}. \tag{5.40}$$

Here we have two real exponential functions, both of which decrease as time goes by (since the coefficients of t in both exponents are negative). In this case, the motion is so damped that it completes no bona fide oscillations. Figure 5.12 shows a typical case in which the oscillator was given a kick from O at $t = 0$; it slid out to a maximum displacement and then slid ever more slowly back again, returning to the origin only in the limit that $t \to \infty$. The first term on the right of (5.40) decreases more slowly than the second, since the coefficient in its exponent is the smaller of the two. Thus the long-term motion is dominated by this first term. In particular, the rate at which the motion dies out can be characterized by the coefficient in the first exponent,

$$\text{(decay parameter)} = \beta - \sqrt{\beta^2 - \omega_\text{o}^2} \qquad \text{[overdamped motion].} \tag{5.41}$$

Careful inspection of (5.41) shows that — contrary to what one might expect — the rate of decay of overdamped motion gets smaller if the damping constant β is made bigger. (See Problem 5.20.)

Figure 5.12   Overdamped motion in which the oscillator is kicked from the origin at $t = 0$. It moves out to a maximum displacement and then moves back toward O asymptotically as $t \to \infty$.

Critical Damping

The boundary between underdamping and overdamping is called **critical damping** and occurs when the damping constant is equal to the natural frequency, $\beta = \omega_\text{o}$. This case has some interesting features, especially from a mathematical point of view. When $\beta = \omega_o$ the two solutions that we found in (5.33) are the same solution, namely

$$x(t) = e^{-\beta t}. \tag{5.42}$$

[This happened because the two solutions of the auxiliary equation (5.30) happen to coincide when $\beta = \omega_o$.] This is the one case where our inspired guess, to seek a solution of the form $x(t) = e^{rt}$, fails to find us two solutions of the equation of motion, and we have to find a second solution by some other method. Fortunately, in this case, it is not hard to spot a second solution: As you can easily check, the function

$$x(t) = te^{-\beta t} \tag{5.43}$$

is also a solution of the equation of motion (5.28) in the special case that $\beta = \omega_o$. (See Problems 5.21 and 5.24.) Thus the general solution for the case of critical damping is

$$x(t) = C_1 e^{-\beta t} + C_2 t\, e^{-\beta t}. \tag{5.44}$$

Notice that both terms contain the same exponential factor $e^{-\beta t}$. Since this factor is what dominates the decay of the oscillations as $t \to \infty$, we can say that both terms decay at about the same rate, with decay parameter

$$\text{(decay parameter)} = \beta = \omega_o \qquad \text{[critical damping]}.$$

It is interesting to compare the rates at which the various types of damped oscillation die out. We have seen that in each case, this rate is determined by a "decay parameter," which is just (minus) the coefficient of t in the exponent of the dominant exponential factor in $x(t)$. Our findings can be summarized as follows:

damping	β	decay parameter
none	$\beta = 0$	0
under	$\beta < \omega_o$	β
critical	$\beta = \omega_o$	β
over	$\beta > \omega_o$	$\beta - \sqrt{\beta^2 - \omega_o^2}$

Figure 5.13 is a plot of the decay parameter as a function of β and shows clearly that the motion dies out most quickly when $\beta = \omega_o$; that is, when the damping is critical. There are situations where one wants any oscillations to die out as quickly as possible. For example, one wants the needle of an analog meter (a voltmeter or pressure gauge, for instance) to settle down rapidly on the correct reading. Similarly, in a car, one wants the oscillations caused by a bumpy road to decay quickly. In such cases one must arrange for the oscillations to be damped (by the shock absorbers in a car), and for the quickest results the damping should be reasonably close to critical.

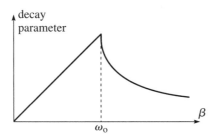

Figure 5.13   The decay parameter for damped oscillations as a function of the damping constant β. The decay parameter is biggest, and the motion dies out most quickly, for critical damping, with $\beta = \omega_0$.

5.5   Driven Damped Oscillations

Any natural oscillator, left to itself, eventually comes to rest, as the inevitable damping forces drain its energy. Thus if one wants the oscillations to continue, one must arrange for some external "driving" force to maintain them. For example, the motion of the pendulum in a grandfather clock is driven by periodic pushes caused by the clock's weights; the motion of a young child on a swing is maintained by periodic pushes from a parent. If we denote the external driving force by $F(t)$ and if we assume as before that the damping force has the form $-bv$, then the net force on the oscillator is $-bv - kx + F(t)$ and the equation of motion can be written as

$$m\ddot{x} + b\dot{x} + kx = F(t). \tag{5.45}$$

Like its counterpart for undriven oscillations, this differential equation crops up in several other areas of physics. A prominent example is the LRC circuit of Figure 5.10. If we want the oscillating current in that circuit to persist, we must apply a driving EMF, $\mathcal{E}(t)$, in which case the equation of motion for the circuit becomes

$$L\ddot{q} + R\dot{q} + \frac{1}{C}q = \mathcal{E}(t) \tag{5.46}$$

in perfect correspondence with (5.45).

As before, we can tidy Equation (5.45) if we divide the equation by m and replace b/m by 2β and k/m by ω_0^2. In addition, I shall denote $F(t)/m$ by

$$f(t) = \frac{F(t)}{m}, \tag{5.47}$$

the force per unit mass. With this notation, (5.45) becomes

$$\ddot{x} + 2\beta\dot{x} + \omega_0^2 x = f(t). \tag{5.48}$$

Linear Differential Operators

Before we discuss how to solve this equation, I would like to streamline our notation. It turns out that it is very helpful to think of the left side of (5.48) as the result of a certain operator acting on the function $x(t)$. Specifically, we define the *differential operator*

$$D = \frac{d^2}{dt^2} + 2\beta \frac{d}{dt} + \omega_o^2. \tag{5.49}$$

The meaning of this definition is simply that when D acts on x it gives the left side of (5.48):

$$Dx \equiv \ddot{x} + 2\beta \dot{x} + \omega_o^2 x.$$

This definition is obviously a notational convenience — the equation (5.48) becomes just

$$Dx = f \tag{5.50}$$

— but it is much more: The notion of an operator like (5.49) proves to be a powerful mathematical tool, with applications throughout physics. For the moment the important thing is that the operator is *linear*: We know from elementary calculus that the derivative of ax (where a is a constant) is just $a\dot{x}$ and that the derivative of $x_1 + x_2$ is just $\dot{x}_1 + \dot{x}_2$. Since this also applies to second derivatives, it applies to the operator D:

$$D(ax) = aDx \quad \text{and} \quad D(x_1 + x_2) = Dx_1 + Dx_2.$$

(Make sure you understand what these two equations mean.) We can combine these into a single equation:

$$D(ax_1 + bx_2) = aDx_1 + bDx_2 \tag{5.51}$$

for any two constants a and b and any two functions $x_1(t)$ and $x_2(t)$. Any operator that satisfies this equation is called a **linear operator**.

We have actually used the property (5.51) of linear operators before. The equation (5.28) for a damped oscillator (not driven) can be written as

$$Dx = 0. \tag{5.52}$$

The superposition principle asserts that if x_1 and x_2 are solutions of this equation, then so is $ax_1 + bx_2$ for any constants a and b. In our new operator notation, the proof is very simple: We are given that $Dx_1 = 0$ and $Dx_2 = 0$, and using (5.51) it immediately follows that

$$D(ax_1 + bx_2) = aDx_1 + bDx_2 = 0 + 0 = 0;$$

that is, $ax_1 + bx_2$ is also a solution.

The equation (5.52), $Dx = 0$, for the undriven oscillator is called a **homogeneous** equation, since every term involves either x or one of its derivatives exactly once. The equation (5.50), $Dx = f$, is called an **inhomogeneous** equation, since it contains the inhomogeneous term f, which does not involve x at all. Our job now is to solve this inhomogeneous equation.

Particular and Homogeneous Solutions

Using our new operator notation, we can find the general solution of Equation (5.48) surprisingly easily; in fact, we've already done most of the work. Suppose first that we have somehow spotted a solution; that is, we've found a function $x_p(t)$ that satisfies

$$Dx_p = f. \tag{5.53}$$

We call this function $x_p(t)$ a **particular solution** of the equation, and the subscript "p" stands for "particular." Next let us consider for a moment the homogeneous equation $Dx = 0$ and suppose we have a solution $x_h(t)$, satisfying

$$Dx_h = 0. \tag{5.54}$$

We'll call this function a **homogeneous solution**.[8] We already know all about the solutions of the homogeneous equation, and we know from (5.32) that x_h must have the form

$$x_h(t) = C_1 e^{r_1 t} + C_2 e^{r_2 t}, \tag{5.55}$$

where both exponentials die out as $t \to \infty$.

We're now ready to prove the crucial result. First, if x_p is a particular solution satisfying (5.53), then $x_p + x_h$ is another solution, for

$$D(x_p + x_h) = Dx_p + Dx_h = f + 0 = f.$$

Given the one particular solution x_p, this gives us a large number of other solutions $x_p + x_h$. And we have in fact found *all* the solutions, since the function x_h contains two arbitrary constants, and we know that the general solution of any second-order equation contains exactly two arbitrary constants. Therefore $x_p + x_h$, with x_h given by (5.55) is the general solution.

This result means that all we have to do is somehow to find a single particular solution $x_p(t)$ of the equation of motion (5.48), and we shall have *every* solution in the form $x(t) = x_p(t) + x_h(t)$.

Complex Solutions for a Sinusoidal Driving Force

I shall now specialize to the case that the driving force $f(t)$ is a sinusoidal function of time,

$$f(t) = f_o \cos(\omega t), \tag{5.56}$$

---

[8] Another common name is the *complementary function*. This has the disadvantage that it's hard to remember which is "particular" and which "complementary."

where f_o denotes the amplitude of the driving force [actually the amplitude divided by the oscillator's mass, since $f(t) = F(t)/m$] and ω is the angular frequency of the driving force. (Be careful to distinguish the *driving frequency* ω from the *natural frequency* ω_o of the oscillator. These are entirely independent frequencies, although we shall see that the oscillator responds most when $\omega \approx \omega_o$.) The driving force for many driven oscillators is at least approximately sinusoidal. For example, even the parent pushing the child on a swing can be crudely approximated by (5.56); the driving EMF induced in your radio's circuits by a broadcast signal is almost perfectly of this form. Probably the chief importance of sinusoidal driving forces is that, according to Fourier's theorem,[9] essentially *any* driving force can be built up as a series of sinusoidal forces.

Let us therefore assume the driving force is given by (5.56), so that the equation of motion (5.48) takes the form

$$\ddot{x} + 2\beta\dot{x} + \omega_o^2 x = f_o \cos(\omega t). \tag{5.57}$$

Solving this equation is greatly simplified by the following trick: For any solution of (5.57), there must be a solution of the same equation, but with the cosine on the right replaced by a sine function. (After all, these two differ only by a shift in the origin of time.) Accordingly, there must also be a function $y(t)$ that satisfies

$$\ddot{y} + 2\beta\dot{y} + \omega_o^2 y = f_o \sin(\omega t). \tag{5.58}$$

Suppose now we define the complex function

$$z(t) = x(t) + i y(t), \tag{5.59}$$

with $x(t)$ as its real part and $y(t)$ as its imaginary part. If we multiply (5.58) by i and add it to (5.57), we find that

$$\ddot{z} + 2\beta\dot{z} + \omega_o^2 z = f_o e^{i\omega t}. \tag{5.60}$$

Although it may not yet look it, Equation (5.60) is a tremendous advance. Because of the simple properties of the exponential function, (5.60) is much easier to solve than either (5.57) or (5.58). And as soon as we find a solution $z(t)$ of (5.60), we have only to take its real part to have a solution of the equation (5.57) whose solutions we actually want.

In seeking a solution of (5.60), we are obviously free to *try* any function we like. In particular, let's see if there is a solution of the form

$$z(t) = C e^{i\omega t}, \tag{5.61}$$

where C is an as yet undetermined constant. If we substitute this guess into the left side of (5.60), we get

$$(-\omega^2 + 2i\beta\omega + \omega_o^2) C e^{i\omega t} = f_o e^{i\omega t}.$$

---

[9] Named for the French mathematician, the Baron Jean Baptiste Joseph Fourier, 1768–1830. See Sections 5.7–5.9.

In other words, our guess (5.61) is a solution of (5.60) if and only if

$$C = \frac{f_0}{\omega_0^2 - \omega^2 + 2i\beta\omega},$$ (5.62)

and we have succeeded in finding a particular solution of the equation of motion.

Before we take the real part of $z(t) = Ce^{i\omega t}$, it is convenient to write the complex coefficient C in the form

$$C = Ae^{-i\delta},$$ (5.63)

where A and δ are real. [Any complex number can be written in this form; the particular notation is chosen to match (5.14).] To identify A and δ, we must compare (5.62) and (5.63). First, taking the absolute value squared of both equations we find that

$$A^2 = CC^* = \frac{f_0}{\omega_0^2 - \omega^2 + 2i\beta\omega} \cdot \frac{f_0}{\omega_0^2 - \omega^2 - 2i\beta\omega}$$

or

$$A^2 = \frac{f_0^2}{(\omega_0^2 - \omega^2)^2 + 4\beta^2\omega^2}.$$ (5.64)

(Make sure you understand this derivation. See Problem 5.35 for some guidance.) We are going to see in a moment that A is just the amplitude of the oscillations caused by the driving force $f(t)$. Thus the result (5.64) is the most important result of this discussion. It shows how the amplitude of oscillations depends on the various parameters. In particular, we see that the amplitude is biggest when $\omega_0 \approx \omega$, so that the denominator is small; in other words, the oscillator responds best when driven at a frequency ω that is close to its natural frequency ω_0, as you would probably have guessed.

Before we continue to discuss the properties of our solution, we need to identify the phase angle δ in (5.63). Comparing (5.63) and (5.62) and rearranging, we see that

$$f_0 e^{i\delta} = A(\omega_0^2 - \omega^2 + 2i\beta\omega).$$

Since f_0 and A are real, this says that the phase angle δ is the same as the phase angle of the complex number $(\omega_0^2 - \omega^2) + 2i\beta\omega$. This relationship is illustrated in Figure 5.14, from which we conclude that

$$\delta = \arctan\left(\frac{2\beta\omega}{\omega_0^2 - \omega^2}\right).$$ (5.65)

Our quest for a particular solution is now complete. The "fictitious" complex solution introduced in (5.59) is

$$z(t) = Ce^{i\omega t} = Ae^{i(\omega t - \delta)}$$

and the real part of this is the solution we are seeking,

$$x(t) = A\cos(\omega t - \delta)$$ (5.66)

where the real constants A and δ are given by (5.64) and (5.65).

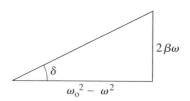

Figure 5.14   The phase angle δ is the angle of this triangle.

The solution (5.66) is just one particular solution of the equation of motion. The general solution is found by adding any solution of the corresponding homogeneous equation, as given by (5.55); that is, the general solution is

$$x(t) = A \cos(\omega t - \delta) + C_1 e^{r_1 t} + C_2 e^{r_2 t}. \tag{5.67}$$

Because the two extra terms in this general solution both die out exponentially as time passes, they are called **transients**. They depend on the initial conditions of the problem but are eventually irrelevant: The long-term behavior of our solution is dominated by the cosine term. Thus the particular solution (5.66) is the solution with which we are usually concerned, and we shall explore its properties in the next section.

Before we discuss an example of the motion (5.67), it is important that you be very clear as to the type of system to which (5.67) applies, namely, any oscillator for which both the restoring force $(-kx)$ and the resistive force $(-b\dot{x})$ are *linear* — a driven, damped *linear* oscillator, whose equation of motion (5.45) is a *linear* differential equation. Because nonlinear differential equations are often hard to solve, most mechanics texts have until recently focussed on linear equations. This created the false impression that linear equations were in some sense the norm, and that the solution (5.67) was the only (or, at least, the only important) way for an oscillator to behave. As we shall see in Chapter 12 on nonlinear mechanics and chaos, an oscillator whose equation of motion is nonlinear can behave in ways that are astonishingly different from (5.67). One important reason for studying the linear oscillator here is to give you a backdrop against which to study the nonlinear oscillator later on.

The details of the motion (5.67) depend on the strength of the damping parameter β. To be specific, let us assume that our oscillator is weakly damped, with β less than the natural frequency ω_o (underdamping). In this case, we know that the two transient terms of (5.67) can be rewritten as in (5.38), so that

$$x(t) = A \cos(\omega t - \delta) + A_{\mathrm{tr}} e^{-\beta t} \cos(\omega_1 t - \delta_{\mathrm{tr}}). \tag{5.68}$$

You need to think very carefully about this potentially confusing formula. The second term on the right is the homogeneous or transient term, and I have added the subscript "tr" to distinguish the constants A_{tr} and δ_{tr} from the A and δ of the first term. The two constants A_{tr} and δ_{tr} are *arbitrary constants;* (5.68) is a possible motion of our system for *any* values of A_{tr} and δ_{tr}, which are determined by the initial conditions.

The factor $e^{-\beta t}$ makes clear that this transient term decays exponentially and is indeed irrelevant to the long-term behavior. As it decays, the transient term oscillates with the angular frequency ω_1 of the undriven (but still damped) oscillator, as in (5.36). The first term is our particular solution and its two constants A and δ are certainly not arbitrary; they are determined by (5.64) and (5.65) in terms of the parameters of the system. This term oscillates with the frequency ω of the driving force and with unchanging amplitude, for as long as the driving force is maintained.

EXAMPLE 5.3   Graphing a Driven Damped Linear Oscillator

Make a plot of $x(t)$ as given by (5.68) for a driven damped linear oscillator which is released from rest at the origin at time $t = 0$, with the following parameters: Drive frequency $\omega = 2\pi$, natural frequency $\omega_o = 5\omega$, decay constant $\beta = \omega_o/20$, and driving amplitude $f_o = 1000$. Show the first five drive cycles.

   The choice of drive frequency equal to 2π means that the drive period is $\tau = 2\pi/\omega = 1$; this means simply that we have chosen to measure time in units of the drive period — often a convenient choice. That $\omega_o = 5\omega$ means that our oscillator has a natural frequency five times the drive frequency; this will let us distinguish easily between the two on a graph. That $\beta = \omega_o/20$ means that the oscillator is rather weakly damped. Finally the choice of $f_o = 1000$ is just a choice of our unit of force; the reason for this odd-seeming choice is that it leads to a conveniently sized amplitude of oscillation (namely A close to 1).

   Our first task is to determine the various constants in (5.68) in terms of the given parameters. In fact this is easier if we rewrite the transient term of (5.68) in the "cosine plus sine" form, so that

$$x(t) = A \cos(\omega t - \delta) + e^{-\beta t}[B_1 \cos(\omega_1 t) + B_2 \sin(\omega_1 t)]. \qquad (5.69)$$

The constants A and δ are determined by (5.64) and (5.65), which, for the given parameters, yield

$$A = 1.06 \quad \text{and} \quad \delta = 0.0208.$$

The frequency ω_1 is

$$\omega_1 = \sqrt{\omega_o^2 - \beta^2} = 9.987\pi,$$

which is very close to ω_o, as we would expect for a weakly damped oscillator. To find B_1 and B_2, we must equate $x(0)$ as given by (5.69) to its given initial value x_o, and likewise the corresponding expression for $\dot{x}_o$ to the initial value v_o. This gives two simultaneous equations for B_1 and B_2, which are easily solved (Problem 5.33) to give

$$B_1 = x_o - A \cos \delta \quad \text{and} \quad B_2 = \frac{1}{\omega_1}(v_o - \omega A \sin \delta + \beta B_1) \qquad (5.70)$$

or, with the numbers, including the initial conditions $x_o = v_o = 0$,

$$B_1 = -1.05 \quad \text{and} \quad B_2 = -0.0572.$$

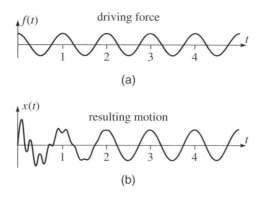

(a)

(b)

Figure 5.15   The response of a damped, linear oscillator to a sinusoidal driving force, with the time t shown in units of the drive
period. **(a)** The driving force is a pure cosine as a function of time.
(b) The resulting motion for the initial conditons $x_o = v_o = 0$. For
the first two or three drive cycles, the transient motion is clearly visible, but after that only the long-term motion remains, oscillating
sinusoidally at exactly the drive frequency. As explained in the text,
the sinusoidal motion after $t \approx 3$ is called an *attractor*.

Putting all of these numbers into (5.69), we can now plot the motion, as
shown in Figure 5.15, where part (a) shows the driving force $f(t) = f_o \cos(\omega t)$
and part (b) the resulting motion $x(t)$ of the oscillator. The driving force is, of
course, perfectly sinusoidal with period 1. The resulting motion is much more
interesting. After about three drive cycles ($t \gtrsim 3$), the motion is indistinguishable
from a pure cosine, oscillating at exactly the drive frequency; that is, the
transients have died out and only the long-term motion remains. Before $t \approx 3$,
however, the effects of the transients are clearly visible. Since they oscillate at
the faster natural frequency ω_o, they show up as a rapid succession of bumps
and dips. In fact, you can easily see that there are five such bumps within the
first drive cycle, indicating that $\omega_o = 5\omega$.

Because the transient motion depends on the initial values x_o and v_o, different
values of x_o and v_o would lead to quite different initial motion. (See Problem 5.36.)
After a short time however (a couple of cycles in this example), the initial differences
disappear and the motion settles down to the same sinusoidal motion of the particular
solution (5.66), *irrespective of the initial conditions*. For this reason, the motion
of (5.66) is sometimes called an **attractor** — the motions corresponding to several
different initial conditions are "attracted" to the particular motion (5.66). For the linear
oscillator discussed here, there is a unique attractor (for a given driving force): Every
possible motion of the system, whatever its initial conditions, is attracted to the same
motion (5.66). We shall see in Chapter 12 that for nonlinear oscillators there can be
several different attractors and that for some values of the parameters the motion of

an attractor can be far more complicated than simple harmonic oscillation at the drive frequency.

The amplitude and phase of the attractor seen in Figure 5.15(b) depend on the parameters of the driving force (but not, of course, on the initial conditions). The dependence of the amplitude and phase on these parameters is the subject of the next section.

5.6   Resonance

In the previous section we considered a damped oscillator that is being driven by a sinusoidal driving force (actually force divided by mass) $f(t) = f_o \cos(\omega t)$ with angular frequency ω. We saw that, apart from transient motions that die out quickly, the system's response is to oscillate sinusoidally at the same frequency, ω:

$$x(t) = A \cos(\omega t - \delta),$$

with amplitude A given by (5.64),

$$A^2 = \frac{f_o{}^2}{(\omega_o^2 - \omega^2)^2 + 4\beta^2\omega^2}, \tag{5.71}$$

and phase shift δ given by (5.65).

The most obvious feature of (5.71) is that the amplitude A of the response is proportional to the amplitude of the driving force, $A \propto f_o$, a result you might have guessed. More interesting is the dependence of A on the frequencies ω_o (the natural frequency of the oscillator) and ω (the frequency of the driver), and on the damping constant β. The most interesting case is when the damping constant β is very small, and this is the case I shall discuss. With β small, the second term in the denominator of (5.71) is small. If ω_o and ω are very different, then the first term in the denominator of (5.71) is large, and the amplitude of the driven oscillations is small. On the other hand, if ω_o is very close to ω, both terms in the denominator are small, and the amplitude A is large. This means that if we vary either ω_o or ω, there can be quite dramatic changes in the amplitude of the oscillator's motion. This is illustrated in Figure 5.16, which shows A^2 as a function of ω_o with ω fixed, for a rather weakly damped system ($\beta = 0.1\omega$). (Note that, because the energy of the system is proportional to A^2, it is usual to make plots of A^2 rather than A.)

Although the behavior illustrated in Figure 5.16 is startlingly dramatic, the qualitative features are what you might have expected. Left to its own devices, the oscillator vibrates at its natural frequency ω_o (or at the slightly lower frequency ω_1 if we allow for the damping). If we try to force it to vibrate at a frequency ω, then for values of ω close to ω_o the oscillator responds very well, but if ω is far from ω_o, it hardly responds at all. We refer to this phenomenon — the dramatically greater response of an oscillator when driven at the right frequency — as **resonance**.

An everyday application of resonance is the reception of radio signals by the LRC circuit in your radio. As we saw, the equation of motion of an LRC circuit is exactly

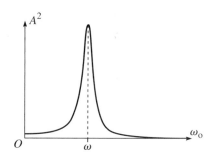

Figure 5.16   The amplitude squared, A^2, of a driven oscillator, shown as a function of the natural frequency ω_{o}, with the driving frequency ω fixed. The response is dramatically largest when ω_{o} and ω are close.

the same as that of a driven oscillator, and LRC circuits show the same phenomenon of resonance. When you tune your radio to receive station KVOD at 90.1 MHz, you are adjusting an LRC circuit in the radio so that its natural frequency is 90.1 MHz. The many radio stations in your neighborhood are all sending out signals, each at its own frequency and each inducing a tiny EMF in the circuit of your radio. But only the signal with the right frequency actually succeeds in driving an appreciable current, which mimics the signal sent out by your favorite KVOD and reproduces its broadcast sounds.

An example of a mechanical resonance of the kind discussed here is the behavior of a car driving on a "washboard" road that has worn into a series of regularly spaced bumps. Each time a wheel crosses a bump, it is given an upward impulse, and the frequency of these impulses depends on the car's speed. There is one speed at which the frequency of these impulses equals the natural frequency of the wheels' vibration on the springs[10] and the wheels resonate, causing an uncomfortable ride. If the car drives slower or faster than this speed, it goes "off resonance" and the ride is much smoother.

Another example occurs when a platoon of soldiers marches across a bridge. A bridge, like almost any mechanical system, has certain natural frequencies, and if the soldiers happened to march with a frequency equal to one of these natural frequencies, the bridge could conceivably resonate sufficiently violently to break. For this reason, soldiers break step when marching across a bridge.

The details of the resonance phenomenon are a bit complicated. For example, the exact location of the maximum response depends on whether we vary ω_{o} with ω fixed or vice versa. The amplitude A is a maximum when the denominator,

$$\text{denominator} = (\omega_{\mathrm{o}}^2 - \omega^2)^2 + 4\beta^2\omega^2 \tag{5.72}$$

---

[10] It is the wheels (plus axles) that exhibit the resonant oscillations; the much heavier body of the car is relatively unaffected.

of (5.71) is a minimum. If we vary ω_0 with ω fixed (as when you tune a radio to pick up your favorite station) this minimum obviously occurs when $\omega_0 = \omega$, making the first term zero. On the other hand, if we vary ω with ω_0 fixed (which is what happens in many applications), then the second term in (5.72) also varies and a straightforward differentiation shows that the maximum occurs when

$$\omega = \omega_2 = \sqrt{\omega_0^2 - 2\beta^2}. \tag{5.73}$$

However, when $\beta \ll \omega_0$ (as is usually the most interesting case), the difference between (5.73) and $\omega = \omega_0$ is negligible.

We have met so many different frequencies in this chapter that it may be worth pausing to review them. First there is ω_0 the natural frequency of the oscillator (in the absence of any damping). Next when we added in a little damping, we found that the same system oscillated sinusoidally with frequency $\omega_1 = \sqrt{\omega_0^2 - \beta^2}$ under an exponentially decaying envelope. Then we added a driving force with frequency ω, which can, in principle, take on any value independently of the previous two. However, the response of the driven oscillator is biggest when $\omega \approx \omega_0$; specifically, if we vary ω with ω_0 fixed, the maximum response comes when $\omega = \omega_2$ as defined by (5.73). To summarize:

$$\omega_0 = \sqrt{k/m} = \text{natural frequency of undamped oscillator}$$

$$\omega_1 = \sqrt{\omega_0^2 - \beta^2} = \text{frequency of damped oscillator}$$

$$\omega = \text{frequency of driving force}$$

$$\omega_2 = \sqrt{\omega_0^2 - 2\beta^2} = \text{value of } \omega \text{ at which response is maximum.}$$

In any case, the maximum amplitude of the driven oscillations is found by putting $\omega_0 \approx \omega$ in (5.71), to give

$$A_{\max} \approx \frac{f_0}{2\beta\omega_0}. \tag{5.74}$$

This shows that smaller values of the damping constant β lead to larger values of the maximum amplitude of oscillation, as illustrated in Figure 5.17.[11]

Width of the Resonance; the Q Factor

You can see clearly from Figure 5.17 that if we make the damping constant β smaller, the resonance peak not only gets higher, but also gets narrower. We can make this idea more precise by defining the **width** (or **full width at half maximum** or **FWHM**) as the interval between the two points where A^2 is equal to half its maximum height. It

---

[11] In this figure I chose to plot A^2 against ω with ω_0 fixed, rather than the other way around as in Figure 5.16. Note that the curves have very similar shapes either way.

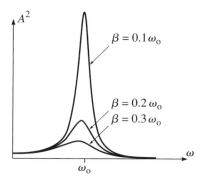

Figure 5.17   The amplitude for driven oscillations as a function of the driving frequency ω for three different values of the damping constant β. Note how as β decreases the resonance peak gets higher and sharper.

is a simple exercise (Problem 5.41) to show that the two half-maximum points are at $\omega \approx \omega_0 \pm \beta$, as in Figure 5.18. Thus the full width at half maximum is

$$\text{FWHM} \approx 2\beta \tag{5.75}$$

or, equivalently, the **half width at half maximum** is

$$\text{HWHM} \approx \beta. \tag{5.76}$$

The sharpness of the resonance peak is indicated by the ratio of its width 2β to its position, ω_0. For many purposes, we want a very sharp resonance, so it is common practice to define a **quality factor** Q as the reciprocal of this ratio:

$$Q = \frac{\omega_0}{2\beta}. \tag{5.77}$$

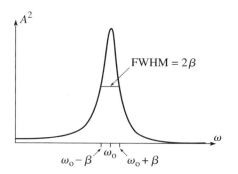

Figure 5.18   The full width at half maximum (FWHM) is the distance between the points where A^2 is half its maximum value.

A large Q indicates a narrow resonance, and vice versa. For example, clocks depend on the resonance in an oscillator (a pendulum, or a quartz crystal, for instance) to regulate the mechanism to move with very well-defined frequency. This requires that the width 2β be very small compared to the natural frequency ω_0. In other words, a good clock needs a high Q. The Q for a typical pendulum may be around 100; that for a quartz crystal around 10,000. Therefore quartz watches keep much better time than a typical grandfather clock.[12]

There is another way to look at the quality factor Q. We saw that, in the absence of any driving force, the oscillations die out in a time of order $1/\beta$,

$$\text{(decay time)} = 1/\beta.$$

(This was actually the time for the amplitude to drop to $1/e$ of its initial value.) The period of a single oscillation is, of course,

$$\text{period} = 2\pi/\omega_0.$$

(Remember we're assuming that $\beta \ll \omega_0$, so we don't need to distinguish between ω_0 and ω_1.) Thus we can rewrite the definition of Q as

$$Q = \frac{\omega_0}{2\beta} = \pi \frac{1/\beta}{2\pi/\omega_0} = \pi \frac{\text{decay time}}{\text{period}}. \tag{5.78}$$

The ratio on the right is just the number of periods in the decay time. Thus, the quality factor Q is π times the number of cycles our oscillator makes in one decay time.[13]

The Phase at Resonance

The phase difference δ by which the oscillator's motion lags behind the driving force is given by (5.65) as

$$\delta = \arctan\left(\frac{2\beta\omega}{\omega_0^2 - \omega^2}\right). \tag{5.79}$$

Let us follow this phase as we vary ω, starting well below a narrow resonance (β small). With $\omega \ll \omega_0$, (5.79) implies that δ is very small; that is, while $\omega \ll \omega_0$ the oscillations are almost perfectly in step with the driving force. (This was the case in Figure 5.15.) As ω is increased toward ω_0, so δ slowly increases. At resonance, where $\omega = \omega_0$, the argument of the arctangent in (5.79) is infinite, so $\delta = \pi/2$ and the oscillations are 90° behind the driving force. Once $\omega > \omega_0$, the argument of the

---

[12] Actually, both quartz watches and grandfather clocks keep much better time than this simple discussion would suggest. A good chronometer keeps the frequency very close to the *center* of the resonance. Thus the variability of the frequency is actually much smaller than the width of the resonance. Nevertheless, the stated conclusion is correct.

[13] Yet another definition (and perhaps the most fundamental) is that $Q = 2\pi$ times the ratio of the energy stored in the oscillator to the energy dissipated in one cycle. See Problem 5.44.

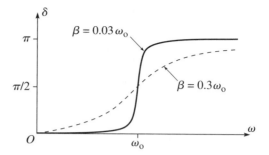

Figure 5.19   The phase shift δ increases from 0 through $\pi/2$ to π as the driving frequency ω passes through resonance. The narrower the resonance, the more suddenly this increase occurs. The solid curve is for a relatively narrow resonance ($\beta = 0.03\omega_o$ or $Q = 16.7$), and the dashed curve is for a wider resonance ($\beta = 0.3\omega_o$ or $Q = 1.67$).

arctangent is negative and approaches 0 as ω increases; thus δ increases beyond $\pi/2$ and eventually approaches π. In particular, once $\omega \gg \omega_o$, the oscillations are almost perfectly out of step with the driving force. All of this behavior is illustrated for two different values of β in Figure 5.19. Notice, in particular, that the narrower the resonance, the more quickly δ jumps from 0 to π.

In the resonances of classical mechanics, the behavior of the phase (as in Figure 5.19) is usually less important than that of the amplitude (as in Figure 5.18).[14] In atomic and nuclear collisions, the phase shift is often the quantity of primary interest. Such collisions are governed by quantum mechanics, but there is a corresponding phenomenon of resonance. A beam of neutrons, for example, can "drive" a target nucleus. When the energy of the beam equals a resonant energy of the system (in quantum mechanics energy plays the role of frequency) a resonance occurs and the phase shift increases rapidly from 0 to π.

5.7   Fourier Series *

Fourier series have broad application in almost every area of modern physics. Nevertheless, we shall not be using them again until Chapter 16. Thus, if you are pressed for time you could omit the last three sections of this chapter on a first reading.

In the last two sections, we have discussed an oscillator that is driven by a sinusoidal driving force $f(t) = f_o \cos(\omega t)$. There are two main reasons for the importance of

---

[14] The behavior of δ can, nevertheless, be observed. Make a simple pendulum from a piece of string and a metal nut, and drive it by holding it at the top and moving your hand from side to side. The most obvious thing is that you will be most successful at driving it when your frequency equals the natural frequency, but you can also see that when you drive more slowly the pendulum moves in step with your hand, whereas when you move more quickly the pendulum moves oppositely to your hand.

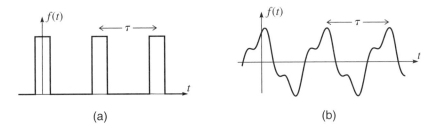

Figure 5.20   Two examples of periodic functions with period τ. **(a)** A rectangular pulse, which could represent a hammer hitting a nail with a constant force at intervals of τ, or a digital signal in a telephone line. **(b)** A smooth periodic signal, which could be the pressure variation of a musical instrument.

sinusoidal driving forces: The first is simply that there are many important systems in which the driving force *is* sinusoidal — the electrical circuit in a radio is a good example. The second is somewhat subtler. It turns out that *any* periodic driving force can be built up from sinusoidal forces using the powerful technique of Fourier series. Thus, in a sense that I shall try to describe, by solving the motion with a sinusoidal driver we have already solved the motion with any periodic driver. Before we can appreciate this wonderful result, we need to review some aspects of Fourier series. In this section I sketch the needed properties of Fourier series;[15] in the next we can apply them to the driven oscillator.

Let us consider a function $f(t)$ that is periodic with period τ; that is, the function repeats itself every time t advances by the period τ:

$$f(t + \tau) = f(t)$$

whatever the value of t. We can describe a function with this property as being τ-periodic. A simple example of a τ-periodic function would be the force exerted on a nail by a hammer that is being swung at intervals of τ, as sketched in Figure 5.20(a). Another could be the pressure exerted on your ear drum by a note played by a musical instrument, as sketched in 5.20(b). It is easy to think up many more periodic functions. In particular, there are lots of sinusoidal functions that are periodic with any given period: The functions

$$\cos(2\pi t/\tau), \ \cos(4\pi t/\tau), \ \cos(6\pi t/\tau), \ \cdots \qquad (5.80)$$

are all τ-periodic, as are the corresponding sine functions. (If t is increased by the amount τ, each of these functions returns to its original value — see Figure 5.21.) We can write these sinusoidal functions a little more compactly if we introduce the angular frequency $\omega = 2\pi/\tau$, in which case all of the functions of (5.80) and the corresponding sines can be written as

$$\cos(n\omega t) \qquad \text{and} \qquad \sin(n\omega t) \qquad [n = 0, 1, 2, \cdots]. \qquad (5.81)$$

---

[15] As usual, I shall try to describe all of the theory that we shall be needing. For more details, see, for example, *Mathematical Methods in the Physical Sciences* by Mary Boas (Wiley, 1983), Ch. 7.

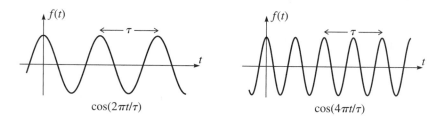

Figure 5.21   Any function of the form $\cos(2n\pi t/\tau)$ (or the corresponding sine) is periodic with period τ if n is an integer. Notice that $\cos(4\pi t/\tau)$ also has the smaller period $\tau/2$, but this doesn't change the fact that it has period τ as well.

(If $n = 0$ the cosine function is just the constant 1 — which is certainly periodic — while the sine is 0 and not at all interesting.)

That the sine and cosine functions (5.81) are all τ-periodic is reasonably obvious. (Be sure you can see this.) What is truly amazing is that, in a sense, these sine and cosine functions define *all possible* τ-periodic functions: In 1807 the French mathematician Jean Baptiste Fourier (1768–1830) realized that *every* τ-periodic function can be written as a linear combination of the sines and cosines of (5.81). That is, if $f(t)$ is any[16] periodic function with period τ then it can be expressed as the sum

$$f(t) = \sum_{n=0}^{\infty} [a_n \cos(n\omega t) + b_n \sin(n\omega t)] \qquad (5.82)$$

where the constants a_n and b_n depend on the function $f(t)$. This extraordinarily useful result is called Fourier's theorem, and the sum (5.82) is called the **Fourier series** for $f(t)$.

It is not hard to see why Fourier's theorem met with considerable surprise, and even skepticism, when he first published it. It claims that a discontinous function, such as the rectangular pulse of Figure 5.20(a), can be built up with sine and cosine functions that are continuous and perfectly smooth. Surprising or not, this turns out to be true, as we shall see by example shortly. Perhaps even more surprising, it is often the case that one gets an excellent approximation by retaining just the first few terms of a Fourier series. Thus, instead of having to handle a fairly disagreeable and possibly discontinuous function, we have only to handle a reasonably small number of sines and cosines. Before we discuss the application of Fourier's theorem to the driven oscillator, we need to look at a few properties of Fourier series.

The proof of Fourier's theorem is difficult — indeed it was many years after Fourier's discovery before a satisfactory proof was found — and I shall simply ask you to accept it. However, once the result is accepted it is easy to learn to use it. In

---

[16] As always with theorems of this kind, there are certain restrictions on the "reasonableness" of the function $f(t)$, but certainly Fourier's theorem is valid for all of the functions we shall have occasion to use.

particular, for any given periodic function $f(t)$ it is easy to find the coefficients a_n and b_n. Problem 5.48 gives you the opportunity to show that these coefficients are given by

$$a_n = \frac{2}{\tau} \int_{-\tau/2}^{\tau/2} f(t) \cos(n\omega t)\, dt \qquad [n \geq 1] \qquad (5.83)$$

and

$$b_n = \frac{2}{\tau} \int_{-\tau/2}^{\tau/2} f(t) \sin(n\omega t)\, dt \qquad [n \geq 1]. \qquad (5.84)$$

Unfortunately the coefficients for $n = 0$ require separate attention. Since the term $\sin n\omega t$ in (5.82) is identically zero for $n = 0$, the coefficient b_0 is irrelevant and we can simply define it to be zero. It is very easy to show (Problem 5.46) that

$$a_0 = \frac{1}{\tau} \int_{-\tau/2}^{\tau/2} f(t)\, dt. \qquad (5.85)$$

Armed with these formulas for the Fourier coefficients, it is easy to find the Fourier series for any given periodic function. In the following example, we do this for the rectangular pulses of Figure 5.20(a).

EXAMPLE 5.4   Fourier Series for the Rectangular Pulse

Find the Fourier series for the periodic rectangular pulse $f(t)$ shown in Figure 5.22 in terms of the period τ, the pulse height $f_{\max}$, and the duration of the pulse $\Delta \tau$. Using the values $\tau = 1$, $f_{\max} = 1$, and $\Delta \tau = 0.25$, plot $f(t)$, as well as the sum of the first three terms of its Fourier series, and the sum of the first eleven terms.

Our first task is to calculate the Fourier coefficients a_n and b_n for the given function. First according to (5.85) the constant term a_0 is

$$a_0 = \frac{1}{\tau} \int_{-\tau/2}^{\tau/2} f(t)\, dt$$

$$= \frac{1}{\tau} \int_{-\Delta\tau/2}^{\Delta\tau/2} f_{\max}\, dt = \frac{f_{\max}\, \Delta\tau}{\tau} \qquad (5.86)$$

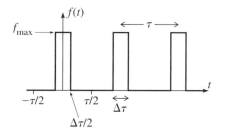

Figure 5.22   A periodic rectangular pulse. The period is τ, the duration of the pulse is $\Delta\tau$, and the pulse height is f_{max}.

where the change in the limits of integration was allowed because the integrand $f(t)$ is zero outside $\pm\Delta\tau/2$. Next, according to (5.83), all the other a coefficients ($n \geq 1$) are given by

$$a_n = \frac{2}{\tau} \int_{-\tau/2}^{\tau/2} f(t) \cos(n\omega t) \, dt$$

$$= \frac{2f_{max}}{\tau} \int_{-\Delta\tau/2}^{\Delta\tau/2} \cos(n\omega t) \, dt$$

$$= \frac{4f_{max}}{\tau} \int_0^{\Delta\tau/2} \cos\left(\frac{2\pi nt}{\tau}\right) \, dt = \frac{2f_{max}}{\pi n} \sin\left(\frac{\pi n \, \Delta\tau}{\tau}\right). \quad (5.87)$$

Notice that in passing from the second to the third line, I used a trick that is often useful in evaluating Fourier coefficients. The integrand on the second line, $\cos(n\omega t)$, is an *even* function; that is, it has the same value at any point t as at $-t$. Therefore we could replace any integral from $-T$ to T by twice the integral from 0 to T.

Finally the b coefficients are all exactly zero, for if you examine the integral (5.84) you will see that (in this case) the integrand is an *odd* function; that is, its value at any point t is the negative of its value at $-t$. [Moving from t to $-t$ leaves $f(t)$ unchanged, but reverses the sign of $\sin(n\omega t)$.] Therefore any integral from $-T$ to T is zero, since the left half exactly cancels the right.

The required Fourier series is therefore

$$f(t) = a_0 + \sum_{n=1}^{\infty} a_n \cos(n\omega t) \quad (5.88)$$

with the constant term a_0 given by (5.86) and all the remaining a coefficients ($n \geq 1$) by (5.87). If we put in the given numbers, these coefficients can all be evaluated and the resulting Fourier series is

$$f(t) = f_{max}\big[0.25 + 0.45\cos(2\pi t) + 0.32\cos(4\pi t) + 0.15\cos(6\pi t)$$

$$+ 0\cos(8\pi t) - 0.09\cos(10\pi t) - 0.11\cos(12\pi t) + \cdots\big] \quad (5.89)$$

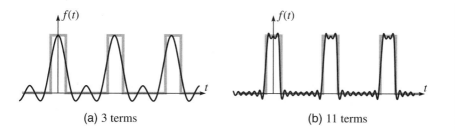

(a) 3 terms   (b) 11 terms

Figure 5.23   (a) The sum of the first three terms of the Fourier series for the rectangular pulse of Figure 5.22. (b) The sum of the first 11 terms.

The practical value of a Fourier series is usually greatest if the series converges rapidly, so that we get a reliable approximation by retaining just the first
few terms of the series. Figure 5.23(a) shows the sum of the first three terms of
the series (5.89) and the rectangular pulse itself. As you might expect, with just
three smooth terms we do not get a sensationally accurate approximation to the
original discontinuous function. Nevertheless, the three terms do a remarkable
job of imitating the general shape. By the time we have included 11 terms, as in
Figure 5.23(b), the fit is amazing.[17] In the next section, we shall use the method
of Fourier series to solve for the motion of an oscillator driven by the periodic
pulses of this example. We shall find the solution as a Fourier series, which converges so quickly that just the first 3 or 4 terms tell us most of what is interesting
to know.

5.8   Fourier Series Solution for the Driven Oscillator*

*This section contains a beautiful application of the method of Fourier series. Important as it
is to understand this method, you can nevertheless omit the section without loss of continuity.*

In this section we shall combine our knowledge of Fourier series (Section 5.7) with
our solution of the sinusoidally driven oscillator (Section 5.5) to solve for the motion
of an oscillator that is driven by an arbitrary periodic driving force. To see how this
works, let us return to the equation of motion (5.48)

$$\ddot{x} + 2\beta\dot{x} + \omega_0^2 x = f$$

where $x = x(t)$ is the position of the oscillator, β is the damping constant, ω_0 is the
natural frequency, and $f = f(t)$ is any periodic driving force (actually force/mass)
with period τ. As before it is convenient to rewrite this in the compact form

$$Dx = f$$

---

[17] Notice, however, that the Fourier series still has a little difficulty in the immediate neighborhood of the discontinuities in $f(t)$. This tendency for the Fourier series to overshoot near a
discontinuity is called the Gibbs phenomenon.

where D stands for the linear differential operator

$$D = \frac{d^2}{dt^2} + 2\beta\frac{d}{dt} + \omega_o^2.$$

The use of Fourier series to solve this problem hinges on the following observation: Suppose that the force $f(t)$ is the sum of two forces, $f(t) = f_1(t) + f_2(t)$, for each of which we have already solved the equation of motion. That is, we already know functions $x_1(t)$ and $x_2(t)$ that satisfy

$$Dx_1 = f_1 \quad \text{and} \quad Dx_2 = f_2.$$

Then the solution[18] to the problem of interest is just the sum $x(t) = x_1(t) + x_2(t)$, as we can easily show:

$$Dx = D(x_1 + x_2) = Dx_1 + Dx_2 = f_1 + f_2 = f$$

where the crucial second step is valid because D is linear. This argument would work equally well however many terms were in the sum for $f(t)$, so we have the conclusion: If the driving force $f(t)$ can be expressed as the sum of any number of terms

$$f(t) = \sum_n f_n(t)$$

and if we know the solutions $x_n(t)$ for each of the individual forces $f_n(t)$, then the solution for the total driving force $f(t)$ is just the sum

$$x(t) = \sum_n x_n(t).$$

This result is ideally suited for use in combination with Fourier's theorem. Any periodic driving force $f(t)$ can be expanded in a Fourier series of sines and cosines, and we already know the solutions for sinusoidal driving forces. Thus, by adding these sinusoidal solutions together, we can find the solution for any periodic driving force. To simplify our writing, let us suppose that the driving force $f(t)$ contains only cosine terms in its Fourier series. [This was the case for the rectangular pulse of Example 5.4, and is true for any even function — satisfying $f(-t) = f(t)$ — because this condition guarantees that the coefficients of the sine terms are all zero.] In this case, the driving force can be written as

$$f(t) = \sum_{n=0}^{\infty} f_n \cos(n\omega t) \tag{5.90}$$

---

[18] Strictly speaking we should not speak of *the* solution, since our second-order differential equation has many solutions. However, we know that the difference between any two solutions is transient — decays to zero — and our main interest is in the long-term behavior, which is therefore essentially unique.

where f_n denotes the nth Fourier coefficient of $f(t)$, and $\omega = 2\pi/\tau$ as usual. Now, each individual term $f_n \cos(n\omega t)$ has the same form (5.56) that we assumed for the sinusoidal driving force in Section 5.5 (except that the amplitude f_o has become f_n and the frequency ω has become $n\omega$). The corresponding solution was given in (5.66),[19]

$$x_n(t) = A_n \cos(n\omega t - \delta_n) \tag{5.91}$$

where

$$A_n = \frac{f_n}{\sqrt{(\omega_0^2 - n^2\omega^2)^2 + 4\beta^2 n^2 \omega^2}} \tag{5.92}$$

from (5.64), and

$$\delta_n = \arctan\left(\frac{2\beta n\omega}{\omega_0^2 - n^2\omega^2}\right) \tag{5.93}$$

from (5.65). Since (5.91) is the solution for the driving force $f_n \cos(n\omega t)$, the solution for the complete force (5.90) is the sum

$$x(t) = \sum_{n=0}^{\infty} A_n \cos(n\omega t - \delta_n). \tag{5.94}$$

This completes the solution for the long-term motion of an oscillator driven by a periodic driving force $f(t)$. To summarize, the steps are:

1. Find the coefficients f_n in the Fourier series (5.90) for the given driving force $f(t)$.
2. Calculate the quantities A_n and δ_n as given by (5.92) and (5.93).
3. Write down the solution $x(t)$ as the Fourier series (5.94).

In practice, one usually needs to include surprisingly few terms of the solution (5.94) to get a satisfactory approximation, as the following example illustrates.[20]

EXAMPLE 5.5   An Oscillator Driven by a Rectangular Pulse

Consider a weakly damped oscillator that is being driven by the periodic rectangular pulses of Example 5.4 (Figure 5.22). Let the natural period of the oscillator be $\tau_0 = 1$, so that the natural frequency is $\omega_0 = 2\pi$, and let the damping constant

---

[19] The $n = 0$, constant term needs separate consideration. It is easy to see that for a constant force f_0 the solution is $x_0 = f_0/\omega_0^2$. This is actually exactly what you get if you just set $n = 0$ in (5.92) and (5.93).

[20] The solution contained in Equations (5.92) to (5.94) can be written more compactly if you don't mind using complex notation. See Problem 5.51.

be $\beta = 0.2$. Let the pulse last for a time $\Delta\tau = 0.25$ and have a height $f_{\max} = 1$. Calculate the first six Fourier coefficients A_n for the long-term motion $x(t)$ of the oscillator, assuming first that the drive period is the same as the natural period, $\tau = \tau_o = 1$. Plot the resulting motion for several complete oscillations. Repeat these exercises for $\tau = 1.5\tau_o$, $2.0\tau_o$, and $2.5\tau_o$.

Before we look at any of these exercises, it is worth thinking of a real system this problem might represent. One simple possibility is a mass hanging from the end of a spring, to which a professor is applying regularly spaced upward taps at intervals τ. An even more familiar example would be a child in a swing, to whom a parent is giving regularly spaced pushes — though in this case we need to be careful to keep the amplitude small to justify the use of Hooke's law. We are told to start by taking $\tau = \tau_o = 1$; that is, the parent is pushing the child at exactly the natural frequency.

The Fourier coefficients f_n of the driving force were already calculated in Equations (5.86) and (5.87) of Example 5.4 (where they were called a_n). If we substitute these into (5.92) for the coefficients A_n and put in the given numbers (including $\tau = \tau_o = 1$), we find for the first six Fourier coefficients $A_0, \cdots, A_5$:

A_0	A_1	A_2	A_3	A_4	A_5
63	1791	27	5	0	-1

(Since the numbers are rather small, I have quoted the values multiplied by 10^4; that is, $A_0 = 63 \times 10^{-4}$, $A_1 = 1791 \times 10^{-4}$, and so on.) Two things stand out about these numbers: First, after A_1, they get rapidly smaller, and for almost all purposes it would be an excellent approximation to ignore all but the first three terms in the Fourier series for $x(t)$. Second, the coefficient A_1 is vastly bigger than all the rest. This is easy to understand if you look at (5.92) for the coefficient A_1: Since $\omega = \omega_o$ (remember the parent is pushing the child at the swing's natural frequency) and $n = 1$, the first term in the denominator is exactly zero, the denominator is anomalously small, and A_1 is anomalously big compared to all the other coefficients. In other words, when the driving frequency is the same as the natural frequency, the $n = 1$ term in the Fourier series for $x(t)$ is at *resonance*, and the oscillator responds especially strongly with frequency ω_o.

Before we can plot $x(t)$ as given by (5.94), we need to calculate the phase shifts δ_n using (5.93). This is easily done, though I shan't waste space displaying the results. We cannot actually plot the infinite series (5.94); instead, we must pick some finite number of terms with which to approximate $x(t)$. In the present case, it seems clear that three terms would be plenty, but to be on the safe side I'll use six. Figure 5.24 shows $x(t)$ as approximated by the sum of the first six terms in (5.94). At the scale shown, this approximate graph is completely indistinguishable from the exact result, which in turn is indistinguishable from a pure cosine[21] with frequency equal to the natural frequency of the oscillator. The strong response at the natural frequency is just what we would expect. For

---

[21] Actually it's a pure *sine*, but this is really $\cos(\omega t - \delta_1)$ with $\delta_1 = \pi/2$ as we should have expected because we're exactly on resonance.

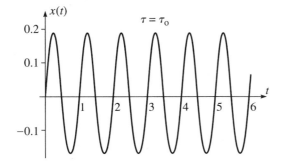

Figure 5.24   The motion of a linear oscillator, driven by periodic rectangular pulses, with the drive period τ equal to the natural period τ_0 of the oscillator (and hence $\omega = \omega_0$). The horizontal axis shows time in units of the natural period τ_0. As expected the motion is almost perfectly sinusoidal, with period equal to the natural period.

instance, anyone who has pushed a child on a swing knows that the most efficient way to get the child swinging high is to administer regularly spaced pushes at intervals of the natural period — that is, $\tau = \tau_0$ — and that the swing will then oscillate vigorously at its natural frequency.

A driving force with any other period τ can be treated in exactly the same way. The Fourier coefficients $A_0, \cdots, A_5$ for all of the values of τ requested above are shown in Table 5.1.

Table 5.1   The first six Fourier coefficients A_n for the motion $x(t)$ of a linear oscillator driven by periodic rectangular pulses, for four different drive periods $\tau = \tau_0$, $1.5\tau_0$, $2.0\tau_0$, and $2.5\tau_0$. All values have been multiplied by 10^4.

	A_0	A_1	A_2	A_3	A_4	A_5
$\tau = 1.0\,\tau_0$	63	1791	27	5	0	−1
$\tau = 1.5\,\tau_0$	42	145	89	18	6	2
$\tau = 2.0\,\tau_0$	32	82	896	40	13	6
$\tau = 2.5\,\tau_0$	25	59	130	97	25	11

The entries in the four rows of this table deserve careful examination. The first row ($\tau = \tau_0$) shows the coefficients already discussed, the most prominent feature of which is that the $n = 1$ coefficient is far the largest, because it is exactly on resonance. In the next row ($\tau = 1.5\tau_0$), the $n = 1$ Fourier component has moved well away from resonance, and A_1 has dropped by a factor of 12 or so. Some of the other coefficients have increased a bit, but the net effect is that the oscillator moves much less than when $\tau = \tau_0$. This is clearly visible in Figures 5.25(a) and (b), which show $x(t)$ (as approximated by the first six terms of its Fourier series) for these two values of the drive period.

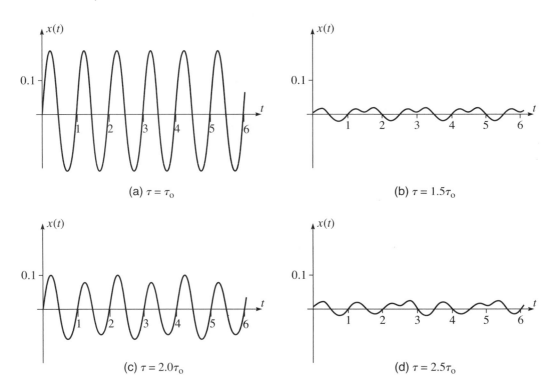

Figure 5.25   The motion of a linear oscillator, driven by periodic rectangular pulses, showing four different values of the drive period τ. **(a)** When the oscillator is driven at its natural frequency ($\tau = \tau_o$), the $n = 1$ term of the Fourier series is at resonance and the oscillator responds energetically. **(b)** When $\tau = 1.5\tau_o$, the response is feeble. **(c)** When $\tau = 2.0\tau_o$, the $n = 2$ term is at resonance and the response is strong again. **(d)** When $\tau = 2.5\tau_o$, the response is again weak.

The third row of Table 5.1 shows the Fourier coefficients for a drive period equal to twice the natural period — the parent is pushing just once every second swing of the child. Now, with $\tau = 2.0\tau_o$, the drive frequency is half the natural frequency ($\omega = \frac{1}{2}\omega_o$). This means that the $n = 2$ Fourier component, with frequency $2\omega = \omega_o$, is exactly on resonance, and the coefficient A_2 is anomalously large. Once again we get a large response, as seen in Figure 5.25(c).

Let us look a little closer at the case of Figure 5.25(c). It is, of course, a matter of experience that a perfectly satisfactory way to get a child swinging is to push once every *two* swings, although this naturally doesn't get quite the result of pushing once for *every* swing. If you look carefully at Figure 5.25(c), you will notice that the swings alternate in size — the even swings are slightly bigger than the odd ones. This too is to be expected: Because the oscillator is damped, the second swing after each push is bound to be a little smaller than the first.

Finally, when $\tau = 2.5\tau_o$, the $n = 2$ Fourier component is well past resonance and A_2 is much smaller again. On the other hand, the $n = 3$ component is approaching resonance so that A_3 is getting bigger. In fact, A_2 and A_3 are roughly the same size, so that $x(t)$ contains two dominant Fourier components and shows the somewhat kinky behavior seen in Figure 5.25(d). (Similar considerations apply to the case $\tau = 1.5\tau_0$, where the coefficients A_1 and A_2 are both fairly large.)

5.9   The RMS Displacement; Parseval's Theorem *

The Parseval relation, which we introduce and apply in this section, is one of the most useful properties of Fourier series. Nevertheless, you could omit this section if pressed for time.

In the last section we studied how the response of an oscillator varied with the frequency of the applied periodic driving force. We did this by solving for the motion $x(t)$, using the method of Fourier series, for each of several interesting applied frequencies. It would be convenient if we could find a single number to measure the oscillator's response and then just plot this number against the driving frequency (or driving period). In fact there are several ways to do this. Perhaps the most obvious thing to try would be the oscillator's average displacement from equilibrium, $\langle x \rangle$. (I am using angle brackets $\langle \, \rangle$ to indicate a time average.) Unfortunately, since the oscillator spends as much time in any region of positive x as in the corresponding region of negative x, the average $\langle x \rangle$ is zero.[22] To get around this difficulty the most convenient quantity to use is the *mean square* displacement $\langle x^2 \rangle$, and to give a quantity with the dimensions of length we usually discuss the **root mean square** or **RMS** displacement

$$x_{\text{rms}} = \sqrt{\langle x^2 \rangle}. \tag{5.95}$$

The definition of the time average needs a little care. The usual practice is to define $\langle \, \rangle$ as the average *over one period* τ. Thus,

$$\langle x^2 \rangle = \frac{1}{\tau} \int_{-\tau/2}^{\tau/2} x^2 \, dt. \tag{5.96}$$

Because the motion is periodic, this is the same as the average over any integer number of periods, and hence also the average over any long time. (If this isn't clear to you, see Problem 5.54.)

To evaluate the average $\langle x^2 \rangle$, we use the Fourier expansion (5.94) of $x(t)$

$$x(t) = \sum_{n=0}^{\infty} A_n \cos(n\omega t - \delta_n). \tag{5.97}$$

(In general, this series will contain sines as well as cosines, but in Example 5.5 the driving force contained only cosines, and for simplicity let us continue to assume this

---

[22] Not quite true. The small constant term A_0 in (5.94) contributes a nonzero average $\langle x \rangle = A_0$, but this does not reflect any *oscillation*, which is what we are trying to characterize.

is the case. For the general case, see Problem 5.56.) Substituting for each of the factors x in (5.96), we get the appalling double sum

$$\langle x^2 \rangle = \frac{1}{\tau} \int_{-\tau/2}^{\tau/2} \sum_n \sum_m A_n \cos(n\omega t - \delta_n) A_m \cos(m\omega t - \delta_m) \, dt. \quad (5.98)$$

Fortunately, this simplifies dramatically. It is fairly easy to show (Problem 5.55) that the integral is just

$$\int_{-\tau/2}^{\tau/2} \cos(n\omega t - \delta_n)\cos(m\omega t - \delta_m)\, dt = \begin{cases} \tau & \text{if } m = n = 0 \\ \tau/2 & \text{if } m = n \neq 0 \\ 0 & \text{if } m \neq n. \end{cases} \quad (5.99)$$

Thus, in the double sum (5.98), only those terms with $m = n$ need to be retained, and we get the surprisingly simple result that

$$\langle x^2 \rangle = A_0^2 + \tfrac{1}{2} \sum_{n=1}^{\infty} A_n^2. \quad (5.100)$$

This relation is called **Parseval's theorem**.[23] It has many important theoretical uses, but for our purposes its main application is this: Since we know how to calculate the coefficients A_n, Parseval's theorem lets us find the response $\langle x^2 \rangle$ of our oscillator. Moreover, by dropping all but some modest finite number of terms in the sum (5.100), we get an excellent and easily calculated approximation for $\langle x^2 \rangle$, as the following example illustrates.

EXAMPLE 5.6   The RMS Displacement for a Driven Oscillator

Consider again the oscillator of Example 5.5, driven by the periodic rectangular pulses of Example 5.4 (Figure 5.22). Find the RMS displacement $x_{\text{rms}} = \sqrt{\langle x^2 \rangle}$ as given by (5.100) for this oscillator. Using the same numerical values as before ($\tau_0 = 1$, $\beta = 0.2$, $f_{\max} = 1$, $\Delta\tau = 0.25$) and approximating (5.100) by its first six terms, make a plot of x_{rms} as a function of the drive period τ for $0.25 < \tau < 5.5$.

We have already done all the calculations needed to write down a formula for $x_{\text{rms}} = \sqrt{\langle x^2 \rangle}$. First, $\langle x^2 \rangle$ is given by (5.100), where the Fourier coefficients A_n are given by (5.92) as

$$A_n = \frac{f_n}{\sqrt{(\omega_0^2 - n^2\omega^2)^2 + 4\beta^2 n^2\omega^2}} \quad (5.101)$$

---

[23] Remember that we made the simplifying assumption that our Fourier series contained only cosine terms. In general, the sum in (5.100) must include contributions B_n^2 from the sine terms as well. See Problem 5.56.

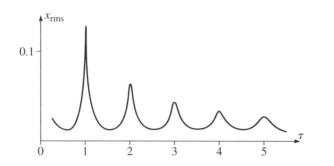

Figure 5.26   The RMS displacement of a linear oscillator, driven by periodic rectangular pulses, as a function of the drive period τ — calculated using the first six terms of the Parseval expression (5.100). The horizontal axis shows τ in units of the natural period τ_o. When τ is an integral multiple of τ_o the response is especially strong.

and the Fourier coefficients f_n of the driving force are given by (5.87) and (5.86) as

$$f_n = \frac{2 f_{max}}{\pi n} \sin\left(\frac{\pi n \Delta \tau}{\tau}\right), \qquad [\text{for } n \geq 1] \qquad (5.102)$$

while $f_0 = f_{max} \Delta \tau / \tau$. Putting all these together gives the desired formula for x_{rms} (which I'll leave you to write down if you want to see it).

   If we now put in the given numbers, we are left with just one independent variable, the period of the driving force τ. (Remember that $\omega = 2\pi/\tau$.) Truncating the infinite series (5.100) after six terms, we arrive at an expression that is easily evaluated with the appropriate software (or even a programmable calculator) and plotted as shown in Figure 5.26. This graph shows clearly and succinctly what we found in the previous example. As we increase the drive period τ, the response of the oscillator varies dramatically. Each time τ passes through an integer multiple of the natural period τ_o (that is, $\tau = n\tau_o$), the response exhibits a sharp maximum, because the nth Fourier component is at resonance. On the other hand, each successive peak is lower than its predecessor, since we elected to fix the width $\Delta \tau$ and height f_{max} of the pulses; thus as the drive period gets longer, the net effect of the force would be expected to get less.

Principal Definitions and Equations of Chapter 5

Hooke's Law

$$F = -kx \iff U = \tfrac{1}{2}kx^2 \qquad [\text{Section 5.1}]$$

Simple Harmonic Motion

$$\ddot{x} = -\omega^2 x \iff x(t) = A\cos(\omega t - \delta), \text{ etc.} \qquad \text{[Section 5.2]}$$

Damped Oscillations

If the oscillator is subject to a damping force $-bv$, then

$$\ddot{x} + 2\beta\dot{x} + \omega_o^2 x = 0 \iff x(t) = Ae^{-\beta t}\cos(\omega_1 t - \delta) \qquad \text{[Eqs. (5.28) \& (5.38)]}$$

where $\beta = b/2m$, $\omega_o = \sqrt{k/m}$, $\omega_1 = \sqrt{\omega_o^2 - \beta^2}$, and the solution given here is for "weak damping" ($\beta < \omega_o$).

Driven Damped Oscillations and Resonance

If the oscillator is also subject to a sinusoidal driving force $F(t) = mf_o\cos(\omega t)$, the long-term motion has the form

$$x(t) = A\cos(\omega t - \delta) \qquad \text{[Eq. (5.66)]}$$

where

$$A^2 = \frac{f_o^2}{(\omega_o^2 - \omega^2)^2 + 4\beta^2\omega^2} \qquad \text{[Eq. (5.64)]}$$

and the phase shift δ is given by (5.65). To this solution can be added a "transient" solution of the corresponding homogeneous equation, but this dies out as time passes. The long-term solution "resonates" (has a sharp maximum) when ω is close to ω_o.

Fourier Series

If the driving force is not sinusoidal, but is still periodic, it can be built up as a Fourier series of sinusoidal terms, as in (5.90), and the resulting motion is the corresponding series of sinusoidal solutions, as in (5.94):

$$x(t) = \sum_{n=0}^{\infty} A_n \cos(n\omega t - \delta_n). \qquad \text{[Eq. (5.94)]}$$

The RMS Displacement

The root mean square displacement

$$x_{\text{rms}} = \sqrt{\frac{1}{\tau} \int_0^\tau x^2 dt} \qquad \text{[Eqs. (5.95) \& (5.96)]}$$

is a good measure of the average response of the oscillator and is given by Parseval's theorem as

$$x_{rms} = \sqrt{A_0^2 + \frac{1}{2} \sum_{n=1}^{\infty} A_n^2}.$$ [Eq. (5.100)]

Problems for Chapter 5

Stars indicate the approximate level of difficulty, from easiest (★) to most difficult (★★★).

SECTION 5.1   Hooke's Law

5.1 ★ A massless spring has unstretched length l_o and force constant k. One end is now attached to the ceiling and a mass m is hung from the other. The equilibrium length of the spring is now l_1. **(a)** Write down the condition that determines l_1. Suppose now the spring is stretched a further distance x beyond its new equilibrium length. Show that the net force (spring plus gravity) on the mass is $F = -kx$. That is, the net force obeys Hooke's law, when x is the distance from the equilibrium position — a very useful result, which lets us treat a mass on a vertical spring just as if it were horizontal. **(b)** Prove the same result by showing that the net potential energy (spring plus gravity) has the form $U(x) = \text{const} + \frac{1}{2}kx^2$.

5.2 ★ The potential energy of two atoms in a molecule can sometimes be approximated by the Morse function,

$$U(r) = A\left[\left(e^{(R-r)/S} - 1\right)^2 - 1\right]$$

where r is the distance between the two atoms and A, R, and S are positive constants with $S \ll R$. Sketch this function for $0 < r < \infty$. Find the equilibrium separation r_o, at which $U(r)$ is minimum. Now write $r = r_o + x$ so that x is the displacement from equilibrium, and show that, for small displacements, U has the approximate form $U = \text{const} + \frac{1}{2}kx^2$. That is, Hooke's law applies. What is the force constant k?

5.3 ★ Write down the potential energy $U(\phi)$ of a simple pendulum (mass m, length l) in terms of the angle ϕ between the pendulum and the vertical. (Choose the zero of U at the bottom.) Show that, for small angles, U has the Hooke's law form $U(\phi) = \frac{1}{2}k\phi^2$, in terms of the coordinate ϕ. What is k?

5.4 ★★ An unusual pendulum is made by fixing a string to a horizontal cylinder of radius R, wrapping the string several times around the cylinder, and then tying a mass m to the loose end. In equilibrium the mass hangs a distance l_o vertically below the edge of the cylinder. Find the potential energy if the pendulum has swung to an angle ϕ from the vertical. Show that for small angles, it can be written in the Hooke's law form $U = \frac{1}{2}k\phi^2$. Comment on the value of k.

SECTION 5.2   Simple Harmonic Motion

5.5 ★ In Section 5.2 we discussed four equivalent ways to represent simple harmonic motion in one dimension:

$$x(t) = C_1 e^{i\omega t} + C_2 e^{-i\omega t} \qquad \text{(I)}$$

$$= B_1 \cos(\omega t) + B_2 \sin(\omega t) \quad \text{(II)}$$

$$= A \cos(\omega t - \delta) \qquad \text{(III)}$$

$$= \text{Re}\, C e^{i\omega t} \qquad \text{(IV)}$$

To make sure you understand all of these, show that they are equivalent by proving the following implications: I $\Rightarrow$ II $\Rightarrow$ III $\Rightarrow$ IV $\Rightarrow$ I. For each form, give an expression for the constants (C_1, C_2, etc.) in terms of the constants of the previous form.

5.6 ⋆ A mass on the end of a spring is oscillating with angular frequency ω. At $t = 0$, its position is $x_o > 0$ and I give it a kick so that it moves back toward the origin and executes simple harmonic motion with amplitude $2x_o$. Find its position as a function of time in the form (III) of Problem 5.5.

5.7 ⋆ **(a)** Solve for the coefficients B_1 and B_2 of the form (II) of Problem 5.5 in terms of the initial position x_o and velocity v_o at $t = 0$. **(b)** If the oscillator's mass is $m = 0.5$ kg and the force constant is $k = 50$ N/m, what is the angular frequency ω? If $x_o = 3.0$ m and $v_o = 50$ m/s, what are B_1 and B_2? Sketch $x(t)$ for a couple of cycles. **(c)** What are the earliest times at which $x = 0$ and at which $\dot{x} = 0$?

5.8 ⋆ **(a)** If a mass $m = 0.2$ kg is tied to one end of a spring whose force constant $k = 80$ N/m and whose other end is held fixed, what are the angular frequency ω, the frequency f, and the period τ of its oscillations? **(b)** If the initial position and velocity are $x_o = 0$ and $v_o = 40$ m/s, what are the constants A and δ in the expression $x(t) = A \cos(\omega t - \delta)$?

5.9 ⋆ The maximum displacement of a mass oscillating about its equilibrium position is 0.2 m, and its maximum speed is 1.2 m/s. What is the period τ of its oscillations?

5.10 ⋆ The force on a mass m at position x on the x axis is $F = -F_o \sinh \alpha x$, where F_o and α are positive constants. Find the potential energy $U(x)$, and give an approximation for $U(x)$ suitable for small oscillations. What is the angular frequency of such oscillations?

5.11 ⋆ You are told that, at the known positions x_1 and x_2, an oscillating mass m has speeds v_1 and v_2. What are the amplitude and the angular frequency of the oscillations?

5.12 ⋆⋆ Consider a simple harmonic oscillator with period τ. Let $\langle f \rangle$ denote the average value of any variable $f(t)$, averaged over one complete cycle:

$$\langle f \rangle = \frac{1}{\tau} \int_0^\tau f(t)\, dt. \qquad (5.103)$$

Prove that $\langle T \rangle = \langle U \rangle = \frac{1}{2} E$ where E is the total energy of the oscillator. [*Hint:* Start by proving the more general, and extremely useful, results that $\langle \sin^2(\omega t - \delta) \rangle = \langle \cos^2(\omega t - \delta) \rangle = \frac{1}{2}$. Explain why these two results are almost obvious, then prove them by using trig identities to rewrite $\sin^2 \theta$ and $\cos^2 \theta$ in terms of $\cos(2\theta)$.]

5.13 ⋆⋆ The potential energy of a one-dimensional mass m at a distance r from the origin is

$$U(r) = U_o \left(\frac{r}{R} + \lambda^2 \frac{R}{r} \right)$$

for $0 < r < \infty$, with U_o, R, and λ all positive constants. Find the equilibrium position r_o. Let x be the distance from equilibrium and show that, for small x, the PE has the form $U = \text{const} + \frac{1}{2} k x^2$. What is the angular frequency of small oscillations?

Figure 5.27   Problem 5.18

SECTION 5.3   Two-Dimensional Oscillators

5.14 ★ Consider a particle in two dimensions, subject to a restoring force of the form (5.21). (The two constants k_x and k_y may or may not be equal; if they are, the oscillator is isotropic.) Prove that its potential energy is (with $U = 0$ at the origin)

$$U = \tfrac{1}{2}(k_x x^2 + k_y y^2). \tag{5.104}$$

5.15 ★ The general solution for a two-dimensional isotropic oscillator is given by (5.19). Show that by changing the origin of time you can cast this in the simpler form (5.20) with $\delta = \delta_y - \delta_x$. [*Hint*: A change of origin of time is a change of variables from t to $t' = t + t_0$. Make this change and choose the constant t_0 appropriately, then rename t' to be t.]

5.16 ★ Consider a two-dimensional isotropic oscillator moving according to Equation (5.20). Show that if the relative phase is $\delta = \pi/2$, the particle moves in an ellipse with semimajor and semiminor axes A_x and A_y.

5.17 ★★ Consider the two-dimensional anisotropic oscillator with motion given by Equation (5.23). **(a)** Prove that if the ratio of frequencies is rational (that is, $\omega_x/\omega_y = p/q$ where p and q are integers) then the motion is periodic. What is the period? **(b)** Prove that if the same ratio is irrational, the motion never repeats itself.

5.18 ★★★ The mass shown from above in Figure 5.27 is resting on a frictionless horizontal table. Each of the two identical springs has force constant k and unstretched length l_0. At equilibrium the mass rests at the origin, and the distances a are not necessarily equal to l_0. (That is, the springs may already be stretched or compressed.) Show that when the mass moves to a position (x, y), with x and y small, the potential energy has the form (5.104) (Problem 5.14) for an anisotropic oscillator. Show that if $a < l_0$ the equilibrium at the origin is unstable and explain why.

5.19 ★★★ Consider the mass attached to four identical springs, as shown in Figure 5.7(b). Each spring has force constant k and unstretched length l_0, and the length of each spring when the mass is at its equilibrium at the origin is a (not necessarily the same as l_0). When the mass is displaced a small distance to the point (x, y), show that its potential energy has the form $\tfrac{1}{2}k'r^2$ appropriate to an isotropic harmonic oscillator. What is the constant k' in terms of k? Give an expression for the corresponding force.

SECTION 5.4   Damped Oscillations

5.20 ★ Verify that the decay parameter $\beta - \sqrt{\beta^2 - \omega_0^2}$ for an overdamped oscillator ($\beta > \omega_0$) *decreases* with increasing β. Sketch its behavior for $\omega_0 < \beta < \infty$.

5.21 ★ Verify that the function (5.43), $x(t) = te^{-\beta t}$, is indeed a second solution of the equation of motion (5.28) for a critically damped oscillator ($\beta = \omega_0$).

5.22 ★ **(a)** Consider a cart on a spring which is critically damped. At time $t = 0$, it is sitting at its equilibrium position and is kicked in the positive direction with velocity v_0. Find its position $x(t)$ for all subsequent times and sketch your answer. **(b)** Do the same for the case that it is released from rest at position $x = x_o$. In this latter case, how far is the cart from equilibrium after a time equal to $\tau_o = 2\pi/\omega_o$, the period in the absence of any damping?

5.23 ★ A damped oscillator satisfies the equation (5.24), where $F_{dmp} = -b\dot{x}$ is the damping force. Find the rate of change of the energy $E = \frac{1}{2}m\dot{x}^2 + \frac{1}{2}kx^2$ (by straightforward differentiation), and, with the help of (5.24), show that dE/dt is (minus) the rate at which energy is dissipated by F_{dmp}.

5.24 ★ In our discussion of critical damping ($\beta = \omega_o$), the second solution (5.43) was rather pulled out of a hat. One can arrive at it in a reasonably systematic way by looking at the solutions for $\beta < \omega_o$ and carefully letting $\beta \to \omega_o$, as follows: For $\beta < \omega_o$, we can write the two solutions as $x_1(t) = e^{-\beta t}\cos(\omega_1 t)$ and $x_2(t) = e^{-\beta t}\sin(\omega_1 t)$. Show that as $\beta \to \omega_o$, the first of these approaches the first solution for critical damping, $x_1(t) = e^{-\beta t}$. Unfortunately, as $\beta \to \omega_o$, the second of them goes to zero. (Check this.) However, as long as $\beta \neq \omega_o$, you can divide $x_2(t)$ by ω_1 and you will still have a perfectly good second solution. Show that as $\beta \to \omega_o$, this new second solution approaches the advertised $te^{-\beta t}$.

5.25 ★★ Consider a damped oscillator with $\beta < \omega_o$. There is a little difficulty defining the "period" τ_1 since the motion (5.38) is not periodic. However, a definition that makes sense is that τ_1 is the time between successive maxima of $x(t)$. **(a)** Make a sketch of $x(t)$ against t and indicate this definition of τ_1 on your graph. Show that $\tau_1 = 2\pi/\omega_1$. **(b)** Show that an equivalent definition is that τ_1 is twice the time between successive zeros of $x(t)$. Show this one on your sketch. **(c)** If $\beta = \omega_o/2$, by what factor does the amplitude shrink in one period?

5.26 ★★ An undamped oscillator has period $\tau_o = 1.000$ s, but I now add a little damping so that its period changes to $\tau_1 = 1.001$ s. What is the damping factor β? By what factor will the amplitude of oscillation decrease after 10 cycles? Which effect of damping would be more noticeable, the change of period or the decrease of the amplitude?

5.27 ★★ As the damping on an oscillator is increased there comes a point when the name "oscillator" seems barely appropriate. **(a)** To illustrate this, prove that a critically damped oscillator can never pass through the origin $x = 0$ more than once. **(b)** Prove the same for an overdamped oscillator.

5.28 ★★ A massless spring is hanging vertically and unloaded, from the ceiling. A mass is attached to the bottom end and released. How close to its final resting position is the mass after 1 second, given that it finally comes to rest 0.5 meters below the point of release and that the motion is critically damped?

5.29 ★★ An undamped oscillator has period $\tau_o = 1$ second. When weak damping is added, it is found that the amplitude of oscillation drops by 50% in one period τ_1. (The period of the damped oscillations is defined as the time between successive maxima, $\tau_1 = 2\pi/\omega_1$. See Problem 5.25.) How big is β compared to ω_o? What is τ_1?

5.30 ★★ The position $x(t)$ of an overdamped oscillator is given by (5.40). **(a)** Find the constants C_1 and C_2 in terms of the initial position x_o and velocity v_o. **(b)** Sketch the behavior of $x(t)$ for the two cases that $v_o = 0$ and that $x_o = 0$. **(c)** To illustrate again how mathematics is sometimes cleverer than we (and check your answer), show that if you let $\beta \to 0$, your solution for $x(t)$ in part (a) approaches the correct solution for undamped motion.

5.31 ★★ [Computer] Consider a cart on a spring with natural frequency $\omega_o = 2\pi$, which is released from rest at $x_o = 1$ and $t = 0$. Using appropriate graphing software, plot the position $x(t)$ for $0 < t < 2$

and for damping constants $\beta = 0, 1, 2, 4, 6, 2\pi, 10$, and 20. [Remember that $x(t)$ is given by different formulas for $\beta < \omega_0$, $\beta = \omega_0$, and $\beta > \omega_0$.]

5.32 ★★ [Computer] Consider an underdamped oscillator (such as a mass on the end of a spring) that is released from rest at position x_0 at time $t = 0$. **(a)** Find the position $x(t)$ at later times in the form

$$x(t) = e^{-\beta t}[B_1 \cos(\omega_1 t) + B_2 \sin(\omega_1 t)].$$

That is, find B_1 and B_2 in terms of x_0. **(b)** Now show that if you let β approach the critical value ω_0, your solution automatically yields the critical solution. **(c)** Using appropriate graphing software, plot the solution for $0 \leq t \leq 20$, with $x_0 = 1$, $\omega_0 = 1$, and $\beta = 0, 0.02, 0.1, 0.3$, and 1.

SECTION 5.5   Driven Damped Oscillations

5.33 ★ The solution for $x(t)$ for a driven, underdamped oscillator is most conveniently found in the form (5.69). Solve that equation and the corresponding expression for $\dot{x}$, to give the coefficients B_1 and B_2 in terms of A, δ, and the initial position and velocity x_0 and v_0. Verify the expressions given in (5.70).

5.34 ★ Suppose that you have found a particular solution $x_p(t)$ of the inhomogeneous equation (5.48) for a driven damped oscillator, so that $Dx_p = f$ in the operator notation of (5.49). Suppose also that $x(t)$ is any other solution, so that $Dx = f$. Prove that the difference $x - x_p$ must satisfy the corresponding homogeneous equation, $D(x - x_p) = 0$. This is an alternative proof that *any* solution x of the inhomogeneous equation can be written as the sum of your particular solution plus a homogeneous solution; that is, $x = x_p + x_h$.

5.35 ★★ This problem is to refresh your memory about some properties of complex numbers needed at several points in this chapter, but especially in deriving the resonance formula (5.64). **(a)** Prove that any complex number $z = x + iy$ (with x and y real) can be written as $z = re^{i\theta}$ where r and θ are the polar coordinates of z in the complex plane. (Remember Euler's formula.) **(b)** Prove that the absolute value of z, defined as $|z| = r$, is also given by $|z|^2 = zz^*$, where z^* denotes the *complex conjugate* of z, defined as $z^* = x - iy$. **(c)** Prove that $z^* = re^{-i\theta}$. **(d)** Prove that $(zw)^* = z^*w^*$ and that $(1/z)^* = 1/z^*$. **(e)** Deduce that if $z = a/(b + ic)$, with a, b, and c real, then $|z|^2 = a^2/(b^2 + c^2)$.

5.36 ★★ [Computer] Repeat the calculations of Example 5.3 (page 185) with all the same parameters, but with the initial conditions $x_0 = 2$ and $v_0 = 0$. Plot $x(t)$ for $0 \leq t \leq 4$ and compare with the plot of Example 5.3. Explain the similarities and differences.

5.37 ★★ [Computer] Repeat the calculations of Example 5.3 (page 185) but with the following parameters

$$\omega = 2\pi, \qquad \omega_0 = 0.25\omega, \qquad \beta = 0.2\omega_0, \qquad f_0 = 1000$$

and with the initial conditions $x_0 = 0$ and $v_0 = 0$. Plot $x(t)$ for $0 \leq t \leq 12$ and compare with the plot of Example 5.3. Explain the similarities and differences. (It will help your explanation if you plot the homogeneous solution as well as the complete solution — homogeneous plus particular.)

5.38 ★★ [Computer] Repeat the calculations of Example 5.3 (page 185) but take the parameters of the system to be $\omega = \omega_0 = 1$, $\beta = 0.1$, and $f_0 = 0.4$, with the initial conditions $x_0 = 0$ and $v_0 = 6$ (all in some apppropriate units). Find A and δ, and then B_1 and B_2, and make a plot of $x(t)$ for the first ten or so periods.

5.39 ★★ [Computer] To get some practice at solving differential equations numerically, repeat the calculations of Example 5.3 (page 185), but instead of finding all the various coefficients just use appropriate software (for example, the NDSolve command of Mathematica) to solve the differential equation (5.48) with the boundary conditions $x_0 = v_0 = 0$. Make sure your graph agrees with Figure 5.15.

SECTION 5.6   Resonance

5.40 ★ Consider a damped oscillator, with fixed natural frequency ω_0 and fixed damping constant β (not too large), that is driven by a sinusoidal force with variable frequency ω. Show that the amplitude of the response, as given by (5.71) is maximum when $\omega = \sqrt{\omega_0^2 - 2\beta^2}$. (Note that so long as the resonance is narrow this implies $\omega \approx \omega_0$.)

5.41 ★ We know that if the driving frequency ω is varied, the maximum response (A^2) of a driven damped oscillator occurs at $\omega \approx \omega_0$ (if the natural frequency is ω_0 and the damping constant $\beta \ll \omega_0$). Show that A^2 is equal to half its maximum value when $\omega \approx \omega_0 \pm \beta$, so that the full width at half maximum is just 2β. [*Hint:* Be careful with your approximations. For instance, it's fine to say $\omega + \omega_0 \approx 2\omega_0$, but you certainly mustn't say $\omega - \omega_0 \approx 0$.]

5.42 ★ A large Foucault pendulum such as hangs in many science museums can swing for many hours before it damps out. Taking the decay time to be about 8 hours and the length to be 30 meters, find the quality factor Q.

5.43 ★★ When a car drives along a "washboard" road, the regular bumps cause the wheels to oscillate on the springs. (What actually oscillates is each axle assembly, comprising the axle and its two wheels.) Find the speed of my car at which this oscillation resonates, given the following information: **(a)** When four 80-kg men climb into my car, the body sinks by a couple of centimeters. Use this to estimate the spring constant k of each of the four springs. **(b)** If an axle assembly (axle plus two wheels) has total mass 50 kg, what is the natural frequency f of the assembly oscillating on its two springs? **(c)** If the bumps on a road are 80 cm apart, at about what speed would these oscillations go into resonance?

5.44 ★★ Another interpretation of the Q of a resonance comes from the following: Consider the motion of a driven damped oscillator after any transients have died out, and suppose that it is being driven close to resonance, so you can set $\omega = \omega_0$. **(a)** Show that the oscillator's total energy (kinetic plus potential) is $E = \frac{1}{2}m\omega^2 A^2$. **(b)** Show that the energy ΔE_{dis} dissipated during one cycle by the damping force F_{dmp} is $2\pi m\beta\omega A^2$. (Remember that the rate at which a force does work is Fv.) **(c)** Hence show that Q is 2π times the ratio $E/\Delta E_{\text{dis}}$.

5.45 ★★★ Consider a damped oscillator, with natural frequency ω_0 and damping constant β both fixed, that is driven by a force $F(t) = F_0 \cos(\omega t)$. **(a)** Find the rate $P(t)$ at which $F(t)$ does work and show that the average rate $\langle P \rangle$ over any number of complete cycles is $m\beta\omega^2 A^2$. **(b)** Verify that this is the same as the average rate at which energy is lost to the resistive force. **(c)** Show that as ω is varied $\langle P \rangle$ is maximum when $\omega = \omega_0$; that is, the resonance of the power occurs at $\omega = \omega_0$ (exactly).

SECTION 5.7   Fourier Series *

5.46 ★ The constant term a_0 in a Fourier series is a bit of a nuisance, always requiring slightly special treatment. At least it has a rather simple interpretation: Show that if $f(t)$ has the standard Fourier series (5.82), then a_0 is equal to the average $\langle f \rangle$ of $f(t)$ taken over one complete cycle.

5.47 ★★ In order to prove the crucial formulas (5.83)–(5.85) for the Fourier coefficients a_n and b_n, you must first prove the following:

$$\int_{-\tau/2}^{\tau/2} \cos(n\omega t) \cos(m\omega t)\, dt = \begin{cases} \tau/2 & \text{if } m = n \neq 0 \\ 0 & \text{if } m \neq n. \end{cases} \tag{5.105}$$

(This integral is obviously τ if $m = n = 0$.) There is an identical result with all cosines replaced by sines, and finally

$$\int_{-\tau/2}^{\tau/2} \cos(n\omega t) \sin(m\omega t)\, dt = 0 \qquad \text{for all integers } n \text{ and } m, \tag{5.106}$$

where as usual $\omega = 2\pi/\tau$. Prove these. [*Hint:* Use trig identities to replace $\cos(\theta)\cos(\phi)$ by terms like $\cos(\theta + \phi)$ and so on.]

5.48 ★★ Use the results (5.105) and (5.106) to prove the formulas (5.83)–(5.85) for the Fourier coefficients a_n and b_n. [*Hint:* Multiply both sides of the Fourier expansion (5.82) by $\cos(m\omega t)$ or $\sin(m\omega t)$ and then integrate from $-\tau/2$ to $\tau/2$.]

5.49 ★★★ [Computer] Find the Fourier coefficients a_n and b_n for the function shown in Figure 5.28(a). Make a plot similar to Figure 5.23, comparing the function itself with the first couple of terms in the Fourier series, and another for the first six or so terms. Take $f_{\text{max}} = 1$.

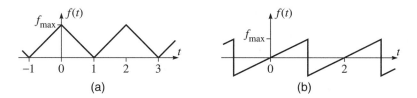

Figure 5.28   **(a)** Problem 5.49. **(b)** Problem 5.50

5.50 ★★★ [Computer] Find the Fourier coefficients a_n and b_n for the function shown in Figure 5.28(b). Make a plot similar to Figure 5.23, comparing the function itself with the sum of the first couple of terms in the Fourier series, and another for the first 10 or so terms. Take $f_{\text{max}} = 1$.

SECTION 5.8   Fourier Series Solution for the Driven Oscillator*

5.51 ★★ You can make the Fourier series solution for a periodically driven oscillator a bit tidier if you don't mind using complex numbers. Obviously the periodic force of Equation (5.90) can be written as $f = \text{Re}(g)$, where the complex function g is

$$g(t) = \sum_{n=0}^{\infty} f_n e^{in\omega t}.$$

Show that the real solution for the oscillator's motion can likewise be written as $x = \text{Re}(z)$, where

$$z(t) = \sum_{n=0}^{\infty} C_n e^{in\omega t}$$

and

$$C_n = \frac{f_n}{\omega_0^2 - n^2\omega^2 + 2i\beta n\omega}.$$

This solution avoids our having to worry about the real amplitude A_n and phase shift δ_n separately. (Of course A_n and δ_n are hidden inside the complex number C_n.)

5.52 ★★★ [Computer] Repeat all the calculations and plots of Example 5.5 (page 199) with all the same parameters except that $\beta = 0.1$. Compare your results with those of the example.

5.53 ★★★ [Computer] An oscillator is driven by the periodic force of Problem 5.49 [Figure 5.28(a)], which has period $\tau = 2$. **(a)** Find the long-term motion $x(t)$, assuming the following parameters: natural period $\tau_0 = 2$ (that is, $\omega_0 = \pi$), damping parameter $\beta = 0.1$, and maximum drive strength $f_{max} = 1$. Find the coefficients in the Fourier series for $x(t)$ and plot the sum of the first four terms in the series for $0 \leq t \leq 6$. **(b)** Repeat, except with natural period equal to 3.

SECTION 5.9   **The RMS Displacement; Parseval's Theorem** *

5.54 ★ Let $f(t)$ be a periodic function with period τ. Explain clearly why the average of f over one period is not necessarily the same as the average over some other time interval. Explain why, on the other hand, the average over a *long* time T approaches the average over one period, as $T \to \infty$.

5.55 ★★ To prove the Parseval relation (5.100), one must first prove the result (5.99) for the integral of a product of cosines. Prove this result, and then use it to prove the Parseval relation.

5.56 ★★ The Parseval relation as stated in (5.100) applies to a function whose Fourier series happens to contain only cosines. Write down the relation and prove it for a function

$$x(t) = \sum_{n=0}^{\infty} [A_n \cos(n\omega t - \delta_n) + B_n \sin(n\omega t - \delta_n)].$$

5.57 ★★ [Computer] Repeat the calculations that led to Figure 5.26, using all the same parameters except taking $\beta = 0.1$. Plot your results and compare your plot with Figure 5.26.

CHAPTER 6

Calculus of Variations

In many problems one needs to use non-Cartesian coordinates. Roughly speaking there are two classes of such problems. First, certain symmetries make it most advantageous to use special coordinates: Problems with spherical symmetry call out for the use of spherical polar coordinates; similarly, problems with axial symmetry are best treated in cylindrical polar coordinates. Second, when particles are constrained in some way, it is usually best to choose an appropriate, and usually non-Cartesian, coordinate system. For example, an object that is constrained to move on the surface of a sphere is probably best treated using spherical polar coordinates; if a bead slides on a curved wire, the best choice of coordinate may be just the distance along the curving wire from some convenient origin.

Unfortunately, as we have seen, the expressions for the components of the acceleration in non-Cartesian coordinates are quite messy, and the situation gets rapidly worse as we move on to more complicated systems. This makes Newton's second law difficult to use in non-Cartesian coordinates. We need an alternative (though ultimately equivalent) equation of motion that works equally well in any coordinates, and the required alternative is provided by Lagrange's equations.

The best way to prove — and to understand the great flexibility of — Lagrange's equations is to use a "variational principle." Variational principles are important in many areas of mathematics and physics. It has proved possible to formulate almost every branch of physics — classical mechanics, quantum mechanics, optics, electromagnetism, and so on — in variational terms. To the beginning student, accustomed to Newton's laws, a reformulation of classical mechanics in terms of a variational principle does not necessarily seem like an improvement. But because they allow a similar formulation of so many different subjects, variational methods have given a unity to physics and have played a crucial role in the recent history of physical theory. For this reason, I would like to introduce variational methods in a reasonably general setting. Therefore this short chapter is a brief introduction to variational problems in general. In the next chapter I shall apply what we learn here to establish the Lagrangian formulation of mechanics. If you are already familiar with the "calculus of variations" you could skip straight to Chapter 7.

6.1   Two Examples

The calculus of variations involves finding the minimum or maximum of a quantity that is expressible as an integral. To see how this can arise, I would like to start with two simple, concrete examples.

The Shortest Path between Two Points

My first example is this problem: Given two points in a plane, what is the shortest path between them? While you certainly know the answer — a straight line — you probably have not seen a proof, unless you have studied the calculus of variations. The problem is illustrated in Figure 6.1, which shows the two given points, (x_1, y_1) and (x_2, y_2), and a path, $y = y(x)$, joining them. Our task is to find the path $y(x)$ that has the shortest length and to show that it is in fact a straight line.

The length of a short segment of the path is $ds = \sqrt{dx^2 + dy^2}$, which, since

$$dy = \frac{dy}{dx} dx \equiv y'(x) \, dx,$$

we can rewrite as

$$ds = \sqrt{dx^2 + dy^2} = \sqrt{1 + y'(x)^2} \, dx. \tag{6.1}$$

Thus the total length of the path between points 1 and 2 is

$$L = \int_1^2 ds = \int_{x_1}^{x_2} \sqrt{1 + y'(x)^2} \, dx. \tag{6.2}$$

This equation puts our problem in mathematical form: The unknown is the *function* $y = y(x)$ that defines the path between points 1 and 2. The problem is to find the function $y(x)$ for which the integral (6.2) is a minimum. It is interesting to contrast this with the standard minimization problem of elementary calculus, where the unknown

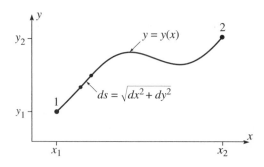

Figure 6.1   A path joining the two points 1 and 2. The length of the short segment is $ds = \sqrt{dx^2 + dy^2}$, and the total length of the path is $L = \int_1^2 ds$.

is the value of a variable x at which a known function $f(x)$ is a minimum. Obviously our new problem is one stage more complicated than this old one.

Before we set up the machinery to solve this new problem, let's consider another example.

Fermat's Principle

A similar problem is to find the path that light will follow between two points. If the refractive index of the medium is constant, then the path is, of course, a straight line, but if the refractive index varies, or if we interpose a mirror or lens, the path is not so obvious. The French mathematician Fermat (1601–1665) discovered that the required path is the path for which the time of travel of the light is minimum. We can illustrate Fermat's principle using Figure 6.1. The time for light to travel a short distance ds is ds/v where v denotes the speed of light in the medium, $v = c/n$ where n is the refractive index. Thus Fermat's principle says that the correct path between points 1 and 2 is the path for which the time

$$\text{(time of travel)} = \int_1^2 dt = \int_1^2 \frac{ds}{v} = \frac{1}{c} \int_1^2 n \, ds$$

is a minimum. If n is constant, then it can be taken outside the integral and the problem reduces to finding the shortest path between points 1 and 2 (and the answer is, of course, a straight line). In general, the refractive index can vary, $n = n(x, y)$, and our problem is to find the path $y(x)$ for which the integral

$$\int_1^2 n(x, y) \, ds = \int_{x_1}^{x_2} n(x, y) \sqrt{1 + y'(x)^2} \, dx \tag{6.3}$$

is minimum. [In writing the last expression, I substituted (6.1) for ds.]

The integral that has to be minimized in connection with Fermat's principle is very similar to the integral (6.2) giving the length of a path; it is just a little more complicated, since the factor $n(x, y)$ introduces an extra dependence on x and y. Similar integrals arise in many other problems. Sometimes we want the path for which an integral is a *maximum*, and sometimes we are interested in both maxima and minima. To get some idea of the possibilities, it is helpful to think again about the problem of finding maxima and minima of functions in elementary calculus. There we know that the necessary condition for a maximum or minimum of a function $f(x)$ is that its derivative vanish, $df/dx = 0$. Unfortunately, this condition is not quite enough to guarantee a maximum or minimum. As you certainly recall from introductory calculus, there are essentially three possibilities, as illustrated in Figure 6.2. A point x_0 where df/dx is zero may be a maximum or a minimum or, if $d^2 f/dx^2$ is also zero, it may be *neither*, as indicated in Figure 6.2(c). When $df/dx = 0$ at a point x_0, but we don't know which of the three possibilities obtains, we say that x_0 is a **stationary point** of the function $f(x)$, since an infinitesimal displacement of x from x_0 leaves $f(x)$ unchanged (because the slope is zero).

The situation for the problems of this chapter is very similar. The method I shall describe in the next section actually finds the path that makes an integral like (6.2) or

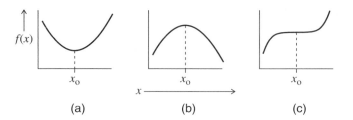

Figure 6.2   If $df/dx = 0$ at x_o, there are three possibilities: **(a)** If the second derivative is positive, then $f(x)$ has a minimum at x_o. **(b)** If the second derivative is negative, then $f(x)$ has a maximum. **(c)** If the second derivative is zero, then there may be a minimum, a maximum, or neither (as shown).

(6.3) **stationary**, in the sense that an infinitesimal variation of the path from its correct course doesn't change the value of the integral concerned. If you need to know that the integral is definitely minimum (or definitely maximum, or perhaps neither), you have to check this separately. Incidentally, we are now ready to explain the name of this chapter: Since our concern is how infinitesimal variations of a path change an integral, the subject is called the **calculus of variations**. For the same reason, the methods we shall develop are called variational methods, and a principle like Fermat's principle is a variational principle.

6.2   The Euler–Lagrange Equation

The two examples of the last section illustrate the general form of the so-called variational problem. We have an integral of the form

$$S = \int_{x_1}^{x_2} f[y(x), y'(x), x]\, dx \tag{6.4}$$

where $y(x)$ is an as-yet unknown curve joining two points (x_1, y_1) and (x_2, y_2) as in Figure 6.1; that is,

$$y(x_1) = y_1 \qquad \text{and} \qquad y(x_2) = y_2. \tag{6.5}$$

Among all the possible curves satisfying (6.5) (that is, joining the points 1 and 2), we have to find the one that makes the integral S a minimum (or maximum or at least stationary). To be definite, I shall suppose that we wish to find a minimum. Notice that the function f in (6.4) is a function of three variables $f = f(y, y', x)$, but because the integral follows the path $y = y(x)$ the integrand $f[y(x), y'(x), x]$ is actually a function of just the one variable x.

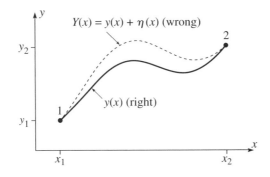

Figure 6.3 The path $y = y(x)$ between points 1 and 2 is the "right" path, the one for which the integral S of (6.4) is a minimum. Any other path $Y(x)$ is "wrong," in that it gives a larger value for S.

Let us denote the correct solution to our problem by $y = y(x)$. Then the integral S in (6.4) evaluated for $y = y(x)$ is less than for any neighboring curve $y = Y(x)$, as sketched in Figure 6.3. It is convenient to write the "wrong" curve $Y(x)$ as

$$Y(x) = y(x) + \eta(x) \tag{6.6}$$

where $\eta(x)$ (Greek "eta") is just the difference between the wrong $Y(x)$ and the right $y(x)$. Since $Y(x)$ must pass through the endpoints 1 and 2, $\eta(x)$ must satisfy

$$\eta(x_1) = \eta(x_2) = 0. \tag{6.7}$$

There are infinitely many choices for the difference $\eta(x)$; for example, we could choose $\eta = (x - x_1)(x_2 - x)$ or $\eta(x) = \sin[\pi(x - x_1)/(x_2 - x_1)]$.

The integral S taken along the wrong curve $Y(x)$ must be larger than that along the right curve $y(x)$, no matter how close the former is to the latter. To express this requirement, I shall introduce a parameter α and redefine $Y(x)$ to be

$$Y(x) = y(x) + \alpha\eta(x). \tag{6.8}$$

The integral S taken along the curve $Y(x)$ now depends on the parameter α, so I shall call it $S(\alpha)$. The right curve $y(x)$ is obtained from (6.8) by setting $\alpha = 0$. Thus the requirement that S is minimum for the right curve $y(x)$ implies that $S(\alpha)$ is a minimum at $\alpha = 0$. With this result, we have converted our problem to the traditional problem from elementary calculus of making sure that an ordinary function [namely $S(\alpha)$] has a minimum at a specified point ($\alpha = 0$). To ensure this, we must just check that the derivative $dS/d\alpha$ is zero when $\alpha = 0$.

If we write out the integral $S(\alpha)$ in detail, it looks like this:

$$S(\alpha) = \int_{x_1}^{x_2} f(Y, Y', x)\, dx$$

$$= \int_{x_1}^{x_2} f(y + \alpha\eta, y' + \alpha\eta', x)\, dx. \tag{6.9}$$

To differentiate (6.9) with respect to α, we note that α appears in the integrand f, so we need to evaluate $\partial f/\partial \alpha$. Since α appears in two of the arguments of f, this gives two terms, namely (using the chain rule)

$$\frac{\partial f(y + \alpha \eta, \, y' + \alpha \eta', \, x)}{\partial \alpha} = \eta \frac{\partial f}{\partial y} + \eta' \frac{\partial f}{\partial y'},$$

and for $dS/d\alpha$ (which has to be zero)

$$\frac{dS}{d\alpha} = \int_{x_1}^{x_2} \frac{\partial f}{\partial \alpha} \, dx = \int_{x_1}^{x_2} \left(\eta \frac{\partial f}{\partial y} + \eta' \frac{\partial f}{\partial y'} \right) dx = 0. \tag{6.10}$$

This condition must be true for any $\eta(x)$ satisfying (6.7); that is, for any choice of the "wrong" path $Y(x) = y(x) + \alpha \eta(x)$.

To take advantage of the condition (6.10), we need to rewrite the second term on the right using integration by parts[1] (remember that η' means $d\eta/dx$):

$$\int_{x_1}^{x_2} \eta'(x) \frac{\partial f}{\partial y'} dx = \left[\eta(x) \frac{\partial f}{\partial y'} \right]_{x_1}^{x_2} - \int_{x_1}^{x_2} \eta(x) \frac{d}{dx} \left(\frac{\partial f}{\partial y'} \right) dx.$$

Because of the condition (6.7), the first term on the right (the "endpoint term") is zero. Thus[2]

$$\int_{x_1}^{x_2} \eta'(x) \frac{\partial f}{\partial y'} dx = - \int_{x_1}^{x_2} \eta(x) \frac{d}{dx} \left(\frac{\partial f}{\partial y'} \right) dx. \tag{6.11}$$

Substituting this identity into (6.10), we find that

$$\int_{x_1}^{x_2} \eta(x) \left(\frac{\partial f}{\partial y} - \frac{d}{dx} \frac{\partial f}{\partial y'} \right) dx = 0. \tag{6.12}$$

This condition must be satisfied for any choice of the function $\eta(x)$. Therefore, as I shall argue in a moment, the factor in large parentheses must be zero:

$$\frac{\partial f}{\partial y} - \frac{d}{dx} \frac{\partial f}{\partial y'} = 0 \qquad \text{(Euler–Lagrange Equation)} \tag{6.13}$$

for all x (in the relevant interval $x_1 \le x \le x_2$). This is the so-called **Euler–Lagrange equation** (named for the Swiss mathematician Leonhard Euler, 1707–1783, and the Italian-French physicist and mathematician Joseph Lagrange, 1736–1813), which lets

---

[1] If you are used to thinking of integration by parts in the form $\int v \, du = [uv] - \int u \, dv$, then you will find it helpful to recognize that another way to say the same thing is: $\int u'v \, dx = [uv] - \int uv' \, dx$. In words: In the integral $\int u'v \, dx$, you can move the prime from the u to the v if you change the sign and add the endpoint contribution $[uv]$.

[2] This is the simple form in which integration by parts often appears in physics: Provided the endpoint term $[uv]$ is zero (as often happens), integration by parts lets you move the differentiation from the u to the v as long as you change the sign; that is, $\int u'v \, dx = - \int uv' dx$.

us find the path for which the integral S is stationary. Before I illustrate its use, I need to discuss the step from (6.12) to (6.13), which is by no means obvious.

Equation (6.12) has the form $\int \eta(x)g(x)\,dx = 0$. I would certainly not claim that this condition alone implies that $g(x) = 0$ for all x. However, (6.12) holds for any choice of the function $\eta(x)$, and if $\int \eta(x)g(x)\,dx = 0$ for *any* $\eta(x)$, then we *can* conclude that $g(x) = 0$ for all x. To prove this, we must assume that all functions concerned are continuous, but, as physicists, we would take for granted that this is the case.[3] Now, to prove the assertion, let us assume the contrary, that $g(x)$ is nonzero in some interval between x_1 and x_2. Then choose a function $\eta(x)$ that has the same sign as $g(x)$ (that is, η is positive where g is positive and η is negative where g is negative). Then the integrand is continuous, satisfies $\eta(x)g(x) \geq 0$, and is nonzero at least in some interval. Under these conditions $\int \eta(x)g(x)\,dx$ cannot be zero. This contradiction implies that $g(x)$ *is* zero for all x.

This completes the proof of the Euler–Lagrange equation. The procedure for using it is this: (1) Set up the problem so that the quantity whose stationary path you seek is expressed as an integral in the standard form

$$S = \int_{x_1}^{x_2} f[y(x), y'(x), x]\,dx, \qquad (6.14)$$

where $f[y(x), y'(x), x]$ is the function appropriate to your problem. (2) Write down the Euler–Lagrange equation (6.13) in terms of the function $f[y(x), y'(x), x]$. (3) Finally, solve (if possible) the differential equation (6.13) for the function $y(x)$ that defines the required stationary path. I shall illustrate this procedure with a couple of examples in the next section.

6.3 Applications of the Euler–Lagrange Equation

Let us start with the problem that began this chapter, finding the shortest path between two points in a plane.

EXAMPLE 6.1 Shortest Path between Two Points

We saw that the length of a path between points 1 and 2 is given by the integral (6.2) as

$$L = \int_1^2 ds = \int_{x_1}^{x_2} \sqrt{1 + y'^2}\,dx.$$

This has the standard form (6.14), with the function f given by

$$f(y, y', x) = (1 + y'^2)^{1/2}. \qquad (6.15)$$

---

[3] The claimed result is clearly false if discontinuous functions are admitted. For instance, if we made $g(x)$ nonzero at just one point, then $\int \eta(x)g(x)\,dx$ would still be zero.

To use the Euler–Lagrange equation (6.13), we must evaluate the two partial derivatives concerned:

$$\frac{\partial f}{\partial y} = 0 \quad \text{and} \quad \frac{\partial f}{\partial y'} = \frac{y'}{(1 + y'^2)^{1/2}}. \tag{6.16}$$

Since $\partial f/\partial y = 0$, (6.13) implies simply that

$$\frac{d}{dx}\frac{\partial f}{\partial y'} = 0.$$

In other words, $\partial f/\partial y'$ is a constant, C. According to (6.16), this implies that

$$y'^2 = C^2(1 + y'^2),$$

or, with a little rearrangement, $y'^2 = $ constant. This implies that $y'(x)$ is a constant, which we could call m. Integrating the equation $y'(x) = m$, we find that $y(x) = mx + b$, and we have proved that the shortest path between two points is a straight line!

A Note on Variables

So far we have considered problems with two variables, which we have called x and y. Of these, x has been the independent variable, and y the dependent, through the relation $y = y(x)$. Unfortunately, we are frequently forced — by convenience or tradition — to name the variables differently. For example, in a simple one-dimensional mechanics problem, the independent variable is the time t and the dependent variable is the position $x = x(t)$. This means you will have to get used to seeing the Euler–Lagrange equation with the variables x and y replaced by an assortment of other variables, such as t and x. In the next example, the two variables are x and y, but the independent variable is y, and the roles of x and y in (6.13) and (6.14) will be exactly reversed.

EXAMPLE 6.2   The Brachistochrone

A famous problem in the calculus of variations is this: Given two points 1 and 2, with 1 higher above the ground, in what shape should we build the track for a frictionless roller coaster so that a car released from point 1 will reach point 2 in the shortest possible time? This problem is called the *brachistochrone* problem, from the Greek words *brachistos* meaning "shortest" and *chronos* meaning "time." The geometry of the problem is sketched in Figure 6.4, where I have taken point 1 as the origin and I have chosen to measure y vertically down.

The time to travel from 1 to 2 is

$$\text{time}(1 \rightarrow 2) = \int_1^2 \frac{ds}{v} \tag{6.17}$$

where the speed at any height y is determined by conservation of energy to be $v = \sqrt{2gy}$. (Problem 6.8.) Because this gives v as a function of y, it is convenient

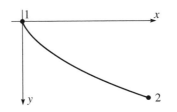

Figure 6.4   The brachistochrone problem is to find the shape of track on which a roller coaster released from point 1 will reach point 2 in the minimum possible time.

to take y as our independent variable. That is, we shall write the unknown path as $x = x(y)$. This means that the distance ds between neighboring points on the path has to be written as

$$ds = \sqrt{dx^2 + dy^2} = \sqrt{x'(y)^2 + 1}\, dy \qquad (6.18)$$

where a prime now denotes differentiation with respect to y; that is, $x'(y) = dx/dy$. Thus according to (6.17) the time of interest is

$$\text{time}(1 \to 2) = \frac{1}{\sqrt{2g}} \int_0^{y_2} \frac{\sqrt{x'(y)^2 + 1}}{\sqrt{y}}\, dy. \qquad (6.19)$$

Equation (6.19) gives the integral whose minimum we have to find. It is of the standard form (6.14), except that the roles of x and y have been interchanged, with the integrand

$$f(x, x', y) = \frac{\sqrt{x'^2 + 1}}{\sqrt{y}}. \qquad (6.20)$$

To find the path that makes the time as small as possible, we have only to apply the Euler–Lagrange equation (again with x and y interchanged) to this function,

$$\frac{\partial f}{\partial x} = \frac{d}{dy}\frac{\partial f}{\partial x'}. \qquad (6.21)$$

The function of (6.20) is independent of x, so the derivative $\partial f/\partial x$ is zero, and (6.21) tells us simply that $\partial f/\partial x'$ is a constant. Evaluating this derivative (and squaring it for convenience) we conclude that

$$\frac{x'^2}{y(1 + x'^2)} = \text{const} = \frac{1}{2a} \qquad (6.22)$$

where I have named the constant $1/2a$ for future convenience. This equation is easily solved for x' to give

$$x' = \sqrt{\frac{y}{2a - y}},$$

whence

$$x = \int \sqrt{\frac{y}{2a - y}} \, dy. \tag{6.23}$$

This integral can be evaluated by the unlikely looking substitution

$$y = a(1 - \cos\theta) \tag{6.24}$$

which gives (as you should check)

$$x = a \int (1 - \cos\theta) \, d\theta$$

$$= a(\theta - \sin\theta) + \text{const.} \tag{6.25}$$

The two equations (6.25) and (6.24) are parametric equations for the required path, giving x and y as functions of the parameter θ. We have chosen the initial point 1 to have $x = y = 0$, so we see from (6.24) that the initial value of θ is zero. This in turn implies that the constant of integration in (6.25) is zero. Thus the final parametric equation for the path is

$$x = a(\theta - \sin\theta) \quad \text{and} \quad y = a(1 - \cos\theta) \tag{6.26}$$

with the constant a chosen so the curve passes through the given point (x_2, y_2).

The curve (6.26) is plotted in Figure 6.5. In that figure I have continued the curve (with dashes) beyond the point 2 to show that the curve that solves the brachistochrone problem happens to be a cycloid — the curve traced out by a point on the rim of a wheel of radius a, rolling along the underside of the x axis (Problem 6.14). Another remarkable feature of this curve is this: If we release the cart from rest at point 2 and let it roll to the bottom of the curve (point 3 in the figure), the time to roll from 2 to 3 is the same whatever the position of 2, anywhere between 1 and 3. This means that the oscillations of a cart rolling back and forth on a cycloid-shaped track are exactly *isochronous* (period perfectly independent of amplitude), in contrast with the oscillations of a simple pendulum, which are only approximately isochronous, to the extent

Figure 6.5 The path for a roller coaster that gives the shortest time between the given points 1 and 2 is part of the cycloid with a vertex at 1 and passing through 2. The cycloid is the curve traced by a point on the rim of a wheel of radius a that rolls along the underside of the x axis. Point 3 is the lowest point on the curve.

that the amplitude is small. (See Problem 6.25.) The isochronous property of the cycloid was actually used in the design of some clocks, one of which can be seen in the Victoria and Albert Museum in London.

Maximum and Minimum vs. Stationary

You have probably noticed that in neither example of this section did I check that the curves that we found actually gave a minimum value to the integral of interest — that the straight line between two points actually makes the path length *minimum*, not a maximum or just stationary. The Euler–Lagrange equation guarantees only to give a path for which the original integral is stationary. The problem of deciding whether we have a minimum or maximum (or a stationary curve that is neither) is generally very difficult. In a few cases, it is easy to see which is the case. For instance, it really is obvious that a straight line gives the *minimum* distance between two points in a plane. In the case of the brachistochrone, it is not at all obvious that the path we found does yield a minimum time, though it is in fact true.

To illustrate the variety of possibilities, consider the problem of finding the shortest path, or **geodesic**, between two points 1 and 2 on the surface of a globe. As you probably know, the answer is the great circle joining the two points.[4] Using the calculus of variations you can prove relatively easily that a great circle does indeed make the distance stationary: Using spherical polar coordinates, every point on the globe can be identified by the two angles θ and ϕ. If you characterize a path as $\phi = \phi(\theta)$ and set up an integral that gives the distance between 1 and 2 along this path, you can show that the Euler–Lagrange equation for $\phi(\theta)$ requires that the path follow a great circle. (See Problem 6.16 for details.) But you have to think a little carefully before deciding that this necessarily gives a minimum distance, since there are *two different* great-circle paths connecting any two points 1 and 2 on the globe: For simplicity consider two towns on the equator, Quito (near the Pacific coast of Ecuador) and Macapá (at the mouth of the Amazon on the Atlantic coast of Brazil). The "right" shortest path between these two is, of course, the great-circle path following the equator for about 2000 miles across South America. But a second possibility, which satisfies the Euler–Lagrange equation just as well, is to head west around the equator from Quito, across the Pacific, the African continent, and the Atlantic, arriving in Macapá some 23,000 miles later. You might guess that this path would be a maximum, but it is in fact neither maximum nor minimum: It is easy to construct nearby paths that are shorter, but it is also easy to find others that are longer. In other words, this second great-circle path gives neither a maximum nor a minimum. This second path is, of course, analogous to the horizontal point of inflection in elementary calculus. In this problem, luckily, it is obvious that the first path gives the true minimum. However, it should be clear that, in general, deciding what sort of stationary path the Euler–Lagrange equation has given us can be tricky.

---

[4] A great circle is the circle in which the globe intersects a plane through the globe's center.

Fortunately for us, these questions are irrelevant for our purposes. We shall find that for the applications in mechanics all that matters is that we have a path which makes a certain integral *stationary*. It simply doesn't matter whether it gives a maximum, minimum, or neither.

6.4   More than Two Variables

So far we have considered only problems with just two variables, the independent variable (usually x) and the dependent (usually y). For most applications in mechanics, we shall find that there are several dependent variables, though fortunately still only one independent variable, which is usually the time t. For a simple example where there are two dependent variables, we can go back to the problem of the shortest path between two points. When we found the shortest path between two points 1 and 2, we assumed that the required path could be written in the form $y = y(x)$. Reasonable as this seems, it is easy to think of paths that cannot be written in this way, such as the path shown in Figure 6.6. If we want to be perfectly sure we have found the shortest path among *all* possible paths, we must find a method that includes these. The way to do this is to write the path in parametric form as

$$x = x(u) \quad \text{and} \quad y = y(u), \tag{6.27}$$

where u is any convenient variable in terms of which the curve can be parameterized (for instance, the distance along the path). The parametric form (6.27) includes all of the curves considered before. [If $y = y(x)$, just use x for the parameter u.] It also includes curves like that of Figure 6.6 and, in fact, all curves of interest.[5]

The length of a small segment of the path (6.27) is

$$ds = \sqrt{dx^2 + dy^2} = \sqrt{x'(u)^2 + y'(u)^2}\, du \tag{6.28}$$

where, as usual, a prime denotes differentiation with respect to the function's argument; that is, $x'(u) = dx/du$ and $y'(u) = dy/du$. Thus the total path length is

$$L = \int_{u_1}^{u_2} \sqrt{x'(u)^2 + y'(u)^2}\, du, \tag{6.29}$$

and our job is to find the two functions $x(u)$ and $y(u)$ for which this integral is minimum.

This problem is more complicated than any we have considered before, because

---

[5] In case you are interested in mathematical niceties, I should say that in what follows I shall assume that all functions concerned are continuous and have continuous second derivatives. This assumption could be weakened a little, for example, by allowing some discontinuities in the derivatives.

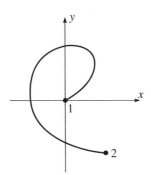

Figure 6.6   This path between the two points 1 and 2 cannot be written as $y = y(x)$ nor as $x = x(y)$. It *can* be written in the parametric form (6.27).

there are now two unknown functions $x(u)$ and $y(u)$. The general problem of this type is this: Given an integral of the form

$$S = \int_{u_1}^{u_2} f[x(u), y(u), x'(u), y'(u), u] \, du \tag{6.30}$$

between two fixed points $[x(u_1), y(u_1)]$ and $[x(u_2), y(u_2)]$, find the path $[x(u), y(u)]$ for which the integral S is stationary. The solution to this problem is very similar to the one-variable case, and I shall just sketch it, leaving you to fill in the details. The upshot is that with two dependent variables, we get two Euler–Lagrange equations. To prove this, we proceed very much as before. Let the correct path be given by

$$x = x(u) \quad \text{and} \quad y = y(u), \tag{6.31}$$

and then consider a neighboring "wrong" path of the form

$$x = x(u) + \alpha \xi(u) \quad \text{and} \quad y = y(u) + \beta \eta(u) \tag{6.32}$$

(where ξ is the Greek letter "xi"). The requirement that the integral S be stationary for the right path (6.31) is equivalent to the requirement that the integral $S(\alpha, \beta)$, taken along the wrong path (6.32), satisfy

$$\frac{\partial S}{\partial \alpha} = 0 \quad \text{and} \quad \frac{\partial S}{\partial \beta} = 0 \tag{6.33}$$

when $\alpha = \beta = 0$. These two conditions are the natural generalization of the condition (6.10) for the one-variable case. By an argument which exactly parallels that leading from (6.10) to (6.13), you can show that these two conditions are equivalent to the *two* Euler–Lagrange equations (see Problem 6.26):

$$\frac{\partial f}{\partial x} = \frac{d}{du} \frac{\partial f}{\partial x'} \quad \text{and} \quad \frac{\partial f}{\partial y} = \frac{d}{du} \frac{\partial f}{\partial y'}. \tag{6.34}$$

These two equations determine a path for which the integral (6.30) is stationary, and, conversely, if the integral is stationary for some path, that path must satisfy these two equations.

EXAMPLE 6.3   The Shortest Path between Two Points Again

We can now solve completely the problem of the shortest path between two points. (That is, solve it including *all* possible paths, such as that in Figure 6.6.) From (6.29), we see that for this problem the integrand f is

$$f(x, x', y, y', u) = \sqrt{x'^2 + y'^2}. \tag{6.35}$$

Since this is independent of x and y, the two derivatives $\partial f/\partial x$ and $\partial f/\partial y$ on the left sides in (6.34) are zero. Therefore, the two Euler–Lagrange equations imply simply that the two derivatives $\partial f/\partial x'$ and $\partial f/\partial y'$ are constants,

$$\frac{\partial f}{\partial x'} = \frac{x'}{\sqrt{x'^2 + y'^2}} = C_1 \quad \text{and} \quad \frac{\partial f}{\partial y'} = \frac{y'}{\sqrt{x'^2 + y'^2}} = C_2. \tag{6.36}$$

If we divide the second equation by the first and recognize that y'/x' is just the derivative dy/dx, we conclude that

$$\frac{dy}{dx} = \frac{y'}{x'} = \frac{C_2}{C_1} = m, \tag{6.37}$$

say. It follows that the required path is a straight line, $y = mx + b$. It is interesting that this proof using a parametric equation is not only better than our previous proof (in that the new proof includes all possible paths), it is also marginally easier.

The generalization of the Euler–Lagrange equation to an arbitrary number of dependent variables is straightforward, and doesn't need to be spelled out in detail. Here I would just like to sketch the way the Euler–Lagrange equations will appear in the Lagrangian formulation of mechanics.

The independent variable in Lagrangian mechanics is the time t. The dependent variables are the coordinates that specify the position, or "configuration," of a system, and are usually denoted by $q_1, q_2, \cdots, q_n$. The number n of coordinates depends on the nature of the system. For a single particle moving unconstrained in three dimensions, n is 3, and the three coordinates q_1, q_2, q_3 could be just the three Cartesian coordinates x, y, z, or they might be the spherical polar coordinates r, θ, ϕ. For N particles moving freely in three dimensions, n is $3N$ and the coordinates $q_1, \cdots, q_n$ could be the $3N$ Cartesian coordinates $x_1, y_1, z_1, \cdots, x_N, y_N, z_N$. For a double pendulum (two simple pendulums, with the second suspended from the bob of the first, as in Figure 6.7), there would be two coordinates q_1, q_2, which could be chosen to be the two angles shown in Figure 6.7. Because the coordinates $q_1, \cdots, q_n$ can take on so many guises, they are often referred to as **generalized coordinates**. It is often useful to think of the n generalized coordinates as defining a point in an n-dimensional **configuration space**, each of whose points labels a unique position, or configuration, of the system.

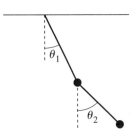

Figure 6.7   A good choice of generalized coordinates to identify
the position of a double pendulum is the pair of angles θ_1 and θ_2
between the pendulums and the vertical.

The ultimate goal in most problems in Lagrangian mechanics is to find how the
coordinates vary with time; that is, to find the n functions $q_1(t), \cdots, q_n(t)$. One can
regard these n functions as defining a path in the n-dimensional configuration space.
This path is, of course, determined by Newton's second law, but we shall find that it
can, equivalently, be characterized as the path for which a certain integral is stationary.
This means that it must satisfy the corresponding Euler–Lagrange equations (called
just Lagrange equations in this context), and it turns out that these Lagrange equations
are usually much easier to write down and use than Newton's second law. In particular,
unlike Newton's second law, Lagrange's equations take exactly the same simple form
in all coordinate systems.

The integral S whose stationary value determines the evolution of the mechanical
system is called the action integral. Its integrand is called the Lagrangian $\mathcal{L}$ and
depends on the n coordinates $q_1, q_2, \cdots, q_n$, their n time derivatives $\dot{q}_1, \dot{q}_2, \cdots, \dot{q}_n$
and the time t,

$$\mathcal{L} = \mathcal{L}(q_1, \dot{q}_1, \cdots, q_n, \dot{q}_n, t). \tag{6.38}$$

Notice that since the independent variable is t, the derivatives of the coordinates q_i
are time derivatives and are denoted, as usual, with dots as $\dot{q}_i$. The requirement that
the action integral

$$S = \int_{t_1}^{t_2} \mathcal{L}(q_1, \dot{q}_1, \cdots, q_n, \dot{q}_n, t) \, dt \tag{6.39}$$

be stationary implies n Euler–Lagrange equations

$$\frac{\partial \mathcal{L}}{\partial q_1} = \frac{d}{dt}\frac{\partial \mathcal{L}}{\partial \dot{q}_1}, \quad \frac{\partial \mathcal{L}}{\partial q_2} = \frac{d}{dt}\frac{\partial \mathcal{L}}{\partial \dot{q}_2}, \quad \cdots, \quad \text{and} \quad \frac{\partial \mathcal{L}}{\partial q_n} = \frac{d}{dt}\frac{\partial \mathcal{L}}{\partial \dot{q}_n}. \tag{6.40}$$

These n equations correspond precisely to the two Euler–Lagrange equations in (6.34)
and are proved in exactly the same way. If these n equations are satisfied, then the
action integral (6.39) is stationary; and if the action integral is stationary, then these n
equations are satisfied. In the next chapter, you will see where these equations come
from and how to use them.

Principal Definitions and Equations of Chapter 6

The Euler–Lagrange Equation

An integral of the form

$$S = \int_{x_1}^{x_2} f[y(x), y'(x), x] \, dx \qquad \text{[Eq. (6.4)]}$$

taken along a path $y = y(x)$ is stationary with respect to variations of that path if and only if $y(x)$ satisfies the **Euler–Lagrange equation**

$$\frac{\partial f}{\partial y} - \frac{d}{dx} \frac{\partial f}{\partial y'} = 0. \qquad \text{[Eq. (6.13)]}$$

Several Variables

If there are n dependent variables in the original integral, there are n Euler–Lagrange equations. For instance, an integral of the form

$$S = \int_{u_1}^{u_2} f[x(u), y(u), x'(u), y'(u), u] \, du,$$

with two dependent variables [$x(u)$ and $y(u)$], is stationary with respect to variations of $x(u)$ and $y(u)$ if and only if these two functions satisfy the two equations

$$\frac{\partial f}{\partial x} = \frac{d}{du} \frac{\partial f}{\partial x'} \qquad \text{and} \qquad \frac{\partial f}{\partial y} = \frac{d}{du} \frac{\partial f}{\partial y'}. \qquad \text{[Eq. (6.34)]}$$

Problems for Chapter 6

Stars indicate the approximate level of difficulty, from easiest ($\star$) to most difficult ($\star\star\star$).

SECTION 6.1   Two Examples

6.1 $\star$ The shortest path between two points on a *curved surface*, such as the surface of a sphere, is called a **geodesic**. To find a geodesic, one has first to set up an integral that gives the length of a path on the surface in question. This will always be similar to the integral (6.2) but may be more complicated (depending on the nature of the surface) and may involve different coordinates than x and y. To illustrate this, use spherical polar coordinates (r, θ, ϕ) to show that the length of a path joining two points on a sphere of radius R is

$$L = R \int_{\theta_1}^{\theta_2} \sqrt{1 + \sin^2 \theta \, \phi'(\theta)^2} \, d\theta \qquad (6.41)$$

if (θ_1, ϕ_1) and (θ_2, ϕ_2) specify the two points and we assume that the path is expressed as $\phi = \phi(\theta)$. (You will find how to minimize this length in Problem 6.16.)

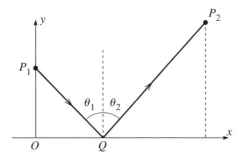

Figure 6.8   Problem 6.3

6.2 ★ Do the same as in Problem 6.1 but find the length L of a path on a cylinder of radius R, using cylindrical polar coordinates (ρ, ϕ, z). Assume that the path is specified in the form $\phi = \phi(z)$.

6.3 ★★ Consider a ray of light traveling in a vacuum from point P_1 to P_2 by way of the point Q on a plane mirror, as in Figure 6.8. Show that Fermat's principle implies that, on the actual path followed, Q lies in the same vertical plane as P_1 and P_2 and obeys the law of reflection, that $\theta_1 = \theta_2$. [*Hints:* Let the mirror lie in the xz plane, and let P_1 lie on the y axis at $(0, y_1, 0)$ and P_2 in the xy plane at $(x_2, y_2, 0)$. Finally let $Q = (x, 0, z)$. Calculate the time for the light to traverse the path $P_1 Q P_2$ and show that it is minimum when Q has $z = 0$ and satisfies the law of reflection.]

6.4 ★★ A ray of light travels from point P_1 in a medium of refractive index n_1 to P_2 in a medium of index n_2, by way of the point Q on the plane interface between the two media, as in Figure 6.9. Show that Fermat's principle implies that, on the actual path followed, Q lies in the same vertical plane as P_1 and P_2 and obeys Snell's law, that $n_1 \sin \theta_1 = n_2 \sin \theta_2$. [*Hints:* Let the interface be the xz plane, and let P_1 lie on the y axis at $(0, h_1, 0)$ and P_2 in the x, y plane at $(x_2, -h_2, 0)$. Finally let $Q = (x, 0, z)$. Calculate the time for the light to traverse the path $P_1 Q P_2$ and show that it is minimum when Q has $z = 0$ and satisfies Snell's law.]

6.5 ★★ Fermat's principle is often stated as "the travel time of a ray of light, moving from point A to B, is minimum along the actual path." Strictly speaking it should say that the time is *stationary*, not minimum. In fact one can construct situations for which the time is maximum along the actual path. Here is one: Consider the concave, hemispherical mirror shown in Figure 6.10, with A and B at opposite ends of a diameter. Consider a ray of light traveling in a vacuum from A to B with one reflection at P, in the same vertical plane as A and B. According to the law of reflection, the actual path goes via point

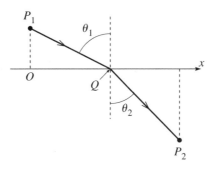

Figure 6.9   Problem 6.4

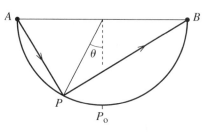

Figure 6.10   Problem 6.5

P_o at the bottom of the hemisphere ($\theta = 0$). Find the time of travel along the path APB as a function of θ and show that it is *maximum* at $P = P_o$. This shows the time is maximum with respect to paths of the form APB with just two straight segments. It is easy to see that it is *minimum* for other kinds of path, so the correct general statement is that it is *stationary* for arbitrary variations of the path.

6.6 ★★ In many problems in the calculus of variations, you need to know the length ds of a short segment of a curve on a surface, as in the expression (6.1). Make a table giving the appropriate expressions for ds in the following eight situations: **(a)** A curve given by $y = y(x)$ in a plane, **(b)** same but $x = x(y)$, **(c)** same but $r = r(\phi)$, **(d)** same but $\phi = \phi(r)$; **(e)** curve given by $\phi = \phi(z)$ on a cylinder of radius R, **(f)** same but $z = z(\phi)$; **(g)** curve given by $\theta = \theta(\phi)$ on a sphere of radius R, **(h)** same but $\phi = \phi(\theta)$.

SECTION 6.3 Applications of the Euler–Lagrange Equation

6.7 ★ Consider a right circular cylinder of radius R centered on the z axis. Find the equation giving ϕ as a function of z for the geodesic (shortest path) on the cylinder between two points with cylindrical polar coordinates (R, ϕ_1, z_1) and (R, ϕ_2, z_2). Describe the geodesic. Is it unique? By imagining the surface of the cylinder unwrapped and laid out flat, explain why the geodesic has the form it does.

6.8 ★ Verify that the speed of the roller coaster car in Example 6.2 (page 222) is $\sqrt{2gy}$. (Assume the wheels have negligible mass and neglect friction.)

6.9 ★ Find the equation of the path joining the origin O to the point $P(1, 1)$ in the xy plane that makes the integral $\int_O^P (y'^2 + yy' + y^2)\, dx$ stationary.

6.10 ★ In general the integrand $f(y, y', x)$ whose integral we wish to minimize depends on y, y', and x. There is a considerable simplification if f happens to be independent of y, that is, $f = f(y', x)$. (This happened in both Examples 6.1 and 6.2, though in the latter the roles of x and y were interchanged.) Prove that when this happens, the Euler–Lagrange equation (6.13) reduces to the statement that

$$\partial f / \partial y' = \text{const.} \qquad (6.42)$$

Since this is a first-order differential equation for $y(x)$, while the Euler–Lagrange equation is generally second order, this is an important simplification and the result (6.42) is sometimes called a **first integral** of the Euler–Lagrange equation. In Lagrangian mechanics we'll see that this simplification arises when a component of momentum is conserved.

6.11 ★★ Find and describe the path $y = y(x)$ for which the integral $\int_{x_1}^{x_2} \sqrt{x}\sqrt{1 + y'^2}\, dx$ is stationary.

6.12 ★★ Show that the path $y = y(x)$ for which the integral $\int_{x_1}^{x_2} x\sqrt{1 - y'^2}\, dx$ is stationary is an arcsinh function.

6.13 ★★ In relativity theory, velocities can be represented by points in a certain "rapidity space" in which the distance between two neighboring points is $ds = [2/(1 - r^2)]\sqrt{dr^2 + r^2\, d\phi^2}$, where r and ϕ are polar coordinates, and we consider just a two-dimensional space. (An expression like this for the distance in a non-Euclidean space is often called the *metric* of the space.) Use the Euler–Lagrange equation to show that the shortest path from the origin to any other point is a straight line.

6.14 ★★ (a) Prove that the brachistochrone curve (6.26) is indeed a cycloid, that is, the curve traced by a point on the circumference of a wheel of radius a rolling along the underside of the x axis. **(b)** Although the cycloid repeats itself indefinitely in a succession of loops, only one loop is relevant to the brachistochrone problem. Sketch a single loop for three different values of a (all with the same starting point 1) and convince yourself that for any point 2 (with positive coordinates x_2, y_2) there is exactly one value of a for which the loop goes through the point 2. **(c)** To find the value of a for a given point x_2, y_2 usually requires solution of a transcendental equation. Here are two cases where you can do it more simply: For $x_2 = \pi b$, $y_2 = 2b$ and again for $x_2 = 2\pi b$, $y_2 = 0$ find the value of a for which the cycloid goes through the point 2 and find the corresponding minimum times.

6.15 ★★ Consider again the brachistochrone problem of Example 6.2 (page 222) but suppose that the car is launched from point 1 with initial speed v_0. Show that the path of minimum time to the fixed point 2 is still a cycloid, but with its cusp (the top point of the curve) a height $v_0^2/2g$ above point 1.

6.16 ★★ Use the result (6.41) of Problem 6.1 to prove that the geodesic (shortest path) between two given points on a sphere is a great circle. [*Hint:* The integrand $f(\phi, \phi', \theta)$ in (6.41) is independent of ϕ, so the Euler–Lagrange equation reduces to $\partial f/\partial \phi' = c$, a constant. This gives you ϕ' as a function of θ. You can avoid doing the final integral by the following trick: There is no loss of generality in choosing your z axis to pass through the point 1. Show that with this choice the constant c is necessarily zero, and describe the corresponding geodesics.]

6.17 ★★ Find the geodesics on the cone whose equation in cylindrical polar coordinates is $z = \lambda\rho$. [Let the required curve have the form $\phi = \phi(\rho)$.] Check your result for the case that $\lambda \to 0$.

6.18 ★★ Show that the shortest path between two given points in a plane is a straight line, using plane polar coordinates.

6.19 ★★ A surface of revolution is generated as follows: Two fixed points (x_1, y_1) and (x_2, y_2) in the x, y plane are joined by a curve $y = y(x)$. [Actually you'll make life easier if you start out writing this as $x = x(y)$.] The whole curve is now rotated about the x axis to generate a surface. Show that the curve for which the area of the surface is stationary has the form $y = y_0 \cosh[(x - x_0)/y_0]$, where x_0 and y_0 are constants. (This is often called the soap-bubble problem, since the resulting surface is usually the shape of a soap bubble held by two coaxial rings of radii y_1 and y_2.)

6.20 ★★ If you haven't done it, take a look at Problem 6.10. Here is a second situation in which you can find a "first integral" of the Euler–Lagrange equation: Argue that if it happens that the integrand $f(y, y', x)$ does not depend explicitly on x, that is, $f = f(y, y')$, then

$$\frac{df}{dx} = \frac{\partial f}{\partial y} y' + \frac{\partial f}{\partial y'} y''.$$

Use the Euler–Lagrange equation to replace $\partial f/\partial y$ on the right, and hence show that

$$\frac{df}{dx} = \frac{d}{dx}\left(y' \frac{\partial f}{\partial y'}\right).$$

This gives you the first integral

$$f - y' \frac{\partial f}{\partial y'} = \text{const.} \tag{6.43}$$

This can simplify several calculations. (See Problems 6.21 and 6.22 for examples.) In Lagrangian mechanics, where the independent variable is the time t, the corresponding result is that if the Lagrangian function is independent of t, then energy is conserved. (See Section 7.8.)

6.21 ★★ In Example 6.2 (page 222) we found the brachistochrone by exchanging the variables x and y. Here is a method that avoids that exchange: Write the time as in Equation (6.19) but using x as the variable of integration. Your integrand should have the form $f(y, y', x) = \sqrt{(y'^2 + 1)/y}$. Since this is independent of x, you can invoke the "first integral" (6.43) of Problem 6.20. Show that this differential equation leads you to the same integral for x as in Equation (6.23) and hence to the same curve as before.

6.22 ★★★ You are given a string of fixed length l with one end fastened at the origin O, and you are to place the string in the xy plane with its other end on the x axis in such a way as to enclose the maximum area between the string and the x axis. Show that the required shape is a semicircle. The area enclosed is of course $\int y\,dx$, but show that you can rewrite this in the form $\int_0^l f\,ds$, where s denotes the distance measured along the string from O, where $f = y\sqrt{1 - y'^2}$, and y' denotes dy/ds. Since f does not involve the independent variable s explicitly, you can exploit the "first integral" (6.43) of Problem 6.20.

6.23 ★★★ An aircraft whose airspeed is v_0 has to fly from town O (at the origin) to town P, which is a distance D due east. There is a steady gentle wind shear, such that $\mathbf{v}_{\text{wind}} = Vy\,\hat{\mathbf{x}}$, where x and y are measured east and north respectively. Find the path, $y = y(x)$, which the plane should follow to minimize its flight time, as follows: **(a)** Find the plane's ground speed in terms of v_0, V, ϕ (the angle by which the plane heads to the north of east), and the plane's position. **(b)** Write down the time of flight as an integral of the form $\int_0^D f\,dx$. Show that if we assume that y' and ϕ both remain small (as is certainly reasonable if the wind speed is not too large), then the integrand f takes the approximate form $f = (1 + \frac{1}{2}y'^2)/(1 + ky)$ (times an uninteresting constant) where $k = V/v_0$. **(c)** Write down the Euler–Lagrange equation that determines the best path. To solve it, make the intelligent guess that $y(x) = \lambda x(D - x)$, which clearly passes through the two towns. Show that it satisfies the Euler–Lagrange equation, provided $\lambda = (\sqrt{4 + 2k^2 D^2} - 2)/(kD^2)$. How far north does this path take the plane, if $D = 2000$ miles, $v_0 = 500$ mph, and the wind shear is $V = 0.5$ mph/mi? How much time does the plane save by following this path? [You'll probably want to use a computer to do this integral.]

6.24 ★★★ Consider a medium in which the refractive index n is inversely proportional to r^2; that is, $n = a/r^2$, where r is the distance from the origin. Use Fermat's principle, that the integral (6.3) is stationary, to find the path of a ray of light travelling in a plane containing the origin. [*Hint:* Use two-dimensional polar coordinates and write the path as $\phi = \phi(r)$. The Fermat integral should have the form $\int f(\phi, \phi', r)\,dr$, where $f(\phi, \phi', r)$ is actually independent of ϕ. The Euler–Lagrange equation therefore reduces to $\partial f/\partial \phi' = \text{const}$. You can solve this for ϕ' and then integrate to give ϕ as a function of r. Rewrite this to give r as a function of ϕ and show that the resulting path is a circle through the origin. Discuss the progress of the light around the circle.]

6.25 ★★★ Consider a single loop of the cycloid (6.26) with a fixed value of a, as shown in Figure 6.11. A car is released from rest at a point P_0 anywhere on the track between O and the lowest point P (that

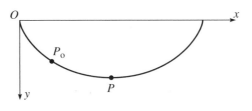

Figure 6.11   Problem 6.25

is, P_o has parameter $0 < \theta_o < \pi$). Show that the time for the cart to roll from P_o to P is given by the integral

$$\text{time}(P_o \rightarrow P) = \sqrt{\frac{a}{g}} \int_{\theta_o}^{\pi} \sqrt{\frac{1 - \cos\theta}{\cos\theta_o - \cos\theta}} \, d\theta$$

and prove that this time is equal to $\pi\sqrt{a/g}$. Since this is *independent of the position of* P_o, the cart takes the same time to roll from P_o to P, whether P_o is at O, or anywhere between O and P, even infinitesimally close to P. Explain qualitatively how this surprising result can possibly be true. [*Hint:* To do the mathematics, you have to make some cunning changes of variables. One route is this: Write $\theta = \pi - 2\alpha$ and then use the relevant trig identities to replace the cosines of θ by sines of α. Now substitute $\sin\alpha = u$ and do the remaining integral.]

SECTION 6.4   **More than Two Variables**

6.26 ★★ Give in detail the argument that leads from the stationary property of the integral (6.30) to the two Euler–Lagrange equations (6.34).

6.27 ★★ Prove that the shortest path between two points in three dimensions is a straight line. Write the path in the parametric form

$$x = x(u), \qquad y = y(u), \qquad \text{and} \qquad z = z(u)$$

and then use the three Euler–Lagrange equations corresponding to (6.34).

7

Lagrange's Equations

The theoretical development of the laws of motion of bodies is a problem of such interest and importance that it has engaged the attention of all the most eminent mathematicians since the invention of dynamics as a mathematical science by Galileo, and especially since the wonderful extension which was given to that science by Newton. Among the successors of those illustrious men, Lagrange has perhaps done more than any other analyst to give extent and harmony to such deductive researches, by showing that the most varied consequences respecting the motions of systems of bodies may be derived from one radical formula; the beauty of the methods so suiting the dignity of the results as to make of his great work a kind of scientific poem.

—William Rowan Hamilton, 1834

Armed with the ideas of the calculus of variations, we are ready to set up the version of mechanics published in 1788 by the Italian–French astronomer and mathematician Lagrange (1736–1813). The Lagrangian formulation has two important advantages over the earlier Newtonian formulation. First, Lagrange's equations, unlike Newton's, take the same form in any coordinate system. Second, in treating constrained systems, such as a bead sliding on a wire, the Lagrangian approach eliminates the forces of constraint (such as the normal force of the wire, which constrains the bead to remain on the wire). This greatly simplifies most problems, since the constraint forces are usually unknown, and this simplification comes at almost no cost, since we usually do not want to know these forces anyway.

In Section 7.1, I prove that Lagrange's equations are equivalent to Newton's second law for a particle moving unconstrained in three dimensions. The extension of this result to N unconstrained particles is surprisingly straightforward, and I leave the details for you to supply (Problem 7.7). In the next few sections, I take up the harder, and more interesting, case of constrained systems. I begin with some simple examples and important definitions (such as degrees of freedom). Then, in Section 7.4, I prove Lagrange's equations for a particle constrained to move on a curved surface (leaving the general case to Problem 7.13). Section 7.5 offers several examples, some of which are distinctly easier to set up in the Lagrangian formulation than in the Newtonian. In

Section 7.6, I introduce the curious terminology of "ignorable coordinates." Finally, after some summarizing remarks in Section 7.7, the chapter concludes with three sections on topics which, although very important, could be omitted on a first reading. In Section 7.8, I discuss how the laws of energy and momentum conservation appear in Lagrangian mechanics. Section 7.9 describes how Lagrange's equations can be extended to include magnetic forces, and Section 7.10 is an introduction to the idea of Lagrange multipliers.

Throughout this chapter, except in Section 7.9, I treat only the case that all nonconstraint forces are conservative or can, at least, be derived from a potential energy function. This restriction can be significantly relaxed, but already includes most of the applications that you are likely to meet in practice.

7.1   Lagrange's Equations for Unconstrained Motion

Consider a particle moving unconstrained in three dimensions, subject to a conservative net force $\mathbf{F}(\mathbf{r})$. The particle's kinetic energy is, of course,

$$T = \tfrac{1}{2}mv^2 = \tfrac{1}{2}m\dot{\mathbf{r}}^2 = \tfrac{1}{2}m(\dot{x}^2 + \dot{y}^2 + \dot{z}^2), \tag{7.1}$$

and its potential energy is

$$U = U(\mathbf{r}) = U(x, y, z). \tag{7.2}$$

The **Lagrangian function**, or just **Lagrangian**, is defined as

$$\mathcal{L} = T - U. \tag{7.3}$$

Notice first that the Lagrangian is the KE *minus* the PE. It is *not* the same as the total energy. You are certainly entitled to ask why the quantity $T - U$ should be of any interest. There seems to be no simple answer to this question except that it is, as we shall see directly. Notice also that I am using a script $\mathcal{L}$ for the Lagrangian[1] (to distinguish it from the angular momentum $\mathbf{L}$ and a length L) and that $\mathcal{L}$ depends on the particle's position (x, y, z) and its velocity $(\dot{x}, \dot{y}, \dot{z})$; that is, $\mathcal{L} = \mathcal{L}(x, y, z, \dot{x}, \dot{y}, \dot{z})$.

Let us consider the two derivatives,

$$\frac{\partial \mathcal{L}}{\partial x} = -\frac{\partial U}{\partial x} = F_x \tag{7.4}$$

and

$$\frac{\partial \mathcal{L}}{\partial \dot{x}} = \frac{\partial T}{\partial \dot{x}} = m\dot{x} = p_x. \tag{7.5}$$

---

[1] This notation gets into difficulty in field theories where the Lagrangian is often denoted by L, and $\mathcal{L}$ is used for the Lagrangian *density*, but this won't be a problem for us.

Differentiating the second equation with respect to time and remembering Newton's second law, $F_x = \dot{p}_x$ (I take for granted that our coordinate frame is inertial), we see that

$$\frac{\partial \mathcal{L}}{\partial x} = \frac{d}{dt} \frac{\partial \mathcal{L}}{\partial \dot{x}}. \tag{7.6}$$

In exactly the same way we can prove corresponding equations in y and z. Thus we have shown that Newton's second law implies the three *Lagrange equations* (in Cartesian coordinates so far):

$$\frac{\partial \mathcal{L}}{\partial x} = \frac{d}{dt} \frac{\partial \mathcal{L}}{\partial \dot{x}}, \qquad \frac{\partial \mathcal{L}}{\partial y} = \frac{d}{dt} \frac{\partial \mathcal{L}}{\partial \dot{y}}, \qquad \text{and} \qquad \frac{\partial \mathcal{L}}{\partial z} = \frac{d}{dt} \frac{\partial \mathcal{L}}{\partial \dot{z}}. \tag{7.7}$$

You can easily check that the argument just given works equally well in reverse, so that (for a single particle in Cartesian coordinates, at least) Newton's second law is exactly equivalent to the three Lagrange equations (7.7). The particle's path as determined by Newton's second law is the same as the path determined by the three Lagrange equations.

Our next step is to recognize that the three equations of (7.7) have exactly the form of the Euler–Lagrange equations (6.40). Therefore, they imply that the integral $S = \int \mathcal{L} \, dt$ is stationary for the path followed by the particle. That this integral, called the action integral, is stationary for the particle's path is called Hamilton's principle[2] (after its inventor, the Irish mathematician, Hamilton, 1805–1865) and can be restated as follows:

Hamilton's Principle

The actual path which a particle follows between two points 1 and 2 in a given time interval, t_1 to t_2, is such that the action integral

$$S = \int_{t_1}^{t_2} \mathcal{L} \, dt \tag{7.8}$$

is stationary when taken along the actual path.

Although we have so far proved this principle only for a single particle and in Cartesian coordinates, we are going to find that it is valid for a huge class of mechanical systems and for almost any choice of coordinates.

So far we have proved for a single particle that the following three statements are exactly equivalent:

---

[2] Try not to be confused by the unlucky circumstance that Hamilton's principle is one possible statement of the Lagrangian formulation of classical mechanics (as opposed to the Hamiltonian formulation).

1. A particle's path is determined by Newton's second law $\mathbf{F} = m\mathbf{a}$.
2. The path is determined by the three Lagrange equations (7.7), at least in Cartesian coordinates.
3. The path is determined by Hamilton's principle.

Hamilton's principle has found generalizations in many fields outside classical mechanics (field theories, for example) and has given a unity to various diverse areas of physics. In the twentieth century it has played an important role in the formulation of quantum theories. However, for our present purposes its great importance is that it lets us prove that Lagrange's equations hold in more-or-less any coordinate system:

Instead of the Cartesian coordinates $\mathbf{r} = (x, y, z)$, suppose that we wish to use some other coordinates. These could be spherical polar coordinates (r, θ, ϕ), or cylindrical polars (ρ, ϕ, z), or any set of "generalized coordinates" q_1, q_2, q_3, with the property that each position $\mathbf{r}$ specifies a unique value of (q_1, q_2, q_3) and vice versa; that is,

$$q_i = q_i(\mathbf{r}) \qquad \text{for } i = 1, 2, \text{ and } 3, \tag{7.9}$$

and

$$\mathbf{r} = \mathbf{r}(q_1, q_2, q_3). \tag{7.10}$$

These two equations guarantee that for any value of $\mathbf{r} = (x, y, z)$ there is a unique (q_1, q_2, q_3) and vice versa. Using (7.10) we can rewrite (x, y, z) and $(\dot{x}, \dot{y}, \dot{z})$ in terms of (q_1, q_2, q_3) and $(\dot{q}_1, \dot{q}_2, \dot{q}_3)$. Next, we can rewrite the Lagrangian $\mathcal{L} = \frac{1}{2}m\dot{\mathbf{r}}^2 - U(\mathbf{r})$ in terms of these new variables as

$$\mathcal{L} = \mathcal{L}(q_1, q_2, q_3, \dot{q}_1, \dot{q}_2, \dot{q}_3)$$

and the action integral as

$$S = \int_{t_1}^{t_2} \mathcal{L}(q_1, q_2, q_3, \dot{q}_1, \dot{q}_2, \dot{q}_3) \, dt.$$

Now, the value of the integral S is unaltered by this change of variables. Therefore, the statement that S is stationary for variations of the path around the correct path must still be true in our new coordinate system, and, by the results of Chapter 6, this means that the correct path must satisfy the three Euler–Lagrange equations,

$$\frac{\partial \mathcal{L}}{\partial q_1} = \frac{d}{dt}\frac{\partial \mathcal{L}}{\partial \dot{q}_1}, \qquad \frac{\partial \mathcal{L}}{\partial q_2} = \frac{d}{dt}\frac{\partial \mathcal{L}}{\partial \dot{q}_2}, \qquad \text{and} \qquad \frac{\partial \mathcal{L}}{\partial q_3} = \frac{d}{dt}\frac{\partial \mathcal{L}}{\partial \dot{q}_3}, \tag{7.11}$$

with respect to the new coordinates q_1, q_2, and q_3. Since these new coordinates are *any* set of generalized coordinates, the qualification "in Cartesian coordinates" can be omitted from the statement (2) above. This result — that Lagrange's equations have the same form for any choice of generalized coordinates — is one of the two main reasons that the Lagrangian formalism is so useful.

There is one point about our derivation of Lagrange's equations that is worth keeping at the back of your mind. A crucial step in our proof was the observation that (7.6) was equivalent to Newton's second law $F_x = \dot{p}_x$, which in turn is true only if the original frame in which we wrote down $\mathcal{L} = T - U$ is inertial. Thus, although

Lagrange's equations are true for any choice of generalized coordinates q_1, q_2, q_3 — and these generalized coordinates may in fact be the coordinates of a noninertial reference frame — we must nevertheless be careful that, when we first write down the Lagrangian $\mathcal{L} = T - U$, we do so in an inertial frame.

We can easily generalize Lagrange's equations to systems of many particles, but let us first look at a couple of simple examples.

EXAMPLE 7.1   One Particle in Two Dimensions; Cartesian Coordinates

Write down Lagrange's equations in Cartesian coordinates for a particle moving in a conservative force field in two dimensions and show that they imply Newton's second law. (Of course, we have already proved this, but it is worth seeing it worked out explicitly.)

The Lagrangian for a single particle in two dimensions is

$$\mathcal{L} = \mathcal{L}(x, y, \dot{x}, \dot{y}) = T - U = \tfrac{1}{2}m(\dot{x}^2 + \dot{y}^2) - U(x, y). \qquad (7.12)$$

To write down the Lagrange equations we need the derivatives

$$\frac{\partial \mathcal{L}}{\partial x} = -\frac{\partial U}{\partial x} = F_x \qquad \text{and} \qquad \frac{\partial \mathcal{L}}{\partial \dot{x}} = \frac{\partial T}{\partial \dot{x}} = m\dot{x}, \qquad (7.13)$$

with corresponding expressions for the y derivatives. Thus the two Lagrange equations can be rewritten as follows:

$$\left.\begin{array}{ccc}
\dfrac{\partial \mathcal{L}}{\partial x} = \dfrac{d}{dt}\dfrac{\partial \mathcal{L}}{\partial \dot{x}} & \Longleftrightarrow & F_x = m\ddot{x} \\[2mm]
\dfrac{\partial \mathcal{L}}{\partial y} = \dfrac{d}{dt}\dfrac{\partial \mathcal{L}}{\partial \dot{y}} & \Longleftrightarrow & F_y = m\ddot{y}
\end{array}\right\} \Longleftrightarrow \mathbf{F} = m\mathbf{a}. \qquad (7.14)$$

Notice how in (7.13) the derivative $\partial \mathcal{L}/\partial x$ is the x component of the force, and $\partial \mathcal{L}/\partial \dot{x}$ is the x component of the momentum (and similarly with the y components). When we use generalized coordinates $q_1, q_2, \cdots, q_n$, we shall find that $\partial \mathcal{L}/\partial q_i$, although not necessarily a force component, plays a role very similar to a force. Similarly, $\partial \mathcal{L}/\partial \dot{q}_i$, although not necessarily a momentum component, acts very like a momentum. For this reason we shall call these derivatives the **generalized force** and **generalized momentum** respectively; that is,

$$\frac{\partial \mathcal{L}}{\partial q_i} = (i\text{th component of generalized force}) \qquad (7.15)$$

and

$$\frac{\partial \mathcal{L}}{\partial \dot{q}_i} = (i\text{th component of generalized momentum}). \qquad (7.16)$$

With these notations, each of the Lagrange equations (7.11)

$$\frac{\partial \mathcal{L}}{\partial q_i} = \frac{d}{dt}\frac{\partial \mathcal{L}}{\partial \dot{q}_i}$$

takes the form

$$\text{(generalized force)} = \text{(rate of change of generalized momentum)} \qquad (7.17)$$

I shall illustrate these ideas in the next example.

EXAMPLE 7.2   One Particle in Two Dimensions; Polar Coordinates

Find Lagrange's equations for the same system, a particle moving in two dimensions, using polar coordinates.

As in all problems in Lagrangian mechanics, our first task is to write down the Lagrangian $\mathcal{L} = T - U$ in terms of the chosen coordinates. In this case we have been told to use polar coordinates, as sketched in Figure 7.1. This means the components of the velocity are $v_r = \dot{r}$ and $v_\phi = r\dot{\phi}$, and the kinetic energy is $T = \frac{1}{2}mv^2 = \frac{1}{2}m(\dot{r}^2 + r^2\dot{\phi}^2)$. Therefore, the Lagrangian is

$$\mathcal{L} = \mathcal{L}(r, \phi, \dot{r}, \dot{\phi}) = T - U = \tfrac{1}{2}m\left(\dot{r}^2 + r^2\dot{\phi}^2\right) - U(r, \phi). \qquad (7.18)$$

Given the Lagrangian, we now have only to write down the two Lagrange equations, one involving derivatives with respect to r and the other derivatives with respect to ϕ.

The r Equation

The equation involving derivatives with respect to r (the r equation) is

$$\frac{\partial \mathcal{L}}{\partial r} = \frac{d}{dt}\frac{\partial \mathcal{L}}{\partial \dot{r}}$$

or

$$mr\dot{\phi}^2 - \frac{\partial U}{\partial r} = \frac{d}{dt}(m\dot{r}) = m\ddot{r}. \qquad (7.19)$$

Since $-\partial U/\partial r$ is just F_r, the radial component of **F**, we can rewrite the r equation as

$$F_r = m(\ddot{r} - r\dot{\phi}^2), \qquad (7.20)$$

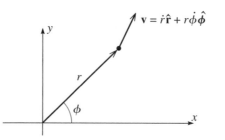

Figure 7.1   The velocity of a particle expressed in two-dimensional polar coordinates.

which you should recognize as $F_r = ma_r$, the r component of $\mathbf{F} = m\mathbf{a}$, first derived in Equation (1.48). (The term $-r\dot{\phi}^2$ is the infamous centripetal acceleration.) That is, when we use polar coordinates (r, ϕ), the Lagrange equation corresponding to r is just the radial component of Newton's second law. (Note, however, that the Lagrangian derivation avoided the tedious calculation of the components of the acceleration.) As we shall see directly, the ϕ equation works a bit differently and illustrates a remarkable feature of the Lagrangian approach.

The ϕ Equation

The Lagrange equation for the coordinate ϕ is

$$\frac{\partial \mathcal{L}}{\partial \phi} = \frac{d}{dt} \frac{\partial \mathcal{L}}{\partial \dot{\phi}} \tag{7.21}$$

or, substituting (7.18) for $\mathcal{L}$,

$$-\frac{\partial U}{\partial \phi} = \frac{d}{dt}(mr^2\dot{\phi}). \tag{7.22}$$

To interpret this equation, we need to relate the left side to the appropriate component of the force $\mathbf{F} = -\nabla U$. This requires that we know the components of ∇U in polar coordinates:

$$\nabla U = \frac{\partial U}{\partial r}\hat{\mathbf{r}} + \frac{1}{r}\frac{\partial U}{\partial \phi}\hat{\boldsymbol{\phi}}. \tag{7.23}$$

(If you don't remember this, see Problem 7.5.) The ϕ component of the force is just the coefficient of $\hat{\boldsymbol{\phi}}$ in $\mathbf{F} = -\nabla U$, that is,

$$F_\phi = -\frac{1}{r}\frac{\partial U}{\partial \phi}.$$

Thus the left side of (7.22) is rF_ϕ, which is simply the *torque* Γ on the particle about the origin. Meanwhile, the quantity $mr^2\dot{\phi}$ on the right can be recognized as the angular momentum L about the origin. Therefore, the ϕ equation (7.22) states that

$$\Gamma = \frac{dL}{dt}, \tag{7.24}$$

the familiar condition from elementary mechanics, that torque equals the rate of change of angular momentum.

The result (7.24) illustrates a wonderful feature of Lagrange's equations, that when we choose an appropriate set of generalized coordinates the corresponding Lagrange equations automatically appear in a corresponding, natural form. When we choose r and ϕ for our coordinates, the ϕ equation turns out to be the equation for angular momentum. In fact, the situation is even better than this. Recall that I introduced the notion of generalized force and generalized momentum in (7.15) and (7.16). In the present case, the ϕ component of the generalized force is just the torque,

$$(\phi \text{ component of generalized force}) = \frac{\partial \mathcal{L}}{\partial \phi} = \Gamma \text{ (torque)} \qquad (7.25)$$

and the corresponding component of the generalized momentum is

$$(\phi \text{ component of generalized momentum}) = \frac{\partial \mathcal{L}}{\partial \dot{\phi}} = L \text{ (angular momentum)}. \qquad (7.26)$$

With the "natural" choice for the coordinates (r and ϕ) the ϕ components of the generalized force and momentum turn out to be the corresponding "natural" quantities, the torque and the angular momentum.

Notice that the generalized "force" does not necessarily have the dimensions of force, nor the generalized "momentum" those of momentum. In the present case, the generalized force (ϕ component) is a torque (that is, force × distance) and the generalized momentum is an angular momentum (momentum × distance).

This example illustrates another feature of Lagrange's equations: The ϕ component $\partial \mathcal{L} / \partial \phi$ of the generalized force turned out to be the torque on the particle. If the torque happens to be zero, then the corresponding generalized momentum $\partial \mathcal{L} / \partial \dot{\phi}$ (the angular momentum, in this case) is conserved. Clearly this is a general result: The ith component of the generalized force is $\partial \mathcal{L} / \partial q_i$. If this happens to be zero, then the Lagrange equation

$$\frac{\partial \mathcal{L}}{\partial q_i} = \frac{d}{dt} \frac{\partial \mathcal{L}}{\partial \dot{q}_i}$$

says simply that the ith component $\partial \mathcal{L} / \partial \dot{q}_i$ of the generalized momentum is constant. That is, if $\mathcal{L}$ is independent of q_i, the ith component of the generalized force is zero, and the corresponding component of the generalized momentum is conserved. In practice, it is often easy to spot that a Lagrangian is independent of a coordinate q_i, and, if you can, then you immediately know a corresponding conservation law. We shall return to this point in Section 7.8.

Several Unconstrained Particles

The extension of the above ideas to a system of N unconstrained particles (a gas of N molecules, for instance) is very straightforward, and I shall leave you to fill in the details (Problems 7.6 and 7.7). Here I shall just sketch the argument for the case of two particles, mainly to show the form of Lagrange's equations for $N > 1$. For two particles, the Lagrangian is defined (exactly as before) as $\mathcal{L} = T - U$, but this now means that

$$\mathcal{L}(\mathbf{r}_1, \mathbf{r}_2, \dot{\mathbf{r}}_1, \dot{\mathbf{r}}_2) = \tfrac{1}{2} m_1 \dot{\mathbf{r}}_1^2 + \tfrac{1}{2} m_2 \dot{\mathbf{r}}_2^2 - U(\mathbf{r}_1, \mathbf{r}_2). \qquad (7.27)$$

As usual, the forces on the two particles are $\mathbf{F}_1 = -\nabla_1 U$ and $\mathbf{F}_2 = -\nabla_2 U$. Newton's second law can be applied to each particle and yields six equations,

$$F_{1x} = \dot{p}_{1x}, \qquad F_{1y} = \dot{p}_{1y}, \qquad \cdots, \qquad F_{2z} = \dot{p}_{2z}.$$

Exactly as in Equation (7.7), each of these six equations is equivalent to a corresponding Lagrange equation

$$\frac{\partial \mathcal{L}}{\partial x_1} = \frac{d}{dt} \frac{\partial \mathcal{L}}{\partial \dot{x}_1}, \qquad \frac{\partial \mathcal{L}}{\partial y_1} = \frac{d}{dt} \frac{\partial \mathcal{L}}{\partial \dot{y}_1}, \qquad \cdots, \qquad \frac{\partial \mathcal{L}}{\partial z_2} = \frac{d}{dt} \frac{\partial \mathcal{L}}{\partial \dot{z}_2}. \qquad (7.28)$$

These six equations imply that the integral $S = \int_{t_1}^{t_2} \mathcal{L} \, dt$ is stationary. Finally, we can change to any other suitable set of six coordinates $q_1, q_2, \cdots, q_6$. The statement that S is stationary must also be true in this new coordinate system, and this implies in turn that Lagrange's equations must be true with respect to the new coordinates:

$$\frac{\partial \mathcal{L}}{\partial q_1} = \frac{d}{dt} \frac{\partial \mathcal{L}}{\partial \dot{q}_1}, \qquad \frac{\partial \mathcal{L}}{\partial q_2} = \frac{d}{dt} \frac{\partial \mathcal{L}}{\partial \dot{q}_2}, \qquad \cdots, \qquad \frac{\partial \mathcal{L}}{\partial q_6} = \frac{d}{dt} \frac{\partial \mathcal{L}}{\partial \dot{q}_6}. \qquad (7.29)$$

An example of a set of six such generalized coordinates that we shall use repeatedly in Chapter 8 is this: In place of the six coordinates of $\mathbf{r}_1$ and $\mathbf{r}_2$, we could use the three coordinates of the CM position $\mathbf{R} = (m_1\mathbf{r}_1 + m_2\mathbf{r}_2)/(m_1 + m_2)$ and the three coordinates of the relative position $\mathbf{r} = \mathbf{r}_1 - \mathbf{r}_2$. We shall find that this choice of coordinates leads to a dramatic simplification. For now, the main point is simply that Lagrange's equations are automatically true in their standard form (7.29) with respect to these new, generalized coordinates.

The extension of these ideas to the case of N unconstrained particles is entirely straightforward, and I leave it for you to check. (See Problem 7.7.) The upshot is that there are $3N$ Lagrange equations

$$\frac{\partial \mathcal{L}}{\partial q_i} = \frac{d}{dt} \frac{\partial \mathcal{L}}{\partial \dot{q}_i}, \qquad [i = 1, 2, \cdots, 3N],$$

valid for any choice of the $3N$ coordinates $q_1, \cdots, q_{3N}$ needed to describe the N particles.

7.2   Constrained Systems; an Example

Perhaps the greatest advantage of the Lagrangian approach is that it can handle systems that are constrained so that they cannot move arbitrarily in the space that they occupy. A familiar example of a constrained system is the bead which is threaded on a wire — the bead can move along the wire, but not anywhere else. Another example of a very constrained system is a rigid body, whose individual atoms can only move in such a way that the distance between any two atoms is fixed. Before I discuss the nature of constraints in general, I shall discuss another simple example, the plane pendulum.

Consider the simple pendulum shown in Figure 7.2. A bob of mass m is fixed to a massless rod, which is pivoted at O and free to swing without friction in the xy plane. The bob moves in both the x and y directions, but it is constrained by the rod so that $\sqrt{x^2 + y^2} = l$ remains constant. In an obvious sense, only one of the coordinates is independent (as x changes, the variation of y is predetermined by the constraint equation), and we say that the system has only one degree of freedom. One way to express this would be to eliminate one of the coordinates, for instance by writing

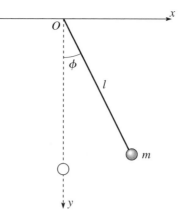

Figure 7.2   A simple pendulum. The bob of mass m is constrained by the rod to remain at distance l from O.

$y = \sqrt{l^2 - x^2}$ and expressing everything in terms of the one coordinate x. Although this is a perfectly legitimate way to proceed, a simpler way is to express both x and y in terms of the single parameter ϕ, the angle between the pendulum and its equilibrium position, as shown in Figure 7.2.

We can express all the quantities of interest in terms of ϕ. The kinetic energy is $T = \frac{1}{2}mv^2 = \frac{1}{2}ml^2\dot{\phi}^2$. The potential energy is $U = mgh$ where h denotes the height of the bob above its equilibrium position and is (as you should check) $h = l(1 - \cos\phi)$. Thus the potential energy is $U = mgl(1 - \cos\phi)$, and the Lagrangian is

$$\mathcal{L} = T - U = \tfrac{1}{2}ml^2\dot{\phi}^2 - mgl(1 - \cos\phi). \tag{7.30}$$

Whichever way we choose to proceed — to write everything in terms of x (or y) or ϕ — the Lagrangian is expressed in terms of a single generalized coordinate q and its time derivative $\dot{q}$, in the form $\mathcal{L} = \mathcal{L}(q, \dot{q})$. Now, it is a fact (which I shall not prove just yet) that once the Lagrangian is written in terms of this one variable (for a system with one degree of freedom), the evolution of the system again satisfies Lagrange's equation (just as we proved for an unconstrained particle in the previous section.) That is,

$$\frac{\partial \mathcal{L}}{\partial q} = \frac{d}{dt}\frac{\partial \mathcal{L}}{\partial \dot{q}}. \tag{7.31}$$

If we choose the angle ϕ as our generalized coordinate, then Lagrange's equation reads

$$\frac{\partial \mathcal{L}}{\partial \phi} = \frac{d}{dt}\frac{\partial \mathcal{L}}{\partial \dot{\phi}}. \tag{7.32}$$

The Lagrangian $\mathcal{L}$ is given by (7.30), and the needed derivatives are easily evaluated to give

$$-mgl\sin\phi = \frac{d}{dt}(ml^2\dot{\phi}) = ml^2\ddot{\phi}. \tag{7.33}$$

Referring to Figure 7.2 you can see that the left side of this equation is just the torque Γ exerted by gravity on the pendulum, while the term ml^2 is the pendulum's moment of inertia I. Since $\ddot{\phi}$ is the angular acceleration α, we see that Lagrange's equation for the simple pendulum simply reproduces the familiar result $\Gamma = I\alpha$.

7.3   Constrained Systems in General

Generalized Coordinates

Consider now an arbitrary system of N particles, $\alpha = 1, \cdots, N$ with positions $\mathbf{r}_\alpha$. We say that the parameters $q_1, \cdots, q_n$ are a set of **generalized coordinates** for the system if each position $\mathbf{r}_\alpha$ can be expressed as a function of $q_1, \cdots, q_n$, and possibly the time t,

$$\mathbf{r}_\alpha = \mathbf{r}_\alpha(q_1, \cdots, q_n, t) \qquad [\alpha = 1, \cdots, N], \qquad (7.34)$$

and conversely each q_i can be expressed in terms of the $\mathbf{r}_\alpha$ and possibly t,

$$q_i = q_i(\mathbf{r}_1, \cdots, \mathbf{r}_N, t) \qquad [i = 1, \cdots, n]. \qquad (7.35)$$

In addition, we require that the number of the generalized coordinates (n) is the smallest number that allows the system to be parametrized in this way. In our three-dimensional world, the number n of generalized coordinates for N particles is certainly no more than $3N$ and, for a constrained system, is usually less — sometimes dramatically so. For example, for a rigid body, the number of particles N may be of order 10^{23}, whereas the number of generalized coordinates n is 6 (three coordinates to give the position of the center of mass and three to give the orientation of the body).

To illustrate the relation (7.34), consider again the simple pendulum of Figure 7.2. There is one particle (the bob) and two Cartesian coordinates (since the pendulum is restricted to two dimensions). As we saw, there is just one generalized coordinate, which we took to be the angle ϕ. The analog of (7.34) is

$$\mathbf{r} \equiv (x, y) = (l \sin \phi, l \cos \phi) \qquad (7.36)$$

and expresses the two Cartesian coordinates x and y in terms of the one generalized coordinate ϕ.

The double pendulum shown in Figure 7.3 has two bobs, both confined to a plane, so it has four Cartesian coordinates, all of which can be expressed in terms of the two generalized coordinates ϕ_1 and ϕ_2. Specifically, if we put our origin at the suspension point of the top pendulum,

$$\mathbf{r}_1 = (l_1 \sin \phi_1, l_1 \cos \phi_1) = \mathbf{r}_1(\phi_1) \qquad (7.37)$$

and

$$\mathbf{r}_2 = (l_1 \sin \phi_1 + l_2 \sin \phi_2, l_1 \cos \phi_1 + l_2 \cos \phi_2) = \mathbf{r}_2(\phi_1, \phi_2). \qquad (7.38)$$

Notice that the components of $\mathbf{r_2}$ depend on ϕ_1 and ϕ_2.

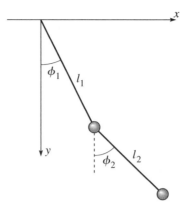

Figure 7.3   The positions of both masses in a double pendulum are uniquely specified by the two generalized coordinates ϕ_1 and ϕ_2, which can themselves be varied independently.

In these two examples, the transformation between the Cartesian and the generalized coordinates did not depend on the time t, but it is easy to think of examples in which it does. Consider the railroad car shown in Figure 7.4, which has a pendulum suspended from its ceiling and is being forced[3] to accelerate with a fixed acceleration a. It is natural to specify the position of the pendulum by the angle ϕ as usual, but we must recognize that, in the first instance, this gives the pendulum's position relative to the accelerating, and hence non-inertial, reference frame of the car. If we wish to specify the bob's position relative to an inertial frame, we can choose a frame fixed relative to the ground, and we can easily express the position relative to this inertial frame in terms of the angle ϕ. The position of the point of suspension relative to the ground is (if we choose our axes and origin properly) just $x_s = \frac{1}{2}at^2$, and the position of the bob is then easily seen to be

$$\mathbf{r} \equiv (x, y) = (l \sin \phi + \tfrac{1}{2}at^2, l \cos \phi) = \mathbf{r}(\phi, t). \tag{7.39}$$

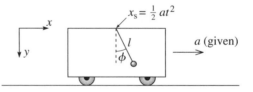

Figure 7.4   A pendulum is suspended from the roof of a railroad car that is being forced to accelerate with a fixed, known acceleration a.

---

[3] The word "forced" is often used to describe a motion that is imposed by some outside agent and is unaffected by the internal motions of the system. In the present example, the "forced" acceleration of the car is assumed to be the same whatever the oscillations of the pendulum.

The relation between $\mathbf{r}$ and the generalized coordinate ϕ depends on the time t, a possibility that I allowed for when writing (7.34).

We shall sometimes describe a set of coordinates $q_1, \cdots, q_n$ as **natural** if the relation (7.34) between the Cartesian coordinates $\mathbf{r}_\alpha$ and the generalized coordinates does *not* involve the time t. We shall find certain convenient properties of natural coordinates that do not generally apply to coordinates for which (7.34) *does* involve the time. Fortunately, as the name implies, there are many problems for which the most convenient choice of coordinates *is* also natural.[4]

Degrees of Freedom

The number of degrees of freedom of a system is the number of coordinates that can be independently varied in a small displacement — the number of independent "directions" in which the system can move from any given initial configuration. For example, the simple pendulum of Figure 7.2 has just one degree of freedom, while the double pendulum of Figure 7.3 has two. A particle that is free to move anywhere in three dimensions has three degrees of freedom, while a gas comprised of N particles has $3N$.

When the number of degrees of freedom of an N-particle system in three dimensions is less than $3N$, we say that the system is *constrained*. (In two dimensions, the corresponding number is $2N$ of course.) The bob of a simple pendulum, with one degree of freedom, is constrained. The two masses of a double pendulum, with two degrees of freedom, are constrained. The N atoms of a rigid body have just six degrees of freedom and are certainly constrained. Other examples are a bead constrained to slide on a fixed wire and a particle constrained to move on a fixed surface in three dimensions.

In all of the examples I have given so far, the number of degrees of freedom was equal to the number of generalized coordinates needed to describe the system's configuration. (The double pendulum has two degrees of freedom and needs two generalized coordinates, and so on.) A system with this natural-seeming property is said to be **holonomic**.[5] That is, a holonomic system has n degrees of freedom and can be described by n generalized coordinates, $q_1, \cdots, q_n$. Holonomic systems are easier to treat than nonholonomic, and in this book I shall restrict myself to holonomic systems.

You might imagine that all systems would be holonomic, or at least that nonholonomic systems would be rare and bizarrely complicated. In fact, there are some quite simple examples of nonholonomic systems. Consider, for instance, a hard rubber ball that is free to roll (but not to slide nor to spin about a vertical axis) on a horizontal table. Starting at any position (x, y) it can move in only two independent directions. Therefore, the ball has two degrees of freedom, and you might well imagine that its

---

[4] Natural coordinates are sometimes called *scleronomous*, and those that are not natural, *rheonomous*. I shall not use these outstandingly forgettable names. Nonnatural coordinates are also sometimes called *forced*, since a time dependence in the relation (7.34) is usually associated with a forced motion, such as the forced acceleration of the car in Figure 7.4.

[5] Many different definitions of "holonomic" can be found, not all of which are exactly equivalent.

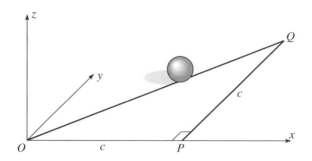

Figure 7.5   The right triangle OPQ lies in the xy plane with sides
OP and PQ of length c. If you roll a ball of circumference c around
OPQ, it will return to its starting point with a changed orientation.

configuration could be uniquely specified by two coordinates, x and y, of its center.
But consider the following: Let us place the ball at the origin O and make a mark on
its top point. Now, carry out the following three moves. (See Figure 7.5.) Roll the ball
along the x axis for a distance equal to the circumference c, to a point P, where the
mark will once again be on the top. Now roll it the same distance c in the y direction to
Q, where the mark is again on top. Finally roll it straight back to the origin along the
hypotenuse of the triangle OPQ. Since this last move has length $\sqrt{2}c$, it brings the ball
back to its starting point, but with the mark no longer on the top. The position (x, y)
has returned to its initial value, but the ball now has a different orientation. Evidently
the two coordinates (x, y) are not enough to specify a unique configuration. In fact,
three more numbers are needed to specify the orientation of the ball, and we need five
coordinates in all to specify the configuration completely. The ball has two degrees
of freedom but needs five generalized coordinates. Evidently it is a nonholonomic
system.

Although nonholonomic systems certainly exist, they are more complicated to analyze than holonomic systems, and I shall not discuss them further. For any holonomic
system with generalized coordinates $q_1, \cdots, q_n$ and potential energy $U(q_1, \cdots, q_n, t)$
(which may depend on the time t as described in Section 4.5), the evolution in time
is determined by the n Lagrange equations

$$\frac{\partial \mathcal{L}}{\partial q_i} = \frac{d}{dt}\frac{\partial \mathcal{L}}{\partial \dot{q}_i} \qquad [i = 1, \cdots, n], \tag{7.40}$$

where the Lagrangian $\mathcal{L}$ is defined as usual to be $\mathcal{L} = T - U$. I shall prove this result
in Section 7.4.

7.4   Proof of Lagrange's Equations with Constraints

We are now ready to prove Lagrange's equations for any holonomic system. To keep
things reasonably simple, I shall treat explicitly the case that there is just one particle.
(The generalization to arbitrary numbers of particles is fairly straightforward — see

Problem 7.13.) To be definite, I shall suppose the particle is constrained to move on a surface.[6] This means that it has two degrees of freedom and can be described by two generalized coordinates q_1 and q_2 that can vary independently.

We must recognize that there are two kinds of forces on the particle (or particles, in the general case). First, there are the forces of constraint: For a bead on a wire the constraining force is the normal force of the wire on the bead; for our particle, constrained to move on a surface, it is the normal force of the surface. For the atoms in a rigid body, the constraining forces are the interatomic forces that hold the atoms in place within the body. In general, the forces of constraint are not necessarily conservative, but this doesn't matter. One of the objectives of the Lagrangian approach is to find equations that do not involve the constraining forces, which we usually don't want to know anyway. (Notice, however, that if the constraining forces are nonconservative, Lagrange's equations in the simple unconstrained form of Section 7.1 certainly do *not* apply.) I shall denote the net constraining force on the particle by $\mathbf{F}_{\text{cstr}}$, which in our case is just the normal force of the surface to which the particle is confined.

Second, there are all the other "nonconstraint" forces on the particle, such as gravity. These are the forces with which we are usually concerned in practice, and I shall denote their resultant by $\mathbf{F}$. I shall assume that the nonconstraint forces all satisfy at least the second condition for conservatism, so that they are derivable from a potential energy, $U(\mathbf{r}, t)$, and

$$\mathbf{F} = -\nabla U(\mathbf{r}, t). \tag{7.41}$$

(If all the nonconstraint forces are actually conservative, then U is independent of t, but we don't need to assume this.) The total force on our particle is $\mathbf{F}_{\text{tot}} = \mathbf{F}_{\text{cstr}} + \mathbf{F}$.

Finally, I shall define the Lagrangian, as usual, to be

$$\mathcal{L} = T - U. \tag{7.42}$$

Since U is the potential energy for the nonconstraint forces only, this definition of $\mathcal{L}$ excludes the constraint forces. This correctly reflects that Lagrange's equations for a constrained system cleverly eliminate the constraint forces, as we shall see.

The Action Integral is Stationary at the Right Path

Consider any two points $\mathbf{r}_1$ and $\mathbf{r}_2$ through which the particle passes at times t_1 and t_2. I shall denote by $\mathbf{r}(t)$ the "right" path, the actual path that our particle follows between the two points, and by $\mathbf{R}(t)$ any neighboring "wrong" path between the same two points. It is convenient to write

$$\mathbf{R}(t) = \mathbf{r}(t) + \boldsymbol{\epsilon}(t), \tag{7.43}$$

---

[6] Actually, it is a bit hard to imagine how to constrain a particle to a single surface so that it can't jump off. If this worries you, you can imagine the particle sandwiched between two parallel surfaces with just enough gap between them to let it slide freely.

which defines $\boldsymbol{\epsilon}(t)$ as the infinitesimal vector pointing from $\mathbf{r}(t)$ on the right path to the corresponding point $\mathbf{R}(t)$ on the wrong path. Since I shall assume that both of the points $\mathbf{r}(t)$ and $\mathbf{R}(t)$ lie in the surface to which the particle is confined, the vector $\boldsymbol{\epsilon}(t)$ is contained in the same surface. Since both $\mathbf{r}(t)$ and $\mathbf{R}(t)$ go through the same endpoints, $\boldsymbol{\epsilon}(t) = 0$ at t_1 and t_2.

Let us denote by S the action integral

$$S = \int_{t_1}^{t_2} \mathcal{L}(\mathbf{R}, \dot{\mathbf{R}}, t)\, dt, \tag{7.44}$$

taken along any path $\mathbf{R}(t)$ lying in the constraining surface, and by S_o the corresponding integral taken along the right path $\mathbf{r}(t)$. As I shall now prove, the integral S is stationary for variations of the path $\mathbf{R}(t)$, when $\mathbf{R}(t) = \mathbf{r}(t)$ or, equivalently, when the difference $\boldsymbol{\epsilon}$ is zero. Another way to say this is that the difference in the action integrals

$$\delta S = S - S_o \tag{7.45}$$

is zero to first order in the distance $\boldsymbol{\epsilon}$ between the paths, and this is what I shall prove.

The difference (7.45) is the integral of the difference between the Lagrangians on the two paths,

$$\delta \mathcal{L} = \mathcal{L}(\mathbf{R}, \dot{\mathbf{R}}, t) - \mathcal{L}(\mathbf{r}, \dot{\mathbf{r}}, t). \tag{7.46}$$

If we substitute $\mathbf{R}(t) = \mathbf{r}(t) + \boldsymbol{\epsilon}(t)$ and

$$\mathcal{L}(\mathbf{r}, \dot{\mathbf{r}}, t) = T - U = \tfrac{1}{2} m \dot{\mathbf{r}}^2 - U(\mathbf{r}, t),$$

this becomes[7]

$$\delta \mathcal{L} = \tfrac{1}{2} m \left[(\dot{\mathbf{r}} + \dot{\boldsymbol{\epsilon}})^2 - \dot{\mathbf{r}}^2 \right] - [U(\mathbf{r} + \boldsymbol{\epsilon}, t) - U(\mathbf{r}, t)]$$

$$= m\, \dot{\mathbf{r}} \cdot \dot{\boldsymbol{\epsilon}} - \boldsymbol{\epsilon} \cdot \nabla U + O(\epsilon^2),$$

where $O(\epsilon^2)$ denotes terms involving squares and higher powers of ϵ and $\dot{\epsilon}$. Returning to the difference (7.45) in the two action integrals, we find that, to first order in ϵ,

$$\delta S = \int_{t_1}^{t_2} \delta \mathcal{L}\, dt = \int_{t_1}^{t_2} [m\, \dot{\mathbf{r}} \cdot \dot{\boldsymbol{\epsilon}} - \boldsymbol{\epsilon} \cdot \nabla U]\, dt. \tag{7.47}$$

The first term in the final integral can be integrated by parts. (Recall that this just means moving the time derivative from one factor to the other and changing the sign.) The difference $\boldsymbol{\epsilon}$ is zero at the two endpoints, so the endpoint contribution is zero, and we get

$$\delta S = -\int_{t_1}^{t_2} \boldsymbol{\epsilon} \cdot [m\ddot{\mathbf{r}} + \nabla U]\, dt. \tag{7.48}$$

---

[7] To understand the second term in the second line, recall that $f(\mathbf{r} + \boldsymbol{\epsilon}) - f(\mathbf{r}) \approx \boldsymbol{\epsilon} \cdot \nabla f$, for any scalar function $f(\mathbf{r})$. See Section 4.3.

Now, the path $\mathbf{r}(t)$ is the "right" path and satisfies Newton's second law. Therefore the term $m\ddot{\mathbf{r}}$ is just the total force on the particle, $\mathbf{F}_{\text{tot}} = \mathbf{F}_{\text{cstr}} + \mathbf{F}$. Meanwhile $\nabla U = -\mathbf{F}$. Therefore, the second term in (7.48) cancels the second piece of the first, and we are left with

$$\delta S = -\int_{t_1}^{t_2} \boldsymbol{\epsilon} \cdot \mathbf{F}_{\text{cstr}} \, dt. \tag{7.49}$$

But the constraint force $\mathbf{F}_{\text{cstr}}$ is normal to the surface in which our particle moves, while $\boldsymbol{\epsilon}$ lies in the surface. Therefore $\boldsymbol{\epsilon} \cdot \mathbf{F}_{\text{cstr}} = 0$, and we have proved that $\delta S = 0$. That is, the action integral is stationary at the right path, as claimed.[8]

The Final Proof

We have proved Hamilton's principle, that the action integral is stationary at the path which the particle actually follows. However, we have proved it, *not* for arbitrary variations of the path, but rather for those variations of path that are *consistent with the constraints* — that is, paths which lie in the surface to which our particle is constrained. This means that we cannot prove Lagrange's equations with respect to the three Cartesian coordinates. On the other hand, we *can* prove them with respect to the appropriate generalized coordinates. We are assuming that our particle is confined by holonomic constraints to move on a surface, that is, a two-dimensional subset of the full three-dimensional world. This means that the particle has two degrees of freedom and can be described by two generalized coordinates, q_1 and q_2, that can be varied independently. *Any* variation of q_1 and q_2 is consistent with the constraints.[9] Accordingly, we can rewrite the action integral in terms of q_1 and q_2 as

$$S = \int_{t_1}^{t_2} \mathcal{L}(q_1, q_2, \dot{q}_1, \dot{q}_2, t) \, dt, \tag{7.50}$$

and this integral is stationary for *any* variations of q_1 and q_2 about the correct path $[q_1(t), q_2(t)]$. Therefore, by the argument of Chapter 6 the correct path must satisfy the two Lagrange equations

$$\frac{\partial \mathcal{L}}{\partial q_1} = \frac{d}{dt}\frac{\partial \mathcal{L}}{\partial \dot{q}_1} \qquad \text{and} \qquad \frac{\partial \mathcal{L}}{\partial q_2} = \frac{d}{dt}\frac{\partial \mathcal{L}}{\partial \dot{q}_2}. \tag{7.51}$$

The proof that I have given here applies directly only to a single particle in three dimensions, constrained to move on a two-dimensional surface, but the main ideas of the general case are all present. The generalization is, for the most part, relatively

---

[8] The observation that the integrand in (7.49) is zero is really the crucial step in our proof. When you consider the generalization of the proof to an arbitrary constrained system (for instance, if you do Problem 7.13), you will find that there is a corresponding step and that the corresponding term is zero, for the same reason: The forces of constraint would do no work in a displacement that is consistent with the constraints. Indeed this is one possible definition of a force of constraint.

[9] For example, if our surface is a sphere, centered at the origin, then the generalized coordinates q_1, q_2 could be the two angles θ, ϕ of spherical polar coordinates. Any variation of θ and ϕ is consistent with the constraint that the particle remain on the sphere.

straightforward (see Problem 7.13), and meanwhile I hope that I have said enough to convince you of the truth of the general result: For any holonomic system, with n degrees of freedom and n generalized coordinates, and with the nonconstraint forces derivable from a potential energy $U(q_1, \cdots, q_n, t)$, the path followed by the system is determined by the n Lagrange equations

$$\frac{\partial \mathcal{L}}{\partial q_i} = \frac{d}{dt}\frac{\partial \mathcal{L}}{\partial \dot{q}_i} \qquad [i = 1, \cdots, n], \qquad (7.52)$$

where $\mathcal{L}$ is the Lagrangian $\mathcal{L} = T - U$ and $U = U(q_1, \cdots, q_n, t)$ is the total potential energy corresponding to all the forces excluding the forces of constraint.

It was essential to our proof of Lagrange's equations that the nonconstraint forces be conservative (or, at a minimum, that they satisfy the second condition for conservatism) so that they are derivable from a potential energy, $\mathbf{F} = -\nabla U$. If this is not true, then Lagrange's equations may not hold, at least in the form (7.52). An obvious example of a force that does not satisfy this condition is sliding friction. Sliding friction cannot be regarded as a force of constraint (it is not normal to the surface) and cannot be derived from a potential energy. Thus, when sliding friction is present, Lagrange's equations do not hold in the form (7.52). Lagrange's equations can be modified to include forces like friction (see Problem 7.12), but the result is clumsy and I shall confine myself to situations where the equations (7.52) do hold.

7.5   Examples of Lagrange's Equations

In this section I present five examples of the use of Lagrange's equations. The first two are sufficiently simple that they can be easily solved within the Newtonian formalism. My main purpose for including them is just to give you experience with using the Lagrangian approach. Nevertheless, even these simple cases show some advantages of the Lagrangian over the Newtonian formalism; in particular, we shall see how the Lagrangian approach obviates any need to consider the forces of constraint. The last three examples are sufficiently complex that solution using the Newtonian approach requires considerable ingenuity; by contrast, the Lagrangian approach lets us write down the equations of motion almost without thinking.

The examples given here illustrate an important point to recognize about Lagrange's equations: The Lagrangian formalism always (or nearly always) gives a straightforward means of writing down the equations of motion. On the other hand, it cannot guarantee that the resulting equations are easy to solve. If we are very lucky, the equations of motion may have an analytic solution, but, even when they do not, they are the essential first step to understanding the solutions and they often suggest a starting point for an approximate solution. The equations of motion can give simple answers to certain subsidiary questions. (For instance, once we have the equations of motion, we can usually find the positions of equilibrium of a system very easily.) And we *can* always solve the equations of motion numerically for given initial conditions.

The Lagrangian method is so important that it certainly deserves more than just five examples. However, the crucial thing is that *you* work through several examples yourself; therefore I have given plenty of problems at the end of the chapter, and it is essential that you work several of these as soon as possible after reading this section.

EXAMPLE 7.3   Atwood's Machine

Consider the Atwood machine first met in Figure 4.15 and shown again in Figure 7.6, in which the two masses m_1 and m_2 are suspended by an inextensible string (length l) which passes over a massless pulley with frictionless bearings and radius R. Write down the Lagrangian $\mathcal{L}$, using the distance x as generalized coordinate, find the Lagrange equation of motion, and solve it for the acceleration $\ddot{x}$. Compare your results with the Newtonian solution.

Because the string has fixed length, the heights x and y of the two masses cannot vary independently. Rather, $x + y + \pi R = l$, the length of the string, so that y can be expressed in terms of x as

$$y = -x + \text{const.} \qquad (7.53)$$

Therefore, we can use x as our one generalized coordinate. From (7.53) we see that $\dot{y} = -\dot{x}$, so that the kinetic energy of the system is

$$T = \tfrac{1}{2}m_1\dot{x}^2 + \tfrac{1}{2}m_2\dot{y}^2 = \tfrac{1}{2}(m_1 + m_2)\dot{x}^2,$$

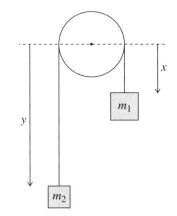

Figure 7.6   An Atwood machine consisting of two masses, m_1 and m_2, suspended by a massless inextensible string that passes over a massless, frictionless pulley of radius R. Because the string's length is fixed, the position of the whole system can be specified by a single variable, which we can take to be the distance x.

while the potential energy is

$$U = -m_1 gx - m_2 gy = -(m_1 - m_2)gx + \text{const.}$$

Combining these, we find the Lagrangian

$$\mathcal{L} = T - U = \tfrac{1}{2}(m_1 + m_2)\dot{x}^2 + (m_1 - m_2)gx, \qquad (7.54)$$

where I have dropped an uninteresting constant.

The Lagrange equation of motion is just

$$\frac{\partial \mathcal{L}}{\partial x} = \frac{d}{dt}\frac{\partial \mathcal{L}}{\partial \dot{x}}$$

or, substituting (7.54) for $\mathcal{L}$,

$$(m_1 - m_2)g = (m_1 + m_2)\ddot{x}, \qquad (7.55)$$

which we can solve at once to give the desired acceleration

$$\ddot{x} = \frac{m_1 - m_2}{m_1 + m_2}g. \qquad (7.56)$$

By choosing m_1 and m_2 fairly close together, one can make this acceleration much less than g, and hence much easier to measure. Therefore, the Atwood machine gave an early and reasonably accurate method for measuring g.

The corresponding Newtonian solution requires us to write down Newton's second law for each of the masses separately. The net force on m_1 is $m_1 g - F_t$ where F_t is the tension in the string. (This is the force of constraint and needed no consideration in the Lagrangian solution.) Thus Newton's second law for m_1 is

$$m_1 g - F_t = m_1 \ddot{x}.$$

In the same way, Newton's second law for m_2 reads

$$F_t - m_2 g = m_2 \ddot{x}.$$

(Remember that the upward acceleration of m_2 is the same as the *downward* acceleration of m_1.) We see that the Newtonian approach has given us two equations for two unknowns, the required acceleration $\ddot{x}$ and the force of constraint F_t. By adding these two equations, we can eliminate F_t and arrive at precisely the equation (7.55) of the Lagrangian method and thence the same value (7.56) for $\ddot{x}$.

The Newtonian solution of the Atwood machine is too simple for us to get very excited by an alternative solution. Nevertheless, this simple example does illustrate how the Lagrangian approach allows us to ignore the unknown (and usually uninteresting) force of constraint and to eliminate at least one step of the Newtonian solution.

EXAMPLE 7.4 A Particle Confined to Move on a Cylinder

Consider a particle of mass m constrained to move on a frictionless cylinder of radius R, given by the equation $\rho = R$ in cylindrical polar coordinates (ρ, ϕ, z), as shown in Figure 7.7. Besides the force of constraint (the normal force of the cylinder), the only force on the mass is a force $\mathbf{F} = -k\mathbf{r}$ directed toward the origin. (This is the three dimensional version of the Hooke's-law force.) Using z and ϕ as generalized coordinates, find the Lagrangian $\mathcal{L}$. Write down and solve Lagrange's equations and describe the motion.

Since the particle's coordinate ρ is fixed at $\rho = R$, we can specify its position by giving just z and ϕ, and since these two coordinates can vary independently the system has two degrees of freedom and we can use (z, ϕ) as generalized coordinates. The velocity has $v_\rho = 0$, $v_\phi = R\dot{\phi}$, and $v_z = \dot{z}$. Therefore the kinetic energy is

$$T = \tfrac{1}{2}mv^2 = \tfrac{1}{2}m(R^2\dot{\phi}^2 + \dot{z}^2).$$

The potential energy for the force $\mathbf{F} = -k\mathbf{r}$ is (Problem 7.25) $U = \tfrac{1}{2}kr^2$, where r is the distance from the origin to the particle, given by $r^2 = R^2 + z^2$ (see Figure 7.7). Therefore

$$U = \tfrac{1}{2}k(R^2 + z^2),$$

and the Lagrangian is

$$\mathcal{L} = \tfrac{1}{2}m(R^2\dot{\phi}^2 + \dot{z}^2) - \tfrac{1}{2}k(R^2 + z^2). \tag{7.57}$$

Since the system has two degrees of freedom, there are two equations of motion. The z equation is

$$\frac{\partial \mathcal{L}}{\partial z} = \frac{d}{dt}\frac{\partial \mathcal{L}}{\partial \dot{z}} \qquad \text{or} \qquad -kz = m\ddot{z}. \tag{7.58}$$

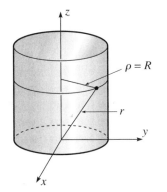

Figure 7.7 A mass m is confined to the surface of the cylinder $\rho = R$ and subject to a Hooke's law force $\mathbf{F} = -k\mathbf{r}$.

The ϕ equation is even simpler. Since $\mathcal{L}$ does not depend on ϕ, it follows that $\partial\mathcal{L}/\partial\phi = 0$ and the ϕ equation is just

$$\frac{\partial\mathcal{L}}{\partial\phi} = \frac{d}{dt}\frac{\partial\mathcal{L}}{\partial\dot\phi} \qquad \text{or} \qquad 0 = \frac{d}{dt}mR^2\dot\phi. \tag{7.59}$$

The z equation (7.58) tells us that the mass executes simple harmonic motion in the z direction, with $z = A\cos(\omega t - \delta)$. The ϕ equation (7.59) tells us that the quantity $mR^2\dot\phi$ is constant, that is, that the z component of the angular momentum is conserved — a result we could have anticipated since there is no torque in this direction. Because ρ is fixed, this implies simply that $\dot\phi$ is constant, and the mass moves around the cylinder with constant angular velocity $\dot\phi$, at the same time that it moves up and down in the z direction in simple harmonic motion.

These two examples illustrate the steps to be followed in solving any problem by the Lagrangian method (provided all constraints are holonomic and the nonconstraint forces are derivable from a potential energy, as we are assuming):

1. Write down the kinetic and potential energies and hence the Lagrangian $\mathcal{L} = T - U$, using any convenient inertial reference frame.
2. Choose a convenient set of n generalized coordinates $q_1, \cdots, q_n$ and find expressions for the original coordinates of step 1 in terms of your chosen generalized coordinates. (Steps 1 and 2 can be done in either order.)
3. Rewrite $\mathcal{L}$ in terms of $q_1, \cdots, q_n$ and $\dot q_1, \cdots, \dot q_n$.
4. Write down the n Lagrange equations (7.52).

As we shall see, these four steps provide an almost infallible route to the equations of motion of any system, no matter how complex. Whether the resulting equations can be easily solved is another matter, but even when they cannot, just having them is a huge step toward understanding a system and an essential step to finding approximate or numerical solutions.

The next two examples illustrate how the Lagrangian approach can give the equations of motion, almost effortlessly, for problems that would require considerable care and ingenuity using Newtonian methods.

EXAMPLE 7.5   A Block Sliding on a Wedge

Consider the block and wedge shown in Figure 7.8. The block (mass m) is free to slide on the wedge, and the wedge (mass M) can slide on the horizontal table, both with negligible friction. The block is released from the top of the wedge, with both initially at rest. If the wedge has angle α and the length of its sloping face is l, how long does the block take to reach the bottom?

The system has two degrees of freedom, and a good choice of the two generalized coordinates is, as shown, the distance q_1 of the block from the top of the wedge and the distance q_2 of the wedge from any convenient fixed point on the table. The quantity we need to find is the acceleration $\ddot q_1$ of the block relative to the wedge, since with this we can quickly find the time required to slide the length of the wedge. Our first task is to write down the Lagrangian, and

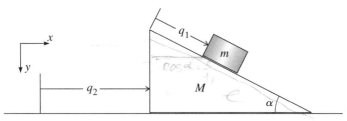

Figure 7.8   A block of mass m slides down a wedge of mass M, which is free to slide over the horizontal table.

it is often safest to do this in Cartesian coordinates, and then rewrite it in terms of the chosen generalized coordinates.

The kinetic energy of the wedge is just $T_M = \frac{1}{2}M\dot{q}_2^2$ but that of the block is more complicated. The block's velocity *relative to the wedge* is $\dot{q}_1$ down the slope, but the wedge itself has a horizontal velocity $\dot{q}_2$ relative to the table. The velocity of the block relative to the inertial frame of the table is the vector sum of these two. Resolving into rectangular components (x to the right, y downward), we find for the velocity of the block relative to the table

$$\mathbf{v} = (v_x, v_y) = (\dot{q}_1 \cos\alpha + \dot{q}_2, \dot{q}_1 \sin\alpha).$$

Thus the kinetic energy of the block is

$$T_m = \tfrac{1}{2}m(v_x^2 + v_y^2) = \tfrac{1}{2}m(\dot{q}_1^2 + \dot{q}_2^2 + 2\dot{q}_1\dot{q}_2 \cos\alpha).$$

(I used the identity $\cos^2\alpha + \sin^2\alpha = 1$ to simplify this.) The total kinetic energy of the system is

$$T = T_M + T_m = \tfrac{1}{2}(M + m)\dot{q}_2^2 + \tfrac{1}{2}m(\dot{q}_1^2 + 2\dot{q}_1\dot{q}_2 \cos\alpha). \qquad (7.60)$$

The potential energy of the wedge is a constant, which we may as well take to be zero. That of the block is $-mgy$, where $y = q_1 \sin\alpha$ is the height of the block measured down from the top of the wedge. Therefore

$$U = -mgq_1 \sin\alpha$$

and the Lagrangian is

$$\mathcal{L} = T - U = \tfrac{1}{2}(M + m)\dot{q}_2^2 + \tfrac{1}{2}m(\dot{q}_1^2 + 2\dot{q}_1\dot{q}_2 \cos\alpha) + mgq_1 \sin\alpha. \qquad (7.61)$$

Once we have found the Lagrangian in terms of the generalized coordinates q_1 and q_2, all we have to do is to write down the two Lagrange equations, one for q_1 and one for q_2, and then solve them. The q_2 equation (which is a little simpler) is

$$\frac{\partial\mathcal{L}}{\partial q_2} = \frac{d}{dt}\frac{\partial\mathcal{L}}{\partial\dot{q}_2} \qquad (7.62)$$

but, since $\mathcal{L}$ in (7.61) is clearly independent of q_2, this just tells us that the generalized momentum $\partial \mathcal{L}/\partial \dot{q}_2$ is constant,

$$M\dot{q}_2 + m(\dot{q}_2 + \dot{q}_1 \cos\alpha) = \text{const} \tag{7.63}$$

— a result you will recognize as conservation of the total momentum in the x direction (and something you could have written down without any help from Lagrange).

The q_1 equation

$$\frac{\partial \mathcal{L}}{\partial q_1} = \frac{d}{dt} \frac{\partial \mathcal{L}}{\partial \dot{q}_1} \tag{7.64}$$

is more complicated, since neither derivative is zero. Sustituting (7.61) for $\mathcal{L}$, we can write this as

$$mg \sin\alpha = \frac{d}{dt} m(\dot{q}_1 + \dot{q}_2 \cos\alpha)$$

$$= m(\ddot{q}_1 + \ddot{q}_2 \cos\alpha). \tag{7.65}$$

Differentiating (7.63) we see that

$$\ddot{q}_2 = -\frac{m}{M+m} \ddot{q}_1 \cos\alpha, \tag{7.66}$$

which lets us eliminate $\ddot{q}_2$ from (7.65) and solve for $\ddot{q}_1$:

$$\ddot{q}_1 = \frac{g \sin\alpha}{1 - \dfrac{m \cos^2\alpha}{M+m}}. \tag{7.67}$$

Armed with this value for $\ddot{q}_1$ we can quickly answer the original question: Since the acceleration down the slope is constant, the distance traveled down the slope in time t is $\frac{1}{2}\ddot{q}_1 t^2$, and the time to travel the length l is just $\sqrt{2l/\ddot{q}_1}$, with $\ddot{q}_1$ given by (7.67). More interesting than this answer is to check that the formula (7.67) for $\ddot{q}_1$ agrees with common sense in various special cases. For example, if $\alpha = 90°$, (7.67) implies that $\ddot{q}_1 = g$, which is clearly right; and, if $M \to \infty$, (7.67) implies that $\ddot{q}_1 \to g \sin\alpha$, which is the well-known acceleration for a block on a fixed incline and clearly makes sense. I leave it as an exercise (Problem 7.19) to check that in the limit that $M \to 0$, our answers agree with what you could have predicted.

EXAMPLE 7.6   A Bead on a Spinning Wire Hoop

A bead of mass m is threaded on a frictionless circular wire hoop of radius R. The hoop lies in a vertical plane, which is forced to rotate about the hoop's vertical diameter with constant angular velocity $\dot{\phi} = \omega$, as shown in Figure 7.9. The bead's position on the hoop is specified by the angle θ measured up from the vertical. Write down the Lagrangian for the system in terms of the

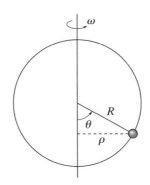

Figure 7.9   A bead is free to move around the frictionless wire hoop, which is spinning at a fixed rate ω about its vertical axis. The bead's position is specified by the angle θ; its distance from the axis of rotation is $\rho = R \sin \theta$.

generalized coordinate θ and find the equation of motion for the bead. Find any equilibrium positions at which the bead can remain with θ constant, and explain their locations in terms of statics and the "centrifugal force" $m\omega^2 \rho$ (where ρ is the bead's distance from the axis). Use the equation of motion to discuss the stability of the equilibrium positions.

Our first task is to write down the Lagrangian. Relative to a nonrotating frame, the bead has velocity $R\dot{\theta}$ tangential to the hoop and $\rho\omega = (R \sin \theta)\omega$ normal to the hoop (the latter due to the spinning of the hoop with angular velocity ω). Thus the kinetic energy is $T = \frac{1}{2}mv^2 = \frac{1}{2}mR^2(\dot{\theta}^2 + \omega^2 \sin^2 \theta)$. The gravitational potential energy is easily seen to be $U = mgR(1 - \cos \theta)$, measured from the bottom of the hoop. Therefore, the Lagrangian is

$$\mathcal{L} = \tfrac{1}{2}mR^2(\dot{\theta}^2 + \omega^2 \sin^2 \theta) - mgR(1 - \cos \theta), \qquad (7.68)$$

and the Lagrange equation is

$$\frac{\partial \mathcal{L}}{\partial \theta} = \frac{d}{dt}\frac{\partial \mathcal{L}}{\partial \dot{\theta}} \qquad \text{or} \qquad mR^2\omega^2 \sin \theta \cos \theta - mgR \sin \theta = mR^2\ddot{\theta}.$$

Dividing through by mR^2, we arrive at the desired equation of motion:

$$\ddot{\theta} = (\omega^2 \cos \theta - g/R) \sin \theta. \qquad (7.69)$$

Although this equation cannot be solved analytically in terms of elementary functions, it can, nevertheless, tell us lots about the system's behavior. To illustrate this, let us use (7.69) to find the equilibrium positions of the bead. An equilibrium point is any value of θ — call it θ_o — satisfying the following condition: If the bead is placed at rest ($\dot{\theta} = 0$) at $\theta = \theta_o$, then it will remain at rest at θ_o. This condition is guaranteed if $\ddot{\theta} = 0$. (To see this, note that if $\ddot{\theta} = 0$, then $\dot{\theta}$ doesn't change and remains zero, which means that θ doesn't change and remains equal to θ_{o}.) Thus to find the equilibrium positions we have only to equate the right side of (7.69) to zero:

$$(\omega^2 \cos\theta - g/R)\sin\theta = 0. \tag{7.70}$$

This equation is satisfied if either of the two factors is zero. The factor $\sin\theta$ is zero if $\theta = 0$ or π. Thus the bead can remain at rest at the bottom or top of the hoop. The first factor in (7.70) vanishes when

$$\cos\theta = \frac{g}{\omega^2 R}.$$

Since $|\cos\theta|$ must be less than or equal to 1, the first factor can vanish only when $\omega^2 \geq g/R$. When this condition is satisfied, there are two more equilibrium positions at

$$\theta_{\mathrm{o}} = \pm \arccos\left(\frac{g}{\omega^2 R}\right). \tag{7.71}$$

We conclude that when the hoop is rotating slowly ($\omega^2 < g/R$), there are just two equilibrium positions, at the bottom and top of the hoop, but when it rotates fast enough ($\omega^2 > g/R$), there are two more, symmetrically placed on either side of the bottom, as given by (7.71).[10]

Perhaps the simplest way to understand the various equilibrium positions is in terms of the "centrifugal force." In most introductory physics courses, the centrifugal force is dismissed as an abomination to be avoided by all right-thinking physicists. As long as we confine our attention to inertial frames, this is a correct (and certainly a safe) point of view. Nevertheless, as we shall see in Chapter 9, from the point of view of a noninertial rotating frame there is a perfectly real centrifugal force $m\omega^2\rho$ (perhaps more familiar as mv^2/ρ), where ρ is the object's distance from the axis of rotation. Thus, taking the point of view of a fly perched on the rotating hoop, we can understand the equilibrium positions as follows: At the bottom or top of the hoop, the bead is on the axis of rotation and $\rho = 0$; therefore, the centrifugal force $m\omega^2\rho$ is zero. Furthermore, the force of gravity is normal to the hoop, so there is no force tending to move the bead along the wire and the bead remains at rest. The other two equilibrium points are a little subtler: At any position off the axis (such as that shown in Figure 7.9) the centrifugal force is nonzero and has a component pushing the bead *outward* along the wire; meanwhile the force of gravity has a component pulling the bead *inward* along the wire (provided the bead is below the halfway marks, $\theta = \pm\pi/2$). At either of the points given by (7.71), these two components are balanced (check this for yourself — Problem 7.28) and the bead can remain at rest.

An equilibrium point θ_{o} is not especially interesting unless it is *stable* — that is, the bead, if nudged a little away from θ_{o}, moves back toward θ_{o}. Using our

---

[10] Notice that when $\omega^2 = g/R$ the two extra positions given by (7.71) have just come into existence and coincide with the first point at the bottom with $\theta = \pm 0$.

equation of motion (7.69), we can easily address this issue, and I'll start with the equilibrium at the bottom, $\theta = 0$. As long as θ remains close to 0, we can set $\cos\theta \approx 1$ and $\sin\theta \approx \theta$ and approximate the equation by

$$\ddot{\theta} = (\omega^2 - g/R)\theta \qquad [\theta \text{ near } 0]. \qquad (7.72)$$

If the hoop is rotating slowly ($\omega^2 < g/R$), this has the form

$$\ddot{\theta} = (\text{negative number})\theta.$$

If we nudge the bead to the right ($\theta > 0$), then since θ is positive $\ddot{\theta}$ is negative, and the bead accelerates to the left, that is, back toward the bottom. If we nudge it to the left, ($\theta < 0$), then $\ddot{\theta}$ becomes positive, and the bead accelerates to the right, which is again back toward the bottom. Either way, the bead returns toward the equilibrium, which is, therefore, *stable*.

If we speed up the rotation of the hoop, so that $\omega^2 > g/R$, then the approximate equation of motion (7.72) takes the form

$$\ddot{\theta} = (\text{positive number})\theta.$$

Now a small displacement to the right makes $\ddot{\theta}$ positive, and the bead accelerates away from the bottom. Similarly a displacement to the left makes $\ddot{\theta}$ negative, and again the bead accelerates away from the bottom. Thus, as we increase ω past the critical value where $\omega^2 = g/R$, the equilibrium at the bottom changes from stable to unstable.

The equilibrium at the top ($\theta = \pi$) is always unstable (see Problem 7.28). This is easy to understand from our discussion of the centrifugal force. Near the top of the hoop, both the centrifugal and gravitational forces tend to push the bead away from the top, so there is no chance of a restoring force to pull it back to the equilibrium position.

The other two equilibrium positions only exist when $\omega^2 > g/R$, and are easily seen to be stable: The equation of motion (7.69) is

$$\ddot{\theta} = (\omega^2 \cos\theta - g/R) \sin\theta. \qquad (7.73)$$

To be definite, let us consider the equilibrium on the right with $0 < \theta < \pi/2$. At the equilibrium point, the term in parenthesis on the right of (7.73) is zero, while $\sin\theta$ is positive. If we increase θ a little (bead moves up and to the right), $\sin\theta$ remains positive, but the term in parenthesis becomes negative. (Remember, $\cos\theta$ is a decreasing function in this quadrant.) Thus $\ddot{\theta}$ becomes negative, and the bead accelerates back toward its equilibrium point. If we decrease θ a little from the equilibrium, then $\ddot{\theta}$ becomes positive, and again the bead accelerates back toward equilibrium. Therefore the equilibrium on the right is stable. As you would expect, a similar analysis shows that the same is true of the equilibrium on the left.

We arrive at the following interesting story: When the hoop is rotating slowly ($\omega^2 < g/R$), there is just one stable equilibrium, at $\theta = 0$. If we speed up the rotation, then as ω passes the critical value where $\omega^2 = g/R$, this original equilibrium becomes unstable, but two new stable equilibrium points appear, emerging from $\theta = 0$ and moving out to the right and left as we increase ω still more. This phenomenon — the disappearance of one stable equilibrium and the simultaneous appearance of two others diverging from the same point — is called a *bifurcation* and will be one of our principal topics in Chapter 12 on chaos theory.

It is interesting to note that the device of this example was used by James Watt (1736–1829) as a governor for his steam engines. The device rotated with the engine, and as the engine sped up the bead rose on the hoop. When the angular velocity ω reached some predetermined maximum allowable value, the bead, arriving at a corresponding height, caused the supply of steam to be shut off.

This example illustrates another strength of the Lagrangian method that was mentioned back in Section 7.1: The generalized coordinates can even be coordinates relative to a noninertial reference frame, just as long as the frame in which the Lagrangian $\mathcal{L} = T - U$ was originally written was inertial. In this example, the angle θ was the polar angle of the bead, measured in the noninertial rotating frame of the hoop, but the Lagrangian (7.68) was defined as $\mathcal{L} = T - U$ with T and U evaluated in the inertial frame relative to which the hoop rotates.[11]

In the next and final example of this section, we pursue the previous example of the bead on the rotating hoop, and obtain approximate solutions of the equation of motion in the neighborhood of the stable equilibrium points.

EXAMPLE 7.7   Oscillations of the Bead near Equilibrium

Consider again the bead of the previous example and use the equation of motion to find the bead's approximate behavior in the neighborhood of the stable equilibrium positions.

When $\omega^2 < g/R$, the only stable equilibrium is at the bottom of the hoop with $\theta = 0$. As long as θ remains small, we can approximate the equation of motion (7.73) by setting $\sin\theta \approx \theta$ and $\cos\theta \approx 1$ to give

$$\ddot{\theta} = -(g/R - \omega^2)\theta \qquad [\theta \text{ near } 0]$$

$$= -\Omega^2 \theta \tag{7.74}$$

where the second line introduces the frequency

$$\Omega = \sqrt{g/R - \omega^2}.$$

As long as $\omega^2 < g/R$, this defines Ω as a real positive number, and we recognize (7.74) as the equation for simple harmonic motion with frequency Ω. We

---

[11] Example 7.5 was another instance: The coordinate q_1 gave the position of the block relative to the accelerating frame of the wedge, but the kinetic energy T was evaluated in the inertial frame of the table. For another example, see Problem 7.30.

conclude that a bead which is displaced a little from the stable equilibrium at $\theta = 0$, executes harmonic motion with frequency Ω,

$$\theta(t) = A\cos(\Omega t - \delta). \tag{7.75}$$

If we speed up the rate of the hoop's rotation until $\omega^2 > g/R$, then Ω becomes pure imaginary, and, since $\cos i\alpha = \cosh\alpha$, our solution (7.75) becomes a hyperbolic cosine, which grows with time, correctly reflecting that the equilibrium at $\theta = 0$ has become unstable.

Once $\omega^2 > g/R$, there are two stable equilibrium positions given by (7.71) and located symmetrically to the right and left of the bottom of the hoop. As you might expect, these behave in the same way, and to be definite I shall focus on the one on the right. Let us denote its position by $\theta = \theta_o$, where, according to (7.70), θ_o satisfies

$$\omega^2 \cos\theta_o - g/R = 0. \tag{7.76}$$

Let us now imagine the bead placed close to θ_o at

$$\theta = \theta_o + \epsilon$$

and investigate the time dependence of the small parameter ϵ. Once again we can approximate the equation of motion (7.73), though this requires more care. If we approximate the factors of $\cos(\theta_o + \epsilon)$ and $\sin(\theta_o + \epsilon)$ by the first two terms of their Taylor series,

$$\cos(\theta_o + \epsilon) \approx \cos\theta_o - \epsilon\sin\theta_o \quad \text{and} \quad \sin(\theta_o + \epsilon) \approx \sin\theta_o + \epsilon\cos\theta_o \tag{7.77}$$

then the equation of motion (7.73) becomes

$$\ddot{\theta} = [\omega^2\cos(\theta_o + \epsilon) - g/R]\sin(\theta_o + \epsilon) \qquad [\theta \text{ near } \theta_o]$$

$$= [\omega^2\cos\theta_o - \epsilon\,\omega^2\sin\theta_o - g/R][\sin\theta_o + \epsilon\cos\theta_o]. \tag{7.78}$$

By (7.76) the first and third terms in the first square bracket cancel, leaving just the middle term $-\epsilon\,\omega^2\sin\theta_o$. To lowest order in ϵ we can drop the second term of the second bracket, and, since $\ddot{\theta}$ is the same as $\ddot{\epsilon}$, we are left with

$$\ddot{\epsilon} = -\epsilon\,\omega^2\sin^2\theta_o = -\Omega'^2\epsilon. \tag{7.79}$$

Here the second equality defines the frequency $\Omega' = \omega\sin\theta_o$, or, using (7.76),

$$\Omega' = \sqrt{\omega^2 - \left(\frac{g}{\omega R}\right)^2}, \tag{7.80}$$

(see Problem 7.26). Equation (7.79) is the equation for simple harmonic motion. Therefore, the parameter ϵ oscillates about zero, and the bead itself oscillates about the equilibrium position θ_o with frequency Ω'.

7.6   Generalized Momenta and Ignorable Coordinates

As I have already mentioned, for any system with n generalized coordinates q_i ($i = 1, \cdots, n$), we refer to the n quantities $\partial \mathcal{L}/\partial q_i = F_i$ as *generalized forces* and $\partial \mathcal{L}/\partial \dot{q}_i = p_i$ as *generalized momenta*. With this terminology, the Lagrange equation,

$$\frac{\partial \mathcal{L}}{\partial q_i} = \frac{d}{dt}\frac{\partial \mathcal{L}}{\partial \dot{q}_i}, \tag{7.81}$$

can be rewritten as

$$F_i = \frac{d}{dt} p_i. \tag{7.82}$$

That is, "generalized force = rate of change of generalized momentum." In particular, if the Lagrangian is independent of a particular coordinate q_i, then $F_i = \partial \mathcal{L}/\partial q_i = 0$ and the corresponding generalized momentum p_i is constant.

Consider, for example, a single projectile subject only to the force of gravity. The potential energy is $U = mgz$ (if we use Cartesian coordinates with z measured vertically up), and the Lagrangian is

$$\mathcal{L} = \mathcal{L}(x, y, z, \dot{x}, \dot{y}, \dot{z}) = \tfrac{1}{2}m(\dot{x}^2 + \dot{y}^2 + \dot{z}^2) - mgz. \tag{7.83}$$

With respect to Cartesian coordinates, the generalized force is just the usual force ($\partial \mathcal{L}/\partial x = -\partial U/\partial x = F_x$, etc.) and the generalized momentum is just the usual momentum ($\partial \mathcal{L}/\partial \dot{x} = m\dot{x} = p_x$, etc.) Because $\mathcal{L}$ is independent of x and y, it immediately follows that the components p_x and p_y are constant, as we already knew.

In general, the generalized forces and momenta are not the same as the usual forces and momenta. For instance, we saw in Equations (7.25) and (7.26) that in two-dimensional polar coordinates the ϕ component of the generalized force is the torque, and that of the generalized momentum is actually the angular momentum. In any case, when the Lagrangian is independent of a coordinate q_i the corresponding generalized momentum is conserved. Thus, if the Lagrangian of a two-dimensional particle is independent of ϕ, then the particle's angular momentum is conserved — another important result (and one that is clear from the Newtonian perspective as well). When the Lagrangian is independent of a coordinate q_i, that coordinate is sometimes said to be **ignorable** or **cyclic**. Obviously it is a good idea, when possible, to choose coordinates so that as many as possible are ignorable and their corresponding momenta are constant. In fact, this is perhaps the main criterion in choosing generalized coordinates for any given problem: Try to find coordinates as many as possible of which are ignorable.

We can rephrase the result of the last three paragraphs by noting that the statement "$\mathcal{L}$ is independent of a coordinate q_i" is equivalent to saying "$\mathcal{L}$ is unchanged, or *invariant*, when q_i varies (with all the other q_j held fixed)." Thus we can say that if $\mathcal{L}$ is invariant under variations of a coordinate q_i then the corresponding generalized momentum p_i is conserved. This connection between invariance of $\mathcal{L}$ and certain conservation laws is the first of several similar results relating invariance under transformations (translations, rotations, and so on) to conservation laws. These results are known collectively as **Noether's theorem**, after the German mathematician Emmy Noether (1882–1935). I shall return to this important theorem in Section 7.8.

7.7   Conclusion

The Lagrangian version of classical mechanics has the two great advantages that, unlike the Newtonian version, it works equally well in all coordinate systems and it can handle constrained systems easily, avoiding any need to discuss the forces of constraint. If the system is constrained, one must choose a suitable set of independent generalized coordinates. Whether or not there are constraints, the next task is to write down the Lagrangian $\mathcal{L}$ in terms of the chosen coordinates. The equations of motion then follow automatically in the standard form

$$\frac{\partial \mathcal{L}}{\partial q_i} = \frac{d}{dt} \frac{\partial \mathcal{L}}{\partial \dot{q}_i} \qquad [i = 1, \cdots, n].$$

There is, of course, no guarantee that the resulting equations will be easy to solve, and in most real problems they are not, requiring numerical solution or at least some approximations before they can be solved analytically.

Even in problems that are only moderately complicated, like the examples of Section 7.5, finding the equations of motion by Lagrange's method is remarkably easier than by using Newton's second law. Indeed, some purists object that the Lagrangian approach makes life *too* easy, removing the need to think about the physics.

The Lagrangian formalism can be extended to include more general systems than those considered so far. One important case is that of magnetic forces, which I take up in Section 7.9. Dissipative forces, such as friction or air resistance, can sometimes be included, but it should be admitted that the Lagrangian formalism is primarily suited to problems where dissipative forces are absent or, at least, negligible.

The final three sections of this chapter treat three advanced topics, all of which are centrally important in Lagrangian mechanics, but all of which could be omitted on a first reading. In Section 7.8, I give two more examples of the remarkable connection between invariance under certain transformations and conservation laws. This connection, known as Noether's theorem, is important in all of modern physics, but especially in quantum physics. Section 7.9 discusses how to include magnetic forces in Lagrangian mechanics, another topic of great importance in quantum theory. Finally, Section 7.10 introduces the method of Lagrange multipliers. This technique appears in many different guises in many areas of physics, but I shall restrict myself to some simple examples in Lagrangian mechanics. These last three sections are arranged to be self-contained and independent. You could study all of them, none of them, or any selection in between.

7.8 More about Conservation Laws*

The material of this section is more advanced than the preceding sections, and you should feel free to omit it on a first reading. Be aware, however, that the material discussed here is needed before you read Section 11.5 and Chapter 13.

In this section I shall discuss how the laws of conservation of momentum and energy fit into the Lagrangian formulation of mechanics. Since we derived the Lagrangian formulation from the Newtonian, anything that we already knew about conservation laws, based on Newtonian mechanics, will naturally still be true in Lagrangian mechanics. Nevertheless, we can gain some new insights by examining the conservation laws from a Lagrangian perspective. Furthermore, much modern work takes the Lagrangian formulation (based on Hamilton's principle, for example) as its starting point. In this context, it is important to know what can be said about conservation laws strictly within the Lagrangian framework.

Conservation of Total Momentum

We already know from Newtonian mechanics that the total momentum of an isolated system of N particles is conserved, but let us examine this important property from the Lagrangian point of view. One of the most prominent features of an isolated system is that it is *translationally invariant;* that is, if we transport all N particles bodily through the same displacement $\boldsymbol{\epsilon}$, nothing physically significant about the system should change. This is illustrated in Figure 7.10, where we see that the effect of moving the whole system through the fixed displacement $\boldsymbol{\epsilon}$ is to replace every position $\mathbf{r}_\alpha$ by $\mathbf{r}_\alpha + \boldsymbol{\epsilon}$,

$$\mathbf{r}_1 \to \mathbf{r}_1 + \boldsymbol{\epsilon}, \qquad \mathbf{r}_2 \to \mathbf{r}_2 + \boldsymbol{\epsilon}, \qquad \cdots, \qquad \mathbf{r}_N \to \mathbf{r}_N + \boldsymbol{\epsilon}. \qquad (7.84)$$

In particular, the potential energy must be unaffected by this displacement, so that

$$U(\mathbf{r}_1 + \boldsymbol{\epsilon}, \cdots, \mathbf{r}_N + \boldsymbol{\epsilon}, t) = U(\mathbf{r}_1, \cdots, \mathbf{r}_N, t) \qquad (7.85)$$

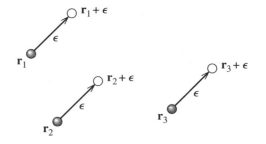

Figure 7.10   An isolated system of N particles is translationally invariant, which means that when every particle is transported through the same displacement $\boldsymbol{\epsilon}$, nothing physically significant changes.

or, more briefly,

$$\delta U = 0$$

where δU denotes the change in U under the translation (7.84). Clearly the velocities are unchanged by the translation (7.84). (Adding a constant ϵ to all the $\mathbf{r}_\alpha$ doesn't change the $\dot{\mathbf{r}}_\alpha$.) Therefore $\delta T = 0$, and hence

$$\delta \mathcal{L} = 0 \qquad (7.86)$$

under the translation (7.84). This result is true for any displacement ϵ. If we choose ϵ to be an infinitesimal displacement in the x direction, then all of the x coordinates $x_1, \cdots, x_N$ increase by ϵ, while the y and z coordinates are unchanged. For this translation, the change in $\mathcal{L}$ is

$$\delta \mathcal{L} = \epsilon \frac{\partial \mathcal{L}}{\partial x_1} + \cdots + \epsilon \frac{\partial \mathcal{L}}{\partial x_N} = 0.$$

This implies that

$$\sum_{\alpha=1}^{N} \frac{\partial \mathcal{L}}{\partial x_\alpha} = 0. \qquad (7.87)$$

Now using Lagrange's equations we can rewrite each derivative as

$$\frac{\partial \mathcal{L}}{\partial x_\alpha} = \frac{d}{dt} \frac{\partial \mathcal{L}}{\partial \dot{x}_\alpha} = \frac{d}{dt} p_{\alpha x}$$

where $p_{\alpha x}$ is the x component of the momentum of particle α. Thus (7.87) becomes

$$\sum_{\alpha=1}^{N} \frac{d}{dt} p_{\alpha x} = \frac{d}{dt} P_x = 0$$

where P_x is the x component of the total momentum $\mathbf{P} = \sum_\alpha \mathbf{p}_\alpha$. By choosing the small displacement ϵ successively in the y and z directions, we can prove the same result for the y and z components, and we reach the conclusion that, provided the Lagrangian is unchanged by the translation (7.84), the total momentum of the N-particle system is conserved. This connection between translational invariance of $\mathcal{L}$ and conservation of total momentum is another example of Noether's theorem.

Conservation of Energy

Finally, I would like to discuss conservation of energy from the Lagrangian point of view. The analysis proves somewhat complicated, but introduces a number of ideas that are important in more advanced work, particularly in the Hamiltonian formulation of mechanics (Chapter 13).

As time advances the function $\mathcal{L}(q_1, \cdots, q_n, \dot{q}_1, \cdots, \dot{q}_n, t)$ changes, both because

t is changing, and because the q's and $\dot{q}$'s change with the evolving system. Thus, by the chain rule,

$$\frac{d}{dt}\mathcal{L}(q_1, \cdots, q_n, \dot{q}_1, \cdots, \dot{q}_n, t) = \sum_i \frac{\partial \mathcal{L}}{\partial q_i}\dot{q}_i + \sum_i \frac{\partial \mathcal{L}}{\partial \dot{q}_i}\ddot{q}_i + \frac{\partial \mathcal{L}}{\partial t}. \qquad (7.88)$$

Now, by Lagrange's equation, I can replace the derivative in the first sum on the right by

$$\frac{\partial \mathcal{L}}{\partial q_i} = \frac{d}{dt}\frac{\partial \mathcal{L}}{\partial \dot{q}_i} = \frac{d}{dt}p_i = \dot{p}_i.$$

Meanwhile, the derivative in the second sum on the right of (7.88) is just the generalized momentum p_i. Thus, I can rewrite (7.88) as

$$\frac{d}{dt}\mathcal{L} = \sum_i \left(\dot{p}_i\dot{q}_i + p_i\ddot{q}_i\right) + \frac{\partial \mathcal{L}}{\partial t}$$

$$= \frac{d}{dt}\sum_i \left(p_i\dot{q}_i\right) + \frac{\partial \mathcal{L}}{\partial t}. \qquad (7.89)$$

Now, for many interesting systems, the Lagrangian does not depend explicitly on time; that is, $\partial \mathcal{L}/\partial t = 0$. When this is the case, the second term on the right of (7.89) vanishes. If we move the left side of (7.89) over to the right, we see that the time derivative of the quantity $\sum p_i\dot{q}_i - \mathcal{L}$ is zero. This quantity is so important that it has its own symbol,

$$\mathcal{H} = \sum_{i=1}^{n} p_i\dot{q}_i - \mathcal{L}, \qquad (7.90)$$

and is called the **Hamiltonian** of the system. With this terminology, we can state the following important conclusion:

> If the Lagrangian $\mathcal{L}$ does not depend explicitly on time (that is, $\partial \mathcal{L}/\partial t = 0$), then the Hamiltonian $\mathcal{H}$ is conserved.

The discovery of any conservation law is a momentous event and is enough to justify saying that the Hamiltonian is an important quantity. In fact, it goes much further than this. As we shall see in Chapter 13, the Hamiltonian $\mathcal{H}$ is the basis of the Hamiltonian formulation of mechanics, in just the same way that $\mathcal{L}$ is the basis of Lagrangian mechanics.

For the moment, the chief importance of our newly discovered Hamiltonian is that in many situations it is in fact just the total energy of the system. Specifically, we shall prove that, *provided the relation between the generalized coordinates and Cartesians is time-independent*,

$$\mathbf{r}_\alpha = \mathbf{r}_\alpha(q_1, \cdots, q_n), \qquad (7.91)$$

the Hamiltonian $\mathcal{H}$ is just the total energy,

$$\mathcal{H} = T + U. \tag{7.92}$$

You may recall that we agreed in Section 7.3 to describe generalized coordinates that satisfy (7.91) as *natural*; thus, we can paraphrase the result (7.92) to say that, provided the generalized coordinates are natural, $\mathcal{H}$ is just the total energy $T + U$. To prove this, let us express the total kinetic energy $T = \frac{1}{2} \sum_\alpha m_\alpha \dot{\mathbf{r}}_\alpha^2$ in terms of the generalized coordinates $q_1, \cdots, q_n$. First, differentiating (7.91) with respect to t and using the chain rule, we find that[12]

$$\dot{\mathbf{r}}_\alpha = \sum_{i=1}^{n} \frac{\partial \mathbf{r}_\alpha}{\partial q_i} \dot{q}_i . \tag{7.93}$$

If we now form the scalar product of this equation with itself, we find

$$\dot{\mathbf{r}}_\alpha^2 = \sum_j \left(\frac{\partial \mathbf{r}_\alpha}{\partial q_j} \dot{q}_j \right) \cdot \sum_k \left(\frac{\partial \mathbf{r}_\alpha}{\partial q_k} \dot{q}_k \right)$$

where I have renamed the summation indices as j and k to avoid future confusion. The kinetic energy is now given as a triple sum, which I can reorganize and write as[13]

$$T = \frac{1}{2} \sum_\alpha m_\alpha \dot{\mathbf{r}}_\alpha^2 = \frac{1}{2} \sum_{j,k} A_{jk} \dot{q}_j \dot{q}_k \tag{7.94}$$

where A_{jk} is shorthand for the sum

$$A_{jk} = A_{jk}(q_1, \cdots, q_n) = \sum_\alpha m_\alpha \left(\frac{\partial \mathbf{r}_\alpha}{\partial q_j} \right) \cdot \left(\frac{\partial \mathbf{r}_\alpha}{\partial q_k} \right) . \tag{7.95}$$

We can now evaluate the generalized momentum p_i by differentiating (7.94) with respect to $\dot{q}_i$ (Problem 7.45),

$$p_i = \frac{\partial \mathcal{L}}{\partial \dot{q}_i} = \frac{\partial T}{\partial \dot{q}_i} = \sum_j A_{ij} \dot{q}_j . \tag{7.96}$$

Returning to Equation (7.90) for the Hamiltonian, we can rewrite the sum on the right as

$$\sum_i p_i \dot{q}_i = \sum_i \left(\sum_j A_{ij} \dot{q}_j \right) \dot{q}_i = \sum_{i,j} A_{ij} \dot{q}_i \dot{q}_j = 2T \tag{7.97}$$

---

[12] If the relation (7.91) were explicitly time-dependent there would be one extra term in this expression for $\dot{\mathbf{r}}_\alpha$, namely $\partial \mathbf{r}_\alpha / \partial t$. This extra term would invalidate the conclusion (7.98) below that $\mathcal{H} = T + U$.

[13] We can restate the result (7.94) to say that, provided the generalized coordinates are natural, the kinetic energy T is a homogeneous quadratic function of the generalized velocities $\dot{q}_i$. This result plays an important role in several later developments. See, for instance, Section 11.5.

where the last step follows from (7.94). Therefore

$$\mathcal{H} = \sum_i p_i \dot{q}_i - \mathcal{L} = 2T - (T - U) = T + U. \tag{7.98}$$

That is, provided the transformation between the Cartesian and generalized coordinates is time-independent, as in (7.91), the Hamiltonian $\mathcal{H}$ is just the total energy of the system.

I have already proved that, provided the Lagrangian is independent of time, the Hamiltonian is conserved. Thus we now see that time independence of the Lagrangian [together with the condition (7.91)] implies conservation of energy. We can rephrase the time independence of $\mathcal{L}$ by saying that $\mathcal{L}$ is unchanged by translations of time, $t \to t + \epsilon$. Thus the result we have just proved is that invariance of $\mathcal{L}$ under time translations is related to energy conservation, in much the same way that invariance of $\mathcal{L}$ under translations of space ($\mathbf{r} \to \mathbf{r} + \boldsymbol{\epsilon}$) is related to conservation of momentum. Both results are manifestations of Noether's famous theorem.

7.9 Lagrange's Equations for Magnetic Forces*

This section requires a knowledge of the scalar and vector potentials of electromagnetism. Although the ideas described here play an important role in the quantum-mechanical treatment of magnetic fields, they will not be used again in this book.

Although I have so far consistently defined the Lagrangian as $\mathcal{L} = T - U$, there are systems, such as a charged particle in a magnetic field, which can be treated by the Lagrangian method, but for which $\mathcal{L}$ is *not* just $T - U$. The natural question to ask is then: What is the definition of the Lagrangian for such systems? This is the first question I address.

Definition and Nonuniqueness of the Lagrangian

Probably the most satisfactory general definition of a Lagrangian for a mechanical system is this:

> ### General Definition of a Lagrangian
> For a given mechanical system with generalized coordinates $q = (q_1, \cdots, q_n)$, a **Lagrangian** $\mathcal{L}$ is a function $\mathcal{L}(q_1, \cdots, q_n, \dot{q}_1, \cdots, \dot{q}_n, t)$ of the coordinates and velocities, such that the correct equations of motion for the system are the Lagrange equations
> $$\frac{\partial \mathcal{L}}{\partial q_i} = \frac{d}{dt}\frac{\partial \mathcal{L}}{\partial \dot{q}_i} \qquad [i = 1, \cdots, n].$$

In other words, a Lagrangian is any function $\mathcal{L}$ for which Lagrange's equations are true for the system under consideration.

Obviously, for the systems that we have discussed so far, the old definition $\mathcal{L} = T - U$ fits this new definition. But the new definition is much more general. In particular, it is easy to see that our new definition does not define a unique Lagrangian function. For example, consider a single particle in one dimension and suppose that we have found a Lagrangian $\mathcal{L}$ for this particle. That is, the equation of motion of the particle is

$$\frac{\partial \mathcal{L}}{\partial x} = \frac{d}{dt}\frac{\partial \mathcal{L}}{\partial \dot{x}}. \tag{7.99}$$

Now let $f(x, \dot{x})$ be any function for which

$$\frac{\partial f}{\partial x} \equiv \frac{d}{dt}\frac{\partial f}{\partial \dot{x}}. \tag{7.100}$$

(It is easy to think up such a function, for instance, $f = x\dot{x}$.) If we replace $\mathcal{L}$ in (7.99) by

$$\mathcal{L}' = \mathcal{L} + f$$

then, by virtue of (7.100), the Lagrangian $\mathcal{L}'$ gives exactly the same equation of motion as $\mathcal{L}$.

The lack of uniqueness of the Lagrangian is similar to, though more radical than, the familiar lack of uniqueness in the potential energy (to which one can add any constant without changing any of the physical predictions). The crucial point is that *any* function $\mathcal{L}$ which gives the right equation of motion has all of the features that we require of a Lagrangian (for instance, that the integral $\int \mathcal{L}\, dt$ is stationary at the right path) and so is just as acceptable as any other such function $\mathcal{L}$. If, for a given system, we can spot a function $\mathcal{L}$ that leads to the right equation of motion, then we don't need to debate whether it is the "right" Lagrangian — if it gives the right equation of motion, then it is just as right as any other conceivable Lagrangian.

Lagrangian for a Charge in a Magnetic Field

Consider now a particle (mass m and charge q) moving in electric and magnetic fields **E** and **B**. The force on the particle is the well-known Lorentz force $\mathbf{F} = q(\mathbf{E} + \mathbf{v} \times \mathbf{B})$, so Newton's second law reads

$$m\ddot{\mathbf{r}} = q(\mathbf{E} + \dot{\mathbf{r}} \times \mathbf{B}). \tag{7.101}$$

To reformulate (7.101) in Lagrangian form, we have only to spot a function $\mathcal{L}$ for which the three Lagrange equations are the same as (7.101). This can be done using the scalar and vector potentials, $V(\mathbf{r}, t)$ and $\mathbf{A}(\mathbf{r}, t)$, in terms of which the two fields can be written[14]

$$\mathbf{E} = -\nabla V - \frac{\partial \mathbf{A}}{\partial t} \quad \text{and} \quad \mathbf{B} = \nabla \times \mathbf{A}. \tag{7.102}$$

---

[14] See, for example, David J. Griffiths, *Introduction to Electrodynamics*, (Prentice-Hall, 1999), p. 416–417.

I now claim that the Lagrangian function[15]

$$\mathcal{L}(\mathbf{r}, \dot{\mathbf{r}}, t) = \tfrac{1}{2}m\dot{\mathbf{r}}^2 - q(V - \dot{\mathbf{r}} \cdot \mathbf{A}) \tag{7.103}$$

$$= \tfrac{1}{2}m(\dot{x}^2 + \dot{y}^2 + \dot{z}^2) - q(V - \dot{x}A_x - \dot{y}A_y - \dot{z}A_z) \tag{7.104}$$

has the desired property, that it reproduces Newton's second law (7.101). To check this, let us examine the first of the three Lagrange equations,

$$\frac{\partial \mathcal{L}}{\partial x} = \frac{d}{dt}\frac{\partial \mathcal{L}}{\partial \dot{x}}. \tag{7.105}$$

To see what this implies we have to evaluate the two derivatives of the proposed Lagrangian (7.104):

$$\frac{\partial \mathcal{L}}{\partial x} = -q\left(\frac{\partial V}{\partial x} - \dot{x}\frac{\partial A_x}{\partial x} - \dot{y}\frac{\partial A_y}{\partial x} - \dot{z}\frac{\partial A_z}{\partial x}\right) \tag{7.106}$$

and

$$\frac{\partial \mathcal{L}}{\partial \dot{x}} = m\dot{x} + qA_x.$$

When we differentiate this with respect to t, we must remember that $A_x = A_x(x, y, z, t)$. As t varies, x, y, z move with the particle and, by the chain rule, we find

$$\frac{d}{dt}\frac{\partial \mathcal{L}}{\partial \dot{x}} = m\ddot{x} + q\left(\dot{x}\frac{\partial A_x}{\partial x} + \dot{y}\frac{\partial A_x}{\partial y} + \dot{z}\frac{\partial A_x}{\partial z} + \frac{\partial A_x}{\partial t}\right). \tag{7.107}$$

Substituting (7.106) and (7.107) into (7.105), cancelling the two terms in $\dot{x}$, and rearranging, we find that Lagrange's equation (the x component, with the proposed Lagrangian) is the same as

$$m\ddot{x} = -q\left(\frac{\partial V}{\partial x} + \frac{\partial A_x}{\partial t}\right) + q\dot{y}\left(\frac{\partial A_y}{\partial x} - \frac{\partial A_x}{\partial y}\right) - q\dot{z}\left(\frac{\partial A_x}{\partial z} - \frac{\partial A_z}{\partial x}\right) \tag{7.108}$$

or, according to (7.102),

$$m\ddot{x} = q\left(E_x + \dot{y}B_z - \dot{z}B_y\right) \tag{7.109}$$

which you will recognize as the x component of Newton's second law (7.101). Since the y and z components work in the same way, we conclude that Lagrange's equations, with the proposed Lagrangian (7.104), are exactly equivalent to Newton's second law for the charged particle. That is, we have successfully recast Newton's second law for a charged particle into Lagrangian form with the Lagrangian (7.104).

Using the Lagrangian (7.104), one can solve various problems involving charged particles in electric and magnetic fields. (See Problem 7.49 for an example.) Theoret-

---

[15] Notice that you can, if you wish, write this as $\mathcal{L} = T - U$, but U is certainly not the usual PE, since it depends on the velocity $\dot{\mathbf{r}}$; U is sometimes called a "velocity-dependent PE," but notice that it is not true that the force on the charge is $-\nabla U$.

ically, the most important conclusion of this analysis emerges when we evaluate the generalized momentum. For example,

$$p_x = \frac{\partial \mathcal{L}}{\partial \dot{x}} = m\dot{x} + qA_x.$$

Since the y and z components work the same way, we conclude that

$$(\text{generalized momentum, } \mathbf{p}) = m\mathbf{v} + q\mathbf{A}. \qquad (7.110)$$

That is, the generalized momentum is the mechanical momentum $m\mathbf{v}$ *plus* a magnetic term $q\mathbf{A}$. This result is at the heart of the quantum theory of a charged particle in a magnetic field, where it turns out that the generalized momentum corresponds to the differential operator $-i\hbar\nabla$ (where $\hbar$ is Planck's constant over 2π), so that the quantum analog of the mechanical momentum $m\mathbf{v}$ is the operator $-i\hbar\nabla - q\mathbf{A}$.

7.10   Lagrange Multipliers and Constraint Forces*

The method of Lagrange multipliers is used in many areas of physics. Nevertheless, we shan't be using it again in this book, and you could omit this section without any loss of continuity.

In this section, we discuss the method of Lagrange multipliers. This powerful method finds application in several areas of physics and takes on quite different appearances in different contexts. Here I shall treat only its application to Lagrangian mechanics,[16] and to keep the analysis simple I shall restrict the discussion to two-dimensional systems with just one degree of freedom.

We have seen that one of the strengths of Lagrangian mechanics is that it can bypass all of the forces of constraint. However, there are situations where one actually needs to know these forces. For example, the designer of a roller coaster needs to know the normal force of the track on the car to know how strong to build the track. In this case, we can still use a modified form of Lagrange's equations, but the procedure is somewhat different: We do not choose generalized coordinates $q_1, \cdots, q_n$ all of which can be independently varied. (Remember that it was the independence of $q_1, \cdots, q_n$ that let us use the standard Lagrange equations without worrying about constraints.) Instead we use a larger number of coordinates and use Lagrange multipliers to handle the constraints.

To illustrate this procedure, we'll consider a system with just two rectangular coordinates x and y, which are restricted by a **constraint equation** of the form[17]

$$f(x, y) = \text{const.} \qquad (7.111)$$

For example, we could consider a simple pendulum with just one degree of freedom (as in Figure 7.2). In treating this by the standard Lagrange approach one would use

---

[16] For applications to other kinds of problems, see, for example, *Mathematical Methods in the Physical Sciences* by Mary Boas (Wiley, 1983), Ch. 4, Section 9 and Ch. 9, Section 6.

[17] We'll see directly by example that some typical constraints can be put in this form. In fact, it is fairly easy to show that any holonomic constraints can be.

the one generalized coordinate ϕ, the angle between the pendulum and the vertical, and avoid any discussion of the constraints. If, instead, we choose to use the original rectangular coordinates x and y, then we must recognize that these coordinates are not independent; they satisfy the constraint equation

$$f(x, y) = \sqrt{x^2 + y^2} = l$$

where l is the length of the pendulum. We shall find that the method of Lagrange multipliers lets us accomodate this constraint, determine the time dependence of x and y, *and* find the tension in the rod. As a second example, consider the Atwood machine of Figure 7.6. In our previous treatment we used the one generalized coordinate x, the position of the mass m_1, but we could instead use both coordinates x and y (the positions of both masses), provided we remember that the constancy of the string's length imposes the constraint that

$$f(x, y) = x + y = \text{const.}$$

Here too, Lagrange multipliers will let us accomodate this constraint, solve for the time dependence of x and y, and find the constraint force, which is here the tension in the string.

To set up our new method we start from Hamilton's principle. Our Lagrangian has the form $\mathcal{L}(x, \dot{x}, y, \dot{y})$. (We could allow it to have an explicit dependence on t as well, but to simplify notation I shall assume it doesn't.) The proof of Hamilton's principle given in Section 7.4 applies even when the coordinates are constrained. (Indeed it was designed to allow for constraints.) Thus we can conclude as before that the action integral

$$S = \int_{t_1}^{t_2} \mathcal{L}(x, \dot{x}, y, \dot{y}) \, dt \tag{7.112}$$

is stationary when taken along the actual path followed. If we denote this "right" path by $x(t)$, $y(t)$ and imagine a small displacement to a neighboring "wrong" path,

$$\left.\begin{array}{l} x(t) \rightarrow x(t) + \delta x(t) \\ y(t) \rightarrow y(t) + \delta y(t) \end{array}\right\} \tag{7.113}$$

then, provided the displacement is consistent with the constraint equation, the action integral (7.112) is unchanged, $\delta S = 0$. To exploit this, we must write δS in terms of the small displacements δx and δy:

$$\delta S = \int \left(\frac{\partial \mathcal{L}}{\partial x} \delta x + \frac{\partial \mathcal{L}}{\partial \dot{x}} \delta \dot{x} + \frac{\partial \mathcal{L}}{\partial y} \delta y + \frac{\partial \mathcal{L}}{\partial \dot{y}} \delta \dot{y} \right) dt. \tag{7.114}$$

The second and fourth terms can be integrated by parts, and we conclude that

$$\delta S = \int \left(\frac{\partial \mathcal{L}}{\partial x} - \frac{d}{dt} \frac{\partial \mathcal{L}}{\partial \dot{x}} \right) \delta x \, dt + \int \left(\frac{\partial \mathcal{L}}{\partial y} - \frac{d}{dt} \frac{\partial \mathcal{L}}{\partial \dot{y}} \right) \delta y \, dt = 0 \tag{7.115}$$

for any displacements δx and δy consistent with the constraints.

If (7.115) were true for *any* displacements, we could immediately prove two separate Lagrange equations, one for x and one for y. (Choosing $\delta y = 0$, we would be left with just the first integral; since this has to vanish for any choice of δx, the factor in parentheses would have to be zero, which implies the usual Lagrange equation with respect to x. And similarly for y.) This is exactly the correct conclusion for the case that there are no constraints.

However, there *are* constraints, and (7.115) is only true for displacements δx and δy consistent with the constraints. Therefore, we proceed as follows: Since all points with which we are concerned satisfy $f(x, y) = \text{const}$, the displacement (7.113) leaves $f(x, y)$ unchanged, so

$$\delta f = \frac{\partial f}{\partial x}\delta x + \frac{\partial f}{\partial y}\delta y = 0 \tag{7.116}$$

for any displacement consistent with the constraints. Since this is zero, we can multiply it by an arbitrary function $\lambda(t)$ — this is the **Lagrange multiplier** — and add it to the integrand in (7.115), without changing the value of the integral (namely zero). Therefore

$$\delta S = \int \left(\frac{\partial \mathcal{L}}{\partial x} + \lambda(t)\frac{\partial f}{\partial x} - \frac{d}{dt}\frac{\partial \mathcal{L}}{\partial \dot{x}} \right) \delta x \, dt$$

$$+ \int \left(\frac{\partial \mathcal{L}}{\partial y} + \lambda(t)\frac{\partial f}{\partial y} - \frac{d}{dt}\frac{\partial \mathcal{L}}{\partial \dot{y}} \right) \delta y \, dt = 0 \tag{7.117}$$

for any displacement consistent with the constraints. Now comes the supreme cunning: So far $\lambda(t)$ is an arbitrary function of t, but we can choose it so that the coefficient of δx in the first integral is zero. That is, by choice of the multiplier $\lambda(t)$, we can arrange that

$$\frac{\partial \mathcal{L}}{\partial x} + \lambda\frac{\partial f}{\partial x} = \frac{d}{dt}\frac{\partial \mathcal{L}}{\partial \dot{x}} \tag{7.118}$$

along the actual path of the system. This is the first of two modified Lagrange equations and differs from the usual equation only by the extra term involving λ on the left. With the multiplier chosen in this way, the whole first integral in (7.117) is zero. Therefore the second integral is also zero (since their sum is), and this is true for *any* choice of δy. (The constraint places no restriction on δx or δy separately — it only fixes δx once δy is chosen, or vice versa.) Therefore the coefficient of δy in this second integral must also be zero and we have a second modified Lagrange equation,

$$\frac{\partial \mathcal{L}}{\partial y} + \lambda\frac{\partial f}{\partial y} = \frac{d}{dt}\frac{\partial \mathcal{L}}{\partial \dot{y}}. \tag{7.119}$$

We now have two modified Lagrange equations for the two unknown functions $x(t)$ and $y(t)$. This elegant result has been bought at the price of introducing a third unknown function, the Lagrange multiplier $\lambda(t)$. To find three unknown functions, we need three equations, but fortunately a third equation is already at hand, the constraint equation

$$f(x, y) = \text{const.} \tag{7.120}$$

The three equations (7.118), (7.119), and (7.120) are sufficient, in principle at least, to determine the coordinates $x(t)$ and $y(t)$ and the multiplier $\lambda(t)$. Before we illustrate this with an example, there is one more bit of theory to develop.

So far the Lagrange multiplier $\lambda(t)$ is just a mathematical artifact, introduced to help us solve our problem. However, it turns out to be closely related to the forces of constraint. To see this, we have only to look more closely at the modified Lagrange equations (7.118) and (7.119). The Lagrangian of our present discussion has the form

$$\mathcal{L} = \tfrac{1}{2}m_1\dot{x}^2 + \tfrac{1}{2}m_2\dot{y}^2 - U(x, y).$$

(In a problem like the simple pendulum, x and y are the two coordinates of a single mass and $m_1 = m_2$. In a problem like the Atwood machine, there are two separate masses and m_1 and m_2 are not necessarily equal.) Inserting this Lagrangian into (7.118), we find

$$-\frac{\partial U}{\partial x} + \lambda\frac{\partial f}{\partial x} = m_1\ddot{x}. \tag{7.121}$$

Now, on the left side $-\partial U/\partial x$ is the x component of the nonconstraint force. (Remember that U was defined as the potential energy of the nonconstraint forces.) On the right, $m_1\ddot{x}$ is the x component of the total force, equal to the sum of the nonconstraint and constraint forces. Thus $m_1\ddot{x} = -\partial U/\partial x + F_x^{\text{cstr}}$. Canceling the term $-\partial U/\partial x$ from both sides of (7.121), we reach the important conclusion that

$$\lambda\frac{\partial f}{\partial x} = F_x^{\text{cstr}} \tag{7.122}$$

with a corresponding result for the y components. This then is the significance of the Lagrange multiplier: Multiplied by the appropriate partial derivatives of the constraint function $f(x, y)$, the Lagrange multiplier $\lambda(t)$ gives the corresponding components of the constraint force.

Let us now see how these ideas work in practice by using the formalism to analyse the example of the Atwood machine.

EXAMPLE 7.8 Atwood's Machine Using a Lagrange Multiplier

Analyze the Atwood machine of Figure 7.6 (shown again here as Figure 7.11) by the method of Lagrange multipliers and using the coordinates x and y of the two masses.

In terms of the given coordinates, the Lagrangian is

$$\mathcal{L} = T - U = \tfrac{1}{2}m_1\dot{x}^2 + \tfrac{1}{2}m_2\dot{y}^2 + m_1 g x + m_2 g y \qquad (7.123)$$

and the constraint equation is

$$f(x, y) = x + y = \text{const.} \qquad (7.124)$$

The modified Lagrange equation (7.118) for x reads

$$\frac{\partial \mathcal{L}}{\partial x} + \lambda \frac{\partial f}{\partial x} = \frac{d}{dt}\frac{\partial \mathcal{L}}{\partial \dot{x}} \qquad \text{or} \qquad m_1 g + \lambda = m_1 \ddot{x} \qquad (7.125)$$

and that for y is

$$\frac{\partial \mathcal{L}}{\partial y} + \lambda \frac{\partial f}{\partial y} = \frac{d}{dt}\frac{\partial \mathcal{L}}{\partial \dot{y}} \qquad \text{or} \qquad m_2 g + \lambda = m_2 \ddot{y}. \qquad (7.126)$$

These two equations, together with the constraint equation (7.124), are easily solved for the unknowns $x(t)$, $y(t)$, and $\lambda(t)$. From (7.124) we see that $\ddot{y} = -\ddot{x}$, and then subtracting (7.126) from (7.125) we can eliminate λ and arrive at the same result as before,

$$\ddot{x} = (m_1 - m_2)g/(m_1 + m_2).$$

To better understand the two modified Lagrange equations (7.125) and (7.126), it is helpful to compare them with the two equations of the Newtonian solution. Newton's second law for m_1 is

$$m_1 g - F_t = m_1 \ddot{x}$$

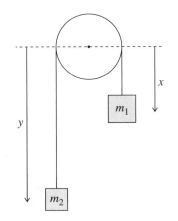

Figure 7.11   The Atwood machine again.

where F_t is the tension in the string, and that for m_2 is

$$m_2 g - F_t = m_2 \ddot{y}.$$

These are precisely the two Lagrange equations (7.125) and (7.126), with the Lagrange multiplier identified as the constraint force

$$\lambda = -F_t.$$

[Two small comments: The minus sign occurs because both coordinates x and y were measured downward, whereas both tension forces are upward. In general, according to (7.122) the constraint force is $\lambda \partial f / \partial x$, but in this simple case, $\partial f / \partial x = 1$.]

You can find some more examples of the use of Lagrange multipliers in Problems 7.50 through 7.52.

Principal Definitions and Equations of Chapter 7

The Lagrangian

The **Lagrangian** $\mathcal{L}$ of a conservative system is defined as

$$\mathcal{L} = T - U, \qquad \text{[Eq. (7.3)]}$$

where T and U are respectively the kinetic and potential energies.

Generalized Coordinates

The n parameters $q_1, \cdots, q_n$ are **generalized coordinates** for an N-particle system if every particle's position $\mathbf{r}_\alpha$ can be expressed as a function of $q_1, \cdots, q_n$ (and possibly the time t) and vice versa, and if n is the smallest number that allows the system to be described in this way.   [Eqs. (7.34) & (7.35)]

If $n < 3N$ (in three dimensions) the system is said to be **constrained**. The coordinates $q_1, \cdots, q_n$ are said to be **natural** if the functional relationships of the $\mathbf{r}_\alpha$ to $q_1, \cdots, q_n$ are independent of time. The number of **degrees of freedom** of a system is the number of coordinates that can be independently varied. If the number of degrees of freedom is equal to the number of generalized coordinates (in some sense the "normal" state of affairs), the system is said to be **holonomic**.   [Section 7.3]

Lagrange's Equations

For any holonomic system, Newton's second law is equivalent to the n **Lagrange equations**

$$\frac{\partial \mathcal{L}}{\partial q_i} = \frac{d}{dt} \frac{\partial \mathcal{L}}{\partial \dot{q}_i} \qquad [i = 1, \cdots, n] \qquad \text{[Sections 7.3 \& 7.4]}$$

and the Lagrange equations are in turn equivalent to Hamilton's principle — a fact we
used only to prove the Lagrange equations.   [Eq. (7.8)]

Generalized Momenta and Ignorable Coordinates

The ith **generalized momentum** p_i is defined to be the derivative

$$p_i = \frac{\partial \mathcal{L}}{\partial \dot{q}_i}.$$

If $\partial \mathcal{L}/\partial q_i = 0$, then we say the coordinate q_i is **ignorable** and the corresponding
generalized momentum is constant.   [Section 7.6]

The Hamiltonian

The **Hamiltonian** is defined as

$$\mathcal{H} = \sum_{i=1}^{n} p_i \dot{q}_i - \mathcal{L}.$$ [Eq. (7.90)]

If $\partial \mathcal{L}/\partial t = 0$, then $\mathcal{H}$ is conserved; if the coordinates $q_1, \cdots, q_n$ are natural, $\mathcal{H}$ is just
the energy of the system.

Lagrangian for a Charge in an Electromagnetic Field

The Lagrangian for a charge q in an electromagnetic field is

$$\mathcal{L}(\mathbf{r}, \dot{\mathbf{r}}, t) = \tfrac{1}{2}m\dot{\mathbf{r}}^2 - q(V - \dot{\mathbf{r}} \cdot \mathbf{A}).$$ [Eq. (7.103)]

Problems for Chapter 7

Stars indicate the approximate level of difficulty, from easiest (⋆) to most difficult (⋆⋆⋆).

SECTION 7.1   Lagrange's Equations for Unconstrained Motion

7.1 ⋆ Write down the Lagrangian for a projectile (subject to no air resistance) in terms of its Cartesian
coordinates (x, y, z), with z measured vertically upward. Find the three Lagrange equations and show
that they are exactly what you would expect for the equations of motion.

7.2 ⋆ Write down the Lagrangian for a one-dimensional particle moving along the x axis and subject
to a force $F = -kx$ (with k positive). Find the Lagrange equation of motion and solve it.

7.3 ⋆ Consider a mass m moving in two dimensions with potential energy $U(x, y) = \tfrac{1}{2}kr^2$, where
$r^2 = x^2 + y^2$. Write down the Lagrangian, using coordinates x and y, and find the two Lagrange
equations of motion. Describe their solutions. [This is the potential energy of an ion in an "ion trap,"
which can be used to study the properties of individual atomic ions.]

7.4 ★ Consider a mass m moving in a frictionless plane that slopes at an angle α with the horizontal. Write down the Lagrangian in terms of coordinates x, measured horizontally across the slope, and y, measured down the slope. (Treat the system as two-dimensional, but include the gravitational potential energy.) Find the two Lagrange equations and show that they are what you should have expected.

7.5 ★ Find the components of $\nabla f(r, \phi)$ in two-dimensional polar coordinates. [*Hint:* Remember that the change in the scalar f as a result of an infinitesimal displacement $d\mathbf{r}$ is $df = \nabla f \cdot d\mathbf{r}$.]

7.6 ★ Consider two particles moving unconstrained in three dimensions, with potential energy $U(\mathbf{r}_1, \mathbf{r}_2)$. **(a)** Write down the six equations of motion obtained by applying Newton's second law to each particle. **(b)** Write down the Lagrangian $\mathcal{L}(\mathbf{r}_1, \mathbf{r}_2, \dot{\mathbf{r}}_1, \dot{\mathbf{r}}_2) = T - U$ and show that the six Lagrange equations are the same as the six Newtonian equations of part (a). This establishes the validity of Lagrange's equations in rectangular coordinates, which in turn establishes Hamilton's principle. Since the latter is independent of coordinates, this proves Lagrange's equations in any coordinate system.

7.7 ★ Do Problem 7.6, but for N particles moving unconstrained in three dimensions (in which case there are $3N$ equations of motion).

7.8 ★★ **(a)** Write down the Lagrangian $\mathcal{L}(x_1, x_2, \dot{x}_1, \dot{x}_2)$ for two particles of equal masses, $m_1 = m_2 = m$, confined to the x axis and connected by a spring with potential energy $U = \frac{1}{2}kx^2$. [Here x is the extension of the spring, $x = (x_1 - x_2 - l)$, where l is the spring's unstretched length, and I assume that mass 1 remains to the right of mass 2 at all times.] **(b)** Rewrite $\mathcal{L}$ in terms of the new variables $X = \frac{1}{2}(x_1 + x_2)$ (the CM position) and x (the extension), and write down the two Lagrange equations for X and x. **(c)** Solve for $X(t)$ and $x(t)$ and describe the motion.

SECTION 7.3   Constrained Systems in General

7.9 ★ Consider a bead that is threaded on a rigid circular hoop of radius R lying in the xy plane with its center at O, and use the angle ϕ of two-dimensional polar coordinates as the one generalized coordinate to describe the bead's position. Write down the equations that give the Cartesian coordinates (x, y) in terms of ϕ and the equation that gives the generalized coordinate ϕ in terms of (x, y).

7.10 ★ A particle is confined to move on the surface of a circular cone with its axis on the z axis, vertex at the origin (pointing down), and half-angle α. The particle's position can be specified by two generalized coordinates, which you can choose to be the coordinates (ρ, ϕ) of cylindrical polar coordinates. Write down the equations that give the three Cartesian coordinates of the particle in terms of the generalized coordinates (ρ, ϕ) and vice versa.

7.11 ★ Consider the pendulum of Figure 7.4, suspended inside a railroad car, but suppose that the car is oscillating back and forth, so that the point of suspension has position $x_s = A \cos \omega t$, $y_s = 0$. Use the angle ϕ as the generalized coordinate and write down the equations that give the Cartesian coordinates of the bob in terms of ϕ and vice versa.

SECTION 7.4   Proof of Lagrange's Equations with Constraints

7.12 ★ Lagrange's equations in the form discussed in this chapter hold only if the forces (at least the nonconstraint forces) are derivable from a potential energy. To get an idea how they can be modified to include forces like friction, consider the following: A single particle in one dimension is subject to various conservative forces (net conservative force $= F = -\partial U / \partial x$) and a nonconservative force (let's

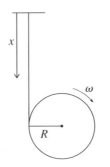

Figure 7.12   Problem 7.14

call it F_{fric}). Define the Lagrangian as $\mathcal{L} = T - U$ and show that the appropriate modification is

$$\frac{\partial \mathcal{L}}{\partial x} + F_{\text{fric}} = \frac{d}{dt}\frac{\partial \mathcal{L}}{\partial \dot{x}}.$$

7.13 ★★ In Section 7.4 [Equations (7.41) through (7.51)], I proved Lagrange's equations for a single particle constrained to move on a two-dimensional surface. Go through the same steps to prove Lagrange's equations for a system consisting of two particles subject to various unspecified constraints. [*Hint:* The net force on particle 1 is the sum of the total constraint force $\mathbf{F}_1^{\text{cstr}}$ and the total nonconstraint force $\mathbf{F}_1$, and likewise for particle 2. The constraint forces come in many guises (the normal force of a surface, the tension force of a string tied between the particles, etc.), but it is always true that the net work done by all constraint forces in any displacement consistent with the constraints is zero — this is the defining property of constraint forces. Meanwhile, we take for granted that the nonconstraint forces are derivable from a potential energy $U(\mathbf{r}_1, \mathbf{r}_2, t)$; that is, $\mathbf{F}_1 = -\nabla_1 U$ and likewise for particle 2. Write down the difference δS between the action integral for the right path given by $\mathbf{r}_1(t)$ and $\mathbf{r}_2(t)$ and any nearby wrong path given by $\mathbf{r}_1(t) + \boldsymbol{\epsilon}_1(t)$ and $\mathbf{r}_2(t) + \boldsymbol{\epsilon}_2(t)$. Paralleling the steps of Section 7.4, you can show that δS is given by an integral analogous to (7.49), and this is zero by the defining property of constraint forces.]

SECTION 7.5   Examples of Lagrange's Equations

7.14 ★ Figure 7.12 shows a crude model of a yoyo. A massless string is suspended vertically from a fixed point and the other end is wrapped several times around a uniform cylinder of mass m and radius R. When the cylinder is released it moves vertically down, rotating as the string unwinds. Write down the Lagrangian, using the distance x as your generalized coordinate. Find the Lagrange equation of motion and show that the cylinder accelerates downward with $\ddot{x} = 2g/3$. [*Hints:* You need to remember from your introductory physics course that the total kinetic energy of a body like the yoyo is $T = \frac{1}{2}mv^2 + \frac{1}{2}I\omega^2$, where v is the velocity of the center of mass, I is the moment of inertia (for a uniform cylinder, $I = \frac{1}{2}mR^2$) and ω is the angular velocity about the CM. You can express ω in terms of $\dot{x}$.]

7.15 ★ A mass m_1 rests on a frictionless horizontal table and is attached to a massless string. The string runs horizontally to the edge of the table, where it passes over a massless, frictionless pulley and then hangs vertically down. A second mass m_2 is now attached to the bottom end of the string. Write down the Lagrangian for the system. Find the Lagrange equation of motion, and solve it for the acceleration of the blocks. For your generalized coordinate, use the distance x of the second mass below the tabletop.

7.16 ★ Write down the Lagrangian for a cylinder (mass m, radius R, and moment of inertia I) that rolls without slipping straight down an inclined plane which is at an angle α from the horizontal. Use as your generalized coordinate the cylinder's distance x measured down the plane from its starting point. Write down the Lagrange equation and solve it for the cylinder's acceleration $\ddot{x}$. Remember that $T = \frac{1}{2}mv^2 + \frac{1}{2}I\omega^2$, where v is the velocity of the center of mass and ω is the angular velocity.

7.17 ★ Use the Lagrangian method to find the acceleration of the Atwood machine of Example 7.3 (page 255) including the effect of the pulley's having moment of inertia I. (The kinetic energy of the pulley is $\frac{1}{2}I\omega^2$, where ω is its angular velocity.)

7.18 ★ A mass m is suspended from a massless string, the other end of which is wrapped several times around a horizontal cylinder of radius R and moment of inertia I, which is free to rotate about a fixed horizontal axle. Using a suitable coordinate, set up the Lagrangian and the Lagrange equation of motion, and find the acceleration of the mass m. [The kinetic energy of the rotating cylinder is $\frac{1}{2}I\omega^2$.]

7.19 ★ In Example 7.5 (page 258) the two accelerations are given by Equations (7.66) and (7.67). Check that the acceleration of the block is given correctly in the limit $M \to 0$. [You need to find the components of this acceleration *relative to the table*.]

7.20 ★ A smooth wire is bent into the shape of a helix, with cylindrical polar coordinates $\rho = R$ and $z = \lambda\phi$, where R and λ are constants and the z axis is vertically up (and gravity vertically down). Using z as your generalized coordinate, write down the Lagrangian for a bead of mass m threaded on the wire. Find the Lagrange equation and hence the bead's vertical acceleration $\ddot{z}$. In the limit that $R \to 0$, what is $\ddot{z}$? Does this make sense?

7.21 ★ The center of a long frictionless rod is pivoted at the origin, and the rod is forced to rotate in a horizontal plane with constant angular velocity ω. Write down the Lagrangian for a bead of mass m threaded on the rod, using r as your generalized coordinate, where r, ϕ are the polar coordinates of the bead. (Notice that ϕ is not an independent variable since it is fixed by the rotation of the rod to be $\phi = \omega t$.) Solve Lagrange's equation for $r(t)$. What happens if the bead is initially at rest at the origin? If it is released from any point $r_0 > 0$, show that $r(t)$ eventually grows exponentially. Explain your results in terms of the centrifugal force $m\omega^2 r$.

7.22 ★ Using the usual angle ϕ as generalized coordinate, write down the Lagrangian for a simple pendulum of length l suspended from the ceiling of an elevator that is accelerating upward with constant acceleration a. (Be careful when writing T; it is probably safest to write the bob's velocity in component form.) Find the Lagrange equation of motion and show that it is the same as that for a normal, nonaccelerating pendulum, except that g has been replaced by $g + a$. In particular, the angular frequency of small oscillations is $\sqrt{(g+a)/l}$.

7.23 ★ A small cart (mass m) is mounted on rails inside a large cart. The two are attached by a spring (force constant k) in such a way that the small cart is in equilibrium at the midpoint of the large. The distance of the small cart from its equilibrium is denoted x and that of the large one from a fixed point on the ground is X, as shown in Figure 7.13. The large cart is now forced to oscillate such that $X = A\cos\omega t$, with both A and ω fixed. Set up the Lagrangian for the motion of the small cart and show that the Lagrange equation has the form

$$\ddot{x} + \omega_0^2 x = B\cos\omega t$$

where ω_0 is the natural frequency $\omega_0 = \sqrt{k/m}$ and B is a constant. This is the form assumed in Section 5.5, Equation (5.57), for driven oscillations (except that we are here ignoring damping). Thus the system

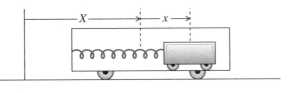

Figure 7.13   Problem 7.23

described here would be one way to realize the motion discussed there. (We could fill the large cart with molasses to provide some damping.)

7.24 ★ We saw in Example 7.3 (page 255) that the acceleration of the Atwood machine is $\ddot{x} = (m_1 - m_2)g/(m_1 + m_2)$. It is sometimes claimed that this result is "obvious" because, it is said, the effective force on the system is $(m_1 - m_2)g$ and the effective mass is $(m_1 + m_2)$. This is not, perhaps, all that obvious, but it does emerge very naturally in the Lagrangian approach. Recall that Lagrange's equation can be thought of as [Equation (7.17)]

$$(\text{generalized force}) = (\text{rate of change of generalized momentum}).$$

Show that for the Atwood machine the generalized force is $(m_1 - m_2)g$ and the generalized momentum $(m_1 + m_2)\dot{x}$. Comment.

7.25 ★ Prove that the potential energy of a central force $\mathbf{F} = -kr^n\hat{\mathbf{r}}$ (with $n \neq -1$) is $U = kr^{n+1}/(n + 1)$ if we choose the zero of U appropriately. In particular, if $n = 1$, then $\mathbf{F} = -k\mathbf{r}$ and $U = \frac{1}{2}kr^2$.

7.26 ★ In Example 7.7 (page 264), we saw that the bead on a spinning hoop can make small oscillations about any of its stable equilibrium points. Verify that the oscillation frequency Ω' defined in (7.79) is equal to $\sqrt{\omega^2 - (g/\omega R)^2}$ as claimed in (7.80).

7.27 ★★ Consider a double Atwood machine constructed as follows: A mass $4m$ is suspended from a string that passes over a massless pulley on frictionless bearings. The other end of this string supports a second similar pulley, over which passes a second string supporting a mass of $3m$ at one end and m at the other. Using two suitable generalized coordinates, set up the Lagrangian and use the Lagrange equations to find the acceleration of the mass $4m$ when the system is released. Explain why the top pulley rotates even though it carries equal weights on each side.

7.28 ★★ A couple of points need checking from Example 7.6 (page 260). **(a)** From the point of view of a noninertial frame rotating with the hoop, the bead is subject to the force of gravity and a centrifugal force $m\omega^2\rho$ (in addition to the constraint force, which is the normal force of the wire). Verify that at the equilibrium points given by (7.71), the tangential components of these two forces balance one another. (A free-body diagram will help.) **(b)** Verify that the equilibrium point at the top ($\theta = \pi$) is unstable. **(c)** Verify that the equilibrium at the second point given by (7.71) (the one on the left, with θ negative) is stable.

7.29 ★★ Figure 7.14 shows a simple pendulum (mass m, length l) whose point of support P is attached to the edge of a wheel (center O, radius R) that is forced to rotate at a fixed angular velocity ω. At $t = 0$, the point P is level with O on the right. Write down the Lagrangian and find the equation of motion for the angle ϕ. [*Hint:* Be careful writing down the kinetic energy T. A safe way to get the velocity right is to write down the position of the bob at time t, and then differentiate.] Check that your answer makes sense in the special case that $\omega = 0$.

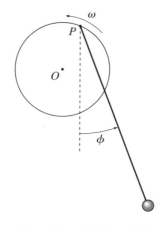

Figure 7.14   Problem 7.29

7.30 ★★ Consider the pendulum of Figure 7.4, suspended inside a railroad car that is being forced to accelerate with a constant acceleration a. **(a)** Write down the Lagrangian for the system and the equation of motion for the angle ϕ. Use a trick similar to the one used in Equation (5.11) to write the combination of $\sin\phi$ and $\cos\phi$ as a multiple of $\sin(\phi + \beta)$. **(b)** Find the equilibrium angle ϕ at which the pendulum can remain fixed (relative to the car) as the car accelerates. Use the equation of motion to show that this equilibrium is stable. What is the frequency of small oscillations about this equilibrium position? (We shall find a much slicker way to solve this problem in Chapter 9, but the Lagrangian method does give a straightforward route to the answer.)

7.31 ★★ A simple pendulum (mass M and length L) is suspended from a cart (mass m) that can oscillate on the end of a spring of force constant k, as shown in Figure 7.15. **(a)** Write the Lagrangian in terms of the two generalized coordinates x and ϕ, where x is the extension of the spring from its equilibrium length. (Read the hint in Problem 7.29.) Find the two Lagrange equations. (Warning: They're pretty ugly!) **(b)** Simplify the equations to the case that both x and ϕ are small. (They're still pretty ugly, and note, in particular, that they are still *coupled*; that is, each equation involves both variables. Nonetheless, we shall see how to solve these equations in Chapter 11 — see particularly Problem 11.19.)

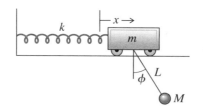

Figure 7.15   Problem 7.31

7.32 ★★ Consider the cube balanced on a cylinder as described in Example 4.7 (page 130). Assuming that $b < r$, use the Lagrangian approach to find the angular frequency of small oscillations about the top. The simplest procedure is to make the small-angle approximations to $\mathcal{L}$ before you differentiate to get Lagrange's equation. As usual, be careful in writing down the kinetic energy; this is $\frac{1}{2}(mv^2 + I\dot{\theta}^2)$, where v is the speed of the CM and I is the moment of inertia about the CM ($2mb^2/3$). The safe way to find v is to write down the coordinates of the CM and then differentiate.

$\frac{1}{2}m\,\dot{x}^2\,\omega^2$

7.33 ★★ A bar of soap (mass m) is at rest on a frictionless rectangular plate that rests on a horizontal table. At time $t = 0$, I start raising one edge of the plate so that the plate pivots about the opposite edge with constant angular velocity ω, and the soap starts to slide toward the downhill edge. Show that the equation of motion for the soap has the form $\ddot{x} - \omega^2 x = -g \sin \omega t$, where x is the soap's distance from the downhill edge. Solve this for $x(t)$, given that $x(0) = x_0$. [You'll need to use the method used to solve Equation (5.48). You can easily solve the homogeneous equation; for a particular solution try $x = A \sin \omega t$ and solve for A.]

7.34 ★★ Consider the well-known problem of a cart of mass m moving along the x axis attached to a spring (force constant k), whose other end is held fixed (Figure 5.2). If we ignore the mass of the spring (as we almost always do) then we know that the cart executes simple harmonic motion with angular frequency $\omega = \sqrt{k/m}$. Using the Lagrangian approach, you can find the effect of the spring's mass M, as follows: **(a)** Assuming that the spring is uniform and stretches uniformly, show that its kinetic energy is $\frac{1}{6}M\dot{x}^2$. (As usual x is the extension of the spring from its equilibrium length.) Write down the Lagrangian for the system of cart plus spring. (*Note:* The potential energy is still $\frac{1}{2}kx^2$.) **(b)** Write down the Lagrange equation and show that the cart still executes SHM but with angular frequency $\omega = \sqrt{k/(m + M/3)}$; that is, the effect of the spring's mass M is just to add $M/3$ to the mass of the cart.

$v = \frac{x}{l}X$ $\frac{M dx}{l}$

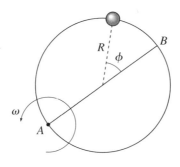

Figure 7.16   Problem 7.35

7.35 ★★ Figure 7.16 is a bird's-eye view of a smooth horizontal wire hoop that is forced to rotate at a fixed angular velocity ω about a vertical axis through the point A. A bead of mass m is threaded on the hoop and is free to move around it, with its position specified by the angle ϕ that it makes at the center with the diameter AB. Find the Lagrangian for this system using ϕ as your generalized coordinate. (Read the hint in Problem 7.29.) Use the Lagrange equation of motion to show that the bead oscillates about the point B exactly like a simple pendulum. What is the frequency of these oscillations if their amplitude is small?

7.36 ★★★ A pendulum is made from a massless spring (force constant k and unstretched length l_0) that is suspended at one end from a fixed pivot O and has a mass m attached to its other end. The spring can stretch and compress but cannot bend, and the whole system is confined to a single vertical plane. **(a)** Write down the Lagrangian for the pendulum, using as generalized coordinates the usual angle ϕ and the length r of the spring. **(b)** Find the two Lagrange equations of the system and interpret them in terms of Newton's second law, as given in Equation (1.48). **(c)** The equations of part (b) cannot be solved analytically in general. However, they *can* be solved for small oscillations. Do this and describe the motion. [*Hint:* Let l denote the equilibrium length of the spring with the mass hanging from it and write $r = l + \epsilon$. "Small oscillations" involve only small values of ϵ and ϕ, so you can use the small-angle approximations and drop from your equations all terms that involve powers of ϵ or ϕ (or their derivatives) higher than the first power (also products of ϵ and ϕ or their derivatives). This dramatically simplifies and uncouples the equations.]

7.37 ★★★ Two equal masses, $m_1 = m_2 = m$, are joined by a massless string of length L that passes through a hole in a frictionless horizontal table. The first mass slides on the table while the second hangs below the table and moves up and down in a vertical line. **(a)** Assuming the string remains taut, write down the Lagrangian for the system in terms of the polar coordinates (r, ϕ) of the mass on the table. **(b)** Find the two Lagrange equations and interpret the ϕ equation in terms of the angular momentum ℓ of the first mass. **(c)** Express $\dot{\phi}$ in terms of ℓ and eliminate $\dot{\phi}$ from the r equation. Now use the r equation to find the value $r = r_0$ at which the first mass can move in a circular path. Interpret your answer in Newtonian terms. **(d)** Suppose the first mass is moving in this circular path and is given a small radial nudge. Write $r(t) = r_0 + \epsilon(t)$ and rewrite the r equation in terms of $\epsilon(t)$ dropping all powers of $\epsilon(t)$ higher than linear. Show that the circular path is stable and that $r(t)$ oscillates sinusoidally about r_0. What is the frequency of its oscillations?

7.38 ★★★ A particle is confined to move on the surface of a circular cone with its axis on the vertical z axis, vertex at the origin (pointing down), and half-angle α. **(a)** Write down the Lagrangian $\mathcal{L}$ in terms of the spherical polar coordinates r and ϕ. **(b)** Find the two equations of motion. Interpret the ϕ equation in terms of the angular momentum ℓ_z, and use it to eliminate $\dot{\phi}$ from the r equation in favor of the constant ℓ_z. Does your r equation make sense in the case that $\ell_z = 0$? Find the value r_0 of r at which the particle can remain in a horizontal circular path. **(c)** Suppose that the particle is given a small radial kick, so that $r(t) = r_0 + \epsilon(t)$, where $\epsilon(t)$ is small. Use the r equation to decide whether the circular path is stable. If so, with what frequency does r oscillate about r_0?

7.39 ★★★ **(a)** Write down the Lagrangian for a particle moving in three dimensions under the influence of a conservative central force with potential energy $U(r)$, using spherical polar coordinates (r, θ, ϕ). **(b)** Write down the three Lagrange equations and explain their significance in terms of radial acceleration, angular momentum, and so forth. (The θ equation is the tricky one, since you will find it implies that the ϕ component of ℓ varies with time, which seems to contradict conservation of angular momentum. Remember, however, that ℓ_ϕ is the component of ℓ in a *variable* direction.) **(c)** Suppose that initially the motion is in the equatorial plane (that is, $\theta_0 = \pi/2$ and $\dot{\theta}_0 = 0$). Describe the subsequent motion. **(d)** Suppose instead that the initial motion is along a line of longitude (that is, $\dot{\phi}_0 = 0$). Describe the subsequent motion.

7.40 ★★★ The "spherical pendulum" is just a simple pendulum that is free to move in any sideways direction. (By contrast a "simple pendulum" — unqualified — is confined to a single vertical plane.) The bob of a spherical pendulum moves on a sphere, centered on the point of support with radius $r = R$, the length of the pendulum. A convenient choice of coordinates is spherical polars, r, θ, ϕ, with the origin at the point of support and the polar axis pointing straight down. The two variables θ and ϕ make a good choice of generalized coordinates. **(a)** Find the Lagrangian and the two Lagrange equations. **(b)** Explain what the ϕ equation tells us about the z component of angular momentum ℓ_z. **(c)** For the special case that $\phi = \text{const}$, describe what the θ equation tells us. **(d)** Use the ϕ equation to replace $\dot{\phi}$ by ℓ_z in the θ equation and discuss the existence of an angle θ_0 at which θ can remain constant. Why is this motion called a conical pendulum? **(e)** Show that if $\theta = \theta_0 + \epsilon$, with ϵ small, then θ oscillates about θ_0 in harmonic motion. Describe the motion of the pendulum's bob.

7.41 ★★★ Consider a bead of mass m sliding without friction on a wire that is bent in the shape of a parabola and is being spun with constant angular velocity ω about its vertical axis, as shown in Figure

7.17. Use cylindrical polar coordinates and let the equation of the parabola be $z = k\rho^2$. Write down the Lagrangian in terms of ρ as the generalized coordinate. Find the equation of motion of the bead and determine whether there are positions of equilibrium, that is, values of ρ at which the bead can remain fixed, without sliding up or down the spinning wire. Discuss the stability of any equilibrium positions you find.

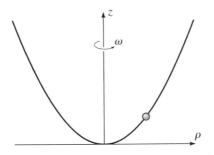

Figure 7.17   Problem 7.41

7.42 ★★★ [Computer] In Example 7.7 (page 264), we saw that the bead on a spinning hoop can make small oscillations about its nonzero stable equilibrium points that are approximately sinusoidal, with frequency $\Omega' = \sqrt{\omega^2 - (g/\omega R)^2}$ as in (7.80). Investigate how good this approximation is by solving the equation of motion (7.73) numerically and then plotting both your numerical solution and the approximate solution $\theta(t) = \theta_0 + A\cos(\Omega' t - \delta)$ on the same graph. Use the following numbers: $g = R = 1$ and $\omega^2 = 2$, and initial conditions $\dot{\theta}(0) = 0$ and $\theta(0) = \theta_0 + \epsilon_0$, where $\epsilon_0 = 1°$. Repeat with $\epsilon_0 = 10°$. Comment on your results.

7.43 ★★★ [Computer] Consider a massless wheel of radius R mounted on a frictionless horizontal axis. A point mass M is glued to the edge, and a massless string is wrapped several times around the perimeter and hangs vertically down with a mass m suspended from its bottom end. (See Figure 4.28.) Initially I am holding the wheel with M vertically below the axle. At $t = 0$, I release the wheel, and m starts to fall vertically down. **(a)** Write down the Lagrangian $\mathcal{L} = T - U$ as a function of the angle ϕ through which the wheel has turned. Find the equation of motion and show that, provided $m < M$, there is one position of stable equilibrium. **(b)** Assuming $m < M$, sketch the potential energy $U(\phi)$ for $-\pi \le \phi \le 4\pi$ and use your graph to explain the equilibrium positions you found. **(c)** Because the equation of motion cannot be solved in terms of elementary functions, you are going to solve it numerically. This requires that you choose numerical values for the various parameters. Take $M = g = R = 1$ (this amounts to a convenient choice of units) and $m = 0.7$. Before solving the equation make a careful plot of $U(\phi)$ against ϕ and predict the kind of motion expected when M is released from rest at $\phi = 0$. Now solve the equation of motion for $0 \le t \le 20$ and verify your prediction. **(d)** Repeat part (c), but with $m = 0.8$.

7.44 ★★★ [Computer] If you haven't already done so, do Problem 7.29. One might expect that the rotation of the wheel would have little effect on the pendulum, provided the wheel is small and rotates slowly. **(a)** Verify this expectation by solving the equation of motion numerically, with the following numbers: Take g and l to be 1. (This means that the natural frequency $\sqrt{g/l}$ of the pendulum is also 1.) Take $\omega = 0.2$, so that the wheel's rotational frequency is small compared to the natural frequency of the pendulum; and take the radius $R = 0.2$, significantly less than the length of the pendulum. As initial conditions take $\phi = 0.2$ and $\dot{\phi} = 0$ at $t = 0$, and make a plot of your solution $\phi(t)$ for $0 < t < 20$. Your graph should look very like the sinusoidal oscillations of an ordinary simple pendulum. Does the period look correct? **(b)** Now plot $\phi(t)$ for $0 < t < 100$ and notice that the rotating support does make a small difference, causing the amplitude of the oscillations to grow and shrink periodically. Comment on the period of these small fluctuations.

SECTION 7.8   More About Conservation Laws *

7.45 ★★ **(a)** Verify that the coefficients A_{ij} in the important expression (7.94) for the kinetic energy of any "natural" system are symmetric; that is, $A_{ij} = A_{ji}$. **(b)** Prove that for any n variables $v_1, \cdots, v_n$

$$\frac{\partial}{\partial v_i} \sum_{j,k} A_{jk} v_j v_k = 2 \sum_j A_{ij} v_j .$$

[*Hint:* Start with the case that $n = 2$, for which you can write out the sums in full. Notice that you need the result of part (a).] This identity is useful in many areas of physics; we needed it to prove the expression (7.96) for the generalized momentum p_i.

7.46 ★★ Noether's theorem asserts a connection between invariance principles and conservation laws. In Section 7.8 we saw that translational invariance of the Lagrangian implies conservation of total linear momentum. Here you will prove that rotational invariance of $\mathcal{L}$ implies conservation of total angular momentum. Suppose that the Lagrangian of an N-particle system is unchanged by rotations about a certain symmetry axis. **(a)** Without loss of generality, take this axis to be the z axis, and show that the Lagrangian is unchanged when all of the particles are simultaneously moved from $(r_\alpha, \theta_\alpha, \phi_\alpha)$ to $(r_\alpha, \theta_\alpha, \phi_\alpha + \epsilon)$ (same ϵ for all particles). Hence show that

$$\sum_{\alpha=1}^{N} \frac{\partial \mathcal{L}}{\partial \phi_\alpha} = 0.$$

(b) Use Lagrange's equations to show that this implies that the total angular momentum L_z about the symmetry axis is constant. In particular, if the Lagrangian is invariant under rotations about all axes, then all components of $\mathbf{L}$ are conserved.

7.47 ★★★ In Chapter 4 (at the end of Section 4.7) I claimed that, for a system with one degree of freedom, positions of stable equilibrium "normally" correspond to minima of the potential energy $U(q)$. Using Lagrangian mechanics, you can now prove this claim. **(a)** Consider a one-degree system of N particles with positions $\mathbf{r}_\alpha = \mathbf{r}_\alpha(q)$, where q is the one generalized coordinate and the transformation between $\mathbf{r}$ and q does not depend on time; that is, q is what we have now agreed to call "natural." (This is the meaning of the qualification "normally" in the statement of the claim. If the transformation depends on time, then the claim is not necessarily true.) Prove that the KE has the form $T = \frac{1}{2} A \dot{q}^2$, where $A = A(q) > 0$ may depend on q but not on $\dot{q}$. [This corresponds exactly to the result (7.94) for n degrees of freedom. If you have trouble with the proof here, review the proof there.] Show that the Lagrange equation of motion has the form

$$A(q)\ddot{q} = -\frac{dU}{dq} - \frac{1}{2}\frac{dA}{dq}\dot{q}^2.$$

(b) A point q_o is an equilibrium point if, when the system is placed at q_o with $\dot{q} = 0$, it remains there. Show that q_o is an equilibrium point if and only if $dU/dq = 0$. **(c)** Show that the equilibrium is stable if and only if U is minimum at q_o. **(d)** If you did Problem 7.30, show that the pendulum of that problem does not satisfy the conditions of this problem and that the result proved here is false for that system.

SECTION 7.9 Lagrange's Equations for Magnetic Forces *

7.48 ★★ Let $F = F(q_1, \cdots, q_n)$ be any function of the generalized coordinates $(q_1, \cdots, q_n)$ of a system with Lagrangian $\mathcal{L}(q_1, \cdots, q_n, \dot{q}_1, \cdots, \dot{q}_n, t)$. Prove that the two Lagrangians $\mathcal{L}$ and $\mathcal{L}' = \mathcal{L} + dF/dt$ give exactly the same equations of motion.

7.49 ★★ Consider a particle of mass m and charge q moving in a uniform constant magnetic field $\mathbf{B}$ in the z direction. **(a)** Prove that $\mathbf{B}$ can be written as $\mathbf{B} = \nabla \times \mathbf{A}$ with $\mathbf{A} = \frac{1}{2}\mathbf{B} \times \mathbf{r}$. Prove equivalently that in cylindrical polar coordinates, $\mathbf{A} = \frac{1}{2}B\rho\,\hat{\boldsymbol{\phi}}$. **(b)** Write the Lagrangian (7.103) in cylindrical polar coordinates and find the three corresponding Lagrange equations. **(c)** Describe in detail those solutions of the Lagrange equations in which ρ is a constant.

SECTION 7.10 Lagrange Multipliers and Constraint Forces *

7.50 ★ A mass m_1 rests on a frictionless horizontal table. Attached to it is a string which runs horizontally to the edge of the table, where it passes over a frictionless, small pulley and down to where it supports a mass m_2. Use as coordinates x and y the distances of m_1 and m_2 from the pulley. These satisfy the constraint equation $f(x, y) = x + y = \text{const}$. Write down the two modified Lagrange equations and solve them (together with the constraint equation) for $\ddot{x}$, $\ddot{y}$, and the Lagrange multiplier λ. Use (7.122) (and the corresponding equation in y) to find the tension forces on the two masses. Verify your answers by solving the problem by the elementary Newtonian approach.

7.51 ★ Write down the Lagrangian for the simple pendulum of Figure 7.2 in terms of the rectangular coordinates x and y. These coordinates are constrained to satisfy the constraint equation $f(x, y) = \sqrt{x^2 + y^2} = l$. **(a)** Write down the two modified Lagrange equations (7.118) and (7.119). Comparing these with the two components of Newton's second law, show that the Lagrange multiplier is (minus) the tension in the rod. Verify Equation (7.122) and the corresponding equation in y. **(b)** The constraint equation can be written in many different ways. For example we could have written $f'(x, y) = x^2 + y^2 = l^2$. Check that using this function would have given the same physical results.

7.52 ★ The method of Lagrange multipliers works perfectly well with non-Cartesian coordinates. Consider a mass m that hangs from a string, the other end of which is wound several times around a wheel (radius R, moment of inertia I) mounted on a frictionless horizontal axle. Use as coordinates for the mass and the wheel x, the distance fallen by the mass, and ϕ, the angle through which the wheel has turned (both measured from some convenient reference position). Write down the modified Lagrange equations for these two variables and solve them (together with the constraint equation) for $\ddot{x}$ and $\ddot{\phi}$ and the Lagrange multiplier. Write down Newton's second law for the mass and wheel, and use them to check your answers for $\ddot{x}$ and $\ddot{\phi}$. Show that $\lambda \partial f/\partial x$ is indeed the tension force on the mass. Comment on the quantity $\lambda \partial f/\partial \phi$.

Two-Body Central-Force Problems

In this chapter, I shall discuss the motion of two bodies each of which exerts a conservative, central force on the other but which are subject to no other, "external," forces. There are many examples of this problem: the two stars of a binary star system, a planet orbiting the sun, the moon orbiting the earth, the electron and proton in a hydrogen atom, the two atoms of a diatomic molecule. In most cases the true situation is more complicated. For example, even if we are interested in just one planet orbiting the sun, we cannot completely neglect the effects of all the other planets; likewise, the moon–earth system is subject to the external force of the sun. Nevertheless, in all cases, it is an excellent starting approximation to treat the two bodies of interest as being isolated from all outside influences.

You may also object that the examples of the hydrogen atom and the diatomic molecule do not belong in classical mechanics, since all such atomic-scale systems must really be treated by quantum mechanics. However, many of the ideas I shall develop in this chapter (the important idea of reduced mass, for instance) play a crucial role in the quantum mechanical two-body problem, and it is probably fair to say that the material covered here is an essential prerequisite for the corresponding quantum material.

8.1 The Problem

Let us consider two objects, with masses m_1 and m_2. For the purposes of this chapter, I shall assume the objects are small enough to be considered as point particles, whose positions (relative to the origin O of some inertial reference frame) I shall denote by $\mathbf{r}_1$ and $\mathbf{r}_2$. The only forces are the forces $\mathbf{F}_{12}$ and $\mathbf{F}_{21}$ of their mutual interaction, which I shall assume is conservative and central. Thus the forces can be derived from a potential energy $U(\mathbf{r}_1, \mathbf{r}_2)$. In the case of two astronomical bodies (the earth and the sun, for instance) the force is the gravitational force $Gm_1m_2/|\mathbf{r}_1 - \mathbf{r}_2|^2$, with the corresponding potential energy (as we saw in Chapter 4)

$$U(\mathbf{r}_1, \mathbf{r}_2) = -\frac{Gm_1m_2}{|\mathbf{r}_1 - \mathbf{r}_2|}. \tag{8.1}$$

For the electron and proton in a hydrogen atom, the potential energy is the Coulomb PE of the two charges (e for the proton and $-e$ for the electron),

$$U(\mathbf{r}_1, \mathbf{r}_2) = -\frac{ke^2}{|\mathbf{r}_1 - \mathbf{r}_2|}, \tag{8.2}$$

where k denotes the Coulomb force constant, $k = 1/4\pi\epsilon_0$.

In both of these examples, U depends only on the difference $(\mathbf{r}_1 - \mathbf{r}_2)$, not on $\mathbf{r}_1$ and $\mathbf{r}_2$ separately. As we saw in Section 4.9, this is no accident: Any isolated system is translationally invariant, and if $U(\mathbf{r}_1, \mathbf{r}_2)$ is translationally invariant it can only depend on $(\mathbf{r}_1 - \mathbf{r}_2)$. In the present case there is a further simplification: As we saw in Section 4.8, if a conservative force is central, then U is independent of the direction of $(\mathbf{r}_1 - \mathbf{r}_2)$. That is, it only depends on the *magnitude* $|\mathbf{r}_1 - \mathbf{r}_2|$, and we can write

$$U(\mathbf{r}_1, \mathbf{r}_2) = U(|\mathbf{r}_1 - \mathbf{r}_2|) \tag{8.3}$$

as is the case in the examples (8.1) and (8.2).

To take advantage of the property (8.3), it is convenient to introduce the new variable

$$\mathbf{r} = \mathbf{r}_1 - \mathbf{r}_2. \tag{8.4}$$

As shown in Figure 8.1, this is just the position of body 1 relative to body 2, and I shall refer to $\mathbf{r}$ as the **relative position**. The result of the previous paragraph can be rephrased to say that the potential energy U depends only on the magnitude r of the relative position $\mathbf{r}$,

$$U = U(r). \tag{8.5}$$

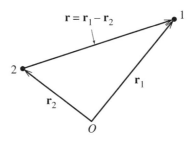

Figure 8.1   The relative position $\mathbf{r} = \mathbf{r}_1 - \mathbf{r}_2$ is the position of body 1 relative to body 2.

We can now state the mathematical problem that we have to solve: We want to find the possible motions of two bodies (the moon and the earth, or an electron and a proton), whose Lagrangian is

$$\mathcal{L} = \tfrac{1}{2}m_1\dot{\mathbf{r}}_1^2 + \tfrac{1}{2}m_2\dot{\mathbf{r}}_2^2 - U(r). \tag{8.6}$$

Of course, I could equally have stated the problem in Newtonian terms, and I shall in fact feel free to move back and forth between the Lagrangian and Newtonian formalisms according to which seems the more convenient. For the present, the Lagrangian formalism is the more transparent.

8.2 CM and Relative Coordinates; Reduced Mass

Our first task is to decide what generalized coordinates to use to solve our problem. There is already a strong suggestion that we should use the relative position $\mathbf{r}$ as one of them (or as three of them, depending on how you count coordinates), because the potential energy $U(r)$ takes such a simple form in terms of $\mathbf{r}$. The question is then, what to choose for the other (vector) variable. The best choice turns out to be the familiar *center of mass* (or CM) position, $\mathbf{R}$, of the two bodies, defined as in Chapter 3 to be

$$\mathbf{R} = \frac{m_1\mathbf{r}_1 + m_2\mathbf{r}_2}{m_1 + m_2} = \frac{m_1\mathbf{r}_1 + m_2\mathbf{r}_2}{M}, \tag{8.7}$$

where as before M denotes the total mass of the two bodies:

$$M = m_1 + m_2.$$

As we saw in Chapter 3, the CM of two particles lies on the line joining them, as shown in Figure 8.2. The distances of the center of mass from the two masses m_2 and m_1 are in the ratio m_1/m_2. In particular, if m_2 is much greater than m_1, then the CM is very close to body 2. (In Figure 8.2, the ratio m_1/m_2 is about 1/3, so the CM is a quarter of the way from m_2 to m_1.)

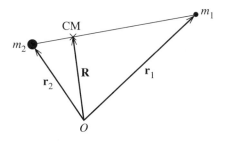

Figure 8.2   The center of mass of the two bodies lies at the position $\mathbf{R} = (m_1\mathbf{r}_1 + m_2\mathbf{r}_2)/M$ on the line joining the two bodies.

We saw in Section 3.3 that the total momentum of the two bodies is the same as if the total mass $M = m_1 + m_2$ were concentrated at the CM and were following the CM as it moves:

$$\mathbf{P} = M\dot{\mathbf{R}}. \tag{8.8}$$

This result has important simplifying consequences: We know, of course, that the total momentum is constant. Therefore, according to (8.8), $\dot{\mathbf{R}}$ is constant; and this means we can choose an inertial reference frame in which the CM is at rest. This **CM frame** is an especially convenient frame in which to analyze the motion, as we shall see.

I am going to use the CM position $\mathbf{R}$ and the relative position $\mathbf{r}$ as generalized coordinates for our discussion of the motion of our two bodies. In terms of these coordinates, we already know that the potential energy takes the simple form $U = U(r)$. To express the kinetic energy in these terms, we need to write the old variables $\mathbf{r}_1$ and $\mathbf{r}_2$ in terms of the new $\mathbf{R}$ and $\mathbf{r}$. It is a straightforward exercise to show that (see Figure 8.2)

$$\mathbf{r}_1 = \mathbf{R} + \frac{m_2}{M}\mathbf{r} \qquad \text{and} \qquad \mathbf{r}_2 = \mathbf{R} - \frac{m_1}{M}\mathbf{r}. \tag{8.9}$$

Thus the kinetic energy is

$$T = \tfrac{1}{2}\left(m_1\dot{\mathbf{r}}_1^2 + m_2\dot{\mathbf{r}}_2^2\right)$$

$$= \tfrac{1}{2}\left(m_1\left[\dot{\mathbf{R}} + \frac{m_2}{M}\dot{\mathbf{r}}\right]^2 + m_2\left[\dot{\mathbf{R}} - \frac{m_1}{M}\dot{\mathbf{r}}\right]^2\right)$$

$$= \tfrac{1}{2}\left(M\dot{\mathbf{R}}^2 + \frac{m_1 m_2}{M}\dot{\mathbf{r}}^2\right). \tag{8.10}$$

The result (8.10) simplifies further if we introduce the parameter

$$\mu = \frac{m_1 m_2}{M} \equiv \frac{m_1 m_2}{m_1 + m_2} \qquad \text{[reduced mass]} \tag{8.11}$$

which has the dimensions of mass and is called the **reduced mass**. You can easily check that μ is always less than both m_1 and m_2 (hence the name). If $m_1 \ll m_2$, then μ is very close to m_1. Thus the reduced mass for the earth–sun system is almost exactly the mass of the earth; the reduced mass of the electron and proton in hydrogen is almost exactly the mass of the electron. On the other hand, if $m_1 = m_2$, then obviously $\mu = \tfrac{1}{2}m_1$.

Returning to (8.10), we can rewrite the kinetic energy in terms of μ as

$$T = \tfrac{1}{2}M\dot{\mathbf{R}}^2 + \tfrac{1}{2}\mu\dot{\mathbf{r}}^2. \tag{8.12}$$

This remarkable result shows that the kinetic energy is the same as that of two "fictitious" particles, one of mass M moving with the speed of the CM, and the other of mass μ (the reduced mass) moving with the speed of the relative position $\mathbf{r}$. Even more significant is the corresponding result for the Lagrangian:

$$\mathcal{L} = T - U = \tfrac{1}{2}M\dot{\mathbf{R}}^2 + \left(\tfrac{1}{2}\mu\dot{\mathbf{r}}^2 - U(r)\right)$$

$$= \mathcal{L}_{\text{cm}} + \mathcal{L}_{\text{rel}}. \tag{8.13}$$

We see that by using the CM and relative positions as our generalized coordinates, we have split the Lagrangian into two separate pieces, one of which involves only the CM coordinate $\mathbf{R}$ and the other only the relative coordinate $\mathbf{r}$. This will mean that we can solve for the motions of $\mathbf{R}$ and $\mathbf{r}$ as two separate problems, which will greatly simplify matters.

8.3   The Equations of Motion

With the Lagrangian (8.13), we can write down the equations of motion of our two-body system. Because $\mathcal{L}$ is independent of $\mathbf{R}$, the $\mathbf{R}$ equation (really three equations, one each for X, Y, and Z) is especially simple,

$$M\ddot{\mathbf{R}} = 0 \qquad \text{or} \qquad \dot{\mathbf{R}} = \text{const.} \tag{8.14}$$

We can explain this result in several ways: First (as we already knew), it is a direct consequence of conservation of total momentum. Alternatively, we can view it as reflecting that $\mathcal{L}$ is independent of $\mathbf{R}$, or, in the terminology introduced in Section 7.6, the CM coordinate $\mathbf{R}$ is "ignorable." More specifically, $\mathcal{L}_{\text{cm}} = \tfrac{1}{2}M\dot{\mathbf{R}}^2$ (which is the only part of $\mathcal{L}$ that involves $\mathbf{R}$) has the form of the Lagrangian of a *free* particle of mass M and position $\mathbf{R}$. Naturally, therefore (Newton's first law), $\mathbf{R}$ moves with constant velocity.

The Lagrange equation for the relative coordinate $\mathbf{r}$ is a little less simple but equally beautiful: $\mathcal{L}_{\text{rel}}$, the only part of $\mathcal{L}$ that involves $\mathbf{r}$, is mathematically indistinguishable from the Lagrangian for a single particle of mass μ and position $\mathbf{r}$, with potential energy $U(r)$. Thus the Lagrange equation corresponding to $\mathbf{r}$ is just (check it and see!)

$$\mu\ddot{\mathbf{r}} = -\nabla U(\mathbf{r}). \tag{8.15}$$

To solve for the relative motion, we have only to solve Newton's second law for a single particle of mass equal to the reduced mass μ and position $\mathbf{r}$, with potential energy $U(r)$.

The CM Reference Frame

Our problem becomes even easier to think about if we make a clever choice of reference frame. Specifically, because $\dot{\mathbf{R}} = \text{const}$, we can choose an inertial reference frame, the so-called **CM frame**, in which the CM is at rest and the total momentum

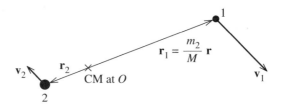

Figure 8.3   In the CM frame the center of mass is stationary at the origin. The relative position $\mathbf{r}$ is the position of particle 1 relative to particle 2; therefore, the position of particle 1 relative to the origin is $\mathbf{r}_1 = (m_2/M)\mathbf{r}$.

is zero. In this frame, $\dot{\mathbf{R}} = 0$ and the CM part of the Lagrangian is zero ($\mathcal{L}_{\text{cm}} = 0$). Thus in the CM frame

$$\mathcal{L} = \mathcal{L}_{\text{rel}} = \tfrac{1}{2}\mu\dot{\mathbf{r}}^2 - U(r) \tag{8.16}$$

and the problem really is reduced to a one-body problem. This dramatic simplification illustrates the curious terminology of the "ignorable coordinate." Recall that a coordinate q_i is said to be ignorable if $\partial\mathcal{L}/\partial q_i = 0$. We see that, in the present case at least, the motion associated with the ignorable coordinate $\mathbf{R}$ really is something that we can ignore.

It is worth taking a moment to consider what the motion looks like in the CM frame, as shown in Figure 8.3. The CM is stationary, and we naturally take it to be the origin. Both particles are moving, but with equal and opposite momenta. If m_2 is much greater than m_1 (as is often the case), the CM is close to m_2 and particle 2 has a small velocity. (In the figure, $m_2 = 3m_1$ and hence $v_2 = \tfrac{1}{3}v_1$.) It is important to note that the relative position $\mathbf{r}$ is the position of particle 1 relative to particle 2, and is not the actual position of either particle. As shown in the picture, the position of particle 1 is actually $\mathbf{r}_1 = (m_2/M)\mathbf{r}$. However, if $m_2 \gg m_1$, then the CM is very close to particle 2, which is almost stationary, and $\mathbf{r}_1 \approx \mathbf{r}$; that is, $\mathbf{r}$ is very nearly the same thing as $\mathbf{r}_1$.

The equation of motion in the CM frame is derived from the Lagrangian $\mathcal{L}_{\text{rel}}$ of (8.16) and is just Equation (8.15). This is precisely the same as the equation for a single particle of mass equal to the reduced mass μ, in the fixed central force field of the potential energy $U(r)$. In the equations of this chapter, the repeated appearance of the mass μ will serve to remind you that the equations apply to the relative motion of two bodies. However, you may find it easier to *visualize* a single body (of mass μ) orbiting about a fixed force center. In particular, if $m_2 \gg m_1$, these two problems are for practical purposes exactly the same. Moreover, if your interest actually is in a single body, of mass m say, orbiting a fixed force center, then you can use all of the same equations, simply replacing μ with m. In any event, any solution for the relative coordinate $\mathbf{r}(t)$ always gives us the motion of particle 1 relative to particle 2. Equivalently, using the relations of Figure 8.3, knowledge of $\mathbf{r}(t)$ tells us the motion of particle 1 (or particle 2) relative to the CM.

Conservation of Angular Momentum

We already know that the total angular momentum of our two particles is conserved. Like so many other things, this condition takes an especially simple form in the CM frame. In any frame, the total angular momentum is

$$\mathbf{L} = \mathbf{r}_1 \times \mathbf{p}_1 + \mathbf{r}_2 \times \mathbf{p}_2$$
$$= m_1 \mathbf{r}_1 \times \dot{\mathbf{r}}_1 + m_2 \mathbf{r}_2 \times \dot{\mathbf{r}}_2. \tag{8.17}$$

In the CM frame, we see from (8.9) (with $\mathbf{R} = 0$) that

$$\mathbf{r}_1 = \frac{m_2}{M}\mathbf{r} \quad \text{and} \quad \mathbf{r}_2 = -\frac{m_1}{M}\mathbf{r}. \tag{8.18}$$

Substituting into (8.17), we see that the angular momentum in the CM frame is

$$\mathbf{L} = \frac{m_1 m_2}{M^2}(m_2 \mathbf{r} \times \dot{\mathbf{r}} + m_1 \mathbf{r} \times \dot{\mathbf{r}})$$
$$= \mathbf{r} \times \mu \dot{\mathbf{r}} \tag{8.19}$$

where I have replaced $m_1 m_2/M$ by the reduced mass μ.

The most remarkable thing about this result is that the total angular momentum in the CM frame is exactly the same as the angular momentum of a single particle with mass μ and position $\mathbf{r}$. For our present purposes the important point is that, because angular momentum is conserved, we see that the vector $\mathbf{r} \times \dot{\mathbf{r}}$ is constant. In particular, the *direction* of $\mathbf{r} \times \dot{\mathbf{r}}$ is constant, which implies that the two vectors $\mathbf{r}$ and $\dot{\mathbf{r}}$ remain in a fixed plane. That is, in the CM frame, the whole motion remains in a fixed plane, which we can take to be the xy plane. In other words, in the CM frame, the two-body problem with central conservative forces is reduced to a two-dimensional problem.

The Two Equations of Motion

To set up the equations of motion for the remaining two-dimensional problem, we need to choose coordinates in the plane of the motion. The obvious choice is to use the polar coordinates r and ϕ, in terms of which the Lagrangian (8.16) is

$$\mathcal{L} = \tfrac{1}{2}\mu(\dot{r}^2 + r^2\dot{\phi}^2) - U(r). \tag{8.20}$$

Since this Lagrangian is independent of ϕ, the coordinate ϕ is ignorable, and the Lagrange equation corresponding to ϕ is just

$$\frac{\partial \mathcal{L}}{\partial \dot{\phi}} = \mu r^2 \dot{\phi} = \text{const} = \ell \qquad [\phi \text{ equation}]. \tag{8.21}$$

Since $\mu r^2 \dot{\phi}$ is the angular momentum ℓ (strictly, the z component ℓ_z), the ϕ equation is just a statement of conservation of angular momentum.

The Lagrange equation corresponding to r (often called the **radial equation**) is

$$\frac{\partial \mathcal{L}}{\partial r} = \frac{d}{dt}\frac{\partial \mathcal{L}}{\partial \dot{r}},$$

or

$$\mu r\dot{\phi}^2 - \frac{dU}{dr} = \mu\ddot{r} \qquad [r \text{ equation}]. \tag{8.22}$$

As we already saw in Example 7.2 [Equations (7.19) and (7.20)], if we move the centripetal term $\mu r\dot{\phi}^2$ over to the right, this is just the radial component of $\mathbf{F} = m\mathbf{a}$ (or rather, $\mathbf{F} = \mu\mathbf{a}$, since μ has replaced m).

8.4   The Equivalent One-Dimensional Problem

The two equations of motion that we have to solve are the ϕ equation (8.21) and radial equation (8.22). The constant ℓ (the angular momentum) in the ϕ equation is determined by the initial conditions, and our main use for the ϕ equation is to solve it for $\dot{\phi}$,

$$\dot{\phi} = \frac{\ell}{\mu r^2}, \tag{8.23}$$

which will let us eliminate $\dot{\phi}$ from the radial equation in favor of the constant ℓ. The radial equation can be rewritten as

$$\mu\ddot{r} = -\frac{dU}{dr} + \mu r\dot{\phi}^2 = -\frac{dU}{dr} + F_{cf} \tag{8.24}$$

which has the form of Newton's second law for a particle in *one* dimension with mass μ and position r, subject to the actual force $-dU/dr$ plus a "fictitious" outward centrifugal force[1]

$$F_{cf} = \mu r\dot{\phi}^2. \tag{8.25}$$

In other words, the particle's radial motion is exactly the same as if the particle were moving in one dimension, subject to the actual force $-dU/dr$ *plus* the centrifugal force F_{cf}.

We have now reduced the problem of the relative motion of two bodies to a single one-dimensional problem, as expressed by (8.24). Before we discuss what the solutions are going to look like, it is helpful to rewrite the centrifugal force, using the ϕ equation (8.23) to eliminate $\dot{\phi}$ in favor of the constant ℓ,

$$F_{cf} = \frac{\ell^2}{\mu r^3}. \tag{8.26}$$

Even better, we can now express the centrifugal force in terms of a centrifugal potential energy,

$$F_{cf} = -\frac{d}{dr}\left(\frac{\ell^2}{2\mu r^2}\right) = -\frac{dU_{cf}}{dr}, \tag{8.27}$$

---

[1] This centrifugal force may be a little more familiar if I write it in terms of the azimuthal velocity $v_\phi = r\dot{\phi}$ as $F_{cf} = \mu v_\phi{}^2/r$.

where the centrifugal potential energy U_{cf} is defined as

$$U_{cf}(r) = \frac{\ell^2}{2\mu r^2}. \tag{8.28}$$

Returning to (8.24), we can now rewrite the radial equation in terms of U_{cf} as

$$\mu\ddot{r} = -\frac{d}{dr}[U(r) + U_{cf}(r)] = -\frac{d}{dr}U_{eff}(r), \tag{8.29}$$

where the **effective potential energy** $U_{eff}(r)$ is the sum of the actual potential energy $U(r)$ and the centrifugal $U_{cf}(r)$:

$$U_{eff}(r) = U(r) + U_{cf}(r) = U(r) + \frac{\ell^2}{2\mu r^2}. \tag{8.30}$$

According to (8.29), the radial motion of the particle is exactly the same as if the particle were moving in one dimension with an effective potential energy $U_{eff} = U + U_{cf}$.

EXAMPLE 8.1 Effective Potential Energy for a Comet

Write down the actual and effective potential energies for a comet (or planet) moving in the gravitational field of the sun. Sketch the three potential energies involved and use the graph of $U_{eff}(r)$ to describe the motion of r. Since planetary motion was first described mathematically by the German astronomer Johannes Kepler, 1571–1630, this problem of the motion of a planet or comet around the sun (or any two bodies interacting via an inverse-square force) is often called the *Kepler problem*.

The actual gravitational potential energy of the comet is given by the well-known formula

$$U(r) = -\frac{Gm_1m_2}{r} \tag{8.31}$$

where G is the universal gravitational constant, and m_1 and m_2 are the masses of the comet and the sun. The centrifugal potential energy is given by (8.28), so the total effective potential energy is

$$U_{eff}(r) = -\frac{Gm_1m_2}{r} + \frac{\ell^2}{2\mu r^2}. \tag{8.32}$$

The general behavior of this effective potential energy is easily seen (Figure 8.4). When r is large, the centrifugal term $\ell^2/2\mu r^2$ is negligible compared to the gravitational term $-Gm_1m_2/r$, and the effective PE, $U_{eff}(r)$, is negative and

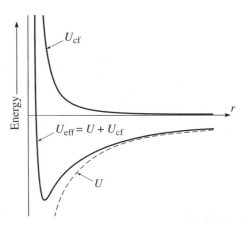

Figure 8.4 The effective potential energy $U_{\text{eff}}(r)$ that governs the radial motion of a comet is the sum of the actual gravitational potential energy $U(r) = -Gm_1m_2/r$ and the centrifugal term $U_{\text{cf}} = \ell^2/2\mu r^2$. For large r, the dominant effect is the attractive gravitational force; for small r, it is the repulsive centrifugal force.

sloping up as r increases. According to (8.29), the acceleration of r is down this slope. [The roller coaster car accelerates down the track defined by $U_{\text{eff}}(r)$.] Thus when a comet is far from the sun, $\ddot{r}$ is always inward.

When r is small, the centrifugal term $\ell^2/2\mu r^2$ dominates the gravitational term $-Gm_1m_2/r$ (unless $\ell = 0$), and near $r = 0$, $U_{\text{eff}}(r)$ is positive and slopes downward. Thus, as a comet gets closer to the sun, $\ddot{r}$ eventually becomes outward, and the comet starts to move away from the sun again. The one exception to this statement is when the angular momentum is exactly zero, $\ell = 0$, in which case (8.23) implies that $\dot{\phi} = 0$; that is, the comet is moving exactly radially, along a line of constant ϕ, and must at some time hit the sun.

Conservation of Energy

To find the details of the orbit we must look more closely at the radial equation (8.29). If we multiply both sides of that equation by $\dot{r}$, we find that

$$\frac{d}{dt}\left(\tfrac{1}{2}\mu\dot{r}^2\right) = -\frac{d}{dt}U_{\text{eff}}(r). \tag{8.33}$$

In other words,

$$\tfrac{1}{2}\mu\dot{r}^2 + U_{\text{eff}}(r) = \text{const.} \tag{8.34}$$

This result is, in fact, just conservation of energy: If we write out U_{eff} as $U + \ell^2/2\mu r^2$ and replace ℓ by $\mu r^2\dot{\phi}$, we see that

$$\tfrac{1}{2}\mu\dot{r}^2 + U_{\text{eff}}(r) = \tfrac{1}{2}\mu\dot{r}^2 + \tfrac{1}{2}\mu r^2\dot{\phi}^2 + U(r)$$

$$= E. \tag{8.35}$$

This completes the rewriting of the two-dimensional problem of the relative motion as an equivalent one-dimensional problem involving just the radial motion. We see that the total energy (which we knew all along is constant) can be thought of as the one-dimensional kinetic energy of the radial motion, plus the effective one-dimensional potential energy U_{eff}, since the latter includes the actual potential energy U *and* the kinetic energy $\frac{1}{2}\mu r^2 \dot{\phi}^2$ of the angular motion. This means that all of our experience with one-dimensional problems, both in terms of forces and in terms of energy, can be immediately transferred to the two-body central-force problem.

EXAMPLE 8.2   Energy Considerations for a Comet or Planet

Examine again the comet (or planet) of Example 8.1 and, by considering its total energy E, find the equation that determines the maximum and minimum distances of the comet from the sun, if $E > 0$ and, again, if $E < 0$.

In the energy equation (8.35) the term $\frac{1}{2}\mu \dot{r}^2$ on the left is always greater than or equal to zero. Therefore, the comet's motion is confined to those regions where $E \geq U_{\text{eff}}$. To see what this implies, I have redrawn in Figure 8.5 the graph of U_{eff} from Figure 8.4. Let us consider first the case that the comet's energy is greater than zero. In the figure I have drawn a dashed horizontal line at height E, labeled $E > 0$. A comet with this energy can move anywhere that this line is above the curve of $U_{\text{eff}}(r)$, but nowhere that the line is below the curve. This means simply that the comet cannot move anywhere inside the turning point labeled r_{min}, determined by the condition

$$U_{\text{eff}}(r_{\text{min}}) = E. \tag{8.36}$$

If the comet is initially moving in, toward the sun, then it will continue to do so until it reaches r_{min}, where $\dot{r} = 0$ instantaneously. It then moves outward, and, since there are no other points at which $\dot{r}$ can vanish, it eventually moves off to infinity, and the orbit is **unbounded**.

If instead $E < 0$, then the line drawn at height E (labeled $E < 0$) meets the curve of $U_{\text{eff}}(r)$ at the two turning points labeled r_{min} and r_{max}, and a comet with

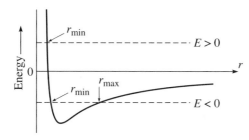

Figure 8.5   Plot of the effective potential energy $U_{\text{eff}}(r)$ against r for a comet. For a given energy E, the comet can only go where $E \geq U_{\text{eff}}(r)$. For $E > 0$ this means it cannot go inside the turning point at r_{min} where $U_{\text{eff}} = E$. For $E < 0$ it is confined between the two turning points labeled r_{min} and r_{max}.

$E < 0$ is trapped between these two values of r. If it is moving away from the sun ($\dot{r} > 0$) it continues to do so until it reaches $r_{\max}$, where $\dot{r}$ vanishes and reverses sign. The comet then moves inward until it reaches $r_{\min}$, where $\dot{r}$ reverses again. Therefore, the comet oscillates in and out between $r_{\min}$ and $r_{\max}$. For obvious reasons, this type of orbit is called a **bounded orbit**.[2]

Finally, if E is equal to the minimum value of $U_{\text{eff}}(r)$ (for a given value of the angular momentum ℓ), the two turning points $r_{\min}$ and $r_{\max}$ coalesce, and the comet is trapped at a fixed radius and moves in a circular orbit.

In this example, I considered just the case of an inverse-square force, but many two-body problems have the same qualitative features. For example, the motion of the two atoms in a diatomic molecule is governed by an effective potential that was sketched in Figure 4.12 and looks very like the gravitational curve of Figure 8.5. Thus all of our qualitative conclusions apply to the diatomic molecule and many other two-body problems.

In thinking about the radial motion of the two-body problem, you must not entirely forget the angular motion. According to (8.23), $\dot{\phi} = \ell/\mu r^2$, and ϕ is always changing, always with the same sign (continually increasing or continually decreasing). For example, as a comet with positive energy approaches the sun, the angle ϕ changes, at a rate that increases as r gets smaller; as the comet moves away, ϕ continues to change in the same direction, but at a rate that decreases as r gets larger. Thus the actual orbit of a positive-energy comet looks something like Figure 8.6. For the case of an inverse-square force (like gravity), the orbit of Figure 8.6 is actually a hyperbola, as we shall prove shortly, but the unbounded orbits (that is, orbits with $E > 0$) are qualitatively similar for many different force laws.

For the bounded orbits ($E < 0$), we have seen that r oscillates between the two extreme values $r_{\min}$ and $r_{\max}$, while ϕ continually increases (or decreases, but let's suppose the comet is orbiting counter-clockwise, so that ϕ is increasing). In the case of the inverse-square force, we shall see that the period of the radial oscillations happens to equal the time for ϕ to make exactly one complete revolution. Therefore, the motion repeats itself exactly once per revolution, as in Figure 8.7(a). (We shall also see that, for any inverse-square force, the bounded orbits are actually ellipses.) For most other force laws, the period of the radial motion is different from the time to make one revolution, and in most cases the orbit is not even closed (that is, it never returns to its initial conditions).[3] Figure 8.7(b) shows an orbit for which r goes from $r_{\min}$ to $r_{\max}$ and back to $r_{\min}$ in the time that the angle ϕ advances by about $330°$, and the orbit certainly does not close on itself after one revolution.

---

[2] If we consider just one comet in orbit around the sun, then energy conservation implies that a bounded orbit ($E < 0$) can never change into an unbounded orbit ($E > 0$), nor vice versa. In reality a comet can occasionally come close enough to another comet or planet to change E, and the orbit can then change from bounded to unbounded or the other way.

[3] Besides the inverse square force, the only important exception is the isotropic harmonic oscillator, for which the orbits are also ellipses, as discussed in Section 5.3.

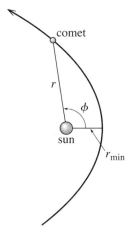

Figure 8.6   Typical unbounded orbit for a positive-energy comet. Initially r decreases from infinity to r_{min} and then goes back out to infinity. Meanwhile the angle ϕ is continually increasing.

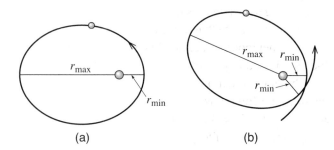

(a)   (b)

Figure 8.7   (a) The bounded orbits for any inverse-square force have the unusual property that r goes from r_{min} to r_{max} and back to r_{min} in exactly the time that ϕ goes from 0 to 360°. Therefore the orbit repeats itself every revolution. (b) For most other force laws, the period of oscillation of r is different from the time in which ϕ advances by 360°, and the orbit does not close on itself after one revolution. In this example, r completes one cycle from r_{min} to r_{max} and back to r_{min} while ϕ advances by about 330°.

8.5   The Equation of the Orbit

The radial equation (8.29) determines r as a function of t, but for many purposes we would like to know r as a function of ϕ. For example, the function $r = r(\phi)$ will tell us the shape of the orbit more directly. Thus we would like to rewrite the radial equation as a differential equation for r in terms of ϕ. There are two tricks for doing this, but let me first write the radial equation in terms of forces:

$$\mu \ddot{r} = F(r) + \frac{\ell^2}{\mu r^3} \qquad (8.37)$$

where $F(r)$ is the actual central force, $F = -dU/dr$, and the second term is the centrifugal force.

The first trick to rewriting this equation in terms of ϕ is to make the substitution

$$u = \frac{1}{r} \qquad \text{or} \qquad r = \frac{1}{u} \qquad (8.38)$$

and the second is to rewrite the differential operator d/dt in terms of $d/d\phi$ using the chain rule:

$$\frac{d}{dt} = \frac{d\phi}{dt}\frac{d}{d\phi} = \dot{\phi}\frac{d}{d\phi} = \frac{\ell}{\mu r^2}\frac{d}{d\phi} = \frac{\ell u^2}{\mu}\frac{d}{d\phi}. \qquad (8.39)$$

(The third equality follows because $\ell = \mu r^2 \dot{\phi}$, and the last results from the change of variables $u = 1/r$.)

Using the identity (8.39) we can rewrite $\ddot{r}$ on the left of the radial equation. First

$$\dot{r} = \frac{d}{dt}(r) = \frac{\ell u^2}{\mu}\frac{d}{d\phi}\left(\frac{1}{u}\right) = -\frac{\ell}{\mu}\frac{du}{d\phi}$$

and hence

$$\ddot{r} = \frac{d}{dt}(\dot{r}) = \frac{\ell u^2}{\mu}\frac{d}{d\phi}\left(-\frac{\ell}{\mu}\frac{du}{d\phi}\right) = -\frac{\ell^2 u^2}{\mu^2}\frac{d^2 u}{d\phi^2}. \qquad (8.40)$$

Substituting back into the radial equation (8.37) we find

$$-\frac{\ell^2 u^2}{\mu}\frac{d^2 u}{d\phi^2} = F + \frac{\ell^2 u^3}{\mu}$$

or

$$u''(\phi) = -u(\phi) - \frac{\mu}{\ell^2 u(\phi)^2}F. \qquad (8.41)$$

For any given central force F, this transformed radial equation is a differential equation for the new variable $u(\phi)$. If we can solve it, then we can immediately write down r as $r = 1/u$. In the next section, we shall solve it for the case of an inverse-square force and show that the resulting orbits are conic sections, that is, ellipses, parabolas, or hyperbolas. First, here is a simpler example.

EXAMPLE 8.3   The Radial Equation for a Free Particle

Solve the transformed radial equation (8.41) for a *free* particle (that is, a particle subject to no forces) and confirm that the resulting orbit is the expected straight line.

This example is probably one of the hardest ways of showing that a free particle moves along a straight line. Nevertheless, it is a nice check that the transformed radial equation makes sense. In the absence of forces, (8.41) is just

$$u''(\phi) = -u(\phi)$$

whose general solution we know to be

$$u(\phi) = A \cos(\phi - \delta), \tag{8.42}$$

where A and δ are arbitrary constants. Therefore, (renaming the constant $A = 1/r_0$)

$$r(\phi) = \frac{1}{u(\phi)} = \frac{r_0}{\cos(\phi - \delta)}. \tag{8.43}$$

This unpromising-looking equation is in fact the equation of a straight line in polar coordinates, as you can see from Figure 8.8. In that picture Q is a fixed point with polar coordinates (r_0, δ), and the line in question is the line through Q perpendicular to OQ. It is easy to see that the point P with polar coordinates (r, ϕ) lies on this line if and only if $r \cos(\phi - \delta) = r_0$. In other words, Equation (8.43) is the equation of this straight line.

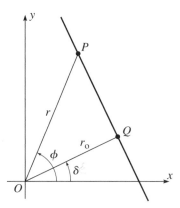

Figure 8.8   The fixed point Q has polar coordinates (r_0, δ) relative to the origin O. The point P with polar coordinates (r, ϕ) lies on the line through Q perpendicular to OQ if and only if $r \cos(\phi - \delta) = r_0$. That is, the equation of this line is (8.43).

In the next section, I shall use the same transformed radial equation (8.41) to solve a much less trivial problem, finding the path of a comet or any other body held in orbit by an inverse-square force.

8.6 The Kepler Orbits

Let us now return to the Kepler problem, the problem of finding the possible orbits of a comet or any other object subject to an inverse-square force. The two important examples of this problem are the motion of comets or planets around the sun (or earth satellites around the earth), in which case the force is the gravitational force $-Gm_1m_2/r^2$, and the orbital motion of two opposite charges q_1 and q_2, in which case the force is the Coulomb force kq_1q_2/r^2. To include both cases and to simplify the equations, I shall write the force as (remember that $u = 1/r$)

$$F(r) = -\frac{\gamma}{r^2} = -\gamma u^2, \tag{8.44}$$

where γ is the "force constant," equal to Gm_1m_2 in the gravitational case.[4]

Thanks to our elaborate preparations, we can now solve the main problem very easily. Inserting the force (8.44) into the transformed radial equation (8.41), we find that $u(\phi)$ must satisfy

$$u''(\phi) = -u(\phi) + \gamma\mu/\ell^2. \tag{8.45}$$

Notice that it is a unique feature of the inverse-square force that the last term in this equation is a constant, since only in this case does the u^2 of the force cancel the $1/u^2$ in (8.41). Because this last term is constant, we can solve (8.45) very easily: If we substitute

$$w(\phi) = u(\phi) - \gamma\mu/\ell^2,$$

the equation becomes

$$w''(\phi) = -w(\phi),$$

which has the general solution

$$w(\phi) = A\cos(\phi - \delta), \tag{8.46}$$

where A is a positive constant and δ is a constant that we can take to be zero by a suitable choice of the direction $\phi = 0$. Thus the general solution for $u(\phi)$ can be written as

$$u(\phi) = \frac{\gamma\mu}{\ell^2} + A\cos\phi = \frac{\gamma\mu}{\ell^2}(1 + \epsilon\cos\phi) \tag{8.47}$$

---

[4] The constant γ is positive for the gravitational force and for the force between two opposite charges. As discussed in Problem 8.31, for two charges of the same sign, γ is negative. For now, we'll assume it is positive.

where ϵ is just a new name for the dimensionless positive constant $A\ell^2/\gamma\mu$. Since $u = 1/r$, the constant $\gamma\mu/\ell^2$ on the right has the dimensions [1/length], and I shall introduce the length

$$c = \frac{\ell^2}{\gamma\mu} \qquad (8.48)$$

in terms of which our solution becomes

$$\frac{1}{r(\phi)} = \frac{1}{c}(1 + \epsilon\cos\phi)$$

or

$$r(\phi) = \frac{c}{1 + \epsilon\cos\phi}. \qquad (8.49)$$

This is our solution for r as a function of ϕ, in terms of the undetermined positive constant ϵ and the length $c = \ell^2/\gamma\mu$ (which is $\ell^2/Gm_1m_2\mu$ in the gravitational problem). I shall now explore its properties, first for the bounded orbits and then for the unbounded.

The Bounded Orbits

The behavior of the orbit $r(\phi)$ in (8.49) is controlled by the as-yet undetermined positive constant ϵ. A glance at (8.49) shows this behavior is very different according as $\epsilon < 1$ or $\epsilon \geq 1$. If $\epsilon < 1$, the denominator of (8.49) never vanishes, and $r(\phi)$ remains bounded for all ϕ. If $\epsilon \geq 1$ the denominator vanishes at some angle, and $r(\phi)$ approaches infinity as ϕ approaches that angle. Evidently the value $\epsilon = 1$ is the boundary between the bounded and unbounded orbits. I shall show shortly that this boundary corresponds exactly to the boundary between $E < 0$ and $E \geq 0$ discussed before. Meanwhile, let us start with the case that the constant ϵ is less than 1. With $\epsilon < 1$, the denominator of $r(\phi)$ in (8.49) oscillates as shown in Figure 8.9 between the values $1 \pm \epsilon$. Therefore, $r(\phi)$ oscillates between

$$r_{\min} = \frac{c}{1 + \epsilon} \qquad \text{and} \qquad r_{\max} = \frac{c}{1 - \epsilon} \qquad (8.50)$$

with $r = r_{\min}$ at the so-called **perihelion** when $\phi = 0$, and $r = r_{\max}$ at the **aphelion** when $\phi = \pi$. Since $r(\phi)$ is obviously periodic in ϕ with period 2π, it follows that $r(2\pi) = r(0)$ and the orbit closes on itself after just one revolution. Thus the general appearance of the orbit is as in Figure 8.10.

While the orbit shown in Figure 8.10 certainly *looks* like an ellipse, I have not yet proved that it really is. However, it is a reasonably easy exercise (see Problem 8.16) to rewrite (8.49) in Cartesian coordinates and cast it in the form

$$\frac{(x + d)^2}{a^2} + \frac{y^2}{b^2} = 1 \qquad (8.51)$$

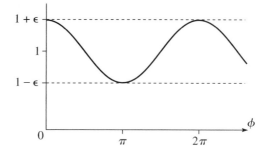

Figure 8.9   The denominator $1 + \epsilon \cos \phi$ in Equation (8.49) for $r(\phi)$ oscillates between $1 + \epsilon$ and $1 - \epsilon$, and is periodic with period 2π.

where (as you can easily check)

$$a = \frac{c}{1 - \epsilon^2}, \qquad b = \frac{c}{\sqrt{1 - \epsilon^2}}, \qquad \text{and} \qquad d = a\epsilon. \qquad (8.52)$$

Equation (8.51) is the standard equation of an ellipse with semimajor and semiminor axes a and b, except that where we expect to see x we have $x + d$. This difference reflects that our origin, the sun, is not at the center of the ellipse, but at a distance d from it, as shown in Figure 8.10.

We can now identify the constant ϵ, which started life as an undetermined constant of integration in (8.47). According to (8.52) the ratio of the major to minor axes is

$$\frac{b}{a} = \sqrt{1 - \epsilon^2}. \qquad (8.53)$$

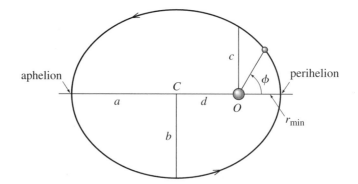

Figure 8.10   The bounded orbits of a comet or planet as given by Equation (8.49) are ellipses. The sun is at the origin O, which is one focus of the ellipse (*not* the center C). The distances a and b are called the semimajor and semiminor axes. The parameter $c = \ell^2/\gamma\mu$ introduced in (8.48) is the value of r when $\phi = 90°$. The points where the comet is closest and farthest from the sun are called the perihelion and aphelion.

Although you almost certainly don't remember it, this equation is the definition (or one possible definition) of the eccentricity of the ellipse. That is, this equation tells us that the constant ϵ is the eccentricity. Notice that if $\epsilon = 0$, then $b = a$ and the ellipse is a circle; if $\epsilon \to 1$, then $b/a \to 0$ and the ellipse becomes very thin and elongated.

Having identified the constant ϵ as the eccentricity, we can now identify the position of the sun in relation to the ellipse. According to (8.52) the distance from the center C to the sun at O is $d = a\epsilon$, and (though again you may not remember it) $a\epsilon$ is the distance from the center to either focus of the ellipse. Thus the position of the sun is actually one of the ellipse's two focuses, and we have now proved **Kepler's first law**, that the planets (and comets whose orbits are bounded) follow orbits that are ellipses with the sun at one focus.

EXAMPLE 8.4   Halley's Comet

Halley's comet, named for the English astronomer Edmund Halley (1656–1742), follows a very eccentric orbit with $\epsilon = 0.967$. At closest approach (the perihelion) the comet is 0.59 AU from the sun, fairly close to the orbit of Mercury. (The AU or astronomical unit is the mean distance of the earth from the sun, about 1.5×10^8 km.) What is the comet's greatest distance from the sun, that is, the distance of the aphelion?

The given distance is $r_{\min} = 0.59$ AU, and, according to (8.50), $r_{\max}/r_{\min} = (1 + \epsilon)/(1 - \epsilon)$. Therefore

$$r_{\max} = \frac{1 + \epsilon}{1 - \epsilon} r_{\min} = \frac{1.967}{0.033} r_{\min} = 60\, r_{\min} = 35 \text{ AU}.$$

This means that at its greatest distance Halley's comet is outside the orbit of Neptune.

The Orbital Period; Kepler's Third Law

We can now find the period of the elliptical orbits of the comets and planets. According to Kepler's second law (Section 3.4), the rate at which a line from the sun to a comet or planet sweeps out area is

$$\frac{dA}{dt} = \frac{\ell}{2\mu}.$$

Since the total area of an ellipse is $A = \pi a b$, the period is

$$\tau = \frac{A}{dA/dt} = \frac{2\pi a b \mu}{\ell}.$$

If we square both sides and use (8.53) to replace b^2 by $a^2(1 - \epsilon^2)$, this becomes

$$\tau^2 = 4\pi^2 \frac{a^4(1 - \epsilon^2)\mu^2}{\ell^2} = 4\pi^2 \frac{a^3 c \mu^2}{\ell^2},$$

where in the last equality I used (8.52) to replace $a(1 - \epsilon^2)$ by c. Since the length c was defined in (8.48) as $\ell^2/\gamma\mu$, this implies that

$$\tau^2 = 4\pi^2 \frac{a^3\mu}{\gamma}. \tag{8.54}$$

Finally, γ is the constant in the inverse-square force law $F = -\gamma/r^2$, and, for the gravitational force, $\gamma = Gm_1m_2 = G\mu M$ where M is the total mass, $M = m_1 + m_2$. (Notice the handy identity that $m_1m_2 = \mu M$.) In our case $m_2 = M_s$, the mass of the sun, which is very much greater than m_1, the mass of the comet or planet. Thus, to an excellent approximation, $M \approx M_s$, and

$$\gamma = Gm_1m_2 \approx G\mu M_s.$$

Therefore, the factor of μ in (8.54) cancels, and we find that

$$\tau^2 = \frac{4\pi^2}{GM_s} a^3. \tag{8.55}$$

This is **Kepler's third law**: Because the mass of the comet (or planet) has canceled out, the law says that for all bodies orbiting the sun, the square of the period is proportional to the cube of the semimajor axis. (For circular orbits, we can replace a^3 by r^3.) The law applies equally to the satellites of any large body. For example, all satellites of the earth, including the moon, obey the same law [with M_s replaced by the earth's mass M_e in (8.55)], and the same applies to all the moons of Jupiter.

EXAMPLE 8.5   Period of a Low-Orbit Earth Satellite

Use Kepler's third law to estimate the period of a satellite in a circular orbit a few tens of miles above the earth's surface.

The period is given by (8.55) with M_s replaced by M_e. Since the orbit is circular, we can replace a by r, and since the orbit is close to the earth's surface, $r \approx R_e$, the radius of the earth. Therefore

$$\tau^2 = \frac{4\pi^2}{GM_e} R_e^3.$$

This simplifies if we recall that $GM_e/R_e^2 = g$, the acceleration of gravity on the earth's surface, and we find that

$$\tau = 2\pi\sqrt{\frac{R_e}{g}} = 2\pi\sqrt{\frac{6.38 \times 10^6 \text{ m}}{9.8 \text{ m/s}^2}} = 5070 \text{ s} \approx 85 \text{ min}, \tag{8.56}$$

in agreement with the well-known observation that low-orbit satellites circle the earth in about one and a half hours.

Relation between Energy and Eccentricity

Finally, we can relate the eccentricity ϵ of the orbit to the energy E of the comet or other orbiting body. The simplest way to do this is to remember that, at its distance of closest approach r_{min}, the comet's energy is equal to the effective potential energy U_{eff} [Equation (8.36)],

$$E = U_{eff}(r_{min}) = -\frac{\gamma}{r_{min}} + \frac{\ell^2}{2\mu r_{min}^2}$$

$$= \frac{1}{2r_{min}}\left(\frac{\ell^2}{\mu r_{min}} - 2\gamma\right). \tag{8.57}$$

Now we know from (8.50) that $r_{min} = c/(1 + \epsilon)$, and from its definition (8.48) that $c = \ell^2/\gamma\mu$. Therefore

$$r_{min} = \frac{\ell^2}{\gamma\mu(1 + \epsilon)}$$

and, substituting into (8.57),

$$E = \frac{\gamma\mu(1 + \epsilon)}{2\ell^2}[\gamma(1 + \epsilon) - 2\gamma]$$

$$= \frac{\gamma^2\mu}{2\ell^2}(\epsilon^2 - 1). \tag{8.58}$$

The calculations leading to (8.58) are equally valid for bounded and unbounded orbits, and they imply the following expected correlations: Negative energies $(E < 0)$ correspond to eccentricities $\epsilon < 1$, which in turn correspond to bounded orbits. Positive energies $(E > 0)$ correspond to eccentricities $\epsilon > 1$, which in turn correspond to unbounded orbits. Equation (8.58) is a useful relation between the mechanical properties E and ℓ and the geometrical property ϵ. It implies some interesting connections. For example, for a given value of the angular momentum ℓ, the orbit of lowest possible energy is the circular orbit with $\epsilon = 0$ (a connection which has an important counterpart in quantum mechanics).

8.7   The Unbounded Kepler Orbits

In the previous section, we found the general Kepler orbit, as given by (8.49),

$$r(\phi) = \frac{c}{1 + \epsilon\cos\phi}, \tag{8.59}$$

and examined in detail the bounded orbits — those for which $\epsilon < 1$ or, equivalently, as we have seen, $E < 0$. In this section, I shall sketch the corresponding analysis of the unbounded orbits, with $\epsilon \geq 1$ and $E \geq 0$.

The boundary between the bounded and unbounded orbits comes when $\epsilon = 1$ or $E = 0$. With $\epsilon = 1$, the denominator of (8.59) vanishes when $\phi = \pm\pi$. Therefore,

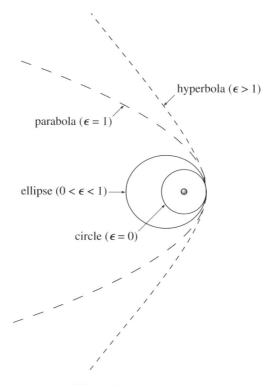

Figure 8.11   Four different Kepler orbits for a comet: a circle, an ellipse, a parabola, and a hyperbola. For clarity, the four orbits were chosen with the same values for r_{min} and with the closest approaches all in the same direction.

$r(\phi) \to \infty$ as $\phi \to \pm\pi$. That is, if $\epsilon = 1$, the orbit is unbounded and goes off to infinity as the comet approaches $\phi = \pm\pi$. Some elementary algebra, parallel to what led to (8.51), shows that with $\epsilon = 1$ the Cartesian version of (8.59) is

$$y^2 = c^2 - 2cx \tag{8.60}$$

which is the equation of a parabola. This orbit is shown (with the long dashes) in Figure 8.11.

If $\epsilon > 1$ (or $E > 0$), the denominator of (8.59) vanishes at a value ϕ_{max} determined by the condition

$$\epsilon \cos(\phi_{max}) = -1.$$

Thus $r(\phi) \to \infty$ when $\phi \to \pm\phi_{max}$ and the orbit is confined to the range of angles $-\phi_{max} < \phi < \phi_{max}$. This gives the orbit the general appearance sketched in Figure 8.6. With $\epsilon > 1$ the Cartesian form of (8.59) is (Problem 8.30)

$$\frac{(x - \delta)^2}{\alpha^2} - \frac{y^2}{\beta^2} = 1, \tag{8.61}$$

where you can easily identify the constants α, β, and δ (Problem 8.30). This is the equation of a hyperbola, and we have proved that, as anticipated, the positive energy Kepler orbits are hyperbolas. One such orbit is shown (with the smaller dashes) in Figure 8.11.

Summary of Kepler Orbits

Our results for the Kepler orbits can be summarized as follows: All of the possible orbits are given by Equation (8.59),

$$r(\phi) = \frac{c}{1 + \epsilon \cos \phi}, \tag{8.62}$$

and are characterized by the two constants of integration[5] ϵ and c. The dimensionless constant ϵ is related to the comet's energy by (8.58),

$$E = \frac{\gamma^2 \mu}{2\ell^2}(\epsilon^2 - 1). \tag{8.63}$$

It is, as we have seen, the eccentricity of the orbit that determines the orbit's shape as follows:

eccentricity	energy	orbit
$\epsilon = 0$	$E < 0$	circle
$0 < \epsilon < 1$	$E < 0$	ellipse
$\epsilon = 1$	$E = 0$	parabola
$\epsilon > 1$	$E > 0$	hyperbola

You can see from (8.62) that the constant c is a scale factor that determines the size of the orbit. It has the dimensions of length and is the distance from sun to comet when $\phi = \pi/2$. It is equal to $\ell^2/\gamma\mu$ or, since γ is the force constant Gm_1m_2,

$$c = \frac{\ell^2}{Gm_1m_2\mu}, \tag{8.64}$$

where m_1 is the mass of the comet, m_2 that of the sun, and μ is the reduced mass $\mu = m_1m_2/(m_1 + m_2)$, which is exceedingly close to m_1 since m_2 is so large.

8.8   Changes of Orbit

In this final section, I shall discuss how a satellite can change from one orbit to another. For example, a spacecraft wishing to visit Venus may want to transfer from a circular

---

[5] Since Newton's second law is a second-order differential equation and the motion is in two dimensions, there are actually four constants of integration in all. The third is the constant δ in (8.46) which we chose to be zero, forcing the axis of the orbit to be the x axis. The fourth is the comet's position on the orbit at time $t = 0$.

orbit close to the earth and centered on the sun to an elliptical orbit that will carry it to the orbit of Venus. Another example, and the one we shall discuss here, is an earth satellite wishing to change from one orbit about the earth to another, perhaps from a circular orbit to an elliptical orbit that will carry it to a higher altitude. The analysis of earth orbits is the same as that of orbits around the sun, except that the mass M_s of the sun must be replaced by the mass M_e of the earth, and the closest and furthest points from earth are called the **perigee** and **apogee** (instead of perihelion and aphelion for the sun). We shall confine attention to bounded, elliptical orbits, for which the most general form is

$$r(\phi) = \frac{c}{1 + \epsilon \cos(\phi - \delta)}. \tag{8.65}$$

(As long as we were interested in just one orbit, we could choose our x axis so that the angle δ was zero. If we're interested in two arbitrary orbits, we cannot get rid of δ in this way — anyway not for both.)

Let us suppose that our spacecraft is initially in a orbit of the form (8.65) with energy E_1, angular momentum ℓ_1 and orbital paramenters c_1, ϵ_1, and δ_1. A common way to change orbits is for the spacecraft to fire its rockets vigorously for a brief time. To a good approximation we can treat this procedure as an impulse that occurs at a unique angle ϕ_0 and causes an instantaneous change of velocity by a known amount. From the known change in velocity, we can calculate the new energy E_2 and angular momentum ℓ_2. From (8.48) we can calculate the new value of c_2, and from (8.58) the new eccentricity ϵ_2. Finally, because the new orbit must join onto the old one at the angle ϕ_0, that is, $r_1(\phi_0) = r_2(\phi_0)$, we can find δ_2 from the equation

$$\frac{c_1}{1 + \epsilon_1 \cos(\phi_0 - \delta_1)} = \frac{c_2}{1 + \epsilon_2 \cos(\phi_0 - \delta_2)}. \tag{8.66}$$

This calculation, though straightforward in principle, is tedious, and not especially illuminating, in practice. To simplify the calculations and to better reveal the important features, I shall treat just one important special case.

A Tangential Thrust at Perigee

Let us consider a satellite that transfers from one orbit to another by firing its rockets in the tangential direction, forward or backward, when it is at the perigee of its initial orbit. By choice of our x axis, we can arrange that this occurs in the direction $\phi = 0$, so that $\phi_0 = 0$ and $\delta_1 = 0$. Moreover, because the rockets are fired in the tangential direction, the velocity just after firing is still in the same direction, which is perpendicular to the radius from earth to the satellite. Therefore, the position at which the rockets are fired is also the perigee for the final orbit,[6] and $\delta_2 = 0$ as well. Thus the equation (8.66) that assures the continuity of the orbit reduces to

$$\frac{c_1}{1 + \epsilon_1} = \frac{c_2}{1 + \epsilon_2}. \tag{8.67}$$

---

[6] Actually, it can be the perigee of the final orbit *or the apogee*, but we can treat both cases at once, as we shall see directly.

Let us denote by λ the ratio of the satellite's speeds just before and just after the firing of the rockets, $v_2 = \lambda v_1$. I shall call λ the **thrust factor**; if $\lambda > 1$, then the thrust was forward and the satellite sped up; if $0 < \lambda < 1$, then the thrust was backward and the satellite slowed down. (In principle, λ could be negative, but this would represent a reversal of direction, an unlikely maneuver that I shan't consider.)

At perigee the angular momentum is just $\ell = \mu r v$. The value of r does not change during the impulse, and I shall assume that the firing of the rockets changes the satellite's mass by a negligible amount. Under these assumptions, the angular momentum changes by the same factor as the speed:

$$\ell_2 = \lambda \ell_1. \tag{8.68}$$

According to (8.48), the parameter c is proportional to ℓ^2. Therefore, the new value of c is

$$c_2 = \lambda^2 c_1. \tag{8.69}$$

Substituting into (8.67) and solving for ϵ_2, we find for the new eccentricity

$$\epsilon_2 = \lambda^2 \epsilon_1 + (\lambda^2 - 1). \tag{8.70}$$

Equation (8.70) contains almost all the interesting information about the new orbit. For example, if $\lambda > 1$ (a forward thrust), it is easy to see that the new orbit has $\epsilon_2 > \epsilon_1$. Thus the new orbit has the same perigee as the old one, but has greater eccentricity and so lies outside the old orbit, as shown by the outer dashed curve in Figure 8.12(a). If we make λ large enough, then the new eccentricity becomes greater than 1; in this case the new orbit is actually a hyperbola, and our spacecraft escapes from the earth.

If we choose the thrust factor $\lambda < 1$ (a backward thrust), then the new eccentricity is less than the old, $\epsilon_2 < \epsilon_1$, and the new orbit lies inside the old, as shown by the inner dashed curve in Figure 8.12(b). As we make λ steadily smaller, eventually ϵ_2 vanishes; that is, if we fire the rockets backward with just the right impulse, we can move the satellite into a circular orbit. If we choose λ still smaller, then ϵ_2 becomes negative. What does this signify? The parameter ϵ started out as a positive constant, but the orbital equation $r = c/(1 + \epsilon \cos \phi)$ makes perfectly good sense with $\epsilon < 0$.

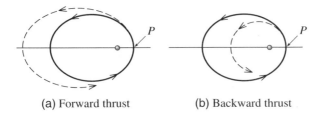

(a) Forward thrust   (b) Backward thrust

Figure 8.12   Changing orbits. The satellite's original orbit is shown as the solid curve, and the rockets are fired when the satellite is at the perigee P. (a) A forward impulse moves the satellite to the larger dashed elliptical orbit. (b) A backward impulse moves the satellite to the smaller dashed elliptical orbit.

The only difference is that the direction $\phi = 0$ is now the direction of maximum r and $\phi = \pi$ is that of minimum r; that is, the apogee and perigee have exchanged places. By administering a large enough backward thrust at P (the old orbit's perigee), we have transferred the satellite to a smaller orbit for which P is now the apogee.

EXAMPLE 8.6 Changing between Circular Orbits

A satellite's crew in a circular orbit of radius R_1 wishes to transfer to a circular orbit of radius $2R_1$. It does this using two successive boosts, as shown in Figure 8.13. First it boosts itself at point P into an elliptical transfer orbit 2, just large enough to take it out to the required radius. Second, on reaching the required radius (at P', the apogee of the transfer orbit) it boosts itself into the desired circular orbit 3. By what factor must it increase its speed in each of these two boosts? That is, what are the required thrust factors λ and λ'? By what factor does the satellite's speed increase as a result of the whole maneuver?

The initial circular orbit has $c_1 = R_1$ and eccentricity $\epsilon_1 = 0$. The final orbit is to have radius $R_3 = 2R_1$. According to (8.69), the transfer orbit has $c_2 = \lambda^2 R_1$ and, according to (8.70), $\epsilon_2 = (\lambda^2 - 1)$, where λ is the thrust factor of the first boost, at P. By the time the satellite reaches the point P', we want it to be at radius R_3. Since P' is the apogee of the transfer orbit, this requires that

$$R_3 = \frac{c_2}{1 - \epsilon_2} = \frac{\lambda^2 R_1}{1 - (\lambda^2 - 1)} = \frac{\lambda^2 R_1}{2 - \lambda^2} \tag{8.71}$$

which is easily solved for λ to give

$$\lambda = \sqrt{\frac{2R_3}{R_1 + R_3}} = \sqrt{\frac{4}{3}} \approx 1.15. \tag{8.72}$$

The satellite must boost its speed by about 15% to move into the required transfer orbit.

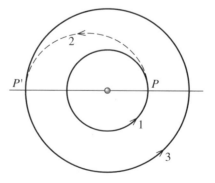

Figure 8.13   Two successive boosts, at P and P', transfer a satellite from the smaller circular orbit 1 to a transfer orbit 2 and thence to the final circular orbit 3.

The second transfer occurs at P', the apogee of the transfer orbit. In Problem 8.33 you can show that the second thrust factor is

$$\lambda' = \sqrt{\frac{R_1 + R_3}{2R_1}} = \sqrt{\frac{3}{2}} \approx 1.22; \qquad (8.73)$$

that is, we need to boost the speed by 22% to move from the transfer orbit to the final circular orbit.

It would be tempting to think that the overall change in speed, moving from the inital to the final orbit, was just the product $\lambda\lambda'$ of the two thrust factors, but this overlooks that the satellite's speed also changes as it moves around the transfer orbit. By conservation of angular momentum, it is easy to see that the speeds at the two ends of the transfer orbit satisfy $v_2(\text{apo})R_3 = v_2(\text{per})R_1$. Therefore, the overall gain in speed is given by

$$v_3 = \lambda' \cdot \frac{v_2(\text{apo})}{v_2(\text{per})} \cdot \lambda \cdot v_1$$

$$= \sqrt{\frac{R_1 + R_3}{2R_1}} \cdot \frac{R_1}{R_3} \cdot \sqrt{\frac{2R_3}{R_1 + R_3}} \cdot v_1 = \sqrt{\frac{R_1}{R_3}} \cdot v_1. \qquad (8.74)$$

In the present case, $R_3 = 2R_1$ and hence $v_3 = v_1/\sqrt{2}$. That is, the final speed is actually less than the initial by a factor of $\sqrt{2}$. This result [and more generally the result (8.74)] could have been anticipated. It is easy to show (Problem 8.32) that for circular orbits $v \propto 1/\sqrt{R}$. Thus doubling the radius necessarily required that the speed be reduced by a factor of $\sqrt{2}$.

Principal Definitions and Equations of Chapter 8

The Relative Coordinate and Reduced Mass

When rewritten in terms of the **relative coordinate**

$$\mathbf{r} = \mathbf{r}_1 - \mathbf{r}_2 \qquad \text{[Eq. (8.4)]}$$

and the CM coordinate $\mathbf{R}$, the two-body problem is reduced to the problem of two independent particles, a free particle with mass $M = m_1 + m_2$ and position $\mathbf{R}$, and a particle with mass equal to the **reduced mass**

$$\mu = \frac{m_1 m_2}{m_1 + m_2}, \qquad \text{[Eq. (8.11)]}$$

position $\mathbf{r}$, and potential energy $U(r)$.

The Equivalent One-Dimensional Problem

The motion of the relative coordinate, with given angular momentum ℓ, is equivalent to the motion of a particle in one (radial) dimension, with mass μ, position r (with $0 < r < \infty$), and **effective potential energy**

$$U_{\text{eff}}(r) = U(r) + U_{\text{cf}}(r) = U(r) + \frac{\ell^2}{2\mu r^2} \qquad \text{[Eq. (8.30)]}$$

where U_{cf} is called the **centrifugal potential energy**.

The Transformed Radial Equation

With the change of variables from r to $u = 1/r$ and elimination of t in favor of ϕ, the equation of the one-dimensional radial motion becomes

$$u''(\phi) = -u(\phi) - \frac{\mu}{\ell^2 u(\phi)^2} F. \qquad \text{[Eq. (8.41)]}$$

The Kepler Orbits

For a planet or comet, the force is $F = Gm_1 m_2/r^2 = \gamma/r^2$, and the solution of (8.41) is

$$r(\phi) = \frac{c}{1 + \epsilon \cos \phi} \qquad \text{[Eq. (8.49)]}$$

where $c = \ell^2/\gamma\mu$ and ϵ is related to the energy by

$$E = \frac{\gamma^2 \mu}{2\ell^2}(\epsilon^2 - 1). \qquad \text{[Eq. (8.58)]}$$

This **Kepler orbit** is an ellipse, parabola, or hyperbola, according as the eccentricity ϵ is less than, equal to, or greater than 1.

Problems for Chapter 8

Stars indicate the approximate level of difficulty, from easiest (⋆) to most difficult (⋆⋆⋆).

SECTION 8.2   CM and Relative Coordinates; Reduced Mass

8.1 ⋆ Verify that the positions of two particles can be written in terms of the CM and relative positions as $\mathbf{r}_1 = \mathbf{R} + m_2\mathbf{r}/M$ and $\mathbf{r}_2 = \mathbf{R} - m_1\mathbf{r}/M$. Hence confirm that the total KE of the two particles can be expressed as $T = \frac{1}{2}M\dot{\mathbf{R}}^2 + \frac{1}{2}\mu\dot{\mathbf{r}}^2$, where μ denotes the reduced mass $\mu = m_1 m_2/M$.

8.2 ⋆⋆ Although the main topic of this chapter is the motion of two particles subject to no external forces, many of the ideas [for example, the splitting of the Lagrangian $\mathcal{L}$ into two independent pieces $\mathcal{L} = \mathcal{L}_{\text{cm}} + \mathcal{L}_{\text{rel}}$ as in Equation (8.13)] extend easily to more general situations. To illustrate this, consider the following: Two masses m_1 and m_2 move in a uniform gravitational field $\mathbf{g}$ and interact via a potential energy $U(r)$. **(a)** Show that the Lagrangian can be decomposed as in (8.13). **(b)** Write down Lagrange's equations for the three CM coordinates X, Y, Z and describe the motion of the CM. Write down the three Lagrange equations for the relative coordinates and show clearly that the motion of $\mathbf{r}$ is the same as that of a single particle of mass equal to the reduced mass μ, with position $\mathbf{r}$ and potential energy $U(r)$.

8.3 ⋆⋆ Two particles of masses m_1 and m_2 are joined by a massless spring of natural length L and force constant k. Initially, m_2 is resting on a table and I am holding m_1 vertically above m_2 at a height L. At time $t = 0$, I project m_1 vertically upward with initial velocity v_o. Find the positions of the two masses at any subsequent time t (before either mass returns to the table) and describe the motion. [*Hints*: See Problem 8.2. Assume that v_o is small enough that the two masses never collide.]

SECTION 8.3 The Equations of Motion

8.4 ⋆ Using the Lagrangian (8.13) write down the three Lagrange equations for the relative coordinates x, y, z and show clearly that the motion of the relative position $\mathbf{r}$ is the same as that of a single particle with position $\mathbf{r}$, potential energy $U(r)$, and mass equal to the reduced mass μ.

8.5 ⋆ The momentum $\mathbf{p}$ conjugate to the relative position $\mathbf{r}$ is defined with components $p_x = \partial \mathcal{L}/\partial \dot{x}$ and so on. Prove that $\mathbf{p} = \mu \dot{\mathbf{r}}$. Prove also that in the CM frame, $\mathbf{p}$ is the same as $\mathbf{p}_1$ the momentum of particle 1 (and also $-\mathbf{p}_2$).

8.6 ⋆ Show that in the CM frame, the angular momentum $\boldsymbol{\ell}_1$ of particle 1 is related to the total angular momentum $\mathbf{L}$ by $\boldsymbol{\ell}_1 = (m_2/M)\mathbf{L}$ and likewise $\boldsymbol{\ell}_2 = (m_1/M)\mathbf{L}$. Since $\mathbf{L}$ is conserved, this shows that the same is true of $\boldsymbol{\ell}_1$ and $\boldsymbol{\ell}_2$ separately in the CM frame.

8.7 ⋆⋆ **(a)** Using elementary Newtonian mechanics find the period of a mass m_1 in a circular orbit of radius r around a *fixed* mass m_2. **(b)** Using the separation into CM and relative motions, find the corresponding period for the case that m_2 is not fixed and the masses circle each other a constant distance r apart. Discuss the limit of this result if $m_2 \to \infty$. **(c)** What would be the orbital period if the earth were replaced by a star of mass equal to the solar mass, in a circular orbit, with the distance between the sun and star equal to the present earth–sun distance? (The mass of the sun is more than 300,000 times that of the earth.)

8.8 ⋆⋆ Two masses m_1 and m_2 move in a plane and interact by a potential energy $U(r) = \frac{1}{2}kr^2$. Write down their Lagrangian in terms of the CM and relative positions $\mathbf{R}$ and $\mathbf{r}$, and find the equations of motion for the coordinates X, Y and x, y. Describe the motion and find the frequency of the relative motion.

8.9 ⋆⋆ Consider two particles of equal masses, $m_1 = m_2$, attached to each other by a light straight spring (force constant k, natural length L) and free to slide over a frictionless horizontal table. **(a)** Write down the Lagrangian in terms of the coordinates $\mathbf{r}_1$ and $\mathbf{r}_2$, and rewrite it in terms of the CM and relative positions, $\mathbf{R}$ and $\mathbf{r}$, using polar coordinates (r, ϕ) for $\mathbf{r}$. **(b)** Write down and solve the Lagrange equations for the CM coordinates X, Y. **(c)** Write down the Lagrange equations for r and ϕ. Solve these for the two special cases that r remains constant and that ϕ remains constant. Describe the corresponding motions. In particular, show that the frequency of oscillations in the second case is $\omega = \sqrt{2k/m_1}$.

8.10 ⋆⋆ Two particles of equal masses $m_1 = m_2$ move on a frictionless horizontal surface in the vicinity of a fixed force center, with potential energies $U_1 = \frac{1}{2}kr_1^2$ and $U_2 = \frac{1}{2}kr_2^2$. In addition, they interact with each other via a potential energy $U_{12} = \frac{1}{2}\alpha kr^2$, where r is the distance between them and α and k are positive constants. **(a)** Find the Lagrangian in terms of the CM position **R** and the relative position $\mathbf{r} = \mathbf{r}_1 - \mathbf{r}_2$. **(b)** Write down and solve the Lagrange equations for the CM and relative coordinates X, Y and x, y. Describe the motion.

8.11 $\star\star$ Consider two particles interacting by a Hooke's law potential energy, $U = \frac{1}{2}kr^2$, where **r** is their relative position $\mathbf{r} = \mathbf{r}_1 - \mathbf{r}_2$, and subject to no external forces. Show that $\mathbf{r}(t)$ describes an ellipse. Hence show that both particles move on similar ellipses around their common CM. [This is surprisingly awkward. Perhaps the simplest procedure is to choose the xy plane as the plane of the orbit and then solve the equation of motion (8.15) for x and y. Your solution will have the form $x = A \cos \omega t + B \sin \omega t$, with a similar expression for y. If you solve these for $\sin \omega t$ and $\cos \omega t$ and remember that $\sin^2 + \cos^2 = 1$, you can put the orbital equation in the form $ax^2 + 2bxy + cy^2 = k$ where k is a positive constant. Now invoke the standard result that if a and c are positive and $ac > b^2$, this equation defines an ellipse.]

SECTION 8.4   The Equivalent One-Dimensional Problem

8.12 $\star\star$ **(a)** By examining the effective potential energy (8.32) find the radius at which a planet (or comet) with angular momentum ℓ can orbit the sun in a circular orbit with fixed radius. [Look at dU_{eff}/dr.] **(b)** Show that this circular orbit is stable, in the sense that a small radial nudge will cause only small radial oscillations. [Look at d^2U_{eff}/dr^2.] Show that the period of these oscillations is equal to the planet's orbital period.

8.13 $\star\star\star$ Two particles whose reduced mass is μ interact via a potential energy $U = \frac{1}{2}kr^2$, where r is the distance between them. **(a)** Make a sketch showing $U(r)$, the centrifugal potential energy $U_{\text{cf}}(r)$, and the effective potential energy $U_{\text{eff}}(r)$. (Treat the angular momentum ℓ as a known, fixed constant.) **(b)** Find the "equilibrium" separation r_0, the distance at which the two particles can circle each other with constant r. [*Hint:* This requires that dU_{eff}/dr be zero.] **(c)** By making a Taylor expansion of $U_{\text{eff}}(r)$ about the equilibrium point r_0 and neglecting all terms in $(r - r_0)^3$ and higher, find the frequency of small oscillations about the circular orbit if the particles are disturbed a little from the separation r_0.

8.14 $\star\star\star$ Consider a particle of reduced mass μ orbiting in a central force with $U = kr^n$ where $kn > 0$. **(a)** Explain what the condition $kn > 0$ tells us about the force. Sketch the effective potential energy U_{eff} for the cases that $n = 2, -1$, and -3. **(b)** Find the radius at which the particle (with given angular momentum ℓ) can orbit at a fixed radius. For what values of n is this circular orbit stable? Do your sketches confirm this conclusion? **(c)** For the stable case, show that the period of small oscillations about the circular orbit is $\tau_{\text{osc}} = \tau_{\text{orb}}/\sqrt{n+2}$. Argue that if $\sqrt{n+2}$ is a rational number, these orbits are closed. Sketch them for the cases that $n = 2, -1$, and 7.

SECTION 8.6   The Kepler Orbits

8.15 $\star$ In deriving Kepler's third law (8.55) we made an approximation based on the fact that the sun's mass M_s is much greater than that of the planet m. Show that the law should actually read $\tau^2 = [4\pi^2/G(M_s + m)]a^3$, and hence that the "constant" of proportionality is actually a little different for different planets. Given that the mass of the heaviest planet (Jupiter) is about 2×10^{27} kg, while M_s is about 2×10^{30} kg (and some planets have masses several orders of magnitude less than Jupiter), by what percent would you expect the "constant" in Kepler's third law to vary among the planets?

8.16 $\star\star$ We have proved in (8.49) that any Kepler orbit can be written in the form $r(\phi) = c/(1 + \epsilon \cos \phi)$, where $c > 0$ and $\epsilon \geq 0$. For the case that $0 \leq \epsilon < 1$, rewrite this equation in rectangular coordinates (x, y) and prove that the equation can be cast in the form (8.51), which is the equation of an ellipse. Verify the values of the constants given in (8.52).

8.17 ★★ If you did Problem 4.41 you met the **virial theorem** for a circular orbit of a particle in a central force with $U = kr^n$. Here is a more general form of the theorem that applies to any periodic orbit of a particle. **(a)** Find the time derivative of the quantity $G = \mathbf{r} \cdot \mathbf{p}$ and, by integrating from time 0 to t, show that

$$\frac{G(t) - G(0)}{t} = 2\langle T \rangle + \langle \mathbf{F} \cdot \mathbf{r} \rangle$$

where $\mathbf{F}$ is the net force on the particle and $\langle f \rangle$ denotes the average over time of any quantity f. **(b)** Explain why, if the particle's orbit is periodic and if we make t sufficiently large, we can make the left-hand side of this equation as small as we please. That is, the left side approaches zero as $t \to \infty$. **(c)** Use this result to prove that if $\mathbf{F}$ comes from the potential energy $U = kr^n$, then $\langle T \rangle = n\langle U \rangle/2$, if now $\langle f \rangle$ denotes the time average over a very long time.

8.18 ★★ An earth satellite is observed at perigee to be 250 km above the earth's surface and traveling at about 8500 m/s. Find the eccentricity of its orbit and its height above the earth at apogee. [*Hint:* The earth's radius is $R_e \approx 6.4 \times 10^6$ m. You will also need to know GM_e, but you can find this if you remember that $GM_e/R_e^2 = g$.]

8.19 ★★ The height of a satellite at perigee is 300 km above the earth's surface and it is 3000 km at apogee. Find the orbit's eccentricity. If we take the orbit to define the xy plane and the major axis in the x direction with the earth at the origin, what is the satellite's height when it crosses the y axis? [See the hint for Problem 8.18.]

8.20 ★★ Consider a comet which passes through its aphelion at a distance r_{max} from the sun. Imagine that, keeping r_{max} fixed, we somehow make the angular momentum ℓ smaller and smaller, though not actually zero; that is, we let $\ell \to 0$. Use equations (8.48) and (8.50) to show that in this limit the eccentricity ϵ of the elliptical orbit approaches 1 and that the distance of closest approach r_{min} approaches zero. Describe the orbit with r_{max} fixed but ℓ very small. What is the semimajor axis a?

8.21 ★★★ **(a)** If you haven't already done so, do Problem 8.20. **(b)** Use Kepler's third law (8.55) to find the period of this orbit in terms of r_{max} (and G and M_s). **(c)** Now consider the extreme case that the comet is released from *rest* at a distance r_{max} from the sun. (In this case ℓ is actually zero.) Use the technique described in connection with (4.58) to find how long the comet takes to reach the sun. (Take the sun's radius to be zero.) **(d)** Assuming the comet can somehow pass freely through the sun, describe its overall motion and find its period. **(e)** Compare your answers in parts (b) and (d).

8.22 ★★★ A particle of mass m moves with angular momentum ℓ about a fixed force center with $F(r) = k/r^3$ where k can be positive or negative. **(a)** Sketch the effective potential energy U_{eff} for various values of k and describe the various possible kinds of orbit. **(b)** Write down and solve the transformed radial equation (8.41), and use your solutions to confirm your predictions in part (a).

8.23 ★★★ A particle of mass m moves with angular momentum ℓ in the field of a fixed force center with

$$F(r) = -\frac{k}{r^2} + \frac{\lambda}{r^3}$$

where k and λ are positive. **(a)** Write down the transformed radial equation (8.41) and prove that the orbit has the form

$$r(\phi) = \frac{c}{1 + \epsilon \cos(\beta\phi)}$$

where c, β, and ϵ are positive constants. **(b)** Find c and β in terms of the given parameters, and describe the orbit for the case that $0 < \epsilon < 1$. **(c)** For what values of β is the orbit closed? What happens to your results as $\lambda \to 0$?

8.24 ★★★ Consider the particle of Problem 8.23, but suppose that the constant λ is negative. Write down the transformed radial equation (8.41) and describe the orbits of low angular momentum (specifically, $\ell^2 < -\lambda m$).

8.25 ★★★ [Computer] Consider a particle with mass m and angular momentum ℓ in the field of a central force $F = -k/r^{5/2}$. To simplify your equations, choose units for which $m = \ell = k = 1$. **(a)** Find the value r_o of r at which U_{eff} is minimum and make a plot of $U_{\text{eff}}(r)$ for $0 < r \le 5r_o$. (Choose your scale so that your plot shows the interesting part of the curve.) **(b)** Assuming now that the particle has energy $E = -0.1$, find an accurate value of r_{min}, the particle's distance of closest approach to the force center. (This will require the use of a computer program to solve the relevant equation numerically.) **(c)** Assuming that the particle is at $r = r_{\text{min}}$ when $\phi = 0$, use a computer program (such as "NDSolve" in Mathematica) to solve the transformed radial equation (8.41) and find the orbit in the form $r = r(\phi)$ for $0 \le \phi \le 7\pi$. Plot the orbit. Does it appear to be closed?

8.26 ★★★ Show that the validity of Kepler's first two laws for any body orbiting the sun implies that the force (assumed conservative) of the sun on any body is central and proportional to $1/r^2$.

8.27 ★★★ At time t_o a comet is observed at radius r_o traveling with speed v_o at an acute angle α to the line from the comet to the sun. Put the sun at the origin O, with the comet on the x axis (at t_o) and its orbit in the xy plane, and then show how you could calculate the parameters of the orbital equation in the form $r = c/[1 + \epsilon \cos(\phi - \delta)]$. Do so for the case that $r_o = 1.0 \times 10^{11}$ m, $v_o = 45$ km/s, and $\alpha = 50$ degrees. [The sun's mass is about 2.0×10^{30} kg, and $G = 6.7 \times 10^{-11}$ N·m²/s².]

SECTION 8.7   The Unbounded Kepler Orbits

8.28 ★ For a given earth satellite with given angular momentum ℓ, show that the distance of closest approach r_{min} on a parabolic orbit is half the radius of the circular orbit.

8.29 ★★ What would become of the earth's orbit (which you may consider to be a circle) if half of the sun's mass were suddenly to disappear? Would the earth remain bound to the sun? [*Hints:* Consider what happens to the earth's KE and PE at the moment of the great disappearance. The virial theorem for the circular orbit (Problem 4.41) helps with this one.] Treat the sun (or what remains of it) as fixed.

8.30 ★★ The general Kepler orbit is given in polar coordinates by (8.49). Rewrite this in Cartesian coordinates for the cases that $\epsilon = 1$ and $\epsilon > 1$. Show that if $\epsilon = 1$, you get the parabola (8.60), and if $\epsilon > 1$, the hyperbola (8.61). For the latter, identify the constants α, β, and δ in terms of c and ϵ.

8.31 ★★★ Consider the motion of two particles subject to a *repulsive* inverse-square force (for example, two positive charges). Show that this system has no states with $E < 0$ (as measured in the CM frame), and that in all states with $E > 0$, the relative motion follows a hyperbola. Sketch a typical orbit. [*Hint:* You can follow closely the analysis of Sections 8.6 and 8.7 except that you must reverse the force; probably the simplest way to do this is to change the sign of γ in (8.44) and all subsequent equations (so that $F(r) = +\gamma/r^2$) and then keep γ itself positive. Assume $\ell \ne 0$.]

SECTION 8.8   Changes of Orbit

8.32 ★ Prove that for circular orbits around a given gravitational force center (such as the sun) the speed of the orbiting body is inversely proportional to the square root of the orbital radius.

8.33 ★★ Figure 8.13 shows a space vehicle boosting from a circular orbit 1 at P to a transfer orbit 2 and then from the transfer orbit at P' to the final circular orbit 3. Example 8.6 derived in detail the thrust factor required for the boost at P. Show similarly that the thrust factor required at P' is $\lambda' = \sqrt{(R_1 + R_3)/2R_1}$. [Your argument should parallel closely that of Example 8.6, but you must account for the fact that P' is the apogee (not perigee) of the transfer orbit. For example, the plus signs in (8.67) should be minus signs here.]

8.34 ★★ Suppose that we decide to send a spacecraft to Neptune, using the simple transfer described in Example 8.6 (page 318). The craft starts in a circular orbit close to the earth (radius 1 AU or astronomical unit) and is to end up in a circular orbit near Neptune (radius about 30 AU). Use Kepler's third law to show that the transfer will take about 31 years. (In practice we can do a lot better than this by arranging that the craft gets a gravitational boost as it passes Jupiter.)

8.35 ★★★ A spacecraft in a circular orbit wishes to transfer to another circular orbit of quarter the radius by means of a tangential thrust to move into an elliptical orbit and a second tangential thrust at the opposite end of the ellipse to move into the desired circular orbit. (The picture looks like Figure 8.13 but run backwards.) Find the thrust factors required and show that the speed in the final orbit is two times *greater* than the initial speed.

Mechanics in Noninertial Frames

We saw in Chapter 1 that Newton's laws are valid only in the special class of *inertial* reference frames — frames that are neither accelerating nor rotating. The natural reaction to this realization is to resolve to treat all mechanics problems using only inertial frames, and this is in fact what we have done so far. Nevertheless, there are situations where it is very desirable to consider a noninertial frame. For example, if you are sitting in a car that is accelerating and you wish to describe the behavior of a coin that you are tossing in the air, it is very natural to want to describe it *as seen by you*, that is, in the accelerating frame of the car in which you are traveling. Another important example is a reference frame fixed to the earth. To an excellent approximation this is an inertial frame. Nevertheless, the earth is rotating on its axis and accelerating in its orbit, and, for both these reasons, a reference frame fixed to the earth is not quite inertial. In the majority of problems, the noninertial character of a frame fixed to the earth is totally negligible, but there are situations, for instance the firing of a long-range rocket, in which the earth's rotation has important consequences. Naturally, we would like to describe the motion relative to the earth-bound frame that we live in, but to do so, we must learn to do mechanics in noninertial frames.

In the first two sections of this chapter, I shall describe the simple case of a reference frame that is accelerating but not rotating. In the remainder of the chapter, I shall discuss the case of a rotating reference frame.

9.1 Acceleration without Rotation

Let us consider an inertial frame $\mathcal{S}_o$ and a second frame $\mathcal{S}$ that is accelerating relative to $\mathcal{S}_o$ with acceleration $\mathbf{A}$, which need not necessarily be constant. Notice that because the noninertial frame is the one we're really interested in, it's the one that I've called $\mathcal{S}$. Also I'm using capital letters for the acceleration (and velocity) of frame $\mathcal{S}$ relative to frame $\mathcal{S}_o$. The inertial frame $\mathcal{S}_o$ could be a frame anchored to the ground. (We'll ignore the tiny acceleration of any earth-bound frame for now.) The frame $\mathcal{S}$ could

327

be a frame fixed in a railroad car that is moving relative to S_o, with velocity $\mathbf{V}$ and acceleration $\mathbf{A} = \dot{\mathbf{V}}$. Suppose now that a passenger in the car is playing catch with a tennis ball of mass m, and let us consider the motion of the ball, first as measured relative to S_o. Because S_o is inertial, we know that Newton's second law holds, so

$$m\ddot{\mathbf{r}}_o = \mathbf{F}, \tag{9.1}$$

where $\mathbf{r}_o$ is the ball's position relative to S_o and $\mathbf{F}$ is the net force on the ball, the vector sum of all forces on the ball (gravity, air resistance, the passenger's hand pushing it, etc.).

Now consider the same ball's motion as measured relative to the accelerating frame S. The ball's position relative to S is $\mathbf{r}$ and, by the vector addition of velocities, its velocity $\dot{\mathbf{r}}$ relative to S is related to $\dot{\mathbf{r}}_o$ by the velocity-addition formula:[1]

$$\dot{\mathbf{r}}_o = \dot{\mathbf{r}} + \mathbf{V} \tag{9.2}$$

that is,

(ball's velocity relative to ground)

= (ball's velocity relative to car) + (car's velocity relative to ground).

Differentiating and rearranging, we find that

$$\ddot{\mathbf{r}} = \ddot{\mathbf{r}}_o - \mathbf{A}. \tag{9.3}$$

If we multiply this equation by m and use (9.1) to replace $m\ddot{\mathbf{r}}_o$ by $\mathbf{F}$, we find that

$$m\ddot{\mathbf{r}} = \mathbf{F} - m\mathbf{A}. \tag{9.4}$$

This equation has exactly the form of Newton's second law, *except* that in addition to $\mathbf{F}$, the sum of all forces identified in the inertial frame, there is an extra term on the right equal to $-m\mathbf{A}$. This means we can continue to use Newton's second law in the noninertial frame S *provided* we agree that in the noninertial frame we must add an extra force-like term, often called the **inertial force**:

$$\mathbf{F}_{\text{inertial}} = -m\mathbf{A}. \tag{9.5}$$

This inertial force experienced in noninertial frames is familiar in several everyday situations: If you sit in an aircraft accelerating rapidly toward takeoff, then, from your point of view, there is a force that pushes you back into your seat. If you are standing in a bus that brakes suddenly ($\mathbf{A}$ backward) the inertial force $-m\mathbf{A}$ is forward and can make you fall on your face if you aren't properly braced. As a car goes rapidly

---

[1] An important discovery of relativity is, of course, that velocities do not combine exactly according to the simple vector addition of (9.2); nevertheless, in the framework of classical mechanics (9.2) is correct. In the same way, I shall take for granted that the times measured in S and S_o are the same, which is correct in classical, though not relativistic, mechanics.

around a sharp curve, the inertial force experienced by the passengers is the so-called centrifugal force that pushes them outward. One can take the view that the inertial force is a "fictitious" force, introduced merely to preserve the form of Newton's second law. Nevertheless, for an observer in an accelerating frame, it is entirely real.

In many problems involving objects in accelerated frames, much the simplest procedure is to go ahead and use Newton's second law in the noninertial frame, always remembering to include the extra inertial force as in (9.4). Here is a simple example:

EXAMPLE 9.1 A Pendulum in an Accelerating Car

Consider a simple pendulum (mass m and length L) mounted inside a railroad car that is accelerating to the right with constant acceleration $\mathbf{A}$ as shown in Figure 9.1. Find the angle ϕ_{eq} at which the pendulum will remain at rest relative to the accelerating car and find the frequency of small oscillations about this equilibrium angle.

As observed in any inertial frame, there are just two forces on the bob, the tension in the string $\mathbf{T}$ and the weight $m\mathbf{g}$; thus the net force (in any inertial frame) is $\mathbf{F} = \mathbf{T} + m\mathbf{g}$. If we choose to work in the noninertial frame of the accelerating car, there is also the inertial force $-m\mathbf{A}$, and the equation of motion (9.4) is

$$m\ddot{\mathbf{r}} = \mathbf{T} + m\mathbf{g} - m\mathbf{A}. \tag{9.6}$$

A remarkable simplification in this problem is that, because the weight $m\mathbf{g}$ and the inertial force $-m\mathbf{A}$ are both proportional to the mass m, we can combine these two terms and write

$$\begin{aligned} m\ddot{\mathbf{r}} &= \mathbf{T} + m(\mathbf{g} - \mathbf{A}) \\ &= \mathbf{T} + m\mathbf{g}_{\mathrm{eff}} \end{aligned} \tag{9.7}$$

where $\mathbf{g}_{\mathrm{eff}} = \mathbf{g} - \mathbf{A}$. We see that the pendulum's equation of motion in the accelerating frame of the car is exactly the same as in an inertial frame, *except*

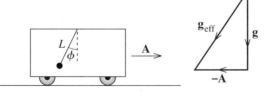

Figure 9.1   A pendulum is suspended from the roof of a railroad car that is accelerating with constant acceleration $\mathbf{A}$. In the noninertial frame of the car, the acceleration manifests itself through the inertial force $-m\mathbf{A}$, which, in turn, is equivalent to the replacement of $\mathbf{g}$ by the effective $\mathbf{g}_{\mathrm{eff}} = \mathbf{g} - \mathbf{A}$.

that **g** *has been replaced by an effective* $\mathbf{g}_{\text{eff}} = \mathbf{g} - \mathbf{A}$, as shown in Figure 9.1.[2] This makes the solution of our problem almost trivially simple.

If the pendulum is to remain at rest (as seen in the car), then **r̈** must be zero, and according to (9.7) **T** must be exactly opposite to $m\mathbf{g}_{\text{eff}}$. In particular, we see from Figure 9.1 that the direction of **T** (and hence of the pendulum) has to be

$$\phi_{\text{eq}} = \arctan(A/g). \tag{9.8}$$

The small-amplitude frequency of a pendulum in an inertial frame is well known to be $\omega = \sqrt{g/L}$. Thus the frequency of our pendulum is obtained by replacing g by $g_{\text{eff}} = \sqrt{g^2 + A^2}$. That is,

$$\omega = \sqrt{\frac{g_{\text{eff}}}{L}} = \sqrt{\frac{\sqrt{g^2 + A^2}}{L}}. \tag{9.9}$$

It is worth comparing this solution with other, more direct methods. First we could certainly have found the equilibrium angle (9.8) working in an inertial frame anchored to the ground. In this frame, there are only two forces on the bob (**T** and $m\mathbf{g}$). If the pendulum is to remain at rest in the railroad car, then (as seen from the ground) it must accelerate at exactly **A**. Therefore the net force $\mathbf{T} + m\mathbf{g}$ must equal $m\mathbf{A}$, and by drawing a triangle of forces you can easily convince yourself that this requires **T** to have the direction (9.8). On the other hand, to find the frequency of oscillations (9.9) directly in the ground-based frame requires considerable ingenuity, and does not give the insight of our noninertial derivation.

We could also have derived both results using the Lagrangian method. This has the distinct advantage that you don't have to think about inertial and non-inertial frames — just write down the Lagrangian $\mathcal{L}$ in terms of the generalized coordinate ϕ and then crank the handle. However, finding the frequency (9.9) this way is quite clumsy, as you discovered if you did Problem 7.30.

9.2   The Tides

A beautiful application of the result (9.4) is the explanation of the tides. As you probably know, the tides are the result of the bulges in the earth's oceans caused by the gravitational attraction of the moon and sun. As the earth rotates, people on the earth's surface move past these bulges and experience a rising and falling of the sea's level. It turns out that the most important contributor to this effect is the moon, and to simplify our discussion I shall, at first, ignore the sun entirely. I shall also assume, at first, that the oceans cover the whole surface of the globe.

---

[2] This result, that the effect of being in an accelerated frame is the same as having an additional gravitational force, is a cornerstone of general relativity, where it is called the *principle of equivalence*.

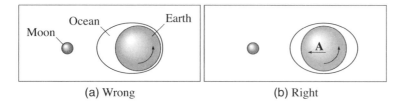

(a) Wrong   (b) Right

Figure 9.2   Views of the earth and moon from high above the North Pole, with the earth rotating counterclockwise about the polar axis. (a) A plausible-sounding, but incorrect, explanation of the tides argues that the attraction of the moon causes the oceans to bulge toward the moon. As the earth rotates once a day, a person fixed on the earth would experience one high tide per day, not the observed two. (b) The correct explanation: The main effect of the moon's attraction is to give the whole earth (including the oceans) a small acceleration **A** toward the moon. As explained in the text, the residual effect is that the oceans bulge both toward and away from the moon as shown. As the earth rotates once a day, these two bulges cause the observed two high tides per day.

A plausible-sounding, though entirely incorrect, explanation of the tides is illustrated in Figure 9.2(a). According to this incorrect argument, the moon's attraction pulls the oceans toward the moon, producing a single bulge toward the moon. The trouble with this argument (beside that it is wrong) is that a single bulge would cause just one high tide per day, instead of the observed two.

The correct explanation, illustrated in Figure 9.2(b), is more complicated. The dominant effect of the moon is to give the whole earth, including the oceans, an acceleration **A** toward the moon. This acceleration is the centripetal acceleration of the earth as the moon and earth circle around their common center of mass and is (almost exactly) the same as if all the masses that make up the earth were concentrated at its center. This centripetal acceleration of any object on earth, as it orbits with the earth, corresponds to the pull of the moon that the object would feel at the earth's center. Now any object on the moon side of the earth is pulled by the moon with a force that is slightly *greater* than it would be at the center. Therefore, as seen from the earth, objects on the side nearest the moon behave as if they felt a slight additional attraction toward the moon. In particular, the ocean surface bulges toward the moon. On the other hand, objects on the far side from the moon are pulled by the moon with a force that is slightly *weaker* than it would be at the center, which means that they move (relative to the earth) as if they were slightly repelled by the moon. This slight repulsion causes the ocean to bulge on the side away from the moon and is responsible for the second high tide of each day.

We can make this argument quantitative (and probably more convincing) if we return to Equation (9.4). The forces on any mass m near the earth's surface are (1) the gravitational pull, $m\mathbf{g}$, of the earth, (2) the gravitational pull, $-GM_{\mathrm{m}}m\hat{\mathbf{d}}/d^2$, of the moon (where M_{m} is the mass of the moon and $\mathbf{d}$ is the position of the object relative to the moon, as in Figure 9.3), and (3) the net non-gravitational force $\mathbf{F}_{\mathrm{ng}}$ (for instance,

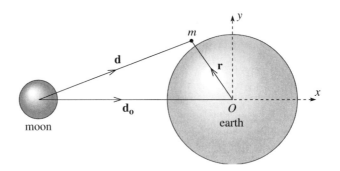

Figure 9.3   A mass m near the earth's surface has position $\mathbf{r}$ relative to
the earth's center, and $\mathbf{d}$ relative to the moon. The vector $\mathbf{d}_\mathrm{o}$ is the position
of the earth's center relative to the moon's.

the buoyant force on a drop of sea water in the ocean). Meanwhile the acceleration of
our origin O, the earth's center, is

$$\mathbf{A} = -GM_\mathrm{m}\frac{\hat{\mathbf{d}}_\mathrm{o}}{d_\mathrm{o}^2}$$

where $\mathbf{d}_\mathrm{o}$ is the position of the earth's center relative to the moon. Putting all this
together we find for (9.4)

$$m\ddot{\mathbf{r}} = \mathbf{F} - m\mathbf{A} \qquad (9.10)$$

$$= \left(m\mathbf{g} - GM_\mathrm{m}m\frac{\hat{\mathbf{d}}}{d^2} + \mathbf{F}_\mathrm{ng}\right) + GM_\mathrm{m}m\frac{\hat{\mathbf{d}}_\mathrm{o}}{d_\mathrm{o}^2} \qquad (9.11)$$

or, if we combine the two terms that involve M_m,

$$m\ddot{\mathbf{r}} = m\mathbf{g} + \mathbf{F}_\mathrm{tid} + \mathbf{F}_\mathrm{ng}$$

where the **tidal force**

$$\mathbf{F}_\mathrm{tid} = -GM_\mathrm{m}m\left(\frac{\hat{\mathbf{d}}}{d^2} - \frac{\hat{\mathbf{d}}_\mathrm{o}}{d_\mathrm{o}^2}\right) \qquad (9.12)$$

is the difference between the actual force of the moon on m and the corresponding
force if m were at the center of the earth.

The entire effect of the moon on the motion (relative to the earth) of any object
near the earth is contained in the tidal force $\mathbf{F}_\mathrm{tid}$ of (9.12). At the point directly facing
the moon, point P in Figure 9.4, the vectors $\mathbf{d} = \overrightarrow{MP}$ and $\mathbf{d}_\mathrm{o} = \overrightarrow{MO}$ point in the same
direction, but $d < d_\mathrm{o}$. Thus the first term in (9.12) dominates, and the tidal force is
toward the moon. At point R, facing directly away from the moon, again $\mathbf{d}$ (equal to
$\overrightarrow{MR}$) and $\mathbf{d}_\mathrm{o}$ point in the same direction, but here $d > d_\mathrm{o}$ and the tidal force points
away from the moon. At the point Q, the vectors $\mathbf{d} = \overrightarrow{MQ}$ and $\mathbf{d}_\mathrm{o} = \overrightarrow{MO}$ point in
different directions; the x components of the two terms in (9.12) cancel almost exactly,
but only the first term has a y component. Thus at Q (and likewise at S) the tidal force

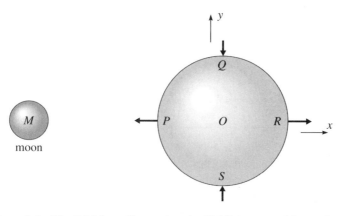

Figure 9.4   The tidal force $\mathbf{F}_{\text{tid}}$ as given by (9.12) is outward (away from the earth's center) at points P and R, but inward (toward the earth's center) at Q and S.

is inward, toward the earth's center, as shown in Figure 9.4. In particular, the effect of the tidal force is to distort the ocean into the shape shown in Figure 9.2(b), with bulges centered on the points P and R, giving the observed two high tides per day.

Magnitude of the Tides

The simplest way to find the height difference between high and low tides is to observe that the surface of the ocean is an equipotential surface — a surface of constant potential energy. To prove this, consider a drop of sea water on the surface of the ocean. The drop is in equilibrium (relative to the earth's reference frame) under the influence of three forces: the earth's gravitational pull $m\mathbf{g}$, the tidal force $\mathbf{F}_{\text{tid}}$, and the pressure force $\mathbf{F}_{\text{p}}$ of the surrounding sea water. Since a static fluid cannot exert any shearing force, the pressure force $\mathbf{F}_{\text{p}}$ must be normal to the surface of the ocean. (It is in fact just the buoyant force of Archimedes' principle.) Since the drop of water is in equilibrium, it follows that $m\mathbf{g} + \mathbf{F}_{\text{tid}}$ must likewise be normal to the surface.

Now, both $m\mathbf{g}$ and $\mathbf{F}_{\text{tid}}$ are conservative, so each can be written as the gradient of a potential energy:

$$m\mathbf{g} = -\nabla U_{\text{eg}} \qquad \text{and} \qquad \mathbf{F}_{\text{tid}} = -\nabla U_{\text{tid}}$$

where U_{eg} is the potential energy due to the earth's gravity and U_{tid} is that of the tidal force, which is, by inspection of (9.12),[3]

$$U_{\text{tid}} = -GM_{\text{m}}m\left(\frac{1}{d} + \frac{x}{d_{\text{o}}^2}\right). \tag{9.13}$$

---

[3] According to (9.12), $\mathbf{F}_{\text{tid}}$ is the sum of two terms. The first is just the usual inverse square force with corresponding potential energy $-GM_{\text{m}}m/d$ and the second is a constant vector pointing in the x direction, which gives the term $-GM_{\text{m}}mx/d_{\text{o}}^2$.

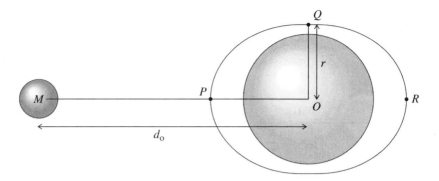

Figure 9.5   The difference h between high and low tides is the difference between the lengths OP and OQ. This difference is much exaggerated here (since h turns out to be of order 1 meter), and both the lengths OP and OQ are very close to the radius of the earth, $R_e \approx 6400$ km. Also, $r \approx R_e$ is actually about 60 times smaller than the earth–moon distance $OM = d_o$.

Thus, the statement that $m\mathbf{g} + \mathbf{F}_{\text{tid}}$ is normal to the surface of the ocean can be rephrased to say that $\nabla(U_{\text{eg}} + U_{\text{tid}})$ is normal to the surface, which in turn implies that $U = (U_{\text{eg}} + U_{\text{tid}})$ is constant on the surface. In other words, the surface of the ocean is an equipotential surface.

Since U is constant on the surface, it follows that

$$U(P) = U(Q)$$

(see Figure 9.5) or

$$U_{\text{eg}}(P) - U_{\text{eg}}(Q) = U_{\text{tid}}(Q) - U_{\text{tid}}(P). \tag{9.14}$$

The left side here is just

$$U_{\text{eg}}(P) - U_{\text{eg}}(Q) = mgh \tag{9.15}$$

where h is the required difference between high and low tides (the difference between the lengths OP and OQ in Figure 9.5). To find the right side of (9.14) we must evaluate the two tidal potential energies, $U_{\text{tid}}(Q)$ and $U_{\text{tid}}(P)$ from the definition (9.13). At point Q, we see that $d = \sqrt{d_o^2 + r^2}$ (with $r \approx R_e$) and $x = 0$. Therefore, from (9.13),

$$U_{\text{tid}}(Q) = -GM_m m \frac{1}{\sqrt{d_o^2 + R_e^2}} .$$

We can rewrite the square root as $\sqrt{d_o^2 + R_e^2} = d_o \sqrt{1 + (R_e/d_o)^2}$. Then, since $R_e/d_o \ll 1$, we can use the binomial approximation $(1 + \epsilon)^{-1/2} \approx 1 - \frac{1}{2}\epsilon$ to give

$$U_{\text{tid}}(Q) \approx -\frac{GM_m m}{d_o}\left(1 - \frac{R_e^2}{2d_o^2}\right). \tag{9.16}$$

At the point P, we see that $d = d_o - R_e$ and $x = -R_e$, and a similar calculation gives (Problem 9.5)

$$U_{\text{tid}}(P) \approx -\frac{G M_m m}{d_o}\left(1 + \frac{R_e^2}{d_o^2}\right). \tag{9.17}$$

(As you can check, one gets the same answer at the point R facing away from the moon, so that, in this approximation, the heights of the two daily tides should be the same.)

Substituting (9.15), (9.16), and (9.17) into (9.14) we get

$$mgh = \frac{G M_m m}{d_o}\frac{3 R_e^2}{2 d_o^2}.$$

If we recall that $g = G M_e / R_e^2$, this implies that

$$h = \frac{3\, M_m\, R_e^4}{2\, M_e\, d_o^3}. \tag{9.18}$$

Putting in the numbers ($M_m = 7.35 \times 10^{22}$ kg, $M_e = 5.98 \times 10^{24}$ kg, $R_e = 6.37 \times 10^6$ m, and $d_o = 3.84 \times 10^8$ m), we find for the height of the tides, due to the moon alone,

$$h = 54 \text{ cm} \qquad \text{[moon alone]}. \tag{9.19}$$

The height of the tides caused by the sun alone is also given by (9.18), but with M_m replaced by the mass of the sun, $M_s = 1.99 \times 10^{30}$ kg, and d_o replaced by the distance from the sun to the earth, 1.495×10^{11} m. This gives

$$h = 25 \text{ cm} \qquad \text{[sun alone]}. \tag{9.20}$$

Although the sun's contribution to the tides is less than the moon's, it is certainly not negligible, and the two effects combine in an interesting way. First, consider a time when the earth, sun, and moon are approximately in line — either with the earth in the middle, as at full moon, or the moon in the middle as at new moon. (See Figure 9.6.) In this case the tidal forces due to the moon and sun reinforce each other (since the two bulges caused by the moon coincide with the two caused by the sun); thus, we would predict large tides (known as *spring tides*) with h given by the sum of (9.19) and (9.20), that is, $h = 54 + 25 = 79$ cm. On the other hand, if the sun, earth, and moon form a right angle, then the two tidal effects will cancel and we would predict smaller tides (known as *neap tides*) with height $h = 54 - 25 = 29$ cm.

Although the theory just presented is basically correct, especially in the middle of the larger oceans, the real situation involves many intriguing complications. Perhaps the most important complication is the effect of the continental land masses. So far I have pretended that the oceans cover the whole world, allowing the tidal forces of the moon and sun to collect the two bulges of Figure 9.5. But the presence of continents can affect our conclusions, leading sometimes to smaller and sometimes to larger tides. A small sea, such as the Black Sea or even the Mediterranean, that is shut off from the main oceans by land will obviously exhibit much smaller tides than we have calculated

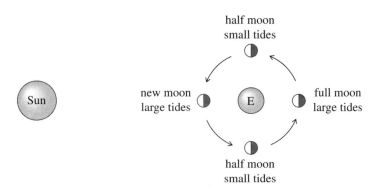

Figure 9.6   Four successive positions of the moon in its monthly orbit around the earth. At new moon and full moon, the tidal effects of the moon and sun reinforce each other and the tides are large ("spring" tides). At half moon, the two effects tend to cancel and the tides are small ("neap" tides).

here. On the other hand, the tides moving across a large ocean can be blocked by the bordering continents and can build up to much greater heights. And tides entering narrow tapering estuaries can cause quite dramatic "tidal bores."

9.3   The Angular Velocity Vector

In the remainder of this chapter, we shall be discussing the motion of objects as seen in reference frames that are *rotating* (relative to inertial frames). Before we begin this discussion, I must introduce some concepts and notation for handling rotations. A detailed study of rotations is actually surprisingly complicated. Fortunately, we do not need many of the details, and some of the properties that are quite hard to prove are reasonably plausible and can be stated without proof.

The rotating axes with which we shall be concerned are almost always axes fixed in a rigid body. The most important example is a set of axes fixed in the rotating earth, but we shall see several other examples in Chapter 10. In discussing the rotation of a rigid body, there are really just two situations that concern us: Sometimes the body is rotating about a point of the body that is *fixed* (in some inertial frame); for example, a wheel that is spinning about a fixed axle, or a pendulum swinging about a fixed pivot. If the rotating body has no fixed point (for example, a baseball that is spinning as it flies through the air), then we usually proceed in two steps: First, we find the motion of the center of mass, and then we analyze the rotational motion of the body relative to its CM. As soon as we restrict attention to the motion relative to the CM, we are in effect examining the motion in a reference frame in which the CM is fixed. Thus either way, our discussion of a rotating body concerns a body with one point effectively fixed.

The crucial result concerning a body rotating about a fixed point is called **Euler's theorem** and states that the most general motion of any body relative to a fixed point O is a rotation about some axis through O. Although this theorem is quite complicated to prove, the result seems very natural and I hope you can accept it without proof.[4] It implies that to specify a rotation about a given point O, we need give only the direction of the axis about which the rotation occurred and the angle through which the body rotated. Here, our concern is more with the *rate* of rotation, or *angular velocity*, and Euler's theorem implies that this can be specified by giving the direction of the axis of rotation and the rate of rotation about this axis. The direction of the axis of rotation can be specified by a unit vector **u** and the rate by the number $\omega = d\theta/dt$. For example, a merry-go-round could be rotating about a vertical axis (**u** points vertically) at a rate of 10 rad/min ($\omega = 10$ rad/min).

It is often convenient to combine the unit vector **u** with ω to form an **angular velocity vector**

$$\boldsymbol{\omega} = \omega \mathbf{u}. \tag{9.21}$$

The single vector $\boldsymbol{\omega}$ specifies both the direction of the axis of rotation (namely, **u**, the direction of $\boldsymbol{\omega}$) and the rate of rotation (namely, ω, the magnitude of $\boldsymbol{\omega}$). Actually we have not yet quite defined a unique vector $\boldsymbol{\omega}$. For example, for the merry-go-round that is rotating about a vertical axis, the vector $\boldsymbol{\omega}$ points vertically, but does it point up or down? We remove this ambiguity using the right-hand rule: We choose the direction of $\boldsymbol{\omega}$ so that when our right thumb points along $\boldsymbol{\omega}$, our right fingers curl in the direction of rotation. Another way to say this is that if you look along the vector $\boldsymbol{\omega}$, you will see the body rotating clockwise.

It is important to recognize that the angular velocity can change with time. If the speed of rotation is changing, then $\boldsymbol{\omega}$ will be changing in magnitude, and if the axis of rotation is changing, then $\boldsymbol{\omega}$ will be changing in direction. For example, the angular velocity of a spacecraft that is tumbling out of control will usually change in both magnitude and direction. In this case $\boldsymbol{\omega} = \boldsymbol{\omega}(t)$ is the instantaneous angular velocity at the time t. On the other hand, there are many interesting situations where $\boldsymbol{\omega}$ is constant (in magnitude and direction); for example, this is true (to an outstanding approximation) of the angular velocity of the earth spinning on its axis.

A Useful Relation

There is a useful relation between the angular velocity of a rigid body and the linear velocity of any point in the body. Consider, for example, the earth, rotating with angular velocity $\boldsymbol{\omega}$ about its center O (which I shall take to be stationary for the present discussion). Next, consider any point P fixed on (or in) the earth, for example the top of Mount Everest, with position **r** relative to O. We can specify **r** by its spherical polar coordinates (r, θ, ϕ) with z axis pointing through the North Pole, so that θ is the **colatitude** — the latitude measured down from the North Pole (instead of up from the equator, as is more usual with geographers). As the earth turns on its axis, the point P is dragged in an easterly direction around a circle of latitude, with radius $\rho = r \sin \theta$, as shown in Figure 9.7. This means that P moves with speed $v = \omega r \sin \theta$, and if you

---

[4] You can find a proof in, for example, *Classical Mechanics* by Herbert Goldstein, Charles Poole, and John Safko (3rd ed., Addison-Wesley, 2002), Section 4-6.

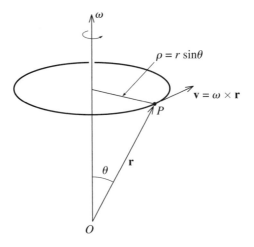

Figure 9.7   The earth's rotation drags the point P on the surface around a circle of latitude (radius $\rho = r\sin\theta$) with speed $v = \omega\rho = \omega r\sin\theta$ and hence velocity $\mathbf{v} = \boldsymbol{\omega} \times \mathbf{r}$.

will check the direction in Figure 9.7, you will see that the vector velocity is $\boldsymbol{\omega} \times \mathbf{r}$. You can easily see that this result is independent of the nature of the rotating body; that is, for any rigid body rotating with angular velocity $\boldsymbol{\omega}$ about an axis through O, the velocity of any point P (position $\mathbf{r}$) fixed on the body is

$$\mathbf{v} = \boldsymbol{\omega} \times \mathbf{r}. \tag{9.22}$$

This useful relation is, of course, a generalization of the familiar relation $v = \omega r$ that you learned in introductory physics for the speed of a point on the perimeter of a wheel of radius r. It is perhaps worth emphasizing that there is a corresponding relation for *any* vector fixed in the rotating body. For example, if $\mathbf{e}$ is a unit vector fixed in the body, then its rate of change, as seen from the non-rotating frame, is

$$\frac{d\mathbf{e}}{dt} = \boldsymbol{\omega} \times \mathbf{e}, \tag{9.23}$$

a result we shall be using shortly.

Addition of Angular Velocities

A final basic property of angular velocities that is worth mentioning is that relative angular velocities add in the same way as relative translational velocities. We know (in the framework of classical mechanics) that if two frames 2 and 1 have relative velocity $\mathbf{v}_{21}$ and if a body 3 has velocity $\mathbf{v}_{32}$ relative to frame 2, then the velocity of 3 relative to frame 1 is just the sum

$$\mathbf{v}_{31} = \mathbf{v}_{32} + \mathbf{v}_{21}. \tag{9.24}$$

Suppose instead that frame 2 is rotating with angular velocity $\boldsymbol{\omega}_{21}$ relative to frame 1 (both frames with the same origin O) and that body 3 is rotating (about O) with angular velocities $\boldsymbol{\omega}_{31}$ and $\boldsymbol{\omega}_{32}$ relative to frames 1 and 2. Now consider any point $\mathbf{r}$ fixed in body 3. Its translational velocities relative to frames 1 and 2 must satisfy (9.24). According to (9.22) this means that

$$\boldsymbol{\omega}_{31} \times \mathbf{r} = (\boldsymbol{\omega}_{32} \times \mathbf{r}) + (\boldsymbol{\omega}_{21} \times \mathbf{r}) = (\boldsymbol{\omega}_{32} + \boldsymbol{\omega}_{21}) \times \mathbf{r}$$

and, since this must be true for any $\mathbf{r}$, it follows that

$$\boldsymbol{\omega}_{31} = \boldsymbol{\omega}_{32} + \boldsymbol{\omega}_{21}. \tag{9.25}$$

That is, angular velocities add in the same way as translational velocities.

Notation for Angular Velocities

In labeling angular velocities, I shall usually observe the following convention: I shall use the lower case letter $\boldsymbol{\omega}$ for the angular velocity of a body (such as a spinning top) whose motion is our primary object of interest. I shall use the capital letter $\boldsymbol{\Omega}$ for the angular velocity of a noninertial, rotating reference frame relative to which we are calculating the motion of one or more objects. This distinction is consistent with the previous two sections, where I used capital $\mathbf{A}$ and $\mathbf{V}$ for the acceleration and velocity of a noninertial frame (relative to an inertial frame). In practice, $\boldsymbol{\omega}$ usually denotes an unknown, while $\boldsymbol{\Omega}$ is usually a given, known angular velocity, such as the angular velocity of the earth as it rotates once a day. In the remainder of this chapter, we shall be concerned with the motion of objects as seen in a rotating reference frame, and, in accordance with this convention, I shall denote the angular velocity of that frame by $\boldsymbol{\Omega}$.

9.4   Time Derivatives in a Rotating Frame

We are now ready to consider the equations of motion for an object that is viewed from a frame S that is rotating with angular velocity $\boldsymbol{\Omega}$ relative to an inertial frame S_o. While our conclusions will apply to any rotating frame, by far the most important example is a frame attached to the rotating earth, and this is the example that you can keep in mind. This being the case, let us pause to calculate the angular velocity of the earth, which rotates on its axis once every 24 hours.[5] Therefore, for a frame attached to the earth

$$\Omega = \frac{2\pi \ \text{rad}}{24 \times 3600 \ \text{s}} \approx 7.3 \times 10^{-5} \ \text{rad/s}. \tag{9.26}$$

---

[5] Strictly speaking the period of rotation about the earth's axis is one *sidereal* day, the time to rotate once relative to the distant stars. This is shorter than the solar day by a factor of 365/366, but the difference is too small to worry about here.

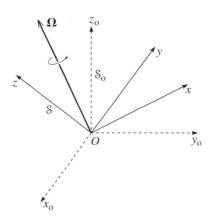

Figure 9.8   The frame S_0 defined by the three dashed axes is inertial.
The frame S defined by the three solid axes shares the same origin
O, but is rotating with angular velocity $\boldsymbol{\Omega}$ relative to S_0.

It is because this angular velocity is so small that we can often ignore it entirely.
Nevertheless, we shall see that the earth's rotation does have measurable effects on
the motion of projectiles, pendulums, and other systems. There are other noninertial
effects (notably the tides), associated with the orbital motion of the earth and moon,
but these are all much less important for the problems that we shall consider here, and
I shall ignore them for now.

I shall assume that the two frames S_0 and S share a common origin O, as shown
in Figure 9.8, so that the only motion of S relative to S_0 is a rotation with angular
velocity $\boldsymbol{\Omega}$. For example, the common origin O could be the center of the earth, S
could be a set of axes fixed in the earth, and S_0 a set of axes with the same origin but
with directions fixed relative to the distant stars. The frame S is convenient for us to
use but is noninertial; the frame S_0 is relatively inconvenient, but is inertial.

Let us now consider an arbitrary vector $\mathbf{Q}$. This could be the velocity or position
of a ball, the net force on an object, or any other vector of interest. Our first task is to
relate the time rate of change of $\mathbf{Q}$ as measured in frame S_0 to the corresponding rate
as measured in S. To distinguish these two rates of change, I shall temporarily use the
following notation:

$$\left(\frac{d\mathbf{Q}}{dt}\right)_{S_0} = \text{(rate of change of vector } \mathbf{Q} \text{ relative to inertial frame } S_0\text{)}$$

and

$$\left(\frac{d\mathbf{Q}}{dt}\right)_{S} = \text{(rate of change of same vector } \mathbf{Q} \text{ relative to rotating frame } S\text{)}.$$

To compare these two rates of change, I shall expand the vector $\mathbf{Q}$ in terms of three
orthogonal unit vectors $\mathbf{e}_1$, $\mathbf{e}_2$, and $\mathbf{e}_3$ that are fixed in the rotating frame S. (For

instance, these three units vectors could point along the three solid axes shown in Figure 9.8.) Thus,[6]

$$\mathbf{Q} = Q_1\mathbf{e}_1 + Q_2\mathbf{e}_2 + Q_3\mathbf{e}_3 = \sum_{i=1}^{3} Q_i\,\mathbf{e}_i\,. \tag{9.27}$$

This expansion is chosen for the convenience of observers in the frame $\mathcal{S}$, since the unit vectors are fixed in that frame. Nonetheless, you should recognize that the expansion is equally valid in both frames. [Whichever frame we use, (9.27) is just an expansion of the one vector $\mathbf{Q}$ in terms of three othogonal vectors $\mathbf{e}_1$, $\mathbf{e}_2$, and $\mathbf{e}_3$.] The only difference is that for observers in $\mathcal{S}$ the vectors $\mathbf{e}_1$, $\mathbf{e}_2$, and $\mathbf{e}_3$ are fixed, but as seen by observers in $\mathcal{S}_o$ the vectors $\mathbf{e}_1$, $\mathbf{e}_2$, and $\mathbf{e}_3$ are rotating.

Let us now differentiate the expansion (9.27) with respect to time. First, as seen in frame $\mathcal{S}$, the vectors $\mathbf{e}_i$ are constant, and we get simply

$$\left(\frac{d\mathbf{Q}}{dt}\right)_{\mathcal{S}} = \sum_i \frac{dQ_i}{dt}\mathbf{e}_i\,. \tag{9.28}$$

[Since the expansion coefficients Q_i in (9.27) are the same in either frame, we don't need to qualify the derivative on the right with a subscript $\mathcal{S}$ or $\mathcal{S}_o$.]

As seen in frame $\mathcal{S}_o$, the vectors $\mathbf{e}_i$ vary with time. Thus, differentiating (9.27) in frame $\mathcal{S}_o$ gives

$$\left(\frac{d\mathbf{Q}}{dt}\right)_{\mathcal{S}_o} = \sum_i \frac{dQ_i}{dt}\mathbf{e}_i + \sum_i Q_i\left(\frac{d\mathbf{e}_i}{dt}\right)_{\mathcal{S}_o}. \tag{9.29}$$

The derivative in the second term on the right is easily evaluated with the help of the "useful relation" (9.23). The vector $\mathbf{e}_i$ is fixed in the frame $\mathcal{S}$, which is rotating with angular velocity $\mathbf{\Omega}$ relative to $\mathcal{S}_o$. Therefore the rate of change of $\mathbf{e}_i$ as seen in $\mathcal{S}_o$ is given by (9.23) as

$$\left(\frac{d\mathbf{e}_i}{dt}\right)_{\mathcal{S}_o} = \mathbf{\Omega} \times \mathbf{e}_i\,.$$

Thus, we can rewrite the second sum in (9.29) as

$$\sum_i Q_i\left(\frac{d\mathbf{e}_i}{dt}\right)_{\mathcal{S}_o} = \sum_i Q_i\,(\mathbf{\Omega} \times \mathbf{e}_i) = \mathbf{\Omega} \times \sum_i Q_i\,\mathbf{e}_i = \mathbf{\Omega} \times \mathbf{Q}.$$

---

[6] Now is one of those moments when, as anticipated in Chapter 1, the notation $\mathbf{e}_i$ with $i = 1, 2, 3$ is more convenient for our three unit vectors than $\hat{\mathbf{x}}$, $\hat{\mathbf{y}}$, and $\hat{\mathbf{z}}$. For now, this is just because it lets us use the summation sign, $\sum$, in sums like (9.27). In Chapter 10, we'll find that the most convenient choice of rotating axes is to use the *principal axes* of the rotating body, and the notation $\mathbf{e}_i$ works very naturally for them. Thus I shall mostly use this notation for the unit vectors fixed in a rotating body, and continue to use $\hat{\mathbf{x}}$, $\hat{\mathbf{y}}$, and $\hat{\mathbf{z}}$ for the nonrotating frame.

Inserting this result into (9.29), and using (9.28) to replace the first sum, we find that

$$\left(\frac{d\mathbf{Q}}{dt}\right)_{\mathcal{S}_o} = \left(\frac{d\mathbf{Q}}{dt}\right)_{\mathcal{S}} + \boldsymbol{\Omega} \times \mathbf{Q}. \tag{9.30}$$

This important identity relates the derivative of any one vector $\mathbf{Q}$ as measured in the inertial frame $\mathcal{S}_o$ to the corresponding derivative in the rotating frame $\mathcal{S}$. In the next section, I shall use it to find the form of Newton's second law in the rotating frame $\mathcal{S}$.

9.5 Newton's Second Law in a Rotating Frame

We are now ready to find the form of Newton's second law in the rotating frame $\mathcal{S}$. To simplify matters, I shall assume that the angular velocity $\boldsymbol{\Omega}$ of $\mathcal{S}$ relative to $\mathcal{S}_o$ is constant, as is the case (to an outstanding approximation) for axes fixed to the earth. A rather surprising aspect of the statement that $\boldsymbol{\Omega}$ is constant is that if it is true in one frame then it is automatically true in the other. This follows immediately from (9.30): Since $\boldsymbol{\Omega} \times \boldsymbol{\Omega} = 0$, the two derivatives of $\boldsymbol{\Omega}$ are always the same; in particular, if one is zero, so is the other.

Consider now a particle of mass m and position $\mathbf{r}$. In the inertial frame $\mathcal{S}_o$, the particle obeys Newton's second law in its normal form,

$$m\left(\frac{d^2\mathbf{r}}{dt^2}\right)_{\mathcal{S}_o} = \mathbf{F} \tag{9.31}$$

where as usual $\mathbf{F}$ denotes the net force on the particle, the vector sum of all forces identified in the inertial frame. The derivative on the left is, of course, the derivative evaluated by observers in the inertial frame $\mathcal{S}_o$. However, we can now use Equation (9.30) to express this derivative in terms of the derivatives evaluated in the rotating frame $\mathcal{S}$. First, according to (9.30)

$$\left(\frac{d\mathbf{r}}{dt}\right)_{\mathcal{S}_o} = \left(\frac{d\mathbf{r}}{dt}\right)_{\mathcal{S}} + \boldsymbol{\Omega} \times \mathbf{r}.$$

Differentiating a second time, we find

$$\left(\frac{d^2\mathbf{r}}{dt^2}\right)_{\mathcal{S}_o} = \left(\frac{d}{dt}\right)_{\mathcal{S}_o}\left(\frac{d\mathbf{r}}{dt}\right)_{\mathcal{S}_o}$$

$$= \left(\frac{d}{dt}\right)_{\mathcal{S}_o}\left[\left(\frac{d\mathbf{r}}{dt}\right)_{\mathcal{S}} + \boldsymbol{\Omega} \times \mathbf{r}\right].$$

Applying (9.30) to the outside derivative on the right, we find

$$\left(\frac{d^2\mathbf{r}}{dt^2}\right)_{\mathcal{S}_o} = \left(\frac{d}{dt}\right)_{\mathcal{S}}\left[\left(\frac{d\mathbf{r}}{dt}\right)_{\mathcal{S}} + \boldsymbol{\Omega} \times \mathbf{r}\right] + \boldsymbol{\Omega} \times \left[\left(\frac{d\mathbf{r}}{dt}\right)_{\mathcal{S}} + \boldsymbol{\Omega} \times \mathbf{r}\right]. \tag{9.32}$$

This rather messy result can be cleaned up. First, since our main concern is going to be with derivatives evaluated in the rotating frame $\mathcal{S}$, we'll revive the "dot" notation for these derivatives. That is, I shall use $\dot{\mathbf{Q}}$ to denote

$$\dot{\mathbf{Q}} \equiv \left(\frac{d\mathbf{Q}}{dt}\right)_{\mathcal{S}},$$

the derivative of any vector $\mathbf{Q}$ in the rotating frame $\mathcal{S}$. If we next note that, since $\boldsymbol{\Omega}$ is constant, its derivative is zero, and we group together two like terms, we can rewrite (9.32) as

$$\left(\frac{d^2\mathbf{r}}{dt^2}\right)_{\mathcal{S}_o} = \ddot{\mathbf{r}} + 2\boldsymbol{\Omega} \times \dot{\mathbf{r}} + \boldsymbol{\Omega} \times (\boldsymbol{\Omega} \times \mathbf{r}) \tag{9.33}$$

where the dots on the right all indicate derivatives evaluated with respect to the rotating frame $\mathcal{S}$.

If we now substitute the result (9.33) into Newton's second law (9.31) for the inertial frame $\mathcal{S}_o$ and move two terms to the right, we find the form of Newton's second law for the rotating frame $\mathcal{S}$ to be

$$m\ddot{\mathbf{r}} = \mathbf{F} + 2m\dot{\mathbf{r}} \times \boldsymbol{\Omega} + m(\boldsymbol{\Omega} \times \mathbf{r}) \times \boldsymbol{\Omega}, \tag{9.34}$$

where, as usual, $\mathbf{F}$ denotes the sum of all the forces as identified in any inertial frame. As with the accelerated frame of Section 9.1, we see that the equation of motion in a rotating reference frame *looks* just like Newton's second law, except that in this case there are two extra terms on the force side of the equation. The first of these extra terms is called the **Coriolis force** (after the French physicist G.G.de Coriolis, 1792–1843, who was the first to explain it),

$$\mathbf{F}_{\text{cor}} = 2m\dot{\mathbf{r}} \times \boldsymbol{\Omega}. \tag{9.35}$$

The second is the so-called **centrifugal force**

$$\mathbf{F}_{\text{cf}} = m(\boldsymbol{\Omega} \times \mathbf{r}) \times \boldsymbol{\Omega}. \tag{9.36}$$

I shall discuss these two terms in the next few sections. For now, the important point is that we can go ahead and use Newton's second law in rotating (and hence noninertial) reference frames, provided we remember always to add these two "fictitious" inertial forces to the net force **F** calculated for an inertial frame. That is, in a rotating frame,[7]

$$m\ddot{\mathbf{r}} = \mathbf{F} + \mathbf{F}_{\text{cor}} + \mathbf{F}_{\text{cf}}. \tag{9.37}$$

9.6 The Centrifugal Force

We have just seen that in order to use Newton's second law in a rotating frame (such as a frame attached to the earth) we must introduce two inertial forces, the centrifugal and the Coriolis forces. To some extent, we can examine the two forces separately. In particular, the Coriolis force on an object is proportional to the object's velocity $\mathbf{v} = \dot{\mathbf{r}}$ relative to the rotating frame. Therefore, the Coriolis force is zero for any object that is at rest in the rotating frame, and it is negligible for objects that are moving sufficiently slowly. For the rest of this chapter, our main concern will be with the rotating frame of the earth, for which we can easily estimate the relative importance of the two inertial forces. Because both forces involve vector products, they depend on the directions of the various vectors, but for an order-of-magnitude estimate we can take

$$F_{\text{cor}} \sim mv\Omega \quad \text{and} \quad F_{\text{cf}} \sim mr\Omega^2,$$

where v is the object's speed relative to the rotating frame of the earth, that is, the speed as observed by us on the earth's surface. Therefore

$$\frac{F_{\text{cor}}}{F_{\text{cf}}} \sim \frac{v}{R\Omega} \sim \frac{v}{V}. \tag{9.38}$$

Here, in the middle expression, I have canceled a common factor of $m\Omega$ and I have replaced r by the earth's radius R. (Remember that the origin is at the earth's center, so for objects near the earth's surface, $r \approx R$.) In the last expression I have replaced $R\Omega$ by V, the speed of a point on the equator as the earth rotates with angular velocity Ω. Since V is approximately 1000 mi/h, (9.38) shows that for projectiles with $v \ll 1000$ mi/h it will be a good starting approximation to ignore the Coriolis force, and this is what I shall do in this section.[8]

The centrifugal force is given by (9.36) as

$$\mathbf{F}_{\text{cf}} = m(\mathbf{\Omega} \times \mathbf{r}) \times \mathbf{\Omega}. \tag{9.39}$$

---

[7] Our derivation of this important result hinged crucially on the relation (9.30) between time derivatives of a given vector in the rotating and nonrotating frames. If you find that relation confusing, you might prefer the alternative derivation based on the Lagrangian formalism and outlined in Problem 9.11.

[8] As we shall see later, even when $v \ll 1000$ mi/h, the Coriolis force can have appreciable effects (for instance, with the Foucault pendulum). Nevertheless, it is certainly true that F_{cor} is small compared to F_{cf}, and it makes sense to ignore the former at first.

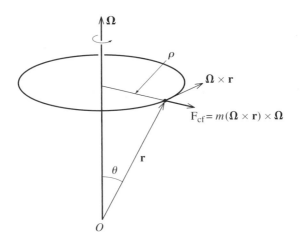

Figure 9.9   The vector $\mathbf{\Omega} \times \mathbf{r}$ is the velocity of an object as it is dragged eastward with speed $\Omega \rho$ by the earth's rotation. Therefore, the centrifugal force, $m(\mathbf{\Omega} \times \mathbf{r}) \times \mathbf{\Omega}$, points radially outward from the axis and has magnitude $m\Omega^2 \rho$.

We can see what this looks like with the help of Figure 9.9, which shows an object on or near the earth's surface at colatitude θ. The earth's rotation carries the object around a circle of latitude, and the vector $\mathbf{\Omega} \times \mathbf{r}$ (which is just the velocity of this circular motion as seen from a nonrotating frame) is tangent to this circle. Thus $(\mathbf{\Omega} \times \mathbf{r}) \times \mathbf{\Omega}$ points radially outward from the axis of rotation in the direction of $\hat{\boldsymbol{\rho}}$, the unit vector in the ρ direction of cylindrical polar coordinates. The magnitude of $(\mathbf{\Omega} \times \mathbf{r}) \times \mathbf{\Omega}$ is easily seen to be $\Omega^2 r \sin \theta = \Omega^2 \rho$. Thus,

$$\mathbf{F}_{cf} = m\Omega^2 \rho \, \hat{\boldsymbol{\rho}}. \qquad (9.40)$$

To summarize, from the point of view of observers rotating with the earth, there is a centrifugal force that is radially outward from the earth's axis and has magnitude $m\Omega^2 \rho$. If we momentarily let $\mathbf{v} = \mathbf{\Omega} \times \mathbf{r}$ denote the velocity associated with the earth's rotation (observed from a nonrotating frame), then its magnitude is $v = \Omega \rho$, and the centrifugal force takes the familiar form mv^2/ρ.

Free-Fall Acceleration

The free-fall acceleration that we call $\mathbf{g}$ is the initial acceleration, relative to the earth, of an object that is released from rest in a vacuum near the earth's surface. We can now see that this is actually a surprisingly complicated notion. The equation of motion (relative to the earth) is[9]

$$m\ddot{\mathbf{r}} = \mathbf{F}_{grav} + \mathbf{F}_{cf}, \qquad (9.41)$$

---

[9] I have defined $\mathbf{g}$ as the *initial* acceleration of a body released from rest to ensure that the Coriolis force is zero. When the object speeds up, we shall find that the Coriolis force eventually becomes important and the acceleration changes (although the effect is usually very small).

where $\mathbf{F}_{cf}$ is given by (9.40) and $\mathbf{F}_{grav}$ is the gravitational force

$$\mathbf{F}_{grav} = -\frac{GMm}{R^2}\hat{\mathbf{r}} = m\mathbf{g}_o. \tag{9.42}$$

Here, M and R are the mass and radius of the earth, and $\hat{\mathbf{r}}$ denotes the unit vector that points radially out from O, the center of the earth.[10] The acceleration $\mathbf{g}_o$ is defined by the second equality and could be called the "true" acceleration of gravity, inasmuch as it is the acceleration we would observe if there were no centrifugal effect.

We see from (9.41) that the initial acceleration of a freely falling object is determined by an effective force which is equal to the sum of two terms,

$$\mathbf{F}_{eff} = \mathbf{F}_{grav} + \mathbf{F}_{cf} = m\mathbf{g}_o + m\Omega^2 R \sin\theta\,\hat{\boldsymbol{\rho}} \tag{9.43}$$

where the last expression for $\mathbf{F}_{cf}$ comes from (9.40) with ρ replaced by $R\sin\theta$. The two forces that make up the effective force are shown in Figure 9.10, from which it is clear that the free-fall acceleration is in general not equal to the true gravitational acceleration, either in magnitude or direction. Specifically, dividing (9.43) by m, we find for the free-fall acceleration $\mathbf{g}$,

$$\mathbf{g} = \mathbf{g}_o + \Omega^2 R \sin\theta\,\hat{\boldsymbol{\rho}}. \tag{9.44}$$

The component of $\mathbf{g}$ in the inward radial direction (the direction of $-\mathbf{r}$)[11] is

$$g_{rad} = g_o - \Omega^2 R \sin^2\theta. \tag{9.45}$$

The second, centrifugal term is zero at the poles ($\theta = 0$ or π) and is largest at the equator, where its magnitude is easily found [using the value of Ω from (9.26)] to be

$$\Omega^2 R = (7.3 \times 10^{-5}\ \mathrm{s}^{-1})^2 \times (6.4 \times 10^6\ \mathrm{m}) \approx 0.034\ \mathrm{m/s^2}. \tag{9.46}$$

Since g_o is about 9.8 m/s², we see that, because of the centrifugal force, the value of g at the equator is about 0.3% less than at the poles.[12] Although this difference is certainly small, it is easily measured with modern gravimeters that can measure g to about 1 part in 10^9.

---

[10] In claiming that the gravitational force is $-(GMm/r^2)\hat{\mathbf{r}}$, I am assuming that the earth is perfectly spherically symmetric, which, although a very good approximation, is not exactly true. Fortunately, all that matters is that $\mathbf{F}_{grav}$ is certainly proportional to m, so it can always be written as $m\mathbf{g}_o$. For almost all purposes we can say that $\mathbf{g}_o$ is in the direction of $-\hat{\mathbf{r}}$, and it is certainly extremely close to that.

[11] Strictly speaking this is the component of $\mathbf{g}$ in the direction of $\mathbf{g}_o$ not $-\mathbf{r}$. By the same token, the factor $\sin^2\theta$ should actually be $\sin\theta\sin\theta'$ where θ' is the angle between the line of $\mathbf{g}_o$ and north (as opposed to θ, the angle between $\mathbf{r}$ and north). However, the difference (which is only because the earth is not perfectly spherically symmetric) is, for most practical purposes, completely negligible.

[12] The actual difference is more like 0.5%, the additional 0.2% being the result of the earth's bulge at the equator.

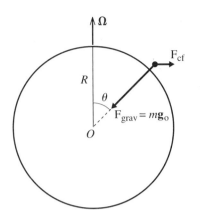

Figure 9.10   The free-fall acceleration (relative to the earth) of an object released from rest close to the earth's surface is the result of two force terms, the true gravitational force $m\mathbf{g}_o$ and the centrifugal, inertial force $\mathbf{F}_{cf}$, which points out from the axis of rotation. (The size of the centrifugal term is much exaggerated in this picture.)

You can see from (9.44) and Figure 9.10 that the tangential component of $\mathbf{g}$ (the component normal to the true gravitational force) comes entirely from the centrifugal force and is

$$g_{\text{tang}} = \Omega^2 R \sin \theta \cos \theta. \tag{9.47}$$

This tangential component of $\mathbf{g}$ is zero at the poles and at the equator, and is maximum at latitude 45°. The most striking feature of a nonzero value for g_{tang} is that it means the free-fall acceleration is not exactly in the direction of the true gravitational force. As you can see from Figure 9.11, the angle between $\mathbf{g}$ and the radial direction is about $\alpha \approx g_{\text{tang}}/g_{\text{rad}}$, and its maximum value (at $\theta = 45°$) is

$$\alpha_{\text{max}} = \frac{\Omega^2 R}{2g_o} \approx \frac{0.034}{2 \times 9.8} \approx 0.0017 \text{ rad} \approx 0.1°. \tag{9.48}$$

This angle α is the angle between the observed free-fall acceleration $\mathbf{g}$ and the true acceleration of gravity $\mathbf{g}_o$ — what we might be tempted to call "vertical." The value of α is actually rather difficult to measure. The direction of the observed $\mathbf{g}$ is easy (in principle, at least). To find the direction of $\mathbf{g}_o$, you might hope to use a plumb line, but a moment's thought should convince you that a plumb line is also subject to the centrifugal force and will hang in the direction of $\mathbf{g}$, not that of $\mathbf{g}_o$. In fact, any attempt to find the direction of $\mathbf{g}_o$ simply and directly winds up finding the direction of $\mathbf{g}$. For this reason, in what follows I shall *define* "vertical" as the direction of a plumb line. Therefore, on those rare occasions when these tiny distinctions matter, "vertical" will mean "in the direction of $\pm\mathbf{g}$." By the same token, "horizontal" will mean "perpendicular to $\mathbf{g}$."

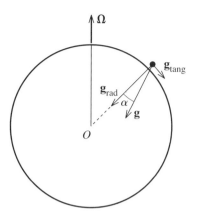

Figure 9.11   Because of the centrifugal force, the free-fall acceleration **g** has a nonzero tangential component (greatly exaggerated here) and **g** deviates from the radial direction by the small angle α.

9.7 The Coriolis Force

When an object is moving, there is a second inertial force that you must include when you want to use Newton's second law in a rotating frame. This is the Coriolis force (9.35)

$$\mathbf{F}_{\text{cor}} = 2m\dot{\mathbf{r}} \times \mathbf{\Omega} = 2m\mathbf{v} \times \mathbf{\Omega} \qquad (9.49)$$

where $\mathbf{v} = \dot{\mathbf{r}}$ is the object's velocity relative to the rotating frame. There is a remarkable parallel between the Coriolis force and the well-known force $q\mathbf{v} \times \mathbf{B}$ on a charge q in a magnetic field **B**. Indeed, if we replace $2m$ by q and $\mathbf{\Omega}$ by **B**, the former becomes exactly the latter. Although this parallel has no deep significance, it can often be a help in visualizing how the Coriolis force is going to affect a particle's motion.

The magnitude of the Coriolis force depends on the magnitudes of **v** and $\mathbf{\Omega}$ as well as their relative orientations. For the case that the rotating frame is the earth, we have seen in (9.26) that $\Omega \approx 7.3 \times 10^{-5} \text{ s}^{-1}$. For an object with $v \approx 50$ m/s (a fast baseball, for example), the maximum acceleration the Coriolis force could produce (acting by itself and with **v** perpendicular to $\mathbf{\Omega}$) would be

$$a_{\text{max}} = 2v\Omega \approx 2 \times (50 \text{ m/s}) \times (7.3 \times 10^{-5} \text{ s}^{-1}) \approx 0.007 \text{ m/s}^2.$$

Compared to the free-fall acceleration $g = 9.8 \text{ m/s}^2$, this is very small, though certainly detectable if we were to take the trouble. Some projectiles, such as rockets and long-range shells, travel much faster than 50 m/s, and for them the Coriolis force is correspondingly more important. In addition, we shall see that there are systems, such as the Foucault pendulum, where the Coriolis force, though very small, can act for a long time and hence produce a large effect.

Direction of the Coriolis Force

Like the magnetic force $q\mathbf{v} \times \mathbf{B}$, the Coriolis force $2m\mathbf{v} \times \mathbf{\Omega}$ is always perpendicular to the velocity of the moving object, with its direction given by the right-hand rule. Figure 9.12 is an overhead view of a horizontal turntable that is rotating counterclockwise relative to the ground. The angular velocity $\mathbf{\Omega}$ points vertically up (out of the page in the figure). If we consider an object sliding or rolling across the turntable, it is easy to see that, whatever the object's position and velocity, the Coriolis force tends to deflect the velocity to the right. Similarly, if the turntable were rotating clockwise, the Coriolis deflection would always be to the left. (Whether the object actually *is* deflected in the specified directions depends on what other forces are acting and how big they are.)

We could imagine Figure 9.12 to be the Northern Hemisphere viewed from above the North Pole. (Since the earth rotates to the east, the angular velocity is directed as shown.) Thus we reach the conclusion that the Coriolis effect due to the earth's rotation tends to deflect moving bodies to the right in the Northern Hemisphere (and, of course, to the left in the Southern Hemisphere).[13] This effect is important to long-range gunners, who must aim to the left of their target in the Northern, and to the right in the Southern, Hemisphere. (See Problem 9.28.) An important example from

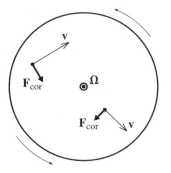

Figure 9.12   Overhead view of a horizontal turntable that is rotating counterclockwise relative to an inertial frame. The turntable's angular velocity $\mathbf{\Omega}$ points up out of the picture. As seen by observers on the turntable, the two objects sliding on the table are subject to Coriolis forces $\mathbf{F}_{\text{cor}} = 2m\mathbf{v} \times \mathbf{\Omega}$. Irrespective of the bodies' positions and velocities, the Coriolis force always tends to deflect the velocity to the right. (If the direction of rotation were clockwise, then $\mathbf{\Omega}$ would be into the page and the Coriolis force would tend to deflect moving objects to the left.)

---

[13] Because the earth is three-dimensional (as opposed to a turntable, which is two-dimensional) the Coriolis effect is actually a little more complicated than this simple statement suggests. However, the statement above is certainly correct for objects moving parallel to the earth's surface and for low trajectory projectiles.

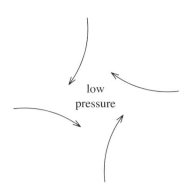

Figure 9.13   A cyclone is the result of air moving into a low-pressure region and being deflected to the right (in the Northern Hemisphere) by the Coriolis effect. This causes a counterclockwise flow with the inward pressure force partially balanced by the outward Coriolis force (and the difference supplying the inward centripetal acceleration).

meteorology is the phenomenon of cyclones. These occur when the air surrounding a region of low pressure moves rapidly inward. Because of the Coriolis effect, the air is deflected to the right, as shown in Figure 9.13, and therefore begins to circulate counterclockwise (in the Northern Hemisphere — clockwise in the Southern). When this happens sufficiently violently, the result is a storm, known variously as a cyclone, hurricane, or typhoon.

It is important to bear in mind that both the Coriolis and centrifugal forces are at root kinematic effects, resulting from our insistence on using a rotating frame of reference. In a few simple cases, it is actually easier (as well as instructive) to analyze the motion in an inertial frame and then transform the results to the rotating frame, as the following example illustrates. Nevertheless, the transformation between the two frames is usually so complicated that it is easier to work all the time in the rotating frame and to live with the "fictitious" Coriolis and centrifugal forces.

EXAMPLE 9.2   Simple Motion on a Turntable

Three observers A, B, and C are standing on a horizontal turntable with A at the center, C at the edge and B halfway between, as shown in Figure 9.14(a). The turntable is rotating counterclockwise (as seen from above) with angular velocity Ω. At time $t = 0$, A kicks a frictionless puck exactly toward B and C, but to his surprise the puck misses both B and C, the latter by an even bigger margin than the former. Explain these events from the points of view both of the observers on the rotating table and of an observer on the ground.

The net force on the puck (as identified in any inertial frame) is zero. Thus in the rotating frame, the only two forces are the centrifugal and Coriolis forces. The former is always radially outward and has no bearing on the puck's deflection. The latter consistently deflects the puck's velocity to the right, just like an upward magnetic field acting on a positive charge. This causes the puck to follow the curved path shown in Figure 9.14(a). At time t_1 when it reaches the

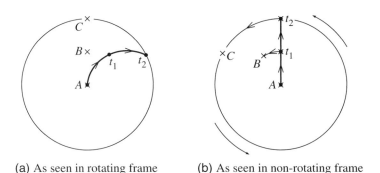

(a) As seen in rotating frame   (b) As seen in non-rotating frame

Figure 9.14   **(a)** Three observers A, B, and C are in a line on a rotating turntable, with A at the center and C at the perimeter. Observer A kicks a puck toward B and C, but because of the Coriolis force the puck veers to the right and misses B and C. **(b)** The same experiment as seen by an observer on the ground. In this frame, the puck travels in a straight line, but by the time t_1 when it reaches the radius of B, observer B has moved to the left. By the time t_2 when the puck reaches the perimeter, C has moved even further to the left.

radius of B, it is a small distance to the right of B, and at time t_2 when it arrives at the turntable's edge, it is even further (about four times, in fact) to the right of C. This explanation is correct and clear, but depends on our understanding the Coriolis force. By analyzing the same experiment in the ground-based frame, we can gain an additional understanding of why the deflection occurs.

In the inertial frame of an observer on the ground, the net force on the puck is zero, and the puck follows a straight path, as shown in Figure 9.14(b). However, by the time t_1, when it should have been hitting B, observer B has moved to the left as the turntable rotates. By time t_2, when it should have been hitting C, observer C has moved even further to the left. As seen by the puck, B and C have moved to the left. Therefore, as seen by B and C, the puck has curved to the right as in Figure 9.14(a).

This simple alternative explanation of the Coriolis effect is actually deceptively simple. In general, the effects of the Coriolis and centrifugal forces are surprisingly complicated and are not nearly so easy to explain by reference to the nonrotating frame. (See Problems 9.20 and 9.24.)

9.8   Free Fall and the Coriolis Force

Let us next consider the effect of the Coriolis force on a freely falling object, that is, an object falling in a vacuum, close to a point **R** on the earth's surface. For this analysis, we must also include the centrifugal force, so the equation of motion is

$$m\ddot{\mathbf{r}} = m\mathbf{g}_o + \mathbf{F}_{cf} + \mathbf{F}_{cor} \tag{9.50}$$

where, as before, $m\mathbf{g}_o$ denotes the true force of the earth's gravity on the object. The centrifugal force is $m(\boldsymbol{\Omega} \times \mathbf{r}) \times \boldsymbol{\Omega}$, (where $\mathbf{r}$ is the object's position relative to the center of the earth), but to an outstanding approximation we can replace $\mathbf{r}$ by $\mathbf{R}$ (the position on the earth's surface where the experiment is being conducted). Thus,

$$\mathbf{F}_{cf} = m(\boldsymbol{\Omega} \times \mathbf{R}) \times \boldsymbol{\Omega}.$$

Returning to the equation of motion (9.50), you will recognize that the sum of the first two terms on the right is just $m\mathbf{g}$, where $\mathbf{g}$ is the observed free-fall acceleration for an object released from rest at position $\mathbf{R}$, as introduced in (9.44). In other words, we can omit the term $\mathbf{F}_{cf}$ from (9.50), if we replace $\mathbf{g}_o$ by the observed $\mathbf{g}$ at the location of our experiment. If we substitute $2m\mathbf{v} \times \boldsymbol{\Omega}$ for $\mathbf{F}_{cor}$, the equation of motion becomes (after cancellation of a factor of m)

$$\ddot{\mathbf{r}} = \mathbf{g} + 2\dot{\mathbf{r}} \times \boldsymbol{\Omega}. \tag{9.51}$$

A simplifying feature of the equation (9.51) is that it does not involve the position $\mathbf{r}$ at all (only its derivatives $\dot{\mathbf{r}}$ and $\ddot{\mathbf{r}}$). This means the equation will not change if we make a change of origin (since a change of origin amounts to adding a constant to $\mathbf{r}$). Accordingly, I shall now choose my origin on the surface of the earth at the position $\mathbf{R}$, as shown in Figure 9.15. With this choice of axes, we can resolve the equation of motion into its three components. The components of $\dot{\mathbf{r}}$ and $\boldsymbol{\Omega}$ are

$$\dot{\mathbf{r}} = (\dot{x}, \dot{y}, \dot{z})$$

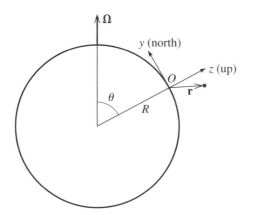

Figure 9.15   Choice of axes for a free-fall experiment. The origin O is on the earth's surface at the experiment's location (position $\mathbf{R}$ relative to the center of the earth). The z axis points vertically up (more precisely, in the direction of $-\mathbf{g}$, where $\mathbf{g}$ is the observed free-fall acceleration), the x and y axes are horizontal (that is, perpendicular to $\mathbf{g}$), with y pointing north, and x due east. The position of the falling object relative to O is $\mathbf{r}$.

and

$$\boldsymbol{\Omega} = (0, \Omega \sin \theta, \Omega \cos \theta).$$

Thus, those of $\dot{\mathbf{r}} \times \boldsymbol{\Omega}$ are

$$\dot{\mathbf{r}} \times \boldsymbol{\Omega} = (\dot{y} \Omega \cos \theta - \dot{z} \Omega \sin \theta, \ -\dot{x} \Omega \cos \theta, \ \dot{x} \Omega \sin \theta) \qquad (9.52)$$

and the equation of motion (9.51) resolves into the following three equations:

$$\ddot{x} = 2\Omega (\dot{y} \cos \theta - \dot{z} \sin \theta)$$
$$\ddot{y} = -2\Omega \dot{x} \cos \theta \qquad\qquad (9.53)$$
$$\ddot{z} = -g + 2\Omega \dot{x} \sin \theta.$$

We shall solve these three equations by making a succession of approximations that depend on the smallness of Ω. First, because Ω is very small, we get a reasonable starting approximation if we ignore Ω entirely. In this approximation, the equations reduce to

$$\ddot{x} = 0, \qquad \ddot{y} = 0, \qquad \text{and} \qquad \ddot{z} = -g, \qquad (9.54)$$

which are the equations of free fall solved in every introductory physics course. If the object is dropped from rest at $x = y = 0$ and $z = h$, then the first two equations imply that $\dot{x}$, $\dot{y}$, x, and y all remain zero, while the last equation implies that $\dot{z} = -gt$ and $z = h - \frac{1}{2} g t^2$. Thus our approximate solution is

$$x = 0, \qquad y = 0, \qquad \text{and} \qquad z = h - \tfrac{1}{2} g t^2, \qquad (9.55)$$

that is, the object falls vertically down with constant acceleration g. This approximation is sometimes called the *zeroth-order* approximation because it involves only the zeroth power of Ω (that is, it is independent of Ω). It is well known to be a very good approximation, but it shows none of the effects of the Coriolis force.

To get the next approximation, we argue as follows: The terms in (9.53) that involve Ω are all small. Thus, it will be safe to evaluate these terms using our zeroth-order approximation for x, y, and z. Substituting (9.55) into the right side of (9.53), we get

$$\ddot{x} = 2\Omega g t \sin \theta, \qquad \ddot{y} = 0, \qquad \text{and} \qquad \ddot{z} = -g. \qquad (9.56)$$

The last two of these are exactly the same as in zeroth order, but the equation for x is new and is easily integrated twice to give

$$x = \tfrac{1}{3} \Omega g t^3 \sin \theta, \qquad (9.57)$$

with y and z the same as in the zeroth approximation (9.55). This result is naturally called the *first-order* approximation (being good through the first power of Ω). We can repeat this process again to get a second-order approximation and so on, but the first-order is good enough for our purposes.

The striking thing about the solution (9.57) is that a freely falling object does not fall straight down. Instead the Coriolis force causes it to curve slightly to the east (positive x direction). To get an idea of the magnitude of the effect, consider an object dropped down a 100-meter-deep mine shaft at the equator, and let us find the total deflection by the time it hits the bottom. The time to reach the bottom is determined by the last of equations (9.55) as $t = \sqrt{2h/g}$, and (9.57) gives for the total easterly deflection (putting $\theta = 90°$ and $g \approx 10$ m/s^2)

$$x = \frac{1}{3}\Omega g \left(\frac{2h}{g}\right)^{3/2}$$

$$\approx \frac{1}{3} \times (7.3 \times 10^{-5}\ \text{s}^{-1}) \times (10\ \text{m/s}^2) \times (20\ \text{s}^2)^{3/2} \approx 2.2\ \text{cm}$$

a small deflection, but certainly detectable. A small easterly deflection of this type was actually predicted by Newton and verified by his rival Robert Hooke (of Hooke's law fame, 1635–1703), although it was not properly explained until the Coriolis effect was understood.

9.9   The Foucault Pendulum

As a final and striking application of the Coriolis effect, let us consider the Foucault pendulum, which can be seen in many science museums around the world and is named for its inventor, the French physicist Jean Foucault (1819–1868). This is a pendulum made of a very heavy mass m suspended by a light wire from a tall ceiling. This arrangement allows the pendulum to swing freely for a very long time and to move in both the east–west and north–south directions. As seen in an inertial frame, there are just two forces on the bob, the tension $\mathbf{T}$ in the wire and the weight $m\mathbf{g}_o$. In the rotating frame of the earth, there are also the centrifugal and Coriolis forces, so the equation of motion in the earth's frame is

$$m\ddot{\mathbf{r}} = \mathbf{T} + m\mathbf{g}_o + m(\boldsymbol{\Omega} \times \mathbf{r}) \times \boldsymbol{\Omega} + 2m\dot{\mathbf{r}} \times \boldsymbol{\Omega}.$$

Exactly as in the previous section, the second and third terms on the right combine to give $m\mathbf{g}$, where $\mathbf{g}$ is the observed free-fall acceleration, and the equation of motion becomes

$$m\ddot{\mathbf{r}} = \mathbf{T} + m\mathbf{g} + 2m\dot{\mathbf{r}} \times \boldsymbol{\Omega}. \tag{9.58}$$

We can now choose our axes as in the previous section, so that x is east, y is north, and z vertically up (direction of $-\mathbf{g}$), and the pendulum is as shown in Figure 9.16.

I shall restrict our discussion to the case of small oscillations, so that the angle β between the pendulum and the vertical is always small. This allows two simplifying approximations: First, the z component of the tension $\mathbf{T}$ is well approximated by the magnitude; that is, $T_z = T\cos\beta \approx T$. Second, it is not hard to see that, for small oscillations, $T_z \approx mg$.[14] Putting these two approximations together, we can write

$$T \approx mg. \tag{9.59}$$

---

[14] Look at the z component of (9.58). In the limit of small oscillations, the term on the left and the last term on the right both approach zero, and you're left with $T_z - mg = 0$.

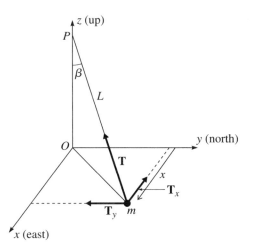

Figure 9.16   A Foucault pendulum comprises a bob of mass m
suspended by a light wire of length L from the point P on a high
ceiling. The tension force on the bob is shown as $\mathbf{T}$ and its x and
y components are T_x and T_y. For small oscillations the angle β
is very small.

We now need to examine the x and y components of the equation of motion (9.58).
This requires that we identify the x and y components of $\mathbf{T}$. If you look at Figure 9.16,
you will see that, by similar triangles, $T_x/T = -x/L$ and similarly for T_y. Combining
this with (9.59), we find that

$$T_x = -mgx/L \quad \text{and} \quad T_y = -mgy/L. \tag{9.60}$$

The x and y components of $\mathbf{g}$ are, of course, zero, and the components of $\dot{\mathbf{r}} \times \boldsymbol{\Omega}$ are
given in (9.52). Putting all of these into (9.58), we find (after canceling a factor of
m and dropping a term involving $\dot{z}$, which is negligible compared to $\dot{x}$ or $\dot{y}$ for small
oscillations)

$$\left.\begin{aligned}
\ddot{x} &= -gx/L + 2\dot{y}\Omega\cos\theta \\
\ddot{y} &= -gy/L - 2\dot{x}\Omega\cos\theta.
\end{aligned}\right\} \tag{9.61}$$

where as usual θ denotes the colatitude of the location of the experiment. The factor
g/L is just ω_o^2, where ω_o is the natural frequency of the pendulum, and $\Omega\cos\theta$ is
just Ω_z, the z component of the earth's angular velocity. Thus these two equations of
motion can be rewritten as

$$\left.\begin{aligned}
\ddot{x} - 2\Omega_z\dot{y} + \omega_o^2 x &= 0 \\
\ddot{y} + 2\Omega_z\dot{x} + \omega_o^2 y &= 0.
\end{aligned}\right\} \tag{9.62}$$

We can solve the coupled equations (9.62) using the trick, introduced in Chapter
2, of defining a complex number

$$\eta = x + iy.$$

Recall that not only does this complex number contain the same information as the position in the xy plane, but a plot of η in the complex plane is an actual bird's eye view of the pendulum's projected position (x, y). If we multiply the second equation of (9.62) by i and add it to the first, we get the single differential equation

$$\ddot{\eta} + 2i\Omega_z \dot{\eta} + \omega_0^2 \eta = 0. \tag{9.63}$$

This is a second-order, linear, homogeneous differential equation and so has exactly two independent solutions. Thus if we can find two independent solutions, we shall know that the most general solution is a linear combination of these two. As often happens, we can find two independent solutions by inspired guesswork: We guess that there is a solution of the form

$$\eta(t) = e^{-i\alpha t} \tag{9.64}$$

for some constant α. Substituting this guess into (9.63), we see immediately that it is a solution if and only if α satisfies

$$\alpha^2 - 2\Omega_z \alpha - \omega_0^2 = 0$$

or

$$\alpha = \Omega_z \pm \sqrt{\Omega_z^2 + \omega_0^2}$$

$$\approx \Omega_z \pm \omega_0 \tag{9.65}$$

where the last line is an extremely good approximation since the earth's angular velocity Ω is so very much smaller than the pendulum's ω_0. This gives us the required two independent solutions, and the general solution to the equation of motion (9.63) is

$$\eta = e^{-i\Omega_z t}\left(C_1 e^{i\omega_0 t} + C_2 e^{-i\omega_0 t}\right). \tag{9.66}$$

To see what this solution looks like, we need to fix the two constants C_1 and C_2 by specifying the initial conditions. Let us suppose that at $t = 0$ the pendulum has been pulled aside in the x direction (east) to a position $x = A$ and $y = 0$, and is released from rest ($v_{xo} = v_{yo} = 0$). With these initial conditions, you can easily check that[15] $C_1 = C_2 = A/2$, and our solution becomes

$$\eta(t) \equiv x(t) + iy(t) = Ae^{-i\Omega_z t} \cos \omega_0 t. \tag{9.67}$$

At $t = 0$ the complex exponential is equal to one, and $x = A$, while $y = 0$. Because $\Omega_z \ll \omega_0$, the cosine factor in (9.67) makes many oscillations before the exponential changes appreciably from one. This implies that, initially, $x(t)$ oscillates with angular frequency ω_0 between $\pm A$, while y remains close to zero. That is, initially, the pendulum swings in simple harmonic motion along the x axis, as indicated in Figure 9.17(a).

---

[15] Actually, there is a small subtlety, in that these simple values depend on the (true) assumption that $\Omega_z \ll \omega_0$, as you will see when you check them.

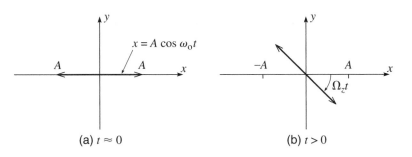

(a) $t \approx 0$ (b) $t > 0$

Figure 9.17   Overhead views of the motion of a Foucault pendulum. (a) For
a while after being released, the pendulum swings back and forth along the
x axis, with amplitude A and frequency ω_0. (b) As time advances, the plane
of its oscillations slowly rotates with angular velocity equal to Ω_z, the z
component of the earth's angular velocity.

However, eventually the complex exponential $e^{-i\Omega_z t}$ begins to change, causing
the complex number $\eta = x + iy$ to rotate through an angle $\Omega_z t$. In the Northern
Hemisphere, where Ω_z is positive, this means that the number $x + iy$ continues
to oscillate sinusoidally (due to the factor $\cos \omega_0 t$), but in a direction that rotates
clockwise. That is, the plane in which the pendulum is swinging rotates slowly
clockwise, with angular velocity Ω_z, as indicated in Figure 9.17(b). In the Southern
Hemisphere, where Ω_z is negative, the corresponding rotation is counterclockwise.

If the Foucault pendulum is located at colatitude θ (latitude $90° - \theta$), then the rate
at which its plane of oscillation rotates is

$$\Omega_z = \Omega \cos \theta. \tag{9.68}$$

At the North Pole ($\theta = 0$), $\Omega_z = \Omega$ and the rate of rotation of the pendulum is the same
as the earth's angular velocity. This result is easy to understand: As seen in an inertial
(nonrotating) frame, a Foucault pendulum at the North Pole would obviously swing
in a fixed plane; meanwhile, as seen in the same inertial frame, the earth is rotating
counterclockwise (as seen from above) with angular velocity Ω. Clearly then, as seen
from the earth, the pendulum's plane of oscillation has to be rotating clockwise with
angular velocity Ω.

At any other latitude, the result is much more complicated from an inertial point of
view, but the rate of rotation of the Foucault pendulum is easily calculated from (9.68).
At the equator ($\theta = 90°$), $\Omega_z = 0$ and the pendulum does not rotate. At a latitude
around $42°$ (the approximate latitude of Boston, Chicago, or Rome),

$$\Omega_z = \Omega \cos 48° \approx \tfrac{2}{3}\Omega.$$

Since Ω equals $360°$/day, $\tfrac{2}{3}\Omega = 240°$/day, and we see that in the course of 6 hours (a
time for which a long, well-built pendulum will certainly continue to swing without
significant damping), the pendulum's plane of motion will rotate through $60°$ — an
easily observable effect.

9.10   Coriolis Force and Coriolis Acceleration

Recall that in Equation (1.48) of Chapter 1 we found the form of Newton's second law in two-dimensional polar coordinates,

$$\mathbf{F} = m\ddot{\mathbf{r}} \qquad \Longleftrightarrow \qquad \begin{cases} F_r = m(\ddot{r} - r\dot{\phi}^2) \\ F_\phi = m(r\ddot{\phi} + 2\dot{r}\dot{\phi}). \end{cases} \tag{9.69}$$

We can now understand the rather ugly last term in each of the two equations on the right in terms of the centrifugal and Coriolis forces.

Consider a particle that is subject to a net force $\mathbf{F}$ and moves in two dimensions. (The exact same analysis works in three dimensions using cylindrical polar coordinates, but for simplicity I shall work in two dimensions.) Relative to any inertial frame $\mathcal{S}$ with origin O, the particle must satisfy (9.69). Now consider a noninertial frame $\mathcal{S}'$ which shares the same origin O and is rotating at constant angular velocity Ω, chosen so that $\Omega = \dot{\phi}$ at one chosen time $t = t_o$. That is, at the chosen instant t_o, the frame $\mathcal{S}'$ and the particle are rotating at the same rate. (For this reason, the frame $\mathcal{S}'$ is sometimes called the *co-rotating frame*.) If the particle has polar coordinates (r', ϕ') relative to $\mathcal{S}'$, then at all times

$$r' = r$$

(since $\mathcal{S}$ and $\mathcal{S}'$ share the same origin), and at the time t_o

$$\dot{\phi}' = 0$$

since the frame $\mathcal{S}'$ and the particle are rotating at the same rate at $t = t_o$. Newton's second law can be applied in the frame $\mathcal{S}'$, provided we include the centrifugal and Coriolis forces. Thus

$$\mathbf{F} + \mathbf{F}_{cf} + \mathbf{F}_{cor} = m\ddot{\mathbf{r}}'. \tag{9.70}$$

Let us write this equation in polar coordinates: The centrifugal force $\mathbf{F}_{cf}$ is purely radial, with radial component $mr\Omega^2$. (Remember that $r' = r$, so it makes no difference whether we write r or r'.) The Coriolis force $\mathbf{F}_{cor}$ is $2m\mathbf{v}' \times \Omega$, and, since $\mathbf{v}'$ is purely radial in the co-rotating frame, $\mathbf{F}_{cor}$ is in the ϕ' direction with ϕ' component $-2m\dot{r}\Omega$. Finally the term $m\ddot{\mathbf{r}}'$ on the right of (9.70) can be replaced by the analog of (9.69), except that in the co-rotating frame $\dot{\phi} = 0$ (at the chosen time t_o), so the terms containing $\dot{\phi}$ will be absent. Putting all this together, we find for the equation of motion of the particle in the co-rotating frame,

$$\mathbf{F} + \mathbf{F}_{cf} + \mathbf{F}_{cor} = m\ddot{\mathbf{r}}' \qquad \Longleftrightarrow \qquad \begin{cases} F_r + mr\Omega^2 = m\ddot{r} \\ F_\phi - 2m\dot{r}\Omega = mr\ddot{\phi}. \end{cases} \tag{9.71}$$

(Because the frame $\mathcal{S}'$ is rotating at a constant rate, I could replace $\ddot{\phi}'$ by $\ddot{\phi}$ since they are equal.)

Let us now compare the equation of motion (9.69) for the inertial frame, with (9.71) for the co-rotating frame. The most important thing to recognize is that, because $\Omega = \dot{\phi}$, they are exactly the same equations for r and ϕ, although certain terms are distributed differently between the two sides. In (9.69) for the nonrotating frame, the only force terms on the left are the real net force, with components F_r and F_ϕ. On the right of (9.69), the acceleration contains the centripetal acceleration $-r\dot{\phi}^2$ in its radial component and the Coriolis acceleration $2\dot{r}\dot{\phi}$ in its ϕ component. In (9.71) for the rotating frame, neither of these additional acceleration terms is present (because we arranged that $\dot{\phi}'$ is zero), but instead they are reincarnated on the force side of the equations (with opposite signs, of course) as the centrifugal force $m\Omega^2 r$ in the radial equation and Coriolis force $-2m\dot{r}\Omega$ in the ϕ equation.

Since the two versions of the equations are the same, it is clear that they are equally correct. In the inertial frame, the forces are simpler (no "fictitious" forces) but the accelerations are more complicated; in the rotating frame, it is the other way round. Which frame one chooses to use is dictated by convenience. In particular, when the observer is anchored to a rotating frame (as we earthlings are), it is generally more convenient to work in the rotating frame and to learn to live with the "fictitious" centrifugal and Coriolis forces.

Principal Definitions and Equations of Chapter 9

Inertial Force in an Accelerating but Nonrotating Frame

The motion of a body, as seen in a frame that has acceleration $\mathbf{A}$ relative to an inertial frame, can be found using Newton's second law in the form $m\ddot{\mathbf{r}} = \mathbf{F} + \mathbf{F}_{\text{inertial}}$, where $\mathbf{F}$ is the net force on the body (as measured in any inertial frame) and $\mathbf{F}_{\text{inertial}}$ is an additional **inertial force**

$$\mathbf{F}_{\text{inertial}} = -m\mathbf{A}. \qquad \text{[Eq. (9.5)]}$$

The Angular Velocity Vector

If a body is rotating about an axis specified by the unit vector $\mathbf{u}$ (direction given by the right-hand rule) at a rate ω (usually measured in radians per second), its **angular velocity vector** is defined as

$$\boldsymbol{\omega} = \omega\mathbf{u}. \qquad \text{[Eq. (9.21)]}$$

The "Useful Relation"

The velocity of a point $\mathbf{r}$ fixed in a rigid body that is rotating with angular velocity $\boldsymbol{\omega}$ is

$$\mathbf{v} = \boldsymbol{\omega} \times \mathbf{r}. \qquad \text{[Eq. (9.22)]}$$

Time Derivatives in a Rotating Frame

If frame S has angular velocity $\mathbf{\Omega}$ relative to frame S_0, then the time derivatives of a single vector $\mathbf{Q}$ as seen in the two frames are related by

$$\left(\frac{d\mathbf{Q}}{dt}\right)_{S_0} = \left(\frac{d\mathbf{Q}}{dt}\right)_{S} + \mathbf{\Omega} \times \mathbf{Q}. \qquad \text{[Eq. (9.30)]}$$

Newton's Second Law in a Rotating Frame

If frame S has angular velocity $\mathbf{\Omega}$ relative to an inertial frame S_0, then Newton's second law in the rotating frame takes the form

$$m\ddot{\mathbf{r}} = \mathbf{F} + \mathbf{F}_{\text{cor}} + \mathbf{F}_{\text{cf}}, \qquad \text{[Eq. (9.37)]}$$

where $\mathbf{F}$ is the net force on the body (as measured in any inertial frame) and the inertial forces $\mathbf{F}_{\text{cor}}$ and $\mathbf{F}_{\text{cf}}$ are the **Coriolis** and **centrifugal forces**,

$$\mathbf{F}_{\text{cor}} = 2m\dot{\mathbf{r}} \times \mathbf{\Omega} \qquad \text{and} \qquad \mathbf{F}_{\text{cf}} = m(\mathbf{\Omega} \times \mathbf{r}) \times \mathbf{\Omega}. \qquad \text{[Eqs. (9.35) \& (9.36)]}$$

Free-Fall Acceleration

The observed free-fall acceleration $\mathbf{g}$ (defined as the initial acceleration, relative to the earth, from rest) includes the "true" gravitational acceleration $\mathbf{g}_0$ and the effect of the centrifugal force

$$\mathbf{g} = \mathbf{g}_0 + (\mathbf{\Omega} \times \mathbf{R}) \times \mathbf{\Omega}. \qquad \text{[Eq. (9.44)]}$$

"Vertical" is defined as the direction of $\mathbf{g}$, and "horizontal" as perpendicular to $\mathbf{g}$.

Problems for Chapter 9

Stars indicate the approximate level of difficulty, from easiest (★) to most difficult (★★★).

SECTION 9.1　Acceleration without Rotation

9.1 ★ Be sure you understand why a pendulum in equilibrium hanging in a car that is accelerating forward tilts backward, and then consider the following: A helium balloon is anchored by a massless string to the floor of a car that is accelerating forward with acceleration A. Explain clearly why the balloon tends to tilt *forward* and find its angle of tilt in equilibrium. [*Hint:* Helium balloons float because of the buoyant Archimedean force, which results from a pressure gradient in the air. What is the relation between the directions of the gravitational field and the buoyant force?]

9.2 ★ A donut-shaped space station (outer radius R) arranges for artificial gravity by spinning on the axis of the donut with angular velocity ω. Sketch the forces on, and accelerations of, an astronaut standing in the station **(a)** as seen from an inertial frame outside the station and **(b)** as seen in the astronaut's personal rest frame (which has a centripetal acceleration $A = \omega^2 R$ as seen in the inertial frame). What angular velocity is needed if $R = 40$ meters and the apparent gravity is to equal the usual value of about 10 m/s^2? **(c)** What is the percentage difference between the perceived g at a six-foot astronaut's feet ($R = 40$ m) and at his head ($R = 38$ m)?

SECTION 9.2 The Tides

9.3 ⋆⋆ **(a)** Consider the tidal force (9.12) on a mass m at the position P of Figure 9.4. Write d as $(d_o - R_e) = d_o(1 - R_e/d_o)$ and use the binomial approximation $(1 - \epsilon)^{-2} \approx 1 + 2\epsilon$ to show that $\mathbf{F}_{\text{tid}} \approx -(2GM_mmR_e/d_o^3)\,\hat{\mathbf{x}}$. Confirm the direction of the force shown in Figure 9.4 and make a numerical comparison of the tidal force with the gravitational force $m\mathbf{g}$ of the earth. **(b)** Do the corresponding calculations for the force at the point R. Compare this force with that of part (a) (magnitude and direction).

9.4 ⋆⋆ Do the same calculations as in Problem 9.3(a) but for the tidal force at the point Q in Figure 9.4. [In this case write $\hat{\mathbf{d}}/d^2 = \mathbf{d}/d^3$ and use the binomial approximation in the form $(1 + \epsilon)^{-3/2} \approx 1 - 3\epsilon/2$.]

9.5 ⋆⋆ Review the derivation of the tidal potential energy (9.16) of a drop of water at the point Q in Figure 9.5 and then give in detail the derivation of (9.17) for the tidal PE at the point P.

9.6 ⋆⋆⋆ Let $h(\theta)$ denote the height of the ocean at any point T on the surface, where $h(\theta)$ is measured up from the level at the point Q of Figure 9.5 and θ is the polar angle TOR of T. Given that the surface of the ocean is an equipotential, show that $h(\theta) = h_o \cos^2 \theta$, where $h_o = 3\,M_m\,R_e^4/(2\,M_e\,d_o^3)$. Sketch and describe the shape of the ocean's surface, bearing in mind that $h_o \ll R_e$. [*Hint:* You will need to evaluate $U_{\text{tid}}(T)$ as given by (9.13), with d equal to the distance MT. To do this you need to find d by the law of cosines and then approximate d^{-1} using the binomial approximation, being very careful to keep *all* terms through order $(R_e/d_o)^2$. Neglect any effects of the sun.]

SECTION 9.4 Time Derivatives in a Rotating Frame

9.7 ⋆ **(a)** Explain the relation (9.30) between the derivatives of a vector $\mathbf{Q}$ in two frames $\mathcal{S}_0$ and $\mathcal{S}$ for the special case that $\mathbf{Q}$ is fixed in the frame $\mathcal{S}$. **(b)** Do the same for a vector $\mathbf{Q}$ that is fixed in the frame $\mathcal{S}_0$ and compare with your answer to part (a).

SECTION 9.5 Newton's Second Law in a Rotating Frame

9.8 ⋆ What are the directions of the centrifugal and Coriolis forces on a person moving **(a)** south near the North Pole, **(b)** east on the equator, and **(c)** south across the equator?

9.9 ⋆ A bullet of mass m is fired with muzzle speed v_o horizontally and due north from a position at colatitude θ. Find the direction and magnitude of the Coriolis force in terms of m, v_o, θ, and the earth's angular velocity Ω. How does the Coriolis force compare with the bullet's weight if $v_o = 1000$ m/s and $\theta = 40$ deg?

9.10 ⋆⋆ The derivation of the equation of motion (9.34) for a rotating frame made the assumption that the angular velocity Ω was constant. Show that if $\dot{\Omega} \neq 0$ then there is a third "fictitious force," sometimes called the *azimuthal force*, on the right side of (9.34) equal to $m\mathbf{r} \times \dot{\Omega}$.

9.11 ⋆⋆⋆ In this problem you will prove the equation of motion (9.34) for a rotating frame using the Lagrangian approach. As usual, the Lagrangian method is in many ways easier than the Newtonian (except that it calls for some slightly tricky vector gymnastics), but is perhaps less insightful. Let $\mathcal{S}$ be a noninertial frame rotating with constant angular velocity Ω relative to the inertial frame $\mathcal{S}_0$. Let both frames have the same origin, $O = O'$. **(a)** Find the Lagrangian $\mathcal{L} = T - U$ in terms of the coordinates $\mathbf{r}$ and $\dot{\mathbf{r}}$ of $\mathcal{S}$. [Remember that you must first evaluate T in the inertial frame. In this connection, recall that $\mathbf{v}_o = \mathbf{v} + \boldsymbol{\Omega} \times \mathbf{r}$.] **(b)** Show that the three Lagrange equations reproduce (9.34) precisely.

SECTION 9.6   The Centrifugal Force

9.12 ★ (a) Show that to design a static structure in a rotating frame (such as a space station) one can use the ordinary rules of statics except that one must include the extra "fictitious" centrifugal force. **(b)** I wish to place a puck on a rotating horizontal turntable (angular velocity Ω) and to have it remain at rest on the table, held by the force of static friction (coefficient μ). What is the maximum distance from the axis of rotation at which I can do this? (Argue from the point of view of an observer in the rotating frame.)

9.13 ★ Show that the angle α between a plumb line and the direction of the earth's center is well approximated by $\tan \alpha = (R_e \Omega^2 \sin 2\theta)/(2g)$, where g is the observed free-fall acceleration and we assume the earth is perfectly spherically symmetric. Estimate the maximum and minimum values of the magnitude of α.

9.14 ★★ I am spinning a bucket of water about its vertical axis with angular velocity Ω. Show that, once the water has settled in equilibrium (relative to the bucket), its surface will be a parabola. (Use cylindrical polar coordinates and remember that the surface is an equipotential under the combined effects of the gravitational and centrifugal forces.)

9.15 ★★ On a certain planet, which is perfectly spherically symmetric, the free-fall acceleration has magnitude $g = g_o$ at the North Pole and $g = \lambda g_o$ at the equator (with $0 \le \lambda \le 1$). Find $g(\theta)$, the free-fall acceleration at colatitude θ as a function of θ.

SECTION 9.7   The Coriolis Force

9.16 ★ The center of a long frictionless rod is pivoted at the origin and the rod is forced to rotate at a constant angular velocity Ω in a horizontal plane. Write down the equation of motion for a bead that is threaded on the rod, using the coordinates x and y of a frame that rotates with the rod (with x along the rod and y perpendicular to it). Solve for $x(t)$. What is the role of the centrifugal force? What of the Coriolis force?

9.17 ★ Consider the bead threaded on a circular hoop of Example 7.6 (page 260), working in a frame that rotates with the hoop. Find the equation of motion of the bead, and check that your result agrees with Equation (7.69). Using a free-body diagram, explain the result (7.71) for the equilibrium positions.

9.18 ★★ A particle of mass m is confined to move, without friction, in a vertical plane, with axes x horizontal and y vertically up. The plane is forced to rotate with constant angular velocity Ω about the y axis. Find the equations of motion for x and y, solve them, and describe the possible motions.

9.19 ★★ I am standing (wearing crampons) on a perfectly frictionless flat merry-go-round, which is rotating counterclockwise with angular velocity Ω about its vertical axis. **(a)** I am holding a puck at rest just above the floor (of the merry-go-round) and release it. Describe the puck's path as seen from above by an observer who is looking down from a nearby tower (fixed to the ground) and also as seen by me on the merry-go-round. In the latter case explain what I see in terms of the centrifugal and Coriolis forces. **(b)** Answer the same questions for a puck which is released from rest by a long-armed spectator who is standing on the ground leaning over the merry-go-round.

9.20 ★★ Consider a frictionless puck on a horizontal turntable that is rotating counterclockwise with angular velocity Ω. **(a)** Write down Newton's second law for the coordinates x and y of the puck as seen by me standing on the turntable. (Be sure to include the centrifugal and Coriolis forces, but ignore the earth's rotation.) **(b)** Solve the two equations by the trick of writing $\eta = x + iy$ and guessing a solution of the form $\eta = e^{-i\alpha t}$. [In this case — as in the case of critically damped SHM discussed in Section 5.4 — you get only one solution this way. The other has the same form (5.43) we found for the second solution in damped SHM.] Write down the general solution. **(c)** At time $t = 0$, I push the puck from position $\mathbf{r}_o = (x_o, 0)$ with velocity $\mathbf{v}_o = (v_{xo}, v_{yo})$ (all as measured by me on the turntable). Show that

$$\left.\begin{array}{l} x(t) = (x_o + v_{xo}t)\cos\Omega t + (v_{yo} + \Omega x_o)t\sin\Omega t \\ y(t) = -(x_o + v_{xo}t)\sin\Omega t + (v_{yo} + \Omega x_o)t\cos\Omega t \end{array}\right\}. \tag{9.72}$$

(d) Describe and sketch the behavior of the puck for large values of t. [*Hint:* When t is large the terms proportional to t dominate (except in the case that both their coefficients are zero). With t large, write (9.72) in the form $x(t) = t(B_1\cos\Omega t + B_2\sin\Omega t)$, with a similar expression for $y(t)$, and use the trick of (5.11) to combine the sine and cosine into a single cosine — or sine, in the case of $y(t)$. By now you can recognize that the path is the same kind of spiral, whatever the initial conditions (with the one exception mentioned).]

9.21 ★★ When a puck slides on a rotating turntable, as in Problems 9.20 and 9.24, it can come instantaneously to rest. Sketch the shape of the path when this happens and explain. If you did Problem 9.24, comment on the relevance of this result to part (d) of that problem.

9.22 ★★ If a negative charge $-q$ (an electron, for example) in an elliptical orbit around a fixed positive charge Q is subjected to a weak uniform magnetic field $\mathbf{B}$, the effect of $\mathbf{B}$ is to make the ellipse precess slowly — an effect known as **Larmor precession**. To prove this, write down the equation of motion of the negative charge in the field of Q and $\mathbf{B}$. Now rewrite it for a frame rotating with angular velocity Ω. [Remember that this changes both $d^2\mathbf{r}/dt^2$ and $d\mathbf{r}/dt$.] Show that by suitable choice of Ω you can arrange that the terms involving $\dot{\mathbf{r}}$ cancel out, but that you are left with one term involving $\mathbf{B} \times (\mathbf{B} \times \mathbf{r})$. If $\mathbf{B}$ is weak enough this term can certainly be neglected. Show that in this case the orbit in the rotating frame is an ellipse (or hyperbola). Describe the appearance of the ellipse as seen in the original nonrotating frame.

9.23 ★★ Here is an unusual way to solve the two-dimensional isotropic oscillator — the motion of a particle subject to a force $-k\mathbf{r}$. Show that by choosing a suitable rotating reference frame, you can arrange that the centrifugal force exactly cancels the force $-k\mathbf{r}$. Recalling the analogy between the Coriolis and magnetic forces, you should be able to write down the general solution for the motion as seen in the rotating frame. If you write your solution in the complex form of Section 2.7, then you can transform back to the nonrotating frame by multiplying by a suitable rotating complex number. Show that the general solution is an ellipse. [See Problem 8.11 for some guidance on this last part.]

9.24 ★★★ [Computer] Use a suitable plotting program (such as ParametricPlot in Mathematica) to plot the orbits (9.72) of the puck of Problem 9.20 on a rotating turntable with $x_o = \Omega = 1$ and the following initial velocities $\mathbf{v}_o$: **(a)** $(0, 1)$, **(b)** $(0, 0)$, **(c)** $(0, -1)$, **(d)** $(-0.5, -0.5)$, **(e)** $(-0.7, -0.7)$, **(f)** $(0, -0.1)$. Comment on any interesting features.

SECTION 9.8 Free Fall and the Coriolis Force

9.25 ★ A high-speed train is traveling at a constant 150 m/s (about 300 mph) on a straight, horizontal track across the South Pole. Find the angle between a plumb line suspended from the ceiling inside the train and another inside a hut on the ground. In what direction is the plumb line on the train deflected?

9.26 ★★ In Section 9.8, we used a method of successive approximations to find the orbit of an object that is dropped from rest, correct to first order in the earth's angular velocity Ω. Show in the same way that if an object is thrown with initial velocity $\mathbf{v}_0$ from a point O on the earth's surface at colatitude θ, then to first order in Ω its orbit is

$$\left.\begin{array}{l} x = v_{xo}t + \Omega(v_{yo}\cos\theta - v_{zo}\sin\theta)t^2 + \tfrac{1}{3}\Omega gt^3\sin\theta \\[4pt] y = v_{yo}t - \Omega(v_{xo}\cos\theta)t^2 \\[4pt] z = v_{zo}t - \tfrac{1}{2}gt^2 + \Omega(v_{xo}\sin\theta)t^2. \end{array}\right\} \tag{9.73}$$

[First solve the equations of motion (9.53) in zeroth order, that is, ignoring Ω entirely. Substitute your zeroth-order solution for $\dot{x}$, $\dot{y}$, and $\dot{z}$ into the right side of equations (9.53) and integrate to give the next approximation. Assume that v_0 is small enough that air resistance is negligible and that $\mathbf{g}$ is a constant throughout the flight.]

9.27 ★★ In Section 9.8, we discussed the path of an object that is dropped from a very tall stepladder above the equator. **(a)** Sketch this path as seen from a tower to the north of the drop and fixed to the earth. Explain why the object lands to the east of its point of release. **(b)** Sketch the same experiment as seen by an inertial observer floating in space to the north of the drop. Explain clearly (from this point of view) why the object lands to the east of its point of release. [*Hint:* The object's angular momentum about the earth's center is conserved. This means that the object's angular velocity $\dot\phi$ changes as it falls.]

9.28 ★★ Use the result (9.73) of Problem 9.26 to do the following: A naval gun shoots a shell at colatitude θ in a direction that is α above the horizontal and due east, with muzzle speed v_0. **(a)** Ignoring the earth's rotation (and air resistance), find how long (t) the shell would be in the air and how far away (R) it would land. If $v_0 = 500$ m/s and $\alpha = 20°$, what are t and R? **(b)** A naval gunner spots an enemy ship due east at the range R of part (a) and, forgetting about the Coriolis effect, aims his gun exactly as in part (a). Find by how far north or south, and in which direction, the shell will miss the target, in terms of Ω, v_0, α, θ, and g. (It will also miss in the east–west direction but this is perhaps less critical.) If the incident occurs at latitude 50° north ($\theta = 40°$), what is this distance? What if the latitude is 50° south? This problem is a serious issue in long-range gunnery: In a battle near the Falkland Islands in World War I, the British navy consistently missed German ships by many tens of yards because they apparently forgot that the Coriolis effect in the southern hemisphere is opposite to that in the north.

9.29 ★★ **(a)** A baseball is thrown vertically up with initial speed v_0 from a point on the ground at colatitude θ. Use the solution (9.73) of Problem 9.26 to show that the ball will return to the ground a distance $(4\Omega v_0{}^3\sin\theta)/(3g^2)$ to the west of its launch point. **(b)** Estimate the size of this effect on the equator if $v_0 = 40$ m/s. **(c)** Sketch the ball's orbit as seen from the north (by an observer fixed to the earth). Compare with the orbit of a ball dropped from a point above the equator, and explain why the Coriolis effect moves the dropped ball to the east, but the thrown ball to the west.

9.30 ★★★ The Coriolis force can produce a torque on a spinning object. To illustrate this, consider a horizontal hoop of mass m and radius r spinning with angular velocity ω about its vertical axis at colatitude θ. Show that the Coriolis force due to the earth's rotation produces a torque of magnitude $m\omega\Omega r^2\sin\theta$ directed to the west, where Ω is the earth's angular velocity. This torque is the basis of the gyrocompass.

9.31 ★★★ The **Compton generator** is a beautiful demonstration of the Coriolis force due to the earth's rotation, invented by the American physicist A. H. Compton (1892–1962, best known as author of the Compton effect) while he was still an undergraduate. A narrow glass tube in the shape of a torus or ring (radius R of the ring $\gg$ radius of the tube) is filled with water, plus some dust particles to let one see any motion of the water. The ring and water are initially stationary and horizontal, but the ring is then spun through 180° about its east–west diameter. Explain why this should cause the water to move around the tube. Show that the speed of the water just after the 180° turn should be $2\Omega R \cos\theta$, where Ω is the earth's angular velocity, and θ is the colatitude of the experiment. What would this speed be if $R \approx 1$ m and $\theta = 40°$? Compton measured this speed with a microscope and got agreement within 3%.

9.32 ★★★ Do all parts of Problem 9.28, but find the distance by which the shell misses its target in both the north–south and east–west directions. [*Hint:* In this case you must recognize that the time of flight is affected by the Coriolis effect.]

SECTION 9.9   The Foucault Pendulum

9.33 ★★ The general solution for the small-amplitude motion of a Foucault pendulum is given by (9.66). If at $t = 0$ the pendulum is at rest with $x = A$ and $y = 0$, find the two coefficients C_1 and C_2, and show that because $\Omega \ll \omega_o$ they are well approximated as $C_1 = C_2 = A/2$, giving the solution (9.67).

9.34 ★★★ At a point P on the earth's surface, an enormous perfectly flat and frictionless platform is built. The platform is exactly horizontal — that is, perpendicular to the local free-fall acceleration $\mathbf{g}_P$. Find the equation of motion for a puck sliding on the platform and show that it has the same form as (9.61) for the Foucault pendulum except that the pendulum's length L is replaced by the earth's radius R. What is the frequency of the puck's oscillations and what is that of its Foucault precession? [*Hints:* Write the puck's position vector, relative to the earth's center O as $\mathbf{R} + \mathbf{r}$, where $\mathbf{R}$ is the position of the point P and $\mathbf{r} = (x, y, 0)$ is the puck's position relative to P. The contribution to the centrifugal force involving $\mathbf{R}$ can be absorbed into $\mathbf{g}_P$ and the contribution involving $\mathbf{r}$ is negligible. The restoring force comes from the variation of $\mathbf{g}$ as the puck moves.] To check the validity of your approximations, compare the approximate size of the gravitational restoring force, the Coriolis force, and the neglected term $m(\mathbf{\Omega} \times \mathbf{r}) \times \mathbf{\Omega}$ in the centrifugal force.

CHAPTER 10

Rotational Motion of Rigid Bodies

A rigid body is a collection of N particles with the property that its shape cannot change — the distance between any two of its constituent particles is fixed. A perfectly rigid body is, of course, an idealization, but an extremely useful one, and one to which many real systems are good approximations. In many ways a rigid body made up of N particles is much simpler than an arbitrary system of N particles: The arbitrary system requires $3N$ coordinates to specify its configuration, three coordinates for each of N particles. The rigid body requires only six coordinates, three to specify the position of the center of mass and three to specify the body's orientation. Further, we shall see that the motion of a rigid body can be divided into two separate simpler problems, the translational motion of the center of mass and the rotation of the body around the CM.

I shall start the chapter with some general results, mostly related to the CM of the body. These results generalize the results we found at the beginning of Chapter 8 for two particles, and most of them apply to *any* system of N particles. However, I shall quickly specialize to the motion of a rigid body. Much the most interesting aspect of the latter is the rotational motion, and this is what will occupy us for most of the chapter.

10.1   Properties of the Center of Mass

Consider a system of N particles $\alpha = 1, \cdots, N$ with masses m_α and positions $\mathbf{r}_\alpha$ measured relative to a chosen origin O. The center of mass of the system was defined in Chapter 3, Equation (3.9), to be the position (relative to the same origin O)

$$\mathbf{R} = \frac{1}{M} \sum_{\alpha=1}^{N} m_\alpha \mathbf{r}_\alpha \qquad \text{or} \qquad \frac{1}{M} \int \mathbf{r}\, dm \qquad (10.1)$$

where M denotes the total mass of all of the particles and the integral form is used when the system can be considered to be a continuous distribution of mass.

The Total Momentum and the CM

Several important parameters of the system's motion can be neatly expressed in terms of the CM. As we saw in Chapter 3, Equation (3.11), the total momentum is

$$\mathbf{P} = \sum_\alpha \mathbf{p}_\alpha = \sum_\alpha m_\alpha \dot{\mathbf{r}}_\alpha = M\dot{\mathbf{R}}. \tag{10.2}$$

That is, the total momentum of the system is exactly the same as that of a single particle of mass equal to the total mass M and velocity equal to that of the CM. If we differentiate this result we see that $\dot{\mathbf{P}} = M\ddot{\mathbf{R}}$ or, since $\dot{\mathbf{P}}$ equals the net external force $\mathbf{F}^{\text{ext}}$ on the system [as we saw in Equation (1.29)],

$$\mathbf{F}^{\text{ext}} = M\ddot{\mathbf{R}}. \tag{10.3}$$

That is, the CM moves just as if it were a single particle of mass M subject to the net external force on the system. This result is the single most important justification for our treating extended objects like baseballs and comets as point particles. To the extent that these nonpoint objects can be represented by their CM, they do move just like point particles.

The Total Angular Momentum

The role of the CM motion in a system's total angular momentum is more complicated, but equally crucial. The following argument does not depend on the system being a rigid body, but to be definite let us consider a rigid body made up of N pieces with masses m_α, as sketched in Figure 10.1, where the body is shown as an ellipsoid. The position of m_α relative to an arbitrary origin O is shown as $\mathbf{r}_\alpha$ and that of the CM relative to O as $\mathbf{R}$. Also shown is the position $\mathbf{r}'_\alpha$ of m_α *relative to the CM*, which satisfies

$$\mathbf{r}_\alpha = \mathbf{R} + \mathbf{r}'_\alpha. \tag{10.4}$$

The angular momentum $\boldsymbol{\ell}_\alpha$ of m_α about the origin O is

$$\boldsymbol{\ell}_\alpha = \mathbf{r}_\alpha \times \mathbf{p}_\alpha = \mathbf{r}_\alpha \times m_\alpha \dot{\mathbf{r}}_\alpha. \tag{10.5}$$

Thus the total angular momentum relative to O is

$$\mathbf{L} = \sum_\alpha \boldsymbol{\ell}_\alpha = \sum_\alpha \mathbf{r}_\alpha \times m_\alpha \dot{\mathbf{r}}_\alpha.$$

If we use (10.4) to rewrite both $\mathbf{r}_\alpha$ and $\dot{\mathbf{r}}_\alpha$, we find that $\mathbf{L}$ is the sum of four terms:

$$\mathbf{L} = \sum \mathbf{R} \times m_\alpha \dot{\mathbf{R}} + \sum \mathbf{R} \times m_\alpha \dot{\mathbf{r}}'_\alpha + \sum \mathbf{r}'_\alpha \times m_\alpha \dot{\mathbf{R}} + \sum \mathbf{r}'_\alpha \times m_\alpha \dot{\mathbf{r}}'_\alpha.$$

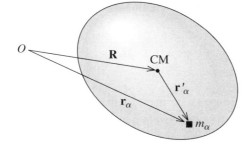

Figure 10.1   A rigid body (shown here as an ellipsoid) is made up of many small pieces, $\alpha = 1, \cdots, N$. The mass of a typical piece is m_α and its position relative to the origin O is $\mathbf{r}_\alpha$. The position of the CM relative to O is $\mathbf{R}$, and $\mathbf{r}'_\alpha$ denotes the position of m_α relative to the CM, so that $\mathbf{r}_\alpha = \mathbf{R} + \mathbf{r}'_\alpha$.

If we factor out of each of these four sums the terms that do not depend on α, we find (remember $\sum m_\alpha = M$)

$$\mathbf{L} = \mathbf{R} \times M\dot{\mathbf{R}} + \mathbf{R} \times \sum m_\alpha \dot{\mathbf{r}}'_\alpha + \left(\sum m_\alpha \mathbf{r}'_\alpha\right) \times \dot{\mathbf{R}} + \sum \mathbf{r}'_\alpha \times m_\alpha \dot{\mathbf{r}}'_\alpha. \quad (10.6)$$

This expression can now be simplified dramatically. Notice first that the sum in parentheses in the third term on the right is the position of the CM *relative to the CM* (times M). This is, of course, zero (Problem 10.1):

$$\sum m_\alpha \mathbf{r}'_\alpha = 0. \quad (10.7)$$

Therefore the third term in (10.6) is zero. Differentiating this relation, we see that the sum in the second term in (10.6) is likewise zero. Thus all that remains of (10.6) is

$$\mathbf{L} = \mathbf{R} \times \mathbf{P} + \sum \mathbf{r}'_\alpha \times m_\alpha \dot{\mathbf{r}}'_\alpha. \quad (10.8)$$

The first term is the angular momentum (relative to O) of the motion of the CM. The second is the angular momentum of the motion relative to the CM. Thus we can re-express (10.8) to say

$$\mathbf{L} = \mathbf{L}(\text{motion of CM}) + \mathbf{L}(\text{motion relative to CM}). \quad (10.9)$$

To illustrate this useful result consider the motion of a planet around the sun (which we can safely treat as fixed because it is so massive). In this case, (10.9) asserts that the total angular momentum of the planet is the angular momentum of the orbital motion of the CM around the sun, plus the angular momentum of its spinning motion around its CM,

$$\mathbf{L} = \mathbf{L}_{\text{orb}} + \mathbf{L}_{\text{spin}}. \tag{10.10}$$

This division of the total angular momentum into its orbital and spin parts is especially useful because it is often true (at least to a good approximation) that the two parts are separately conserved. To see this, note first that since $\mathbf{L}_{\text{orb}} = \mathbf{R} \times \mathbf{P}$,

$$\dot{\mathbf{L}}_{\text{orb}} = \dot{\mathbf{R}} \times \mathbf{P} + \mathbf{R} \times \dot{\mathbf{P}} = \mathbf{R} \times \mathbf{F}^{\text{ext}} \tag{10.11}$$

(since the first cross product is zero and $\dot{\mathbf{P}} = \mathbf{F}^{\text{ext}}$). That is, $\mathbf{L}_{\text{orb}}$ evolves just as if the planet were a point particle, with all its mass concentrated at its CM. In particular, if the force of the sun on the planet were perfectly central ($\mathbf{F}^{\text{ext}}$ exactly collinear with $\mathbf{R}$), then $\mathbf{L}_{\text{orb}}$ would be constant. In practice the force is not exactly central (since planets are not perfectly spherical and the sun's gravitational field is not perfectly uniform), but it is true to an excellent approximation.

To find $\dot{\mathbf{L}}_{\text{spin}}$, we can write $\mathbf{L}_{\text{spin}} = \mathbf{L} - \mathbf{L}_{\text{orb}}$. We already know that $\dot{\mathbf{L}} = \boldsymbol{\Gamma}^{\text{ext}}$, so

$$\dot{\mathbf{L}} = \sum \mathbf{r}_\alpha \times \mathbf{F}_\alpha^{\text{ext}} = \sum (\mathbf{r}'_\alpha + \mathbf{R}) \times \mathbf{F}_\alpha^{\text{ext}} = \sum \mathbf{r}'_\alpha \times \mathbf{F}_\alpha^{\text{ext}} + \mathbf{R} \times \mathbf{F}^{\text{ext}}. \tag{10.12}$$

Subtracting (10.11) from (10.12) gives $\dot{\mathbf{L}}_{\text{spin}}$,

$$\dot{\mathbf{L}}_{\text{spin}} = \dot{\mathbf{L}} - \dot{\mathbf{L}}_{\text{orb}} = \sum \mathbf{r}'_\alpha \times \mathbf{F}_\alpha^{\text{ext}} = \boldsymbol{\Gamma}^{\text{ext}} (\text{about CM});$$

that is, the rate of change of $\mathbf{L}_{\text{spin}}$, the angular momentum about the CM, is just the net external torque, *measured relative to the CM*. [This natural-seeming result was mentioned without proof in Equation (3.28). What makes it a little surprising is that a reference frame attached to the CM is generally *not* an inertial frame. Surprising or not, the result is true and very useful.] Since the torque of the sun about the CM of any planet is very small, $\mathbf{L}_{\text{spin}}$ is very nearly constant. Nevertheless, this useful conclusion, although an excellent approximation, is not exact. For instance, because of our own earth's equatorial bulge, there is a small torque on the earth due to the sun (and moon), and $\mathbf{L}_{\text{spin}}$ is not quite constant. The slow changing of $\mathbf{L}_{\text{spin}}$ is responsible for the effect known as the *precession of the equinoxes*, the rotation of the earth's axis relative to the stars by some 50 arcseconds per year.

There is a corresponding (though not exactly analogous) division of angular momentum into its orbital and spin parts in quantum mechanics. For example, the angular momentum of the electron orbiting around the proton in a hydrogen atom is made up of two terms as in (10.10), and for much the same reasons each separate kind of angular momentum is almost perfectly conserved. Here too, this useful result is only approximately true: In this case there is a weak magnetic torque on the electron and neither the spin nor the orbital angular momentum is exactly conserved (although the total angular momentum is).

Kinetic Energy

The total kinetic energy of N particles is

$$T = \sum_{\alpha=1}^{N} \tfrac{1}{2} m_\alpha \dot{\mathbf{r}}_\alpha^{\,2}. \tag{10.13}$$

As before, we can use (10.4) to replace $\mathbf{r}_\alpha$ by $\mathbf{R} + \mathbf{r}'_\alpha$, which gives

$$\dot{\mathbf{r}}_\alpha^{\,2} = (\dot{\mathbf{R}} + \dot{\mathbf{r}}'_\alpha)^2 = \dot{\mathbf{R}}^2 + \dot{\mathbf{r}}'^{\,2}_\alpha + 2\dot{\mathbf{R}} \cdot \dot{\mathbf{r}}'_\alpha$$

and hence

$$T = \tfrac{1}{2} \sum m_\alpha \dot{\mathbf{R}}^2 + \tfrac{1}{2} \sum m_\alpha \dot{\mathbf{r}}'^{\,2}_\alpha + \dot{\mathbf{R}} \cdot \sum m_\alpha \dot{\mathbf{r}}'_\alpha . \tag{10.14}$$

The sum in the last term on the right is zero by (10.7) and we find that

$$T = \tfrac{1}{2} M \dot{\mathbf{R}}^2 + \tfrac{1}{2} \sum m_\alpha \dot{\mathbf{r}}'^{\,2}_\alpha \tag{10.15}$$

or

$$T = T(\text{motion of CM}) + T(\text{motion relative to CM}). \tag{10.16}$$

For a rigid body, the only possible motion relative to the CM is rotation. Thus we can rephrase this result to say that

$$T = T(\text{motion of CM}) + T(\text{rotation about CM}). \tag{10.17}$$

This useful result says, for example, that the kinetic energy of a wheel rolling down the road is the translational energy of the CM plus the energy of the rotation about the axle.

From (10.14) we can derive an alternative and sometimes useful expression for the total kinetic energy. The derivation of (10.14) did not depend on $\mathbf{R}$ being the CM position, and (10.14) is actually valid for any point $\mathbf{R}$ fixed in the body. In particular, suppose we choose $\mathbf{R}$ to be a point of the body that happens to be at rest (even just instantaneously at rest). In this case the first and third terms on the right of (10.14) are both zero, and we find that

$$T = \tfrac{1}{2} \sum m_\alpha \dot{\mathbf{r}}'^{\,2}_\alpha . \tag{10.18}$$

This says that the total kinetic energy of a rigid body is just the rotational energy of the body relative to any point of the body that is instantaneously at rest. For example the kinetic energy of a rolling wheel can be evaluated as the energy of rotation about the point of contact with the road, since this point is instantaneously at rest.

Potential Energy of a Rigid Body

If all the forces on and within an N-particle rigid body are conservative, then, as we saw in Section 4.10, the total potential energy can be written as

$$U = U^{\text{ext}} + U^{\text{int}} \tag{10.19}$$

where U^{ext} is the sum of all potential energies due to any external forces. (For example, if the body of interest is a baseball, U^{ext} could be the gravitational energy of all the particles that comprise the ball in the field of the "external" earth.) The internal potential energy U^{int} is the sum of the potential energies for all pairs of particles,

$$U^{\text{int}} = \sum_{\alpha < \beta} U_{\alpha\beta}(r_{\alpha\beta}) \tag{10.20}$$

where $r_{\alpha\beta}$ is the distance[1] between particles α and β. However, in a rigid body all of the interparticle distances $r_{\alpha\beta}$ are fixed. Therefore, the internal potential energy is a constant and may as well be ignored. In other words, in discussing the motion of a rigid body we have to consider only the *external* forces and their corresponding potential energies.

10.2   Rotation about a Fixed Axis

The results of the previous section show the importance of rotational motion. For example, the kinetic energy of any extended body flying through the air (you might think of a drum major's twirling baton) is the sum of two terms: the translational energy of the CM and the rotational energy of its spinning about the CM. The former we understand rather completely, but the latter we need to study. In most of the remainder of this chapter, we shall be focussing on rotational motion.

In this section, we'll start with the special case of a body that is rotating about a fixed axis, such as the piece of wood shown in Figure 10.2 spinning on a fixed rod, and first calculate its angular momentum.

Because the axis of rotation is fixed, we can agree to call it the z axis, with the origin O somewhere on the axis of rotation. As usual we imagine the body divided into many small pieces with masses m_α ($\alpha = 1, \cdots, N$) and the angular momentum is given by the usual formula

$$\mathbf{L} = \sum \boldsymbol{\ell}_\alpha = \sum \mathbf{r}_\alpha \times m_\alpha \mathbf{v}_\alpha \tag{10.21}$$

---

[1] Throughout this chapter I shall assume that all the internal forces are central. This guarantees that $U_{\alpha\beta}$ depends only on the magnitude of $\mathbf{r}_{\alpha\beta}$, not on its direction. It also ensures that the internal forces never contribute to changes in the total angular momentum.

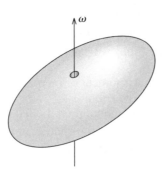

Figure 10.2   An egg-shaped block of wood with a hole drilled through it is threaded on a rod fixed on the z axis. The block is spinning with angular velocity $\boldsymbol{\omega}$.

where the velocities $\mathbf{v}_\alpha$ are the velocities with which the pieces of the body are being carried in circles by the body's rotational velocity $\boldsymbol{\omega}$. We saw in (9.22) that these are just $\mathbf{v}_\alpha = \boldsymbol{\omega} \times \mathbf{r}_\alpha$. With our z axis along $\boldsymbol{\omega}$, the components of $\boldsymbol{\omega}$ are

$$\boldsymbol{\omega} = (0, 0, \omega)$$

and

$$\mathbf{r}_\alpha = (x_\alpha, y_\alpha, z_\alpha).$$

Thus $\mathbf{v}_\alpha = \boldsymbol{\omega} \times \mathbf{r}_\alpha$ has components

$$\mathbf{v}_\alpha = \boldsymbol{\omega} \times \mathbf{r}_\alpha = (-\omega y_\alpha, \omega x_\alpha, 0)$$

and finally

$$\boldsymbol{\ell}_\alpha = m_\alpha \mathbf{r}_\alpha \times \mathbf{v}_\alpha = m_\alpha \omega (-z_\alpha x_\alpha, -z_\alpha y_\alpha, x_\alpha^2 + y_\alpha^2). \tag{10.22}$$

At last we are ready to calculate the total angular momentum of our spinning solid, and I shall start with the z component. If we put the z component of (10.22) into (10.21), we find that

$$L_z = \sum m_\alpha (x_\alpha^2 + y_\alpha^2) \omega. \tag{10.23}$$

Now, the quantity $(x_\alpha^2 + y_\alpha^2)$ is the same as ρ_α^2, where, as usual, $\rho = \sqrt{x^2 + y^2}$ denotes the distance of any point (x, y, z) from the z axis. Therefore,

$$L_z = \sum m_\alpha \rho_\alpha^2 \omega = I_z \omega, \tag{10.24}$$

where

$$I_z = \sum m_\alpha \rho_\alpha^2 \tag{10.25}$$

is the familiar **moment of inertia about the** z **axis**, as defined in every introductory physics course — the sum of all the constituent masses, each multiplied by the square of its distance from the z axis.[2] Thus we have proved the familiar result that

(angular momentum) = (moment of inertia) × (angular velocity).

Note, however, that the angular momentum on the left of (10.24) is actually L_z, and the moment of inertia is, of course, that for rotation about the z axis.

To reinforce this gratifying result, let us calculate the kinetic energy of our rotating body. This is

$$T = \tfrac{1}{2} \sum m_\alpha v_\alpha^2,$$

or, since the speed of m_α as it is carried in a circle around the z axis with angular velocity ω is $v_\alpha = \rho_\alpha \omega$,

$$T = \tfrac{1}{2} \sum m_\alpha \rho_\alpha^2 \omega^2 = \tfrac{1}{2} I_z \omega^2, \tag{10.26}$$

another familiar result from introductory physics.

So far we have met no surprises, but when we calculate the x and y components of **L** we find something unexpected: Substituting the x and y components of (10.22) into (10.21), we find for the x and y components of **L**:

$$L_x = -\sum m_\alpha x_\alpha z_\alpha \omega \qquad \text{and} \qquad L_y = -\sum m_\alpha y_\alpha z_\alpha \omega. \tag{10.27}$$

As we shall see in a moment the sums here are in general *not* zero, and we have the following surprising conclusion: The angular velocity $\boldsymbol{\omega}$ points in the z direction (the body rotates about the z axis), but, since L_x and L_y can be nonzero, the angular momentum **L** may be in a *different* direction. That is, the angular momentum may not be in the same direction as the angular velocity, and the relation $\mathbf{L} = I\boldsymbol{\omega}$ that you may have learned in introductory physics is generally not true!

To better understand this rather unexpected conclusion, consider a rigid body that consists of a single mass m on the end of a massless rod, pivoted about the z axis at a fixed angle α, as in Figure 10.3. As this body rotates about the z axis, it is easy to see that the mass m has velocity **v** into the page (negative x direction) and hence that $\mathbf{L} = \mathbf{r} \times m\mathbf{v}$ is in the direction shown, at an angle $(90° - \alpha)$ with the z axis. Clearly L_y is not equal to zero, and, even though the body is rotating about the z axis, the angular momentum is not in that direction. In other words, **L** is not parallel to $\boldsymbol{\omega}$.

This example is worth pursuing a little further. It is clear from the picture that as the body rotates steadily about the z axis, the direction of **L** changes. (Specifically, **L** itself sweeps around the z axis.) Therefore, $\dot{\mathbf{L}} \neq 0$, and a torque is required simply to keep the body rotating steadily. This conclusion, at first rather surprising, is actually easy to understand: The required torque is in the direction of $\dot{\mathbf{L}}$, which is out of the page (positive x direction) in Figure 10.3; that is, the torque must be counterclockwise. The easiest way to understand this is to put yourself in a frame that is rotating with

---

[2] You may recall that, when actually calculating moments of inertia, we often replace the sum in (10.25) by an integral. For now, however, I shall continue to write moments of inertia as sums.

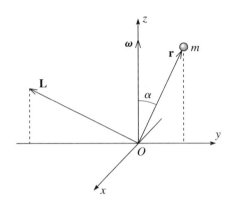

Figure 10.3   A rigid rotating body comprising a single mass m anchored to the z axis by a massless rod at a fixed angle α, shown at a moment when m happens to lie in the yz plane. As the body rotates about the z axis, m has velocity, and hence momentum, into the page (in the negative x direction) at the moment shown. Therefore the angular momentum $\mathbf{L} = \mathbf{r} \times \mathbf{p}$ is directed as shown and is certainly not parallel to the angular velocity $\boldsymbol{\omega}$.

the body. In this frame, the mass m experiences a centrifugal force out from the z axis (to the right in the picture). Therefore a counterclockwise torque is required to prevent the rod that holds m from bending or breaking away from its anchor on the axis.

When a body, such as the wheel of your car coasting along the highway, is rotating steadily about a fixed direction, we usually do not want to have to exert any torque on it. This means the body must be designed so that its angular momentum is parallel to its angular velocity. In the case of your car, this is guaranteed by the process of dynamical balancing of the wheels. If the wheels are not properly balanced, you are made quickly aware of it by a disagreeable vibration of the car. More generally, the question whether or not $\mathbf{L}$ and $\boldsymbol{\omega}$ are parallel is an important issue throughout the study of rotating bodies and leads us to the important concept of *principal axes*, as we shall discuss in Section 10.4.

The Products of Inertia

We need to collect together our results for the angular momentum of a body rotating about the z axis and to streamline our notation. It is clear from (10.27) that L_x and L_y are proportional to ω, and the constants of proportionality are generally denoted by I_{xz} and I_{yz}. Thus I shall rewrite (10.27) as

$$L_x = I_{xz}\omega \quad \text{and} \quad L_y = I_{yz}\omega \tag{10.28}$$

where

$$I_{xz} = -\sum m_\alpha x_\alpha z_\alpha \quad \text{and} \quad I_{yz} = -\sum m_\alpha y_\alpha z_\alpha. \qquad (10.29)$$

The two coefficients I_{xz} and I_{yz} are called the **products of inertia** of the body. The rationale for this new notation is that I_{xz} tells us the x component of $\mathbf{L}$ when $\boldsymbol{\omega}$ is in the z direction (and likewise for I_{yz}). To conform with this notation, we have to rename I_z, the old-fashioned moment of inertia about the z axis and call it I_{zz},

$$I_{zz} = \sum m_\alpha \rho_\alpha^2 = \sum m_\alpha (x_\alpha^2 + y_\alpha^2). \qquad (10.30)$$

With this notation, we can say that, for a body rotating about the z axis, the angular momentum is

$$\mathbf{L} = (I_{xz}\omega, \ I_{yz}\omega, \ I_{zz}\omega). \qquad (10.31)$$

Obviously, it is important to be able to calculate the coefficients I_{xz}, I_{yz}, and I_{zz}, for bodies of different shapes. I shall work out some examples in the remainder of this chapter and there are plenty more in the problems. Here are three simple examples for a start.

EXAMPLE 10.1   Calculating Some Simple Moments and Products of Inertia

Calculate the moment and products of inertia for rotation about the z axis of the following rigid bodies: **(a)** A single mass m located at the position $(0, y_o, z_o)$ as shown in Figure 10.3. **(b)** The same as in part (a) but with a second equal mass placed symmetrically below the xy plane, as in Figure 10.4(a). **(c)** A uniform ring of mass M and radius ρ_o centered on the z axis and parallel to the xy plane, as in Figure 10.4(b).

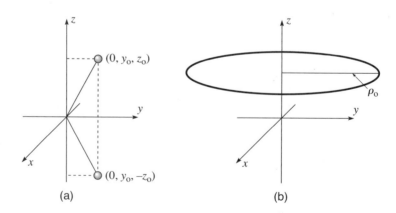

(a)   (b)

Figure 10.4   **(a)** A rigid body comprising two equal masses m held at equal distances above and below the xy plane and rotating about the z axis (shown at an instant when the two masses lie in the yz plane). **(b)** A uniform continuous ring of total mass M and radius ρ_o, centered on the z axis and parallel to the xy plane. (Example 10.1.)

(a) For the single mass of Figure 10.3, the sums in (10.29) and (10.30) each reduce to a single term and we find

$$I_{xz} = 0, \qquad I_{yz} = -my_o z_o, \qquad \text{and} \qquad I_{zz} = my_o^2.$$

That I_{yz} is nonzero confirms that $\mathbf{L}$ has a nonzero y component and hence that the angular momentum $\mathbf{L}$ is not in the same direction as the axis of rotation. The remaining parts (b) and (c) illustrate how when a rigid body has certain symmetries, the products of inertia can turn out to be zero.

(b) For the two masses of Figure 10.4(a) each of the sums in (10.29) and (10.30) contains two terms and we find

$$I_{xz} = - \sum m_\alpha x_\alpha z_\alpha = 0 \tag{10.32}$$

because both masses have $x_\alpha = 0$,

$$I_{yz} = - \sum m_\alpha y_\alpha z_\alpha = -m[y_o z_o + y_o(-z_o)] = 0, \tag{10.33}$$

and

$$I_{zz} = \sum m_\alpha (x_\alpha^2 + y_\alpha^2) = m(0 + y_o^2 + 0 + y_o^2) = 2my_o^2.$$

The interesting case here is the product of inertia I_{yz}, which is zero because the contribution of the first mass is exactly cancelled by that of the second mass, at the "mirror image" point with the opposite sign of z_α. Clearly this will happen for any body that has reflection symmetry in the plane $z = 0$;[3] both of the products of inertia I_{xz} and I_{yz} will be zero because every term in each of the sums of (10.32) and (10.33) will be cancelled by another term with the opposite sign of z_α. With $I_{xz} = I_{yz} = 0$, we see from (10.31) that when the body rotates about the z axis, the angular momentum $\mathbf{L}$ is also along the z axis.

(c) Because the body in Figure 10.4(b) is a continuously distributed mass, we should in general evaluate the products and moment of inertia as integrals, but in this case we can see the answers without actually doing any integrals. Consider, first, the product of inertia I_{xz} as given by the sum in (10.32). Referring to Figure 10.4(b), it is easy to see that this sum is zero: Each contribution to the sum from a small mass m_α at $(x_\alpha, y_\alpha, z_\alpha)$ can be paired with the contribution from an equal mass diametrically across the circle at $(-x_\alpha, -y_\alpha, z_\alpha)$. This second contribution, with the same value of z but the opposite value for x, exactly cancels the first, and we conclude that the whole sum I_{xz} is zero. By exactly the same argument, $I_{yz} = 0$. The moment I_{zz} is most easily evaluated in the form of the first sum in (10.30). Since all terms in the sum have the same value of ρ_α (namely ρ_o), we can factor ρ_α^2 out of the sum and are left with

$$I_{zz} = \left(\sum m_\alpha \right) \rho_o^2 = M\rho_o^2.$$

---

[3] We say that a body has reflection symmetry in the plane $z = 0$ if the mass density is the same at any point (x, y, z) as at the point $(x, y, -z)$, the reflection of (x, y, z) in a mirror located in the plane $z = 0$.

In this example, the two products of inertia were zero because the body was axially symmetric about its axis of rotation,[4] and it is easy to see that the same result holds quite generally: If a rigid body is axially symmetric about a certain axis (like a well-balanced wheel about its axle or a circular cone about its center line), then the two products of inertia for rotation about that axis are automatically zero, since the terms of the sums in (10.32) and (10.33) cancel in pairs. In particular, if a body is axially symmetric about a certain axis and is rotating about this symmetry axis, its angular momentum will be in the same direction.

10.3   Rotation about Any Axis; the Inertia Tensor

So far we have considered only a body that is rotating about the z axis. In a certain sense this is quite general: Whatever axis a body is rotating about, we can choose to call it the z axis. Unfortunately, although this statement is true, it does not tell the whole story. First, we are often interested in bodies that are free to rotate about any axis — a gyroscope's bearings allow it to rotate about any axis, and a projectile (such as a baseball or a drum major's baton thrown in the air) has the same freedom. When this is the case, the axis about which the body rotates may *change with time*. If this happens, then we can certainly choose as our z axis the axis of rotation at one instant, but a moment later our chosen z axis is almost certainly *not* the axis of rotation. For this reason alone, we must clearly examine the form of the angular momentum when a body is spinning about an arbitrary axis.

The second reason that we need to consider rotation about an arbitrary axis is subtler, and I shall return to it later, but let me mention it briefly here. We have seen that in general the direction of the angular momentum of a spinning body is not the same as the axis of rotation. On the other hand, it sometimes happens that these two directions *are* the same. (For instance, we have seen that this is the case for an axially symmetric body rotating about its axis of symmetry.) When this is true, we say that the axis in question is a *principal axis*. We shall find for any given body, rotating about any given point, that there are three mutually perpendicular principal axes. Because much of our discussion of rotations is much easier when referred to these principal axes, we often wish to choose the principal axes as our coordinate axes. If we do this, then we are no longer at liberty to choose our z axis to coincide with an arbitrary axis of rotation. Again, we must learn to allow *any* axis to be the axis of rotation, and our first order of business is to calculate the angular momentum corresponding to such rotation.

---

[4] We say that a body is axially, or rotationally, symmetric about an axis if the mass density is the same at all points on any circle centered on, and perpendicular to, the axis. In terms of cylindrical polar coordinates (ρ, ϕ, z) centered on the axis in question, the mass density is independent of ϕ. Alternatively, the mass distribution is unchanged by any rotation about the axis of symmetry.

Angular Momentum for an Arbitrary Angular Velocity

Let us then consider a rigid body rotating about an arbitrary axis with angular velocity

$$\boldsymbol{\omega} = (\omega_x, \omega_y, \omega_z).$$

Before we launch into the calculation of the angular momentum, let us pause to consider the kind of situation to which our calculations will apply. There are in fact two important cases that you could keep in mind: First, it sometimes happens that a rigid body has one fixed point, so that its only possible motion is a rotation about that fixed point. For example, the Foucault pendulum is fixed at its point of support in the ceiling, and its only possible motion is rotation about that fixed point. Again, a top spinning on a table can get its tip caught in a small dent in the table, and, from then on, it can only rotate about the fixed position of its tip. In either case, the magnitude and direction of the angular velocity $\boldsymbol{\omega}$ can change, but the rotation will always be about the fixed point, which we shall naturally take to be our origin.

The second case that you could bear in mind is that of an object that has been thrown in the air. In this case there is certainly not any fixed point, but we have seen that we can analyze the motion in terms of the motion of the CM and the rotational motion relative to the CM. In this case, the motion that we are now analyzing is the motion relative to the CM, which we shall naturally take to be the origin.

With these examples in mind, let us calculate the body's angular momentum,

$$\mathbf{L} = \sum m_\alpha \mathbf{r}_\alpha \times \mathbf{v}_\alpha$$

$$= \sum m_\alpha \mathbf{r}_\alpha \times (\boldsymbol{\omega} \times \mathbf{r}_\alpha). \tag{10.34}$$

We can evaluate this almost exactly as we did in the previous section for the case that $\boldsymbol{\omega}$ was along the z axis: For any position $\mathbf{r}$ we can write down the components of $\boldsymbol{\omega} \times \mathbf{r}$ and then $\mathbf{r} \times (\boldsymbol{\omega} \times \mathbf{r})$, with the rather ugly result (there are several ways to do this, one of which is the so-called $BAC - CAB$ rule — see Problem 10.19)

$$\mathbf{r} \times (\boldsymbol{\omega} \times \mathbf{r}) = \big((y^2 + z^2)\omega_x - xy\omega_y - xz\omega_z,$$

$$- yx\omega_x + (z^2 + x^2)\omega_y - yz\omega_z,$$

$$- zx\omega_x - zy\omega_y + (x^2 + y^2)\omega_z\big). \tag{10.35}$$

Substituting (10.35) into (10.34), we can write down the three components of $\mathbf{L}$ as follows:

$$\left.\begin{array}{ccc}
L_x & = & I_{xx}\omega_x + I_{xy}\omega_y + I_{xz}\omega_z \\
L_y & = & I_{yx}\omega_x + I_{yy}\omega_y + I_{yz}\omega_z \\
L_z & = & I_{zx}\omega_x + I_{zy}\omega_y + I_{zz}\omega_z
\end{array}\right\} \tag{10.36}$$

where the three moments of inertia, I_{xx}, I_{yy}, I_{zz}, and the six products of inertia, $I_{xy}, \cdots$, are defined in exact parallel with the definitions (10.29) and (10.30) of the previous section. For example,

$$I_{xx} = \sum m_\alpha (y_\alpha^2 + z_\alpha^2) \qquad (10.37)$$

and similarly for I_{yy} and I_{zz}, and

$$I_{xy} = -\sum m_\alpha x_\alpha y_\alpha \qquad (10.38)$$

and so on.

The rather clumsy result (10.36) can be streamlined in a couple of ways: If you don't mind replacing the subscripts x, y, and z with $i = 1, 2, 3$, then (10.36) takes the compact form

$$L_i = \sum_{j=1}^{3} I_{ij}\omega_j . \qquad (10.39)$$

This suggests another way to think about (10.36) since, as you may recognize, (10.39) is the rule for matrix multiplication. Thus, (10.36) can be rewritten in matrix form. First, we introduce the 3×3 matrix

$$\mathbf{I} = \begin{bmatrix} I_{xx} & I_{xy} & I_{xz} \\ I_{yx} & I_{yy} & I_{yz} \\ I_{zx} & I_{zy} & I_{zz} \end{bmatrix} \qquad (10.40)$$

which is called the **moment of inertia tensor**[5] or just **inertia tensor**. In addition, let us agree temporarily to think of three-dimensional vectors as 3×1 columns made up of their three components; that is, we write

$$\mathbf{L} = \begin{bmatrix} L_x \\ L_y \\ L_z \end{bmatrix} \qquad \text{and} \qquad \boldsymbol{\omega} = \begin{bmatrix} \omega_x \\ \omega_y \\ \omega_z \end{bmatrix} . \qquad (10.41)$$

(Notice that I am now using boldface for two kinds of matrices — square 3×3 matrices like $\mathbf{I}$, and 3×1 column matrices like $\mathbf{L}$ and $\boldsymbol{\omega}$ that are really just vectors. You will quickly learn to distinguish the two kinds of matrix from the context.) With these notations, Equation (10.36) takes the very compact matrix form

$$\mathbf{L} = \mathbf{I}\boldsymbol{\omega} \qquad (10.42)$$

_____

[5] The full definition of a tensor involves the transformation properties of its elements when we rotate our coordinate axes (see Section 15.17), but for our present purposes it is sufficient to say that a three-dimensional tensor is a set of nine numbers arranged as a 3×3 matrix as in (10.40).

where the product on the right is the standard product of two matrices, the first a 3×3 square matrix and the second a 3×1 column.

This beautiful result is our first example of the great usefulness of matrix algebra in mechanics. In many areas of physics — perhaps most especially in quantum mechanics, but also certainly in classical mechanics — the formulation of many problems in matrix notation is so much simpler than in any other way, that it is absolutely essential that you become familiar with the basics of matrix algebra.[6]

An important property of the moment of inertia tensor (10.40) is that it is a *symmetric* matrix; that is, its elements satisfy

$$I_{ij} = I_{ji}. \tag{10.43}$$

Another way to say this is that the matrix (10.40) is unchanged if we reflect it in the *main diagonal* — the diagonal running from top left to bottom right. Each element above the diagonal (for instance I_{xy}) is equal to its mirror image (I_{yx}) below the diagonal. To prove this property, you have only to look at the definition (10.38) to see that I_{xy} is the same as I_{yx} and similarly with all the off-diagonal elements I_{ij} (the elements with $i \neq j$). Yet another way to state this property is to define the *transpose* of any matrix $\mathbf{A}$ as the matrix $\tilde{\mathbf{A}}$ obtained by reflecting $\mathbf{A}$ in its main diagonal — the ij element of $\tilde{\mathbf{A}}$ is the ji element of $\mathbf{A}$. Thus the result (10.43) means that the matrix $\mathbf{I}$ is equal to its own transpose,

$$\mathbf{I} = \tilde{\mathbf{I}}. \tag{10.44}$$

This property — that the matrix $\mathbf{I}$ is symmetric — plays a key role in the mathematical theory of the moment of inertia tensor.

EXAMPLE 10.2 Inertia Tensor for a Solid Cube

Find the moment of inertia tensor for **(a)** a uniform solid cube, of mass M and side a, rotating about a corner (Figure 10.5) and **(b)** the same cube rotating about its center. Use axes parallel to the cube's edges. For both cases, find the angular momentum when the axis of rotation is parallel to $\hat{\mathbf{x}}$ [that is, $\boldsymbol{\omega} = (\omega, 0, 0)$] and also when $\boldsymbol{\omega}$ is along the body diagonal in the direction $(1, 1, 1)$.

(a) Because the mass is continuously distributed we need to replace the sums in the definitions (10.37) and (10.38) by integrals. Thus (10.37) becomes

$$I_{xx} = \int_0^a dx \int_0^a dy \int_0^a dz \, \varrho \, (y^2 + z^2) \tag{10.45}$$

where $\varrho = M/a^3$ denotes the mass density of the cube. This is the sum of two triple integrals, each of which can be factored into three single integrals. For example,

---

[6] The matrix operations that I shall assume you already know — matrix addition, multiplication, transposition, determinants, and a few more — can be found in Chapter 3 of Mary Boas, *Mathematical Methods in the Physical Sciences* (Wiley, 1983). Some of the ideas I shall be developing in this and the next chapter are discussed in more detail in Chapter 10 of the same book.

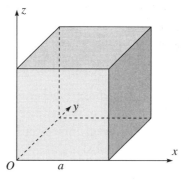

Figure 10.5   A uniform solid cube of side a that is
free to rotate about the corner O.

$$\int_0^a dx \int_0^a dy \int_0^a dz \, \varrho \, y^2 = \varrho \left(\int_0^a dx \right) \left(\int_0^a y^2 \, dy \right) \left(\int_0^a dz \right)$$

$$= \tfrac{1}{3} \varrho a^5 = \tfrac{1}{3} M a^2. \tag{10.46}$$

The second term in (10.45) has the same value, and we conclude that

$$I_{xx} = \tfrac{2}{3} M a^2 \tag{10.47}$$

with (by symmetry) the same values for I_{yy} and I_{zz}.

The integral form of (10.38) for the off-diagonal elements of $\mathbf{I}$ is

$$I_{xy} = -\int_0^a dx \int_0^a dy \int_0^a dz \, \varrho xy$$

$$= -\varrho \left(\int_0^a x \, dx \right) \left(\int_0^a y \, dy \right) \left(\int_0^a dz \right) \tag{10.48}$$

$$= -\tfrac{1}{4} \varrho a^5 = -\tfrac{1}{4} M a^2$$

with (again by symmetry) the same answer for all the other off-diagonal elements.

Putting all of these results together, we find for the moment of inertia tensor of a cube rotating about its corner

$$\mathbf{I} = \begin{bmatrix} \tfrac{2}{3} M a^2 & -\tfrac{1}{4} M a^2 & -\tfrac{1}{4} M a^2 \\ -\tfrac{1}{4} M a^2 & \tfrac{2}{3} M a^2 & -\tfrac{1}{4} M a^2 \\ -\tfrac{1}{4} M a^2 & -\tfrac{1}{4} M a^2 & \tfrac{2}{3} M a^2 \end{bmatrix} = \frac{M a^2}{12} \begin{bmatrix} 8 & -3 & -3 \\ -3 & 8 & -3 \\ -3 & -3 & 8 \end{bmatrix}$$

$$\text{[about corner]} \tag{10.49}$$

where the second, more compact, form follows from the rules for multiplying a matrix by a number. (Notice that, as expected, $\mathbf{I}$ is a symmetric matrix.)

According to (10.42) the angular momentum $\mathbf{L}$ corresponding to an angular velocity $\boldsymbol{\omega}$ is given by the matrix product $\mathbf{L} = \mathbf{I}\boldsymbol{\omega}$, where the vectors $\mathbf{L}$ and $\boldsymbol{\omega}$ are understood as 3×1 columns made up of the three components of the vector concerned. Thus if the cube is rotating about the x axis,

$$\mathbf{L} = \mathbf{I}\boldsymbol{\omega} = \frac{Ma^2}{12} \begin{bmatrix} 8 & -3 & -3 \\ -3 & 8 & -3 \\ -3 & -3 & 8 \end{bmatrix} \begin{bmatrix} \omega \\ 0 \\ 0 \end{bmatrix} = \frac{Ma^2}{12} \begin{bmatrix} 8\omega \\ -3\omega \\ -3\omega \end{bmatrix} \quad (10.50)$$

or, reverting to more standard vector notation,

$$\mathbf{L} = Ma^2\omega \left(\tfrac{2}{3}, -\tfrac{1}{4}, -\tfrac{1}{4} \right). \quad (10.51)$$

As we have come to recognize is possible, we see that in this case $\mathbf{L}$ is not in the same direction as the angular velocity $\boldsymbol{\omega} = (\omega, 0, 0)$.

If the cube is rotating about its main diagonal, then the unit vector in the direction of rotation is $\mathbf{u} = (1/\sqrt{3})(1, 1, 1)$ and the angular velocity vector is $\boldsymbol{\omega} = \omega\mathbf{u} = (\omega/\sqrt{3})(1, 1, 1)$. Thus according to (10.42), the angular momentum for this case is

$$\mathbf{L} = \mathbf{I}\boldsymbol{\omega} = \frac{Ma^2}{12} \frac{\omega}{\sqrt{3}} \begin{bmatrix} 8 & -3 & -3 \\ -3 & 8 & -3 \\ -3 & -3 & 8 \end{bmatrix} \begin{bmatrix} 1 \\ 1 \\ 1 \end{bmatrix}$$

$$= \frac{Ma^2}{12} \frac{\omega}{\sqrt{3}} \begin{bmatrix} 2 \\ 2 \\ 2 \end{bmatrix} = \frac{Ma^2}{6} \boldsymbol{\omega}. \quad (10.52)$$

In this case, rotation about the main diagonal of the cube, we see that the angular momentum is in the same direction as the angular velocity.

(b) If the cube is rotating about its center, then in Figure 10.5 we must move the origin O to the center of the cube. This means that all of the integrals in (10.45) and (10.48) run from $-a/2$ to $a/2$ instead of 0 to a. Evaluating (10.45) for I_{xx} [as in (10.46)] we find

$$I_{xx} = \tfrac{1}{6}Ma^2 \quad (10.53)$$

and likewise I_{yy} and I_{zz}. When the limits in (10.48) are replaced by $-a/2$ and $a/2$, both of the first two integrals are zero, and we conclude that

$$I_{xy} = 0.$$

As you can easily check, all of the off-diagonal elements of $\mathbf{I}$ work in the same way and are zero. We could actually have anticipated this vanishing of the off-diagonal elements of $\mathbf{I}$ based on Example 10.1. We saw there that if the plane $z = 0$ is a plane of reflection symmetry, then both I_{xz} and I_{yz} are automatically zero. (Every contribution from above the plane $z = 0$ was canceled by the contribution from the corresponding point below the plane.) Therefore, if two of the three coordinate planes $x = 0$, $y = 0$, and $z = 0$ are planes of reflection symmetry, this guarantees that *all* of the products of inertia are zero. For the cube (with O at its center) all three of the coordinate planes are planes of reflection symmetry, so it was inevitable that the off-diagonal elements of $\mathbf{I}$ turned out to be zero.

Collecting results, we conclude that for a cube rotating about its center, the moment of inertia tensor is

$$\mathbf{I} = \tfrac{1}{6}Ma^2 \begin{bmatrix} 1 & 0 & 0 \\ 0 & 1 & 0 \\ 0 & 0 & 1 \end{bmatrix} = \tfrac{1}{6}Ma^2\mathbf{1} \qquad \text{[about CM]}, \qquad (10.54)$$

where $\mathbf{1}$ denotes the 3×3 **unit matrix**,

$$\mathbf{1} = \begin{bmatrix} 1 & 0 & 0 \\ 0 & 1 & 0 \\ 0 & 0 & 1 \end{bmatrix}. \qquad (10.55)$$

We shall see that, because the moment of inertia tensor for rotation about the center of a cube is a multiple of the unit matrix, rotations about the center of a cube are especially easy to analyze. In particular, the angular momentum of the cube (rotating about its center) is

$$\mathbf{L} = \mathbf{I}\boldsymbol{\omega} = \tfrac{1}{6}Ma^2\mathbf{1}\boldsymbol{\omega} = \tfrac{1}{6}Ma^2\boldsymbol{\omega}, \qquad (10.56)$$

which implies that the angular momentum $\mathbf{L}$ is in the same direction as the angular velocity $\boldsymbol{\omega}$, *whatever the direction of* $\boldsymbol{\omega}$. This simple result is a consequence of the high degree of symmetry of the cube relative to its center.

You will notice that (10.56) for rotation about any axis through the center of the cube agrees with (10.52) for rotation about the main diagonal through the corner of the cube. This is not an accident: The main diagonal through the center of the cube is exactly the same as the main diagonal through the corner. Therefore the angular momenta for rotations about these two axes *have* to agree.

This example illustrated many of the features of a typical calculation of the moment of inertia tensor $\mathbf{I}$, and you should certainly try some of the problems at the end of this chapter to get used to doing such calculations yourself. Meanwhile, here is one more example to illustrate the special features of a body that is axially symmetric.

EXAMPLE 10.3   Inertia Tensor for a Solid Cone

Find the moment of inertia tensor $\mathbf{I}$ for a spinning top that is a uniform solid cone (mass M, height h, and base radius R) spinning about its tip. Choose the z axis along the axis of symmetry of the cone, as shown in Figure 10.6. For an arbitrary angular velocity $\boldsymbol{\omega}$, what is the top's angular momentum $\mathbf{L}$?

The moment of inertia about the z axis, I_{zz}, is given by the integral

$$I_{zz} = \int_V dV \varrho(x^2 + y^2) \qquad (10.57)$$

where the subscript V on the integral indicates that the integral runs over the volume of the body, dV is an element of volume, and ϱ is the constant mass density $\varrho = M/V = 3M/(\pi R^2 h)$. This integral is easily evaluated in cylindrical

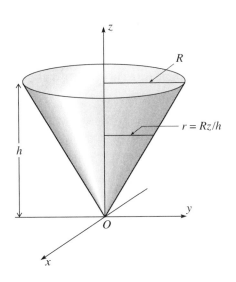

Figure 10.6   A top comprising a uniform solid cone, of mass M, height h, and base radius R, spins about its tip. The radius of the cone at height z is $r = Rz/h$. (The top is not necessarily vertical, but, whatever its orientation, we choose the z axis along the axis of symmetry.)

polar coordinates (ρ, ϕ, z) since $x^2 + y^2 = \rho^2$. Thus the integral can be written as[7]

$$I_{zz} = \varrho \int_V dV\, \rho^2 = \varrho \int_0^h dz \int_0^{2\pi} d\phi \int_0^r \rho\, d\rho\, \rho^2 \qquad (10.58)$$

where the upper limit r of the ρ integral is the radius of the cone at height z, as shown in Figure 10.6. These integrals are easily carried out and give (Problem 10.26)

$$I_{zz} = \tfrac{3}{10} M R^2. \qquad (10.59)$$

Because of the top's rotational symmetry about the z axis, the other two moments of inertia, I_{xx} and I_{yy}, are equal. (A rotation through 90° about the z axis leaves the body unchanged but interchanges I_{xx} and I_{yy}. Therefore, $I_{xx} = I_{yy}$.) To evaluate I_{xx} we write

$$I_{xx} = \int_V dV \varrho(y^2 + z^2) = \int_V dV \varrho y^2 + \int_V dV \varrho z^2. \qquad (10.60)$$

---

[7] Here is one of those horrible moments when the traditional use of the Greek letter "rho" for density collides with its use for the cylindrical coordinate equal to the distance from the z axis. Notice that I am using two different versions of the letter, ϱ for density and ρ for the coordinate. Fortunately, this unlucky collision happens only very occasionally. If you need to refresh your memory about volume integrals in cylindrical coordinates, see Problem 10.26.

The first integral here is the same as the second term in (10.57), and, by the rotatonal symmetry, the two terms in (10.57) are equal. Therefore, the first term of (10.60) is just $I_{zz}/2$ or $\frac{3}{20} M R^2$. The second integral in (10.60) can be evaluated using cylindrical polar coordinates, as in (10.58), and gives $\frac{3}{5} M h^2$. Therefore

$$I_{xx} = I_{yy} = \tfrac{3}{20} M (R^2 + 4h^2). \tag{10.61}$$

This leaves the off-diagonal products of inertia, $I_{xy}, \cdots$, to calculate, but we can easily see that all of these are zero. The point is that because of the axial symmetry about the z axis, both of the planes $x = 0$ and $y = 0$ are planes of reflection symmetry. (See Figure 10.6.) By the argument given in Example 10.1, symmetry about the plane $x = 0$ implies that $I_{xy} = I_{xz} = 0$. Similarly, symmetry about $y = 0$ implies that $I_{yz} = I_{yx} = 0$. Thus, symmetry about any two coordinate planes guarantees that *all* of the products of inertia are zero.

Collecting results, we find that the moment of inertia tensor for a uniform cone (relative to its tip) is

$$\mathbf{I} = \tfrac{3}{20} M \begin{bmatrix} R^2 + 4h^2 & 0 & 0 \\ 0 & R^2 + 4h^2 & 0 \\ 0 & 0 & 2R^2 \end{bmatrix} = \begin{bmatrix} \lambda_1 & 0 & 0 \\ 0 & \lambda_2 & 0 \\ 0 & 0 & \lambda_3 \end{bmatrix}, \tag{10.62}$$

where the last form is just for convenience of discussion (and in the present case, it happens that $\lambda_1 = \lambda_2$). The most striking thing about this matrix is that its off-diagonal elements are all zero. A matrix with this property is called a **diagonal matrix**. [The inertia tensor (10.54) for a cube about its center was also diagonal, but since all its diagonal elements were equal, it was actually a multiple of the unit matrix **1**, making it an even more special case.] The important consequence of **I** being diagonal emerges when we evaluate the angular momentum **L** for an arbitrary angular velocity $\boldsymbol{\omega} = (\omega_x, \omega_y, \omega_z)$:

$$\mathbf{L} = \mathbf{I}\boldsymbol{\omega} = (\lambda_1 \omega_x, \lambda_2 \omega_y, \lambda_3 \omega_z). \tag{10.63}$$

While this may not look remarkable, notice that if the angular velocity $\boldsymbol{\omega}$ points along one of the coordinate axes, then the same is true of the angular momentum **L**. For example, if $\boldsymbol{\omega}$ points along the x axis, then $\omega_y = \omega_z = 0$ and (10.63) implies that

$$\mathbf{L} = \mathbf{I}\boldsymbol{\omega} = (\lambda_1 \omega_x, 0, 0) \tag{10.64}$$

and **L** also points along the x axis. Obviously, the same thing happens if $\boldsymbol{\omega}$ points along the y or z axes, and we see quite generally that if the inertia tensor **I** is diagonal, then **L** will be parallel to $\boldsymbol{\omega}$ whenever $\boldsymbol{\omega}$ points along one of the three coordinate axes.

10.4   Principal Axes of Inertia

We have seen that in general the angular momentum of a body spinning about a point O is not in the same direction as the axis of rotation; that is, $\mathbf{L}$ is not parallel to $\boldsymbol{\omega}$. We have also seen that, for certain bodies at least, there can be certain axes for which $\mathbf{L}$ and $\boldsymbol{\omega}$ *are* parallel. When this happens, we say that the axis in question is a **principal axis**. To express this definition mathematically, note that two nonzero vectors $\mathbf{a}$ and $\mathbf{b}$ are parallel if and only if $\mathbf{a} = \lambda \mathbf{b}$ for some real number λ. Thus we can define a principal axis of a body (about an origin O) as any axis through O with the property that if $\boldsymbol{\omega}$ points along the axis, then

$$\mathbf{L} = \lambda \boldsymbol{\omega} \tag{10.65}$$

for some real number λ. To see the significance of the number λ in this equation, we can temporarily choose the direction of $\boldsymbol{\omega}$ to be the z direction, in which case $\mathbf{L}$ is given by (10.31) as $\mathbf{L} = (I_{xz}\omega,\ I_{yz}\omega,\ I_{zz}\omega)$. Since $\mathbf{L}$ is parallel to $\boldsymbol{\omega}$, the first two components are zero, and we conclude that $\mathbf{L} = (0, 0, I_{zz}\omega) = I_{zz}\boldsymbol{\omega}$. Comparing with (10.65), we see that the number λ in (10.65) is just the body's moment of inertia about the axis in question. To summarize, if the angular velocity $\boldsymbol{\omega}$ points along a principal axis, then $\mathbf{L} = \lambda\boldsymbol{\omega}$, where λ is the moment of inertia about the axis in question.

Let us review what we already know about the existence of principal axes. We saw at the end of the last section that if the inertia tensor $\mathbf{I}$, with respect to a chosen set of axes, turns out to be diagonal,

$$\mathbf{I} = \begin{bmatrix} \lambda_1 & 0 & 0 \\ 0 & \lambda_2 & 0 \\ 0 & 0 & \lambda_3 \end{bmatrix}, \tag{10.66}$$

then the chosen x, y, and z axes are principal axes. Conversely, if the x, y, and z axes are principal axes, then it is easy to see (Problem 10.29) that $\mathbf{I}$ must be diagonal, as in (10.66). The three numbers that I have denoted by λ_1, λ_2, and λ_3 are the moments of inertia about the three principal axes and are called the **principal moments**.

If a body has a symmetry axis through O (like the spinning top of Example 10.3), then that axis is a principal axis. Furthermore, any two axes perpendicular to the symmetry axis (like the x and y axes in Example 10.3) are also principal axes, since the inertia tensor with respect to these axes is diagonal. Again, if a body has two perpendicular planes of reflection symmetry through O (like the cube spinning about its center[8] in Example 10.2), then the three perpendicular axes defined by these two planes and O are principal axes.

So far all our examples of principal axes have involved bodies with special symmetries, and you may very reasonably be thinking that the existence of principal axes is somehow tied to a body's having some symmetry. In fact however, this is most

---

[8] Of course the cube has *three* planes of reflection symmetry, but two are enough to guarantee the claim made here. For example, if the cone of Figure 10.6 were an elliptical cone, the z axis would no longer be an axis of rotational symmetry, but the planes $x = 0$ and $y = 0$ would still provide two planes of reflection symmetry, and the three coordinate axes would still be principal axes.

emphatically not so. Although symmetries of a body make it much easier to spot the principal axes, it turns out that *any rigid body rotating about any point has three principal axes:*

Existence of Principal Axes

For any rigid body and any point O there are three perpendicular principal axes through O. That is, there are three perpendicular axes through O with respect to which the inertia tensor $\mathbf{I}$ is diagonal and, hence, with the property that when the angular velocity $\boldsymbol{\omega}$ points along any one of these axes the same is true of the angular momentum $\mathbf{L}$.

This surprising result is the consequence of an important mathematical theorem which states that if $\mathbf{I}$ is any real symmetric matrix (such as the inertia tensor of some body with respect to any chosen set of orthogonal axes) then there exists another set of orthogonal axes (with the same origin) such that the corresponding matrix (call it $\mathbf{I}'$), evaluated with respect to the new axes, has the diagonal form (10.66). This result, which is proved in the appendix, is extremely useful, since the discussion of rotational motion is much simpler if it can be referred to a set of principal axes, and the result guarantees that this can always be done. (It may be worth mentioning right away, however, that the principal axes of a rigid body are naturally fixed in the body. Thus when we choose the principal axes as our coordinate axes, we are committing ourselves to using a rotating set of axes.)

While it is not essential to see a proof of the existence of principal axes, we certainly do need to know how to *find* the principal axes, and this is what I take up in the next section.

Kinetic Energy of a Rotating Body

It is naturally important to be able to write down the kinetic energy of a rotating body. I shall leave it as a challenging exercise (Problem 10.33) to show that

$$T = \tfrac{1}{2}\boldsymbol{\omega} \cdot \mathbf{L}. \tag{10.67}$$

In particular, if we use a set of principal axes for our coordinate system, then $\mathbf{L} = (\lambda_1\omega_1, \lambda_2\omega_2, \lambda_3\omega_3)$ and

$$T = \tfrac{1}{2}(\lambda_1\omega_1^2 + \lambda_2\omega_2^2 + \lambda_3\omega_3^2). \tag{10.68}$$

This important result is a natural generalization of Equation (10.26), $T = \tfrac{1}{2}I_{zz}\omega^2$, for rotation about a fixed z axis. We shall use the result in writing the Lagrangian for a spinning body in Section 10.9.

10.5   Finding the Principal Axes; Eigenvalue Equations

Suppose that we wish to find the principal axes of a rigid body rotating about a point O. Suppose further that, using some given set of axes, we have calculated the inertia tensor $\mathbf{I}$ for the body. If $\mathbf{I}$ is diagonal, then our axes already are the principal axes of the body, and there is nothing more to do. Suppose, however, that $\mathbf{I}$ is *not* diagonal. How are we to find the principal axes? The essential clue is Equation (10.65): If $\boldsymbol{\omega}$ points along a principal axis, then $\mathbf{L}$ must equal $\lambda\boldsymbol{\omega}$ (for some number λ). Since $\mathbf{L} = \mathbf{I}\boldsymbol{\omega}$, it follows that $\boldsymbol{\omega}$ must satisfy the equation

$$\mathbf{I}\boldsymbol{\omega} = \lambda\boldsymbol{\omega} \tag{10.69}$$

for some (as yet unknown) number λ. The equation (10.69) has the form

$$(\text{matrix}) \times (\text{vector}) = (\text{number}) \times (\text{same vector})$$

and is called an **eigenvalue equation**. Eigenvalue equations are among the most important equations of modern physics and arise in many different areas. They always express the same idea, that some mathematical operation performed on a vector ($\boldsymbol{\omega}$ in our case) produces a second vector ($\mathbf{I}\boldsymbol{\omega}$ here) that has the same direction as the first. A vector $\boldsymbol{\omega}$ that satisfies (10.69) is called an **eigenvector** and the corresponding number λ, the corresponding **eigenvalue**.

There are actually two parts to solving the eigenvalue problem of (10.69). Usually we want to know the directions of $\boldsymbol{\omega}$ for which (10.69) is satisfied (namely the directions of the principal axes), and in most cases we also want to know the corresponding eigenvalues λ (namely, the moments of inertia about the principal axes). In practice, we usually solve these two parts of the problem in the opposite order — first find the possible eigenvalues λ and then the corresponding directions of $\boldsymbol{\omega}$.

The first step in solving the matrix equation (10.69) is to rewrite it. Since $\boldsymbol{\omega} = \mathbf{1}\boldsymbol{\omega}$ (where $\mathbf{1}$ is the 3×3 unit matrix), Equation (10.69) is the same as $\mathbf{I}\boldsymbol{\omega} = \lambda\mathbf{1}\boldsymbol{\omega}$, or, moving the right side over to the left,

$$(\mathbf{I} - \lambda\mathbf{1})\boldsymbol{\omega} = 0. \tag{10.70}$$

This is a matrix equation of the form $\mathbf{A}\boldsymbol{\omega} = 0$, where $\mathbf{A}$ is a 3×3 matrix and $\boldsymbol{\omega}$ is a vector, that is, a 3×1 column of numbers ω_x, ω_y, and ω_z. The matrix equation $\mathbf{A}\boldsymbol{\omega} = 0$ is really three simultaneous equations for the three numbers ω_x, ω_y, and ω_z, and it is a well-known property of such equations that they have a nonzero solution if and only if the determinant, $\det(\mathbf{A})$, is zero.[9] Therefore, the eigenvalue equation (10.70) has a nonzero solution if and only if

$$\det(\mathbf{I} - \lambda\mathbf{1}) = 0. \tag{10.71}$$

This equation is called the **characteristic equation** (or *secular equation*) for the matrix $\mathbf{I}$. The determinant involved is a cubic in the number λ. Therefore, the equation

---

[9] See, for example, Mary Boas, *Mathematical Methods in the Physical Sciences* (Wiley, 1983), page 133.

is a cubic equation for the eigenvalues λ and will, in general, have three solutions, λ_1, λ_2, and λ_3, the three principal moments. For each of these values of λ, the equation (10.70) can be solved to give the corresponding vector $\boldsymbol{\omega}$, whose direction tells us the direction of one of the three principal axes of the rigid body under consideration.

If you have never seen this procedure for finding the principal axes (or, more generally, for solving eigenvalue problems), you will certainly want to see an example and to work through some yourself. Here is one as a start.

EXAMPLE 10.4   Principal Axes of a Cube about a Corner

Find the principal axes and corresponding moments for the cube of Example 10.2, rotating about its corner. What is the form of the inertia tensor evaluated with respect to the principal axes?

Using axes parallel to the edges of the cube, we found the inertia tensor to be [Equation (10.49)]

$$\mathbf{I} = \mu \begin{bmatrix} 8 & -3 & -3 \\ -3 & 8 & -3 \\ -3 & -3 & 8 \end{bmatrix} \tag{10.72}$$

where I have introduced the abbreviation μ for the constant $\mu = Ma^2/12$, which has the dimensions of moment of inertia. Since $\mathbf{I}$ is not diagonal, it is clear that our original chosen axes (parallel to the edges of the cube) are not the principal axes. To find the principal axes, we must find the directions of $\boldsymbol{\omega}$ that satisfy the eigenvalue equation $\mathbf{I}\boldsymbol{\omega} = \lambda\boldsymbol{\omega}$.

Our first step is to find the values of λ (the eigenvalues) that satisfy the characteristic equation $\det(\mathbf{I} - \lambda\mathbf{1}) = 0$. Substituting (10.72) for $\mathbf{I}$, we find that

$$\mathbf{I} - \lambda\mathbf{1} = \begin{bmatrix} 8\mu & -3\mu & -3\mu \\ -3\mu & 8\mu & -3\mu \\ -3\mu & -3\mu & 8\mu \end{bmatrix} - \begin{bmatrix} \lambda & 0 & 0 \\ 0 & \lambda & 0 \\ 0 & 0 & \lambda \end{bmatrix}$$

$$= \begin{bmatrix} 8\mu - \lambda & -3\mu & -3\mu \\ -3\mu & 8\mu - \lambda & -3\mu \\ -3\mu & -3\mu & 8\mu - \lambda \end{bmatrix}.$$

The determinant of this matrix is straightforward to evaluate and is

$$\det(\mathbf{I} - \lambda\mathbf{1}) = (2\mu - \lambda)(11\mu - \lambda)^2. \tag{10.73}$$

Thus the three roots of the equation $\det(\mathbf{I} - \lambda\mathbf{1}) = 0$ (the eigenvalues) are

$$\lambda_1 = 2\mu \quad \text{and} \quad \lambda_2 = \lambda_3 = 11\mu. \tag{10.74}$$

Notice that in this case two of the three roots of the cubic (10.73) happen to be equal.

Armed with the eigenvalues, we can now find the eigenvectors, that is, the directions of the three principal axes of our cube rotating about its corner. These are determined by Equation (10.70), which we must examine for each of the eigenvalues λ_1, λ_2, and λ_3 in turn (though in this case the last two are equal). Let us start with λ_1.

With $\lambda = \lambda_1 = 2\mu$, Equation (10.70) becomes

$$(\mathbf{I} - \lambda\mathbf{1})\boldsymbol{\omega} = \mu \begin{bmatrix} 6 & -3 & -3 \\ -3 & 6 & -3 \\ -3 & -3 & 6 \end{bmatrix} \begin{bmatrix} \omega_x \\ \omega_y \\ \omega_z \end{bmatrix} = 0. \qquad (10.75)$$

This gives three equations for the components of $\boldsymbol{\omega}$,

$$\begin{aligned} 2\omega_x - \omega_y - \omega_z &= 0 \\ -\omega_x + 2\omega_y - \omega_z &= 0 \\ -\omega_x - \omega_y + 2\omega_z &= 0. \end{aligned} \qquad (10.76)$$

Subtracting the second equation from the first we see that $\omega_x = \omega_y$, and the first then tells us that $\omega_x = \omega_z$. Therefore, $\omega_x = \omega_y = \omega_z$, and we conclude that the first principal axis is in the direction $(1, 1, 1)$ along the principal diagonal of the cube. If we define a unit vector $\mathbf{e}_1$ in this direction,

$$\mathbf{e}_1 = \tfrac{1}{\sqrt{3}}(1, 1, 1), \qquad (10.77)$$

then $\mathbf{e}_1$ specifies the direction of our first principal axis. If $\boldsymbol{\omega}$ points along $\mathbf{e}_1$, then $\mathbf{L} = \mathbf{I}\boldsymbol{\omega} = \lambda_1\boldsymbol{\omega}$. This says simply that the moment of inertia about this principal axis is $\lambda_1 = 2\mu = \tfrac{1}{6}Ma^2$. Thus our analysis of the first eigenvalue has produced this conclusion: One of the principal axes of a cube, rotating about its corner O, is the principal diagonal through O (direction $\mathbf{e}_1$), and the moment of inertia for that axis is the corresponding eigenvalue $\tfrac{1}{6}Ma^2$.

The other two eigenvalues are equal ($\lambda_2 = \lambda_3 = 11\mu$), so there is just one more case to consider. With $\lambda = 11\mu$, the eigenvalue equation (10.70) reads

$$(\mathbf{I} - \lambda\mathbf{1})\boldsymbol{\omega} = \mu \begin{bmatrix} -3 & -3 & -3 \\ -3 & -3 & -3 \\ -3 & -3 & -3 \end{bmatrix} \begin{bmatrix} \omega_x \\ \omega_y \\ \omega_z \end{bmatrix} = 0.$$

This gives three equations for the components of $\boldsymbol{\omega}$, but all three equations are actually the same equation, namely

$$\omega_x + \omega_y + \omega_z = 0. \qquad (10.78)$$

This equation does not uniquely determine the direction of $\boldsymbol{\omega}$. To see what it does imply, notice that $\omega_x + \omega_y + \omega_z$ can be viewed as the scalar product of $\boldsymbol{\omega}$ with the vector $(1, 1, 1)$. Thus Equation (10.78) states simply that $\boldsymbol{\omega} \cdot \mathbf{e}_1 = 0$; that is, $\boldsymbol{\omega}$ needs only to be orthogonal to our first principal axis $\mathbf{e}_1$. In other words, any two orthogonal directions $\mathbf{e}_2$ and $\mathbf{e}_3$ that are perpendicular to $\mathbf{e}_1$ will serve as the other two principal axes, both with moment of inertia $\lambda_2 = \lambda_3 = 11\mu = \tfrac{11}{12}Ma^2$. This freedom in choosing the last two principal axes is directly related to the circumstance that the last two eigenvalues λ_2 and λ_3 are equal; when all three eigenvalues are different, each one leads to a unique direction for the corresponding principal axis.

Finally, if we were to re-evaluate the inertia tensor with respect to new axes in the directions $\mathbf{e}_1$, $\mathbf{e}_2$, and $\mathbf{e}_3$, then the new matrix $\mathbf{I}'$ would be diagonal, with the principal moments down the diagonal,

$$\mathbf{I}' = \begin{bmatrix} \lambda_1 & 0 & 0 \\ 0 & \lambda_2 & 0 \\ 0 & 0 & \lambda_3 \end{bmatrix} = \tfrac{1}{12}Ma^2 \begin{bmatrix} 2 & 0 & 0 \\ 0 & 11 & 0 \\ 0 & 0 & 11 \end{bmatrix}.$$

For this reason the process of finding the principal axes of a body is described as **diagonalization of the inertia tensor**.

The last paragraph of this example illustrates a useful point: By the time we have found the principal axes of a body with the corresponding principal moments, there is no need to re-evaluate the inertia tensor with respect to the new axes. We *know* that with respect to the principal axes it is bound to be diagonal,

$$\mathbf{I}' = \begin{bmatrix} \lambda_1 & 0 & 0 \\ 0 & \lambda_2 & 0 \\ 0 & 0 & \lambda_3 \end{bmatrix}, \tag{10.79}$$

with the principal moments λ_1, λ_2, and λ_3 down the diagonal. In general the three principal moments will all be different, in which case the directions of the three principal axes are uniquely determined and are automatically orthogonal (see Problem 10.38). As we saw in Example 10.4, it can happen that two of the principal moments are equal, in which case the corresponding two principal axes can have any direction that is orthogonal to the third axis. (This is what happened in the Example 10.4, and also what happens with any body that has rotational symmetry about an axis through O.) If all three principal moments are the same (as with a cube or sphere about its center) then, in fact, *any* axis is a principal axis. For proofs of these statements about the uniqueness or otherwise of the principal axes, see Problem 10.38.

10.6   Precession of a Top due to a Weak Torque

We now know enough about the angular momentum of a rigid body to start solving some interesting problems. We'll start with the phenomenon of precession of a spinning top subject to a weak torque.

Consider a symmetric top as shown in Figure 10.7. The axes labeled x, y, and z are fixed to the ground with the z axis vertically up. The top is pivoted freely at its tip O, and it makes an angle θ with the vertical. Because of the top's axial symmetry, its inertia tensor is diagonal, with the form

$$\mathbf{I} = \begin{bmatrix} \lambda_1 & 0 & 0 \\ 0 & \lambda_1 & 0 \\ 0 & 0 & \lambda_3 \end{bmatrix}, \tag{10.80}$$

relative to the top's principal axes (namely, an axis along $\mathbf{e}_3$, the symmetry axis, and any two orthogonal axes perpendicular to $\mathbf{e}_3$).

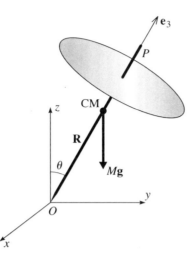

Figure 10.7   A spinning top is made of a rod OP fastened normally through the center of a uniform circular disc and freely pivoted at O. The total mass is M, and $\mathbf{R}$ denotes the CM position relative to O. The principal axes of the top are in the directions of the unit vector $\mathbf{e}_3$ along the symmetry axis of the top, and any two orthogonal unit vectors $\mathbf{e}_1$ and $\mathbf{e}_2$ perpendicular to $\mathbf{e}_3$. The top's weight $M\mathbf{g}$ produces a torque, which causes the angular momentum $\mathbf{L}$ to change.

Let us suppose first that gravity has been switched off and that the top is spinning about its symmetry axis, with angular velocity $\boldsymbol{\omega} = \omega\mathbf{e}_3$ and angular momentum

$$\mathbf{L} = \lambda_3\boldsymbol{\omega} = \lambda_3\omega\mathbf{e}_3. \tag{10.81}$$

Since there is no net torque on the top, $\mathbf{L}$ is constant. Therefore, the top will continue indefinitely to spin about the same axis with the same angular velocity.[10]

Now let us switch gravity back on, causing a torque $\boldsymbol{\Gamma} = \mathbf{R} \times M\mathbf{g}$, with magnitude $RMg \sin\theta$ and a direction which is perpendicular to both the vertical z axis and the axis of the top. Let us suppose further that this torque is small. (We can ensure this by arranging that any or all of the parameters R, M, or g are small compared with the other relevant parameters of the system.) The existence of a torque implies that the angular momentum starts to change, since $\boldsymbol{\Gamma} = \dot{\mathbf{L}}$.

The changing of $\mathbf{L}$ implies that $\boldsymbol{\omega}$ starts to change and that the components ω_1 and ω_2 cease to be zero. However, to the extent that the torque is small, we can expect ω_1 and ω_2 to remain small.[11] This means that Equation (10.81) remains a good approximation. (That is, the main contribution to $\mathbf{L}$ continues to be the spinning about

---

[10] That the angular velocity remains constant is actually not completely obvious. However, we shall prove it later on, so please accept it as a reasonable claim for now.

[11] Again, this plausible statement should be (and will be) proved, but let us accept it for now.

$\mathbf{e}_3$.) In this approximation, the torque $\boldsymbol{\Gamma}$ is perpendicular to $\mathbf{L}$ (since $\boldsymbol{\Gamma}$ is perpendicular to $\mathbf{e}_3$), which means that $\mathbf{L}$ changes in direction but not in magnitude. From (10.81) we see that $\mathbf{e}_3$ starts changing in direction, while ω remains constant. Specifically, the equation $\dot{\mathbf{L}} = \boldsymbol{\Gamma}$ becomes

$$\lambda_3 \omega \dot{\mathbf{e}}_3 = \mathbf{R} \times M\mathbf{g}$$

or, substituting $\mathbf{R} = R\mathbf{e}_3$ and $\mathbf{g} = -g\hat{\mathbf{z}}$ (where $\hat{\mathbf{z}}$ is the unit vector that points vertically upward),

$$\dot{\mathbf{e}}_3 = \frac{MgR}{\lambda_3 \omega}\, \hat{\mathbf{z}} \times \mathbf{e}_3 = \boldsymbol{\Omega} \times \mathbf{e}_3 \tag{10.82}$$

where

$$\boldsymbol{\Omega} = \frac{MgR}{\lambda_3 \omega}\, \hat{\mathbf{z}}. \tag{10.83}$$

You will recognize (10.82) as saying that the axis of the top, $\mathbf{e}_3$, rotates with angular velocity $\boldsymbol{\Omega}$ about the vertical direction $\hat{\mathbf{z}}$.

Our conclusion is that the torque exerted by gravity causes the top's axis to *precess*, that is, to move slowly around a vertical cone, with fixed angle θ and with angular frequency $\Omega = RMg/\lambda_3\omega$.[12] This precession, although surprising at first glance, can be understood in elementary terms: In the view of Figure 10.7, the gravitational torque is clockwise, and the torque vector $\boldsymbol{\Gamma}$ is into the page. Since $\dot{\mathbf{L}} = \boldsymbol{\Gamma}$, this requires that the change in $\mathbf{L}$ be into the page, which is exactly the direction of the predicted precession.

This precession of the axis of a spinning body is an effect that you have almost certainly observed when playing with a child's top. The same effect shows up in several other situations. For example, the earth spins on its axis, much like a spinning top, and the axis of spin is inclined at an angle $\theta = 23°$ from the normal to the earth's orbit around the sun. Because of the earth's equatorial bulge, the sun and moon exert small torques on the earth and these torques cause the earth's axis to precess slowly (one complete turn in 26,000 years), tracing out a cone of half-angle $23°$ around the normal to the orbital plane — a phenomenon known as the **precession of the equinoxes**. This means that in another 13,000 years the pole star will be some $46°$ away from true north.

10.7   Euler's Equations

We are now ready to set up the equations of motion (or at least one form of the equations of motion) for a rotating rigid body. The two situations to which our discussion will principally apply are these: (1) A body that is pivoted about one fixed point, like the spinning top of Section 10.6, and (2) a body without any fixed point, like

---

[12] I should perhaps emphasize again that the discussion here is an approximation, the criterion for its validity being that $\Omega \ll \omega$. An exact analysis shows that, if launched as described here, the top will also make very small oscillations called nutations in the θ direction, although in practice these are quickly damped out by friction.

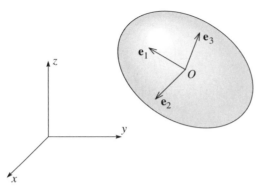

Figure 10.8 Axes used to derive Euler's equations for a rotating rigid body (shown here as an egg). The axes labeled x, y, and z define an inertial frame, often called the *space frame*. The unit vectors, $\mathbf{e}_1$, $\mathbf{e}_2$, and $\mathbf{e}_3$ point along the principal axes of the body and define the rotating, noninertial *body frame*. If the body has no fixed point (like an egg thrown in the air) then O is normally chosen to be the CM of the body. If the body has a fixed pivot, then O is that fixed point, and we would generally choose O to be the origin of both reference frames.

a drum major's flying baton, whose rotational motion about the CM we have chosen to examine. The equations that I shall derive are called Euler's equations (named for the same mathematician as the Euler–Lagrange equations of Chapter 6) and can be regarded as the rotational version of Newton's second law in the form $m\mathbf{a} = \mathbf{F}$. As we shall see, there are some problems that can be easily solved using Euler's equations. However, many problems are more easily solved using the Lagrangian approach, which we'll take up in Section 10.10.

Before we launch into the derivation of Euler's equations, there is a complication we must now face up to. To take advantage of our understanding of the inertia tensor, and particularly the principal axes, we naturally want to use the principal axes of the body as our coordinate axes. However, because the principal axes are fixed in the rotating body, this inevitably involves us in using axes that rotate. We shall, therefore, need to use the machinery of Chapter 9 for handling rotating reference frames. The notation that I shall use is illustrated in Figure 10.8. For an inertial frame, relative to which Newton's laws hold in their simple form, we'll use the axes labeled x, y, and z. This frame is traditionally called the **space frame**, presumably because it is fixed in space — that is, inertial. The rotating frame is defined by the three unit vectors $\mathbf{e}_1$, $\mathbf{e}_2$, and $\mathbf{e}_3$, fixed in the body and pointing along the principal axes of the body. This frame, fixed in the body, is called the **body frame**.

If the angular velocity of the body is $\boldsymbol{\omega}$, and the principal moments of the body are λ_1, λ_2, and λ_3, then the angular momentum, as measured in the body frame, is

$$\mathbf{L} = (\lambda_1\omega_1, \lambda_2\omega_2, \lambda_3\omega_3), \qquad \text{[in the body frame]}. \qquad (10.84)$$

Now, if $\mathbf{\Gamma}$ is the torque acting on the body, we know that *as seen in the space frame*

$$\left(\frac{d\mathbf{L}}{dt}\right)_{\text{space}} = \mathbf{\Gamma}. \tag{10.85}$$

We saw in Chapter 9 that the rates of change of any vector as seen in the two frames are related by (9.30)

$$\left(\frac{d\mathbf{L}}{dt}\right)_{\text{space}} = \left(\frac{d\mathbf{L}}{dt}\right)_{\text{body}} + \boldsymbol{\omega} \times \mathbf{L}$$

$$= \dot{\mathbf{L}} + \boldsymbol{\omega} \times \mathbf{L} \tag{10.86}$$

where in the second line I have reintroduced the convention that a dot represents a time derivative evaluated in the rotating body frame (whose angular velocity is $\boldsymbol{\omega}$, the angular velocity of the body itself). Substituting (10.86) into (10.85), we arrive at the equation of motion for the rotating body frame:

$$\dot{\mathbf{L}} + \boldsymbol{\omega} \times \mathbf{L} = \mathbf{\Gamma}. \tag{10.87}$$

This equation is called **Euler's equation**. Using (10.84), we can resolve Euler's equation into its three components:

$$\left. \begin{aligned} \lambda_1 \dot{\omega}_1 - (\lambda_2 - \lambda_3)\omega_2\omega_3 &= \Gamma_1 \\ \lambda_2 \dot{\omega}_2 - (\lambda_3 - \lambda_1)\omega_3\omega_1 &= \Gamma_2 \\ \lambda_3 \dot{\omega}_3 - (\lambda_1 - \lambda_2)\omega_1\omega_2 &= \Gamma_3 \end{aligned} \right\} \quad \text{[Euler's equations]} \quad (10.88)$$

which are often referred to as Euler's equations.

The three Euler equations determine the evolution of $\boldsymbol{\omega}$ as seen in a frame fixed in the body. In general, they are difficult to use because the components Γ_1, Γ_2, and Γ_3 of the applied torque as seen in the rotating body frame are complicated (and unknown) functions of time. In fact, the main use of Euler's equations is in the case that the applied torque is zero, as I shall discuss in the next section. However, there are a few other cases where the torque is simple enough that we can get useful information from Euler's equations. For example, consider again the spinning top of Section 10.6. As we saw there, the gravitational torque on the top is always perpendicular to the axis $\mathbf{e}_3$, so Γ_3 is always zero. Furthermore, because of the top's axial symmetry, the two moments of inertia λ_1 and λ_2 are equal. Thus the third of the Euler equations (10.88) reduces to

$$\lambda_3 \dot{\omega}_3 = 0.$$

That is, the component of $\boldsymbol{\omega}$ along the symmetry axis is constant, a result that I stated as reasonable, but did not prove, in the discussion of Section 10.6.[13]

---

[13] The other main unproved assertion of that discussion — that the components ω_1 and ω_2 remain small for all times — can also be understood using Euler's equations. The components Γ_1 and Γ_2

10.8   Euler's Equations with Zero Torque

Let us now consider a rotating body subject to zero torque. In this case, Euler's equations (10.88) take the simple form

$$\left.\begin{array}{l} \lambda_1 \dot{\omega}_1 = (\lambda_2 - \lambda_3)\omega_2\omega_3 \\ \lambda_2 \dot{\omega}_2 = (\lambda_3 - \lambda_1)\omega_3\omega_1 \\ \lambda_3 \dot{\omega}_3 = (\lambda_1 - \lambda_2)\omega_1\omega_2 \end{array}\right\}. \tag{10.89}$$

I shall discuss these equations, first for the case that the three principal moments λ_1, λ_2, and λ_3 are all different, and then for the case that $\lambda_1 = \lambda_2 \neq \lambda_3$ (as with a spinning top).

A Body with Three Different Principal Moments

Let us suppose first that the principal moments of the body under consideration are all different. If at time $t = 0$ the body happens to be rotating about one of its principal axes ($\mathbf{e}_3$ say), then $\omega_1 = \omega_2 = 0$ at $t = 0$. Now, with $\omega_1 = \omega_2 = 0$, the right sides of all three Euler equations (10.89) are zero. This shows that as long as ω_1 and ω_2 are zero, all three components of $\boldsymbol{\omega}$ remain constant. That is, ω_1 and ω_2 remain zero and ω_3 is constant. In other words, if the body starts out rotating about one of its principal axes, it will continue to do so, with constant angular velocity $\boldsymbol{\omega}$. This statement applies, in the first instance, to the angular velocity as measured in the rotating body frame. However, with $\boldsymbol{\omega}$ along $\mathbf{e}_3$, the angular momentum is $\mathbf{L} = \lambda_3\boldsymbol{\omega}$, and we know that $\mathbf{L}$ is constant as seen in any inertial frame. Thus our result applies equally in any inertial frame: If a body that is subject to no torque is spinning initially about any principal axis, it will continue to do so indefinitely with constant angular velocity.

The converse of this result is also true. If at $t = 0$, the angular velocity is *not* along a principal axis, then $\boldsymbol{\omega}$ is not constant. To see this, notice first that if $\boldsymbol{\omega}$ is not along a principal axis, then at least two of its components are nonzero. If you look at the Euler equations (10.89) you will see that, with two components of $\boldsymbol{\omega}$ nonzero, at least one component of $\dot{\boldsymbol{\omega}}$ must be nonzero. (For example, if ω_1 and ω_2 are nonzero, then $\dot{\omega}_3 \neq 0$.) Therefore, with two of its components nonzero, $\boldsymbol{\omega}$ cannot be constant.

We conclude that the only way a body with three different principal moments can rotate freely with constant angular velocity is by rotating about one of its principal axes. It is interesting to know if this kind of rotation is stable. That is, if a body is rotating about one of its principal axes and is given a very small kick, will it continue to rotate close to its original axis, or will its motion change completely? Let us suppose that the body is rotating about the axis $\mathbf{e}_3$, with $\omega_1 = \omega_2 = 0$. If now we give it a tiny kick, then ω_1 and ω_2 will pick up nonzero values that are, at least initially, small. The question is whether the values of ω_1 and ω_2 will remain small or start to grow bigger.

---

of the torque are nonzero, which is what drives ω_1 and ω_2. Nonetheless, they are both small and *oscillate rapidly* as the top spins. It is reasonably easy to see that, under these conditions, the first two Euler equations (10.88) imply that ω_1 and ω_2 also oscillate rapidly and with *small amplitude*. This is the required result.

To answer this, we note from the third of the Euler equations (10.89) that as long as both ω_1 and ω_2 are small, $\dot{\omega}_3$ remains very small ("small $\times$ small"). Thus initially, at least, it is a good approximation to take ω_3 to be constant. In this case, the first two Euler equations read

$$\left.\begin{array}{l} \lambda_1\dot{\omega}_1 = [(\lambda_2 - \lambda_3)\omega_3]\omega_2 \\ \lambda_2\dot{\omega}_2 = [(\lambda_3 - \lambda_1)\omega_3]\omega_1 \end{array}\right\} \tag{10.90}$$

where the coefficients in square brackets are (approximately) constant. These coupled first-order equations for ω_1 and ω_2 are easily solved. Perhaps the simplest method is to differentiate the first equation once and then substitute the second, to give

$$\ddot{\omega}_1 = -\left[\frac{(\lambda_3 - \lambda_2)(\lambda_3 - \lambda_1)}{\lambda_1\lambda_2}\omega_3^2\right]\omega_1. \tag{10.91}$$

If the coefficient in square brackets is positive, the solution for ω_1 is a sine or cosine in time, and ω_1 undergoes small oscillations, returning repeatedly to zero. According to the first equation of (10.90), ω_2 is proportional to $\dot{\omega}_1$, so, under the same conditions, ω_2 undergoes similar small oscillations. To see what these conditions imply, notice that the coefficient in (10.91) is positive if λ_3 is larger than both λ_1 and λ_2 or smaller than both λ_1 and λ_2. Therefore, we have shown that if the body is spinning about either the principal axis with largest moment or that with smallest moment, the motion is stable against small disturbances.

On the other hand if λ_3 lies between λ_1 and λ_2, the coefficient in brackets in (10.91) is negative, and the solution for ω_1 is a real exponential, which moves rapidly away from zero.[14] Since $\omega_2 \propto \dot{\omega}_1$, the same is true of ω_2, and we reach the following conclusion: For a freely spinning body, rotation about the principal axis with the intermediate moment (λ_3 less than λ_1 but more than λ_2 or vice versa) is unstable. You can test this interesting claim with a book, held shut with a rubber band. If you throw it up giving it a spin about either the maximum or minimum axis, it will continue to spin stably about that axis. If you give it a spin about the intermediate axis, it will tumble wildly (and be much harder to catch).

Motion of a Body with Two Equal Moments: Free Precession

The complete solution of the Euler equations (10.89) for a freely rotating body with three different principal moments is possible, but complicated and unilluminating. If two of the three principal moments are equal (as with a spinning top), the corresponding problem is much easier and quite interesting. Let us, therefore, consider this case,

---

[14] There is a small subtlety that may be worth mentioning: The solution for ω_1 is a combination of two exponentials, $e^{\pm\alpha t}$, and you might think that the special case of a pure decaying exponential, $e^{-\alpha t}$, was an exception to my claim. However, this solution is excluded by the initial conditions: If the body is given a small kick from $\omega_1 = 0$, then ω_1 and $\dot{\omega}_1$ must have the same sign, whereas for the pure decaying exponential they have opposite signs.

and suppose that the first two moments are equal, $\lambda_1 = \lambda_2$. The crucial simplifying feature of this case is that the third of the Euler equations (10.89) becomes

$$\dot{\omega}_3 = 0.$$

That is, ω_3, the third component of the angular velocity (as measured in the body frame), is constant,

$$\omega_3 = \text{const.}$$

Knowing that ω_3 is constant, we can now rewrite the first two Euler equations as

$$\left.\begin{array}{l} \dot{\omega}_1 = \dfrac{(\lambda_1 - \lambda_3)\omega_3}{\lambda_1}\,\omega_2 = \Omega_b\omega_2 \\[4mm] \dot{\omega}_2 = -\dfrac{(\lambda_1 - \lambda_3)\omega_3}{\lambda_1}\,\omega_1 = -\Omega_b\omega_1 \end{array}\right\} \qquad (10.92)$$

where I have defined the constant frequency

$$\Omega_b = \frac{\lambda_1 - \lambda_3}{\lambda_1}\,\omega_3. \qquad (10.93)$$

(The subscript "b" stands for body, for reasons we'll see in a moment.) The coupled equations (10.92) for ω_1 and ω_2 are easily solved by the now familiar trick of setting $\omega_1 + i\omega_2 = \eta$, which reduces (10.92) to

$$\dot{\eta} = -i\Omega_b\eta,$$

whence

$$\eta = \eta_0 e^{-i\Omega_b t}.$$

If we choose our axes so that $\omega_1 = \omega_0$ and $\omega_2 = 0$ at $t = 0$, then $\eta_0 = \omega_0$ and, taking real and imaginary parts of η, we find for the complete solution

$$\boldsymbol{\omega} = \left(\omega_0 \cos \Omega_b t,\ -\omega_0 \sin \Omega_b t,\ \omega_3\right) \qquad (10.94)$$

with ω_0 and ω_3 both constant. The two components ω_1 and ω_2 rotate with angular velocity Ω_b, while ω_3 remains constant. Since ω_0 and ω_3 are constant, so is the angle α between $\boldsymbol{\omega}$ and $\mathbf{e}_3$. Therefore, as seen in the body frame, $\boldsymbol{\omega}$ moves steadily around a cone, called the **body cone** with angular frequency Ω_b given by (10.93), as indicated in Figure 10.9(a).

The angular momentum $\mathbf{L}$ is given by

$$\mathbf{L} = (\lambda_1\omega_1,\ \lambda_1\omega_2,\ \lambda_3\omega_3)$$
$$= (\lambda_1\omega_0 \cos \Omega_b t,\ -\lambda_1\omega_0 \sin \Omega_b t,\ \lambda_3\omega_3). \qquad (10.95)$$

Comparison of (10.94) and (10.95) should convince you that the three vectors, $\boldsymbol{\omega}$, $\mathbf{L}$, and $\mathbf{e}_3$ lie in a single plane, with the angles between any two being constant in time. (We've shown this in the body frame, but this means that it's true in any frame.) Thus, as seen in the body frame, both $\boldsymbol{\omega}$ and $\mathbf{L}$ precess around $\mathbf{e}_3$ at the same rate Ω_b.

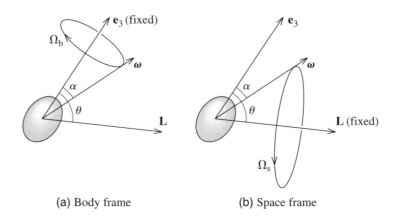

(a) Body frame   (b) Space frame

Figure 10.9   An axially symmetric body (shown here as a prolate spheroid or "egg-shaped" solid) is rotating with angular velocity $\boldsymbol{\omega}$, not in the direction of any of the principal axes. **(a)** As seen in the body frame, both $\boldsymbol{\omega}$ and $\mathbf{L}$ precess about the symmetry axis, $\mathbf{e}_3$, with angular frequency Ω_b given by (10.93). **(b)** As seen in the space frame, $\mathbf{L}$ is fixed, and both $\boldsymbol{\omega}$ and $\mathbf{e}_3$ precess about $\mathbf{L}$ with frequency Ω_s given by (10.96).

To find what happens as seen in the space frame, note that in any inertial frame the vector $\mathbf{L}$ is constant. Therefore, as seen in the space frame the plane containing $\boldsymbol{\omega}$, $\mathbf{L}$, and $\mathbf{e}_3$ must rotate about $\mathbf{L}$, and the two vectors $\boldsymbol{\omega}$ and $\mathbf{e}_3$ precess about $\mathbf{L}$, as shown in Figure 10.9(b). In particular, in the space frame, $\boldsymbol{\omega}$ traces out a cone, called the **space cone**, around which the body cone rolls. I leave it as a fairly difficult exercise (Problem 10.46) to show that the rate of precession of $\boldsymbol{\omega}$ around the space cone can be expressed in several ways, of which the simplest is this:

$$\Omega_s = \frac{L}{\lambda_1}. \tag{10.96}$$

Note well that the **free precession** derived here has nothing to do with any external torques on the spinning body. On the contrary, we have derived it for a body subject to *no* external torque. An interesting example of this precession is provided by the earth. I have already mentioned that the sun and moon produce a small torque on the earth's equatorial bulge, and that this torque causes the precession of the equinoxes, the precession of the polar axis with a period of 26,000 years. But the equatorial bulge also means that the earth's moment of inertia about the polar axis is larger than the other two principal moments by about 1 part in 300. According to (10.93) this should imply a precession (unless the earth's rotation is *perfectly* aligned with the principal axis) with frequency $\Omega_b = \omega_3/300$. Since ω_3 represents one revolution per day, Ω_b should correspond to one revolution every 300 days. A tiny wobbling of the polar axis (by less than a second of arc) was discovered by the American amateur astronomer Seth Chandler (1846–1913). This **Chandler wobble** is apparently due to the free-body precession discussed here, although its period is more like 400 days, supposedly because the earth is not a perfectly rigid body.

10.9 Euler Angles*

*Sections marked with an asterisk contain material that is usually somewhat more advanced and could be omitted if you are pressed for time.

The trouble with the Euler equations (10.88) is that they refer to axes fixed in the body, and, except in a few very simple problems, such axes are hopelessly awkward to work with. We need to come up with equations of motion relative to a nonrotating space frame, and before we can do that we need a set of coordinates that specify the orientation of our body relative to such a frame. There are several ways to do this, all of them surprisingly cumbersome, but by far the most popular and useful is due to Euler (yet again!) and specifies the orientation of a rigid body by three *Euler angles*. In many applications, the body of interest rotates about a fixed point, and in this case (the only case we'll consider in detail) we naturally choose the fixed point to be the origin O of both the space and the body axes. As before, the basis vectors of the space frame will be called $\hat{\mathbf{x}}$, $\hat{\mathbf{y}}$, and $\hat{\mathbf{z}}$. For the body frame, we'll use the principal axes of the body, with directions $\mathbf{e}_1$, $\mathbf{e}_2$, and $\mathbf{e}_3$. If two of the principal moments are equal, we'll take them to be numbers 1 and 2, so that $\lambda_1 = \lambda_2$, and we'll refer to the third direction $\mathbf{e}_3$ as the axis of symmetry. Let us imagine the body oriented initially with its three axes lying along the corresponding space axes ($\mathbf{e}_1$ along $\hat{\mathbf{x}}$ and so on). We're going to see that by a sequence of three rotations, through angles θ, ϕ, and ψ about three different axes, we can bring the body into any assigned orientation and that the angles (θ, ϕ, ψ) specify a unique orientation of the body. In particular, the angles θ and ϕ will be just the polar angles of the axis $\mathbf{e}_3$ relative to the space frame.

Step (a). Starting with the body axes aligned with the space axes, we first rotate the body through an angle ϕ about the axis $\hat{\mathbf{z}}$, as illustrated in the first frame of Figure 10.10. This rotates the first and second body axes in the xy plane. In particular, the second body axis now points in the direction labeled $\mathbf{e}_2'$.

Step (b). Next we rotate the body through an angle θ about the new axis $\mathbf{e}_2'$. This moves the body axis $\mathbf{e}_3$ to the direction whose polar angles are θ and ϕ. Evidently our

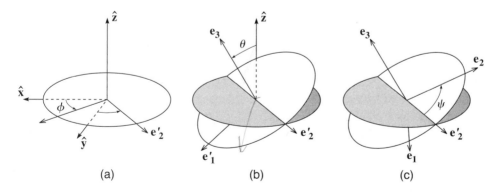

(a)   (b)   (c)

Figure 10.10 Definition of the Euler angles θ, ϕ, and ψ. Starting with the body axes $\mathbf{e}_1$, $\mathbf{e}_2$, $\mathbf{e}_3$ and space axes $\hat{\mathbf{x}}$, $\hat{\mathbf{y}}$, $\hat{\mathbf{z}}$ aligned, the three successive rotations bring the body axes to any prescribed orientation.

first two steps can bring the body axis $\mathbf{e}_3$ to any assigned orientation, and, with $\mathbf{e}_3$ in position, the only remaining freedom is a rotation about $\mathbf{e}_3$.

Step (c). Finally, we'll rotate the body about $\mathbf{e}_3$ through whatever angle ψ is needed to bring the body axes $\mathbf{e}_2$ and $\mathbf{e}_1$ into their assigned directions, as shown in the third frame of Figure 10.10.

The three angles (θ, ϕ, ψ) are the **Euler angles**[15] that specify the body's orientation. Before we can use them, we must calculate a few parameters in terms of them, starting with the angular velocity $\boldsymbol{\omega}$. To find $\boldsymbol{\omega}$, notice that we can regard the steps of Figure 10.10 as defining a sequence of four reference frames, starting with the space axes defined by $\hat{\mathbf{x}}$, $\hat{\mathbf{y}}$, and $\hat{\mathbf{z}}$, and moving via two intermediate frames to wind up with the body axes defined by $\mathbf{e}_1$, $\mathbf{e}_2$, and $\mathbf{e}_3$. To find the angular velocity of the body axes relative to the space axes, we have only to find the velocity of each of these frames relative to its predecessor and form their vector sum. [Remember that relative angular velocities add, as we observed in (9.25).] As ϕ varies, the frame defined by step (a) rotates with angular velocity $\boldsymbol{\omega}_a = \dot{\phi}\hat{\mathbf{z}}$ relative to the space axes. Similarly the angular velocity of the frame defined by step (b) relative to its predecessor is $\boldsymbol{\omega}_b = \dot{\theta}\mathbf{e}_2'$, and that of the body frame in step (c) is $\boldsymbol{\omega}_c = \dot{\psi}\mathbf{e}_3$. Therefore, the required angular velocity of the body frame relative to the space frame is

$$\boldsymbol{\omega} = \boldsymbol{\omega}_a + \boldsymbol{\omega}_b + \boldsymbol{\omega}_c = \dot{\phi}\hat{\mathbf{z}} + \dot{\theta}\mathbf{e}_2' + \dot{\psi}\mathbf{e}_3. \tag{10.97}$$

This expresses $\boldsymbol{\omega}$ in terms of a rather messy mixture of unit vectors, but it is a simple matter to rewrite these vectors in terms of the unit vectors of any one frame (Problem 10.48).

To find the angular momentum or kinetic energy, we need in general to find the components of $\boldsymbol{\omega}$ relative to the principal axes $\mathbf{e}_1$, $\mathbf{e}_2$, and $\mathbf{e}_3$. However, if we are content to consider just the symmetric case that $\lambda_1 = \lambda_2$, we don't even have to do this. The point is that with $\lambda_1 = \lambda_2$ any two perpendicular axes in the plane of $\mathbf{e}_1$ and $\mathbf{e}_2$ are also principal axes. Thus, instead of $\mathbf{e}_1$, $\mathbf{e}_2$, and $\mathbf{e}_3$ we can use the axes $\mathbf{e}_1'$, $\mathbf{e}_2'$, and $\mathbf{e}_3$ where $\mathbf{e}_1'$ and $\mathbf{e}_2'$ are both shown in the second frame of Figure 10.10. This choice has the advantage that the last two terms in (10.97) need no conversion and, since (as you should check)

$$\hat{\mathbf{z}} = (\cos\theta)\mathbf{e}_3 - (\sin\theta)\mathbf{e}_1', \tag{10.98}$$

we find that

$$\boldsymbol{\omega} = (-\dot{\phi}\sin\theta)\mathbf{e}_1' + \dot{\theta}\mathbf{e}_2' + (\dot{\psi} + \dot{\phi}\cos\theta)\mathbf{e}_3. \tag{10.99}$$

---

[15] Beware of the many different conventions used in defining Euler's angles. The convention followed here is most popular in quantum mechanics but less so among authors of classical mechanics. It has the great advantage that θ and ϕ are precisely the polar angles of the body axis $\mathbf{e}_3$.

Knowing the angular velocity $\boldsymbol{\omega}$ with respect to a set of principal axes, we can immediately write down the angular momentum $\mathbf{L}$ and the kinetic energy T. The angular momentum is $\mathbf{L} = (\lambda_1 \omega_1, \lambda_2 \omega_2, \lambda_3 \omega_3)$ (relative to any set of principal axes). Thus in our case

$$\mathbf{L} = (-\lambda_1 \dot{\phi} \sin\theta)\mathbf{e}_1' + \lambda_1 \dot{\theta}\mathbf{e}_2' + \lambda_3(\dot{\psi} + \dot{\phi}\cos\theta)\mathbf{e}_3. \qquad (10.100)$$

For future reference, note that the component of $\mathbf{L}$ along the body axis $\mathbf{e}_3$ is just

$$L_3 = \lambda_3 \omega_3 = \lambda_3(\dot{\psi} + \dot{\phi}\cos\theta). \qquad (10.101)$$

Also, as you can check (Problem 10.49), the component along the space axis $\hat{\mathbf{z}}$ is

$$L_z = \lambda_1 \dot{\phi}\sin^2\theta + \lambda_3(\dot{\psi} + \dot{\phi}\cos\theta)\cos\theta \qquad (10.102)$$

$$= \lambda_1 \dot{\phi}\sin^2\theta + L_3\cos\theta \qquad (10.103)$$

where in the second line I have used (10.101). Also for future reference, note that we can solve (10.103) for $\dot{\phi}$ in terms of θ, L_z and L_3:

$$\dot{\phi} = \frac{L_z - L_3\cos\theta}{\lambda_1 \sin^2\theta}. \qquad (10.104)$$

We saw in (10.68) that the kinetic energy is $T = \frac{1}{2}(\lambda_1 \omega_1^2 + \lambda_2 \omega_2^2 + \lambda_3 \omega_3^2)$. Thus, for a body whose first two principal moments are equal, (10.99) gives

$$T = \tfrac{1}{2}\lambda_1(\dot{\phi}^2 \sin^2\theta + \dot{\theta}^2) + \tfrac{1}{2}\lambda_3(\dot{\psi} + \dot{\phi}\cos\theta)^2. \qquad (10.105)$$

We shall use this result in a moment to write down the Lagrangian for a spinning top.

10.10 Motion of a Spinning Top *

Lagrange's Equations

To illustrate the use of Euler angles, let us return to the symmetric spinning top discussed in Section 10.6 and shown in Figure 10.7. The motion of this system is most easily solved using the Lagrangian approach, so we'll start by writing down the Lagrangian $\mathcal{L} = T - U$. The kinetic energy is given by (10.105), while the potential energy is $U = MgR\cos\theta$. Therefore the Lagrangian of the top is

$$\mathcal{L} = \tfrac{1}{2}\lambda_1(\dot{\phi}^2 \sin^2\theta + \dot{\theta}^2) + \tfrac{1}{2}\lambda_3(\dot{\psi} + \dot{\phi}\cos\theta)^2 - MgR\cos\theta. \qquad (10.106)$$

With three generalized coordinates, there are three Lagrange equations. The θ equation is

$$\lambda_1 \ddot{\theta} = \lambda_1 \dot{\phi}^2 \sin\theta \cos\theta - \lambda_3(\dot{\psi} + \dot{\phi}\cos\theta)\dot{\phi}\sin\theta + MgR\sin\theta \qquad [\theta \text{ equation}].$$

(10.107)

The ϕ and ψ equations are simpler because neither ϕ nor ψ appears in $\mathcal{L}$, so that both ϕ and ψ are ignorable, and the corresponding generalized momenta are constants. For p_ϕ this gives

$$p_\phi = \frac{\partial \mathcal{L}}{\partial \dot{\phi}} = \lambda_1 \dot{\phi} \sin^2\theta + \lambda_3(\dot{\psi} + \dot{\phi}\cos\theta)\cos\theta = \text{const} \qquad [\phi \text{ equation}]. \quad (10.108)$$

Comparing with (10.102), we see that the generalized momentum p_ϕ is just the z component L_z of the angular momentum, and the constancy of p_ϕ is just a statement that L_z is conserved — a result we could have anticipated since there is no torque about the z axis. Similarly, for p_ψ we get

$$p_\psi = \frac{\partial \mathcal{L}}{\partial \dot{\psi}} = \lambda_3(\dot{\psi} + \dot{\phi}\cos\theta) = \text{const} \qquad [\psi \text{ equation}]. \qquad (10.109)$$

Comparing this with (10.101), we see that p_ψ is the component of $\mathbf{L}$ along the body's symmetry axis $\mathbf{e}_3$, and the constancy of p_ψ tells us that L_3 is conserved. An important consequence is that, since $L_3 = \lambda_3 \omega_3$, the component ω_3 of the angular velocity is also constant (a result we proved back at the end of Section 10.7 using Euler's equations).

Steady Precession

As a first application of the Lagrange equations, let us investigate whether the top can exhibit a precession in which the top's axis moves around the z axis tracing a cone with constant angle θ. From (10.104) we see that if θ is constant, then so is $\dot{\phi}$. That is, if the top's axis is to move around a cone of fixed angle θ, it has to do so at a constant angular velocity $\dot{\phi} = \Omega$, say. Looking next at the ψ equation (10.109), we see that, with $\dot{\phi}$ and θ fixed, $\dot{\psi}$ must also be constant.

The rate Ω at which the top precesses is determined by the θ equation (10.107). If θ is constant, the left side is zero, and, replacing $\dot{\phi}$ by Ω, and $(\dot{\psi} + \dot{\phi}\cos\theta)$ by ω_3, we find that Ω must satisfy

$$\lambda_1 \Omega^2 \cos\theta - \lambda_3 \omega_3 \Omega + MgR = 0. \qquad (10.110)$$

This is a quadratic equation for Ω. Thus, as long as the roots are real, there are two different rates Ω at which the top can precess, for a given tilt θ and given rate of spin ω_3. We can write down these two values of Ω for any values of the other parameters, but the most interesting case is when the top is spinning rapidly and ω_3 is large. In this case, it is easy to see that the two roots *are* real and that one root is much smaller than the other. The small root is (Problem 10.53)

$$\Omega \approx \frac{MgR}{\lambda_3 \omega_3}. \qquad (10.111)$$

This slow precession is precisely the motion predicted in Section 10.6 with Ω given by (10.83).[16]

The second, larger root of (10.110) (again assuming that ω_3 is very large) is

$$\Omega \approx \frac{\lambda_3 \omega_3}{\lambda_1 \cos \theta}. \tag{10.112}$$

(See Problem 10.53.) Notice that this faster precession does not depend on g, so we would expect to observe this one even in the absence of gravity. In fact, this precession is precisely the free precession predicted in Section 10.8 for a symmetric body moving in the absence of any torques. As you can check, the value of Ω predicted here is the same as what was called Ω_s in (10.96). (See Problem 10.52.)

Nutation

In general, as a top precesses around the vertical axis (ϕ changing), the angle θ varies as well. Thus as the axis swings around the vertical it also nods up or down in a motion called **nutation**, from a Latin word meaning to nod repeatedly. We can investigate the variation of θ using the θ equation (10.107). The first step is to use the ϕ and ψ equations to eliminate the variables $\dot{\phi}$ and $\dot{\psi}$ in favor of the constants $p_\phi = L_z$ and $p_\psi = L_3$. This gives a second-order, ordinary differential equation for θ, which can, at least in principle, be solved to give the time dependence of θ.

To get a qualitative picture of the motion, a simpler procedure is to look at the total energy, $E = T + U$, with T given by (10.105) and $U = MgR\cos\theta$. In T, we can replace the variables $\dot{\phi}$ and $\dot{\psi}$ in favor of the constants L_z and L_3, and we find (Problem 10.51)

$$E = \tfrac{1}{2}\lambda_1 \dot{\theta}^2 + U_{\text{eff}}(\theta) \tag{10.113}$$

where the effective potential energy $U_{\text{eff}}(\theta)$ is

$$U_{\text{eff}}(\theta) = \frac{(L_z - L_3 \cos\theta)^2}{2\lambda_1 \sin^2\theta} + \frac{L_3^2}{2\lambda_3} + MgR\cos\theta. \tag{10.114}$$

Equation (10.113) serves to emphasize that our problem has been reduced to a one-dimensional problem, involving just the coordinate θ. We can predict the qualitative behavior of θ by looking at a graph of $U_{\text{eff}}(\theta)$. The coordinate θ ranges from 0 to π, and, because of the factor of $\sin^2\theta$ in the denominator of the first term, $U_{\text{eff}}(\theta)$ approaches $+\infty$ at the two ends, $\theta = 0$ and π. It is not hard to convince oneself that the graph of $U_{\text{eff}}(\theta)$ has the "U" shape shown in Figure 10.11. From (10.113), it is clear that $E \geq U_{\text{eff}}(\theta)$, so θ is confined between the two turning points, θ_1 and θ_2, where $E = U_{\text{eff}}(\theta)$ as shown in the figure. Evidently, θ oscillates periodically, or "nutates," between θ_1 and θ_2 at the same time that the top's axis precesses around the vertical.

---

[16] Here the denominator has a factor ω_3 where (10.83) has ω, but this difference is irrelevant since both discussions assume that ω_3 is very large, so that $\omega_3 \approx \omega$.

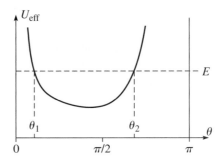

Figure 10.11   The effective potential energy (10.114) that determines the time dependence of θ for the symmetric top. Since E
can be no less than $U_{\text{eff}}(\theta)$, the motion is confined to the interval
$\theta_1 \leq \theta \leq \theta_2$.

The details of the motion depend on just how ϕ varies. According to (10.104),

$$\dot{\phi} = \frac{L_z - L_3 \cos \theta}{\lambda_1 \sin^2 \theta}, \tag{10.115}$$

which shows that there are two main possibilities: If L_z is larger than L_3 (in magnitude), then $\dot{\phi}$ cannot vanish. Thus, although $\dot{\phi}$ may vary, it can never change sign, so
ϕ changes in the same direction all the time (always increasing or always decreasing).
Thus, the top precesses in a single direction, while its angle of tilt θ oscillates between
θ_1 and θ_2, producing the motion sketched in Figure 10.12(a). If L_z is smaller than L_3,
then $\dot{\phi}$ would vanish at an angle θ_0, such that $L_z - L_3 \cos \theta_0 = 0$. If this angle lies out-

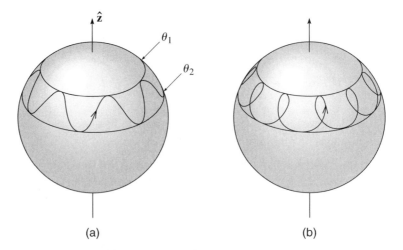

(a)   (b)

Figure 10.12   Nutation of a top. The top end of the spinning top moves on a
sphere centered on the pivot at the bottom end. (a) Here $\dot{\phi}$ never vanishes, and
ϕ always moves steadily in one direction while θ oscillates between θ_1 and θ_2.
(b) If $\dot{\phi}$ changes sign, then ϕ moves first forward and then backward while θ
oscillates.

side the range between θ_1 and θ_2 to which the motion is confined, then $\dot{\phi}$ still cannot vanish and the motion is still as sketched in Figure 10.12(a). On the other hand, if the angle θ_o lies between θ_1 and θ_2, then $\dot{\phi}$ will change sign twice in each oscillation of θ. In this case, the precession moves first in one direction, then in the other, and the overall motion is as sketched in Figure 10.12(b).

Principal Definitions and Equations of Chapter 10

CM and Relative Motions

$$\mathbf{L} = \mathbf{L}(\text{motion of CM}) + \mathbf{L} \text{ (motion relative to CM)}. \qquad [\text{Eq. (10.9)}]$$

and

$$T = T(\text{motion of CM}) + T(\text{motion relative to CM}). \qquad [\text{Eq. (10.16)}]$$

The Moment of Inertia Tensor

The angular momentum $\mathbf{L}$ and angular velocity $\boldsymbol{\omega}$ of a rigid body are related by

$$\mathbf{L} = \mathbf{I}\boldsymbol{\omega} \qquad [\text{Eq.(10.42)}]$$

where $\mathbf{L}$ and $\boldsymbol{\omega}$ must be seen as 3×1 columns and $\mathbf{I}$ is the 3×3 **moment of inertia tensor**, whose diagonal and off-diagonal elements are defined as

$$I_{xx} = \sum_\alpha m_\alpha(y_\alpha^2 + z_\alpha^2), \text{ etc.} \quad \text{and} \quad I_{xy} = -\sum_\alpha m_\alpha x_\alpha y_\alpha, \text{ etc.}$$

$$[\text{Eqs. (10.37) \& (10.38)}]$$

Principal Axes

A **principal axis** of a body (about a point O) is any axis through O with the property that if $\boldsymbol{\omega}$ points along the axis, then $\mathbf{L}$ is parallel to $\boldsymbol{\omega}$; that is,

$$\mathbf{L} = \lambda\boldsymbol{\omega} \qquad [\text{Eq. (10.65)}]$$

for some real number λ. For any body and any point O, there are three perpendicular principal axes through O.   [Section 10.4 and Appendix]

Evaluated with respect to its principal axes, the inertia tensor has the **diagonal form**

$$\mathbf{I}' = \begin{bmatrix} \lambda_1 & 0 & 0 \\ 0 & \lambda_2 & 0 \\ 0 & 0 & \lambda_3 \end{bmatrix}. \qquad [\text{Eq. (10.79)}]$$

Euler's Equations

If $\dot{\mathbf{L}}$ denotes the rate of change of a body's angular momentum as seen in a frame fixed in the body (body frame), then it satisfies **Euler's equations**

$$\dot{\mathbf{L}} + \boldsymbol{\omega} \times \mathbf{L} = \boldsymbol{\Gamma}. \qquad \text{[Eqs. (10.87) \& (10.88)]}$$

Euler's Angles

The orientation of a rigid body can be specified by the three **Euler angles** θ, ϕ, ψ defined in Figure 10.10.   [Sections 10.9 & 10.10]

   The Lagrangian for a rigid body spinning about a fixed pivot is

$$\mathcal{L} = \tfrac{1}{2}\lambda_1(\dot{\phi}^2 \sin^2\theta + \dot{\theta}^2) + \tfrac{1}{2}\lambda_3(\dot{\psi} + \dot{\phi}\cos\theta)^2 - MgR\cos\theta. \qquad \text{[Eq. (10.106)]}$$

Problems for Chapter 10

Stars indicate the approximate level of difficulty, from easiest (★) to most difficult (★★★).

SECTION 10.1   Properties of the Center of Mass

10.1 ★ The result (10.7), that $\sum m_\alpha \mathbf{r}'_\alpha = 0$, can be paraphrased to say that the position vector of the CM relative to the CM is zero, and, in this form, is nearly obvious. Nevertheless, to be sure you understand the result, prove it by solving (10.4) for $\mathbf{r}'_\alpha$ and substituting into the sum concerned.

10.2 ★ To illustrate the result (10.18), that the total KE of a body is just the rotational KE relative to any point that is instantaneously at rest, do the following: Write down the KE of a uniform wheel (mass M, radius R) rolling with speed v along a flat road, as the sum of the energies of the CM motion and the rotation about the CM. Now write it as the energy of the rotation about the instantaneous point of contact with the road and show that you get the same answer. (The energy of rotation is $\tfrac{1}{2}I\omega^2$. The moment of inertia of a uniform wheel about its center is $I = \tfrac{1}{2}MR^2$. That about a point on the rim is $I' = \tfrac{3}{2}MR^2$.)

10.3 ★ Five equal point masses are placed at the five corners of a square pyramid whose square base is centered on the origin in the xy plane, with side L, and whose apex is on the z axis at a height H above the origin. Find the CM of the five-mass system.

10.4 ★★ The calculation of centers of mass or moments of inertia usually involves doing an integral, most often a volume integral, and such integrals are often best done in spherical polar coordinates (defined back in Figure 4.16). Prove that

$$\int dV f(\mathbf{r}) = \int r^2 dr \int \sin\theta \, d\theta \int d\phi \, f(r, \theta, \phi).$$

[Think about the small volume dV enclosed between r and $r + dr$, θ and $\theta + d\theta$, and ϕ and $\phi + d\phi$.] If the volume integral on the left runs over all space, what are the limits of the three integrals on the right?

10.5 ★★ A uniform solid hemisphere of radius R has its flat base in the xy plane, with its center at the origin. Use the result of Problem 10.4 to find the center of mass. [Comment: This and the next two problems are intended to reactivate your skills at finding centers of mass by integration. In all cases, you will need to use the integral form of the definition (10.1) of the CM. If the mass is distributed through a volume (as here), the integral will be a volume integral with $dm = \varrho \, dV$.]

10.6 ★★ **(a)** Find the CM of a uniform hemispherical shell of inner and outer radii a and b and mass M positioned as in Problem 10.5. [See the comment to Problem 10.5 and use the result of Problem 10.4.] **(b)** What becomes of your answer when $a = 0$? **(c)** What if $b \to a$?

10.7 ★★ A "rounded cone" is made by cutting out of a uniform sphere of radius R the volume with $\theta \le \theta_\mathrm{o}$, where θ is the usual angle measured from the polar axis and θ_o is a constant between 0 and π. **(a)** Describe this cone and use the result of Problem 10.4 to find its volume. **(b)** Find its CM and comment on your results for the cases that $\theta_\mathrm{o} = \pi$ and $\theta_\mathrm{o} \to 0$.

10.8 ★★ A uniform thin wire lies along the y axis between $y = \pm L/2$. It is now bent toward the left into an arc of a circle with radius R, leaving the midpoint at the origin and tangent to the y axis. Find the CM. [See the comment to Problem 10.5. In this case the integral is a one-dimensional integral.] Comment on your answer for the cases that $R \to \infty$ and that $2\pi R = L$.

SECTION 10.2 Rotation about a Fixed Axis

10.9 ★ The moment of inertia of a continuous mass distribution with density ϱ is obtained by converting the sum of (10.25) into the volume integral $\int \rho^2 \, dm = \int \rho^2 \varrho \, dV$. (Note the two forms of the Greek "rho": ρ = distance from z axis, ϱ = mass density.) Find the moment of inertia of a uniform circular cylinder of radius R and mass M for rotation about its axis. Explain why the products of inertia are zero.

10.10 ★ **(a)** A thin uniform rod of mass M and length L lies on the x axis with one end at the origin. Find its moment of inertia for rotation about the z axis. [Here the sum of (10.25) must be replaced by an integral of the form $\int x^2 \mu \, dx$ where μ is the linear mass density, mass/length.] **(b)** What if the rod's center is at the origin?

10.11 ★★ **(a)** Use the result of Problem 10.4 to find the moment of inertia of a uniform solid sphere (mass M, radius R) for rotation about a diameter. **(b)** Do the same for a uniform hollow sphere whose inner and outer radii are a and b. [One slick way to do this is to think of the hollow sphere as a solid sphere of radius b from which you have removed a sphere of the same density but radius a.]

10.12 ★★ A triangular prism (like a box of Toblerone) of mass M, whose two ends are equilateral triangles parallel to the xy plane with side $2a$, is centered on the origin with its axis along the z axis. Find its moment of inertia for rotation about the z axis. Without doing any integrals write down and explain its two products of inertia for rotation about the z axis.

10.13 ★★ A thin rod (of width zero, but not necessarily uniform) is pivoted freely at one end about the horizontal z axis, being free to swing in the xy plane (x horizontal, y vertically down). Its mass is m, its CM is a distance a from the pivot, and its moment of inertia (about the z axis) is I. **(a)** Write down the equation of motion $\dot{L}_z = \Gamma_z$ and, assuming the motion is confined to small angles (measured from the downward vertical), find the period of this compound pendulum. ("Compound pendulum" is traditionally used to mean any pendulum whose mass is distributed — as contrasted with a "simple pendulum," whose mass is concentrated at a single point on a massless arm.) **(b)** What is the length of the "equivalent" simple pendulum, that is, the simple pendulum with the same period?

10.14 ★★ A stationary space station can be approximated as a hollow spherical shell of mass 6 tonnes (6000 kg) and inner and outer radii of 5 m and 6 m. To change its orientation, a uniform flywheel (radius

10 cm, mass 10 kg) at the center is spun up quickly from rest to 1000 rpm. **(a)** How long will it take the station to rotate by 10 degrees? **(b)** What energy was needed for the whole operation? [To find the necessary moment of inertia, you could do Problem 10.11.]

10.15 ★★ **(a)** Write down the integral (as in Problem 10.9) for the moment of inertia of a uniform cube of side a and mass M, rotating about an edge, and show that it is equal to $\frac{2}{3}Ma^2$. **(b)** If I balance the cube on an edge in unstable equilibrium on a rough table, it will eventually topple and rotate until it hits the table. By considering the energy of the cube, find its angular velocity just before it hits the table. (Assume the edge does not slide on the table.)

10.16 ★★ Find the moment of inertia for a uniform cube of mass M and edge a as in Problem 10.15 and then do the following: The cube is sliding with velocity $\mathbf{v}$ across a flat horizontal frictionless table when it hits a straight very low step perpendicular to $\mathbf{v}$, and the leading lower edge comes abruptly to rest. **(a)** By considering what quantities are conserved before, during, and after the brief collision, find the cube's angular velocity just after the collision. **(b)** Find the minimum speed v for which the cube rolls over after hitting the step.

10.17 ★★ Write down the integral for the moment of inertia of a uniform ellipsoid (mass M) with surface $(x/a)^2 + (y/b)^2 + (z/c)^2 = 1$ for rotation about the z axis. One simple way to do the integral is to make a change of variables to $\xi = x/a$, $\eta = y/b$, and $\zeta = z/c$. Each of the two resulting integrals can be related to the corresponding integrals for a sphere (as in Problem 10.11). Do this. Check your answer for the case $a = b = c$.

10.18 ★★★ Consider the rod of Problem 10.13. The rod is struck sharply with a horizontal force F which delivers an impulse $F\,\Delta t = \xi$ a distance b below the pivot. **(a)** Find the rod's angular momentum about the pivot, and hence momentum, just after the impulse. **(b)** Find the impulse η delivered to the pivot. **(c)** For what value of b (call it b_o) is $\eta = 0$? (The distance b_o defines the so-called "sweet spot." If the rod were a tennis racket and the pivot your hand, then if the ball hits the sweet spot, your hand would experience no impulse.)

SECTION 10.3   Rotation about Any Axis; the Inertia Tensor

10.19 ★ Verify that the components of the vector $\mathbf{r} \times (\boldsymbol{\omega} \times \mathbf{r})$ are given correctly by Equation (10.35). Do this both by working with components and by using the so-called $BAC - CAB$ rule, that is $\mathbf{A} \times (\mathbf{B} \times \mathbf{C}) = \mathbf{B}(\mathbf{A} \cdot \mathbf{C}) - \mathbf{C}(\mathbf{A} \cdot \mathbf{B})$.

10.20 ★ Show that the inertia tensor is additive, in this sense: Suppose a body A is made up of two parts B and C. (For instance, a hammer is made up of a wooden handle wedged into a metal head.) Then $\mathbf{I}_A = \mathbf{I}_B + \mathbf{I}_C$. Similarly, if A can be thought of as the result of removing C from B (as a hollow spherical shell is the result of removing a small sphere from inside a larger sphere), then $\mathbf{I}_A = \mathbf{I}_B - \mathbf{I}_C$.

10.21 ★★ The definition of the inertia tensor in Equations (10.37) and (10.38) has the rather ugly feature that the diagonal and off-diagonal elements are defined by completely different equations. Show that the two definitions can be combined into the single equation (which is slightly less messy in integral form)

$$I_{ij} = \int \varrho(r^2\delta_{ij} - r_i r_j)\,dV$$

where δ_{ij} is the **Kronecker delta symbol**

$$\delta_{ij} = \begin{cases} 1 & i = j \\ 0 & i \neq j. \end{cases} \tag{10.116}$$

10.22 ⋆⋆ A rigid body comprises 8 equal masses m at the corners of a cube of side a, held together by massless struts. **(a)** Use the definitions (10.37) and (10.38) to find the moment of inertia tensor **I** for rotation about a corner O of the cube. (Use axes along the three edges through O.) **(b)** Find the inertia tensor of the same body but for rotation about the center of the cube. (Again use axes parallel to the edges.) Explain why in this case certain elements of **I** could be expected to be zero.

10.23 ⋆⋆ Consider a rigid plane body or "lamina," such as a flat piece of sheet metal, rotating about a point O in the body. If we choose axes so that the lamina lies in the xy plane, which elements of the inertia tensor **I** are automatically zero? Prove that $I_{zz} = I_{xx} + I_{yy}$.

10.24 ⋆⋆ **(a)** If $\mathbf{I}^{\text{cm}}$ denotes the moment of inertia tensor of a rigid body (mass M) about its CM, and **I** the corresponding tensor about a point P displaced from the CM by $\boldsymbol{\Delta} = (\xi, \eta, \zeta)$, prove that

$$I_{xx} = I_{xx}^{\text{cm}} + M(\eta^2 + \zeta^2) \qquad \text{and} \qquad I_{yz} = I_{yz}^{\text{cm}} - M\eta\zeta, \tag{10.117}$$

and so forth. (These results, which generalize the parallel-axis theorem that you probably learned in introductory physics, mean that once you know the inertia tensor for rotation about the CM, calculating it for any other origin is trivially easy.) **(b)** Confirm that the results of Example 10.2 (page 381) fulfill the identities (10.117) [so that the calculations of part (a) of the example were actually unnecessary].

10.25 ⋆⋆ **(a)** Find all nine elements of the moment of inertia tensor with respect to the CM of a uniform cuboid (a rectangular brick shape) whose sides are $2a$, $2b$, and $2c$ in the x, y, and z directions and whose mass is M. Explain clearly why you could write down the off-diagonal elements without doing any integration. **(b)** Combine the results of part (a) and Problem 10.24 to find the moment of inertia tensor of the same cuboid with respect to the corner A at (a, b, c). **(c)** What is the angular momentum about A if the cuboid is spinning with angular velocity ω around the edge through A and parallel to the x axis?

10.26 ⋆⋆ **(a)** Prove that in cylindrical polar coordinates a volume integral takes the form

$$\int dV \, f(\mathbf{r}) = \int \rho \, d\rho \int d\phi \int dz \, f(\rho, \phi, z).$$

(b) Show that the moment of inertia of the cone in Figure 10.6 pivoted at its tip and rotating about its axis is given by the integral (10.58), explaining clearly the limits of integration. Show that the integral evaluates to $\frac{3}{10}MR^2$. **(c)** Prove also that $I_{xx} = \frac{3}{20}M(R^2 + 4h^2)$ as in Equation (10.61).

10.27 ⋆⋆⋆ Find the inertia tensor for a uniform, thin hollow cone, such as an ice-cream cone, of mass M, height h, and base radius R, spinning about its pointed end.

10.28 ⋆⋆⋆ Find the moment of inertia tensor **I** for the triangular prism of Problem 10.12 with height h. (If you did Problem 10.12, you've already done about half the work.) Your result should show that **I** has the form we've found for an axisymmetric body. This suggests what is true, that three-fold symmetry about an axis (symmetry under rotations of 120 degrees) is enough to ensure this form.

SECTION 10.4   Principal Axes of Inertia

10.29 ★ Prove that if the axes Ox, Oy, and Oz are principal axes of a certain rigid body, then the inertia tensor (with respect to these axes) is diagonal with the principal moments down the diagonal as in (10.66).

10.30 ★ Consider a lamina, such as a flat piece of sheet metal, rotating about a point O in the body. Prove that the axis through O and perpendicular to the plane is a principal axis. [*Hint:* See Problem 10.23.]

10.31 ★★ Consider an arbitrary rigid body with an axis of rotational symmetry, which we'll call $\hat{z}$. **(a)** Prove that the axis of symmetry is a principal axis. **(b)** Prove that any two directions $\hat{x}$ and $\hat{y}$ perpendicular to $\hat{z}$ and each other are also principal axes. **(c)** Prove that the principal moments corresponding to these two axes are equal: $\lambda_1 = \lambda_2$.

10.32 ★★ **(a)** Show that the principal moments of any rigid body satisfy $\lambda_3 \leq \lambda_1 + \lambda_2$. [*Hint:* Look at the integrals that define these moments.] In particular, if $\lambda_1 = \lambda_2$, then $\lambda_3 \leq 2\lambda_1$. **(b)** For what shape of body is $\lambda_3 = \lambda_1 + \lambda_2$?

10.33 ★★★ Here is a good exercise in vector identities and matrices, leading to some important general results: **(a)** For a rigid body made up of particles of mass m_α, spinning about an axis through the origin with angular velocity ω, prove that its total kinetic energy can be written as

$$T = \tfrac{1}{2} \sum m_\alpha [(\omega r_\alpha)^2 - (\boldsymbol{\omega} \cdot \mathbf{r}_\alpha)^2].$$

Remember that $\mathbf{v}_\alpha = \boldsymbol{\omega} \times \mathbf{r}_\alpha$. You may find the following vector identity useful: For any two vectors $\mathbf{a}$ and $\mathbf{b}$,

$$(\mathbf{a} \times \mathbf{b})^2 = a^2 b^2 - (\mathbf{a} \cdot \mathbf{b})^2.$$

(If you use the identity, please prove it.) **(b)** Prove that the angular momentum $\mathbf{L}$ of the body can be written as

$$\mathbf{L} = \sum m_\alpha [\omega r_\alpha{}^2 - \mathbf{r}_\alpha (\boldsymbol{\omega} \cdot \mathbf{r}_\alpha)].$$

For this you will need the so-called $BAC - CAB$ rule, that $\mathbf{A} \times (\mathbf{B} \times \mathbf{C}) = \mathbf{B}(\mathbf{A} \cdot \mathbf{C}) - \mathbf{C}(\mathbf{A} \cdot \mathbf{B})$. **(c)** Combine the results of parts (a) and (b) to prove that

$$T = \tfrac{1}{2}\boldsymbol{\omega} \cdot \mathbf{L} = \tfrac{1}{2}\tilde{\omega}\mathbf{I}\omega.$$

Prove both equalities. The last expression is a matrix product; ω denotes the 3×1 column of numbers $\omega_x, \omega_y, \omega_z$, the tilde on $\tilde{\omega}$ denotes the matrix transpose (in this case a row), and $\mathbf{I}$ is the moment of inertia tensor. This result is actually quite important; it corresponds to the much more obvious result that for a particle, $T = \tfrac{1}{2}\mathbf{v} \cdot \mathbf{p}$. **(d)** Show that with respect to the principal axes, $T = \tfrac{1}{2}(\lambda_1 \omega_1^2 + \lambda_2 \omega_2^2 + \lambda_3 \omega_3^2)$, as in Equation (10.68).

SECTION 10.5   Finding the Principal Axes; Eigenvalue Equations

10.34 ★ The inertia tensor $\mathbf{I}$ for a solid cube is given by (10.72). Verify that $\det(\mathbf{I} - \lambda\mathbf{1})$ is as given in (10.73).

10.35 ★★ A rigid body consists of three masses fastened as follows: m at $(a, 0, 0)$, $2m$ at $(0, a, a)$, and $3m$ at $(0, a, -a)$. **(a)** Find the inertia tensor **I**. **(b)** Find the principal moments and a set of orthogonal principal axes.

10.36 ★★ A rigid body consists of three equal masses (m) fastened at the positions $(a, 0, 0)$, $(0, a, 2a)$, and $(0, 2a, a)$. **(a)** Find the inertia tensor **I**. **(b)** Find the principal moments and a set of orthogonal principal axes.

10.37 ★★★ A thin, flat, uniform metal triangle lies in the xy plane with its corners at $(1, 0, 0)$, $(0, 1, 0)$, and the origin. Its surface density (mass/area) is $\sigma = 24$. (Distances and masses are measured in unspecified units, and the number 24 was chosen to make the answer come out nicely.) **(a)** Find the triangle's inertia tensor **I**. **(b)** What are its principal moments and the corresponding axes?

10.38 ★★★ Suppose that you have found three independent principal axes (directions $\mathbf{e}_1$, $\mathbf{e}_2$, $\mathbf{e}_3$) and corresponding principal moments λ_1, λ_2, λ_3 of a rigid body whose moment of inertia tensor **I** (not diagonal) you had calculated. (You may assume, what is actually fairly easy to prove, that all of the quantities concerned are real.) **(a)** Prove that if $\lambda_i \neq \lambda_j$ then it is automatically the case that $\mathbf{e}_i \cdot \mathbf{e}_j = 0$. (It may help to introduce a notation that distinguishes between vectors and matrices. For example, you could use an underline to indicate a matrix, so that $\underline{\mathbf{a}}$ is the 3×1 matrix that represents the vector $\mathbf{a}$, and the vector scalar product $\mathbf{a} \cdot \mathbf{b}$ is the same as the matrix product $\tilde{\underline{\mathbf{a}}}\,\underline{\mathbf{b}}$ or $\tilde{\underline{\mathbf{b}}}\,\underline{\mathbf{a}}$. Then consider the number $\tilde{\underline{\mathbf{e}}}_i \underline{\mathbf{I}}\,\underline{\mathbf{e}}_j$, which can be evaluated in two ways using the fact that both $\mathbf{e}_i$ and $\mathbf{e}_j$ are eigenvectors of **I**.) **(b)** Use the result of part (a) to show that if the three principal moments are all different, then the directions of three principal axes are uniquely determined. **(c)** Prove that if two of the principal moments are equal, $\lambda_1 = \lambda_2$ say, then any direction in the plane of $\mathbf{e}_1$ and $\mathbf{e}_2$ is also a principal axis with the same principal moment. In other words, when $\lambda_1 = \lambda_2$ the corresponding principal axes are not uniquely determined. **(d)** Prove that if all three principal moments are equal, then *any* axis is a principal axis with the same principal moment.

SECTION 10.6   Precession of a Top Due to a Weak Torque

10.39 ★ Consider a top consisting of a uniform cone spinning freely about its tip at 1800 rpm. If its height is 10 cm and its base radius 2.5 cm, at what angular velocity will it precess?

SECTION 10.7   Euler's Equations

10.40 ★★ **(a)** A rigid body is rotating freely, subject to zero torque. Use Euler's equations (10.88) to prove that the magnitude of the angular momentum **L** is constant. (Multiply the ith equation by $L_i = \lambda_i \omega_i$ and add the three equations.) **(b)** In much the same way, show that the kinetic energy of rotation $T_{\text{rot}} = \frac{1}{2}(\lambda_1 \omega_1^2 + \lambda_2 \omega_2^2 + \lambda_3 \omega_3^2)$, as in (10.68), is constant.

10.41 ★★ Consider a lamina rotating freely (no torques) about a point O of the lamina. Use Euler's equations to show that the component of $\boldsymbol{\omega}$ in the plane of the lamina has constant magnitude. [*Hint:* Use the results of Problems 10.23 and 10.30. According to Problem 10.30, if you choose the direction $\mathbf{e}_3$ normal to the plane of the lamina, $\mathbf{e}_3$ points along a principal axis. Then what you have to prove is that the time derivative of $\omega_1^2 + \omega_2^2$ is zero.]

SECTION 10.8   Euler's Equations with Zero Torque

10.42 ★ I take a book that is 30 cm × 20 cm × 3 cm and is held shut by a rubber band, and I throw it into the air spinning about an axis that is close to the book's shortest symmetry axis at 180 rpm. What is the angular frequency of the small oscillations of its axis of rotation? What if I spin it about an axis close to the longest symmetry axis?

10.43 ★★ I throw a thin uniform circular disc (think of a frisbee) into the air so that it spins with angular velocity $\boldsymbol{\omega}$ about an axis which makes an angle α with the axis of the disc. **(a)** Show that the magnitude of $\boldsymbol{\omega}$ is constant. [Look at Equation (10.94).] **(b)** Show that as seen by me, the disc's axis precesses around the fixed direction of the angular momentum with angular velocity $\Omega_s = \omega\sqrt{4 - 3\sin^2\alpha}$. (The results of Problems 10.23 and 10.46 will be useful.)

10.44 ★★ An axially symmetric space station (principal axis $\mathbf{e}_3$, and $\lambda_1 = \lambda_2$) is floating in free space. It has rockets mounted symmetrically on either side that are firing and exert a constant torque Γ about the symmetry axis. Solve Euler's equations exactly for $\boldsymbol{\omega}$ (relative to the body axis) and describe the motion. At $t = 0$ take $\boldsymbol{\omega} = (\omega_{10}, 0, \omega_{30})$.

10.45 ★★ Because of the earth's equatorial bulge, its moment about the polar axis is slightly greater that the other two moments, $\lambda_3 = 1.00327\lambda_1$ (but $\lambda_1 = \lambda_2$). **(a)** Show that the free precession described in Section 10.8 should have period 305 days. (As described in the text, the period of this "Chandler wobble" is actually more like 400 days.) **(b)** The angle between the polar axis and $\boldsymbol{\omega}$ is about 0.2 arc seconds. Use Equation (10.118) from Problem 10.46 to show that as seen from the space frame the period of this wobble should be about a day.

10.46 ★★★ We saw in Section 10.8 that in the free precession of an axially symmetric body the three vectors $\mathbf{e}_3$ (the body axis), $\boldsymbol{\omega}$, and $\mathbf{L}$ lie in a plane. As seen in the body frame, $\mathbf{e}_3$ is fixed, and $\boldsymbol{\omega}$ and $\mathbf{L}$ precess around $\mathbf{e}_3$ with angular velocity $\Omega_b = \omega_3(\lambda_1 - \lambda_3)/\lambda_1$. As seen in the space frame $\mathbf{L}$ is fixed and $\boldsymbol{\omega}$ and $\mathbf{e}_3$ precess around $\mathbf{L}$ with angular velocity Ω_s. In this problem you will find three equivalent expressions for Ω_s. **(a)** Argue that $\boldsymbol{\Omega}_s = \boldsymbol{\Omega}_b + \boldsymbol{\omega}$. [Remember that relative angular velocities add like vectors.] **(b)** Bearing in mind that $\boldsymbol{\Omega}_b$ is parallel to $\mathbf{e}_3$ prove that $\Omega_s = \omega\sin\alpha/\sin\theta$ where α is the angle between $\mathbf{e}_3$ and $\boldsymbol{\omega}$ and θ is that between $\mathbf{e}_3$ and $\mathbf{L}$ (see Figure 10.9). **(c)** Thence prove that

$$\Omega_s = \omega\frac{\sin\alpha}{\sin\theta} = \frac{L}{\lambda_1} = \omega\frac{\sqrt{\lambda_3^2 + (\lambda_1^2 - \lambda_3^2)\sin^2\alpha}}{\lambda_1}. \tag{10.118}$$

10.47 ★★★ Imagine that this world is perfectly rigid, uniform, and spherical and is spinning about its usual axis at its usual rate. A huge mountain of mass 10^{-8} earth masses is now added at colatitude $60°$, causing the earth to begin the free precession described in Section 10.8. How long will it take the North Pole (defined as the northern end of the diameter along $\boldsymbol{\omega}$) to move 100 miles from its current position? [Take the earth's radius to be 4000 miles.]

SECTION 10.9   Euler Angles *

10.48 ★★ Equation (10.97) gives the angular velocity of a body in terms of an unholy mixture of unit vectors. **(a)** Find $\boldsymbol{\omega}$ in terms of $\hat{\mathbf{x}}$, $\hat{\mathbf{y}}$, and $\hat{\mathbf{z}}$. **(b)** Do the same in terms of $\mathbf{e}_1$, $\mathbf{e}_2$, and $\mathbf{e}_3$.

10.49 ★★ Starting from Equation (10.100) for $\mathbf{L}$, verify that L_z is correctly given by Equations (10.102) and (10.103).

10.50 ★★ Equation (10.105) gives the kinetic energy in terms of Euler angles for a body with $\lambda_1 = \lambda_2$. Find the corresponding expression for a body whose three principal moments are all different.

SECTION 10.10 Motion of a Spinning Top *

10.51 ⋆ Verify that the energy of a symmetric top can be written as $E = \frac{1}{2}\lambda_1\dot\theta^2 + U_{\text{eff}}(\theta)$, where the effective potential energy is as given in (10.114).

10.52 ⋆⋆ Consider the rapid steady precession of a symmetric top predicted in connection with (10.112). **(a)** Show that in this motion the angular momentum **L** must be very close to the vertical. [*Hint:* Use (10.100) to write down the horizontal component L_{hor} of **L**. Show that if $\dot\phi$ is given by the right side of (10.112), L_{hor} is exactly zero.] **(b)** Use this result to show that the rate of precession Ω given in (10.112) agrees with the free precession rate Ω_s found in (10.96).

10.53 ⋆⋆ In the discussion of steady precession of a top in Section 10.10, the rates Ω at which steady precession can occur were determined by the quadratic equation (10.110). In particular, we examined this equation for the case that ω_3 is very large. In this case you can write the equation as $a\Omega^2 + b\Omega + c = 0$ where b is very large. **(a)** Verify that when b is very large, the two solutions of this equation are approximately $-c/b$ (which is small) and $-b/a$ (which is large). What precisely does the condition "b is very large" entail? **(b)** Verify that these give the two solutions claimed in (10.111) and (10.112).

10.54 ⋆⋆⋆ [Computer] The nutation of a top is controlled by the effective potential energy (10.114). Make a plot of $U_{\text{eff}}(\theta)$ as follows: **(a)** First, since the second term of $U_{\text{eff}}(\theta)$ is a constant, you can ignore it. Next, by choice of your units, you can take $MgR = 1 = \lambda_1$. The remaining parameters L_z and L_3 are genuinely independent parameters. To be definite set $L_z = 10$ and $L_3 = 8$ and plot $U_{\text{eff}}(\theta)$ as a function of θ. **(b)** Explain clearly how you could use your graph to determine the angle θ_o at which the top could precess steadily with $\theta = $ constant. Find θ_o to three significant figures. **(c)** Find the rate of this steady precession, $\Omega = \dot\phi$, as given by (10.115). Compare with the approximate value of Ω given by (10.112).

10.55 ⋆⋆⋆ The analysis of the free precession of a symmetric body in Section 10.8 was based on Euler's equations. Obtain the same results using Euler's angles as follows: Since **L** is constant you may as well choose the space axis $\hat{z}$ so that $\mathbf{L} = L\hat{z}$. **(a)** Use Equation (10.98) for $\hat{z}$ to write **L** in terms of the unit vectors $\mathbf{e}'_1$, $\mathbf{e}'_2$, and $\mathbf{e}_3$. **(b)** By comparing this expression with (10.100), obtain three equations for $\dot\theta$, $\dot\phi$, and $\dot\psi$. **(c)** Hence show that θ and $\dot\phi$ are constant, and that the rate of precession of the body axis about the space axis $\hat{z}$ is $\Omega_s = L/\lambda_1$ as in (10.96). **(d)** Using (10.99) show that the angle between $\boldsymbol{\omega}$ and $\mathbf{e}_3$ is constant and that the three vectors **L**, $\boldsymbol{\omega}$, and $\mathbf{e}_3$ are always coplanar.

10.56 ⋆⋆⋆ An important special case of the motion of a symmetric top occurs when it spins about a vertical axis. Analyze this motion as follows: **(a)** By inspecting the effective PE (10.114), show that if at any time $\theta = 0$, then L_3 and L_z must be equal. **(b)** Set $L_z = L_3 = \lambda_3\omega_3$ and then make a Taylor expansion of $U_{\text{eff}}(\theta)$ about $\theta = 0$ to terms of order θ^2. **(c)** Show that if $\omega_3 > \omega_{\text{min}} = 2\sqrt{MgR\lambda_1/\lambda_3^2}$, then the position $\theta = 0$ is stable, but if $\omega_3 < \omega_{\text{min}}$ it is unstable. (In practice, friction slows the top's spinning. Thus with ω_3 sufficiently fast, the vertical top is stable, but as it slows down the top will eventually lurch away from the vertical when ω_3 reaches ω_{min}.)

10.57 ⋆⋆⋆ **(a)** Find the Lagrangian for a symmetric top whose tip is free to slide on a frictionless, horizontal table. For generalized coordinates use the Euler angles (θ, ϕ, ψ) plus X and Y, where (X, Y, Z) is the CM position relative to a fixed point on the table. (Note that the vertical position Z is not an independent coordinate, since $Z = R\cos\theta$.) **(b)** Show that the CM motion of (X, Y) separates completely from the rotational motion. **(c)** Consider the two possible rates of steady precession (10.111) and (10.112) (for given θ and ω_3). How do these differ in the present case from their corresponding values when the tip is held at a fixed pivot?

11

Coupled Oscillators and Normal Modes

In Chapter 5 we discussed the oscillations of a single body, such as a mass on the end of a fixed spring. I now want to take up the oscillations of several bodies, such as the atoms that make up a molecule like CO_2, which we can imagine as a system of masses connected to one another by springs. If each mass were attached to a separate fixed spring, with no connections between the masses, then each would oscillate independently, as described in Chapter 5, and there would be nothing more to say. Thus our interest here is a system of masses that can oscillate and are connected to one another in some way — a system of **coupled oscillators**. A single oscillator has a single natural frequency, at which (in the absence of damping or driving forces) it will oscillate for ever. We shall find that two or more coupled oscillators have several natural (or "normal") frequencies and that the general motion is a combination of vibrations at all the different natural frequencies.

Like the theory of rotating bodies in Chapter 10, the theory of coupled oscillators makes essential use of matrices, and many of the ideas you learned in Chapter 10 play an important role here. The most obvious applications of the ideas of this chapter are to the study of molecules, but there are many others, including acoustics, the vibrations of structures like bridges and buildings, and coupled electrical circuits.

Throughout this chapter I shall assume that all of the forces with which we are concerned obey Hooke's law and hence that the equations of motion are all linear. While this is a special case, it is a very important special case, with many applications in mechanics and throughout physics. Nevertheless, you should bear in mind that the systems discussed here are a special case; we shall see in Chapter 12 how startlingly more complicated the motion of nonlinear oscillators can be.

11.1 Two Masses and Three Springs

As a simple first example of coupled oscillators, consider the two carts shown in Figure 11.1. The carts move without friction on a horizontal track, between two fixed

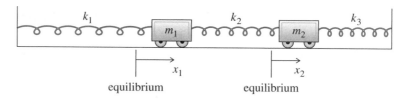

Figure 11.1   Two carts attached to fixed walls by the springs labeled k_1 and k_3, and to each other by k_2. The carts' positions x_1 and x_2 are measured from their respective equilibrium positions.

walls. Each is attached to its adjacent wall by a spring (force constants k_1 and k_3), and the carts are attached to each other by a spring with force constant k_2. In the absence of the spring 2, the two carts would oscillate independently of each other. Thus it is spring 2 that "couples" the two oscillators. In fact, spring 2 makes it impossible that either cart move without the other moving as well: For example, if cart 1 is stationary and cart 2 moves, the length of spring 2 will change, which will produce a changing force on cart 1, causing it to move as well.[1]

It is easy to find the equations of motion for the two carts, using either Newton's second law or Lagrange's equations. In general, Lagrange's equations are easier to write down, but in the present simple case, Newton's law may be a little more instructive. Suppose the two carts have moved distances x_1 and x_2 (measured to the right) from their equilibrium positions. Spring 1 is now stretched by an amount x_1 and so exerts a force $k_1 x_1$ to the left on cart 1. Spring 2 is more complicated since it is affected by the positions of both carts, but you can easily convince yourself that it is stretched by the amount $x_2 - x_1$ and exerts a force $k_2(x_2 - x_1)$ to the right on cart 1. Thus the net force on cart 1 is

$$\text{(net force on cart 1)} = -k_1 x_1 + k_2(x_2 - x_1)$$
$$= -(k_1 + k_2)x_1 + k_2 x_2 \tag{11.1}$$

where the second line is just to show more clearly the dependence on the two variables x_1 and x_2. You can find the net force on cart 2 in the same way, and the two equations of motion are

$$\left. \begin{array}{l} m_1 \ddot{x}_1 = -(k_1 + k_2)x_1 + k_2 x_2 \\ m_2 \ddot{x}_2 = k_2 x_1 - (k_2 + k_3)x_2. \end{array} \right\} \tag{11.2}$$

Before we try to solve these two coupled equations, notice that they can be written in the beautifully compact matrix form

$$\mathbf{M\ddot{x}} = -\mathbf{Kx}. \tag{11.3}$$

---

[1] In the following discussion, it is simplest to assume that when the two carts are at their equilibrium positions the three springs are neither stretched nor compressed. (Their lengths are equal to their natural, unstretched lengths.) However, depending on the distance between the two walls, it could be that all three springs are compressed or all three are stretched. Fortunately, as you can easily check (Problem 11.1), none of the results of the next three sections is affected by these possibilities.

Here I have introduced the (2×1) column matrix (or "column vector")

$$\mathbf{x} = \begin{bmatrix} x_1 \\ x_2 \end{bmatrix} \tag{11.4}$$

which labels the configuration of our system. (It has 2 elements because the system has 2 degrees of freedom; for a system with n degrees of freedom it would have n elements.) I have also defined two square matrices,

$$\mathbf{M} = \begin{bmatrix} m_1 & 0 \\ 0 & m_2 \end{bmatrix} \quad \text{and} \quad \mathbf{K} = \begin{bmatrix} k_1 + k_2 & -k_2 \\ -k_2 & k_2 + k_3 \end{bmatrix}. \tag{11.5}$$

The "mass matrix" $\mathbf{M}$ is (in this simple case, at least) a diagonal matrix, with the masses m_1 and m_2 down the diagonal. The "spring-constant matrix" $\mathbf{K}$ has nonzero off-diagonal elements, reflecting that the right sides of the two equations (11.2) couple x_1 and x_2. Notice that the matrix equation (11.3) is a very natural generalization of the equation of motion of a single cart on a single spring: With just one degree of freedom, all three matrices $\mathbf{x}$, $\mathbf{M}$, and $\mathbf{K}$ are just (1×1) matrices, that is, ordinary numbers. The configuration $\mathbf{x}$ is the cart's position x, the mass matrix $\mathbf{M}$ is the cart's mass m, and $\mathbf{K}$ is the spring constant k. And the equation of motion (11.3) is just the familiar $m\ddot{x} = -kx$. Notice also that both matrices $\mathbf{M}$ and $\mathbf{K}$ are symmetric, as will be true of all the corresponding matrices in this chapter. Although the symmetry of $\mathbf{M}$ and $\mathbf{K}$ does not play a very obvious role in the discussions here, it is in fact a key property of the underlying mathematics, as we shall see in the appendix.

In trying to solve the equation of motion (11.3) we might reasonably guess that there could be solutions in which both carts oscillate sinusoidally with the same angular frequency ω; that is,

$$x_1(t) = \alpha_1 \cos(\omega t - \delta_1) \tag{11.6}$$

and

$$x_2(t) = \alpha_2 \cos(\omega t - \delta_2). \tag{11.7}$$

In any event, there is nothing to stop us *trying* to find solutions of this form. (And we shall, in fact, succeed!) If there is a solution of this form, then there will certainly also be a solution of the same form but with the cosines replaced by sines:

$$y_1(t) = \alpha_1 \sin(\omega t - \delta_1)$$

and

$$y_2(t) = \alpha_2 \sin(\omega t - \delta_2)$$

and there is nothing to stop me combining these two solutions into a single complex solution

$$z_1(t) = x_1(t) + iy_1(t) = \alpha_1 e^{i(\omega t - \delta_1)} = \alpha_1 e^{-i\delta_1} e^{i\omega t} = a_1 e^{i\omega t}, \tag{11.8}$$

where $a_1 = \alpha_1 e^{-i\delta_1}$, and, likewise,

$$z_2(t) = x_2(t) + iy_2(t) = \alpha_2 e^{i(\omega t - \delta_2)} = \alpha_2 e^{-i\delta_2} e^{i\omega t} = a_2 e^{i\omega t}. \tag{11.9}$$

This trick of introducing a "fictitious" complex solution to the equation of motion is the same trick introduced in Section 5.5. I am not, of course, claiming that these complex numbers represent the actual motion of the two carts. The actual motion is given by the two real numbers (11.6) and (11.7). Nevertheless, for the right choices of a_1, a_2, and ω, the two complex numbers (11.8) and (11.9) are (as we shall see) solutions of the equation of motion, and their real parts describe the actual motion of our system. The great advantage of the complex numbers is that, as you can see from the right sides of (11.8) and (11.9), they both have the same time dependence, given by the common factor $e^{i\omega t}$. This lets us combine the two complex solutions into a single (2×1) matrix solution of the form

$$\mathbf{z}(t) = \begin{bmatrix} z_1(t) \\ z_2(t) \end{bmatrix} = \begin{bmatrix} a_1 \\ a_2 \end{bmatrix} e^{i\omega t} = \mathbf{a} e^{i\omega t} \tag{11.10}$$

where the column $\mathbf{a}$ is a constant, made up of two complex numbers,

$$\mathbf{a} = \begin{bmatrix} a_1 \\ a_2 \end{bmatrix} = \begin{bmatrix} \alpha_1 e^{-i\delta_1} \\ \alpha_2 e^{-i\delta_2} \end{bmatrix}.$$

In seeking solutions of the equation of motion (11.3), we shall accordingly try for solutions $\mathbf{z}(t)$ of the complex form (11.10), bearing in mind that when we find such solutions the actual motion $\mathbf{x}(t)$ is equal to the real part of $\mathbf{z}(t)$,

$$\mathbf{x}(t) = \mathrm{Re}\,\mathbf{z}(t).$$

When we substitute the form (11.10) into Equation (11.3), $\mathbf{M\ddot{x}} = -\mathbf{Kx}$, we obtain the equation

$$-\omega^2 \mathbf{M a}\, e^{i\omega t} = -\mathbf{K a}\, e^{i\omega t},$$

or, cancelling the common exponential factor and rearranging,

$$(\mathbf{K} - \omega^2 \mathbf{M})\mathbf{a} = 0. \tag{11.11}$$

This equation is a generalization of the eigenvalue equation studied in Section 10.5. (In the usual eigenvalue equation, what we are calling ω^2 is the eigenvalue, and where we have the matrix $\mathbf{M}$ the ordinary eigenvalue equation has the unit matrix $\mathbf{1}$.) It can be solved in almost exactly the same way. If the matrix $(\mathbf{K} - \omega^2 \mathbf{M})$ has nonzero determinant, then the only solution of (11.11) is the trivial solution $\mathbf{a} = 0$, corresponding to no motion at all. On the other hand, if

$$\det(\mathbf{K} - \omega^2 \mathbf{M}) = 0, \tag{11.12}$$

then there certainly is a nontrivial solution of (11.11) and hence a solution of the equations of motion with our assumed sinusoidal form (11.10). In the present case, the matrices $\mathbf{K}$ and $\mathbf{M}$ are (2×2) matrices, so the equation (11.12) is a quadratic equation for ω^2 and has (in general) two solutions for ω^2. This implies that there are two frequencies ω at which the carts can oscillate in pure sinusoidal motion as in (11.10) [or, rather, (11.6) and (11.7) for the actual real motion].[2]

The two frequencies at which our system can oscillate sinusoidally (the so-called **normal frequencies**) are determined by the quadratic equation (11.12) for ω^2. The details of this equation depend on the values of the three spring constants and the two masses. While the general case is perfectly straightforward, it is not especially illuminating, and I shall discuss instead two special cases where one can understand more easily what is going on. I shall start with the case that the three springs are identical, and likewise the two masses.

11.2   Identical Springs and Equal Masses

Let us continue to examine the two carts of Figure 11.1, but suppose now that the two masses are equal, $m_1 = m_2 = m$, and similarly the three spring constants, $k_1 = k_2 = k_3 = k$. In this case, the matrices $\mathbf{M}$ and $\mathbf{K}$ defined in (11.5) reduce to

$$\mathbf{M} = \begin{bmatrix} m & 0 \\ 0 & m \end{bmatrix} \quad \text{and} \quad \mathbf{K} = \begin{bmatrix} 2k & -k \\ -k & 2k \end{bmatrix}. \tag{11.13}$$

The matrix $(\mathbf{K} - \omega^2\mathbf{M})$ of the generalized[3] eigenvalue equation (11.11) becomes

$$(\mathbf{K} - \omega^2\mathbf{M}) = \begin{bmatrix} 2k - m\omega^2 & -k \\ -k & 2k - m\omega^2 \end{bmatrix} \tag{11.14}$$

and its determinant is

$$\det(\mathbf{K} - \omega^2\mathbf{M}) = (2k - m\omega^2)^2 - k^2 = (k - m\omega^2)(3k - m\omega^2).$$

The two normal frequencies are determined by the condition that this determinant be zero and are therefore

$$\omega = \sqrt{\frac{k}{m}} = \omega_1 \quad \text{and} \quad \omega = \sqrt{\frac{3k}{m}} = \omega_2. \tag{11.15}$$

These two normal frequencies are the frequencies at which our two carts can oscillate in purely sinusoidal motion. Notice that the first one, ω_1, is precisely the frequency of a single mass m on a single spring k. We shall see the reason for this apparent coincidence in a moment.

---

[2] Since there are two solutions for ω^2, you might think this would give four solutions for $\omega = \pm\sqrt{\omega^2}$. However, a glance at Equations (11.6) and (11.7) will convince you that $+\omega$ and $-\omega$ constitute the *same* frequency for the real motion.

[3] From now on, I shall refer to (11.11) as the eigenvalue equation, omitting the "generalized."

Equation (11.15) tells us the two possible frequencies of our system, but we have not yet described the corresponding motions. Recall that the actual motion is given by the column of real numbers $\mathbf{x}(t) = \text{Re } \mathbf{z}(t)$ where the complex column $\mathbf{z}(t) = \mathbf{a}e^{i\omega t}$, and $\mathbf{a}$ is made up of two fixed numbers,

$$\mathbf{a} = \begin{bmatrix} a_1 \\ a_2 \end{bmatrix},$$

which must satisfy the eigenvalue equation

$$(\mathbf{K} - \omega^2\mathbf{M})\mathbf{a} = 0. \tag{11.16}$$

Now that we know the possible normal frequencies, we must solve this equation for the vector $\mathbf{a}$ for each normal frequency in turn. The sinusoidal motion with any one of the normal frequencies is called a **normal mode**, and I shall start with the first normal mode.

The First Normal Mode

If we choose ω equal to the first normal frequency, $\omega_1 = \sqrt{k/m}$, then the matrix $(\mathbf{K} - \omega^2\mathbf{M})$ of (11.14) becomes

$$(\mathbf{K} - \omega_1^2\mathbf{M}) = \begin{bmatrix} k & -k \\ -k & k \end{bmatrix}. \tag{11.17}$$

(Notice that this matrix has determinant 0, as it should.) Therefore, for this case, the eigenvalue equation (11.16) reads

$$\begin{bmatrix} 1 & -1 \\ -1 & 1 \end{bmatrix}\begin{bmatrix} a_1 \\ a_2 \end{bmatrix} = 0$$

which is equivalent to the two equations

$$a_1 - a_2 = 0$$
$$-a_1 + a_2 = 0.$$

Notice that these two equations are actually the same equation, and either one implies that $a_1 = a_2 = Ae^{-i\delta}$, say. The complex column $\mathbf{z}(t)$ is therefore

$$\mathbf{z}(t) = \begin{bmatrix} a_1 \\ a_2 \end{bmatrix}e^{i\omega_1 t} = \begin{bmatrix} A \\ A \end{bmatrix}e^{i(\omega_1 t - \delta)}$$

and the corresponding actual motion is given by the real column $\mathbf{x}(t) = \text{Re } \mathbf{z}(t)$ or

$$\mathbf{x}(t) = \begin{bmatrix} x_1(t) \\ x_2(t) \end{bmatrix} = \begin{bmatrix} A \\ A \end{bmatrix}\cos(\omega_1 t - \delta).$$

That is,

$$\left.\begin{aligned} x_1(t) &= A\cos(\omega_1 t - \delta) \\ x_2(t) &= A\cos(\omega_1 t - \delta) \end{aligned}\right\} \qquad \text{[first normal mode]}. \tag{11.18}$$

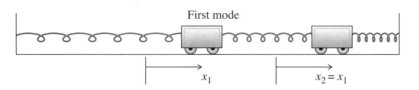

Figure 11.2 The first normal mode for two equal-mass carts with three identical springs. The two carts oscillate back and forth with equal amplitudes and exactly in phase, so that $x_1(t) = x_2(t)$, and the middle spring remains at its equilibrium length all the time.

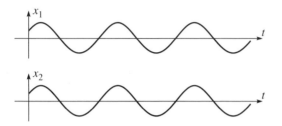

Figure 11.3 In the first mode, the two positions oscillate sinusoidally, with equal amplitudes and in phase.

We see that in the first normal mode the two carts oscillate in phase and with the same amplitude A, as shown in Figure 11.2.

A striking feature of Figure 11.2 is that, because $x_1(t) = x_2(t)$, the middle spring is neither stretched nor compressed during the oscillations. This means that, for the first normal mode, the middle spring is actually irrelevant, and each cart oscillates just as if it were attached to a single spring. This explains why the first normal frequency $\omega_1 = \sqrt{k/m}$ is the same as for a single cart on a single spring.

Another way to illustrate the motion in the first normal mode is just to plot the two positions x_1 and x_2 as functions of t. This is shown in Figure 11.3.

The Second Normal Mode

The second normal frequency at which our system can oscillate sinusoidally is given by (11.15) as $\omega_2 = \sqrt{3k/m}$, which, when substituted into (11.14), gives

$$(\mathbf{K} - \omega_2^2\mathbf{M}) = \begin{bmatrix} -k & -k \\ -k & -k \end{bmatrix}. \tag{11.19}$$

Thus, for this normal mode, the eigenvalue equation $(\mathbf{K} - \omega_2^2\mathbf{M})\mathbf{a} = 0$ implies that

$$\begin{bmatrix} 1 & 1 \\ 1 & 1 \end{bmatrix}\begin{bmatrix} a_1 \\ a_2 \end{bmatrix} = 0$$

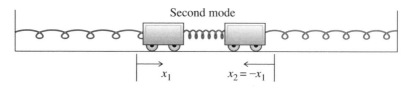

Figure 11.4   The second normal mode for two equal-mass carts with three identical springs. The two carts oscillate back and forth with equal amplitudes but exactly out of phase, so that $x_2(t) = -x_1(t)$ at all times.

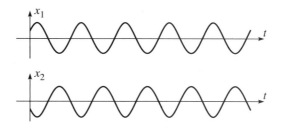

Figure 11.5   In the second mode, the two positions oscillate sinusoidally, with equal amplitudes but exactly out of phase.

which implies that $a_1 + a_2 = 0$, or $a_1 = -a_2 = Ae^{-i\delta}$, say. The complex column $\mathbf{z}(t)$ is therefore

$$\mathbf{z}(t) = \begin{bmatrix} a_1 \\ a_2 \end{bmatrix} e^{i\omega_2 t} = \begin{bmatrix} A \\ -A \end{bmatrix} e^{i(\omega_2 t - \delta)}$$

and the corresponding actual motion is given by the real column $\mathbf{x}(t) = \text{Re}\,\mathbf{z}(t)$ or

$$\mathbf{x}(t) = \begin{bmatrix} x_1(t) \\ x_2(t) \end{bmatrix} = \begin{bmatrix} A \\ -A \end{bmatrix} \cos(\omega_2 t - \delta).$$

That is,

$$\left.\begin{aligned} x_1(t) &= A\cos(\omega_2 t - \delta) \\ x_2(t) &= -A\cos(\omega_2 t - \delta) \end{aligned}\right\} \qquad \text{[second normal mode].} \qquad (11.20)$$

We see that in the second normal mode the two carts oscillate with the same amplitude A but exactly out of phase, as shown in the picture of Figure 11.4 and the graphs of Figure 11.5.

Notice that in the second normal mode, when cart 1 is displaced to the right, cart 2 is displaced an equal distance to the left, and vice versa. This means that when the outer two springs are stretched (as in Figure 11.4), the middle spring is compressed by twice as much. Thus, for example, when the left spring is pulling cart 1 to the left, the middle spring is pushing cart 1, also to the left, with a force that is twice as large. This means that each cart moves as if it were attached to a single spring with force constant $3k$. In particular, the second normal frequency is $\omega_2 = \sqrt{3k/m}$.

The General Solution

We have now found two normal-mode solutions, which we can rewrite as

$$\mathbf{x}(t) = A_1 \begin{bmatrix} 1 \\ 1 \end{bmatrix} \cos(\omega_1 t - \delta_1) \qquad \text{and} \qquad \mathbf{x}(t) = A_2 \begin{bmatrix} 1 \\ -1 \end{bmatrix} \cos(\omega_2 t - \delta_2)$$

where ω_1 and ω_2 are the normal frequencies (11.15). Both of these solutions satisfy the equation of motion $\mathbf{M\ddot{x}} = -\mathbf{Kx}$ for any values of the four real constants A_1, δ_1, A_2, and δ_2. Because the equation of motion is linear and homogeneous, the sum of these two solutions is also a solution:

$$\mathbf{x}(t) = A_1 \begin{bmatrix} 1 \\ 1 \end{bmatrix} \cos(\omega_1 t - \delta_1) + A_2 \begin{bmatrix} 1 \\ -1 \end{bmatrix} \cos(\omega_2 t - \delta_2). \qquad (11.21)$$

Because the equation of motion is really two second-order differential equations for the two variables $x_1(t)$ and $x_2(t)$, its general solution has four constants of integration. Therefore the solution (11.21), with its four arbitrary constants, is in fact the general solution. *Any* solution can be written in the form (11.21), with the constants A_1, A_2, δ_1, and δ_2 determined by the initial conditions.

The general solution (11.21) is hard to visualize and describe. The motion of each cart is a mixture of the two frequencies, ω_1 and ω_2. Since $\omega_2 = \sqrt{3}\omega_1$ the motion never repeats itself, except in the special case that one of the constants A_1 or A_2 is zero (which gives us back one of the normal modes). Figure 11.6 shows graphs of the two positions in a typical nonnormal mode (with $A_1 = 1$, $A_2 = 0.7$, $\delta_1 = 0$, and $\delta_2 = \pi/2$). About the only simple thing one can say about these graphs is that they certainly are not very simple!

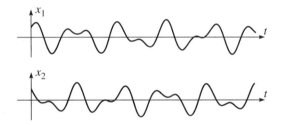

Figure 11.6   In the general solution, both $x_1(t)$ and $x_2(t)$ oscillate with *both* of the normal frequencies, producing a quite complicated non-periodic motion.

Normal Coordinates

We have seen that in any possible motion of our two-cart system, both of the co-ordinates $x_1(t)$ and $x_2(t)$ vary with time. In the normal modes, their time dependence is simple (sinusoidal), but it is still true that both vary, reflecting that the two carts are coupled and that one cart cannot move without the other. It is possible to introduce alternative, so-called **normal coordinates** which, although less physically transparent, have the convenient property that each can vary independently of the other. This statement is true for any system of coupled oscillators, but is especially easy to see in the present case of two equal masses joined by three identical springs.

In place of the coordinates x_1 and x_2, we can characterize the positions of the two carts by the two *normal coordinates*

$$\xi_1 = \tfrac{1}{2}(x_1 + x_2) \tag{11.22}$$

and

$$\xi_2 = \tfrac{1}{2}(x_1 - x_2). \tag{11.23}$$

The physical significance of the original variables x_1 and x_2 (as the positions of the two carts) is obviously more transparent, but ξ_1 and ξ_2 serve just as well to label the configuration of the system. Moreover, if you refer back to (11.18) for the first normal mode, you will see that in the first mode the new variables are given by

$$\left.\begin{array}{rcl}
\xi_1(t) & = & A\cos(\omega_1 t - \delta) \\
\xi_2(t) & = & 0
\end{array}\right\} \qquad \text{[first normal mode],} \tag{11.24}$$

whereas in the second mode, we see from (11.20) that

$$\left.\begin{array}{rcl}
\xi_1(t) & = & 0 \\
\xi_2(t) & = & A\cos(\omega_2 t - \delta)
\end{array}\right\} \qquad \text{[second normal mode].} \tag{11.25}$$

In the first normal mode the new variable ξ_1 oscillates, but ξ_2 remains zero. In the second mode it is the other way round. In this sense, the new coordinates are independent — either can oscillate without the other. The general motion of our system is a superposition of both modes, and in this case both ξ_1 and ξ_2 oscillate, but ξ_1 oscillates at the frequency ω_1 only, and ξ_2 at the frequency ω_2 only. In some more complicated problems, these new normal coordinates represent a considerable simplification. (See Problems 11.9, 11.10, and 11.11 for some examples and Section 11.7 for further discussion.)

11.3   Two Weakly Coupled Oscillators

In the last section we discussed the oscillations of two equal masses joined by three equal springs. For this system, the two normal modes were easy to understand and to visualize, but the nonnormal oscillations were much less so. A system where some of the nonnormal oscillations are readily visualized is a pair of oscillators which have the same natural frequency and which are *weakly coupled*. As an example of such a system, consider the two identical carts shown in Figure 11.7, which are attached to their adjacent walls by identical springs (force constants k) and to each other by a much weaker spring (force constant $k_2 \ll k$).

We can quickly solve for the normal modes of this system. The mass matrix $\mathbf{M}$ is the same as before. The spring matrix $\mathbf{K}$ and the crucial combination $(\mathbf{K} - \omega^2\mathbf{M})$

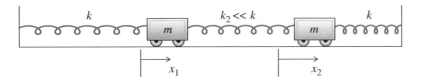

Figure 11.7   Two weakly coupled carts. The middle spring which couples the two carts is much weaker than the outer two springs.

that determines the eigenvalue problem are easily written down [starting with (11.5) for $\mathbf{K}$]:

$$\mathbf{K} = \begin{bmatrix} k + k_2 & -k_2 \\ -k_2 & k + k_2 \end{bmatrix}$$

and

$$(\mathbf{K} - \omega^2 \mathbf{M}) = \begin{bmatrix} k + k_2 - m\omega^2 & -k_2 \\ -k_2 & k + k_2 - m\omega^2 \end{bmatrix}. \tag{11.26}$$

The determinant of $(\mathbf{K} - \omega^2 \mathbf{M})$ is $(k - m\omega^2)(k + 2k_2 - m\omega^2)$, and we conclude that the two normal frequencies are

$$\omega_1 = \sqrt{\frac{k}{m}} \qquad \text{and} \qquad \omega_2 = \sqrt{\frac{k + 2k_2}{m}}. \tag{11.27}$$

The first frequency is exactly the same as in the previous example, and we can easily see why. The motion in this first mode is, as you can check, the same motion as shown in Figure 11.2 for the first mode of the equal-spring case. The important point is that in this mode the two carts move together in such a way that the middle spring is undisturbed and hence irrelevant. Naturally we get the same frequency for this mode whatever the strength of the middle spring.

In the second mode also, the motion is the same as for the corresponding mode of the equal-spring example — namely, the two carts oscillating exactly out of phase, both moving inward or both moving outward at any one time, as in Figure 11.4. But in this mode, the strength of the middle spring is, of course, relevant, and the second normal frequency ω_2 as given by (11.27) depends on k_2. In the present case, ω_2 is very close to ω_1, since $k_2 \ll k$. To take advantage of this closeness, it is convenient to define ω_o to be the average of the two normal frequencies

$$\omega_o = \frac{\omega_1 + \omega_2}{2}.$$

Since ω_1 and ω_2 are very close to each other, ω_o is very close to either, and for most purposes we can think of ω_o as essentially the same as $\omega_1 = \sqrt{k/m}$. To show the small difference between ω_1 and ω_2, I shall write

$$\omega_1 = \omega_o - \epsilon \qquad \text{and} \qquad \omega_2 = \omega_o + \epsilon.$$

That is, the small number ϵ is half the difference between the two normal frequencies.

The two normal modes of the weakly coupled carts can now be written as

$$\mathbf{z}(t) = C_1 \begin{bmatrix} 1 \\ 1 \end{bmatrix} e^{i(\omega_0 - \epsilon)t} \qquad \text{and} \qquad \mathbf{z}(t) = C_2 \begin{bmatrix} 1 \\ -1 \end{bmatrix} e^{i(\omega_0 + \epsilon)t}.$$

Both of these satisfy the equation of motion for any values of the two complex numbers C_1 and C_2. (It is convenient to continue to work with the "fictitious" complex solutions for a bit longer.) The sum of these two solutions is also a solution,

$$\mathbf{z}(t) = C_1 \begin{bmatrix} 1 \\ 1 \end{bmatrix} e^{i(\omega_0 - \epsilon)t} + C_2 \begin{bmatrix} 1 \\ -1 \end{bmatrix} e^{i(\omega_0 + \epsilon)t}, \qquad (11.28)$$

and, since it contains four arbitrary real constants (the two complex constants C_1 and C_2 are equivalent to four real constants), it is the general solution. The constants C_1 and C_2 in (11.28) are determined by the initial conditions — the positions and velocities of the two carts at $t = 0$.

To see some general features of the solution (11.28), it is helpful to factor it as

$$\mathbf{z}(t) = \left\{ C_1 \begin{bmatrix} 1 \\ 1 \end{bmatrix} e^{-i\epsilon t} + C_2 \begin{bmatrix} 1 \\ -1 \end{bmatrix} e^{i\epsilon t} \right\} e^{i\omega_0 t}. \qquad (11.29)$$

This expresses our solution as a product of two terms. The term in braces, $\{\cdots\}$, is a (2×1) column matrix which depends on t. But because ϵ is very small, this column varies very slowly compared to the second factor $e^{i\omega_0 t}$. Over any reasonably short time interval, the first factor is essentially constant and our solution behaves like $\mathbf{z}(t) = \mathbf{a}e^{i\omega_0 t}$, with $\mathbf{a}$ constant. That is, over any short time interval, the two carts will oscillate sinusoidally with angular frequency ω_0. But if we wait long enough, the "constant" $\mathbf{a}$ will vary slowly, and the details of the two carts' motion will change. I shall illustrate this behavior in detail in a moment.

Let us now examine the behavior of (11.29) for some simple values of the constants C_1 and C_2. First, if either C_1 or C_2 is zero, the solution (11.29) reverts to one of the normal modes. (For instance, if $C_1 = 0$, the solution is the second normal mode.) A more interesting case is that C_1 and C_2 are equal in magnitude, and, to simplify the discussion, I shall suppose C_1 and C_2 are equal and real,

$$C_1 = C_2 = A/2,$$

say. (The 2 is just for future convenience.) In this case (11.29) becomes

$$\mathbf{z}(t) = \frac{A}{2} \begin{bmatrix} e^{-i\epsilon t} + e^{i\epsilon t} \\ e^{-i\epsilon t} - e^{i\epsilon t} \end{bmatrix} e^{i\omega_0 t} = A \begin{bmatrix} \cos \epsilon t \\ -i \sin \epsilon t \end{bmatrix} e^{i\omega_0 t}. \qquad (11.30)$$

To find the actual motion of the two carts, we must take the real part of this matrix, $\mathbf{x}(t) = \text{Re } \mathbf{z}(t)$, whose two elements are the two positions,

$$\left. \begin{array}{rcl} x_1(t) & = & A \cos \epsilon t \cos \omega_0 t \\ x_2(t) & = & A \sin \epsilon t \sin \omega_0 t. \end{array} \right\} \qquad (11.31)$$

The solution (11.31) has a simple and elegant interpretation. Notice first that at time zero, $x_1 = A$, whereas $\dot{x}_1 = x_2 = \dot{x}_2 = 0$. That is, our solution describes the motion when cart 1 is pulled a distance A to the right and released at $t = 0$, with cart 2 stationary at its equilibrium position. Because ϵ is very small, there is an appreciable interval (namely, $0 \leq t \ll 1/\epsilon$) during which the functions in (11.31) that involve ϵt remain essentially unchanged, that is, $\cos \epsilon t \approx 1$ and $\sin \epsilon t \approx 0$. During this initial interval, the two positions, as given by (11.31), are just

$$\left. \begin{array}{l} x_1(t) \approx A \cos \omega_0 t \\ x_2(t) \approx 0 \end{array} \right\} \qquad [t \approx 0]. \tag{11.32}$$

Initially, cart 1 oscillates with amplitude A and frequency ω_0, while cart 2 remains stationary.

This simple state of affairs cannot last for ever. As soon as cart 1 begins to move, it starts to flex the weak middle spring, which starts to push and pull on cart 2. Although the force exerted by the middle spring is weak, it eventually starts to make cart 2 oscillate. This can be seen in (11.31), where the factor $\sin \epsilon t$ eventually becomes appreciable, and cart 2 starts to oscillate, also at the frequency ω_0. Notice that as the factor $\sin \epsilon t$ in $x_2(t)$ grows toward 1, the factor $\cos \epsilon t$ in $x_1(t)$ shrinks toward zero, as it has to do to keep the total energy of the two oscillating carts constant. Eventually, when $t = \pi/2\epsilon$, the factor $\sin \epsilon t$ reaches 1 (and $\cos \epsilon t$ reaches zero), and there is an interval when[4]

$$\left. \begin{array}{l} x_1(t) \approx 0 \\ x_2(t) \approx A \sin \omega_0 t \end{array} \right\} \qquad [t \approx \pi/2\epsilon]. \tag{11.33}$$

Now that cart 2 is oscillating at maximum amplitude and cart 1 not at all, cart 2 starts to drive cart 1. Cart 1 begins to oscillate with increasing amplitude, and the amplitude of cart 2's oscillations begins to diminish again. This process, in which the two carts pass the energy back and forth from one to the other, continues indefinitely (or until dissipative forces — which we are ignoring — have removed all the energy). It is illustrated in Figure 11.8, which shows $x_1(t)$ and $x_2(t)$ as given by (11.31) as functions of t for a couple of cycles of passing the energy from cart 1 to 2 and back to 1 again.

If you have studied the phenomenon of beats, you have probably noticed the similarity of either graph in Figure 11.8 to a graph of beats. Beats are the result of superposing two waves — sound waves, for example — with nearly equal frequencies. Because of the small difference in frequencies, the two waves move regularly in and out of phase (at any one location). This means that the resulting interference of the waves is alternately constructive and destructive, and a graph of the resultant signal looks just like either one of the graphs in Figure 11.8. To understand what is beating in the case of our two carts, we need to consider again the two normal coordinates of Equations (11.22) and (11.23), $\xi_1 = \frac{1}{2}(x_1 + x_2)$ and $\xi_2 = \frac{1}{2}(x_1 - x_2)$. For the present solution

---

[4] Notice that we can interpret what has happened in terms of the discussion of Section 5.6 on resonance. The weak spring has been driving cart 2 at the resonant frequency ω_0, so cart 2 should have responded by oscillating $\pi/2$ behind the driver. This is exactly what we see from (11.32) and (11.33), since $\sin \omega_0 t$ is indeed $\pi/2$ behind $\cos \omega_0 t$.

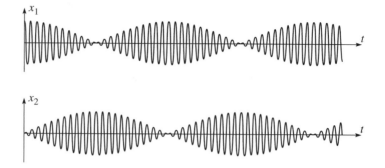

Figure 11.8   The positions $x_1(t)$ and $x_2(t)$ of two weakly coupled oscillating carts if cart 1 is released from rest at $x_1 = A > 0$ and cart 2 at $x_2 = 0$.

(11.31) these are [remember the trig identity $\cos\theta\cos\phi + \sin\theta\sin\phi = \cos(\theta - \phi)$]

$$\left.\begin{array}{rcl} \xi_1(t) &=& \frac{1}{2}A\cos(\omega_0 - \epsilon)t = \frac{1}{2}A\cos\omega_1 t \\ \xi_2(t) &=& \frac{1}{2}A\cos(\omega_0 + \epsilon)t = \frac{1}{2}A\cos\omega_2 t. \end{array}\right\} \qquad (11.34)$$

That is, the two normal coordinates oscillate with equal amplitudes, the first at the frequency ω_1 and the second at the nearby frequency ω_2. Since $x_1(t) = \xi_1(t) + \xi_2(t)$, we see that $x_1(t)$ is the superposition of $\xi_1(t)$ and $\xi_2(t)$, and the waxing and waning of $x_1(t)$ is the result of beats between these two signals of nearly equal frequencies. The same applies to $x_2(t)$, except that, because $x_2(t) = \xi_1(t) - \xi_2(t)$, the moments of *constructive* interference for $x_1(t)$ are moments of *destructive* interference for $x_2(t)$ and vice versa, as is clearly seen in Figure 11.8.

11.4   Lagrangian Approach: The Double Pendulum

The analysis of the two oscillating carts in the last three sections was based on Newton's second law. We could equally have derived the equations of motion using the Lagrangian formalism, although there is no particular advantage to doing so. However, we shall find that as we study systems of increasing complexity, the advantages of the Lagrangian approach rapidly become overwhelming. I shall start this section by rederiving the equations for the familiar two carts from their Lagrangian. I shall then do the same for another simple system with two degrees of freedom, the double pendulum. These two examples will pave the way for the general discussion in the next section.

Lagrangian Approach for Two Carts on Three Springs

Let us consider once more the two carts of Figure 11.1. We could write down the equations of motion (11.2) from Newton's second law as soon as we had identified the forces on each of the carts. To do the same thing with Lagrange's equations, we have first to write down the kinetic and potential energies, T and U, and then the Lagrangian $\mathcal{L} = T - U$. The kinetic energy is just

$$T = \tfrac{1}{2}m_1\dot{x}_1^2 + \tfrac{1}{2}m_2\dot{x}_2^2. \tag{11.35}$$

To write down the potential energy, we must identify the extensions of the three springs as x_1, $x_2 - x_1$, and $-x_2$, from which it immediately follows that the potential energy is

$$
\begin{aligned}
U &= \tfrac{1}{2}k_1x_1^2 + \tfrac{1}{2}k_2(x_1 - x_2)^2 + \tfrac{1}{2}k_3x_2^2 \\
&= \tfrac{1}{2}(k_1 + k_2)x_1^2 - k_2x_1x_2 + \tfrac{1}{2}(k_2 + k_3)x_2^2.
\end{aligned} \tag{11.36}
$$

These results immediately give us the Lagrangian $\mathcal{L} = T - U$ and thence the two Lagrange equations of motion:

$$\frac{d}{dt}\frac{\partial \mathcal{L}}{\partial \dot{x}_1} = \frac{\partial \mathcal{L}}{\partial x_1} \qquad \text{or} \qquad m_1\ddot{x}_1 = -(k_1 + k_2)x_1 + k_2x_2$$

and

$$\frac{d}{dt}\frac{\partial \mathcal{L}}{\partial \dot{x}_2} = \frac{\partial \mathcal{L}}{\partial x_2} \qquad \text{or} \qquad m_2\ddot{x}_2 = k_2x_1 - (k_2 + k_3)x_2.$$

These are precisely the two equations of motion (11.2), which we rewrote in the compact matrix form as $\mathbf{M\ddot{x}} = -\mathbf{Kx}$. This alternative derivation of the same equations has no particular advantage for this simple system. Here is a second system, which is still very simple, but for which the Lagrangian approach is already distinctly more straightforward than the Newtonian.

The Double Pendulum

Consider a double pendulum, comprising a mass m_1 suspended by a massless rod of length L_1 from a fixed pivot, and a second mass m_2 suspended by a massless rod of length L_2 from m_1, as shown in Figure 11.9. It is a straightforward matter to write down the Lagrangian $\mathcal{L}$ as a function of the two generalized coordinates ϕ_1 and ϕ_2 shown. When the angle ϕ_1 increases from 0, the mass m_1 rises by an amount $L_1(1 - \cos\phi_1)$ and gains a potential energy

$$U_1 = m_1gL_1(1 - \cos\phi_1).$$

Similarly, as ϕ_2 increases from 0, the second mass rises by $L_2(1 - \cos\phi_2)$ but, in addition, its point of support (m_1) has risen by $L_1(1 - \cos\phi_1)$. Thus

$$U_2 = m_2g[L_1(1 - \cos\phi_1) + L_2(1 - \cos\phi_2)].$$

The total potential energy is therefore

$$U(\phi_1, \phi_2) = (m_1 + m_2)gL_1(1 - \cos\phi_1) + m_2gL_2(1 - \cos\phi_2). \tag{11.37}$$

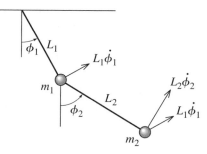

Figure 11.9   A double pendulum. The velocity of m_2 is the vector sum of the two velocities shown, separated by an angle $\phi_2 - \phi_1$.

The velocity of m_1 is just $L_1\dot\phi_1$ in the tangential direction, as shown in Figure 11.9, so its kinetic energy is

$$T_1 = \tfrac{1}{2}m_1 L_1^2 \dot\phi_1^2.$$

The velocity of m_2 is the vector sum of two velocities, as I have indicated in Figure 11.9 — the velocity $L_2\dot\phi_2$ of m_2 relative to its support m_1 plus the velocity $L_1\dot\phi_1$ of its support. The angle between these two velocities is $(\phi_2 - \phi_1)$, so the kinetic energy of m_2 is

$$T_2 = \tfrac{1}{2}m_2[L_1^2\dot\phi_1^2 + 2L_1 L_2 \dot\phi_1 \dot\phi_2 \cos(\phi_1 - \phi_2) + L_2^2\dot\phi_2^2]$$

and the total kinetic energy is

$$T = \tfrac{1}{2}(m_1 + m_2)L_1^2\dot\phi_1^2 + m_2 L_1 L_2 \dot\phi_1 \dot\phi_2 \cos(\phi_1 - \phi_2) + \tfrac{1}{2}m_2 L_2^2\dot\phi_2^2. \quad (11.38)$$

From (11.38) and (11.37) we can write down the Lagrangian $\mathcal{L} = T - U$ and then the two Lagrange equations for ϕ_1 and ϕ_2. However, the resulting equations are too complicated to be especially illuminating and can certainly not be solved analytically. This situation is reminiscent of the simple pendulum, whose equation of motion ($L\ddot\phi = -g\sin\phi$) is also unsolvable analytically, forcing us to solve it numerically or to make a suitable approximation. In that case, and in the present case, the simplest useful approximation is the small-angle approximation, which reduces the simple pendulum's equation to the solvable $L\ddot\phi = -g\phi$. We are going to see that for almost all coupled oscillating systems, the exact equations are not analytically solvable, but that if we confine attention to *small oscillations* (certainly an important special case), the equations reduce to a standard form that *is* solvable.

Returning to the equations of the double pendulum, let us assume that both angles ϕ_1 and ϕ_2 and the corresponding velocities $\dot\phi_1$ and $\dot\phi_2$ remain small at all times. This lets us simplify the expressions for T and U by Taylor expanding them and dropping all terms that are of third power or higher in the four small quantities. In (11.38) for the kinetic energy, the only term that needs attention is the middle one. Since the factor $\cos(\phi_1 - \phi_2)$ is already multiplied by the doubly small product $\dot\phi_1\dot\phi_2$, we can approximate the cosine by 1, to give

$$T = \tfrac{1}{2}(m_1 + m_2)L_1^2\dot\phi_1^2 + m_2L_1L_2\dot\phi_1\dot\phi_2 + \tfrac{1}{2}m_2L_2^2\dot\phi_2^2. \tag{11.39}$$

In (11.37) for the potential energy, we must handle the cosines more carefully (since they are not already multiplied by any small quantities). The Taylor series for $\cos\phi$ gives the approximation $\cos\phi \approx 1 - \phi^2/2$, which reduces (11.37) to

$$U = \tfrac{1}{2}(m_1 + m_2)gL_1\phi_1^2 + \tfrac{1}{2}m_2gL_2\phi_2^2. \tag{11.40}$$

Before we use these simplified expressions for T and U to give us the equations of motion, let us pause to examine what our assumption of small oscillations has achieved. The exact expression (11.38) for T was a transcendental function of the coordinates ϕ_1, ϕ_2 and velocities $\dot\phi_1$, $\dot\phi_2$; the small-angle approximation reduced this to a homogeneous quadratic function[5] of the two velocities only. The exact expression (11.37) for U was a transcendental function of ϕ_1 and ϕ_2; the small-angle approximation reduced this to a homogeneous quadratic function of ϕ_1 and ϕ_2. We shall see that the same simplifications occur for a wide class of oscillating systems: The assumption that all oscillations are small reduces T to a homogeneous quadratic function of the velocities and U to a homogeneous quadratic function of the coordinates.[6] The simplifying feature of these homogeneous quadratic forms for T and U is that when we differentiate them to get Lagrange's equations they reduce to homogeneous linear functions, yielding equations of motion that can always be easily solved.

We can now substitute the approximate expressions (11.39) and (11.40) for T and U into the Lagrangian $\mathcal{L} = T - U$ and write down the two Lagrange equations of motion for ϕ_1 and ϕ_2:

$$\frac{d}{dt}\frac{\partial\mathcal{L}}{\partial\dot\phi_1} = \frac{\partial\mathcal{L}}{\partial\phi_1} \quad \text{or} \quad (m_1 + m_2)L_1^2\ddot\phi_1 + m_2L_1L_2\ddot\phi_2 = -(m_1 + m_2)gL_1\phi_1 \tag{11.41}$$

and

$$\frac{d}{dt}\frac{\partial\mathcal{L}}{\partial\dot\phi_2} = \frac{\partial\mathcal{L}}{\partial\phi_2} \quad \text{or} \quad m_2L_1L_2\ddot\phi_1 + m_2L_2^2\ddot\phi_2 = -m_2gL_2\phi_2. \tag{11.42}$$

These two equations for ϕ_1 and ϕ_2 can be rewritten as a single matrix equation

$$\mathbf{M}\ddot{\boldsymbol{\phi}} = -\mathbf{K}\boldsymbol{\phi} \tag{11.43}$$

---

[5] A homogeneous quadratic function contains only second powers of its arguments — no first powers or constant terms, and no powers higher than two.

[6] It is an almost unique feature of systems of masses connected by Hooke's-law springs that the exact expressions for T and U [as in (11.35) and (11.36)] already are in these simple forms, without our having to make any approximations.

if we introduce the (2×1) column of coordinates

$$\boldsymbol{\phi} = \begin{bmatrix} \phi_1 \\ \phi_2 \end{bmatrix}$$

and the two (2×2) matrices

$$\mathbf{M} = \begin{bmatrix} (m_1 + m_2)L_1^2 & m_2 L_1 L_2 \\ m_2 L_1 L_2 & m_2 L_2^2 \end{bmatrix} \quad \text{and} \quad \mathbf{K} = \begin{bmatrix} (m_1 + m_2)g L_1 & 0 \\ 0 & m_2 g L_2 \end{bmatrix}.$$

$$(11.44)$$

The matrix equation (11.43) is exactly analogous to (11.3) for the two carts on springs. In the present case, the "mass" matrix $\mathbf{M}$ is not actually made up of masses, but it still plays the role of inertia in the equation of motion (11.43). (That is, it multiplies the second derivatives of the coordinates.) Similarly the "spring-constant" matrix $\mathbf{K}$ is not actually made up of spring constants, but it plays the analogous role in the equation of motion.

The procedure for solving the equations of motion (11.43) is exactly the same as for the two carts of Section 11.1. We first try to find solutions — normal modes — in which the two coordinates ϕ_1 and ϕ_2 vary sinusoidally with the same angular frequency ω. Exactly as before, any such solution $\boldsymbol{\phi}(t)$ can be written as the real part of a complex solution $\mathbf{z}(t)$ whose time dependence is just $e^{i\omega t}$; that is,

$$\boldsymbol{\phi}(t) = \operatorname{Re} \mathbf{z}(t) \qquad \text{where} \qquad \mathbf{z}(t) = \mathbf{a} e^{i\omega t} = \begin{bmatrix} a_1 \\ a_2 \end{bmatrix} e^{i\omega t},$$

and the two components a_1, a_2 of $\mathbf{a}$ are constants. Exactly as before, a function of this form satisfies the equation of motion (11.43) if and only if the frequency ω and the column $\mathbf{a}$ satisfy the eigenvalue equation $(\mathbf{K} - \omega^2 \mathbf{M})\mathbf{a} = 0$. This equation for $\mathbf{a}$ has a solution if and only if $\det(\mathbf{K} - \omega^2 \mathbf{M}) = 0$, a quadratic equation for ω^2, which determines the two normal frequencies of the double pendulum. Knowing these two frequencies, we can go back and find the corresponding columns $\mathbf{a}$, and we then know the two normal modes. Finally, the general motion of the system is just an arbitrary superposition of these two normal modes.

Equal Lengths and Masses

To simplify the discussion, let us now restrict our attention to the case that our double pendulum has equal masses, $m_1 = m_2 = m$, and equal lengths, $L_1 = L_2 = L$, say. The equations tidy up appreciably if we recognize that $\sqrt{g/L}$ is the frequency of a single pendulum of the same length L. If we call this frequency ω_o, then we can replace g everywhere by $L\omega_o^2$ and the two matrices $\mathbf{M}$ and $\mathbf{K}$ of (11.44) become (as you can check)

$$\mathbf{M} = mL^2 \begin{bmatrix} 2 & 1 \\ 1 & 1 \end{bmatrix} \quad \text{and} \quad \mathbf{K} = mL^2 \begin{bmatrix} 2\omega_o^2 & 0 \\ 0 & \omega_o^2 \end{bmatrix}. \qquad (11.45)$$

The matrix $(\mathbf{K} - \omega^2\mathbf{M})$ of the eigenvalue equation is therefore

$$(\mathbf{K} - \omega^2\mathbf{M}) = mL^2 \begin{bmatrix} 2(\omega_o^2 - \omega^2) & -\omega^2 \\ -\omega^2 & (\omega_o^2 - \omega^2) \end{bmatrix}. \qquad (11.46)$$

The normal frequencies are determined by the condition $\det(\mathbf{K} - \omega^2\mathbf{M}) = 0$, which gives

$$2(\omega_o^2 - \omega^2)^2 - \omega^4 = \omega^4 - 4\omega_o^2\omega^2 + 2\omega_o^4 = 0$$

with the two solutions $\omega^2 = (2 \pm \sqrt{2})\omega_o^2$. That is, the two normal frequencies are given by

$$\omega_1^2 = (2 - \sqrt{2})\omega_o^2 \qquad \text{and} \qquad \omega_2^2 = (2 + \sqrt{2})\omega_o^2 \qquad (11.47)$$

(or $\omega_1 \approx 0.77\omega_o$ and $\omega_2 \approx 1.85\omega_o$) where $\omega_o = \sqrt{g/L}$ is the frequency of a single pendulum of length L.

Knowing the two normal frequencies we can now find the motion of the double pendulum in the corresponding normal modes, by solving the equation $(\mathbf{K} - \omega^2\mathbf{M})\mathbf{a} = 0$ with $\omega = \omega_1$ and ω_2 in turn. If we substitute $\omega = \omega_1$, as given by (11.47), into (11.46), we get

$$(\mathbf{K} - \omega_1^2\mathbf{M}) = mL^2\omega_o^2(\sqrt{2} - 1)\begin{bmatrix} 2 & -\sqrt{2} \\ -\sqrt{2} & 1 \end{bmatrix}.$$

Therefore, the equation $(\mathbf{K} - \omega_1^2\mathbf{M})\mathbf{a} = 0$ implies that $a_2 = \sqrt{2}a_1$, and if we write $a_1 = A_1 e^{-i\delta_1}$, the two coordinates are

$$\boldsymbol{\phi}(t) = \begin{bmatrix} \phi_1(t) \\ \phi_2(t) \end{bmatrix} = \operatorname{Re}\mathbf{a}e^{i\omega_1 t} = A_1\begin{bmatrix} 1 \\ \sqrt{2} \end{bmatrix}\cos(\omega_1 t - \delta_1) \qquad \text{[first mode].} \qquad (11.48)$$

We see that in the first normal mode the two pendulums oscillate exactly in phase, with the amplitude of the lower pendulum $\sqrt{2}$ times that of the upper pendulum, as shown in Figure 11.10.

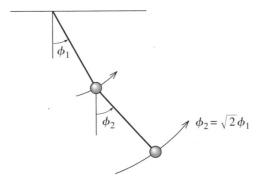

Figure 11.10   The first normal mode for a double pendulum with equal masses and equal lengths. The two angles ϕ_1 and ϕ_2 oscillate in phase, with the amplitude for ϕ_2 larger by a factor of $\sqrt{2}$.

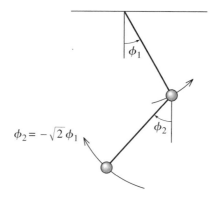

Figure 11.11   The second normal mode for a double pendulum with equal masses and equal lengths. The two angles ϕ_1 and ϕ_2 oscillate exactly out of phase, with the amplitude for ϕ_2 larger by a factor of $\sqrt{2}$.

Turning to the second mode, we find from (11.46) and (11.47) that

$$(\mathbf{K} - \omega_2^2 \mathbf{M}) = -mL^2\omega_o^2(\sqrt{2} + 1) \begin{bmatrix} 2 & \sqrt{2} \\ \sqrt{2} & 1 \end{bmatrix}.$$

The equation $(\mathbf{K} - \omega_2^2 \mathbf{M})\mathbf{a} = 0$ implies that $a_2 = -\sqrt{2}a_1$, and if we write $a_1 = A_2 e^{-i\delta_2}$, the two coordinates are

$$\boldsymbol{\phi}(t) = \begin{bmatrix} \phi_1(t) \\ \phi_2(t) \end{bmatrix} = \mathrm{Re}\,\mathbf{a}e^{i\omega_2 t} = A_2 \begin{bmatrix} 1 \\ -\sqrt{2} \end{bmatrix} \cos(\omega_2 t - \delta_2) \qquad \text{[second mode]}.$$

$$(11.49)$$

In this second mode, $\phi_2(t)$ oscillates exactly out of phase with $\phi_1(t)$, again with an amplitude that is $\sqrt{2}$ times bigger than that of $\phi_1(t)$, as shown in Figure 11.11. The general solution is an arbitrary linear combination of the two normal modes (11.48) and (11.49).

11.5   The General Case

We have now studied in great detail the normal modes of two systems — a pair of carts attached to three springs and a double pendulum — and are ready to discuss the general case of a system with n degrees of freedom that is oscillating about a point of stable equilibrium. Since the system has n degrees of freedom, its configuration can be specified by n generalized coordinates,[7] $q_1, \cdots, q_n$. To avoid too much notational clutter, I shall now abbreviate the set of all n coordinates by a single boldface $\mathbf{q}$,

$$\mathbf{q} = (q_1, \cdots, q_n).$$

---

[7] I take for granted that the system is holonomic, so that the number of degrees of freedom equals the number of generalized coordinates, as discussed in Section 7.3.

[Thus for the two carts of Section 11.1, $\mathbf{q}$ would denote the two displacements $\mathbf{q} = (x_1, x_2)$, and for the double pendulum, $\mathbf{q} = (\phi_1, \phi_2)$.] (Note well that $\mathbf{q}$ is not, in general, a three-dimensional vector; it is a vector in the n-dimensional space of the generalized coordinates $q_1, \cdots, q_n$.)

I shall assume that the system is conservative, so that it has a potential energy

$$U(q_1, \cdots, q_n) = U(\mathbf{q})$$

and Lagrangian $\mathcal{L} = T - U$. The kinetic energy is, of course, $T = \sum_\alpha \frac{1}{2} m_\alpha \dot{\mathbf{r}}_\alpha{}^2$, where the sum runs over all the particles, $\alpha = 1, \cdots, N$, that comprise the system. This has to be rewritten in terms of the generalized coordinates $\mathbf{q} = (q_1, \cdots, q_n)$ using the relation between the Cartesian coordinates $\mathbf{r}_\alpha$ and the generalized coordinates

$$\mathbf{r}_\alpha = \mathbf{r}_\alpha(q_1, \cdots, q_n) \tag{11.50}$$

where I shall take for granted that this relation does not involve the time t explicitly. [Recall that, in the terminology of Section 7.3, generalized coordinates for which (11.50) does not involve the time are called "natural."] We saw in detail in Section 7.8 that if we differentiate (11.50) with respect to t and substitute into the kinetic energy, we find that [compare Equation (7.94)]

$$T = T(\mathbf{q}, \dot{\mathbf{q}}) = \frac{1}{2} \sum_{j,k} A_{jk}(\mathbf{q}) \, \dot{q}_j \, \dot{q}_k \tag{11.51}$$

where the coefficients $A_{jk}(\mathbf{q})$ may depend on the coordinates $\mathbf{q}$. [Compare (11.38) for the case of the double pendulum.] Under our present assumptions, the Lagrangian has the general form $\mathcal{L}(\mathbf{q}, \dot{\mathbf{q}}) = T(\mathbf{q}, \dot{\mathbf{q}}) - U(\mathbf{q})$, where $T(\mathbf{q}, \dot{\mathbf{q}})$ is given by (11.51), and $U(\mathbf{q})$ is an as-yet unspecified function of the coordinates $\mathbf{q}$.

Our final assumption on the system is that it is making small oscillations about a configuration of stable equilibrium. By redefining the coordinates if necessary, we can arrange that the equilibrium position is $\mathbf{q} = 0$ (that is, $q_1 = \cdots = q_n = 0$). Then, since we are interested only in small oscillations, we have only to concern ourselves with small values of the coordinates $\mathbf{q}$, and we can use Taylor expansions of U and T about the equilibrium point $\mathbf{q} = 0$. For U this gives

$$U(\mathbf{q}) = U(0) + \sum_j \frac{\partial U}{\partial q_j} q_j + \frac{1}{2} \sum_{j,k} \frac{\partial^2 U}{\partial q_j \partial q_k} q_j q_k + \cdots \tag{11.52}$$

where all derivatives are evaluated at $\mathbf{q} = 0$. This can be much simplified. First, since $U(0)$ is a constant we can simply drop it, by redefining the zero of potential energy. Second, since $\mathbf{q} = 0$ is an equilibrium point, all of the first derivatives $\partial U/\partial q_j$ are zero. I shall rename the second derivatives as $\partial^2 U/\partial q_j \partial q_k = K_{jk}$ (which satisfy $K_{jk} = K_{kj}$ since it makes no difference in which order we evaluate second derivatives). Finally, since the oscillations are small, I shall neglect all terms higher than second order in the small quantities $\mathbf{q}$ or $\dot{\mathbf{q}}$. This reduces U to

$$U = U(\mathbf{q}) = \frac{1}{2} \sum_{j,k} K_{jk} q_j q_k . \tag{11.53}$$

The kinetic energy is even simpler. Every term in (11.51) contains a factor $\dot{q}_j \dot{q}_k$ which is already second order in small quantities. Therefore, we can ignore everything but the constant term in the expansion of $A_{jk}(\mathbf{q})$. If we call this constant term $A_{jk}(0) = M_{jk}$, this reduces the kinetic energy to

$$T = T(\dot{\mathbf{q}}) = \tfrac{1}{2} \sum_{j,k} M_{jk}\, \dot{q}_j\, \dot{q}_k \qquad (11.54)$$

and the Lagrangian to

$$\mathcal{L}(\mathbf{q}, \dot{\mathbf{q}}) = T(\dot{\mathbf{q}}) - U(\mathbf{q}) \qquad (11.55)$$

with $T(\dot{\mathbf{q}})$ given by (11.54) and $U(\mathbf{q})$ by (11.53). Notice that the approximate forms (11.54) and (11.53) correspond to the approximations (11.39) and (11.40) for the double pendulum. Just like the latter approximations, they reduce the kinetic energy to a homogeneous quadratic function of the velocities $\dot{\mathbf{q}}$ and the potential energy to a homogeneous quadratic function of the coordinates $\mathbf{q}$. Just as with the double pendulum, this will guarantee that the equations of motion are solvable linear equations, but before we take up the equations, let's see one more simple example of this dramatic simplification of T and U that results from the assumption of small oscillations.

EXAMPLE 11.1   A Bead on a Wire

A bead of mass m is threaded on a frictionless wire that lies in the xy plane (y vertically up), bent in the shape $y = f(x)$ with a minimum at the origin, as shown in Figure 11.12. Write down the potential and kinetic energies and their simplified forms appropriate for small oscillations about O.

This system has just one degree of freedom, and the natural choice of generalized coordinate is just x. With this choice, the potential energy is simply $U = mgy = mgf(x)$. When we confine ourselves to small oscillations, we can Taylor expand $f(x)$. Since $f(0) = f'(0) = 0$, this gives

$$U(x) = mgf(x) \approx \tfrac{1}{2} mgf''(0)x^2.$$

The kinetic energy is $T = \tfrac{1}{2} m(\dot{x}^2 + \dot{y}^2)$, where, by the chain rule, $\dot{y} = f'(x)\dot{x}$. Therefore, $T = \tfrac{1}{2} m[1 + f'(x)^2]\dot{x}^2$. Notice that the exact expression for T depends on x as well as $\dot{x}$. However, since T already contains the factor $\dot{x}^2$, when

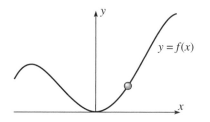

Figure 11.12   A bead threaded on a frictionless wire in the shape $y = f(x)$.

we make the small-oscillation approximation we can simply replace the term $f'(x)$ by its value at $x = 0$ (namely zero) and we get

$$T(x, \dot{x}) = \tfrac{1}{2}m[1 + f'(x)^2]\dot{x}^2 \approx \tfrac{1}{2}m\dot{x}^2.$$

As expected, the small-oscillation approximation has reduced both U and T to homogeneous quadratic functions of x (for U) or $\dot{x}$ (for T).

The Equation of Motion

Returning to the approximate Lagrangian (11.55) for the general system, we can easily write down the equations of motion. Since there are n generalized coordinates q_i, $(i = 1, \cdots, n)$, there are n corresponding equations

$$\frac{d}{dt}\frac{\partial \mathcal{L}}{\partial \dot{q}_i} = \frac{\partial \mathcal{L}}{\partial q_i} \qquad [i = 1, \cdots, n]. \tag{11.56}$$

To write these equations explicitly, we must differentiate the expressions (11.54) and (11.53) for T and U. If you have never tried differentiating sums like these, it may help to write them out explicitly at first. For example, for a system with just two degrees of freedom ($n = 2$), Equation (11.53) for U reads

$$U = \tfrac{1}{2}\sum_{j,k=1}^{2} K_{jk}\, q_j\, q_k = \tfrac{1}{2}(K_{11}q_1^2 + K_{12}q_1q_2 + K_{21}q_2q_1 + K_{22}q_2^2)$$

$$= \tfrac{1}{2}(K_{11}q_1^2 + 2K_{12}q_1q_2 + K_{22}q_2^2) \tag{11.57}$$

where the second line follows because $K_{12} = K_{21}$. In this form we can easily differentiate with respect to either q_1 or q_2. For example,

$$\frac{\partial U}{\partial q_1} = K_{11}q_1 + K_{12}q_2$$

with a corresponding expression for $\partial U/\partial q_2$, and quite generally (however many degrees of freedom there are)

$$\frac{\partial U}{\partial q_i} = \sum_{j} K_{ij}q_j \qquad [i = 1, \cdots, n]. \tag{11.58}$$

Since differentiation of the kinetic energy (11.54) works in exactly the same way, we can write down the n Lagrange equations (11.56):

$$\sum_{j} M_{ij}\ddot{q}_j = -\sum_{j} K_{ij}q_j \qquad [i = 1, \cdots, n]. \tag{11.59}$$

These n equations can be immediately grouped into a single matrix equation

$$\mathbf{M\ddot{q}} = -\mathbf{Kq} \tag{11.60}$$

where $\mathbf{q}$ is the $(n \times 1)$ column

$$\mathbf{q} = \begin{bmatrix} q_1 \\ \vdots \\ q_n \end{bmatrix}$$

and $\mathbf{M}$ and $\mathbf{K}$ are the $(n \times n)$ "mass" and "spring-constant" matrices comprised of the numbers M_{ij} and K_{ij} respectively.[8]

The matrix equation (11.60) is, of course, the n-dimensional equivalent of the two-dimensional equations (11.3) for the pair of carts and (11.43) for the double pendulum, and it is solved in exactly the same way. We first seek normal modes with the now-familiar form

$$\mathbf{q}(t) = \mathrm{Re}\,\mathbf{z}(t), \qquad \text{where} \qquad \mathbf{z}(t) = \mathbf{a}e^{i\omega t} \tag{11.61}$$

and $\mathbf{a}$ is a constant $(n \times 1)$ column. These lead us to the eigenvalue equation

$$(\mathbf{K} - \omega^2 \mathbf{M})\mathbf{a} = 0, \tag{11.62}$$

which has solutions if and only if ω satisfies the **characteristic** or **secular equation**

$$\det(\mathbf{K} - \omega^2 \mathbf{M}) = 0. \tag{11.63}$$

This determinant is an nth degree polynomial in ω^2, so equation (11.63) has n solutions, which tell us the n normal frequencies of the system.[9] With ω set equal to each of the normal frequencies in turn, Equation (11.62) determines the motion of the system in the corresponding normal mode. Finally, the general motion of the system is given by an arbitrary sum of the normal mode solutions (11.61).

The general procedure outlined in the last three paragraphs is just what we have already discussed in detail for the examples of the two carts and the double pendulum. I shall go through one more example, this one with three degrees of freedom, in the next section, and you should certainly work some of the examples in the problems at the end of the chapter.

---

[8] From this point in the calculation all we need is the two matrices $\mathbf{M}$ and $\mathbf{K}$. Thus, in practice, there is actually no need to write down the Lagrangian, nor the Lagrange equations, since the matrices $\mathbf{M}$ and $\mathbf{K}$ can be read off directly from the approximate expressions (11.54) and (11.53) for T and U.

[9] Two subtleties: First, it may happen that some of the roots of (11.63) are equal; this simply means that some of the normal modes have equal frequencies and presents no serious problem. Second — and much deeper — we need the n solutions of (11.63) for ω^2 to be real and positive, in order that the normal frequencies be real. That this is actually so follows from properties of the matrices $\mathbf{K}$ and $\mathbf{M}$ as I shall discuss in the appendix.

11.6   Three Coupled Pendulums

Consider three identical pendulums coupled by two identical springs as shown in Figure 11.13. As generalized coordinates it is natural to use the three angles shown as ϕ_1, ϕ_2, and ϕ_3, with the equilibrium position at $\phi_1 = \phi_2 = \phi_3 = 0$. Our first task is to write down the Lagrangian for the system, at least for small displacements. The systematic, and perhaps the safest, procedure would be to write down the exact expressions for T and U and then make the small-angle approximations. In practice, finding the exact expressions can be very tedious. (In the present case, the potential energy of the springs depends on their extensions, and the exact expressions for these, good for any angles, are very cumbersome. See Problem 11.22.) It often happens that with care one can write down the small-angle approximations for T and U directly and save a lot of trouble, and this is what I shall do here.

The kinetic energy of the three pendulums is easily seen to be

$$T = \tfrac{1}{2}mL^2(\dot{\phi}_1^2 + \dot{\phi}_2^2 + \dot{\phi}_3^2) \tag{11.64}$$

which does not require any approximation. The gravitational potential energy of each pendulum has the form $mgL(1 - \cos\phi) \approx \tfrac{1}{2}mgL\phi^2$, where the last expression is the well-known small-angle approximation. Thus the total gravitational potential energy is

$$U_{\text{grav}} = \tfrac{1}{2}mgL(\phi_1^2 + \phi_2^2 + \phi_3^2). \tag{11.65}$$

To find the potential energy of the two springs, we have to find how much each is stretched. For arbitrary values of the angles ϕ this is a fairly messy affair, but for small angles the only appreciable stretching comes from the horizontal displacements of the pendulum bobs, each of which moves a distance of approximately $L\phi$ to the

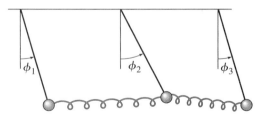

Figure 11.13   Three identical pendulums of lengths L and masses m are coupled by two identical springs with spring constants k. The generalized coordinates are the three angles ϕ_1, ϕ_2, and ϕ_3. The springs' natural lengths are equal to the separation of the supports of the pendulums, so the equilibrium position is $\phi_1 = \phi_2 = \phi_3 = 0$, with all three pendulums hanging vertically.

right. Thus, for example, the left spring is stretched by about $L(\phi_2 - \phi_1)$, and the total spring potential energy is

$$U_{\mathrm{spr}} = \tfrac{1}{2}kL^2 \left[(\phi_2 - \phi_1)^2 + (\phi_3 - \phi_2)^2 \right]$$

$$= \tfrac{1}{2}kL^2(\phi_1^2 + 2\phi_2^2 + \phi_3^2 - 2\phi_1\phi_2 - 2\phi_2\phi_3). \qquad (11.66)$$

Before we combine these expressions for T and U to get the Lagrangian and the equations of motion, there is a useful device I would like to introduce. The equations we are going to derive involve several fixed parameters, (m, L, g, k), some of which are not especially interesting. The repeated writing of these parameters is, at the very least, an annoying chore and can easily lead to careless mistakes. Thus it is helpful to find some way to get rid of uninteresting parameters before we do any more calculation. A radical way to do this that is very popular with theoretical physicists is to choose a system of units such that the uninteresting parameters have the value 1 — a process sometimes described as choosing **natural units**. In the present problem, for example, we can choose m to be the unit of mass and L to be the unit of length. With this choice m and L naturally have the value 1, and they disappear from all subsequent work. This trick materially simplifies the trivial details of our calculations, reduces the danger of errors, and helps us to see the truly interesting features.

The only serious disadvantage of using natural units is this: Once our calculations are complete, we sometimes want to know how our answers depend on the values of the parameters that have been suppressed. (What is the frequency of a certain normal mode, if $L = 1.5$ m?) To answer this kind of question, we have to put the banished parameters back into our answers. Although this process (of restoring the banished parameters) seems daunting to the beginner, it is usually fairly easy. For example, with $m = L = 1$, we are going to find that one of the normal frequencies is given by $\omega^2 = g$. This implies that (in our system of units) the quantity g/ω^2 has the value 1. But you will recognize that g/ω^2 has the dimensions of a length, and to say that a length has the value 1 in our units, is the same as saying that it has the value L in any system of units. Therefore, $g/\omega^2 = L$ in general, and our answer is that $\omega^2 = g/L$ whatever units we use. This puts the "L" back into our answer and lets us find ω for any given value of L.

Let us then choose units with $m = L = 1$, so that the kinetic and potential energies obtained from (11.64), (11.65), and (11.66) become

$$T = \tfrac{1}{2}(\dot{\phi}_1^2 + \dot{\phi}_2^2 + \dot{\phi}_3^2) \qquad (11.67)$$

and

$$U = \tfrac{1}{2}g(\phi_1^2 + \phi_2^2 + \phi_3^2) + \tfrac{1}{2}k(\phi_1^2 + 2\phi_2^2 + \phi_3^2 - 2\phi_1\phi_2 - 2\phi_2\phi_3). \qquad (11.68)$$

We could now write down the Lagrangian and then the equations of motion, but there is actually no need to do this. We already know that the result will be the now-familiar matrix equation

$$\mathbf{M}\ddot{\boldsymbol{\phi}} = -\mathbf{K}\boldsymbol{\phi} \qquad (11.69)$$

where, in this case, $\boldsymbol{\phi}$ is the (3×1) column comprising the three angles ϕ_1, ϕ_2, and ϕ_3. The elements of the (3×3) matrices $\mathbf{M}$ and $\mathbf{K}$ can be read directly from (11.67) and (11.68) to give[10]

$$\mathbf{M} = \begin{bmatrix} 1 & 0 & 0 \\ 0 & 1 & 0 \\ 0 & 0 & 1 \end{bmatrix} \quad \text{and} \quad \mathbf{K} = \begin{bmatrix} g+k & -k & 0 \\ -k & g+2k & -k \\ 0 & -k & g+k \end{bmatrix}. \quad (11.70)$$

The normal modes of our system have the familiar form $\boldsymbol{\phi}(t) = \operatorname{Re} \mathbf{z}(t) = \operatorname{Re} \mathbf{a} e^{i\omega t}$, where $\mathbf{a}$ and ω are determined by the eigenvalue equation

$$(\mathbf{K} - \omega^2 \mathbf{M})\mathbf{a} = 0. \quad (11.71)$$

Our first step is to find the possible normal frequencies from the characteristic equation $\det(\mathbf{K} - \omega^2 \mathbf{M}) = 0$, to which end we need to write down the matrix $(\mathbf{K} - \omega^2 \mathbf{M})$,

$$(\mathbf{K} - \omega^2 \mathbf{M}) = \begin{bmatrix} g+k-\omega^2 & -k & 0 \\ -k & g+2k-\omega^2 & -k \\ 0 & -k & g+k-\omega^2 \end{bmatrix}. \quad (11.72)$$

The determinant is easily evaluated as

$$\det(\mathbf{K} - \omega^2 \mathbf{M}) = (g - \omega^2)(g + k - \omega^2)(g + 3k - \omega^2),$$

so that the three normal frequencies are given by

$$\omega_1^2 = g, \qquad \omega_2^2 = g + k, \qquad \text{and} \qquad \omega_3^2 = g + 3k. \quad (11.73)$$

Knowing the three normal frequencies, we can now find the three corresponding normal modes in turn. The first normal frequency has $\omega_1 = \sqrt{g}$. (This is in our units, where $L = 1$. As I have already mentioned, in arbitrary units it is $\omega_1 = \sqrt{g/L}$, which is the frequency for a single pendulum of length L. We'll see the reason for this coincidence in a moment.) If we substitute ω_1 into Equation (11.72) for $(\mathbf{K} - \omega^2 \mathbf{M})$, then the eigenvalue equation (11.71) implies (as you should check) that

$$a_1 = a_2 = a_3 = A e^{-i\delta}, \qquad \text{[first mode]}$$

say. That is, in the first mode,

$$\phi_1(t) = \phi_2(t) = \phi_3(t) = A \cos(\omega_1 t - \delta)$$

and the three pendulums oscillate in unison (equal amplitudes and phases), as shown in Figure 11.14(a). In this mode, the springs are neither compressed nor stretched, and their presence is irrelevant. Therefore, each pendulum oscillates just like a single pendulum, with frequency $\omega_1 = \sqrt{g/L}$ (or $\sqrt{g}$ in our units).

---

[10] One has to think a little when reading off the elements of these matrices. The rule is this: If you ignore the factor $\frac{1}{2}$ in front of (11.68), for example, then the diagonal element K_{ii} is just the coefficient of ϕ_i^2, while the off-diagonal element K_{ij} is *half* the coefficient of $\phi_i \phi_j$. To understand this, look at (11.57).

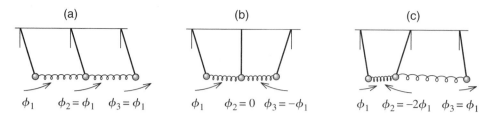

Figure 11.14   The three normal modes for three coupled pendulums. **(a)** In the first mode, the three pendulums oscillate in unison. Since neither spring is stretched or compressed, the frequency is just that for a single pendulum of the same length. **(b)** In the second mode, the outer two pendulums oscillate exactly out of phase, while the middle one doesn't move at all. **(c)** In the third mode, the outer two pendulums oscillate in unison, but the middle one is exactly out of phase and has twice the amplitude.

If we substitute $\omega = \omega_2$, the eigenvalue equation (11.71) implies that

$$a_1 = -a_3 = Ae^{-i\delta}, \qquad \text{but} \qquad a_2 = 0 \qquad \text{[second mode]}.$$

Therefore, the outer two pendulums oscillate exactly out of phase, while the middle one sits at rest, as shown in Figure 11.14(b). Finally, substituting $\omega = \omega_3$ into the eigenvalue equation (11.71) leads to the result

$$a_1 = -\tfrac{1}{2}a_2 = a_3 = Ae^{-i\delta}, \qquad \text{[third mode]}$$

say. Thus, in the third mode, the outer two pendulums oscillate in unison, but the middle one oscillates with twice the amplitude and exactly out of phase, as shown in Figure 11.14(c). The general solution is an arbitrary linear combination of all three normal modes.

11.7   Normal Coordinates *

This section could be skipped if you are pressed for time.

In Section 11.2, we found the normal modes for two equal masses joined by three identical springs. At the end of that section, I mentioned that one can replace the two coordinates x_1 and x_2 by two "normal coordinates"

$$\xi_1 = \tfrac{1}{2}(x_1 + x_2) \qquad \text{and} \qquad \xi_2 = \tfrac{1}{2}(x_1 - x_2). \qquad (11.74)$$

These new coordinates have the property that each always oscillates at just one of the two normal frequencies — ξ_1 at the frequency ω_1 and ξ_2 at the frequency ω_2. In this section I will show that we can do the same thing for any system oscillating about a stable equilibrium point. If the system has n degrees of freedom, then it is described by n generalized coordinates $q_1, \cdots, q_n$ and has n normal modes with frequencies $\omega_1, \cdots, \omega_n$. What I shall show is that we can introduce n new, normal, coordinates $\xi_1, \cdots, \xi_n$, such that each normal coordinate ξ_i oscillates at just one frequency, namely the normal frequency ω_i. Before I show this, I need to review and extend the discussion of Section 11.2 for the two carts.

Normal Coordinates for Two Carts on Springs

The equations of motion for the positions x_1 and x_2 of the two carts were given in (11.2), which I rewrite here (for the case of equal masses and equal spring constants) as

$$\left.\begin{array}{rcl} m\ddot{x}_1 &=& -2kx_1 + kx_2 \\ m\ddot{x}_2 &=& kx_1 - 2kx_2 . \end{array}\right\} \tag{11.75}$$

A moment's inspection should convince you that if we add these two equations, we get an equation for $\xi_1 = \frac{1}{2}(x_1 + x_2)$ alone, and if we subtract them, we get an equation for $\xi_2 = \frac{1}{2}(x_1 - x_2)$ alone:

$$\left.\begin{array}{rcl} m\ddot{\xi}_1 &=& -k\xi_1 \\ m\ddot{\xi}_2 &=& -3k\xi_2 . \end{array}\right\} \tag{11.76}$$

These two equations are *uncoupled*, and show that each normal coordinate oscillates, as claimed, at a single frequency — ξ_1 at frequency $\omega_1 = \sqrt{k/m}$ and ξ_2 at $\omega_2 = \sqrt{3k/m}$. In other words, the normal coordinates behave just like the coordinates of two *uncoupled* oscillators — by going over to the normal coordinates, we have "uncoupled" the oscillations.

Just as the equations (11.75) for x_1 and x_2 can be rewritten as a single matrix equation $\mathbf{M\ddot{x}} = -\mathbf{Kx}$, so the two equations (11.76) for ξ_1 and ξ_2 can be written as $\mathbf{M'\ddot{\xi}} = -\mathbf{K'\xi}$, with the important difference that the two matrices $\mathbf{M'}$ and $\mathbf{K'}$ are both diagonal:

$$\mathbf{M'} = \begin{bmatrix} m & 0 \\ 0 & m \end{bmatrix} \quad \text{and} \quad \mathbf{K'} = \begin{bmatrix} k & 0 \\ 0 & 3k \end{bmatrix} . \tag{11.77}$$

The transition from the original coordinates (x_1, x_2) to the normal coordinates (ξ_1, ξ_2) is said to *diagonalize* the matrices $\mathbf{M}$ and $\mathbf{K}$. That the new matrices are diagonal is precisely equivalent to the statement that the equations (11.76) for ξ_1 and ξ_2 are uncoupled and that ξ_1 and ξ_2 oscillate independently.

We can define the two normal coordinates ξ_1 and ξ_2 differently, and more generally, in terms of the eigenvectors $\mathbf{a}$ that describe the motion of the normal modes and are determined by the eigenvalue equation $(\mathbf{K} - \omega^2 \mathbf{M})\mathbf{a} = 0$. We saw in Section 11.2 that these two (2×1) columns are (for our two carts)

$$\mathbf{a}_{(1)} = \begin{bmatrix} 1 \\ 1 \end{bmatrix} \quad \text{and} \quad \mathbf{a}_{(2)} = \begin{bmatrix} 1 \\ -1 \end{bmatrix} . \tag{11.78}$$

[Two important points: Each of these vectors contains an arbitrary multiplier A, but I now want to fix this, and the simplest choice here is to make $A = 1$; another, and sometimes better, choice is to normalize the vectors by putting in a factor of $1/\sqrt{2}$. Second, each column $\mathbf{a}$ is made up of two numbers, which I have been calling a_1 and a_2. But I am now discussing two *different columns*, $\mathbf{a}_{(1)}$ and $\mathbf{a}_{(2)}$, one for each normal mode, and I am using the parentheses in the subscripts to emphasize this distinction.] Now, it is easy to see that *any* (2×1) column vector can be written as a combination of the two vectors $\mathbf{a}_{(1)}$ and $\mathbf{a}_{(2)}$. (See Problem 11.33.) In particular, I can expand the column $\mathbf{x}$ in this way as

$$\mathbf{x} = \xi_1 \mathbf{a}_{(1)} + \xi_2 \mathbf{a}_{(2)} = \begin{bmatrix} \xi_1 + \xi_2 \\ \xi_1 - \xi_2 \end{bmatrix}. \tag{11.79}$$

The first equality defines ξ_1 and ξ_2 as the coefficients in this expansion of $\mathbf{x}$ in terms of the eigenvectors $\mathbf{a}_{(1)}$ and $\mathbf{a}_{(2)}$, but inspection of the last expression in (11.79) should convince you that it defines ξ_1 and ξ_2 to be precisely the normal coordinates of (11.74). That is, the normal coordinates, with the property that each oscillates independently at one of the normal frequencies, can be defined as the coefficients in the expansion of $\mathbf{x}$ in terms of the eigenvectors $\mathbf{a}_{(1)}$ and $\mathbf{a}_{(2)}$. We shall see that this definition carries over naturally to the general case of oscillations of any system with n degrees of freedom.

The General Case

We can now easily introduce normal coordinates for an arbitrary oscillating system with n generalized coordinates $q_1, \cdots, q_n$. We know that such a system has n normal modes. In mode i, the column vector $\mathbf{q}$ oscillates sinusoidally,

$$\mathbf{q}(t) = \mathbf{a}_{(i)} \cos(\omega_i t - \delta_i)$$

where the fixed column $\mathbf{a}_{(i)}$ satisfies the eigenvalue equation

$$\mathbf{K}\mathbf{a}_{(i)} = \omega_i^2 \mathbf{M}\mathbf{a}_{(i)}. \tag{11.80}$$

The columns $\mathbf{a}_{(1)}, \cdots, \mathbf{a}_{(n)}$ are n independent, real[11] $(n \times 1)$ columns, and *any* $(n \times 1)$ column can be expanded in terms of them; that is, the vectors $\mathbf{a}_{(1)}, \cdots, \mathbf{a}_{(n)}$ are a **basis** or **complete set** for the space of all $(n \times 1)$ vectors. (For a proof of these properties, see the appendix.) Thus any solution of the equations of motion $\mathbf{q}(t)$ can be expanded as

$$\mathbf{q}(t) = \sum_{i=1}^{n} \xi_i(t)\, \mathbf{a}_{(i)}. \tag{11.81}$$

This definition of the normal coordinates ξ_i exactly parallels the new definition (11.79) for the system of two carts on springs, and we can now prove that it has the desired property that the different $\xi_i(t)$ oscillate independently, as follows:

The column $\mathbf{q}(t)$ satisfies the equation of motion

$$\mathbf{M}\ddot{\mathbf{q}} = -\mathbf{K}\mathbf{q}.$$

---

[11] That the vectors $\mathbf{a}_{(i)}$ are real is not obvious. In the case of the two carts, you can see from (11.78) that they are. In the general case, they are determined by the real eigenvalue equation (11.80) and it is fairly easy to prove that either they are real or they can be redefined so that they are. See the appendix.

If we replace $\mathbf{q}$ by its expansion (11.81), this equation becomes

$$\sum_{i=1}^{n} \ddot{\xi}_i(t)\, \mathbf{Ma}_{(i)} = -\sum_{i=1}^{n} \xi_i(t)\, \mathbf{Ka}_{(i)} = -\sum_{i=1}^{n} \xi_i(t)\, \omega_i^2 \mathbf{Ma}_{(i)} \qquad (11.82)$$

where the last equality follows from the eigenvalue equation (11.80). Now, the n column vectors $\mathbf{a}_{(1)}, \cdots, \mathbf{a}_{(n)}$ are independent, and this property is unchanged when they are multiplied by the matrix $\mathbf{M}$. Therefore, the n vectors $\mathbf{Ma}_{(1)}, \cdots, \mathbf{Ma}_{(n)}$ are also independent, and the equality (11.82) can only hold if all corresponding coefficients on each side are equal.[12] That is,

$$\ddot{\xi}_i(t) = -\omega_i^2\, \xi_i(t).$$

This establishes that the normal coordinates ξ_i defined by (11.81) do indeed oscillate independently at the advertised frequencies.

Principal Definitions and Equations of Chapter 11

The Equations of Motion in Matrix Form

The configuration of a system with n degrees of freedom can be specified by an $n \times 1$ column matrix $\mathbf{q}$, comprising the n generalized coordinates $q_1, \cdots, q_n$. The equation of motion for small oscillations about a stable equilibrium (with the coordinates chosen so that $\mathbf{q} = 0$ at equilibrium) has the matrix form

$$\mathbf{M\ddot{q}} = -\mathbf{Kq} \qquad \text{[Eq. (11.60)]}$$

where $\mathbf{M}$ and $\mathbf{K}$ are the $n \times n$ **"mass"** and **"spring-constant" matrices**. One way to find these matrices is to write the KE and PE of the system in the forms

$$T = \tfrac{1}{2}\sum_{j,k} M_{jk}\, \dot{q}_j\, \dot{q}_k \quad \text{and} \quad U = \tfrac{1}{2}\sum_{j,k} K_{jk}\, q_j\, q_k. \qquad \text{[Eqs. (11.54) \& (11.53)]}$$

Normal Modes

A **normal mode** is any motion in which all n coordinates oscillate sinusoidally with the same frequency ω (a **normal frequency**) and can be written as

$$\mathbf{q}(t) = \mathrm{Re}\!\left(\mathbf{a}\, e^{i\omega t}\right) \qquad \text{[Eq. (11.61)]}$$

where the constant $n \times 1$ column $\mathbf{a}$ must satisfy the generalized eigenvalue equation

$$(\mathbf{K} - \omega^2 \mathbf{M})\mathbf{a} = 0. \qquad \text{[Eq. (11.62)]}$$

---

[12] The result I am using here is the analog of the familiar result in three-dimensional space that the equality $\sum_1^3 \lambda_i \mathbf{e}_i = \sum_1^3 \mu_i \mathbf{e}_i$ is only possible if $\lambda_i = \mu_i$ for all i. For a proof of the independence of the vectors $\mathbf{a}_{(1)}, \cdots, \mathbf{a}_{(n)}$ see the appendix.

For any system with n degrees of freedom and a stable equilibrium at $\mathbf{q} = 0$, there are n normal frequencies $\omega_1, \cdots, \omega_n$ (some of which may be equal) and n independent corresponding eigenvectors $\mathbf{a}_{(1)}, \cdots, \mathbf{a}_{(n)}$. Any $n \times 1$ column can be expanded in terms of these eigenvectors, and any solution of the equations of motion can be expanded in terms of the normal modes.

Normal Coordinates

When any solution is expanded in terms of the normal modes, the expansion coefficients $\xi_i(t)$ are called **normal coordinates**, and each oscillates at the corresponding frequency ω_i.

[Section 11.7]

Problems for Chapter 11

Stars indicate the approximate level of difficulty, from easiest (⋆) to most difficult (⋆⋆⋆).

SECTION 11.1   Two Masses and Three Springs

11.1 ⋆ In discussing the two carts of Figure 11.1, I mentioned that it is simplest to assume that when the two carts are in equilibrium the lengths L_1, L_2, L_3 of the three springs are equal to their natural, unstretched lengths l_1, l_2, l_3. However, this assumption is not needed, and the three springs could all be in tension (or compression) at the equilibrium position. **(a)** Find the relations among these six lengths (and the three spring constants k_1, k_2, k_3) required for the two carts to be in equilibrium. **(b)** Show that the net force on either cart is exactly as given in Equation (11.2), irrespective of how L_1, L_2, L_3 compare with l_1, l_2, l_3, just as long as x_1 and x_2 are measured from the carts' equilibrium positions.

11.2 ⋆⋆ A massless spring (force constant k_1) is suspended from the ceiling, with a mass m_1 hanging from its lower end. A second massless spring (force constant k_2) is suspended from m_1, and a second mass m_2 is suspended from the second spring's lower end. Assuming that the masses move only in a vertical direction and using coordinates y_1 and y_2 measured from the masses' equilibrium positions, show that the equations of motion can be written in the matrix form $\mathbf{M}\ddot{\mathbf{y}} = -\mathbf{K}\mathbf{y}$, where $\mathbf{y}$ is the 2×1 column made up of y_1 and y_2. Find the 2×2 matrices $\mathbf{M}$ and $\mathbf{K}$.

SECTION 11.2   Identical Springs and Equal Masses

11.3 ⋆ Find the normal frequencies for the system of two carts and three springs shown in Figure 11.1, for arbitrary values of m_1 and m_2 and of k_1, k_2, and k_3. Check that your answer is correct for the case that $m_1 = m_2$ and $k_1 = k_2 = k_3$.

11.4 ⋆⋆ **(a)** Find the normal frequencies for the system of two carts and three springs shown in Figure 11.1, for the case that $m_1 = m_2$ and $k_1 = k_3$, (but k_2 may be different). Check that your answer is correct for the case that $k_1 = k_2$ as well. **(b)** Find and describe the motion in each of the two normal modes in turn. Compare with the motion found for the case that $k_1 = k_2$ in Section 11.2. Explain any similarities.

11.5 ⋆⋆ **(a)** Find the normal frequencies, ω_1 and ω_2, for the two carts shown in Figure 11.15, assuming that $m_1 = m_2$ and $k_1 = k_2$. **(b)** Find and describe the motion for each of the normal modes in turn.

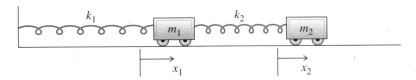

Figure 11.15   Problems 11.5 and 11.6.

11.6 ★★ Answer the same questions as in Problem 11.5 but for the case that $m_1 = m_2$ and $k_1 = 3k_2/2$. (Write $k_1 = 3k$ and $k_2 = 2k$.) Explain the motion in the two normal modes.

11.7 ★★ [Computer] The most general motion of the two carts of Section 11.2 is given by (11.21), with the constants A_1, A_2, δ_1, and δ_2 determined by the initial conditions. **(a)** Show that (11.21) can be rewritten as

$$\mathbf{x}(t) = (B_1 \cos \omega_1 t + C_1 \sin \omega_1 t) \begin{bmatrix} 1 \\ 1 \end{bmatrix} + (B_2 \cos \omega_2 t + C_2 \sin \omega_2 t) \begin{bmatrix} 1 \\ -1 \end{bmatrix}.$$

This form is usually a little more convenient for matching to given initial conditions. **(b)** If the carts are released from rest at positions $x_1(0) = x_2(0) = A$, find the coefficients B_1, B_2, C_1, and C_2 and plot $x_1(t)$ and $x_2(t)$. Take $A = \omega_1 = 1$ and $0 \le t \le 30$ for your plots. **(c)** Same as part (b), except that $x_1(0) = A$ but $x_2(0) = 0$.

11.8 ★★ [Computer] Same as Problem 11.7 but in part (b) the carts are at their equilibrium positions at $t = 0$ and are kicked away from each other, each with speed v_0. In part (c), the carts start out at their equilibrium positions and cart 2 has speed v_0 to the right but cart 1 has initial speed 0. Take $v_0 = \omega_1 = 1$ and $0 \le t \le 30$ for your plots.

11.9 ★★ **(a)** Write down the equations of motion (11.2) for the equal-mass carts of Section 11.2 with three identical springs. Show that the change of variables to the normal coordinates $\xi_1 = \frac{1}{2}(x_1 + x_2)$ and $\xi_2 = \frac{1}{2}(x_1 - x_2)$ leads to uncoupled equations for ξ_1 and ξ_2. **(b)** Solve for ξ_1 and ξ_2 and hence write down the general solution for x_1 and x_2. (Notice how very simple this procedure is, once you have guessed what the normal coordinates are. For a simple symmetric system like this, you can sometimes guess the form of ξ_1 and ξ_2 by considering the symmetry — especially once you have some experience working with normal modes.)

11.10 ★★★ [Computer] In general, the analysis of coupled oscillators with dissipative forces is much more complicated than the conservative case considered in this chapter. However, there are a few cases where the same methods still work, as the following problem illustrates: **(a)** Write down the equations of motion corresponding to (11.2) for the equal-mass carts of Section 11.2 with three identical springs, but with each cart subject to a linear resistive force $-b\mathbf{v}$ (same coefficient b for both carts). **(b)** Show that if you change variables to the normal coordinates $\xi_1 = \frac{1}{2}(x_1 + x_2)$ and $\xi_2 = \frac{1}{2}(x_1 - x_2)$, the equations of motion for ξ_1 and ξ_2 are uncoupled. **(c)** Write down the general solutions for the normal coordinates and hence for x_1 and x_2. (Assume that b is small, so that the oscillations are underdamped.) **(d)** Find $x_1(t)$ and $x_2(t)$ for the initial conditions $x_1(0) = A$ and $x_2(0) = v_1(0) = v_2(0) = 0$, and plot them for $0 \le t \le 10\pi$ using the values $A = k = m = 1$, and $b = 0.1$.

11.11 ★★★ **(a)** Write down the equations of motion corresponding to (11.2) for the equal-mass carts of Section 11.2 with three identical springs, but with each cart subject to a linear resistive force $-b\mathbf{v}$ (same coefficient b for both carts) and with a driving force $F(t) = F_0 \cos \omega t$ applied to cart 1. **(b)** Show that if you change variables to the normal coordinates $\xi_1 = \frac{1}{2}(x_1 + x_2)$ and $\xi_2 = \frac{1}{2}(x_1 - x_2)$, the equations of motion for ξ_1 and ξ_2 are uncoupled. **(c)** Using the methods of Section 5.5, write down the general solutions. **(d)** Assuming that $\beta = b/2m \ll \omega_0$, show that ξ_1 resonates when $\omega \approx \omega_0 = \sqrt{k/m}$ and likewise ξ_2 when $\omega \approx \sqrt{3}\omega_0$. **(e)** Prove, on the other hand, that if both carts are driven in phase with the same force $F_0 \cos \omega t$, only ξ_1 shows a resonance. Explain.

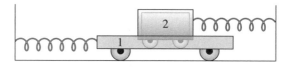

Figure 11.16   Problem 11.12

11.12 ★★★ Here is a different way to couple two oscillators. The two carts in Figure 11.16 have equal masses m (though different shapes). They are joined by identical but separate springs (force constant k) to separate walls. Cart 2 rides in cart 1, as shown, and cart 1 is filled with molasses, whose viscous drag supplies the coupling between the carts.

(a) Assuming that the drag force has magnitude $\beta m v$ where $\mathbf{v}$ is the relative velocity of the two carts, write down the equations of motion of the two carts using as coordinates x_1 and x_2, the displacements of the carts from their equilibrium positions. Show that they can be written in matrix form as $\ddot{\mathbf{x}} + \beta \mathbf{D}\dot{\mathbf{x}} + \omega_0^2 \mathbf{x} = 0$, where $\mathbf{x}$ is the 2×1 column made up of x_1 and x_2, $\omega_0 = \sqrt{k/m}$, and $\mathbf{D}$ is a certain 2×2 square matrix. **(b)** There is nothing to stop you from seeking a solution of the form $\mathbf{x}(t) = \operatorname{Re} \mathbf{z}(t)$, with $\mathbf{z}(t) = \mathbf{a}e^{rt}$. Show that you do indeed get two solutions of this form with $r = i\omega_0$ or $r = -\beta + i\omega_1$ where $\omega_1 = \sqrt{\omega_0^2 - \beta^2}$. (Assume that the viscous force is weak, so that $\beta < \omega_0$.) **(c)** Describe the corresponding motions. Explain why one of these modes is damped but the other is not.

SECTION 11.3   Two Weakly Coupled Oscillators

11.13 ★★★ [Computer] Consider the two carts of Section 11.3, coupled by a weak spring and subject to a resistive force $-bv$ (same force for each cart). **(a)** Write down the equations of motion for x_1 and x_2 in the form (11.2) and show that if you change to the normal coordinates $\xi_1 = \frac{1}{2}(x_1 + x_2)$ and $\xi_2 = \frac{1}{2}(x_1 - x_2)$ the equations of motion for ξ_1 and ξ_2 are uncoupled. **(b)** Solve for $\xi_1(t)$ and $\xi_2(t)$ assuming that the dissipative coefficient b is small ("underdamped motion," as in Section 5.4), and hence write down the general solution for $x_1(t)$ and $x_2(t)$. **(c)** Suppose that cart 1 is held at $x_1 = A$ and cart 2 at $x_2 = 0$, and they are released from rest at $t = 0$. Find and plot the two positions as functions of t for $0 \le t \le 80$, using the values $A = k = m = 1$, $k_2 = 0.2$, and $b = 0.04$. (In matching the initial conditions, take advantage of the fact that $b \ll 1$, and use a suitable computer program to make the plot.) Comment on your plots.

SECTION 11.4   Lagrangian Approach: The Double Pendulum

11.14 ★★ Consider two identical plane pendulums (each of length L and mass m) that are joined by a massless spring (force constant k) as shown in Figure 11.17. The pendulums' positions are specified by the angles ϕ_1 and ϕ_2 shown. The natural length of the spring is equal to the distance between the two supports, so the equilibrium position is at $\phi_1 = \phi_2 = 0$ with the two pendulums vertical. **(a)** Write down the total kinetic energy and the gravitational and spring potential energies. [Assume that both angles remain small at all times. This means that the extension of the spring is well approximated by

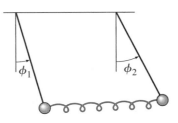

Figure 11.17   Problem 11.14

$L(\phi_2 - \phi_1)$.] Write down the Lagrange equations of motion. **(b)** Find and describe the normal modes for these two coupled pendulums.

11.15 ⋆⋆ Write down the exact Lagrangian (good for all angles) for the double pendulum of Figure 11.9 and find the corresponding equations of motion. Show that they reduce to Equations (11.41) and (11.42) if both angles are small.

11.16 ⋆⋆ **(a)** Find the normal frequencies for small oscillations of the double pendulum of Figure 11.9 for arbitrary values of the masses and lengths. **(b)** Check that your answers are correct for the special case that $m_1 = m_2$ and $L_1 = L_2$. **(c)** Discuss the limit that $m_2 \to 0$.

11.17 ⋆⋆ **(a)** Find the normal frequencies and modes of the double pendulum of Figure 11.9, given that $m_1 = 8m$, $m_2 = m$, and $L_1 = L_2 = L$. **(b)** Find the actual motion $[\phi_1(t), \phi_2(t)]$ if the pendulum is released from rest with $\phi_1 = 0$ and $\phi_2 = \alpha$. Is this motion periodic?

11.18 ⋆⋆ Two equal masses m are constrained to move without friction, one on the positive x axis and one on the positive y axis. They are attached to two identical springs (force constant k) whose other ends are attached to the origin. In addition, the two masses are connected to each other by a third spring of force constant k'. The springs are chosen so that the system is in equilibrium with all three springs relaxed (length equal to unstretched length). What are the normal frequencies? Find and describe the normal modes. Consider only small displacements from equilibrium.

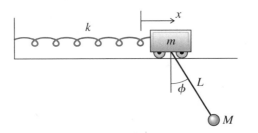

Figure 11.18   Problem 11.19

11.19 ⋆⋆⋆ A simple pendulum (mass M and length L) is suspended from a cart (mass m) that can oscillate on the end of a spring of force constant k, as shown in Figure 11.18. **(a)** Assuming that the angle ϕ remains small, write down the system's Lagrangian and the equations of motion for x and ϕ. **(b)** Assuming that $m = M = L = g = 1$ and $k = 2$ (all in appropriate units) find the normal frequencies, and for each normal frequency find and describe the motion of the corresponding normal mode.

11.20 ★★★ A thin uniform rod of length $2b$ is suspended by two vertical light strings, both of fixed length l, fastened to the ceiling. Assuming only small displacements from equilibrium, find the Lagrangian of the system and the normal frequencies. Find and describe the normal modes. [*Hint:* A possible choice of generalized coordinates would be x, the longitudinal displacement of the rod, and y_1 and y_2, the sideways displacements of the rod's two ends. You'll need to find how high the two ends are above their equilibrium height and what angle the rod has turned through.]

SECTIONS 11.5 and 11.6   The General Case & Three Coupled Pendulums

11.21 ★ Verify that if $U = \frac{1}{2}\sum_j \sum_k K_{jk}q_j q_k$, where the coefficients K_{jk} are all constant and satisfy $K_{ij} = K_{ji}$, then $\partial U/\partial q_i = \sum_j K_{ij}q_j$, as claimed in Equation (11.58).

11.22 ★ Write down the exact potential energy of the three pendulums of Figure 11.13, good for all angles, small or large, and show that your answer reduces to (11.68) if all three angles are small.

11.23 ★★ Equation (11.73) gives the three normal frequencies of three coupled pendulums in natural units with $L = m = 1$. We have already seen that the value of ω_1^2 in arbitrary units is g/L. Find the values of ω_2^2 and ω_3^2 in arbitrary units. [*Hint:* Start by considering the quantity $\omega_2^2 - \omega_1^2$.]

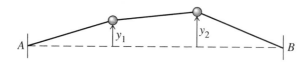

Figure 11.19   Problem 11.24

11.24 ★★ Two equal masses m move on a frictionless horizontal table. They are held by three identical taut strings (each of length L, tension T), as shown in Figure 11.19, so that their equilibrium position is a straight line between the anchors at A and B. The two masses move in the transverse (y) direction, but not in the longitudinal (x) direction. Write down the Lagrangian for small displacements, and find and describe the motion in the corresponding normal modes. [*Hint:* "Small" displacements have y_1 and y_2 much less than L, which means that you can treat the tensions as constant. Therefore the PE of each string is just Td, where d is the amount by which its length has increased from equilibrium.]

11.25 ★★ Consider a system of carts and springs like that in Figure 11.1 except that there are *three* equal-mass carts and *four* identical springs. Solve for the three normal frequencies, and find and describe the motion in the corresponding normal modes.

11.26 ★★ A bead of mass m is threaded on a frictionless circular wire hoop of radius R and mass m (same mass). The hoop is suspended at the point A and is free to swing in its own vertical plane as shown in Figure 11.20. Using the angles ϕ_1 and ϕ_2 as generalized coordinates, solve for the normal frequencies of small oscillations, and find and describe the motion in the corresponding normal modes. [*Hint:* The KE of the hoop is $\frac{1}{2}I\dot{\phi}_1^2$, where I is its moment of inertia about A and can be found using the parallel axis theorem.]

11.27 ★★ Consider two carts of equal mass m on a horizontal, frictionless track. The carts are connected to each other by a single spring of force constant k, but are otherwise free to move freely along the track. **(a)** Write down the Lagrangian and find the normal frequencies of the system. Show that one of the

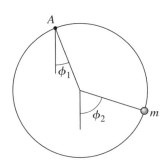

Figure 11.20   Problem 11.26

normal frequencies is zero. **(b)** Find and describe the motion in the normal mode whose frequency is nonzero. **(c)** Do the same for the mode with zero frequency. [*Hint:* This one requires some thought. It isn't immediately clear what oscillations of zero frequency are. Notice that the eigenvalue equation $(\mathbf{K} - \omega^2\mathbf{M})\mathbf{a} = 0$ reduces to $\mathbf{Ka} = 0$ in this case. Consider a solution $\mathbf{x}(t) = \mathbf{a}f(t)$, where $f(t)$ is an undetermined function of t and use the equation of motion, $\mathbf{M}\ddot{\mathbf{x}} = -\mathbf{Kx}$, to show that this solution represents motion of the whole system with constant velocity. Explain why this kind of motion is possible here but not in the previous examples.]

11.28 ★★ A simple pendulum (mass M and length L) is suspended from a cart of mass m that moves freely along a horizontal track. (See Figure 11.18, but imagine the spring removed.) **(a)** What are the normal frequencies? **(b)** Find and describe the corresponding normal modes. [See the hint for Problem 11.27].

11.29 ★★★ A thin rod of length $2b$ and mass m is suspended by its two ends with two identical vertical springs (force constant k) that are attached to the horizontal ceiling. Assuming that the whole system is constrained to move in just the one vertical plane, find the normal frequencies and normal modes of small oscillations. Describe and explain the normal modes. [*Hint:* It is crucial to make a wise choice of generalized coordinates. One possibility would be r, ϕ, and α, where r and ϕ specify the position of the rod's CM relative to an origin half way between the springs on the ceiling, and α is the angle of tilt of the rod. Be careful when writing down the potential energy.]

11.30 ★★★ [Computer] Consider a system of carts and springs like that in Figure 11.1 except that there are *four* equal-mass carts and *five* identical springs. Solve for the four normal frequencies, and find and describe the motion in the corresponding normal modes. [This can be solved without the aid of a computer, but it would probably be worth your while to learn how to evaluate determinants and to solve the characteristic equation using suitable computer software.]

11.31 ★★★ Consider a frictionless rigid horizontal hoop of radius R. Onto this hoop I thread three beads with masses $2m$, m, and m, and, between the beads, three identical springs, each with force constant k. Solve for the three normal frequencies and find and describe the three normal modes.

11.32 ★★★ As a model of a linear triatomic molecule (such as CO_2), consider the system shown in Figure 11.21, with two identical atoms each of mass m connected by two identical springs to a single atom of mass M. To simplify matters, assume that the system is confined to move in one dimension. **(a)** Write down the Lagrangian and find the normal frequencies of the system. Show that one of the

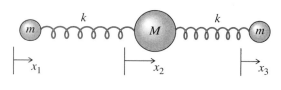

Figure 11.21   Problem 11.32

normal frequencies is zero. **(b)** Find and describe the motion in the normal modes whose frequencies are nonzero. **(c)** Do the same for the mode with zero frequency. [*Hint:* See the comments at the end of Problem 11.27.]

SECTION 11.7   **Normal Coordinates** *

11.33 ★ The eigenvectors $a_{(1)}$ and $a_{(2)}$ that describe the motion in the two normal modes of the two carts of Section 11.2 are given in (11.78). Prove that *any* (2×1) column x can be written as a linear combination of these two eigenvectors; that is, $a_{(1)}$ and $a_{(2)}$ are a *basis* of the space of (2×1) columns.

11.34 ★★ It is a crucial property of the eigenvectors, $a_{(1)}, \cdots, a_{(n)}$, describing the motion in the normal modes of an oscillating system that *any* $(n \times 1)$ column x can be written as a linear combination of the n eigenvectors; that is, the eigenvectors are a *basis* of the space of $(n \times 1)$ columns. This is proved in the appendix, but to illustrate it, do the following: **(a)** Write down the three eigenvectors $a_{(1)}$, $a_{(2)}$, and $a_{(3)}$ for the coupled pendulums of Section 11.6. (Each of these contains an arbitrary overall factor, which you can choose at your convenience.) Prove that they have the property that any (3×1) column x can be expanded in terms of them. **(b)** The expansion coefficients in this expansion are the normal coordinates ξ_1, ξ_2, ξ_3. Find the normal coordinates for the three coupled pendulums, and explain the sense in which they describe the three normal modes of Figure 11.14.

11.35 ★★ Consider the two coupled pendulums of Problem 11.14. **(a)** What would be a natural choice for the normal coordinates ξ_1 and ξ_2? **(b)** Show that even if both pendulums are subject to a resisitive force of magnitude bv (with b small), the equations of motion for ξ_1 and ξ_2 are still uncoupled. **(c)** Find and describe the motion of the two pendulums for the two modes.

Further Topics

Part I of this book contains the material that is in some sense essential. Part II contains mostly more advanced topics that could be considered optional — although all are of the greatest importance in modern physics. While the chapters of Part I were designed to be read pretty much in sequence, I have tried to make the five chapters of Part II mutually independent, so that you could pick and choose according to your tastes and available time. If you understand most of the material of Part I, you are ready to launch into any of the chapters of Part II. This is not to say that you have *to have studied all of Part I before reading any of Part II. For example, you could read Chapter 12 on chaos as soon as you have studied Chapter 5 on oscillations. Similarly, you could read Chapter 13 on Hamiltonian mechanics immediately after Chapter 7 on Lagrangian mechanics, and likewise, Chapter 14 on collision theory immediately after Chapter 8 on two-body motion.*

CHAPTER 12

Nonlinear Mechanics and Chaos

One of the most fascinating and exciting discoveries of the last few decades has been the recognition that most systems whose equations of motion are nonlinear can exhibit *chaos*. This startling phenomenon, which shows up in many different areas — oscillating mechanical systems, chemical reactions, fluid flow, lasers, population growth, the spread of diseases, and many more — means that although a system obeys deterministic equations of motion (such as Newton's laws) its detailed future behavior may, as a practical matter, be unpredictable.

The behavior of a chaotic mechanical system can be very complicated, and the need to describe this behavior has spawned a whole array of new ways to view the motion of such systems — state-space orbits, Poincaré sections, bifurcation diagrams. Fortunately, there are systems that are complicated enough to exhibit chaos, but still simple enough not to need all these new tools for their description. In particular, a driven damped pendulum, whose equation of motion is nonlinear, can exhibit chaos but can be described in reasonably elementary terms. For this reason, I shall mostly focus on this system here. After a brief review of the difference between linear and nonlinear equations and of the properties of a damped *linear* oscillator, I shall describe in some detail the motion of a driven damped pendulum, using just straightforward graphs of its position ϕ against time. Once you are familiar with the basic phenomena, I shall give an introduction to some of the more sophisticated tools that you will need if you want to explore the rapidly expanding literature on chaos. To conclude this chapter I shall describe another system that can exhibit chaos — the so-called logistic map. Although this is not strictly part of mechanics, it shows many strong parallels with mechanical systems, and has the great advantage of a simplicity that allows an easy understanding of several aspects of its behavior. It also serves to illustrate some of the amazing universality of chaos.

In planning this chapter, I felt strongly that it was important to convey a good understanding of a few topics, rather than a superficial glimpse of the whole field. In particular, chaos theory is divided into two broad areas — dissipative systems, such as the damped pendulum, and nondissipative, or "Hamiltonian" systems — and I decided to restrict myself entirely to the former. I know that some readers will question this decision. Many important applications of chaos theory (in astronomy and statistical mechanics, for instance) concern nondissipative systems, as did the pioneering work of Poincaré. Nevertheless, it is my experience that the most accessible topics for the beginner concern dissipative systems, and, to keep the length of this chapter reasonably finite, I decided to treat only these. Please see this chapter as a sampler of the good things in chaos theory; it certainly makes no claim to tell the whole story.

This chapter is very different from all the other chapters of this book. The theory of chaos is new and not at all elementary. (And parts of the theory have yet to be discovered!) It requires a much deeper understanding of differential equations than I am assuming in this book, and a proper exposition of chaos theory requires a whole book rather than one chapter.[1] Therefore, I shall restrict myself here to simply *describing* the fascinating main properties of chaotic motion, without much attempt to prove that the motion is as I claim. This is actually a reasonably satisfactory situation. Before you try to read any of the more advanced books, it is almost certainly a good thing to have some idea of what chaos involves and some familiarity with the tools used to describe it, and these are what I hope to communicate.

12.1 Linearity and Nonlinearity

For a system to exhibit chaos its equations of motion must be *nonlinear*. We have noted examples of linear and nonlinear equations from time to time in this book, but let us review the two concepts now. A differential equation is linear if it involves the dependent variable or variables and their derivatives only linearly. The equation of motion of a cart (mass m) on a spring (force constant k),

$$m\ddot{x} = -kx, \tag{12.1}$$

is a linear differential equation for the cart's position x. Similarly, the equations of motion for the two carts discussed in Chapter 11 [Equations (11.2) for example] are linear equations for the two carts' positions x_1 and x_2. If we apply a driving force $F(t)$ to the cart of Equation (12.1), the resulting equation,

$$m\ddot{x} = -kx + F(t), \tag{12.2}$$

is still linear [though no longer homogeneous, since the "inhomogeneous" term $F(t)$ does not involve the dependent variable x at all]. By contrast, the equation of motion for a simple pendulum (mass m, length L) is $I\ddot{\phi} = \Gamma$ or

$$mL^2\ddot{\phi} = -mgL \sin\phi, \tag{12.3}$$

which is a *nonlinear* equation for ϕ, since $\sin\phi$ is not linear in ϕ. (If the oscillations are small, then $\sin\phi \approx \phi$, and the equation is well approximated by a linear equation;

---

[1] Several such books exist, of which my favorite is *Nonlinear Dynamics and Chaos* by Steven H. Strogatz, Addison-Wesley, Reading, MA (1994), but be warned, it takes eight chapters of mathematical preliminaries to get to the chaos.

in general, however, the equation for the simple pendulum is definitely nonlinear.) Another example is the equation of motion of a single planet in the field of the sun,

$$m\ddot{\mathbf{r}} = -GmM\hat{\mathbf{r}}/r^2, \tag{12.4}$$

which is a nonlinear equation for the variables $\mathbf{r} = (x, y, z)$ because the force term is nonlinear in x, y, and z. These two examples show that nonlinear equations are not especially unusual. On the contrary, many perfectly everyday systems have equations of motion that are nonlinear.

So far in this book the main difference between linear and nonlinear differential equations has been that the former have been easily solved analytically, whereas most of the latter have been *impossible* to solve analytically. In fact, our experience in this regard reflects the true state of affairs: Almost all of the linear equations of mechanics *are* analytically solvable, and almost none of the nonlinear ones are.[2] This circumstance is largely to blame for the failure until recently of scientists to recognize that chaos is an important and widespread phenomenon. Because nonlinear equations are so intractable, textbooks always focussed on linear problems. When nonlinear problems *had* to be addressed, they were often solved using approximations that reduced then to linear problems. In this way, the astonishingly rich variety of complications that occur for nonlinear systems went almost completely unrecognized. The first person to notice some of the symptoms of chaos was the French mathematician Poincaré (1854–1912) in his studies of the gravitational three-body problem — the motion of three bodies (such as the sun, earth, and moon) interacting via the gravitational force. The equation of motion for this system is nonlinear, like its two-body counterpart (12.4), and Poincaré observed that it exhibits the phenomenon now called *sensitivity to initial conditions* that is one of the characteristics of chaotic motion, as we shall see.

That Poincaré's observation of chaos went nearly unnoticed by physicists until the 1970s is probably due to several factors. The discoveries of relativity (1905) and then of quantum mechanics (around 1925) diverted most physicists' attention away from classical mechanics. And the difficulty of solving nonlinear equations without the aid of computers certainly discouraged the pursuit of nonlinear problems. In any case it was only in the 1970s that computer solutions of various nonlinear problems[3] drew the attention of significant numbers of scientists (physicists and many others) to the phenomenon that we now call chaos.

Nonlinearity is essential for chaos — if a system's equations of motion are linear, it cannot exhibit chaos. But nonlinearity does not guarantee chaos. For example, the equation (12.3) for a simple pendulum is nonlinear, but even when the amplitude is large (and the linear approximation is definitely not good) the simple pendulum never

---

[2] One of the rare examples of a solvable nonlinear equation is (12.4) for a planet, whose orbit we found in Chapter 8. But notice that we did this by a cunning change of variables that reduced the nonlinear equation (8.37) for r to the linear equation (8.45) for u.

[3] The first such calculation, of atmospheric convection, was made by the meteorologist Edward Lorenz at MIT in 1963, but this work did not attract widespread attention for another decade. For an exhaustive, but very readable, history of chaos theory see *Chaos, Making a New Science* by James Gleick, Viking-Penguin, New York (1987).

exhibits chaos. On the other hand, if we add in a damping force $-bv = -bL\dot\phi$ and a driving force $F(t)$, (12.3) becomes the equation of the *driven, damped pendulum*:

$$mL^2\ddot\phi = -mgL\sin\phi - bL^2\dot\phi + LF(t) \qquad (12.5)$$

and this equation *does* lead to chaos for some values of the parameters. Loosely speaking, the requirement for chaos is that the equations of motion be nonlinear and *somewhat complicated*. Equation (12.3) for the simple pendulum is not sufficiently complicated, but Equation (12.5) for the driven damped pendulum is. Unfortunately, a discussion of precisely what is "sufficiently complicated" to produce chaos would be well beyond the scope of this book.[4]

Another relatively simple example of a nonlinear system that exhibits chaos is the double pendulum of Section 11.4. In the small angle approximation, the equations of motion for the double pendulum are linear equations for the two angles ϕ_1 and ϕ_2 [see (11.41) and (11.42)], but in general they are nonlinear (see Problem 11.15), and they are sufficiently complicated to produce chaos. The driven damped pendulum and the double pendulum are two of the simplest mechanical systems that exhibit chaos. The driven damped pendulum has just one degree of freedom (one coordinate, ϕ, needed to specify its configuration), whereas the double pendulum has two (two coordinates, ϕ_1 and ϕ_2, needed). For this reason the driven damped pendulum is the simpler one to analyze and will be the main focus of our discussion here.

What's Special about Nonlinearity?

In the enormous set of all possible differential equations, the linear equations form a miniscule subset, with many simple properties that are not shared by the general non-linear equation. Thus it is really the linear equations which are "special." Nevertheless, for the reasons already mentioned, many physicists are much more conversant with the linear case, and they are sometimes tempted to assume that familiar properties of linear equations will carry over to nonlinear equations. This dangerous assumption is frequently wrong. In particular, the main message of this chapter is that chaos, which never appears in linear systems, is a common occurrence in nonlinear systems. Unfor-tunately, the underlying theory of this particular difference is beyond the scope of this book, and we shall have to be content with seeing some simple examples of chaotic motion, without a detailed understanding of *why* chaos occurs. Here, I would like to mention just one huge difference between linear and nonlinear equations that high-

---

[4] For the record, the criterion is this: As we shall see in Chapter 13, a set of second-order differential equations (like Newton's second law) for n variables can usually be rewritten as a set of first-order equations for N variables, $\xi_1, \cdots, \xi_N$ where $N > n$, with the general form $\dot\xi_i = f_i(\xi_1, \cdots, \xi_N)$ for $i = 1, \cdots, N$. For instance, if we write $\dot\phi = \omega$, the one equation (12.3) for the angle ϕ of the simple pendulum becomes two first-order equations, one for ϕ and one for ω, namely $\dot\phi = \omega$ and $\dot\omega = -(g/L)\sin\phi$. When the right-hand sides of these equations are independent of t (as they are here) the equations are said to be *autonomous*. For a dissipative system to exhibit chaos, its equations of motion, when put in this standard autonomous form, must be nonlinear and have N variables with $N \geq 3$. Nondissipative systems need nonlinearity and $N \geq 4$.

lights the importance of not letting the linear prejudice that many of us share mislead us in our study of nonlinear equations.

Nonlinear Equations Don't Obey the Superposition Principle

We saw in Chapter 5 that linear homogeneous equations satisfy the superposition principle — that any linear combination of solutions gives us another solution. We have used this result several times, particularly in Chapters 5 and 11, but let me refresh your memory with the example of a second-order equation of the form

$$p(t)\ddot{x}(t) + q(t)\dot{x}(t) + r(t)x(t) = 0, \tag{12.6}$$

where $x(t)$ is the unknown and $p(t), q(t)$, and $r(t)$ are known fixed functions. [An example of such an equation is (12.1) for a cart on a spring.] Notice first that, because every term in this equation is linear in $x(t)$ (or its derivatives), we can multiply through by any constant a and see at once that if $x(t)$ is a solution, then so is $ax(t)$. Second, if $x_1(t)$ and $x_2(t)$ are both solutions, then we can add the two corresponding equations, one for $x_1(t)$ and one for $x_2(t)$, and conclude that $x_1(t) + x_2(t)$ is also a solution. Thus any linear combination

$$x(t) = a_1 x_1(t) + a_2 x_2(t)$$

is also a solution of (12.6) — the result called the superposition principle. On the other hand, it is easy to see that neither of the arguments just given works if the equation is nonlinear. [Make sure you can see this. Suppose, for example, the last term in (12.6) was $r(t)\sqrt{x(t)}$; see Problem 12.3.] Therefore, the superposition principle does not apply to nonlinear equations.

   An important consequence of the superposition principle that we have used repeatedly is this : To find all the solutions of (12.6) we have only to find two independent solutions $x_1(t)$ and $x_2(t)$; then *every* solution can be expressed as a linear combination of $x_1(t)$ and $x_2(t)$. More generally, to find all the solutions of an nth order homogeneous linear differential equation, we have only to find n independent solutions and then every solution can be expressed as a linear combination of these n solutions. Since the superposition principle does not apply to nonlinear equations, this dramatic simplification does not apply to nonlinear equations.

   There is a corresponding situation for inhomogeneous equations, such as (12.2) and (12.5), again as we saw in Chapter 5. If $x_p(t)$ is any one particular solution of a *linear* nth order inhomogeneous equation, then *every* solution can be written as $x_p(t)$ plus a linear combination of n independent solutions of the corresponding homogeneous equation. For nonlinear equations, there is no corresponding result (Problem 12.4). Thus every solution of any nth order linear equation (homogeneous or inhomogeneous) can be expressed simply in terms of n independent functions, but for nonlinear equations there is no such simple expression.

   With these general observations on nonlinear equations, let us take up the one nonlinear equation that we shall discuss in detail, Equation (12.5) for a driven damped pendulum. I shall first describe some features of its motion that we would expect (or at least that are not wholly unexpected), and I shall then take up the surprising features associated with the pendulum's chaotic motion.

12.2 The Driven Damped Pendulum DDP

The equation of motion for the **driven damped pendulum** (or **DDP**) was given in (12.5). Since this equation is going to occupy us for several sections to come, I would like to make quite sure that you are clear where it came from and to tidy it up. The pendulum is sketched in Figure 12.1. The equation of motion is just $I\ddot{\phi} = \Gamma$, where I is the moment of inertia and Γ is the net torque about the pivot. In this case $I = mL^2$, and the torque arises from the three forces shown in Figure 12.1. The resistive force has magnitude bv and hence exerts a torque $-Lbv = -bL^2\dot{\phi}$. The torque of the weight is $-mgL\sin\phi$, and that of the driving force is $LF(t)$. Thus the equation of motion $I\ddot{\phi} = \Gamma$ is

$$mL^2\ddot{\phi} = -bL^2\dot{\phi} - mgL\sin\phi + LF(t) \tag{12.7}$$

exactly as in (12.5).

Throughout this chapter I shall assume that the driving force $F(t)$ is sinusoidal, specifically that

$$F(t) = F_\text{o}\cos(\omega t) \tag{12.8}$$

where F_o is the *drive amplitude* (the amplitude of the driving force) and ω the **drive frequency**. As I argued in Chapter 5, several real and interesting driving forces approximate this sinusoidal form quite closely, and it has proved possible to reproduce such sinusoidal forces with remarkable precision for experiments on chaos. Substituting into (12.7) and reorganizing a little, we find that

$$\ddot{\phi} + \frac{b}{m}\dot{\phi} + \frac{g}{L}\sin\phi = \frac{F_\text{o}}{mL}\cos\omega t. \tag{12.9}$$

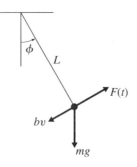

Figure 12.1   The three important forces on the driven damped pendulum are the resistive force with magnitude bv, the weight mg, and the driving force $F(t)$. (There is also a reaction force from the pivot at the top, but this contributes nothing to the torque.)

In this equation, you will recognize the coefficient b/m as the constant that we renamed as 2β in Chapter 5,

$$\frac{b}{m} = 2\beta,$$

where β was called the **damping constant**. Similarly the coefficient g/L is just ω_o^2,

$$\frac{g}{L} = \omega_o^2,$$

where ω_o is the **natural frequency** of the pendulum. Finally, the coefficient F_o/mL must have the dimensions of (time)$^{-2}$; that is, F_o/mL has the same units as ω_o^2. It is convenient to rewrite this coefficient as $F_o/mL = \gamma\omega_o^2$. That is, we introduce a dimensionless parameter

$$\gamma = \frac{F_o}{mL\omega_o^2} = \frac{F_o}{mg}, \tag{12.10}$$

which I shall call the **drive strength** and is just the ratio of the drive amplitude F_o to the weight mg. This parameter γ is a dimensionless measure of the strength of the driving force. When $\gamma < 1$, the drive force is less than the weight and we would expect it to produce a relatively small motion. (For instance, the drive force is insufficient to hold the pendulum out at $\phi = 90°$.) Conversely, if $\gamma \geq 1$, the drive force exceeds the pendulum's weight, and we should anticipate that it will produce larger scale motions (for instance, motion in which the pendulum is pushed all the way over the top at $\phi = \pi$).

Making all these substitutions, we get our final form of the equation of motion (12.9) for a driven damped pendulum

$$\ddot{\phi} + 2\beta\dot{\phi} + \omega_o^2 \sin\phi = \gamma\omega_o^2 \cos\omega t. \tag{12.11}$$

This is the equation whose solutions we shall be studying for the next several sections.

12.3   Some Expected Features of the DDP

Properties of the Linear Oscillator

To appreciate the extraordinary richness of the chaotic motion of our driven damped pendulum, we must first review what sort of behavior we might *expect*, based on our experiences with linear oscillators. Specifically, if we release the pendulum near the equilibrium position $\phi = 0$ with a small initial velocity and if the drive strength is small, $\gamma \ll 1$, we would expect ϕ to remain small at all times. Thus we should be

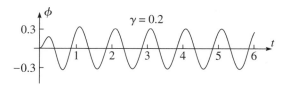

Figure 12.2   The motion of a DDP with a relatively weak drive strength of $\gamma = 0.2$. The drive period was chosen to be $\tau = 1$, so that the horizontal axis gives the time in units of the drive period. You can see clearly that after about two cycles the motion has settled down to a perfectly sinusoidal motion with period equal to the drive period.

able to approximate the term $\sin\phi$ in (12.11) as ϕ, and the equation of motion would become the linear equation

$$\ddot{\phi} + 2\beta\dot{\phi} + \omega_o^2\phi = \gamma\omega_o^2\cos\omega t \qquad (12.12)$$

which has exactly the form of (5.57) for the linear oscillator of Chapter 5. Thus the "expected" behavior of the driven damped pendulum, at least for a weak enough driving force, is just the behavior described in Section 5.5. This behavior can be quickly summarized: The initial behavior of the pendulum depends on the initial conditions, but any differences (or "transients") due to the initial conditions die out rapidly, and the motion approaches a unique "attractor," in which the pendulum oscillates sinusoidally with exactly the frequency of the driving force:

$$\phi(t) = A\cos(\omega t - \delta). \qquad (12.13)$$

These predictions are nicely illustrated in Figure 12.2, which shows the actual motion of the driven damped pendulum for a fairly weak drive strength of $\gamma = 0.2$. [Since the exact equation of motion (12.11) cannot be solved analytically, this and all subsequent plots of the motion of the DDP were made from numerical solutions of (12.11).[5]] The drive frequency was chosen to be $\omega = 2\pi$, so that the drive period is $\tau = 2\pi/\omega = 1$. This means that the horizontal axis shows time in units of the drive period. The natural frequency was chosen to be $\omega_o = 1.5\omega$, so that the system is fairly close to resonance, since this is where chaotic motion is usually easiest to find.[6] The most striking feature of this plot is that after about two cycles the motion has settled down to a purely sinusoidal motion with exactly the period of the driving force, $\tau = 1$. The initial conditions chosen for this plot were that $\phi = \dot{\phi} = 0$ at $t = 0$. It is a fact (though not one that our one plot can show) that, whatever initial conditions we were to choose, the motion of a linear oscillator would always approach the same unique attractor as the initial transients die out.

---

[5] All of these plots were made using Mathematica's numerical solver NDSolve. For many plots the default precision of 15 digits was more than sufficient, but where there was any reason for doubt, the precision was increased in integer steps until two successive calculations were indistinguishable.

[6] The other parameters used were as follows: Damping constant $\beta = \omega_o/4$, and initial conditions $\phi = \dot{\phi} = 0$ at $t = 0$.

To summarize, for a linear damped oscillator, with a sinusoidal driving force: **(1)** There is a unique attractor which the motion approaches, irrespective of the chosen initial conditions. **(2)** The motion of this attractor is itself sinusoidal with frequency exactly equal to the drive frequency.

Nearly Linear Oscillations of the DDP

Let us now imagine increasing the drive strength so that the amplitude of oscillation increases to a value where the approximation

$$\sin \phi \approx \phi$$

is no longer satisfactory. As long as the amplitude is not too large, we would expect to get a satisfactory approximation by including just one more term in the Taylor series for $\sin \phi$ and writing

$$\sin \phi \approx \phi - \tfrac{1}{6}\phi^3.$$

If we use this approximation in the exact equation of motion (12.11), we get the approximate equation

$$\ddot{\phi} + 2\beta\dot{\phi} + \omega_o^2 \left(\phi - \tfrac{1}{6}\phi^3 \right) = \gamma \omega_o^2 \cos \omega t. \tag{12.14}$$

To the extent that the new nonlinear term involving ϕ^3 is small, we can anticipate that the solution of this equation will still be reasonably approximated (once the transients have died out) by an expression of the same form as before,

$$\phi(t) \approx A \cos(\omega t - \delta).$$

When this is put into (12.14), the small term involving ϕ^3 contributes a term proportional to $\cos^3(\omega t - \delta)$. Since

$$\cos^3 x = \tfrac{1}{4}(\cos 3x + 3 \cos x) \tag{12.15}$$

(see Problem 12.5) there is now a small term on the left side of (12.14) proportional to $\cos 3(\omega t - \delta)$. Since the right side contains no terms with this time dependence, it follows that at least one of the terms on the left (ϕ, $\dot{\phi}$, or $\ddot{\phi}$, and in fact all three) must. That is, a more exact expression for $\phi(t)$ must have the form

$$\phi(t) = A \cos(\omega t - \delta) + B \cos 3(\omega t - \delta), \tag{12.16}$$

with B much smaller than A. Therefore, we must anticipate that, as we increase the driving force and the amplitude increases, the solution will pick up a small term that oscillates with frequency 3ω.

We can repeat this argument: If we substitute the improved solution (12.16) back into (12.14), then the term in ϕ^3 will give even smaller terms of the form $\cos n(\omega t - \delta)$, with n an integer greater than 3. Therefore we must expect smaller corrections to (12.16) with frequencies $n\omega$, with n equal to various integers. Any term oscillating

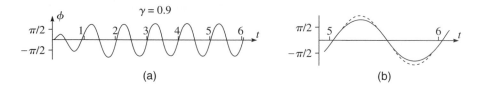

Figure 12.3   **(a)** The motion of a DDP with drive strength $\gamma = 0.9$ (and all other parameters the same as in Figure 12.2). After two or three drive cycles, the motion settles down to a regular oscillation, which has period equal to the drive period and looks at least approximately sinusoidal. **(b)** The solid curve is an enlargement of a single cycle of part (a), from $t = 5$ to 6. The dashed curve, which is a pure cosine with the same frequency, phase, and slope where they cross the axis, shows clearly that the actual motion is no longer perfectly sinusoidal; it is appreciably flatter at the extremes.

with frequency equal to an integer multiple of ω is called a **harmonic** of the drive frequency. Thus our conclusion is that, as the drive strength is increased and the nonlinearity becomes more important, the pendulum's motion will pick up various harmonics of the drive frequency ω, the most important being the $n = 3$ harmonic already included in (12.16).

The nth harmonic, with frequency $n\omega$, is periodic with period $\tau_n = 2\pi/n\omega = \tau/n$, where $\tau = 2\pi/\omega$ is the drive period. Thus in one drive period, the nth harmonic repeats itself n times. In particular, in one drive period, every harmonic will have cycled back to its original value, and a motion that is made up of various harmonics will still be periodic with the same period as the driving force.

The main difference between the motion implied by (12.16) (possibly with other harmonics included) and the motion (12.13) of the linear oscillator is that, with its extra term (or terms), (12.16) is no longer given by a single cosine function. We should be able to see this in a graph of $\phi(t)$ against t, which must deviate slightly from a pure sinusoidal shape. However, in the regime we are considering, the coefficient B in (12.16) and the coefficients of any higher harmonics are all much smaller than A, and the difference between the actual motion and a pure cosine is quite hard to see. Figure 12.3(a) shows the motion of a damped driven pendulum with drive strength $\gamma = 0.9$ (just below our rough boundary at $\gamma = 1$ between weak and strong drive strengths). Just as in Figure 12.2, the motion has quickly settled down to steady oscillation with exactly the period of the driving force. At first glance the curve (after about $t = 2$) appears to be a pure cosine, but on closer examination you may be able to convince yourself that it is a little too flat at the crests and troughs. Figure 12.3(b) is an enlargement of one cycle of the motion (solid curve), with a superposed pure cosine with the same period and phase (dashed curve). This comparison shows clearly that the actual motion is no longer a single pure cosine.[7]

_____

[7] The flattened shape at the extremes is nicely consistent with (12.16): Provided B and A have opposite signs, the second term in (12.16) reduces $\phi(t)$ at the crests and troughs and increases it near where it crosses the axis. The behavior is also easy to understand physically: At the extremes, the restoring torque of gravity ($mgL \sin\phi$) is weaker than the linear approximation ($mgL\phi$), and the actual motion is less sharply curved.

The behavior of the DDP in the linear and nearly linear regimes can be quickly summarized: As the initial transients die out, the motion approaches a unique attractor, in which the pendulum oscillates with the same period as the driver. In the linear regime (driving strength $\gamma \ll 1$) this limiting motion is given by a simple cosine function with frequency equal to the drive frequency ω. In the not-quite-linear regime (γ somewhat larger, but definitely not much greater than 1), the limiting motion is still periodic, with the same period, but it picks up some harmonics and is a sum of cosines with frequencies $n\omega$ as in (12.16). As we shall see in the next section, we have only to increase the drive strength a little above $\gamma = 1$, to encounter some dramatically different behavior.

12.4 The DDP: Approach to Chaos

Let us now continue increasing the strength of our DDP's driver. Figure 12.4 shows the motion (ϕ against t) for all the same parameters and initial conditions as in the last two figures, except that I have increased the drive strength to $\gamma = 1.06$, just a little above the rough boundary at $\gamma = 1$ between weak and strong driving. The most striking thing about this plot is the dramatic oscillation of the initial transient motion. In the first three drive cycles, the pendulum swings from $\phi = 0$ to nearly 5π; that is, it makes nearly two and a half counterclockwise rotations. In the next two cycles it swings back nearly to $\phi = \pi$ and eventually settles down to more-or-less sinusoidal oscillations around $\phi \approx 2\pi$. (The position $\phi = 2\pi$ is, of course, the same as $\phi = 0$, but the statement that ϕ eventually centers on 2π is nonetheless meaningful, indicating that the pendulum has made one net counterclockwise rotation since $t = 0$.)

It is impossible to be completely sure, based on a graph such as Figure 12.4, that the eventual motion really is exactly periodic. One way to examine this question more closely is to print out the positions $\phi(t)$ at successive one-cycle intervals, $t = t_0, t_0 + 1, t_0 + 2, t_0 + 3, \cdots$. The larger we choose t_0, the more closely these positions should agree with one another (if, the eventual motion is indeed periodic). For example, starting at $t = 34$ the positions $\phi(t)$ (as given by the same numerical solution on which Figure 12.4 was based) are

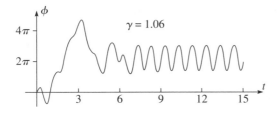

Figure 12.4   The motion of a DDP with drive strength $\gamma = 1.06$. The initial, rather wild, transients die out after about 9 drive cycles, and the motion settles down to an attractor with the same period as the driver.

t	$\phi(t)$
34	6.0366
35	6.0367
36	6.0366
37	6.0366
38	6.0366
39	6.0366

with all subsequent values equal to 6.0366. Evidently, to five significant figures, the motion has settled down to be perfectly periodic after 35 drive cycles.[8] Of course, it is *possible* that the motion does something nonperiodic in between the integer times shown, and certainly no one would accept our data as mathematical proof. Nevertheless, the evidence is overwhelming that for $\gamma = 1.06$ (and with the initial conditions used) $\phi(t)$ does approach an attractor that is periodic with the same period as the driver. In this respect, the motion shown for $\gamma = 1.06$ is not much different from that for $\gamma = 0.9$ as shown in Figure 12.3. However, the dramatic initial swings in Figure 12.4 are harbingers of interesting developments to come.

Period Two

Figure 12.5(a) corresponds exactly to the previous figure except that I have now increased the drive strength to $\gamma = 1.073$. Again the most obvious feature is the wild initial oscillation, which now lasts for nearly 20 drive cycles before the motion settles down to steady oscillations that are at least approximately sinusoidal. However, if you look closely at these oscillations, you will notice that the crests and troughs (especially the troughs) are not all of the same height. Figure 12.5(b) is a many-fold enlargement of the same troughs between $t = 20$ and 30, and you can see clearly that the troughs alternate between two distinct heights. You might wonder if this alternation is itself a transient that will disappear after enough cycles, but this is not in fact so. A plot of the oscillations for $990 \le t \le 1000$ looks exactly the same as that for $20 \le t \le 30$. Another way to show this is to print out the numerical values of $\phi(t)$ at one-cycle intervals. Starting at $t = 30$, this yields

t	$\phi(t)$
30	−6.6438
31	−6.4090
32	−6.6438
33	−6.4090
34	−6.6438
35	−6.4090

---

[8] Naturally, the motion takes longer to settle down to a constant if we insist on more significant figures. For example, it is not until 46 cycles have passed that $\phi(t)$ starts repeating to six significant figures, after which $\phi(t) = 6.03662$ for $t = 46, 47, 48, \cdots$.

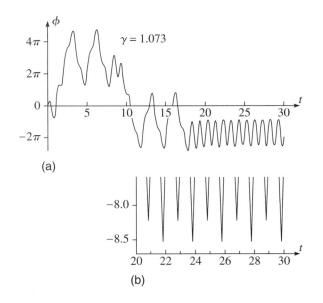

(a)

(b)

Figure 12.5   (a) The first 30 cycles of a DDP with drive strength $\gamma = 1.073$. The wild initial oscillations persist for nearly 20 drive cycles, after which the motion settles down to an attractor that is approximately sinusoidal. However, closer inspection shows that the crests and troughs of this attractor are not all of the same height. (b) An enlargement of the attractor for $20 \le t \le 30$ showing just the troughs of part (a). The troughs alternate in height, repeating themselves once every *two* drive cycles.

a pattern that repeats precisely forever. Evidently, by $t = 30$, the motion has settled down so that $\phi(t)$ has the value -6.6438 (to 5 significant figures) for all even values of t and has the distinct value -6.4090 for all odd values of t.

This behavior means that the motion no longer repeats itself with the frequency of the driver. Rather, the motion is periodic with period equal to *twice the drive period*, and we say that the motion has **period two**. (In our units this last statement is literally true; in general it means that the period of the motion is two times the drive period.) It is important to recognise that this development is quite different from the appearance of the harmonics that we noticed in the case of nearly linear motion. A harmonic has frequency $n\omega$, an integer multiple of the drive frequency, and hence period equal to an integer *submultiple* of the drive period. What we have now found has period equal to an integer *multiple* of the drive period, and hence frequency ω/n, which can be described as a **subharmonic** of the drive frequency. Looking at Figure 12.5(a) you can see that the motion is still very nearly sinusoidal with the period of the driver (period 1). Thus the dominant term in $\phi(t)$ is still of the form $A \cos(\omega t - \delta)$; nevertheless, $\phi(t)$ definitely contains a small subharmonic term with period 2.

Period Three

Although the attractor shown in Figure 12.5 has period two, the dominant behavior is still clearly of period one; that is, the new $n = 2$ subharmonic contributes only a small amount to the solution. If we increase the drive strength a little further, we find an

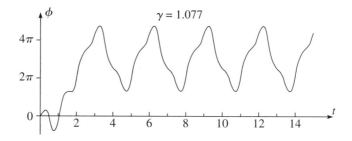

Figure 12.6   The motion of a DDP with drive strength $\gamma = 1.077$. After little more than two drive cycles, the motion has settled down to a periodic attractor which repeats itself every three drive cycles (for example, the troughs come just before $t = 5$, 8, 11, 14, and so on); therefore, the attractor has period three.

attractor for which the subharmonic is the dominant term. Figure 12.6 shows the first 15 cycles of the motion of our DDP with the drive strength increased to $\gamma = 1.077$ (and all other parameters the same as before). In this case it is obvious at just a glance that the motion settles down to an attractor that repeats itself every *three* drive cycles and hence has period three. While it would be hard to question that this graph has period three, we can reinforce the conclusion by looking at the values of $\phi(t)$ at one-cycle intervals. Starting from $t = 30$ these are as follows:

t	$\phi(t)$
30	13.81225
31	7.75854
32	6.87265
33	13.81225
34	7.75854
35	6.87265
36	13.81225
37	7.75854
38	6.87265

with exactly the same pattern, repeating once every three drive cycles, continuing indefinitely. Evidently the solution has picked up a period-three term, and this term dominates the solution.

More than One Attractor

For a linear oscillator, with a given set of parameters, we proved in Section 5.5 that there is a unique attractor; that is, whatever the initial values of ϕ and $\dot{\phi}$, the eventual motion will always be the same, once the transients have died out. For a nonlinear oscillator, this is not the case, and the DDP with the drive strength $\gamma = 1.077$ of Figure 12.6 furnishes a clear example. In Figure 12.7, I have shown the motion for a DDP with the same parameters as in Figure 12.6 (including the same drive strength), but with two different sets of initial conditions. The dashed curve is the same solution as

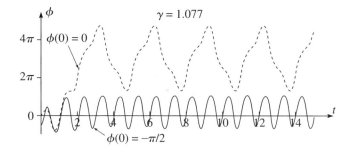

Figure 12.7   Two solutions for the same DDP, with the same drive strengths, but different initial conditions [$\phi(0) = \dot{\phi}(0) = 0$ for the dashed curve, but $\phi(0) = -\pi/2$ and $\dot{\phi}(0) = 0$ for the solid curve]. Even after the transients have died out, the two motions are totally different.

in Figure 12.6, with the same initial conditions as we have used for every graph up to now, $\phi(0) = \dot{\phi}(0) = 0$. The solid curve shows the motion of the exact same DDP, also with $\dot{\phi}(0) = 0$, but with $\phi(0) = -\pi/2$; that is, for the solid curve, the pendulum was released from 90° on the left. As you can clearly see, the two attractors (the curves to which the actual motions converge as the initial transients die out) are totally different. For the dashed curve, the attractor has period three, for the solid curve the eventual period is (as you can see if you look closely) actually two, with alternate troughs (and alternate crests) having slightly different heights. Evidently, for a nonlinear oscillator, different initial conditions can lead to totally different attractors.

A Period-Doubling Cascade

Having recognized that different initial conditions can lead to different attractors, we must anticipate that the evolution of the oscillations as we vary γ may depend on the initial conditions that we choose. In the sequence of Figures 12.2 through 12.6, I used the initial conditions $\phi(0) = 0$ and $\dot{\phi}(0) = 0$ for all five pictures. It turns out that the new initial conditions $\phi(0) = -\pi/2$ and $\dot{\phi}(0) = 0$, introduced in Figure 12.7, lead to a quite different and very interesting evolution. In Figure 12.8, I have shown the motion of the DDP for four successively larger values of γ, all with these new initial conditions. The left-hand pictures show $\phi(t)$ as a function of t for the first ten cycles of the driver. The first graph is for $\gamma = 1.06$, the same value as was used in Figure 12.4, and, as in Figure 12.4, the motion settles down to a steady oscillation with period equal to the drive period; that is, the attractor has period one. To confirm this conclusion, the right-hand picture shows the same motion, but for $28 < t < 40$ (by which time any initial transients have completely disappeared at the scale shown), with the vertical scale magnified to show clearly that successive oscillations are all of equal amplitude.

For the second pair of graphs, the drive strength was increased to $\gamma = 1.078$. At first glance, the motion looks very similar to that for $\gamma = 1.06$, but on closer inspection you can see that the maxima and minima are not all of the same height. This is very visible

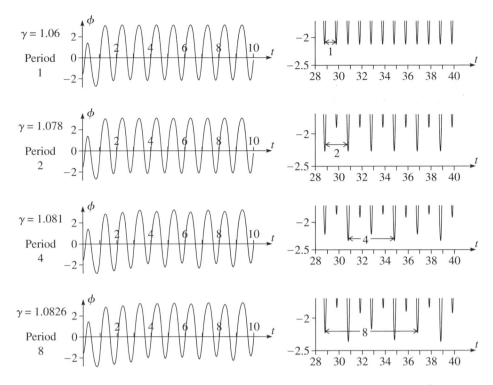

Figure 12.8   A period-doubling cascade. The left-hand pictures show the first ten drive cycles of a DDP with successively larger drive strengths, as indicated on the left. All other parameters, including the initial conditions $\phi(0) = -\pi/2$ and $\dot{\phi}(0) = 0$, are the same in all pictures. In each picture on the right I have enlarged the bottom of the corresponding motion on the left, to show more clearly the differences in extent of successive oscillations; these enlargements show 12 drive cycles, starting from $t = 28$, by which time the motion has settled down to a perfectly periodic attractor (at least at the scale shown). Each double-headed arrow shows one complete cycle of the corresponding motion; the periods of the four attractors are clearly seen to be 1, 2, 4, and 8, as indicated.

in the enlargement on the right where you can see easily that the minima alternate between two distinct, fixed heights, so that the attractor now has period two.

With $\gamma = 1.081$, as in the third pair, the graph on the left again looks pretty much like its two predecessors, and it is hard to be sure just what is going on. One of the reasons is that we can't be sure that ten drive cycles (the number shown on the left) are long enough for all transients to have disappeared, but in the right-hand enlargement it is quite clear that the minima are alternating among four different values. That is, the period has doubled again to period four.

In the last pair of pictures, with $\gamma = 1.0826$, it is even harder to be sure what is happening in the left-hand picture, but the enlargement on the right makes clear that the motion eventually repeats once every eight drive cycles. That is, the attractor has period eight. The **period-doubling cascade** seen in these four pairs of pictures

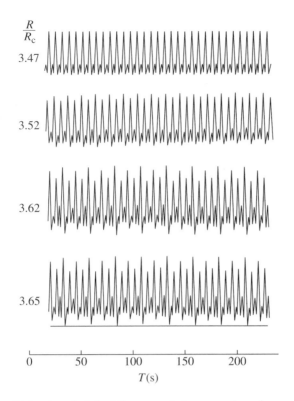

Figure 12.9   A period-doubling cascade in convection of mercury in
a small convection cell. The plots show the temperature at one fixed
point in the cell as a function of time, for four successively larger
temperature gradients as given by the parameter R/R_c.

continues. If we increase the drive strength further, we find motion with period 16,
then 32, and so on to infinity.

The period-doubling cascade of Figure 12.8 is a very striking phenomenon, but
the quantitative differences between the four successive unenlarged graphs are quite
small. You might guess that to build a driven damped pendulum sufficiently precise to
observe these subtle differences would be very hard, and this is indeed the case. Nevertheless, such pendulums have been constructed and have been used to observe all
of the effects described in this chapter, with amazing agreement between theory and
experiment.[9] Perhaps even more remarkable is that the phenomenon of period doubling is found in many completely different nonlinear systems — electrical circuits,
chemical reactions, balls bouncing on oscillating surfaces, and many more. In each of
these systems, there is a "control parameter" that can be varied (the driving strength
of a DDP, a voltage in an electrical circuit, a flow rate in a chemical reaction). The

---

[9] For a description of three of the commercially available "chaotic pendulums" see J. A. Blackburn and G. L. Baker, "A Comparison of Commercial Chaotic Pendulums," *American Journal of
Physics*, Vol. 66, p. 821, (1998). The Daedalon pendulum described there was used to get the data
shown in Figure 12.32.

behavior of the system is monitored as this parameter is varied, and it is found that the behavior exhibits period-doubling cascades. Figure 12.9 shows a cascade observed by Libchaber et al.[10] in the convection of mercury in a small box whose bottom is maintained at a slightly higher temperature than its top. This temperature difference is the control parameter and is measured by a number R, called the Rayleigh number. When R is very small, heat is conducted up with no convection. Then at a critical temperature difference, R_c, steady convection sets in, and, as R is increased still further, the convection becomes oscillatory. These oscillations can be observed by measuring the temperature at any fixed point in the cell, and Figure 12.9 shows four plots of the observed temperature (at one fixed point) against time, for four successively larger values of the control parameter R. The period doublings from 1 to 2, from 2 to 4, and from 4 to 8 are beautifully clear.

Not only are period-doubling cascades observed in numerous different systems. In a sense that I shall describe directly, the cascades occur *in the same way*, a circumstance referred to as "universality."

The Feigenbaum Number and Universality

Returning to the period-doublings of the DDP, you can see from the values of the drive strength γ shown in Figure 12.8 that the doublings occur faster and faster as we increase γ. To make this idea quantitative, we need to examine the **threshold values** of γ at which the period actually doubles. For example, looking at the numbers in Figure 12.8, it seems clear that somewhere between $\gamma = 1.06$ and 1.078, there must be a value γ_1 where the period changes from 1 to 2. Finding where this threshold (or **"bifurcation point"**) actually occurs is surprisingly hard, but it turns out that (to 5 significant figures) $\gamma_1 = 1.0663$. Similarly, at $\gamma_2 = 1.0793$, the period changes from 2 to 4. If we let γ_n denote the threshold at which the period changes from 2^{n-1} to 2^n, then the first few thresholds γ_n are as shown in Table 12.1. In the last column of the table, I have shown the distances $\gamma_n - \gamma_{n-1}$ between successive thresholds, which, as you can see, shrink geometrically,[11] each interval being about one fifth of its predecessor.

In the late 1970's, the physicist Mitchell Feigenbaum (born 1944) showed not only that many different nonlinear systems undergo similar period-doubling cascades but that the cascades all show the same geometric acceleration; specifically, the intervals between the thresholds for the control parameter (the drive strength, in our case) satisfy

$$(\gamma_{n+1} - \gamma_n) \approx \frac{1}{\delta}(\gamma_n - \gamma_{n-1}) \tag{12.17}$$

---

[10] Reproduced with permission from A. Libchaber, C. Laroche, and S. Fauve, *Journal de Physique-Lettres*, vol. 43, p. 211 (1982).

[11] A sequence of numbers, $a_1, a_2, \cdots$, is geometric if $a_{n+1} = k a_n$ for some fixed number k. If $k < 1$, the geometric sequence goes to zero as $n \to \infty$.

Table 12.1   The first four thresholds γ_n at which the period of the DDP [with the initial conditions $\phi(0) = -\pi/2$ and $\dot{\phi}(0) = 0$] doubles from 1 to 2, 2 to 4, 4 to 8, and 8 to 16. The last column shows the widths of the intervals between successive thresholds.

n	period	γ_n	interval
1	$1 \to 2$	1.0663	
			0.0130
2	$2 \to 4$	1.0793	
			0.0028
3	$4 \to 8$	1.0821	
			0.0006
4	$8 \to 16$	1.0827	

where the constant δ has the same value

$$\delta = 4.6692016 \tag{12.18}$$

for all such systems and is called the **Feigenbaum number**.[12] It is the widespread occurrence of period doubling and the fact that δ has the same value for so many different systems that has led to the phenomenon of period doubling being characterized as **universal**. I have written the Feigenbaum relation (12.17) with an "approximately equal" sign, because, strictly speaking, the relation holds only in the limit that $n \to \infty$. For many systems, however, the relation is a very good approximation for *all* values of n. (See Problems 12.11 and 12.29.)

The Feigenbaum relation (12.17) implies that the intervals between successive thresholds approach zero rapidly, and hence that the thresholds themselves approach a finite limit γ_c,

$$\gamma_n \to \gamma_c \qquad (\text{as } n \to \infty). \tag{12.19}$$

Therefore, the sequence of thresholds γ_n satisfies

$$\gamma_1 < \gamma_2 < \cdots < \gamma_n < \cdots < \gamma_c$$

with infinitely many thresholds squeezed in the rapidly narrowing gap between γ_n and γ_c. For our DDP, the limit γ_c is found to be

$$\gamma_c = 1.0829. \tag{12.20}$$

We shall see that beyond the critical value γ_c, chaos sets in, so the period-doubling cascade is called a **route to chaos**. However, I should emphasize that there are systems that exhibit chaos without first going through a period-doubling cascade; that is, the period-doubling cascade is just one of several possible routes to chaos.

---

[12] Actually there are two Feigenbaum numbers, and this one is often called Feigenbaum's delta.

12.5   Chaos and Sensitivity to Initial Conditions

If we increase the drive strength γ beyond the critical value $\gamma_c = 1.0829$, then our DDP begins to exhibit the behavior that has come to be called "chaos." Figure 12.10 shows the first thirty drive cycles of the DDP with $\gamma = 1.105$. The pendulum is obviously "trying" to oscillate with the period of the driver. Nevertheless, the actual oscillations wander around erratically and never repeat themselves exactly. Of course you might wonder if I have not given the oscillations time to settle down; perhaps at some later time they would converge to a periodic motion. In fact, however, a graph of any time interval is just as erratic, but never an exact repetition of any other interval. Even though the driving force is perfectly periodic and even after the transients have all died out, the long-term motion is definitely nonperiodic. This erratic, nonperiodic long-term behavior is one of the defining characteristics of chaos. The other defining characteristic is the phenomenon called sensitivity to initial conditions.

Sensitivity to Initial Conditions

The issue of sensitivity to initial conditions arises in connection with the following questions: Imagine two identical DDP's, with all parameters the same, but launched at $t = 0$ with slightly different initial conditions. [Perhaps the initial angles $\phi(0)$ differ by a fraction of a degree.] As time goes by, do the motions of the two pendulums remain nearly the same? Do they perhaps get closer to one another? Or do they diverge and become more and more different?

To make these questions more precise, let us denote the positions of the two pendulums by $\phi_1(t)$ and $\phi_2(t)$. These two functions satisfy exactly the same equation of motion, but have slightly different initial conditions. If now $\Delta\phi(t)$ denotes the difference between our two solutions,

$$\Delta\phi(t) = \phi_2(t) - \phi_1(t), \tag{12.21}$$

the issue is the time dependence of $\Delta\phi(t)$. Does $\Delta\phi(t)$ stay more-or-less constant? Does it decrease as time goes by? Or does it increase?

For the linear oscillator discussed in Chapter 5, the answer is that $\Delta\phi(t)$ goes to zero, since we proved that all solutions of the equation of motion approach the

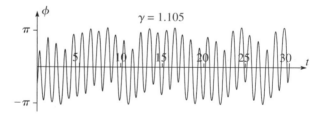

Figure 12.10   Chaos. The first 30 drive cycles of the DDP with $\gamma = 1.105$ are erratic and show no signs of periodicity. In fact the oscillations never do settle down to a regular periodic motion, and this erratic, nonperiodic long-term motion is one of the defining characteristics of chaos.

same attractor as $t \to \infty$. Therefore the *difference* of any two solutions must approach zero. Further, the difference must approach zero *exponentially*. To see this, recall from Equation (5.67) that any solution has the form

$$\phi(t) = A \cos(\omega t - \delta) + C_1 e^{r_1 t} + C_2 e^{r_2 t} \tag{12.22}$$

where the cosine term is the same for all solutions, whereas the two decaying exponential terms have coefficients C_1 and C_2 that depend on the initial conditions. This implies that when we take the difference of two solutions the cosine term drops out and we are left with

$$\Delta\phi(t) = B_1 e^{r_1 t} + B_2 e^{r_2 t}, \tag{12.23}$$

where the constants B_1 and B_2 depend on the two sets of initial conditions. The precise behavior of this difference depends on the relative sizes of the damping constant β and the natural frequency ω_o. In all of the examples so far in this chapter I have chosen $\beta = 0.25\omega_o$, so that $\beta < \omega_o$ (the situation called underdamping). In this case, we saw in Section 5.4 that the coefficients r_1 and r_2 have the form $-\beta \pm i\omega_1$. Some simple algebra then puts (12.23) in the form [compare Equation (5.38)]

$$\Delta\phi(t) = De^{-\beta t} \cos(\omega_1 t - \delta). \tag{12.24}$$

That is, $\Delta\phi(t)$ is the exponential $e^{-\beta t}$ times an oscillatory cosine.

There is a problem in trying to display the time dependence of a function like (12.24). The exponential factor decays so fast that one cannot easily show its range of values on a conventional graph. For example, with the values I have been using, $\beta = 0.25\omega_o = 0.75\pi = 2.356$, after just one drive cycle ($t = 1$) the exponential factor is $e^{-\beta t} = e^{-2.356} \approx 0.09$, and $\Delta\phi(t)$ has diminished by an order of magnitude. If we wanted to plot $\Delta\phi(t)$ against t over 10 cycles, say, then $\Delta\phi(t)$ would shrink by about ten orders of magnitude — a range that cannot possibly be shown on a simple linear plot of $\Delta\phi(t)$ against t.

As you probably know, the solution to this problem is to make a logarithmic plot; that is, we plot the *log* of $\Delta\phi(t)$ against t. Actually since $\Delta\phi(t)$ can be negative, we must plot $\ln |\Delta\phi(t)|$ against t. According to (12.24) this should obey

$$\ln |\Delta\phi(t)| = \ln D - \beta t + \ln |\cos(\omega_1 t - \delta)|. \tag{12.25}$$

The first term on the right is a constant, and the second is linear in t with slope $-\beta$. The third is a little complicated: Since $|\cos(\omega_1 t - \delta)|$ oscillates between 1 and 0, its natural log oscillates between 0 and $-\infty$. Thus a graph of $\ln |\Delta\phi(t)|$ against t should bounce up and down (going to $-\infty$ each time the cosine term vanishes), underneath an envelope that decreases linearly with slope $-\beta$. This is clearly visible in Figure 12.11, which shows a plot of $\log |\Delta\phi(t)|$ against t for the relatively weak driving strength $\gamma = 0.1$, for which the linear approximation is certainly good. [I plotted the log to base 10 rather than the natural log, because the former is easier to interpret on a graph. Since $\log(x)$ is just the constant $\log(e)$ times $\ln(x)$, this changes none of our qualitative predictions.] To plot this graph, I gave the first pendulum the same initial conditions as in Figures 12.8 and 12.10; the second pendulum was released with its

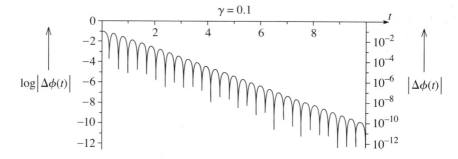

Figure 12.11   Logarithmic plot of $\Delta\phi(t)$, the separation of two identical DDP's, with a weak drive strength $\gamma = 0.1$, that were released with initial positions that differ by 0.1 radians (or about 6°). The vertical axis on the left shows $\log|\Delta\phi(t)|$, while that on the right shows $|\Delta\phi(t)|$ itself. The picture shows clearly that the maxima of $\log|\Delta\phi(t)|$ decrease linearly, and hence that $\Delta\phi(t)$ decays exponentially.

initial position 0.1 radians lower, so that the initial difference was $\Delta\phi(0) = 0.1$ rad, or about 6 degrees. The most important feature of the plot is that the successive maxima of $\log|\Delta\phi(t)|$ decrease perfectly linearly, confirming that $\Delta\phi(t)$ decays exponentially, dropping by about 10 orders of magnitude in the first ten drive cycles (as you can easily check from the graph).[13]

So far we have proved that, in the linear regime, the separation $\Delta\phi(t)$ of two identical DDP's, launched with different initial conditions, decreases exponentially. This has an important practical consequence: In practice, we cannot possibly know the initial conditions of any system *exactly*. Therefore, when we try to predict the future behavior of our DDP we must recognize that the initial conditions we use may differ a little from the true initial conditions. This means that our predicted motion for $t > 0$ may differ from the true motion. But because $\Delta\phi(t)$ goes to zero exponentially, we can be sure that our error will never be worse than the initial error and will, in fact, rapidly approach zero. We can say that the linear oscillator is *insensitive to its initial conditions*. To achieve any prescribed accuracy in our predictions, we have only to ascertain the initial condition to this same accuracy.

What happens as we increase the drive strength γ out of the linear regime? Naturally, we can no longer depend on our proofs for the linear oscillator. However, we can reasonably expect that the difference $\Delta\phi(t)$ will continue to decay exponentially for at least some range of drive strengths. The question is "How large is this range?" and the answer is quite surprising: Provided the difference in initial conditions is sufficiently small, the difference $\Delta\phi(t)$ continues to decay exponentially for all values of γ up to the critical value γ_c at which chaos sets in. For example, if $\gamma = 1.07$ we know

---

[13] A second noteworthy feature is that the points where $|\Delta\phi(t)|$ vanishes (and, hence, $\log|\Delta\phi(t)|$ goes to $-\infty$) show up on the logarithmic plot as sharp downward spikes. This is because the plotting program can only sample a finite number of points and naturally misses the points where $\log|\Delta\phi(t)| = -\infty$. Instead it can only detect that there are points where $\log|\Delta\phi(t)|$ has a precipitous minimum, and this is what it shows.

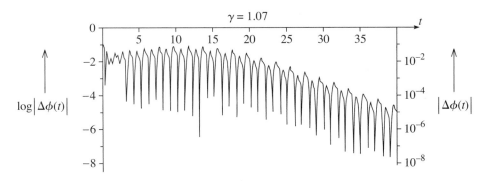

Figure 12.12   Logarithmic plot of $\Delta\phi(t)$, the separation of two identical DDP's, with drive strength $\gamma = 1.07$, that were released with initial positions that differ by 0.1 radians (or about 6°). For the first 15 or so drive cycles, $\Delta\phi(t)$ holds fairly constant in amplitude, but then the maxima of $\log|\Delta\phi(t)|$ decrease linearly, implying that $\Delta\phi(t)$ decays exponentially.

from Table 12.1 that the motion has period 2 (anyway for the initial conditions of that table), and the motion is distinctly nonlinear; nonetheless, the difference $\Delta\phi(t)$ still decays exponentially, as is clearly visible in Figure 12.12, which shows the difference $\Delta\phi(t)$ for two solutions with the same initial conditions as in Figure 12.11. In this case, $\Delta\phi(t)$ remains pretty well constant in amplitude for the first 15 or 20 drive cycles, but then the crests of $\log|\Delta\phi(t)|$ drop perfectly linearly, indicating that $\Delta\phi(t)$ decays exponentially as $t \to \infty$. Notice, however, that the exponential decay is considerably slower than in the linear case: Here the amplitude drops by about 4 orders of magnitude in the last 25 cycles; in the linear case of Figure 12.11, it dropped by 10 orders of magnitude in just 10 cycles. Nevertheless, the main point is that $\Delta\phi(t)$ goes to zero exponentially, and, as in the linear regime, we can predict the future behavior of our DDP, confident that any uncertainties in our predictions will be not much larger (and usually much smaller) than our uncertainty in the initial conditions.

If we now increase the drive strength past $\gamma_c = 1.0829$ into the chaotic regime, the picture changes completely. Figure 12.13 shows $\Delta\phi(t)$ for the same DDP as in Figures 12.11 and 12.12, except that the drive strength is now $\gamma = 1.105$, the same value used in our first plot of chaotic motion in Figure 12.10. The most obvious feature of this graph is that $\Delta\phi(t)$ clearly *grows* with time. In fact, you will notice that to highlight the growth of $\Delta\phi(t)$ I chose the initial difference to be just $\Delta\phi(0) = 0.0001$ radians. Starting from this tiny value, $|\Delta\phi|$ has increased in 16 drive cycles by more than 4 orders of magnitude to about $|\Delta\phi| \approx 3.5$.

From $t = 1$ through $t = 16$ (where the graph is leveling off), the maxima in Figure 12.13 grow almost perfectly linearly, implying that $\Delta\phi(t)$ grows exponentially.[14] This exponential growth spells disaster for any attempt at accurate prediction of the DDP's

---

[14] The eventual leveling of the curve is easily understood. You can see from Figure 12.10 that the angle $\phi(t)$ [actually, $\phi_1(t)$, but the same applies to $\phi_2(t)$] oscillates between about $\pm\pi$. That is, neither $\phi_1(t)$ nor $\phi_2(t)$ ever exceeds magnitude π. Therefore their difference $\Delta\phi(t)$ can never exceed 2π. Thus the curve *has* to level off before $\Delta\phi(t)$ reaches 2π.

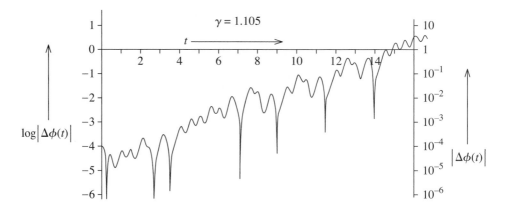

Figure 12.13   The separation $\Delta\phi(t)$ of two identical pendulums, both with drive strength $\gamma = 1.105$ and with an initial separation $\Delta\phi(0) = 10^{-4}$ rad. After a small initial drop, the crests of $\log|\Delta\phi(t)|$ *increase linearly*, showing that $\Delta\phi(t)$ itself *grows exponentially*.

long-term motion. For the present case, an error as small as 10^{-4} radians in our initial conditions will have grown in 16 cycles to an error of about 3.5, or more than π radians. Thus an uncertainty of $\pm 10^{-4}$ radians in the initial conditions grows to an uncertainty of $\pm\pi$, and an uncertainty of $\pm\pi$ in the angle of a pendulum means that we have *no idea at all* where the pendulum is! I chose this example because it is especially dramatic. Nevertheless, in any chaotic motion, $\Delta\phi(t)$ grows exponentially for a while at least. Even if this growth levels out before $\Delta\phi(t)$ reaches π, the exponential growth means that a tiny uncertainty in the initial conditions quickly grows into a large uncertainty in the predicted motion. It is in this sense that we say chaos exhibits **extreme sensitivity to initial conditions**, and this sensitivity is what can make the reliable prediction of chaotic motion a practical impossibility.

The Liapunov Exponent

What we have seen in the preceding three examples can be rephrased to say that the difference $\Delta\phi(t)$ between two identical DDP's released with slightly different initial conditions behaves exponentially:

$$|\Delta\phi(t)| \sim K e^{\lambda t} \tag{12.26}$$

(where the symbol "$\sim$" signifies that $\Delta\phi(t)$ may oscillate underneath an envelope with the advertized behavior, and K is a positive constant). The coefficient λ in the exponent is called the **Liapunov exponent**.[15] If the long-term motion is nonchaotic (settles down to periodic oscillation) the Liapunov exponent is negative; if the long-term motion is chaotic (erratic and nonperiodic) the Liapunov exponent is positive.

---

[15] Strictly speaking there are several Liapunov exponents, of which the one discussed here is the largest.

Higher Values of γ

So far we have seen that as we increased the drive strength γ, the motion of our DDP became more and more complicated — from the linear regime, with its pure sinusoidal response, to the nearly linear regime, with the addition of harmonics, to the appearance of subharmonics and (for certain initial conditions at least) a period-doubling cascade, and finally to chaos. You might naturally anticipate that, if we were to increase γ still further, the chaos would continue and intensify, but as usual our nonlinear system defies predictions. As γ increases, the DDP actually alternates between intervals of chaos separated by intervals of periodic, nonchaotic motion. I shall illustrate this with just two examples.

We saw that with $\gamma = 1.105$, the DDP exhibits chaotic behavior (Figure 12.10) and exponential divergence of neighboring solutions (Figure 12.13). We have only to increase the drive strength to $\gamma = 1.13$ to enter a narrow "window" of *nonchaotic* period-3 oscillation with exponential *convergence* of neighboring solutions, as shown in Figure 12.14. In part (a), you can see that within three drive cycles the motion settles into regular period-3 oscillations. Part (b) shows the separation $\Delta\phi(t)$ of two pendulums released with an initial difference $\Delta\phi(0)$ of 0.001 radians; in the first eight drive cycles, $\Delta\phi(t)$ actually increases, but from then on it decreases exponentially to zero, dropping by six orders of magnitude in the next twelve cycles.

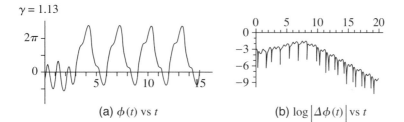

(a) $\phi(t)$ vs t (b) $\log|\Delta\phi(t)|$ vs t

Figure 12.14   Motion of a DDP with $\gamma = 1.13$. **(a)** The graph of $\phi(t)$ quickly settles down to oscillations of period 3 (same initial conditions as in Figures 12.8 and 12.10). **(b)** Logarithmic plot of the separation of two identical pendulums, the first with the same initial conditions as in part (a), the second with its initial angle 0.001 radians lower. After an initial modest increase, $\Delta\phi(t)$ goes to 0 exponentially.

Figure 12.15 shows the corresponding two graphs for a drive strength of $\gamma = 1.503$, where the motion has returned to being chaotic.[16] In part (a) we see a new kind of chaotic motion. The driving force is now strong enough to keep the pendulum rolling right over the top, and in the first 18 drive cycles the pendulum makes 13 complete clockwise rotations [$\phi(t)$ decreases by 26π]. The motion here can be seen as a steady rotation at about one revolution per drive cycle, with an erratic

---

[16] As we shall see, between the values $\gamma = 1.13$ and 1.503, shown in Figures 12.14 and 12.15, the DDP has passed through several intervals of chaotic and nonchaotic motion, but I omit the details for now.

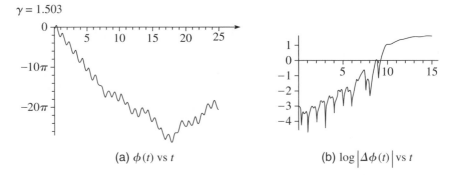

(a) $\phi(t)$ vs t (b) $\log|\Delta\phi(t)|$ vs t

Figure 12.15   Motion of a DDP with $\gamma = 1.503$. (Same initial conditions as in Figure 12.14.) **(a)** The graph of $\phi(t)$ against t oscillates erratically. In the first 18 cycles, it plunges to about -26π; that is, it makes about 13 complete clockwise rotations. It then starts to climb back but never actually repeats itself. **(b)** Logarthmic plot of the distance between two identical pendulums with an initial separation of 0.001 radians. For the first nine or ten cycles, $\Delta\phi(t)$ grows exponentially and then levels out.

oscillation superposed.[17] At $t = 18$, the motion reverses to a more-or-less steady counterclockwise rotation with erratic oscillations superposed, and, as the picture suggests, the motion never settles down to be periodic.

The logarithmic plot of Figure 12.15(b) shows the divergence of two pendulums with the same drive strength $\gamma = 1.503$, but an initial separation of 0.001 radians. The separation of the two pendulums increases exponentially for the first 9 or 10 cycles and levels out by about $t = 15$. A dramatic feature of this divergence is that it is big enough to be seen in the conventional linear plot of Figure 12.16, which shows the actual positions $\phi_1(t)$ and $\phi_2(t)$ of the two pendulums. At first sight it is perhaps surprising that for the first 8.5 cycles the two curves are completely indistinguishable, but that the difference is then so abundantly visible. However, you can understand this striking behavior by reference to Figure 12.15(b), where you can see that until $t \approx 8.5$ the separation $\Delta\phi(t)$, although growing rapidly, is nevertheless always less than about 1/3 radian — too small to be seen on the scale of Figure 12.16. By the time $t \approx 9.5$, $\Delta\phi(t)$ has reached about 3 — which is easily visible on the linear plot — and is still climbing rapidly. Thus from $t \approx 9.5$ the two curves are completely distinct.

The main morals to be drawn from these last two examples are these: **(1)** Once the drive strength γ of our DDP is past the critical value $\gamma_c = 1.0829$, there are intervals where the motion is chaotic and others where it is not. These intervals are often quite narrow, so that the chaotic motion comes and goes with startling rapidity. **(2)** The chaos can take on several different forms, such as the erratic "rolling" motion of Figure

---

[17] If you look closely you can see that for the first 7 cycles, the motion is very close to being a steady rotation of -2π per cycle, with a *regular* period-1 oscillation superposed. This type of motion is actually periodic, since a change of -2π brings the pendulum back to the same place. For some values of the drive strength, the long-term motion settles down exactly this way, a phenomenon called phase locking. (See Problem 12.17.)

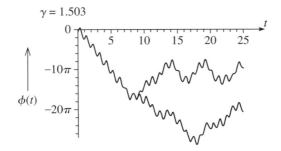

Figure 12.16   Linear plot of the positions of the same two identical DDP's whose separation $\Delta\phi(t)$ was shown in Figure 12.15(b) [$\Delta\phi(0) = 0.001$ rad]. For the first eight and a half drive cycles, the two curves are indistinguishable; after this the difference is dramatically apparent.

12.15(a). **(3)** The erratic motion of chaos always goes along with the sensitivity to initial conditions associated with the exponential divergence of neighboring solutions of the equation of motion.

12.6   Bifurcation Diagrams

So far, each of our pictures of the motion of the driven damped pendulum has shown the motion for one particular value of the drive strength γ. To observe the evolution of the motion as γ changes, we have had to draw several different plots, one for each value of γ. One would like to construct a single plot that somehow displayed the whole story, with its changing periods and its alternating periodicity and chaos as γ varies. This is the purpose of the bifurcation diagram.

A **bifurcation diagram** is a cunningly constructed plot of $\phi(t)$ against γ as in Figure 12.17. Perhaps the best way to explain what this plot shows is just to describe in detail how it was made. Having decided on a range of values of γ to display (from $\gamma = 1.06$ to 1.087 in Figure 12.17) one must first choose a large number of values of γ, evenly spaced across the chosen range. For Figure 12.17, I chose 271 values of γ, spaced at intervals of 0.0001,

$$\gamma = 1.0600, \ 1.0601, \ 1.0602, \ \cdots, \ 1.0869, \ 1.0870.$$

For each chosen value of γ, the next step is to solve numerically the equation of motion (12.11) from $t = 0$ to a time $t_{\max}$ picked so that all transients have long since died out. To make Figure 12.17, I chose the same initial conditions as in the last few pictures, namely $\phi(0) = -\pi/2$ and $\dot{\phi}(0) = 0$.[18]

---

[18] Some authors like to superpose the plots for several different initial conditions. This gives a more complete picture of the many possible motions, but makes the plot harder to interpret. For simplicity, I chose to use just one set of initial conditions.

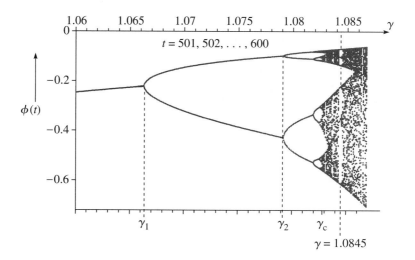

Figure 12.17 Bifurcation diagram for the driven damped pendulum for drive strengths $1.060 \leq \gamma \leq 1.087$. The period-doubling cascade is clearly visible: At $\gamma_1 = 1.0663$ the period changes from 1 to 2, and at $\gamma_2 = 1.0793$ from 2 to 4. The next bifurcation, from period 4 to 8, is easily seen at $\gamma_3 = 1.0821$, and that from 8 to 16 is just discernable at $\gamma_4 = 1.0827$. To the right of the critical value, $\gamma_c = 1.0829$, the motion is mostly chaotic, although at $\gamma = 1.0845$ you can just make out a brief interval of period-6 motion.

To understand our next move, recall that a good way to check for periodicity (or non-periodicity) is to examine the values

$$\phi(t_o), \ \phi(t_o + 1), \ \phi(t_o + 2), \ \cdots$$

of $\phi(t)$ for a large number of times at one-cycle intervals. If the motion is periodic with period n, these will repeat themselves after n cycles, otherwise not. Therefore, our next step is to use our solutions for $\phi(t)$ to find the values of $\phi(t)$ over a range of times at integer intervals from some chosen $t_{\min}$ to $t_{\max}$ (with $t_{\min}$ large enough that all transients have died out). For Figure 12.17, I found $\phi(t)$ for 100 times,

$$t = 501, \ 502, \ \cdots, \ 600.$$

(Since this had to be done for 271 different values of γ, there were in all 271×100 or nearly 30,000 calculations to do, and the whole process took several hours.) Finally, for each value of γ, these hundred values of $\phi(t)$ were drawn as dots on the plot of ϕ against γ. To see what this accomplishes, consider first a value of γ such as $\gamma = 1.065$ where we know that the motion has period 1. With the period equal to 1, the hundred successive values of $\phi(t)$ are all the same, and the 100 dots all land at the same place in the plot of ϕ against γ. Thus what we see at any γ for which the period is 1 is a *single* dot. From $\gamma = 1.06$ till the threshold value $\gamma_1 = 1.0663$ where the period doubles, our plot is therefore a single curve.

At the threshold $\gamma_1 = 1.0663$, the period changes to 2, and the positions

$$\phi(501), \ \phi(502), \ \phi(503), \ \cdots, \ \phi(600)$$

now alternate between *two different* values. Therefore, these 100 points actually create exactly two distinct dots on the plot, and the single curve *bifurcates* at γ_1 into *two* curves. At $\gamma_2 = 1.0793$ the period doubles again, to period 4, and each of the two curves bifurcates, giving four curves in all. The next doubling, to period 8, is easily seen (though I have not actually indicated it on the picture), and if you look closely you can just pick out some of the bifurcations to period 16. After this, the graph becomes a nearly solid confusion of points, and it is impossible to tell (from the graph, at least) the exact value γ_c where chaos begins, though it is clearly somewhere just below $\gamma = 1.083$. Beyond this point, for the remainder of Figure 12.17, the motion is mostly chaotic, though you can see a small window at $\gamma = 1.0845$, indicated by a vertical dashed line. (The window is especially noticeable in the upper section of the plot, where the dots are otherwise denser.) If you hold a ruler to this vertical line, you will see that at this particular value of γ there are just six distinct points. That is, at $\gamma = 1.0845$, the motion has returned briefly to being periodic, this time with period 6.

A Larger View

Figure 12.17 shows a rather small range of drive strengths ($1.06 \leq \gamma \leq 1.087$) in great detail. Before we examine a larger ranges of drive strengths, we must cope with one small complication. As γ increases, we have seen that the pendulum can start a "rolling" motion in which it makes many complete revolutions. In some cases, it can continue to "roll" indefinitely, so that $\phi(t)$ eventually approaches $\pm\infty$. Even if this rolling motion is perfectly periodic, the successive values

$$\phi(t_o),\ \phi(t_o + 1),\ \phi(t_o + 2),\ \cdots$$

never repeat themselves, since they increase by a multiple of 2π in each cycle. This renders a bifurcation plot, drawn exactly as in Figure 12.17, useless. The most obvious way to get around this difficulty is to redefine ϕ so that it always lies in the range

$$-\pi < \phi \leq \pi.$$

Each time ϕ increases past π, we subtract 2π, and each time it decreases past $-\pi$, we add 2π. With this modification, we can now draw a bifurcation diagram as before. However, keeping ϕ between $\pm\pi$ in this way has the disadvantage that it introduces a meaningless discontinuous jump into $\phi(t)$, each time it passes $\pm\pi$.

A second, and sometimes simpler, way around the problem of the 2π-ambiguity in ϕ is to plot the values of the *angular velocity*

$$\dot{\phi}(t_o),\ \dot{\phi}(t_o + 1),\ \dot{\phi}(t_o + 2),\ \cdots \tag{12.27}$$

instead of the angular position $\phi(t_o),\ \cdots$. The angular velocity $\dot{\phi}$ is immune to the 2π-ambiguity of ϕ (since $\dot{\phi}$ is unaffected by the addition of any multiple of 2π to ϕ). Thus, if the motion is periodic with period n, then the values (12.27) will repeat themselves after n cycles, and otherwise not. Therefore, a bifurcation diagram drawn using the values of $\dot{\phi}$ instead of ϕ will work just like Figure 12.17, even if the pendulum undergoes a rolling motion.

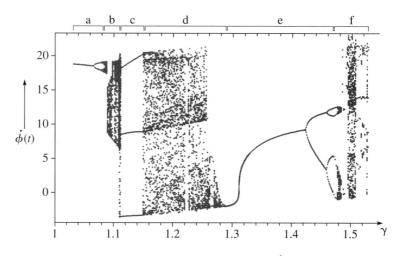

Figure 12.18 Bifurcation diagram showing values of $\dot\phi$ for the DDP with drive strengths $1.03 \le \gamma \le 1.53$. The intervals labelled $a, b, \cdots, f$ across the top are as follows: **(a)** This interval is the same as was shown in much greater detail in Figure 12.17. It starts with period 1, followed by a period-doubling cascade leading to chaos. **(b)** Mostly chaos. **(c)** Period 3. **(d)** Mostly chaos. **(e)** Period 1, followed by another period-doubling cascade. **(f)** Mostly chaos.

Figure 12.18 is a bifurcation diagram drawn using values of $\dot\phi$ over a range from just above $\gamma = 1.0$ to just above $\gamma = 1.5$. The first part of this picture, labelled (a) at the top, is the interval that was shown in much greater detail in Figure 12.17, with a period-doubling cascade that starts from period 1 and ends in chaos. Section (b) is mostly chaos, although we already know that it contains some narrow windows of periodicity (most of which are completely hidden at the scale used here). Section (c) is very clearly period 3 and includes the value $\gamma = 1.13$ that was shown in Figure 12.14. Section (d) is mostly chaos, while (e) starts with a long stretch of period 1, followed by another period-doubling cascade. Finally, section (f) is mostly chaos, although you can just pick out some windows of periodicity.

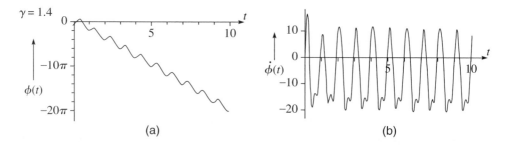

Figure 12.19 Motion of the DDP with drive strength $\gamma = 1.4$. **(a)** The plot of $\phi(t)$ against t shows a periodic rolling motion in which ϕ decreases by 2π in each drive cycle. **(b)** The plot of angular velocity $\dot\phi(t)$ against t shows even more clearly that after about two drive cycles the motion becomes periodic, with $\dot\phi(t)$ returning to the same value once each cycle.

A striking feature of Figure 12.18 is the long interval of period-1 motion, from just below $\gamma = 1.3$ to just above $\gamma = 1.4$. This period-1 motion is actually a rolling motion, as you can see in Figure 12.19 which shows the motion for $\gamma = 1.4$. In part (a), which shows $\phi(t)$ as a function of t, you can see that the pendulum is rolling clockwise at a rate of one complete revolution per drive cycle (ϕ decreases by 20π in 10 cycles). That the motion really is periodic is even more evident in part (b), which shows $\dot{\phi}(t)$ as a function of t. After about two drive cycles, $\dot{\phi}(t)$ is clearly periodic with period 1.

12.7 State-Space Orbits

In the next two sections I give a brief introduction to the *Poincaré section*, which is an important alternative way to view the motion of chaotic (and nonchaotic) systems. The Poincaré section is a simplification of the so-called *state-space orbit*. This simplification is especially helpful for complicated multidimensional systems but can be introduced in the context of our one-dimensional driven damped pendulum. Thus I shall start in this section by describing state-space orbits for the DDP.

In our discussion of the DDP we have focussed almost exclusively on the position $\phi(t)$ as a function of t. It turns out, however, that it is sometimes an advantage to follow *both* the position $\phi(t)$ *and* the angular velocity $\dot{\phi}(t)$ as time evolves. In principle, if one knows $\phi(t)$ for all t, then one can calculate $\dot{\phi}(t)$ by straightforward differentiation. Thus to follow $\dot{\phi}(t)$ as well as $\phi(t)$ is, in this sense, redundant. Nevertheless, following both variables can provide new insights into the motion, and this is what we shall now discuss.

There is an immediate problem in plotting the two variables $\phi(t)$ and $\dot{\phi}(t)$ as functions of the third variable t, since this requires a three-dimensional plot — something which is hard to make, and not especially illuminating when made. The usual procedure is to draw the pair of values $[\phi(t), \dot{\phi}(t)]$ as a point in a two-dimensional plane where the horizontal axis labels ϕ and the vertical axis $\dot{\phi}$. (For reasons I'll discuss in a moment, this plane with coordinates ϕ and $\dot{\phi}$ is called *state space*.) As time passes, the point $[\phi(t), \dot{\phi}(t)]$ moves in this two-dimensional space and traces out a curve, which is called a **state-space orbit** (or phase-space trajectory). Once you get used to interpreting these state-space orbits, you will find that they give a rather clear picture of the system's motion.

As a first example, let us consider a DDP with $\gamma = 0.6$ (a drive strength for which the linear approximation is still fairly good) and with our favorite initial conditions $\phi(0) = -\pi/2$ and $\dot{\phi}(0) = 0$. Figure 12.20 shows a conventional plot of $\phi(t)$ against t for this case. To interpret this picture one has to know (as you certainly do) that the changing position $\phi(t)$ is shown by the vertical displacement of the graph while the time t advances from left to right. With this understood, you can clearly see the motion starting at $t = 0$ with $\phi(0) = -\pi/2$ and quickly approaching the expected sinusoidal attractor, with $\phi(t)$ of the form $\phi(t) = A\cos(\omega t - \delta)$.

Figure 12.21 shows the state-space orbit for the same DDP with the same initial conditions. Part (a) shows the first twenty cycles, $0 \le t \le 20$. To interpret this picture,

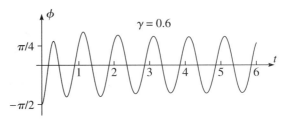

Figure 12.20   Conventional plot of $\phi(t)$ against t for a DDP with drive strength $\gamma = 0.6$. The motion quickly settles down to almost perfectly sinusoidal oscillation.

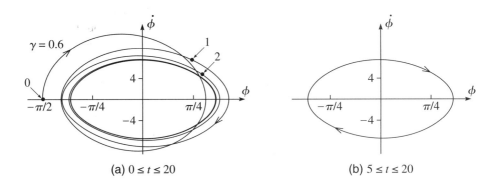

(a) $0 \le t \le 20$ (b) $5 \le t \le 20$

Figure 12.21   State-space orbit for a DDP with drive strength $\gamma = 0.6$. State space is the two-dimensional plane with coordinates ϕ and $\dot{\phi}$; the state-space orbit is just the path traced by the point $[\phi(t), \dot{\phi}(t)]$ as time passes. **(a)** The first 20 cycles, starting from the initial values $\phi(0) = -\pi/2$ and $\dot{\phi}(0) = 0$. The three dots labelled 0, 1, and 2 show the positions of $[\phi(t), \dot{\phi}(t)]$ at $t = 0$, 1, and 2. The orbit spirals inward and rapidly approaches the period-one attractor, which appears as an ellipse in state space. **(b)** The same as (a) but with the first 5 cycles omitted so that only the elliptical attractor is seen. Between the times $5 \le t \le 20$, the point $[\phi(t), \dot{\phi}(t)]$ moves 15 times around the same elliptical path.

one has to know that as t advances the curve is traced, in the direction of the arrows, by the pair $[\phi(t), \dot{\phi}(t)]$. With this understood, you can see clearly that the orbit starts out from $\phi(0) = -\pi/2$ and $\dot{\phi}(0) = 0$. Since the initial acceleration $\ddot{\phi}$ is positive,[19] $\dot{\phi}(t)$ increases from the outset, and $\phi(t)$ begins to increase as soon as $\dot{\phi}$ is nonzero. Thus the point $[\phi(t), \dot{\phi}(t)]$ moves up initially, curving to the right. The oscillation of $\phi(t)$ is evidenced by the back and forth, left-right, motion of the orbit; the oscillation of $\dot{\phi}(t)$ by the up and down vertical motion. Eventually, as the transients die out, the motion approaches its long-term attractor, in which (in the linear approximation) we know that $\phi(t)$ has the form

$$\phi(t) = A\cos(\omega t - \delta). \tag{12.28}$$

---

[19] As you can easily check, with the given initial conditions, both gravity and the drive force give the pendulum a positive acceleration at first.

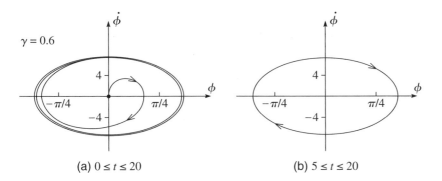

γ = 0.6

(a) $0 \le t \le 20$ (b) $5 \le t \le 20$

Figure 12.22   State-space orbit for a DDP with drive strength $\gamma = 0.6$ and initial conditions $\phi(0) = \dot{\phi}(0) = 0$. **(a)** The first 20 cycles, starting from the origin and spiralling out toward the elliptical attractor. **(b)** In the 15 cycles $5 \le t \le 20$, the orbit moves 15 times around the elliptical attractor to give exactly the same picture as in Figure 12.21(b).

This implies that the angular velocity $\dot{\phi}(t)$ approaches the form

$$\dot{\phi}(t) = -\omega A \sin(\omega t - \delta). \tag{12.29}$$

The two equations (12.28) and (12.29) are the parametric equations for an ellipse drawn clockwise in the plane of $(\phi, \dot{\phi})$, with semimajor and semiminor axes A and ωA. Thus, once the transients have died out, the point $[\phi(t), \dot{\phi}(t)]$ moves around this ellipse with angular frequency equal to the drive frequency ω; that is, the state-space orbit completes one revolution per drive cycle. In Figure 12.21(a), the state-space orbit spirals in toward this ellipse, merging with it after about three cycles. [This already illustrates one small advantage of the state-space orbit over the conventional plot of $\phi(t)$ against t: In the conventional Figure 12.20 the actual motion has become indistinguishable from the limiting sinusoidal motion after little more than 1 cycle; in the state-space plot of Figure 12.21(a) the actual and limiting orbits can be told apart for some three cycles. Thus the state-space orbit gives a more sensitive picture of the approach to the attractor.] Figure 12.21(b) is the same as part (a), except that I have omitted the first 5 cycles; that is, in part (b), $5 \le t \le 20$ and only the elliptical attractor shows up. Since our main interest is usually in the limiting motion, state-space plots are usually drawn as in part (b), with enough initial cycles omitted so that only the limiting motion is visible.

Figure 12.22 shows the state-space orbit for exactly the same DDP as in Figure 12.21, but with initial conditions $\phi(0) = \dot{\phi}(0) = 0$. In part (a) you can easily see that the orbit starts out with the stated initial conditions, and spirals outward, completing some 2.5 cycles before merging with the elliptical attractor. Part (b) shows the 15 cycles starting from $t = 5$, by which time the orbit is indistinguishable from its long-term attractor. In particular, Figure 12.22(b) is exactly the same as Figure 12.21(b), because for $\gamma = 0.6$ all initial conditions lead to the same attractor.

State Space

I shall give a detailed discussion of state space in Chapter 13, but here is a brief explanation of the terminology. For our pendulum, **state space** (also called *phase space*) is the two-dimensional plane defined by the two variables ϕ, the angular position, and $\dot{\phi}$, the angular velocity. This is to be contrasted with the one-dimensional **configuration space** defined by the one variable ϕ that gives the position, or *configuration* of the system. More generally, the configuration space of an n-dimensional mechanical system is the n-dimensional space of its n position coordinates $q_1, \cdots, q_n$, whereas state space is the $2n$-dimensional space of the coordinates $q_1, \cdots, q_n$ *and* velocities $\dot{q}_1, \cdots, \dot{q}_n$. I shall discuss several properties and uses of state space in Chapter 13. Here I mention just one important feature: The "**state**" (or "state of motion" in full) of a mechanical system is often used to mean a specification of the motion (at any chosen time t_o) that is complete enough to determine uniquely the motion at all later times. That is, the state of a system defines the *initial conditions* needed to specify a unique solution of the equation of motion. For our pendulum, specification of the position ϕ at time t_o is not sufficient to determine a unique solution, but specification of ϕ and $\dot{\phi}$ *is*. That is, the two variables ϕ and $\dot{\phi}$ define the state of the pendulum, and the space of all pairs $(\phi, \dot{\phi})$ is naturally called *state space*.

A **state-space orbit** is simply the path traced in state space by the pair $[\phi(t), \dot{\phi}(t)]$ as time evolves. Natural as this name is, you must recognise that a state-space orbit is very different from the orbit of, say, a planet in ordinary space with coordinates $\mathbf{r} = (x, y, z)$. For example, a planet can have many different orbits passing through a single point $\mathbf{r}$ at a given time t_o. On the other hand, from what was just said about initial conditions, it follows that for any "point" $(\phi, \dot{\phi})$ in state space, our pendulum has exactly one state-space orbit passing through $(\phi, \dot{\phi})$ at any given t_o. Another curious feature of state-space orbits concerns their direction of flow: Since the vertical axis represents the velocity $\dot{\phi}$, the motion at any point above the horizontal axis ($\dot{\phi} > 0$) is always to the right (increasing ϕ), as seen in Figure 12.22. Similarly, the motion at any point below the horizontal axis has to be to the left. If an orbit crosses the horizontal axis then, since $\dot{\phi} = 0$, the orbit must be moving exactly vertically (ϕ not changing). All of these properties are illustrated in Figure 12.22. They imply that any closed state-space orbit, such as the elliptical attractor of Figure 12.22(b), is always traced in a clockwise direction.

More State-Space Orbits

As we increase the drive strength γ, we know that the motion of our DDP undergoes various dramatic changes, some of which show up very nicely in plots of the state-space orbits. For instance, Figure 12.23 shows the state-space orbits for $\gamma = 1.078$ and $\gamma = 1.081$, both in the middle of the period-doubling cascade first shown in Figure 12.8. Both plots show forty cycles starting from $t = 20$, by which time all initial transients have completely died out. That is, both plots show the limiting, long-term motion and are to be compared with Figure 12.22(b). As in that picture, these new orbits move around the origin in more-or-less elliptical loops, but in both cases the

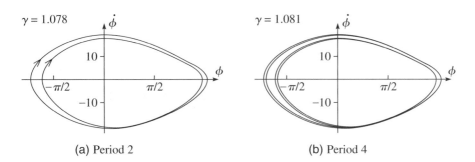

Figure 12.23   State-space orbits showing the periodic attractors for **(a)** $\gamma =$ 1.078 with period 2 and **(b)** $\gamma = 1.081$ with period 4. Both plots show the forty cycles from $t = 20$ to 60. In part (a), the orbit traces just two distinct loops twenty times each; in part (b) it traces four loops ten times each. Compare Figure 12.8 (middle two lines).

orbit makes more than one loop before closing on itself. In part (a) there are two distinct loops, each of which lasts for one drive cycle, so that the motion repeats itself once every two cycles — that is, it has period two. In part (b) there are four distinct loops, indicating very clearly that the period has doubled again to period four. It is important to understand that it makes no difference how many cycles we plot in these two figures [as long as we start after the transients have died out, and plot at least two cycles in part (a) and four in part (b)]. I could have plotted from $t = 20$ to 100 or from 20 to 1000, and part (a) would still have shown the same two loops and part (b) the same four loops.

Chaos

If we increase the drive strength γ a little further, we enter a region of chaos. Figure 12.24 shows the state-space orbit for drive strength $\gamma = 1.105$, whose chaotic character was shown in Figures 12.10 and 12.13. Part (a) shows seven cycles from $t = 14$ to $t = 21$, and you can see clearly that in seven cycles the orbit fails to repeat or to close on itself. Thus if the motion is periodic, its period must be greater than 7. To decide whether it *is* periodic, we need to plot more cycles. In part (b), I have plotted from $t = 14$ to 200, and the plot has become an almost solid swath of black but has still not repeated itself. [The evidence for this last claim is that in a plot out to $t = 400$ (not shown) the curve moves into several of the remaining gaps of part (b); thus it has certainly not begun to repeat by $t = 200$.] Therefore, Figure 12.24 adds strong support to our conclusion that the motion never repeats itself and is in fact chaotic.

The black swath of Figure 12.24(b) is very striking, but is too full of information to be of much use. We need a way to extract from this picture a smaller amount of information that might actually tell us more. The technique for doing this is the so-called Poincaré section, but before we take this up, I want to give two more examples of state-space orbits.

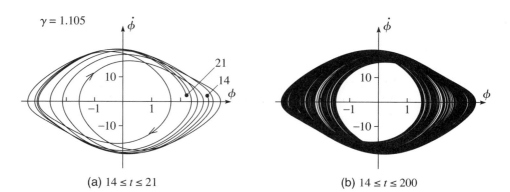

(a) $14 \leq t \leq 21$ (b) $14 \leq t \leq 200$

Figure 12.24   State-space orbits for a DDP with $\gamma = 1.105$ showing the chaotic attractor. **(a)** In the seven cycles from $t = 14$ to 21 the orbit does not close on itself. **(b)** The same is true in the 186 cycles from $t = 14$ to 200, and by now it is pretty clear that the motion is never going to repeat itself and is in fact chaotic.

State-Space Orbits for Rolling Motion

We have already seen that for $\gamma = 1.4$ our DDP executes a "rolling motion," making a complete clockwise rotation once each drive cycle (Figure 12.19). The state-space orbit for this motion is shown in Figure 12.25. In this plot you can see clearly how, after a couple of cycles, the pendulum settles down to a periodic motion in which ϕ decreases by 2π and the pendulum makes a complete clockwise rotation once per cycle.

The plot of Figure 12.25 is a very satisfactory way of showing the state-space orbit over a small number of cycles. Sometimes, however, (if the motion is chaotic, for instance) one would like to show the orbit over a very long time interval — several hundred cycles, perhaps — and in this case ϕ may range over many hundreds of complete revolutions. To show this, in the format of Figure 12.25, we would be forced to compress the scale on the ϕ axis to the point where the motion would be completely indecipherable. The usual way around this difficulty is to redefine ϕ so that it always

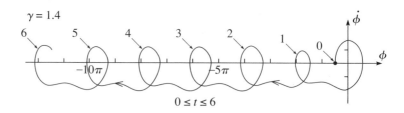

$0 \leq t \leq 6$

Figure 12.25   First six cycles of the state-space orbit for a DDP with $\gamma = 1.4$, showing the periodic rolling motion, in which ϕ decreases by 2π in each cycle. The numbers $0, 1, \cdots, 6$ indicate the state-space "position" $(\phi, \dot{\phi})$ at the times $t = 0, 1, \cdots, 6$.

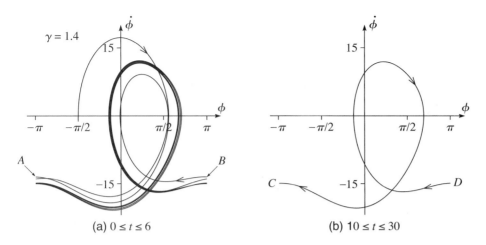

Figure 12.26   **(a)** The exact same orbit as in Figure 12.25 but with ϕ redefined so that it remains between $-\pi$ and π. Each time ϕ decreases to $-\pi$ (at A for example), the orbit disappears and reappears at $+\pi$ (at B for example). **(b)** By the time $t = 10$ the orbit has settled down to perfectly periodic motion, with each successive cycle lying exactly (on this scale) on top of its predecessor.

lies between $-\pi$ and π: Each time ϕ decreases past $-\pi$ we add on 2π and each time it increases past π we subtract 2π. (This is acceptable since any two values of ϕ that differ by a multiple of 2π represent the same position of the pendulum.) With ϕ redefined in this way, the state-space orbit of Figure 12.25 looks as shown in Figure 12.26(a). This new plot is not an obvious improvement on Figure 12.25 (though we shall see that it does have some advantages), but you should study it carefully to understand the relationship of the two kinds of picture. You can think of the new picture as being obtained from Figure 12.25 by cutting apart the intervals $-3\pi < \phi < -\pi$, and $-5\pi < \phi < -3\pi$, and so on, and pasting them all back on top of the interval $-\pi < \phi < \pi$. In the resulting picture, ϕ makes a discontinuous jump each time it arrives at $\phi = \pm\pi$. For example, at about $t = 0.7$, ϕ decreases to $-\pi$ at the point A and jumps to the point B.

An advantage of Figure 12.26(a) over Figure 12.25 is that the new picture gives a more incisive test of the periodicity of the orbit. In the new picture, you can see that the orbit is approaching a periodic attractor, but it is also clear that in the interval $0 \le t \le 6$ the orbit has definitely not *reached* the periodic attractor. (In fact you can just about see that there are 6 distinct loops.) On the other hand, by the time $t = 10$ the successive cycles are indistinguishable on the scale of these pictures. The twenty cycles shown in Figure 12.26(b) all disappear on the left at the same point C, reappear at D, and follow the exact same path back to C twenty times over.

A disadvantage of either plot in Figure 12.26 is the spurious discontinuity each time ϕ jumps from $-\pi$ to π, as at points A and B, for example. We can get rid of these discontinuities (at least in our minds) if we imagine the plot cut out and rolled into a cylinder with the vertical lines $\phi = \pm\pi$ glued together. In this way point A becomes the same as point B, and the state-space orbit moves continuously around the vertical cylinder.

More Chaos

As a final example of a state-space orbit, I have shown the orbit for a DDP with $\gamma = 1.5$ in Figure 12.27. We already know that the motion is chaotic for this value of γ, though for this picture I chose a smaller damping constant, $\beta = \omega_0/8$ instead of the value $\beta = \omega_0/4$ that I used for all previous pictures in this chapter. As it turns out, this smaller damping makes the chaotic motion more wild and produces a Poincaré section that is even more interesting and elegant (as I describe in the next section). With these parameters, the pendulum undergoes an erratic rolling motion, making many complete revolutions, first in one direction and then in the other. Thus we are forced to make a plot with ϕ confined between $-\pi$ and π as in Figure 12.26 — but with dramatically different results. The motion does not repeat itself in the 190 cycles shown, with $10 \leq t \leq 200$. (The evidence for this claim is that in a plot for $10 \leq t \leq 250$ — not shown — the orbit moves into some of the unvisited regions of Figure 12.27. If it had begun to repeat itself before $t = 200$, it could not visit new ground after $t = 200$.)

The dense tangle of threads running through Figure 12.27 lend strong support to the claim that for these parameters the motion is chaotic. Unfortunately, one could not claim that the picture sheds much light on the nature of chaotic motion. It is just too densely packed with information to convey any useful message. In the next section I describe the Poincaré section, which is a technique for culling out of pictures like Figure 12.27 enough information to allow an interesting pattern to emerge.

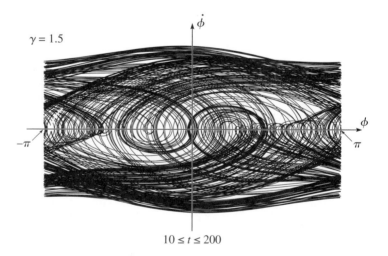

Figure 12.27   The chaotic state-space orbit for a DDP with $\gamma = 1.5$ and $\beta = \omega_0/8$. In the 190 cycles shown, the motion does not repeat itself, and, in fact, it never does.

12.8   Poincaré Sections

For the periodic motion of a DDP, a state-space orbit is a descriptive way of viewing the pendulum's history. For chaotic motion, a state-space orbit conveys a sense of the dramatic nature of chaos, but is too full of information to be of much serious use. One way around this difficulty is a trick which we have used earlier and was suggested by Poincaré: Instead of following the motion as a function of the continuous variable t, we look at the position just once per cycle at times $t = t_o, t_o + 1, t_o + 2, \cdots$. The **Poincaré section** for a DDP is just a plot showing the pendulum's "position" $[\phi(t), \dot{\phi}(t)]$ in state space at one-cycle intervals

$$t = t_o, t_o + 1, t_o + 2, \cdots , \tag{12.30}$$

with t_o usually chosen so that the initial transients have died out.[20] To illustrate this, consider the state-space orbit shown in Figure 12.28(a) for a DDP with $\gamma = 1.078$ (and the damping constant restored to our usual value $\beta = \omega_o/4$). The two loops of this orbit indicate that (as we already knew) the long-term motion has period two. To emphasize this I have drawn dots to show the position $[\phi(t), \dot{\phi}(t)]$ at one-cycle intervals, $t = 20, 21, 22, \cdots$. Since the motion has period two, these alternate between just two distinct positions and show up as just two dots. In a Poincaré section one dispenses with the orbit and draws just the dots at one-cycle intervals as in Figure 12.28(b). When the motion is periodic, there is no particular advantage to the Poincaré section over the complete state-space orbit of Figure 12.28(a), although the Poincaré section does show the period very clearly. (A Poincaré section with four dots would show period-four motion, and so on.) On the other hand, when the motion is chaotic,

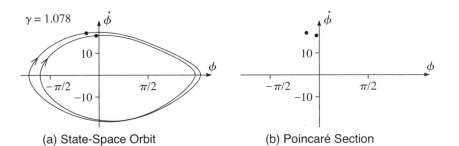

(a) State-Space Orbit   (b) Poincaré Section

Figure 12.28   **(a)** State-space orbit of a DDP with $\gamma = 1.078$ for $20 \leq t \leq 60$. The dots show the positions at $t = 20, 21, 22, \cdots$, but, since the motion has period two, these alternate between just two fixed points. The right point shows the positions for $t = 20, 22, \cdots$; the left one, those for $t = 21, 23, \cdots$. **(b)** In the corresponding Poincaré section, one omits the orbit and draws only the dots showing the positions at $t = 20, 21, 22, \cdots$. The presence of just two dots is a clear indication of period-two motion.

---

[20] For a multidimensional system, the Poincaré section involves taking a two-dimensional slice, or *section*, through the multidimensional state space. Hence the word "section."

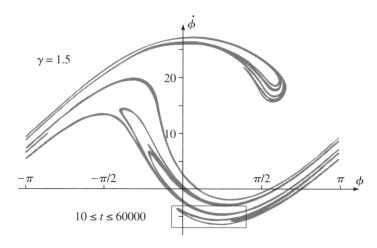

Figure 12.29   Poincaré section for a pendulum with $\gamma = 1.5$ and damping constant $\beta = \omega_o/8$ for times $10 \leq t \leq 60000$. This figure is made of nearly 60000 points showing the "position" $[\phi(t), \dot{\phi}(t)]$ at one-cycle intervals, $t = 10, 11, \cdots, 60000$. The rectangular box indicates the region that is enlarged in Figure 12.30.

no two cycles of the motion are the same, and the state-space orbit can be a real mess, as we saw in Figure 12.27. In this case, the Poincaré section reveals some totally unexpected structures.

To illustrate a Poincaré section for chaotic motion, I chose the pendulum whose chaotic state-space orbit was shown in Figure 12.27. It is clear that, since this motion never repeats itself, the Poincaré section will contain infinitely many points, and these infinitely many points will comprise a subset of the points of the full orbit. It is probably fair to say that no one could ever *guess* what this subset would look like, but with the aid of a high-speed computer we can find out. The result is shown in Figure 12.29. Although it is certainly not obvious exactly what this elegant figure signifies, it certainly *is* obvious that it signifies something. By selecting from Figure 12.27 just those points at one-cycle intervals, we have reduced the dense, and nearly solid, tangle of Figure 12.27 to the elegant curve of Figure 12.29. Actually, while Figure 12.29 *looks* like a relatively simple curve, it is not a curve at all, but rather a **fractal**. A fractal can be defined in various ways, but a characteristic feature of fractals is that when one enlarges the scale and zooms in on a portion of the picture, one uncovers further structures that are in some ways similar to the original picture (somewhat like a photograph of a person holding a photograph of a person holding a photograph . . . and so on). To illustrate this property of our Poincaré section, I have zoomed in on the region indicated by the rectangular box at the bottom of Figure 12.29. Notice that (at the scale of Figure 12.29) this region comprises a prominent "tongue" pointing to the left near the left of the box, with a second tongue inside the first near the right of the box. Figure 12.30 is a fourfold enlargement of this box. This enlargement makes clear that the apparently single tongue on the right of the box of 12.29 is actually four tongues, while that on the left is actually at least five.

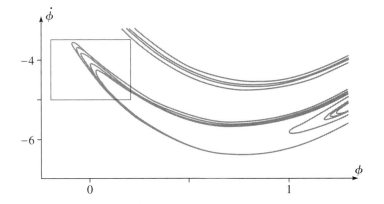

Figure 12.30   Enlargement of the small box at the bottom of the Poincaré section of Figure 12.29. Each of the two tongues in the box of 12.29 is seen to be made up of several tongues. The box on the left is the region that is further enlarged in Figure 12.31.

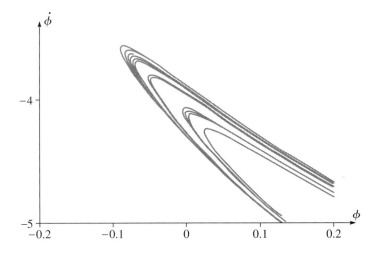

Figure 12.31   A further enlargement of the box at the left of the enlargement of Figure 12.30. Each of the five tongues in the box of 12.30 (except perhaps the innermost one) is seen to be made up of several tongues.

This process of zooming in on successively smaller regions of the Poincaré section can, at least in principle, be continued indefinitely. Figure 12.31 is a further fourfold enlargement of the region shown by the gray rectangle on the left of Figure 12.30. In this enlargement, we see that each of the five tongues on the left of Figure 12.30 (except perhaps the fifth one) actually consists of several separate tongues. This so-called **self similarity** of the figure is one of the characteristic features of a fractal.

When the Poincaré section of the motion of a chaotic system is a fractal, the long-term motion is said to be a **strange attractor**. It would unfortunately be well beyond the scope of this book to explain what it signifies that the Poincaré section of a chaotic attractor is fractal, and indeed there is still much about this phenomenon that is not

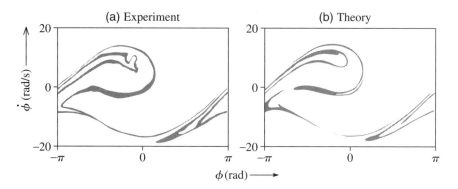

Figure 12.32   Poincaré section for a DDP. **(a)** Experimental results using the Daedalon Chaotic Pendulum. **(b)** Theoretical prediction using the same parameters as in part (a). Courtesy of Professors H.J.T. Smith and James Blackburn and the Daedalon Corporation.

understood. Nevertheless, it is undeniably fascinating that the strange geometrical structure of the fractal appears in our study of the long-term behavior of chaotic systems. This discovery has stimulated much research on both the physics of chaotic systems and the mathematics of fractals.

To observe a strange attractor with a real pendulum would obviously be challenging, but once again the experimentalists have risen to the challenge. Figure 12.32 shows a Poincaré section made with the Daedalon chaotic pendulum. Part (a) shows the experimental results and part (b) the theoretical prediction (that is, a numerical solution of the equation of motion using the experimental values of the parameters). Considering the great subtlety of these graphs, the agreement is outstanding.[21]

12.9   The Logistic Map

As I have repeatedly emphasized, the phenomenon of chaos appears in many different situations. In particular, there are certain systems that can exhibit chaos, but whose equations of motion — called *maps* — are simpler than the equations of any mechanical system. Although these systems are not strictly part of classical mechanics, they are worth mentioning here, for several reasons: Because their equations of motion are simple, several aspects of their motion can be understood using quite elementary methods. Any understanding of chaos that we get from studying these simpler systems can shed light on the corresponding behavior of mechanical systems. In particular, there is an intimate connection between these "maps" and the Poincaré sections of mechanical systems. Finally, a discussion of chaos in this new context highlights the diversity of systems that exhibit the phenomenon.

---

[21] There are, nevertheless, differences. One possible cause is the difficulty of making a drive motor that is perfectly sinusoidal.

Discrete Time and Maps

In almost all problems of mechanics, one is concerned with the evolution of a system as time advances continuously. However, there are systems for which time is a discrete variable. The history of any event that occurs just once a year, such as the Super Bowl, is an example. The score at Super Bowl games is defined only at a sequence of discrete times starting in 1967 and spaced at one-year intervals:

$$t = 1967, 1968, 1969, \cdots. \tag{12.31}$$

The attendance at your weekly lunch group is defined only for discrete times spaced a week apart. The total rainfall in the annual Indian monsoon is defined only for discrete times spaced a year apart.

Even when a variable is defined as a function of continuous time, we may find that we need its values only at certain discrete times. For example, entymologists studying the population of a particular bug may have no interest in the population's day-by-day evolution; rather they may need to record the bug population just once a year, immediately after the year's new arrivals have hatched. Another example of this situation is the Poincaré section for a mechanical system, as described in Section 12.8. Our ultimate interest is to know the state of the system for all (continuous) times, but we saw that it is sometimes useful to record just the state at discrete, one-cycle intervals. To the extent that we are prepared to settle for this smaller amount of information, the Poincaré section reduces the problem of the pendulum (or whatever) to a discrete-time problem, and anything we can learn about discrete-time systems should shed light on the possible behavior of Poincaré sections.

In the case of the driven damped pendulum, we know that the state of the system, as given by the pair $[\phi(t), \dot{\phi}(t)]$, at any time t determines uniquely the state at any later time. In particular, it determines the state $[\phi(t + 1), \dot{\phi}(t + 1)]$ one cycle later. This means that there exists a function f (which we don't know, but which certainly exists) that, acting on any chosen pair $[\phi(t), \dot{\phi}(t)]$, gives the corresponding pair $[\phi(t + 1), \dot{\phi}(t + 1)]$. That is,

$$[\phi(t + 1), \dot{\phi}(t + 1)] = f\left([\phi(t), \dot{\phi}(t)]\right). \tag{12.32}$$

In the same way, we could imagine a bug species with the property that the population n_{t+1} in year $t + 1$ is uniquely determined[22] by the population n_t in the preceding year t. Again this would imply the existence of a function f that carries any n_t onto its corresponding n_{t+1}:

$$n_{t+1} = f(n_t). \tag{12.33}$$

---

[22] This is, of course, a very simplified model. In the real world, the population n_{t+1} certainly depends on n_t, but also on many other factors, such as the rainfall in year t, the supply of bug food, and the population of birds that like to eat the bug. Nevertheless, we can imagine a temperate island, with a constant supply of bug food and no bug predators, on which n_{t+1} is uniquely determined by n_t alone.

We can call an equation of this form the **growth equation** for the population concerned.

EXAMPLE 12.1   Exponential Population Growth

The simplest example of a growth equation of the type (12.33) is the case where n_{t+1} is proportional to n_t:

$$n_{t+1} = f(n_t) = rn_t. \tag{12.34}$$

That is, the function $f(n)$ that gives next year's population in terms of this year's is just

$$f(n) = rn \tag{12.35}$$

where the positive constant r could be called the **growth rate** or **growth parameter** of the population. [For example if every bug alive this spring dies before next spring but leaves two surviving offspring, then the population would satisfy (12.34) with $r = 2$.] Solve the equation (12.34) for n_t in terms of n_0 and discuss the long-term behavior of n_t.

The solution to (12.34) is easily seen by inspection. Observe that

$$n_1 = f(n_0) = rn_0$$

and

$$n_2 = f(n_1) = f(f(n_0)) = r^2 n_0$$

from which it is clear that

$$n_t = f(n_{t-1}) = \overbrace{f(f(\cdots f(n_0)\cdots))}^{t \text{ terms}} = r^t n_0. \tag{12.36}$$

We see that if $r > 1$, the population n_t grows exponentially, approaching infinity as $t \to \infty$. If $r = 1$, the population stays constant, and if $r < 1$ it decreases exponentially to zero.

Before we discuss a more interesting growth equation, I need to introduce some terminology. In mathematics, the words "function" and "map" are used as almost exact synonyms. Thus we can say that Equation (12.33) defines n_{t+1} as a *function* of n_t. Or we can say that (12.33) is a **map** in which f carries n_t onto the corresponding[23] n_{t+1}, a relationship that we can represent thus:

$$n_t \xrightarrow{f} n_{t+1} = f(n_t). \tag{12.37}$$

---

[23] The origin of this rather strange usage seems to be in cartography. A cartographer's map of the US, for example, establishes a correspondence between each actual point of the US and the corresponding point on a piece of paper, in somewhat the same way the function $y = f(x)$ establishes a correspondence between each value x and the corresponding value $y = f(x)$.

The whole sequence of numbers, $n_0, n_1, n_2, \cdots$ can be written similarly as

$$n_0 \xrightarrow{f} n_1 \xrightarrow{f} n_2 \xrightarrow{f} n_3 \xrightarrow{f} \cdots \tag{12.38}$$

and is naturally described as an **iterated map**, or just **map** for short. For some reason the word "map" (as opposed to "function") is used almost universally to describe relationships like (12.37) between successive values of any discrete-time variable. The map (12.37) is a *one-dimensional map* since it carries the single number n_t onto the single number n_{t+1}. The corresponding relationship (12.32) for the Poincaré section of a DDP defines a *two-dimensional map*, since it carries the pair of numbers $[\phi(t), \dot{\phi}(t)]$ onto the pair $[\phi(t+1), \dot{\phi}(t+1)]$.

The Logistic Map

The exponential map of Example 12.1 with $r > 1$ is a reasonably realistic model for the initial growth of many populations, but no real population can grow exponentially for ever. Something — overcrowding or shortage of food, for example — eventually slows down the growth. There are many ways to modify the map (12.34) to give a more realistic model of population growth. One of the simplest is to replace the function $f(n) = rn$ of (12.34) by

$$f(n) = rn(1 - n/N) \tag{12.39}$$

where N is a large positive constant, whose significance we shall see directly. That is, we replace the exponential map (12.34) with the so-called **logistic map**

$$n_{t+1} = f(n_t) = rn_t(1 - n_t/N). \tag{12.40}$$

As long as the population remains small compared to N, the term n_t/N in (12.40) is unimportant, and our new map produces the same exponential evolution as the exponential map. But if n_t grows toward N, the term n_t/N becomes important, and the parenthesis $(1 - n_t/N)$ begins to diminish and "kill off" some of the excess growth. Thus this "mortality factor" $(1 - n/N)$ in (12.39) produces exactly the expected slowing of the population growth as n becomes large, and overcrowding or starvation become important. In particular, if n_t were to reach the value N, the parenthesis $(1 - n_t/N)$ in (12.40) would vanish, and next year's population n_{t+1} would be zero. If n_t were greater than N, then n_{t+1} would be negative — which is impossible. In other words, a population governed by the logistic map (12.40) can never exceed the value N, the maximum or **carrying capacity** of the model. Notice that, because of the term involving n in the mortality factor, the logistic map (12.39) (unlike the exponential map) is *nonlinear*. It is this nonlinearity that makes possible the chaotic behavior of the logistic map.

Figure 12.33 compares exponential and logistic growth, for a growth parameter $r = 2$ and an initial population $n_0 = 4$. The upper curve (gray dots) shows the unending doubling of the exponential case; the lower curve (black dots) shows the growth predicted by the logistic map (12.40), with the same growth parameter $r = 2$ and with a carrying capacity $N = 1000$. As long as n remains small (much less than 1000), the

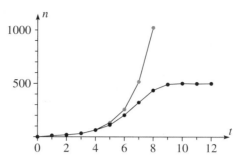

Figure 12.33   Exponential and logistic growth, both with growth parameter $r = 2$. The gray dots show the exponential growth, increasing without limit; the black dots show the logistic growth, which eventually slows down and approaches an equilibrium at $n = 500$. The lines joining the dots are just to guide the eye.

logistic growth is indistinguishable from pure exponential growth, but once n reaches 100 or so, the mortality factor visibly slows the logistic growth, which eventually levels off at around $n = 500$.

Before we discuss the logistic map in any detail it is convenient to simplify it by changing variables from the population n to the *relative population*,

$$x = n/N, \tag{12.41}$$

the ratio of the actual population n to its maximum possible value N. Dividing both sides of (12.40) by N, we see that x_t obeys the growth equation

$$x_{t+1} = f(x_t) = rx_t(1 - x_t) \tag{12.42}$$

where I have redefined the map f as a function of x to be

$$f(x) = rx(1 - x). \tag{12.43}$$

Since the population n is confined to the range $0 \leq n \leq N$, the relative population $x = n/N$ is restricted to

$$0 \leq x \leq 1. \tag{12.44}$$

Within this range, the function $x(1 - x)$ has a maximum of 1/4 (at $x = 1/2$). Thus, to guarantee that x_{t+1}, as given by (12.42), never exceeds 1, we must limit the growth factor to $0 \leq r \leq 4$. Therefore, we shall be studying the map (12.42) in the ranges $0 \leq x \leq 1$ and $0 \leq r \leq 4$.

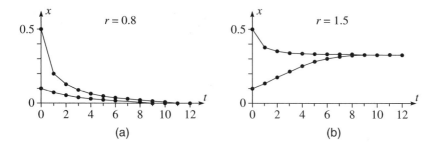

Figure 12.34   The relative population $x_t = n_t/N$ for the logisitic map (12.42), with two different initial conditions for each of two different growth rates. **(a)** With the growth parameter $r = 0.8$, the population rapidly approaches zero whether $x_0 = 0.1$ or $x_0 = 0.5$. **(b)** With $r = 1.5$ and the same two initial conditions, the population approaches the fixed value 0.33.

Before we look at some of the exotic aspects of the logistic map, let us first look at a couple of cases where it behaves just as one might expect. Figure 12.34(a) shows the logistic population for a growth parameter $r = 0.8$ and for two different initial values, $x_0 = 0.1$ and $x_0 = 0.5$. You can see that in either case $x_t \to 0$ as $t \to \infty$. In fact it is easy to see that as long as $r < 1$, the population eventually goes to zero whatever its initial value: From (12.42) we see that $x_t \le rx_{t-1}$ and hence that $x_t \le r^t x_0$; therefore, if $r < 1$, we conclude that $x_t \to 0$ as $t \to \infty$.

Figure 12.34(b) shows the logistic population for a growth parameter $r = 1.5$ and for the same two initial conditions. For $x_0 = 0.1$ we see that the population increases at first. On the other hand, for the larger initial value $x_0 = 0.5$ the mortality factor causes the population to *decrease* at first. In either case, it eventually levels out at $x = 0.33$.

Fixed Points

In both the cases shown in Figure 12.34 we can say that the logistic map has a constant **attractor** towards which the population eventually moves, namely $x = 0$ for any $r < 1$, and $x = 0.33$ for $r = 1.5$. If the population happens to start out equal to such a constant attractor, $x_0 = x^*$ say, then it simply remains fixed there for all time; that is, $x_t = x^*$ for all t. This obviously happens if and only if

$$f(x^*) = x^*. \tag{12.45}$$

Any value x^* which satisfies this equation is called a **fixed point** of the map f. These fixed points are analogous to the equilibrium points of a mechanical system, in that a system which starts at a fixed point remains there for ever.

For a given map, we can solve the equation (12.45) to find the map's fixed points. For example, the fixed points of the logistic map must satisfy

$$rx^*(1 - x^*) = x^*, \tag{12.46}$$

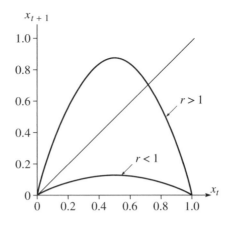

Figure 12.35   Graphs of x against x (the 45° line) and of the logistic function $f(x) = rx(1 - x)$ (the two curves) for two choices of r, one less than and one greater than 1. The fixed points of the logistic map lie at the intersections of the line with the curve. When $r < 1$ there is just one intersection, at $x^* = 0$; when $r > 1$ there are two intersections, one still at $x^* = 0$ and the other at an $x^* > 0$.

which is easily solved to give

$$x^* = 0 \qquad \text{or} \qquad x^* = \frac{r - 1}{r}. \tag{12.47}$$

The first solution is the fixed point $x^* = 0$ that we have already noted. The second solution depends on the value of r. For $r < 1$ it is negative and hence irrelevant. For $r = 1$ it coincides with the first solution $x^* = 0$, but for $r > 1$ it is a distinct, second fixed point. For example, for $r = 1.5$ it gives the fixed point we have already noted at $x^* = 1/3$.

It is a fortunate circumstance that we can actually solve the equation (12.45) analytically to find the fixed points of the logistic map, but it is also instructive to examine the equation graphically, since graphical considerations give additional insight and can be applied to many different maps, some of which cannot necessarilly be solved analytically. To solve Equation (12.45) graphically, we just plot the two functions x and $f(x)$ against x as in Figure 12.35 and read off the fixed points as those values x^* where the two graphs intersect. When r is small, the curve of $f(x)$ lies below the 45° line of x against x, and the only intersection is at $x = 0$; that is, the only fixed point is $x^* = 0$. When r is large the curve of $f(x)$ bulges above the 45° line and there are two fixed points. The boundary between the two cases is easily found by noting that the slope of $f(x)$ at $x = 0$ is just r. Thus as we increase r, the curve moves across the 45°line (whose slope is 1) when $r = 1$. Thus for $r < 1$ there is just one fixed point at $x^* = 0$; but when $r > 1$ there are two fixed points, one at $x^* = 0$ and the other at an $x^* > 0$. The advantage of this graphical argument is that it works equally well for any similar function $f(x)$ as long as it is a single concave-down arch. [For example, $f(x)$ could be the function $f(x) = r \sin(\pi x)$ of Problem 12.23.]

A Test for Stability

That x^* is a fixed point (that is, an equilibrium value) for the logistic map guarantees that when the population starts out at x^* it will stay there. By itself this is not enough to ensure that the value x^* is an attractor for the map. We need to check in addition that x^* is a **stable fixed point**, that is, that if the population starts out *close* to x^* it will evolve towards x^* not away from it. (This issue has an exact parallel in the study of equilibrium points of a mechanical system: If a system starts out exactly at an equilibrium point then it will — in principle — stay there indefinitely. But only if the equilibrium is stable will the system move back to equilibrium if disturbed a little away.)

There is a simple test for stability, which we can derive as follows: If x_t is close to a fixed point x^*, we naturally write

$$x_t = x^* + \epsilon_t. \tag{12.48}$$

That is, we define ϵ_t as the difference between x_t and the fixed point x^*. If ϵ_t is small, this lets us evaluate x_{t+1} as

$$x_{t+1} = f(x_t) = f(x^* + \epsilon_t)$$
$$\approx f(x^*) + f'(x^*)\epsilon_t = x^* + \lambda\epsilon_t \tag{12.49}$$

where in the last expression I have used the fact that x^* is a fixed point [so that $f(x^*) = x^*$] and I have introduced the notation λ for the derivative of $f(x)$ at x^*,

$$\lambda = f'(x^*). \tag{12.50}$$

Now, according to (12.48), $x_{t+1} = x^* + \epsilon_{t+1}$. Comparing this with the last expression of (12.49) we see that

$$\epsilon_{t+1} \approx \lambda\epsilon_t. \tag{12.51}$$

Because of this simple relation, the number $\lambda = f'(x^*)$ is called the **multiplier** or **eigenvalue** of the fixed point. It shows that if $|\lambda| < 1$, then once x_t is close to x^*, successive values get closer and closer to x^*. On the other hand, if $|\lambda| > 1$, then when x_t is close to x^*, the succeeding values move *away* from x^*. This is our required test for stability:

Stability of Fixed Points

Let x^* be a fixed point of the map $x_{t+1} = f(x_t)$; that is, $f(x^*) = x^*$. If $|f'(x^*)| < 1$, then x^* is stable and acts as an attractor. If $|f'(x^*)| > 1$, then x^* is unstable and acts as a repeller.

We can immediately apply this test to the two fixed points of the logistic map: Since $f(x) = rx(1 - x)$, its derivative is

$$f'(x) = r(1 - 2x).$$

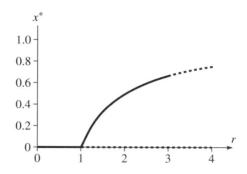

Figure 12.36   The fixed points x^* of the logistic map as functions of the growth parameter r. The solid curves show the stable fixed points and the dashed curves the unstable. Note how the fixed point $x^* = 0$ becomes unstable at precisely the place ($r = 1$) where the nonzero fixed point first appears.

At the fixed point $x^* = 0$, this means the crucial derivative is

$$f'(x^*) = r \ ;$$

therefore, the fixed point $x^* = 0$ is stable for $r < 1$ but unstable for $r > 1$. At the fixed point $x^* = (r - 1)/r$, the derivative is

$$f'(x^*) = 2 - r \ ,$$

so that this fixed point is stable for $1 < r < 3$, but unstable for $r > 3$. These results are summarised in Figure 12.36, which shows the fixed-point values x^* as functions of the growth parameter r with the stable fixed points shown as solid curves and the unstable as dashed curves.

The arguments just given show exactly when each of the two fixed points becomes unstable and that the second one appears exactly when the first becomes unstable. On the other hand, it would be nice to have an argument that made clearer *why* the fixed points behave the way they do and showed more clearly (what is true) that the same qualitative conclusions apply to any other one-dimensional map with the same general features. Such an argument can be found by examining the graph of Figure 12.35. In that figure, we saw that the fixed points of the map correspond to intersections of the curve of $f(x)$ against x with the 45° line (x against x). When r is small, it was clear that there is only one such intersection at $x^* = 0$. As r increased the curve bulged up more and eventually crossed the 45° line producing a second intersection and hence a second, nonzero fixed point. We saw that this second intersection appears when the slope $f'(0)$ of the curve at $x = 0$ is exactly 1 (that is, when it is tangent to the 45° line). In light of our test for stability, this means the second fixed point has to appear at exactly the moment when the first one at $x^* = 0$ changes from stable to unstable.

The First Period Doubling

Figure 12.36 tells the whole story of the fixed points of the logistic map. The solid curves show the stable fixed points, which are the constant attractors. We see that for $r < 1$, the logistic population approaches 0 as $t \to \infty$; for $1 < r < 3$, it approaches the other fixed point $x^* = (r - 1)/r$ as $t \to \infty$. But what happens — and this proves the most interesting question — when $r > 3$? In the computer age this is easily answered. Figure 12.37 shows the first 30 cycles of the logistic population with a growth parameter of $r = 3.2$. The striking feature of this graph is that it no longer settles down to a single constant value. Instead, it bounces back and forth between the two fixed values shown as x_a and x_b, repeating itself once every two cycles. In the language developed for the driven damped pendulum, we can say that the period has doubled to period two, and this period-two limiting motion is called a **two-cycle**.

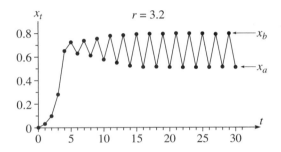

Figure 12.37   A logistic population with growth parameter $r = 3.2$ and initial value $x_0 = 0.01$. The population never settles down to a constant value; rather, it oscillates between two values, repeating itself once every *two* cycles. In other words, it has doubled its period to period two.

We can understand the doubling of the period of the logistic map with the graphical methods already developed, although the argument is a bit more complicated. The essential observations are these: First, neither of the two limiting values x_a and x_b is a fixed point of the map $f(x)$. Instead,

$$f(x_a) = x_b \qquad \text{and} \qquad f(x_b) = x_a. \tag{12.52}$$

Let us, however, consider the **double map** (or **second iterate map**)

$$g(x) = f(f(x)), \tag{12.53}$$

which carries the population x_t onto the population two years hence,

$$x_{t+2} = g(x_t). \tag{12.54}$$

It is clear from Figure 12.37 or Equations (12.52) that both x_a and x_b are fixed points of the double map $g(x)$,[24]

$$g(x_a) = x_a \qquad \text{and} \qquad g(x_b) = x_b. \tag{12.55}$$

---

[24] These points are also called second-order fixed points.

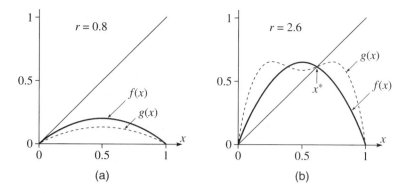

Figure 12.38   The logistic map $f(x)$ and its second iterate $g(x) = f(f(x))$. **(a)** For a growth parameter $r = 0.8$, each function is a single arch which intersects the 45° line just once, at the origin. **(b)** When $r = 2.6$, $f(x)$ is higher than before, but still just a single arch; $g(x)$ has developed two maxima with a valley in between. Both functions have acquired a second intersection with the 45° line, at the same point marked x^*.

Thus to study the two-cycles of the map $f(x)$ we have only to examine the fixed points of the double map $g(x) = f(f(x))$, and for this we can use our understanding of fixed points. Before we do this, it is worth noticing that any fixed point of $f(x)$ is automatically also a fixed point of $g(x)$. (If $x_{t+1} = x_t$ for all t, then certainly $x_{t+2} = x_t$.) Therefore, the two-cycles of $f(x)$ correspond to those fixed points of $g(x)$ that are not also fixed points of $f(x)$.

Since $f(x)$ is a quadratic function of x, it follows that $g(x) = f(f(x))$ is a quartic function, whose explicit form can be written down and studied. However, we can gain a better understanding by considering its graph. When r is small, we know that $f(x)$ is a single low arch (as in Figure 12.35) and you can easily convince yourself that $g(x)$ is an even lower arch, as sketched in Figure 12.38(a), which shows both functions for a growth parameter $r = 0.8$. As r increases, both arches rise, and by the time $r = 2.6$ the function $g(x)$ has developed two maxima as seen in Figure 12.38(b). (You can explore the reason for this development in Problem 12.26.) Also, both curves now intersect the 45° line twice, once at the origin and once at the fixed point $x^* = (r - 1)/r$. That both curves intersect the 45° line at the same points shows two things: First, as we already knew, every fixed point of $f(x)$ is also a fixed point of $g(x)$, and, second, (for the growth parameters shown in this figure) every fixed point of $g(x)$ is also a fixed point of $f(x)$; that is, there are no two-cycles yet.

As we increase the growth parameter r still further, the two crests of the double map $g(x)$ continue to rise, while the valley between them gets lower. (Again see Problem 12.26 for the reason.) Figure 12.39 shows the curves of $f(x)$ and $g(x)$ for growth parameters $r = 2.8$, 3.0, and 3.4. With $r = 2.8$ [part (a)], the double map $g(x)$ still has just the same two fixed points as $f(x)$, at $x = 0$ and at the point indicated as x^*. By the time $r = 3.4$ [part (c)], the double map has developed two additional fixed points, shown as x_a and x_b; that is, the logistic map now has a two-cycle. The threshold value at which the two-cycle appears is clearly the value for which the curve $g(x)$ is *tangent* to the 45° line — namely $r = 3$ for the logistic map, as in part (b) of the figure. If

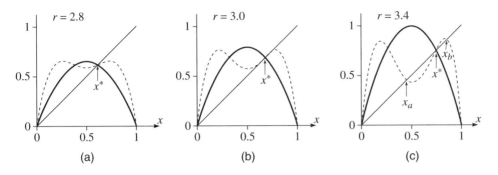

Figure 12.39   The logistic map $f(x)$ (solid curves) and its second iterate $g(x) = f(f(x))$ (dashed) for $r = 2.8$, 3.0, and 3.4. **(a)** With $r = 2.8$, the map $g(x)$ has the same two fixed points as $f(x)$, namely $x = 0$ and $x = x^*$ as shown. **(b)** When r reaches 3.0, the curve of $g(x)$ is tangent to the 45° line, and for any larger value of r, as in **(c)**, the map $g(x)$ has two extra fixed points labelled x_a and x_b.

$r < 3$, the curve $g(x)$ crosses the 45° line just once, at x^*; if $r > 3$ it crosses three times, once at x^* and two more times, at x_a and x_b (one above and one below x^*).

We already know that $r = 3$ is the threshold at which the fixed point x^* becomes unstable. Thus, Figure 12.39 shows that the two-cycle appears at the moment when the "one-cycle" (that is, the fixed point) becomes unstable. Happily, we are now in a position to see *why* this has to be. We have already noted that the two-cycle appears precisely when the curve $g(x)$ is tangent to the 45° line at the point x^*, that is, when

$$g'(x^*) = 1. \tag{12.56}$$

To see what this implies, let us evaluate the derivative $g'(x)$ of the double map $g(x)$ at either of the two-cycle fixed points, x_a say,

$$g'(x_a) = \frac{d}{dx}g(x)\bigg|_{x_a} = \frac{d}{dx}f(f(x))\bigg|_{x_a} = f'(f(x)) \cdot f'(x)\bigg|_{x_a}$$

$$= f'(x_b)f'(x_a). \tag{12.57}$$

Here in the last expression of the first line I have used the chain rule, and in the second line I have used the fact that $f(x_a) = x_b$. Let us apply this result to the birth of the two-cycle in Figure 12.39(b). At the moment of birth, $x^* = x_a = x_b$, and we can combine (12.56) and (12.57) to give

$$[f'(x^*)]^2 = 1.$$

This means that $|f'(x^*)| = 1$, and, by our test for stability, we see that the moment when the two-cycle is born is precisely the moment when the fixed point x^* becomes unstable.

We can use these same techniques to explore what happens as we increase r still further. For example, one can show (Problem 12.28) that the two-cycle that we have just seen appearing at $r = 3$ becomes unstable at $r = 1 + \sqrt{6} = 3.449$ and is succeeded by a stable four-cycle. However, to keep this long chapter from growing totally out of

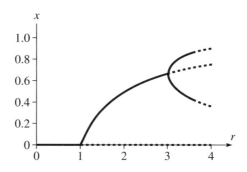

Figure 12.40   The fixed-points and two-cycles of the logistic map as functions of the growth parameter r. The solid curves are stable and the dashed curves unstable.

bounds, I shall just sketch briefly some highlights that can be found by exploring the logistic map numerically.

Bifurcation Diagrams

In Figure 12.36 we saw the complete history of the fixed points of the logistic map itself. If we add onto that picture the fixed points x_a and x_b of the double map $g(x) = f(f(x))$ (that is, the two-cycles of the logistic map) we get the graphs shown in Figure 12.40. These curves are reminiscent of the beginnings of the bifurcation diagram, Figure 12.17, for the driven damped pendulum. In fact, we can redraw Figure 12.40 using the same procedure as was used for Figure 12.17: First one picks a large number of equally spaced values of the growth parameter in the range of interest. (I chose the range $2.8 \le r \le 4$, since this is where the excitement lies, and chose 1200 equally spaced values in this range.) Then for each value of r one calculates the populations $x_0, x_1, x_2, \cdots, x_{t\,max}$, where t_{max} is some very large time. (I chose $t_{max} = 1000$.) Next one chooses a time t_{min} large enough to let all transients die out. (I chose $t_{min} = 900$.) Finally, in a plot of x against r, one shows the values of x_t for $t_{min} \le t \le t_{max}$ as dots above the corresponding value of r. The resulting bifurcation diagram for the logistic map is shown in Figure 12.41.

The similarity of the bifurcation diagram of Figure 12.41 for the logistic map and Figure 12.17 for the driven pendulum is striking indeed. The interpretation of the two pictures is also similar. On the left of Figure 12.41 we see the period-one attractor for $r \le 3$, followed by the first period doubling at $r = 3$. This is followed by the second doubling at $r = 3.449$, and a whole cascade of doublings that end in chaos near $r = 3.570$. With the help of this diagram we can predict the long-term behavior of the logistic population for any particular choice of r (though there is clearly much fine detail that cannot be distinguished at the scale of Figure 12.41). For example at $r = 3.5$, it is clear that the population should have period four, a claim that is borne out in Figure 12.42(a), which shows the twenty cycles from $t = 100$ to 120 for this value of r. Similarly, around $r = 3.84$, sandwiched between wide intervals of chaos, you can see a narrow window that appears to have period three, an observation borne

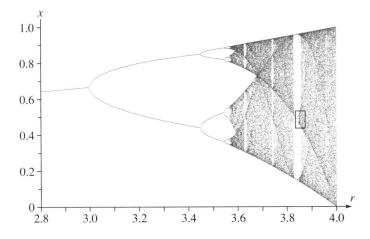

Figure 12.41   Bifurcation diagram for the logistic map. A period-doubling cascade is clearly visible starting at $r_1 = 3$, with a second doubling at $r_2 = 3.449$, and
ending in chaos at $r_c = 3.570$. Several windows of periodicity stand out clearly
amongst the chaos, especially the period-three window near $r = 3.84$. The tiny
rectangle near $r = 3.84$ is the region that is enlarged in Figure 12.44.

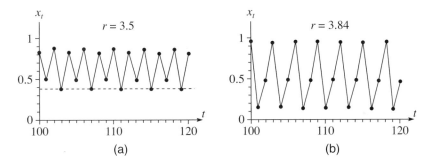

Figure 12.42   Long-term evolution of logistic populations with growth parameters $r = 3.5$ and 3.84. **(a)** Period four. With $r = 3.5$, the twenty cycles $100 \leq$
$t \leq 120$ take on just four distinct values at intervals of four cycles. (The dashed
horizontal line is just to highlight the constancy of every fourth dot.) **(b)** Period
three. With $r = 3.84$, the population repeats every three cycles.

out in Figure 12.42(b). As an example of chaos, Figure 12.43 shows the evolution
of a population with $r = 3.7$; in the eighty cycles shown, there is no evidence of any
repetition.

From Figure 12.41 (and careful enlargements), one can read off the threshold
values of r at which the period doublings occur. If we denote by r_n the threshold
at which the cycle of period 2^n appears, these are found to be

$$r_1 = 3, \quad r_2 = 3.4495, \quad r_3 = 3.5441, \quad r_4 = 3.5644,$$

$$r_5 = 3.5688, \quad r_6 = 3.5697, \quad r_7 = 3.5699. \tag{12.58}$$

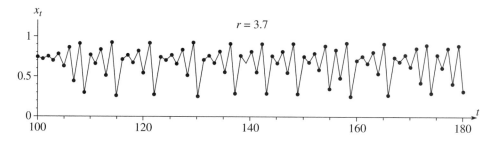

Figure 12.43   Chaos. With $r = 3.7$, the eighty cycles, $100 \leq t \leq 180$, of the logistic map show no tendency to repeat themselves.

As you can check, the separation between successive thresholds shrinks geometrically, just like the corresponding intervals for the period doubling of the driven pendulum. In fact, (see Problem 12.29) the numbers (12.58) give a remarkable fit to the Feigenbaum relation (12.17) that we first met in connection with the DDP, with the same Feigenbaum constant (12.18). Another striking parallel with the driven pendulum is this (Problem 12.30): For those r for which the evolution is non-chaotic, if two populations start out sufficiently close to one another, their difference will converge exponentially to zero as $t \to \infty$. For those r for which the evolution is chaotic, the same difference *diverges* exponentially as $t \to \infty$. That is, the chaotic evolution of the logistic map shows the same sensitive dependence on initial conditions that we found for the driven pendulum.

Perhaps the most striking feature of the logistic bifurcation diagram is that, when one zooms in on certain parts of the diagram, a perfect self similarity emerges. The small rectangle near $r = 3.84$ in Figure 12.41 has been enlarged many fold in Figure 12.44. Apart from the facts that this new picture is upside down and its scale is vastly different, it is a perfect copy of the whole original diagram of which it is a part. This is a striking example of the *self similarity* which appears in many places in the study of chaos and which we met in connection with the Poincaré section of the DDP shown in Figures 12.29, 12.30, and 12.31.

There are many other features of the logistic map and more parallels with the DDP, all worth exploring and some treated in the problems at the end of this chapter. Here, however, I shall leave the logistic map and close this chapter with the hope that you feel at home with some of the main features of chaos and the tools used to explore them. I hope too that your appetite has been whetted to explore this fascinating subject further.[25]

---

[25] For a comprehensive history, with very little mathematics, see *Chaos, Making a New Science* by James Gleick, Viking-Penguin, New York (1987). For a quite mathematical, but highly readable, account of chaos in many different fields see *Nonlinear Dynamics and Chaos* by Steven H. Strogatz, Addison-Wesley, Reading, MA (1994). Two books which focus mostly on chaos in physical systems are *Chaotic Dynamics: An Introduction* by G.L. Baker and J.P.Gollub, Cambridge Universtiy Press, Cambridge (1996) and *Chaos and Nonlinear Dynamics* by Robert C. Hilborn, Oxford University Press, New York (2000).

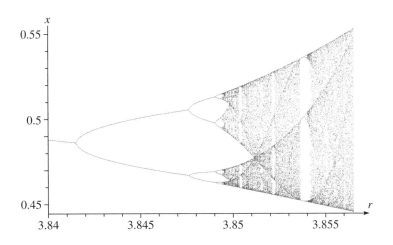

Figure 12.44   A many-fold enlargement of the small rectangle in the logistic bifurcation diagram of Figure 12.41. This tiny section of the original diagram is a perfect, upside-down copy of the whole original. Note that this section is just one of three strands in the original; thus, although this diagram starts out *looking* just like period 1 doubling to period 2, it is actually period 3 doubling to period 6 and so on.

Principal Definitions and Equations of Chapter 12

The Driven Damped Pendulum

A damped pendulum that is driven by a sinusoidal force $F(t) = F_{\mathrm{o}} \cos(\omega t)$ satisfies the nonlinear equation

$$\ddot{\phi} + 2\beta\dot{\phi} + \omega_{\mathrm{o}}^2 \sin\phi = \gamma\omega_{\mathrm{o}}^2 \cos\omega t \qquad \text{[Eq. (12.11)]}$$

where $\gamma = F_{\mathrm{o}}/mg$ is called the **drive strength** and is the ratio of the drive amplitude to the weight.

Period Doubling

For small drive strengths, ($\gamma \lesssim 1$) the long-term response, or **attractor**, of the pendulum has the same period as the drive force. But if γ is increased past $\gamma_1 = 1.0663$, for certain initial conditions and drive frequencies, the attractor undergoes a **period-doubling cascade**, in which the period repeatedly doubles, approaching infinity as $\gamma \to \gamma_{\mathrm{c}} = 1.0829$. [Section 12.4]

Chaos

If the drive strength is increased beyond γ_{c}, at least for certain choices of drive frequency and initial conditions, the long-term motion becomes nonperiodic, and we say that **chaos** has set in. As γ is increased still further, the long-term motion varies, sometimes chaotic, sometimes periodic. [Sections 12.5 & 12.6]

Sensitivity to Initial Conditions

Chaotic motion is **extremely sensitive to initial conditions**. If two identical chaotic pendulums with identical drive forces are launched with slightly different initial conditions, their separation increases exponentially with time, however small the initial difference.   [Section 12.5]

Bifurcation Diagrams

A **bifurcation diagram** is a plot of the system's position at discrete times, $t_o, t_o + 1, t_o + 2, \cdots$ (more generally $t_o, t_o + \tau, t_o + 2\tau, \cdots$) as a function of the drive strength (more generally the appropriate control parameter).   [Figures 12.17 & 12.18]

State-Space Orbits and Poincaré Sections

The **state space** for a system with n degrees of freedom is the $2n$-dimensional space comprising the n generalized coordinates and the n generalized velocities. For the DDP, the points in state space have the form $(\phi, \dot{\phi})$. A **state-space orbit** is just the path traced in state space by a system as t evolves. A **Poincaré section** is a state-space orbit restricted to discrete times $t_o, t_o + 1, t_o + 2, \cdots$ (and, when $n \geq 2$, to a subspace of fewer dimensions).   [Sections 12.7 & 12.8]

The Logistic Map

The **logistic map** is a function (or "map") that gives a number x_t at regular discrete intervals (for example, the relative population of a certain bug once each year) as

$$x_{t+1} = rx_t(1 - x_t). \qquad \text{[Eq. (12.42)]}$$

Although this is not a mechanical system, it exhibits many of the features (period doubling, chaos, sensitivity to initial conditions) of nonlinear mechanical systems.
                                                                          [Section 12.9]

Problems for Chapter 12

Stars indicate the approximate level of difficulty, from easiest ($\star$*) to most difficult (*$\star\star\star$*).*
Warning: Even when the motion is nonchaotic, it can be very sensitive to tiny errors. In several of the computer problems you may need to increase your working precision to get satisfactory results.

SECTION 12.1   Linearity and Nonlinearity

12.1 $\star$ Consider the nonlinear first-order equation $\dot{x} = 2\sqrt{x-1}$. **(a)** By separating variables, find a solution $x_1(t)$. **(b)** Your solution should contain one constant of integration k, so you might reasonably expect it to be the general solution. Show, however, that there is another solution, $x_2(t) = 1$, that is not of the form of $x_1(t)$ whatever the value of k. **(c)** Show that although $x_1(t)$ and $x_2(t)$ are solutions, neither $Ax_1(t)$, nor $Bx_2(t)$, nor $x_1(t) + x_2(t)$ are solutions. (That is, the superposition principle does not apply to this equation.)

12.2 ★ Here is a different example of the disagreeable things that can happen with nonlinear equations. Consider the nonlinear equation $\dot{x} = 2\sqrt{x}$. Since this is first-order, one would expect that specification of $x(0)$ would determine a unique solution. Show that for this equation there are two different solutions, both satisfying the initial condition $x(0) = 0$. [*Hint:* Find one solution $x_1(t)$ by separating variables, but note that $x_2(t) = 0$ is another. Fortunately none of the equations normally encountered in classical mechanics suffer from this disagreeable ambiguity.]

12.3 ★ Consider a second-order linear homogeneous equation of the form (12.6). **(a)** Write out a detailed proof of the superposition principle, that if $x_1(t)$ and $x_2(t)$ are solutions of this equation, then so is any linear combination $a_1x_1(t) + a_2x_2(t)$, where a_1 and a_2 are any two constants. **(b)** Consider now the nonlinear equation in which the third term of (12.6) is replaced by $r(t)\sqrt{x(t)}$. Explain clearly why the superposition principle does not hold for this equation.

12.4 ★ Consider an *inhomogeneous* second-order linear equation of the form

$$p(t)\ddot{x}(t) + q(t)\dot{x}(t) + r(t)x(t) = f(t). \tag{12.59}$$

Let $x_p(t)$ denote a solution (a "*particular*" solution) of this equation and prove that *any* solution $x(t)$ can be written as

$$x(t) = x_p(t) + a_1x_1(t) + a_2x_2(t) \tag{12.60}$$

where $x_1(t)$ and $x_2(t)$ are two independent solutions of the corresponding homogeneous equation — that is, (12.59) with $f(t)$ deleted. [*Hint:* Write down the equations for $x(t)$ and $x_p(t)$ and subtract.] This result shows that to find *all* solutions of (12.59), we have only to find one particular solution and two independent solutions of the corresponding homogeneous equation. **(b)** Explain clearly why the result you proved in part (a) is not, in general, true for a nonlinear equation such as

$$p(t)\ddot{x}(t) + q(t)\dot{x}(t) + r(t)\sqrt{x(t)} = f(t).$$

SECTION 12.3   Some Expected Features of the DDP

12.5 ★ Use Euler's relation and the corresponding expression for $\cos\phi$ (inside the front cover) to prove the identity (12.15).

12.6 ★★ [Computer] **(a)** Use appropriate sofware to solve the equation (12.11) numerically, for a DDP with the following parameters: drive strength $\gamma = 0.9$, drive frequency $\omega = 2\pi$, natural frequency $\omega_o = 1.5\omega$, damping constant $\beta = \omega_o/4$, and initial conditions $\phi(0) = \dot{\phi}(0) = 0$. Solve the equation and plot your solution for six cycles, $0 \le t \le 6$, and verify that you get the result shown in Figure 12.3. **(b)** and **(c)** Solve the same equation twice more with the two different initial conditions $\phi(0) = \pm\pi/2$ — both with $\dot{\phi}(0) = 0$ — and plot all three solutions on the same picture. Do your results bear out the claim that, for this drive strength, all solutions (whatever their initial conditions) approach the same periodic attractor?

12.7 ★★ [Computer] Do all the same calculations as in Problem 12.6 but with a drive strength $\gamma = 1.06$ and for $0 \leq t \leq 10$. In part (a) verify that your results agree with Figure 12.4. Do your results suggest that there is still a unique attractor to which all solutions (whatever their initial conditions) converge? (At first sight the answer may appear to be "No," but remember that values of ϕ that differ by 2π should be considered to be the same.)

12.8 ★★ [Computer] Use a computer to find a numerical solution of the equation of motion (12.11) for a DDP with the following parameters: drive strength $\gamma = 1.073$, drive frequency $\omega = 2\pi$, natural frequency $\omega_0 = 1.5\omega$, damping constant $\beta = \omega_0/4$, and initial conditions $\phi(0) = \pi/2$ and $\dot{\phi}(0) = 0$. **(a)** Solve for $0 \leq t \leq 50$, and then plot the first ten cycles, $0 \leq t \leq 10$. **(b)** To be sure that the initial transients have died out, plot the ten cycles $40 \leq t \leq 50$. What is the period of the long-term motion (the attractor)?

12.9 ★★ [Computer] Do the same calculations as in Problem 12.8 with all the same parameters except that $\phi(0) = 0$. Plot the first 30 cycles $0 \leq t \leq 30$ and check that you agree with Figure 12.5. Plot the ten cycles $40 \leq t \leq 50$ and find the period of the long-term motion.

12.10 ★★ [Computer] Explore the behavior of the DDP with the same parameters as in Problem 12.8, but with several different initial conditions. For example, you might keep $\dot{\phi}(0) = 0$ but try various different values for $\phi(0)$ between $-\pi$ and π. You will find that the initial behavior varies quite a lot according to the initial conditions, but the long-term motion is the same in all cases (as long as you remember that values of ϕ that differ by multiples of 2π represent the same position).

12.11 ★★ Test how well the values of the thresholds γ_n given in Table 12.1 fit the Feigenbaum relation (12.17) as follows: **(a)** Assuming that the Feigenbaum relation is exactly true use it to prove that $(\gamma_{n+1} - \gamma_n) = (1/\delta)^{n-1}(\gamma_2 - \gamma_1)$ and, hence, that a plot of $\ln(\gamma_{n+1} - \gamma_n)$ against n should be a straight line with slope $-\ln \delta$. **(b)** Make this plot for the three differences of Table 12.1. How well do your points seem to bear out our prediction? Fit a line to your plot (either graphically or using a least squares fit) and find the slope and hence the Feigenbaum number δ. [You would not expect to get very good agreement with the known value (12.18) for two reasons: You have only three points to plot and the Feigenbaum relation (12.17) is only approximate, except in the limit of large n. Under the circumstances, you will find the agreement is remarkable.]

12.12 ★★ Here is another way to look at the Feigenbaum relation (12.17): **(a)** Assuming that (12.17) is exactly true, prove that the thresholds γ_n approach a finite limit γ_c and then prove that $\gamma_n = \gamma_c - K/\delta^n$, where K is a constant. This means that a plot of γ_n against δ^{-n} should be a straight line. **(b)** Using the known value (12.18) of δ and the four values of γ_n in Table 12.1 make this plot. Does it seem to fit our prediction? The vertical intercept of your graph should be γ_c. What is your value and how well does it agree with (12.20)?

12.13 ★ You can see in Figure 12.13 that for $\gamma = 1.105$, the separation of two identical pendulums with slightly different initial conditions increases exponentially. Specifically $\Delta\phi$ starts out at 10^{-4} and by $t = 14.5$ it has reached about 1. Use this to estimate the Liapunov exponent λ as defined in (12.26), $\Delta\phi(t) \sim Ke^{\lambda t}$. Your answer should confirm that $\lambda > 0$ for chaotic motion.

12.14 ★★ [Computer] Numerically solve the equation of motion (12.11) for a DDP with drive strength $\gamma = 1.084$, and the following other parameters: drive frequency $\omega = 2\pi$, natural frequency $\omega_0 = 1.5\omega$, damping constant $\beta = \omega_\mathrm{o}/4$, and initial conditions $\phi(0) = \dot{\phi}(0) = 0$. Solve for the first seven drive cycles ($0 \leq t \leq 7$) and call your solution $\phi_1(t)$. Solve again for all the same parameters except that $\phi(0) = 0.00001$ and call this solution $\phi_2(t)$. Let $\Delta\phi(t) = \phi_2(t) - \phi_1(t)$ and make a plot of $\log|\Delta\phi(t)|$ against t. With this drive strength the motion is chaotic. Does your plot confirm this? In what sense?

12.15 ★★ [Computer] Numerically solve the equation of motion (12.11) for a DDP with the following parameters: drive strength $\gamma = 0.3$, drive frequency $\omega = 2\pi$, natural frequency $\omega_\mathrm{o} = 1.5\omega$, damping constant $\beta = \omega_\mathrm{o}/4$, and initial conditions $\phi(0) = \dot{\phi}(0) = 0$. Solve for the first five drive cycles ($0 \leq t \leq 5$) and call your solution $\phi_1(t)$. Solve again for all the same parameters except that $\phi(0) = 1$ (that is, the initial angle is one radian) and call this solution $\phi_2(t)$. Let $\Delta\phi(t) = \phi_2(t) - \phi_1(t)$ and make a plot of $\log|\Delta\phi(t)|$ against t. Does your plot confirm that $\Delta\phi(t)$ goes to zero exponentially? Note: The exponential decay continues indefinitely, but $\Delta\phi(t)$ eventually gets so small that it is smaller than the rounding errors, and the exponential decay cannot be seen. If you want to go further, you will probably need to crank up your precision.

12.16 ★★ Consider the chaotic motion of a DDP for which the Liapunov exponent is $\lambda = 1$, with time measured in units of the drive period as usual. (This is very roughly the value found in Problem 12.13.) **(a)** Suppose that you need to predict $\phi(t)$ with an accuracy of $1/100$ rad and that you know the initial value $\phi(0)$ within 10^{-6} rad. What is the maximum time $t_{\max}$ for which you can predict $\phi(t)$ within the required accuracy? This $t_{\max}$ is sometimes called the **time horizon** for prediction within a specified accuracy. **(b)** Suppose that, with a vast expenditure of money and labor, you manage to improve the accuracy of your initial value to 10^{-9} radians (a thousand-fold improvement). What is the time horizon now (for the same required accuracy of prediction)? By what factor has $t_{\max}$ improved? Your results illustrate the difficulty of making accurate long-term predictions for chaotic motion.

12.17 ★★ [Computer] In Figure 12.15, you can see that for $\gamma = 1.503$ the DDP "tries" to execute a steady rolling motion changing by 2π once each cycle, but that there is superposed an erratic wobbling and that the direction of the rolling reverses itself from time to time. For other values of γ, the pendulum actually does approach a steady, periodic rolling. **(a)** Solve the equation of motion (12.11) for a drive strength $\gamma = 1.3$ and all other parameters as in the first part of Problem 12.14, for $0 \leq t \leq 8$. Call your solution $\phi_1(t)$ and plot it as a function of t. Describe the motion. **(b)** It is hard to be sure that the motion is periodic based on this graph, because of the steady rolling through -2π each cycle. As a better check, plot $\phi_1(t) + 2\pi t$ against time. Describe what this shows. This kind of periodic rolling motion is sometimes described as **phase-locked**.

12.18 ★★ [Computer] Since the rolling motion of Problem 12.17 is periodic (and hence not chaotic) we would expect the difference $\Delta\phi(t)$ between neighboring solutions (solutions of the same equation, but with slightly different initial conditions) to decrease exponentially. To illustrate this, do part (a) of Problem 12.17 and then find the solution of the same problem except that $\phi(0) = 1$. Call this second solution $\phi_2(t)$ and let $\Delta\phi(t) = \phi_2(t) - \phi_1(t)$. Make a plot of $\log|\Delta\phi(t)|$ against t and comment.

SECTION 12.7   **State-Space Orbits**

12.19 ★ Consider an undamped, undriven simple harmonic oscillator — a mass m on the end of a spring whose force constant is k. **(a)** Write down the general solution $x(t)$ for the position as a function of time t. Use this to sketch the state-space orbit, showing the motion of the point $[x(t), \dot{x}(t)]$ in the two-dimensional state space with coordinates $(x, \dot{x})$. Explain the direction in which the orbit is traced as time advances. **(b)** Write down the total energy of the system and use conservation of energy to prove that the state-space orbit is an ellipse.

12.20 ★ Consider a weakly-damped, but undriven oscillator as described by Equation (5.28). The general motion is given by (5.38). **(a)** Use this to sketch the state-space orbit, showing the movement of $(x, \dot{x})$ for $0 \le t \le 10$ with the parameters $\delta = 0$, $\omega_1 = 2\pi$, and $\beta = 0.5$. **(b)** This system has a unique stationary attractor. What is it? Explain it, both with reference to your sketch and in terms of energy.

SECTION 12.9   The Logistic Map

12.21 ★★ Here is an iterated map that is easily studied with the help of your calculator: Let $x_{t+1} = f(x_t)$ where $f(x) = \cos(x)$. If you choose any value for x_0, you can find $x_1, x_2, x_3, \cdots$ by simply pressing the cosine button on your calculator over and over again. (Be sure the calculator is in radians mode.) **(a)** Try this for several different choices of x_0, finding the first 30 or so values of x_t. Describe what happens. **(b)** You should have found that there seems to be a single fixed attractor. What is it? Explain it, by examining (graphically, for instance) the equation for a fixed point $f(x^*) = x^*$ and applying our test for stability [namely, that a fixed point x^* is stable if $|f'(x^*)| < 1$].

12.22 ★★ Consider the iterated map $x_{t+1} = f(x_t)$ where $f(x) = x^2$. **(a)** Show that it has exactly two fixed points of which just one is stable. What are they? **(b)** Show that x_t approaches the stable fixed point if and only if $-1 < x_0 < 1$. The interval $-1 < x_0 < 1$ is called a **basin of attraction** since all sequences $x_0, x_1, \cdots$ that start in the "basin" are attracted to the same attractor. **(c)** Show that $x_t \to \infty$ if and only if $|x_0| > 1$. (Thus we could say the map has a second stable fixed point at $x = \infty$ and the basin of attraction for this fixed point is the set $|x_0| > 1$.) For chaotic systems, the basins of attraction can be much more complicated than these examples and are often fractals.

12.23 ★★ [Computer] Consider the **sine map** $x_{t+1} = f(x_t)$ where $f(x) = r \sin(\pi x)$. The interesting behavior of this map is for $0 \le x \le 1$ and $0 \le r \le 1$, so restrict your attention to these ranges. **(a)** Using a plot analogous to Figure 12.35, discuss the fixed points of this map. Show that the map has either one or two fixed points, depending on the value of r. Show that when r is small there is just one fixed point, which is stable. **(b)** At what value of r (call it r_0) does the second fixed point appear? Show that r_0 is also the value of r at which the first fixed point becomes unstable. **(c)** As r increases, the second fixed point eventually becomes unstable. Find numerically the value r_1 at which this occurs.

12.24 ★★ [Computer] Consider the sine map of Problem 12.23. Using a programmable calculator (or a computer) you can easily find the first ten or twenty values of x_t for any chosen initial value x_0. Taking $x_0 = 0.3$, calculate the first 10 values of x_t for each of the following values of the parameter r: **(a)** $r = 0.1$, **(b)** $r = 0.5$, **(c)** $r = 0.78$. In each case plot your results (x_t against t) and describe the long-term attractor. If you did Problem 12.23, are your results here consistent with what you proved there?

12.25 ★★ [Computer] The sine map of Problem 12.23 exhibits period-doubling cascades just like the logistic map. To illustrate this, take $x_0 = 0.8$ and find the first twenty values x_t for each of the following values of the parameter r: **(a)** $r = 0.60$, **(b)** $r = 0.79$, **(c)** $r = 0.85$, and **(d)** $r = 0.865$. Plot your results (as four separate plots) and comment.

12.26 ★★ The appearance of a two-cycle of the logistic map at the exact moment when the one-cycle becomes unstable follows directly from the behavior of the graphs of $f(x)$ and $g(x) = f(f(x))$ as shown in Figures 12.38 and 12.39. The crucial point is that the function $f(x)$ is a simple arch that gets steadily higher as we increase the control parameter r from 0; at the same time, $f(f(x))$ starts out as a simple arch which is lower than $f(x)$, but developes two maxima [higher than $f(x)$] with a minimum [lower than $f(x)$] in between. Explain clearly in words why this behavior of $f(f(x))$ follows inevitably from that of $f(x)$. [*Hint:* Since $f(x)$ is symmetric about $x = 0.5$, you need only consider its behavior as x runs from 0 to 0.5. The advantage of this argument is that it applies to any map of the form $f(x) = r\phi(x)$ where $\phi(x)$ has a single symmetric arch (such as the sine map of Problem 12.23).]

12.27 ★★ The two-cycles of the map $f(x)$ correspond to fixed points of the second-iterate $g(x) = f(f(x))$. Thus the two values shown as x_a and x_b in Figure 12.37 are roots of the equation $f(f(x)) = x$. In the case of the logistic map this is a quartic equation, which is not too hard to solve: **(a)** Verify that for the logistic function $f(x)$

$$x - f(f(x)) = rx \left(x - \frac{r-1}{r} \right) \left[r^2 x^2 - r(r+1)x + r + 1 \right].$$ (12.61)

Thus the fixed-point equation $x - f(f(x)) = 0$ has four roots. The first two are $x = 0$ and $x = (r-1)/r$. Explain these and show that the other two roots are

$$x_a, x_b = \frac{r + 1 \pm \sqrt{(r+1)(r-3)}}{2r}.$$ (12.62)

Explain how you know that these are the two points of a two-cycle. **(b)** Show that for $r < 3$ these roots are complex and hence that there is no real two-cycle. **(c)** For $r \geq 3$ these two roots are real and there is a real two-cycle. Find the values of x_a and x_b for the case $r = 3.2$ and verify the values shown in Figure 12.37.

12.28 ★★ Equation (12.62) in Problem 12.27 gives the two fixed points of the two-cycle of the logistic map. One can observe this two-cycle only if it is stable, which will happen if $|g'(x)| < 1$, where $g(x)$ is the double map $g(x) = f(f(x))$ and x is either of the values x_a or x_b. **(a)** Combine (12.62) with Equation (12.57) to find $g'(x_a)$. [Notice that because (12.57) is symmetric in x_a and x_b, you will get the same result whether you use x_a or x_b; that is, the two points necessarily become stable or unstable at the same time.] **(b)** Show that the two-cycle is stable for $3 < r < 1 + \sqrt{6}$. This establishes that the threshold at which period 2 is replaced by period 4 is $r_2 = 1 + \sqrt{6} = 3.449$.

12.29 ★★ [Computer] The thresholds r_n for period doubling of the logistic map are given by Equation (12.58). These should satisfy the Feigenbaum relation (12.17), at least in the limit that $n \to \infty$ (with γ replaced by r, of course). Test this claim as follows: **(a)** If you have not done Problem 12.11 prove that the Feigenbaum relation (if exactly true) implies that $(r_{n+1} - r_n) = K/\delta^n$. **(b)** Make a plot of $\ln(r_{n+1} - r_n)$ against n. Find the best-fit straight line to the data and from its slope predict the Feigenbaum constant. How does your answer compare with the accepted value $\delta = 4.67$?

12.30 ★★ [Computer] The chaotic evolution of the logistic map shows the same sensitivity to initial conditions that we met in the DDP. To illustrate this do the following: **(a)** Using a growth rate $r = 2.6$, calculate x_t for $1 \leq t \leq 40$ starting from $x_0 = 0.4$. Repeat but with the initial condition $x_0' = 0.5$ (the prime is just to distinguish this second solution from the first — it does not denote differentiation) and then plot $\log|x_t' - x_t|$ against t. Describe the behavior of the difference $x_t' - x_t$. **(b)** Repeat part (a) but with $r = 3.3$. In this case, the long term evolution has period 2. Again describe the behavior of the difference $x_t' - x_t$. **(c)** Repeat parts (a) and (b), but with $r = 3.6$. In this case, the evolution is chaotic, and we expect the difference to *grow* exponentially; therefore, it is more interesting to take the two initial values much closer together. To be definite take $x_0 = 0.4$ and $x_0' = 0.400001$. How does the difference behave?

12.31 ★★★ [Computer] When the evolution of the logistic map is non-chaotic, two solutions with the same r that start out sufficiently close, will converge exponentially. (This was illustrated in Problem 12.30.) This does not mean that *any* two solutions with the same r will converge. **(a)** Repeat Problem 12.30(a), with all the same parameters except $r = 3.5$ (a value for which we know the long-term motion has period 4). Does $x'_t - x_t$ approach zero? **(b)** Now do the same exercise but with the initial conditions $x_0 = 0.45$ and $x'_0 = 0.5$. Comment. Can you explain why, if the period is greater than 1, it is impossible that the difference $x'_t - x_t$ go to zero for all choices of initial conditions?

12.32 ★★★ [Computer] Make a bifurcation diagram for the logistic map, in the style of Figure 12.41 but for the range $0 \leq r \leq 3.55$. Take $x_0 = 0.1$. Comment on its main features. [*Hint:* Start by using a very small number of points, perhaps just r going from 0 to 3.5 in steps of 0.5 and t going from 51 to 54. This will let you calculate for each of the values of r individually, and get the feel of how things work. To make a good diagram, you will then need to increase the number of points (r going from 0 to 3.55 in steps of 0.025, and t from 51 to 60, perhaps), and you will certainly need to automate the calculation of the large number of points.]

12.33 ★★★ [Computer] Reproduce the logistic bifurcation diagram of Figure 12.41 for the range $2.8 \leq r \leq 4$. Take $x_0 = 0.1$. [*Hint:* To make Figure 12.41 I used about 50,000 points, but you certainly don't need to use that many. In any case, start by using a very small number of points, perhaps just r going from 2.8 to 3.4 in steps of 0.2 and t going from 51 to 54. This will let you calculate for each of the values of r individually, and get the feel of how things work. To make a good diagram, you will then need to increase the number of points (r going from 2.8 to 4 in steps of 0.025, and t from 500 to 600, perhaps), and you will certainly need to automate the calculation of the large number of points.]

12.34 ★★★ [Computer] Make a bifurcation diagram for the sine map of Problems 12.23 and 12.25. This should resemble Figure 12.41 but for the range $0.6 \leq r \leq 1$. Take $x_0 = 0.1$. Comment on its main features. [*Hint:* Start by using a very small number of points, perhaps just r going from 0.6 to 0.8 in steps of 0.05 and t going from 51 to 54. This will let you calculate for each of the values of r individually, and get the feel of how things work. To make a good diagram, you will then need to increase the number of points (r going from 0.6 to 1 in steps of 0.005, and t from 400 to 500, perhaps), and you will certainly need to automate the calculation of the large number of points.]

CHAPTER 13

Hamiltonian Mechanics

In the first six chapters of this book, we worked entirely with the Newtonian form of mechanics, which describes the world in terms of forces and accelerations (as related by the second law) and is primarily suited for use in Cartesian coordinate systems. In Chapter 7, we met the Lagrangian formulation. This second formulation is entirely equivalent to Newton's, in the sense that either one can be derived from the other, but the Lagrangian form is considerably more flexible with regard to choice of coordinates. The n Cartesian coordinates that describe a system in Newtonian terms are replaced by a set of n generalized coordinates $q_1, q_2, \cdots, q_n$, and Lagrange's equations are equally valid for essentially any choice of $q_1, q_2, \cdots, q_n$. As we have seen on many occasions, this versatility allows one to solve many problems much more easily using Lagrange's formulation. The Lagrangian approach also has the advantage of eliminating the forces of constraint. On the other hand, the Lagrangian method is at a disadvantage when applied to dissipative systems (for example, systems with friction). By now, I hope you feel comfortable with both formulations and are familiar with the advantages and disadvantages of each.

Newtonian mechanics was first expounded by Newton in his *Principia Mathematica*, published in 1687. Lagrange published his formulation in his book *Méchanique Analytique* in 1788. In the early nineteenth century, various physicists, including Lagrange, developed yet a third formulation of mechanics, which was put into a complete form in 1834 by the Irish mathematician William Hamilton (1805–1865) and has come to be called Hamiltonian mechanics. It is this third formulation of mechanics that is the subject of this chapter.

Like the Lagrangian version, Hamiltonian mechanics is equivalent to Newtonian but is considerably more flexible in its choice of coordinates. In fact, in this respect it is even more flexible than the Lagrangian approach. Where the Lagrangian formalism centers on the Lagrangian function $\mathcal{L}$, the Hamiltonian approach is based on the Hamiltonian function $\mathcal{H}$ (which we met briefly in Chapter 7). For most of the systems we shall meet, $\mathcal{H}$ is just the total energy. Thus one advantage of Hamilton's formalism is that it is based on a function, $\mathcal{H}$, which (unlike the Lagrangian $\mathcal{L}$) has a clear physical significance and is frequently conserved. The Hamiltonian approach is also especially well suited for handling other conserved quantities and implementing various approximation schemes. It has been generalized to various different branches of physics; in particular, Hamiltonian mechanics leads very naturally from classical mechanics into quantum mechanics. For all these reasons, Hamilton's formulation plays an important role in many branches of modern physics, including astrophysics, plasma physics, and the design of particle accelerators. Unfortunately, at the level of this book, it is hard to demonstrate many of the advantages of Hamilton's version over Lagrange's, and in this chapter I shall have to ask you to be content with learning the former as just an alternative to the latter — an alternative several of whose advantages I can mention but not explore in depth. If you go on to take a more advanced course in classical mechanics or to study quantum mechanics, you will certainly meet many of this chapter's ideas again.

13.1 The Basic Variables

Because the Hamiltonian version of mechanics is closer to the Lagrangian than to the Newtonian and arises naturally from the Lagrangian, let us start by reviewing the main features of the latter, which centers on the Lagrangian function $\mathcal{L}$. For most systems of interest, $\mathcal{L}$ is just the difference of the kinetic and potential energies, $\mathcal{L} = T - U$, and in this chapter we shall confine attention to systems for which this is the case. The Lagrangian $\mathcal{L}$ is a function of the n generalized coordinates $q_1, \cdots, q_n$, their n time derivatives (or generalized velocities) $\dot{q}_1, \cdots, \dot{q}_n$, and, perhaps, the time:

$$\mathcal{L} = \mathcal{L}(q_1, \cdots, q_n, \dot{q}_1, \cdots, \dot{q}_n, t) = T - U. \tag{13.1}$$

The n coordinates $(q_1, \cdots, q_n)$ specify a position or "configuration" of the system, and can be thought of as defining a point in an n-dimensional **configuration space**. The $2n$ coordinates $(q_1, \cdots, q_n, \dot{q}_1, \cdots, \dot{q}_n)$ define a point in **state space**, and specify a set of initial conditions (at any chosen time t_o) that determine a unique solution of the n second-order differential equations of motion, Lagrange's equations,

$$\frac{\partial \mathcal{L}}{\partial q_i} = \frac{d}{dt} \frac{\partial \mathcal{L}}{\partial \dot{q}_i} \qquad [i = 1, \cdots, n]. \tag{13.2}$$

For each set of initial conditions, these equations of motion determine a unique path or "orbit" through state space.

You may also recall that we defined a **generalized momentum** given by

$$p_i = \frac{\partial \mathcal{L}}{\partial \dot{q}_i}. \tag{13.3}$$

If the coordinates $(q_1, \cdots, q_n)$ are in fact Cartesian coordinates, the generalized momenta p_i are the corresponding components of the usual momenta; in general, p_i is not actually a momentum, but does, as we have seen, play an analogous role. The generalized momentum p_i is also called the **canonical momentum** or the **momentum conjugate** to q_i.

In the Hamiltonian approach, the central role of the Lagrangian $\mathcal{L}$ is taken over by the **Hamiltonian function**, or just **Hamiltonian**, $\mathcal{H}$ defined as

$$\mathcal{H} = \sum_{i=1}^{n} p_i \dot{q}_i - \mathcal{L}. \tag{13.4}$$

The equations of motion, which we shall derive in the next two sections, involve derivatives of $\mathcal{H}$ rather than $\mathcal{L}$ as in Lagrange's equations. We met the Hamiltonian function briefly in Section 7.8, where we proved that, provided the generalized coordinates $(q_1, \cdots, q_n)$ are "natural" (that is, the relation between the q's and the underlying Cartesian coordinates is time independent), $\mathcal{H}$ is just the total energy of the system and is, therefore, familiar and easy to visualize.

There is a second important difference between the Lagrangian and Hamiltonian formalisms. In the former we label the state of the system by the $2n$ coordinates

$$(q_1, \cdots, q_n, \dot{q}_1, \cdots, \dot{q}_n), \qquad \text{[Lagrange]} \tag{13.5}$$

whereas in the latter we shall use the coordinates

$$(q_1, \cdots, q_n, p_1, \cdots, p_n), \qquad \text{[Hamilton]} \tag{13.6}$$

consisting of the n generalized positions and the n generalized *momenta* (instead of the generalized velocities). This choice of coordinates has several advantages, a few of which I shall sketch and some of which you will have to take on faith.

Just as we can regard the $2n$ coordinates (13.5) of the Lagrange approach as defining a point in a $2n$-dimensional state space, so we can regard the $2n$ coordinates (13.6) of the Hamiltonian approach as defining a point in a $2n$-dimensional space, which is usually called **phase space**.[1] Just as the Lagrange equations of motion (13.2) determine a unique path in state space starting from any initial point (13.5), so (we shall see) Hamilton's equations determine a unique path in phase space starting from any initial point (13.6). A succinct way to state some of the advantages of the Hamiltonian formalism is that phase space has certain geometrical properties that make it more convenient than state space.

Like Lagrange's approach, Hamilton's is best suited to systems that are subject to no frictional forces. Accordingly, I shall assume throughout this chapter that all the forces of interest are conservative or can at least be derived from a potential energy function. Although this restriction excludes many interesting mechanical systems, it still includes a huge number of important problems, especially in astrophysics and at the microscopic — atomic and molecular — level.

---

[1] Many authors use the names "state space" and "phase space" interchangeably, but it is convenient to have different names for the different spaces, and I shall reserve "state space" for the space of positions and generalized velocities, and "phase space" for that of positions and generalized momenta.

13.2   Hamilton's Equations for One-Dimensional Systems

To minimize notational complications, I shall first derive Hamilton's equations of motion for a conservative, one-dimensional system with a single "natural" generalized coordinate q. For example, you could think of a simple, plane pendulum, in which case q could be the usual angle ϕ, or a bead on a stationary wire, in which case q could be the horizontal distance x along the wire. For any such system, the Lagrangian is a function of q and $\dot{q}$, that is,

$$\mathcal{L} = \mathcal{L}(q, \dot{q}) = T(q, \dot{q}) - U(q). \tag{13.7}$$

Recall that, in general, the kinetic energy can depend on q as well as $\dot{q}$, whereas, for conservative systems, the potential energy depends only on q. For example, for the simple pendulum (mass m and length L),

$$\mathcal{L} = \mathcal{L}(\phi, \dot{\phi}) = \tfrac{1}{2}mL^2\dot{\phi}^2 - mgL(1 - \cos\phi), \tag{13.8}$$

where, in this case, the kinetic energy involves only $\dot{\phi}$, not ϕ. For a bead sliding on a frictionless wire of variable height $y = f(x)$, we saw in Example 11.1 (page 438) that

$$\mathcal{L} = \mathcal{L}(x, \dot{x}) = T - U = \tfrac{1}{2}m[1 + f'(x)^2]\dot{x}^2 - mgf(x). \tag{13.9}$$

Here the x dependence of the kinetic energy came about when we rewrote the v in $\tfrac{1}{2}mv^2$ in terms of the horizontal distance x. The two examples (13.8) and (13.9) illustrate a general result that we proved in Section 7.8 that the Lagrangian for a conservative system with "natural" coordinates (and in one dimension here) has the general form

$$\mathcal{L} = \mathcal{L}(q, \dot{q}) = T - U = \tfrac{1}{2}A(q)\dot{q}^2 - U(q). \tag{13.10}$$

Notice that, while the kinetic energy can depend on q in a complicated way, through the function $A(q)$, its dependence on $\dot{q}$ is just through the simple quadratic factor $\dot{q}^2$. As you can easily check by writing it down, Lagrange's equation for this Lagrangian is automatically a second-order differential equation for q.

The Hamiltonian is defined by (13.4), which in one dimension reduces to

$$\mathcal{H} = p\dot{q} - \mathcal{L}. \tag{13.11}$$

In the discussion of Section 7.8, I offered some reason why one might perhaps expect a function defined in this way to be an interesting function to study. For now, let us just accept the definition as an inspired suggestion by Hamilton — a suggestion whose merit will appear as we proceed.[2] Given the form (13.10) of $\mathcal{L}$, we can calculate the

---

[2] Actually, the change from $\mathcal{L}$ to $\mathcal{H}$ as the object of primary interest is an example of a mathematical maneuver, called a *Legendre transformation*, which plays an important role in several fields, most notably thermodynamics. For example, the change from the thermodynamic internal

generalized momentum p as

$$p = \frac{\partial \mathcal{L}}{\partial \dot{q}} = A(q)\dot{q} \tag{13.12}$$

so that $p\dot{q} = A(q)\dot{q}^2 = 2T$. Substituting into (13.11), we find that

$$\mathcal{H} = p\dot{q} - \mathcal{L} = 2T - (T - U) = T + U.$$

That is, the Hamiltonian $\mathcal{H}$ for the "natural" system considered here is precisely the total energy — the same result we proved for any "natural" system in any number of dimensions in Section 7.8.

The next step in setting up the Hamiltonian formalism is perhaps the most subtle. In the Lagrangian approach, we think of $\mathcal{L}$ as a function of q and $\dot{q}$, as is indicated explicitly in (13.10). Similarly, (13.12) gives the generalized momentum p in terms of q and $\dot{q}$. However, we can solve (13.12) for $\dot{q}$ in terms of q and p:

$$\dot{q} = p/A(q) = \dot{q}(q, p), \tag{13.13}$$

say. With $\dot{q}$ expressed as a function of q and p, let us now look at the Hamiltonian. Wherever $\dot{q}$ appears in $\mathcal{H}$, we can replace it by $\dot{q}(q, p)$, and $\mathcal{H}$ becomes a function of q and p. In all its horrible detail, (13.11) becomes

$$\mathcal{H}(q, p) = p\,\dot{q}(q, p) - \mathcal{L}(q, \dot{q}(q, p)). \tag{13.14}$$

Our final step is to get Hamilton's equations of motion. To find these, we just evaluate the derivatives of $\mathcal{H}(q, p)$ with respect to q and p. First, using the chain rule, we differentiate (13.14) with respect to q:

$$\frac{\partial \mathcal{H}}{\partial q} = p\frac{\partial \dot{q}}{\partial q} - \left[\frac{\partial \mathcal{L}}{\partial q} + \frac{\partial \mathcal{L}}{\partial \dot{q}}\frac{\partial \dot{q}}{\partial q}\right].$$

Now, in the third term on the right, you will recognize that $\partial \mathcal{L}/\partial \dot{q} = p$. Thus, the first and third terms on the right cancel one another, leaving just

$$\frac{\partial \mathcal{H}}{\partial q} = -\frac{\partial \mathcal{L}}{\partial q} = -\frac{d}{dt}\frac{\partial \mathcal{L}}{\partial \dot{q}} = -\frac{d}{dt}p = -\dot{p} \tag{13.15}$$

where the second equality follows from the Lagrange equation (13.2). This equation gives the time derivative of p (that is, $\dot{p}$) in terms of the Hamiltonian $\mathcal{H}$ and is the first of the two Hamiltonian equations of motion. Before we discuss it, let's derive the second one.

Differentiating (13.14) with respect to p and using the chain rule, we find

$$\frac{\partial \mathcal{H}}{\partial p} = \left[\dot{q} + p\frac{\partial \dot{q}}{\partial p}\right] - \frac{\partial \mathcal{L}}{\partial \dot{q}}\frac{\partial \dot{q}}{\partial p} = \dot{q} \tag{13.16}$$

---

energy U to enthalpy H is a Legendre transformation, closely analogous to Hamilton's change from $\mathcal{L}$ to $\mathcal{H}$.

since the second and third terms in the middle expression cancel exactly. This is the second of the Hamiltonian equations of motion and gives $\dot{q}$ in terms of the Hamiltonian $\mathcal{H}$. Collecting them together (and reordering a bit), we have **Hamilton's equations** for a one-dimensional system:

$$\dot{q} = \frac{\partial \mathcal{H}}{\partial p} \quad \text{and} \quad \dot{p} = -\frac{\partial \mathcal{H}}{\partial q}. \tag{13.17}$$

In the Lagrangian formalism, the equation of motion of a one-dimensional system is a *single second-order* differential equation for q. In the Hamiltonian approach, there are *two first-order* equations, one for q and one for p. Before we extend this result to more general systems or discuss any advantages the new formalism may have, let us look at a couple of simple examples.

EXAMPLE 13.1   A Bead on a Straight Wire

Consider a bead sliding on a frictionless rigid straight wire lying along the x axis, as shown in Figure 13.1. The bead has mass m and is subject to a conservative force, with corresponding potential energy $U(x)$. Write down the Lagrangian and Lagrange's equation of motion. Find the Hamiltonian and Hamilton's equations, and compare the two approaches.

Naturally, we take as our generalized coordinate q the Cartesian x. The Lagrangian is then

$$\mathcal{L}(x, \dot{x}) = T - U = \tfrac{1}{2}m\dot{x}^2 - U(x).$$

The corresponding Lagrange equation is

$$\frac{\partial \mathcal{L}}{\partial x} = \frac{d}{dt}\frac{\partial \mathcal{L}}{\partial \dot{x}} \quad \text{or} \quad -\frac{dU}{dx} = m\ddot{x} \tag{13.18}$$

which is just Newton's $F = ma$, as we would expect.

To set up the Hamiltonian formalism, we must first find the generalized momentum,

$$p = \frac{\partial \mathcal{L}}{\partial \dot{x}} = m\dot{x}.$$

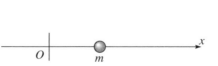

Figure 13.1   A bead of mass m sliding on a frictionless straight wire.

As expected, this is just the conventional mv momentum. This equation can be solved to give $\dot{x} = p/m$, which can then be substituted into the Hamiltonian to give

$$\mathcal{H} = p\dot{x} - \mathcal{L} = \frac{p^2}{m} - \left[\frac{p^2}{2m} - U(x)\right] = \frac{p^2}{2m} + U(x)$$

which you will recognize as the total energy, with the kinetic term $\frac{1}{2}m\dot{x}^2$ rewritten in terms of momentum as $p^2/(2m)$. Finally, the two Hamilton equations (13.17) are

$$\dot{x} = \frac{\partial \mathcal{H}}{\partial p} = \frac{p}{m} \quad \text{and} \quad \dot{p} = -\frac{\partial \mathcal{H}}{\partial x} = -\frac{dU}{dx}.$$

The first of these is, from a Newtonian point of view, just the traditional definition of momentum, and, when we substitute this definition into the second equation, it gives us back $m\ddot{x} = -dU/dx$ again. As had to be the case, Newton, Lagrange, and Hamilton all lead us to the same familiar equation. In this very simple example, neither Lagrange nor Hamilton has any visible advantage over Newton.

EXAMPLE 13.2   Atwood's Machine

Set up the Hamiltonian formalism for the Atwood machine, first shown as Figure 4.15 and shown again here as Figure 13.2. Use the height x of m_1 measured downward as the one generalized coordinate.

The Lagrangian is $\mathcal{L} = T - U$, where, as we saw in Example 7.3 (page 255),

$$T = \tfrac{1}{2}(m_1 + m_2)\dot{x}^2 \quad \text{and} \quad U = -(m_1 - m_2)gx. \quad (13.19)$$

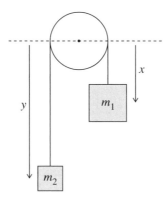

Figure 13.2   An Atwood machine consisting of two masses, m_1 and m_2, suspended by a massless inextensible string that passes over a massless, frictionless pulley. Because the string's length is fixed, the position of the whole system is specified by the distance x of m_1 below any convenient fixed level.

We can calculate the Hamiltonian $\mathcal{H}$ either as $\mathcal{H} = p\dot{x} - \mathcal{L}$ or, what is usually a little quicker (provided it is true[3]), as $\mathcal{H} = T + U$. Either way, we must first find the generalized momentum $p = \partial\mathcal{L}/\partial\dot{x}$ or, since U does not involve $\dot{x}$,

$$p = \frac{\partial T}{\partial \dot{x}} = (m_1 + m_2)\dot{x}.$$

We solve this to give $\dot{x}$ in terms of p as $\dot{x} = p/(m_1 + m_2)$, which we substitute into $\mathcal{H}$ to give $\mathcal{H}$ as a function of x and p:

$$\mathcal{H} = T + U = \frac{p^2}{2(m_1 + m_2)} - (m_1 - m_2)gx. \tag{13.20}$$

We can now write down the two Hamilton equations of motion (13.17) as

$$\dot{x} = \frac{\partial\mathcal{H}}{\partial p} = \frac{p}{m_1 + m_2} \quad \text{and} \quad \dot{p} = -\frac{\partial\mathcal{H}}{\partial x} = (m_1 - m_2)g.$$

Again, the first of these is just a restatement of the definition of the generalized momentum and, when we combine this with the second, we get the well-known result for the acceleration of the Atwood machine,

$$\ddot{x} = \frac{m_1 - m_2}{m_1 + m_2}g.$$

These two examples illustrate several of the general features of the Hamiltonian approach: Our first task is alway to write down the Hamiltonian $\mathcal{H}$ (just as in the Lagrangian approach the first task is to write down $\mathcal{L}$). In the Hamiltonian approach there are usually a couple of extra steps, which are to write down the generalized momentum, to solve the resulting equation for the generalized velocity, and to express $\mathcal{H}$ as a function of position and momentum. Once this is done, one can just turn the handle and crank out Hamilton's equations. In general, there is no guarantee that the resulting equations will be easy to solve, but it is a wonderful property of Hamilton's approach (like Lagrange's) that it provides an almost infallible way to find the equations of motion.

13.3   Hamilton's Equations in Several Dimensions

Our derivation of Hamilton's equations for a one-dimensional system is easily extended to multidimensional systems. The only real problem is that the equations can become badly cluttered with indices, so, to minimize the clutter, I shall use the abbreviation introduced in Section 11.5: The configuration of an n-dimensional system

---

[3] Remember that this second expression is true provided the generalized coordinate is "natural," that is, the relation between the generalized coordinate and the underlying Cartesian coordinates is independent of time — a condition which is certainly met here (Problem 13.4).

is given by n generalized coordinates $q_1, \cdots, q_n$, which I shall represent by a single bold-face $\mathbf{q}$:

$$\mathbf{q} = (q_1, \cdots, q_n).$$

Similarly, the generalized velocities become $\dot{\mathbf{q}} = (\dot{q}_1, \cdots, \dot{q}_n)$, and the generalized momenta are

$$\mathbf{p} = (p_1, \cdots, p_n).$$

It is important to remember that for now a bold-face $\mathbf{q}$ or $\mathbf{p}$ is not necessarily a three-dimensional vector. Rather $\mathbf{q}$ and $\mathbf{p}$ are n-dimensional vectors in the space of generalized positions or generalized momenta.

Hamilton's equations follow directly from Lagrange's equations in their standard form. Thus to prove the former, we have only to assume the truth of the latter. To be specific, however, I shall make the same assumptions that we used in Chapter 7: I shall assume that any constraints are holonomic; that is, the number of degrees of freedom is equal to the number of generalized coordinates. I shall also assume that the nonconstraint forces can be derived from a potential energy function, though it is not essential that they be conservative (that is, the potential energy is allowed to depend on t). The equations that relate the N underlying Cartesian coordinates $\mathbf{r}_1, \cdots, \mathbf{r}_N$ to the n generalized coordinates $q_1, \cdots, q_n$ *can* depend on time; that is, it is not essential that the generalized coordinates be "natural." These assumptions are enough to guarantee that the standard Lagrangian formalism applies, and will let us derive from it the Hamiltonian one. Thus our starting point is that there is a Lagrangian

$$\mathcal{L} = \mathcal{L}(\mathbf{q}, \dot{\mathbf{q}}, t) = T - U$$

and that the evolution of our system is governed by the n Lagrange equations,

$$\frac{\partial \mathcal{L}}{\partial q_i} = \frac{d}{dt} \frac{\partial \mathcal{L}}{\partial \dot{q}_i} \qquad [i = 1, \cdots, n]. \tag{13.21}$$

We shall define the Hamiltonian function as in (13.4),

$$\mathcal{H} = \sum_{i=1}^{n} p_i \dot{q}_i - \mathcal{L}, \tag{13.22}$$

where the generalized momenta are defined by

$$p_i = \frac{\partial \mathcal{L}(\mathbf{q}, \dot{\mathbf{q}}, t)}{\partial \dot{q}_i} \qquad [i = 1, \cdots, n] \tag{13.23}$$

as in (13.3). Just as in the one-dimensional case, our next step is to express the Hamiltonian as a function of the $2n$ variables $\mathbf{q}$ and $\mathbf{p}$. To this end, note that we can view the equations (13.23) as n simultaneous equations for the n generalized velocities $\dot{\mathbf{q}}$. We can in principle solve these equations to give the generalized velocities in terms of the variables $\mathbf{p}$, $\mathbf{q}$, and t:

$$\dot{q}_i = \dot{q}_i(q_1, \cdots, q_n, p_1, \cdots, p_n, t) \qquad [i = 1, \cdots, n]$$

or, more succinctly

$$\dot{\mathbf{q}} = \dot{\mathbf{q}}(\mathbf{q}, \mathbf{p}, t).$$

We can now eliminate the generalized velocities from our definition of the Hamiltonian to give (again in agonizing detail)

$$\mathcal{H} = \mathcal{H}(\mathbf{q}, \mathbf{p}, t) = \sum_{i=1}^{n} p_i \dot{q}_i(\mathbf{q}, \mathbf{p}, t) - \mathcal{L}(\mathbf{q}, \dot{\mathbf{q}}(\mathbf{q}, \mathbf{p}, t), t). \tag{13.24}$$

The derivation of Hamilton's equations now proceeds very much as in one dimension, and I shall leave you to fill in the details (Problem 13.15). Following the same steps as led from (13.14) to (13.17), we differentiate $\mathcal{H}$ with respect to q_i and then p_i, and this leads to **Hamilton's equations**:

$$\dot{q}_i = \frac{\partial \mathcal{H}}{\partial p_i} \quad \text{and} \quad \dot{p}_i = -\frac{\partial \mathcal{H}}{\partial q_i} \quad [i = 1, \cdots, n]. \tag{13.25}$$

Notice that for a system with n degrees of freedom, the Hamiltonian approach gives us $2n$ first-order differential equations, instead of the n second-order equations of Lagrange.

Before we discuss an example of Hamilton's equations, there is one more derivative of $\mathcal{H}$ to consider, its derivative with respect to time. This is actually quite subtle. The function $\mathcal{H}(\mathbf{q}, \mathbf{p}, t)$ could vary with time for two reasons: First, as the motion proceeds, the $2n$ coordinates $(\mathbf{q}, \mathbf{p})$ vary, and this could cause $\mathcal{H}(\mathbf{q}, \mathbf{p}, t)$ to change; in addition, $\mathcal{H}(\mathbf{q}, \mathbf{p}, t)$ may have an explicit time dependence, as indicated by the final argument t, and this also can make $\mathcal{H}$ vary with time. Mathematically, this means that $d\mathcal{H}/dt$ contains $2n + 1$ terms, as follows:

$$\frac{d\mathcal{H}}{dt} = \sum_{i=1}^{n} \left[\frac{\partial \mathcal{H}}{\partial q_i} \dot{q}_i + \frac{\partial \mathcal{H}}{\partial p_i} \dot{p}_i \right] + \frac{\partial \mathcal{H}}{\partial t}. \tag{13.26}$$

It is important to understand the difference between the two derivatives of $\mathcal{H}$ in this equation. The derivative on the left, $d\mathcal{H}/dt$ (sometimes called the total derivative), is the actual rate of change of $\mathcal{H}$ as the motion proceeds, with all the coordinates $q_1, \cdots, q_n, p_1, \cdots, p_n$ changing as t advances. That on the right, $\partial \mathcal{H}/\partial t$, is the partial derivative, which is the rate of change of $\mathcal{H}$ if we vary its last argument t holding all the other arguments fixed. In particular, if $\mathcal{H}$ does not depend explicitly on t, this partial derivative will be zero. Now it is easy to see that, because of Hamilton's equations (13.25), each pair of terms in the sum of (13.26) is exactly zero, so that we have the simple result

$$\frac{d\mathcal{H}}{dt} = \frac{\partial \mathcal{H}}{\partial t}.$$

That is, $\mathcal{H}$ varies with time only to the extent that it is explicitly time dependent. In particular, if $\mathcal{H}$ does not depend explicitly on t (as is often the case), then $\mathcal{H}$ is a constant in time; that is, the quantity $\mathcal{H}$ is conserved. This is the same result we derived in Section 7.8.[4]

In Section 7.8, we proved a second result regarding time dependence: If the relation of the generalized coordinates $q_1, \cdots, q_n$ to the underlying rectangular coordinates is independent of t (that is, our generalized coordinates are "natural"), then the Hamiltonian $\mathcal{H}$ is just the total energy, $\mathcal{H} = T + U$. In the remainder of this chapter, I shall consider only the case that the generalized coordinates *are* "natural" *and* that $\mathcal{H}$ is *not* explicitly time-dependent. Thus it will be true from now on that $\mathcal{H}$ is the total energy and that total energy is conserved.

Let us now work out an example of the Hamiltonian formalism for a system in two spatial dimensions. Unfortunately, like all reasonably simple examples, this does not exhibit any significant advantages of the Hamiltonian over the Lagrangian approach; rather, in this example, the Hamiltonian approach is just an alternative route to the same final equation of motion.

**EXAMPLE 13.3 Hamilton's Equations for a Particle
in a Central Force Field**

Set up Hamilton's equations for a particle of mass m subject to a conservative central force field with potential energy $U(r)$, using as generalized coordinates the usual polar coordinates r and ϕ.

By conservation of angular momentum, we know that the motion is confined to a fixed plane, in which we can define the polar coordinates r and ϕ. The kinetic energy is given in terms of these generalized coordinates by the familiar expression

$$T = \tfrac{1}{2}m(\dot{r}^2 + r^2\dot{\phi}^2). \tag{13.27}$$

Since the equations relating (r, ϕ) to (x, y) are time-independent, we know that $\mathcal{H} = T + U$, which we must express in terms of r and ϕ and the corresponding generalized momenta p_r and p_ϕ. These latter are defined by the relation[5] $p_i = \partial \mathcal{L}/\partial \dot{q}_i = \partial T/\partial \dot{q}_i$, which gives

$$p_r = \partial T/\partial \dot{r} = m\dot{r} \quad \text{and} \quad p_\phi = \partial T/\partial \dot{\phi} = mr^2\dot{\phi}. \tag{13.28}$$

The momentum p_r conjugate to r is just the radial component of the ordinary momentum $m\mathbf{v}$, but, as we first saw in Section 7.1 [Equation (7.26)], the

---

[4] We proved there, in Equations (7.89) and (7.90), that $\mathcal{H}$ is conserved if and only if $\mathcal{L}$ does not depend explicitly on the time. These two conditions ($\mathcal{H}$ not explicitly time dependent or $\mathcal{L}$ likewise) are equivalent, for, as you can easily check, $\partial \mathcal{H}/\partial t = -\partial \mathcal{L}/\partial t$. See Problem 13.16.

[5] In any problem where the potential energy $U = U(\mathbf{q})$ is independent of the velocities $\dot{\mathbf{q}}$ (as we are certainly assuming here), there is this small simplification that we can replace $\mathcal{L}$ by T in the definition of p_i.

momentum p_ϕ conjugate to ϕ is the *angular* momentum. We must next solve the two equations (13.28) to give the velocities $\dot{r}$ and $\dot{\phi}$ in terms of the momenta p_r and p_ϕ:

$$\dot{r} = \frac{p_r}{m} \qquad \text{and} \qquad \dot{\phi} = \frac{p_\phi}{mr^2}.$$

We can now substitute these into (13.27) and we arrive at the Hamiltonian, expressed as a function of the proper variables,

$$\mathcal{H} = T + U = \frac{1}{2m}\left(p_r^2 + \frac{p_\phi^2}{r^2}\right) + U(r). \tag{13.29}$$

We can now write down the four Hamilton equations (13.25). The two radial equations are

$$\dot{r} = \frac{\partial \mathcal{H}}{\partial p_r} = \frac{p_r}{m} \qquad \text{and} \qquad \dot{p}_r = -\frac{\partial \mathcal{H}}{\partial r} = \frac{p_\phi^2}{mr^3} - \frac{dU}{dr}. \tag{13.30}$$

The first of these reproduces the definition of the radial momentum. If we substitute the first into the second, we obtain the familiar result that $m\ddot{r}$ is the sum of the actual radial force $(-dU/dr)$ plus the centrifugal force p_ϕ^2/mr^3. [See Equations (8.24) and (8.26).] The two ϕ equations are

$$\dot{\phi} = \frac{\partial \mathcal{H}}{\partial p_\phi} = \frac{p_\phi}{mr^2} \qquad \text{and} \qquad \dot{p}_\phi = -\frac{\partial \mathcal{H}}{\partial \phi} = 0. \tag{13.31}$$

The first of these reproduces the definition of p_ϕ. The second tells us what we already knew, that the angular momentum is conserved. As in the previous two examples, we see that the Hamiltonian formalism provides an alternative route to the same final equations of motion as we could find using either the Newtonian or Lagrangian approaches.

This example illustrates the general procedure to be followed in setting up Hamilton's equations for any given system:

1. Choose suitable generalized coordinates, $q_1, \cdots, q_n$.
2. Write down the kinetic and potential energies, T and U, in terms of the q's and $\dot{q}$'s.
3. Find the generalized momenta $p_1, \cdots, p_n$. (We are now assuming our system is conservative, so U is independent of $\dot{q}_i$ and we can use $p_i = \partial T/\partial \dot{q}_i$. In general, one must use $p_i = \partial \mathcal{L}/\partial \dot{q}_i$.)
4. Solve for the $\dot{q}$'s in terms of the p's and q's.
5. Write down the Hamiltonian $\mathcal{H}$ as a function of the p's and q's. [Provided our coordinates are "natural" (relation between generalized coordinates and underlying Cartesians is independent of time), $\mathcal{H}$ is just the total energy $\mathcal{H} = T + U$, but when in doubt, use $\mathcal{H} = \sum p_i \dot{q}_i - \mathcal{L}$. See Problems 13.11 and 13.12.]
6. Write down Hamilton's equations (13.25).

If you look back over the last example, you will see that the solution followed all of these six steps, and the same will be true of all later examples and problems. Before we do another example, let us compare these six steps with the corresponding steps in the Lagrangian approach. To set up Lagrange's equations, we follow the same first two steps (choose generalized coordinates and write down T and U). Steps (3) and (4) are unnecessary, since we don't have to know the generalized momenta, nor to eliminate the $\dot{q}$'s in favor of the p's. Finally, one must carry out the analogs of (5) and (6); namely, write down the Lagrangian and Lagrange's equations. Evidently, setting up the Hamiltonian approach involves two small extra steps [Steps (3) and (4) above] as compared to the Lagrangian. Although both steps are usually quite straightforward, this is undeniably a small disadvantage of Hamilton's formalism. Now, here is another example.

EXAMPLE 13.4   Hamilton's Equations for a Mass on a Cone

Consider a mass m which is constrained to move on the frictionless surface of a vertical cone $\rho = cz$ (in cylindrical polar coordinates ρ, ϕ, z with $z > 0$) in a uniform gravitational field g vertically down (Figure 13.3). Set up Hamilton's equations using z and ϕ as generalized coordinates. Show that for any given solution there are maximum and minimum heights $z_{\max}$ and $z_{\min}$ between which the motion is confined. Use this result to describe the motion of the mass on the cone. Show that for any given value of $z > 0$ there is a solution in which the mass moves in a circular path at fixed height z.

Our generalized coordinates are z and ϕ, with ρ determined by the constraint that the mass remain on the cone, $\rho = cz$. The kinetic energy is therefore

$$T = \tfrac{1}{2}m\left[\dot{\rho}^2 + (\rho\dot{\phi})^2 + \dot{z}^2\right] = \tfrac{1}{2}m\left[(c^2 + 1)\dot{z}^2 + (cz\dot{\phi})^2\right].$$

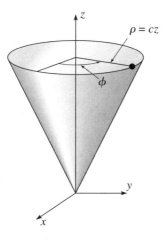

Figure 13.3   A mass m is constrained to move on the surface of the cone shown. For clarity, the cone is shown truncated at the height of the mass, although it actually continues upward indefinitely.

The potential energy is of course $U = mgz$. The generalized momenta are

$$p_z = \frac{\partial T}{\partial \dot{z}} = m(c^2 + 1)\dot{z} \qquad \text{and} \qquad p_\phi = \frac{\partial T}{\partial \dot{\phi}} = mc^2z^2\dot{\phi}. \quad (13.32)$$

These are trivially solved for $\dot{z}$ and $\dot{\phi}$, and we can write down the Hamiltonian,

$$\mathcal{H} = T + U = \frac{1}{2m}\left[\frac{p_z^2}{(c^2 + 1)} + \frac{p_\phi^2}{c^2z^2}\right] + mgz. \quad (13.33)$$

Hamilton's equations are now easily found: The two z equations are

$$\dot{z} = \frac{\partial \mathcal{H}}{\partial p_z} = \frac{p_z}{m(c^2 + 1)} \qquad \text{and} \qquad \dot{p}_z = -\frac{\partial \mathcal{H}}{\partial z} = \frac{p_\phi^2}{mc^2z^3} - mg. \quad (13.34)$$

The two ϕ equations are

$$\dot{\phi} = \frac{\partial \mathcal{H}}{\partial p_\phi} = \frac{p_\phi}{mc^2z^2} \qquad \text{and} \qquad \dot{p}_\phi = -\frac{\partial \mathcal{H}}{\partial \phi} = 0. \quad (13.35)$$

The last of these tells us, what we could well have guessed, that p_ϕ, which is just the z component of angular momentum, is constant.

The easiest way to see that, for any given solution, z is confined between two bounds, $z_{\min}$ and $z_{\max}$, is to remember that the Hamiltonian function (13.33) is equal to the total energy, and that energy is conserved. Thus, for any given solution, (13.33) is equal to a fixed constant E. Now, the function $\mathcal{H}$ in (13.33) is the sum of three positive terms, and as $z \to \infty$ the last term tends to infinity. Since $\mathcal{H}$ must equal the fixed constant E, there must be a $z_{\max}$ which z cannot exceed. In the same way, the second term in (13.33) approaches infinity as $z \to 0$; so there has to be a $z_{\min} > 0$ below which z cannot go. In particular, this means that the mass can never fall all the way into the bottom of the cone at $z = 0$.[6] The motion of the mass on the cone is now easy to describe. It moves around the z axis with constant angular momentum $p_\phi = mc^2z^2\dot{\phi}$. Since p_ϕ is constant, the angular velocity $\dot{\phi}$ varies — increasing as z gets smaller and decreasing as z gets bigger. At the same time, the mass's height z oscillates up and down between $z_{\min}$ and $z_{\max}$. (See Problems 13.14 and 13.17 for more details.)

To investigate the possibility of a solution in which the mass stays at a fixed height z, notice that this requires that $\dot{z} = 0$ for all time. This in turn requires that $p_z = 0$ for all time, and hence $\dot{p}_z = 0$. From the second of the z equations (13.34), we see that $\dot{p}_z = 0$ if and only if

$$p_\phi = \pm\sqrt{m^2c^2gz^3}. \quad (13.36)$$

---

[6] Two comments: It is easy to see that the second term in (13.33) is related to the centrifugal force; thus, we can say that the mass is held away from the bottom of the cone by the centrifugal force. Second, the one exception to this statement is if the angular momentum $p_\phi = 0$; in this case, the mass moves up and down the cone in the radial direction (ϕ constant) and *will* eventually fall to the bottom.

If, for any chosen initial height z, we launch the mass with $p_z = 0$ and p_ϕ equal to one of these two values (going either clockwise or counterclockwise), then since $\dot{p}_z = 0$, p_z and hence $\dot{z}$ both remain zero, and the mass continues to move at its initial height around a horizontal circle.

13.4 Ignorable Coordinates

So far we have set up Hamilton's formalism and have seen that it is valid whenever Lagrange's is. The former enjoys almost all the advantages and disadvantages of the latter, when either is compared with the Newtonian formalism. But it is not yet clear that there are any significant advantages to using Hamilton instead of Lagrange, or vice versa. As I have already mentioned, in more advanced theoretical work, Hamilton's approach has some distinct advantages, and, I shall try to give some feeling for a few of these advantages in the next four sections.

In Chapter 7, we saw that if the Lagrangian $\mathcal{L}$ happens to be independent of a coordinate q_i, then the corresponding generalized momentum p_i is constant. [This followed at once from the Lagrange equation $\partial\mathcal{L}/\partial q_i = (d/dt)\partial\mathcal{L}/\partial\dot{q}_i$, which can be rewritten as $\partial\mathcal{L}/\partial q_i = \dot{p}_i$. Therefore, if $\partial\mathcal{L}/\partial q_i = 0$, it immediately follows that $\dot{p}_i = 0$.] When this happens, we say that the coordinate q_i is **ignorable**.

In the same way, if the Hamiltonian $\mathcal{H}$ is independent of q_i, it follows from the Hamilton equation $\dot{p}_i = -\partial\mathcal{H}/\partial q_i$ that its conjugate momentum p_i is a constant. We saw this in Equation (13.35) of the last example, where $\mathcal{H}$ was independent of ϕ, and the conjugate momentum p_ϕ (actually the angular momentum) was constant. The results of this and the last paragraph are in fact the *same* result, since it is easy to prove that $\partial\mathcal{L}/\partial q_i = -\partial\mathcal{H}/\partial q_i$ (Problem 13.22). Thus $\mathcal{L}$ is independent of q_i if and only if $\mathcal{H}$ is independent of q_i. If a coordinate q_i is ignorable in the Lagrangian then it is also ignorable in the Hamiltonian and vice versa.

It is nevertheless true that the Hamiltonian formalism is more convenient for handling ignorable coordinates. To see this, let us consider a system with just two degrees of freedom and suppose that the Hamiltonian is independent of q_2. This means that the Hamiltonian depends on only three variables,

$$\mathcal{H} = \mathcal{H}(q_1, p_1, p_2). \tag{13.37}$$

For example, the Hamiltonian (13.29) for the central force problem has this property, being independent of the coordinate ϕ. This means that $p_2 = k$, a constant that is determined by the initial conditions. Substitution of this constant into the Hamiltonian leaves

$$\mathcal{H} = \mathcal{H}(q_1, p_1, k)$$

which is a function of just the two variables q_1 and p_1, and the solution of the motion is reduced to a one-dimensional problem with this effectively one-dimensional Hamiltonian. More generally, if a system with n degrees of freedom has an ignorable coordinate q_i, then solution of the motion in the Hamiltonian framework is exactly equivalent to a problem with $(n - 1)$ degrees of freedom in which q_i and p_i can be entirely ignored. If there are several ignorable coordinates, the problem is correspondingly further simplified.

In the Lagrangian formalism, it is of course true that if q_i is ignorable, then p_i is constant, but this does not lead to the same elegant simplification: Supposing again that the system has two degrees of freedom and that q_2 is ignorable, then corresponding to (13.37) we would have

$$\mathcal{L} = \mathcal{L}(q_1, \dot{q}_1, \dot{q}_2).$$

Now, even though q_2 is ignorable and p_2 is a constant, it is not necessarily true that $\dot{q}_2$ is constant. Thus the Lagrangian does not reduce cleanly to a one-dimensional function that depends only on q_1 and $\dot{q}_1$. [For example, in the central-force problem, the Lagrangian has the form $\mathcal{L}(r, \dot{r}, \dot{\phi})$, but, even though ϕ is ignorable, $\dot{\phi}$ still varies as the motion proceeds, and the problem does not automatically reduce to a problem with one degree of freedom.[7]]

13.5   Lagrange's Equations vs. Hamilton's Equations

For a system with n degrees of freedom, the Lagrangian formalism presents us with n second-order differential equations for the n variables $q_1, \cdots, q_n$. For the same system, the Hamiltonian formalism gives us $2n$ *first-order* differential equations for the $2n$ variables $q_1, \cdots, q_n, p_1, \cdots, p_n$. That Hamilton could recast n second-order equations into $2n$ first-order equations is no particular surprise. In fact it is easy to see that *any* set of n second-order equations can be recast in this way: For simplicity, let us consider the case that there is just one degree of freedom, so that Lagrange's approach gives just one second-order equation for the one coordinate q. This equation can be written as

$$f(\ddot{q}, \dot{q}, q) = 0 \tag{13.38}$$

where f is some function of its three arguments. Let us now define a second variable

$$s = \dot{q}. \tag{13.39}$$

In terms of this second variable, $\ddot{q} = \dot{s}$ and our original differential equation (13.38) becomes

$$f(\dot{s}, s, q) = 0. \tag{13.40}$$

We have now replaced the one second-order equation (13.38) for q with the two first-order equations (13.39) and (13.40) for q and s.

---

[7] In this particular case, the difficulty is fairly easy to circumvent. In the discussion of Section 8.4, we wrote the Lagrange radial equation (8.24) in terms of the variable $\dot{\phi}$ and then rewrote it eliminating $\dot{\phi}$ in favor of $p_\phi = \ell$, which is constant.

Evidently, the fact that Hamilton's equations are first-order where Lagrange's are second-order is no particular improvement. However, the specific form of Hamilton's equations is in fact a big improvement. To see this, let us rewrite Hamilton's equations in a more streamlined form. First we can rewrite the first n of Equations (13.25) as

$$\dot{q}_i = \frac{\partial \mathcal{H}}{\partial p_i} = f_i(\mathbf{q}, \mathbf{p}) \qquad [i = 1, \cdots, n] \qquad (13.41)$$

where each f_i is some function of $\mathbf{q}$ and $\mathbf{p}$, and we can combine these n equations into a single n-dimensional equation

$$\dot{\mathbf{q}} = \mathbf{f}(\mathbf{q}, \mathbf{p}) \qquad (13.42)$$

where the boldface $\mathbf{f}$ stands for the vector comprised of the n functions $f_i = \partial \mathcal{H} / \partial p_i$. In the same way we can rewrite the n equations for the $\dot{p}_i$ in a similar form:

$$\dot{\mathbf{p}} = \mathbf{g}(\mathbf{q}, \mathbf{p}) \qquad (13.43)$$

where the boldface $\mathbf{g}$ stands for the vector comprised of the n functions $g_i = -\partial \mathcal{H} / \partial q_i$. Finally, we can introduce a $2n$-dimensional vector

$$\mathbf{z} = (\mathbf{q}, \mathbf{p}) = (q_1, \cdots, q_n, p_1, \cdots, p_n). \qquad (13.44)$$

This **phase-space vector** or **phase point z** comprises all of the generalized coordinates *and* all of their conjugate momenta. Each value of $\mathbf{z}$ labels a unique point in phase space and identifies a unique set of initial conditions for our system. With this new notation, we can combine the two equations (13.42) and (13.43) into a single grand, $2n$-component equation of motion

$$\dot{\mathbf{z}} = \mathbf{h}(\mathbf{z}) \qquad (13.45)$$

where the function $\mathbf{h}$ is a vector comprising the $2n$ functions $f_1, \cdots, f_n$ and $g_1, \cdots, g_n$ of Equations (13.42) and (13.43).

Equation (13.45) expresses Hamilton's equations as a first-order differential equation for the phase-space vector $\mathbf{z}$. Furthermore, it is a first-order equation with the especially simple form:[8]

(first derivative of $\mathbf{z}$) = (function of $\mathbf{z}$).

A large part of the mathematical literature on differential equations is devoted to equations with this standard form, and it is a distinct advantage of the Hamiltonian formalism that Hamilton's equations — unlike Lagrange's — are automatically of this form.

Our combining of the n position coordinates $\mathbf{q}$ with the n momenta $\mathbf{p}$ to form a single phase space vector $\mathbf{z} = (\mathbf{q}, \mathbf{p})$ suggests a certain equality between the position and momentum coordinates in phase-space, and this suggestion proves correct. We

---

[8] If the Hamiltonian was explicitly time dependent, then (13.45) would take the (still very simple) form $\dot{\mathbf{z}} = \mathbf{h}(\mathbf{z}, t)$. The form (13.45) without any explicit time dependence is said to be *autonomous*.

have known since Chapter 7 that one of the strengths of the Lagrangian formalism is its great flexibility with respect to coordinates: Any set of generalized coordinates $\mathbf{q} = (q_1, \cdots, q_n)$ can be replaced by a second set $\mathbf{Q} = (Q_1, \cdots, Q_n)$, where each of the new Q_i is a function of the original $(q_1, \cdots, q_n)$,

$$\mathbf{Q} = \mathbf{Q}(\mathbf{q}), \tag{13.46}$$

and Lagrange's equations will be just as valid with respect to the new coordinates $\mathbf{Q}$ as they were with respect to the old $\mathbf{q}$.[9] We can paraphrase this to say that Lagrange's equations are unchanged (or invariant) under any coordinate change in the n-dimensional configuration space defined by $\mathbf{q} = (q_1, \cdots, q_n)$. The Hamiltonian formalism shares this same flexibility — Hamilton's equations are invariant under any coordinate change (13.46) in configuration space. However, the Hamiltonian formalism actually has a much greater flexibility and allows for certain coordinate changes in the $2n$-dimensional phase space. We can consider changes of coordinates of the form

$$\mathbf{Q} = \mathbf{Q}(\mathbf{q}, \mathbf{p}) \qquad \text{and} \qquad \mathbf{P} = \mathbf{P}(\mathbf{q}, \mathbf{p}), \tag{13.47}$$

that is, coordinate changes in which both the q's and the p's are intermingled. If the equations (13.47) satisfy certain conditions, this change of coordinates is called a **canonical transformation**, and it turns out that Hamilton's equations are invariant under these canonical transformations. Any further discussion of canonical transformations would carry us beyond the scope of this book, but you should be aware of their existence and that they are one of the properties that make the Hamiltonian approach such a powerful theoretical tool.[10] Problems 13.24 and 13.25 offer two examples of canonical transformations.

13.6 Phase-Space Orbits

One can view the phase-space vector $\mathbf{z} = (\mathbf{q}, \mathbf{p})$ of (13.44) as defining the system's "position" in phase space. Any point $\mathbf{z}_0$ defines a possible initial condition (at any chosen time t_0), and Hamilton's equations (13.45) define a unique **phase-space orbit** or **trajectory** which starts from $\mathbf{z}_0$ at t_0 and which the system follows as time progresses. Since phase space has $2n$ dimensions, the visualization of these orbits presents some challenges unless $n = 1$. For example, for a single unconstrained particle in three dimensions, $n = 3$, and the phase space is six-dimensional — not something that most of us can visualize easily. There are various techniques, such as the Poincaré section

---

[9] Of course, this statement has to be qualified a little: The coordinates $\mathbf{Q}$ must be "reasonable" in the sense that each set $\mathbf{Q}$ determines a unique set $\mathbf{q}$ and vice versa, and the function $\mathbf{Q}(\mathbf{q})$ has to be suitably differentiable.

[10] I should emphasize that there is no corresponding transformation in Lagrangian mechanics, which operates in the state space defined by the $2n$-dimensional vector $(\mathbf{q}, \dot{\mathbf{q}})$. Since $\dot{\mathbf{q}}$ is defined as the time derivative of $\mathbf{q}$, there is no analog to (13.47) in which the q's and $\dot{q}$'s get intermingled.

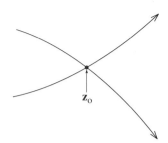

Figure 13.4   One can imagine two orbits passing through the same point $\mathbf{z}_0$ in phase space. However, Hamilton's equations guarantee that for any given point $\mathbf{z}_0$, there is a unique orbit passing through $\mathbf{z}_0$, so the two orbits must in fact be the same.

described in Section 12.8, for viewing phase-space orbits in a subspace of fewer dimensions than the full phase space, but here I shall give just two examples of systems for which $n = 1$, and the phase space is therefore only two-dimensional.

Before we look at these examples, there is an important property of phase-space orbits that deserves mention right away: It is easy to see that no two different phase-space orbits can pass through the same point in phase space; that is, no two phase-space orbits can cross one another. For suppose two orbits pass through one and the same point $\mathbf{z}_0$, as in Figure 13.4.

Now, from Hamilton's equations (13.45), it follows that for any point $\mathbf{z}_0$ there can be only one distinct orbit passing through $\mathbf{z}_0$. Therefore the two orbits passing through $\mathbf{z}_0$ have to be the same. Notice that this result excludes different orbits from passing through the same point *even at different times*: If one orbit passes through $\mathbf{z}_0$ today, then no different orbit can pass through $\mathbf{z}_0$ today, yesterday, or tomorrow.[11] This result — that no two phase-space orbits can cross — places severe restrictions on the way in which these orbits are traced out in phase space. It has important consequences in, for example, the analysis of chaotic motion of Hamiltonian systems.

EXAMPLE 13.5   A One-Dimensional Harmonic Oscillator

Set up Hamilton's equations for a one-dimensional simple-harmonic oscillator with mass m and force constant k, and describe the possible orbits in the phase space defined by the coordinates (x, p).

The kinetic energy is $T = \frac{1}{2}m\dot{x}^2$ and the potential energy is $U = \frac{1}{2}kx^2 = \frac{1}{2}m\omega^2 x^2$, if we introduce the natural frequency $\omega = \sqrt{k/m}$. The generalized momentum is $p = \partial T/\partial \dot{x} = m\dot{x}$, and the Hamiltonian (written as a function of x and p) is

$$\mathcal{H} = T + U = \frac{p^2}{2m} + \frac{1}{2}m\omega^2 x^2. \qquad (13.48)$$

---

[11] This is because our Hamiltonian is time-independent. If $\mathcal{H}$ is explicitly time-dependent, then we can assert only that no two orbits can pass through one point at the same time.

Thus, Hamilton's equations give

$$\dot{x} = \frac{\partial \mathcal{H}}{\partial p} = \frac{p}{m} \quad \text{and} \quad \dot{p} = -\frac{\partial \mathcal{H}}{\partial x} = -m\omega^2 x.$$

The simplest way to solve these two equations is to eliminate p and get the familiar second-order equation $\ddot{x} = -\omega^2 x$, with the equally familiar solution

$$x = A\cos(\omega t - \delta) \quad \text{and hence} \quad p = m\dot{x} = -m\omega A \sin(\omega t - \delta). \quad (13.49)$$

Phase space for the one-dimensional oscillator is the two-dimensional space with coordinates (x, p). In this space, the solution (13.49) is the parametric form for an ellipse, traced in the clockwise direction, as in Figure 13.5, which shows two phase-space orbits for the cases that the oscillator started out from rest at $x = A$ (solid curve) and $x = A/2$ (dashed curve). That the orbits *have* to be ellipses follows from conservation of energy: The total energy is given by the Hamiltonian (13.48), whose initial value (for the solid curve with $x = A$ and $p = 0$) is $\frac{1}{2}m\omega^2 A^2$. Thus conservation of energy implies that

$$\frac{p^2}{2m} + \frac{1}{2}m\omega^2 x^2 = \frac{1}{2}m\omega^2 A^2$$

or

$$\frac{x^2}{A^2} + \frac{p^2}{(m\omega A)^2} = 1. \quad (13.50)$$

This is the equation for an ellipse with semimajor and semiminor axes A and $m\omega A$, in agreement with (13.49).

It is perhaps worth following one of the phase-space orbits of Figure 13.5 in detail. For the solid curve, the motion starts from rest with x at its maximum, at the point $x = A$, $p = 0$, shown as a dot in the figure. The restoring force causes m to accelerate back toward $x = 0$, so that x gets smaller while p becomes

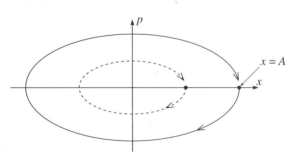

Figure 13.5   The phase space for a one-dimensional harmonic oscillator is a plane, with axes labelled by x (the position) and p (the momentum), in which the point representing the state of motion traces a clockwise ellipse. There is a unique orbit through each phase point (x, p). The outer orbit (solid curve) started from rest with $x = A$ and $p = 0$; the inner one started from $x = A/2$ and $p = 0$.

increasingly negative. By the time x has reached 0, p has reached its largest negative value $p = -m\omega A$. The oscillator now overshoots the equilibrium point, so x becomes increasingly negative, while p is still negative but getting less so. By the time x reaches $-A$, p is again zero, and so on, until the oscillator gets back to its starting point $x = A$, $p = 0$ and the cycle starts over again. Notice that, in agreement with our general argument, the two orbits shown in Figure 13.5 do not cross one another. In fact, it is easy to see that no two ellipses with the form (13.50) can have any point in common unless they have the same value of A (in which case they are the *same* ellipse).

In this simple one-dimensional case, the phase-space plot doesn't tell us anything we couldn't learn from the simple solution $x = A \cos(\omega t - \delta)$, but I hope you will agree that it does show some details of the motion rather clearly.

If you have read Chapter 12, you will recognize that a phase-space orbit is closely related to the state-space orbit described in Section 12.7. The only difference is that the former traces the system's evolution in the (x, p) plane whereas the latter uses the $(x, \dot{x})$ plane. In the present case there is almost no difference, since for one-dimensional motion along the x axis, p is proportional to $\dot{x}$ (specifically, $p = m\dot{x}$). Thus the two kinds of plot are identical except that the former is stretched by a factor of m in the vertical direction. Nevertheless, in the general $2n$-dimensional case, the spaces defined by $(\mathbf{q}, \mathbf{p})$ and by $(\mathbf{q}, \dot{\mathbf{q}})$ can be very different. As we shall see in the next section, the phase-space plot has some elegant properties not shared by the state-space one.

It often happens that one needs to follow not just one orbit but several different orbits through phase space. For example, in the study of chaos we saw that it is of great interest to follow the evolution of two identical systems that are launched with slightly different initial conditions. If the motion is nonchaotic, then the two systems remain close together in phase space, but if the motion is chaotic they move apart so rapidly that their detailed motion is effectively unpredictable. In the next example, we look at four neighboring phase-space orbits of a particle falling under the influence of gravity.

EXAMPLE 13.6   A Falling Body

Set up the Hamiltonian formalism for a mass m constrained to move in a vertical line, subject only to the force of gravity. Use the coordinate x, measured downward from a convenient origin, and its conjugate momentum. Describe the phase-space orbits and in particular sketch the orbits from time 0 to a later time t for the following four different initial conditions at $t = 0$:

(a) $x_0 = p_0 = 0$ (that is, the mass is released from rest at $x = 0$);
(b) $x_0 = X$, but $p_0 = 0$ (the mass is released from rest at $x = X$);
(c) $x_0 = X$ and $p_0 = P$ (the mass is thrown from $x = X$ with initial momentum $p = P$);

(d) $x_o = 0$ and $p_o = P$ (the mass is thrown from $x = 0$ with initial momentum $p = P$).

The kinetic and potential energies are $T = \frac{1}{2}m\dot{x}^2$ and $U = -mgx$. (Remember that x is measured downward.) The conjugate momentum is, of course, $p = m\dot{x}$ and the Hamiltonian is

$$\mathcal{H} = T + U = \frac{p^2}{2m} - mgx. \tag{13.51}$$

To find the shape of the phase-space orbits, we don't have to solve the equations of motion, since conservation of energy requires that they satisfy $\mathcal{H} = $ const. For the Hamiltonian (13.51), this defines a parabola with the form $x = kp^2 + $ const, with its symmetry axis on the x axis.

To draw the four orbits asked for, it is helpful to solve the equations of motion

$$\dot{x} = \frac{\partial \mathcal{H}}{\partial p} = \frac{p}{m} \quad \text{and} \quad \dot{p} = -\frac{\partial \mathcal{H}}{\partial x} = mg.$$

The second of these gives

$$p = p_o + mgt$$

and the first then gives

$$x = x_o + \frac{p_o}{m}t + \frac{1}{2}gt^2$$

— both results that are very familiar from elementary mechanics. Putting in the given initial conditions, one gets the four curves shown in Figure 13.6. As expected, the four orbits $A_oA, \cdots, D_oD$ are parabolas, no one of which crosses any other. You can see that the initial rectangle $A_oB_oC_oD_o$ has evolved

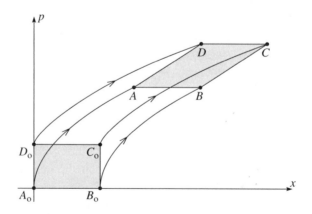

Figure 13.6   Four different phase-space orbits for a body moving vertically under the influence of gravity, with position x (measured vertically down) and momentum p. The four different initial states at time 0 are shown by the dots labeled A_o, B_o, C_o, and D_o, with corresponding final states at a later time t labeled A, B, C, and D.

into a parallelogram $ABCD$. However, it is easy to show that the area of the parallelogram is the same as that of the original rectangle. (Same base, $A_oB_o = AB$, and same height. See Problem 13.27.) We shall see in the next section that these two properties (changing shape but unchanging area) are general properties of all phase-space orbits. In particular, that the area does not change is an example of the important result known as Liouville's theorem.

13.7 Liouville's Theorem *

This section contains material that is more advanced than most other material in this book. In particular, it uses the divergence operator and divergence theorem of vector calculus. If you haven't met these ideas before, you could, if you wish, omit this section.

In Example 13.6 we saw that one can use a plot of phase space to track the motion of several identical systems evolving from various different initial conditions. In many problems, especially those of statistical mechanics, one has occasion to track the motion of an enormous number of identical systems. The state of each system can be labelled by a dot in phase space, and, if the number of these dots is large enough, we can view the resulting swarm of dots as a kind of fluid, with a density ρ measured in dots per volume of phase space. For example, in the statistical mechanics of an ideal gas, one wants to follow the motion of some 10^{23} identical molecules as they move inside a container. Each molecule is governed by the same Hamiltonian and moves in the same six-dimensional phase space with coordinates (x, y, z, p_x, p_y, p_z). Thus the state of the system at any one time can be specified by giving the positions of 10^{23} dots in this phase space; these 10^{23} dots form a swarm whose motion can (for many purposes) be treated like that of a fluid. For most of this section, you could bear this example in mind, though in specific cases I shall often specialize to a system with just one spatial dimension and hence two dimensions of phase space.

The tracking of the motion of many identical systems by means of a cloud of dots in their phase space is illustrated in Figure 13.7. This picture could be viewed as a schematic representation of a multidimensional phase space, but let us, for simplicity, think of it as the two-dimensional phase space, with the coordinates $\mathbf{z} = (q, p)$ of a system with one degree of freedom. Each dot in the lower cloud represents the initial state of one system (for example, one molecule in a gas) by giving its position $\mathbf{z} = (q, p)$ in phase space at time t_o. Hamilton's equations determine the "velocity" with which each dot moves through phase space:

$$\text{(phase-space velocity)} = \dot{\mathbf{z}} = (\dot{q}, \dot{p}) = \left(\frac{\partial \mathcal{H}}{\partial p}, -\frac{\partial \mathcal{H}}{\partial q} \right). \qquad (13.52)$$

For each dot $\mathbf{z}$ in the initial cloud, there is a unique velocity $\dot{\mathbf{z}}$, and each dot moves off with its assigned velocity. In general, different dots will have different velocities, and the cloud can change its shape and orientation, as shown. On the other hand, as we shall prove shortly, the volume occupied by the cloud cannot change — the result called Liouville's theorem.

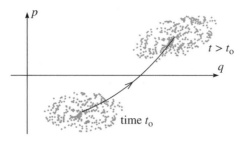

Figure 13.7   The dots in the lower cloud label the states of some 300 identical systems at time t_0. As time progresses each dot moves through phase space in accordance with Hamilton's equations, and the whole cloud moves to the upper position.

To make this last idea more precise, we need to consider a closed surface in phase space such as that shown at the lower left of Figure 13.8. Each point on this closed surface defines a unique set of initial conditions and moves along the corresponding phase-space orbit as time advances. Thus the whole surface moves through phase space. It is easy to see that any point that is initially *inside* the surface must remain inside for all time: For suppose that such a point could move outside. Then at a certain moment it would have to cross the moving surface. At this moment we would have two distinct orbits crossing one another, which we know is impossible. By the same argument, any point that is initially outside the surface remains outside for all time. Thus the number of dots (representing molecules, for example) inside the surface is a constant in time. The main result of this section — Liouville's theorem — is that the volume of the moving closed surface in Figure 13.8 is constant in time. To prove this, we need to know the relationship between the rate of change of this kind of volume and the so-called *divergence* of the velocity vector.

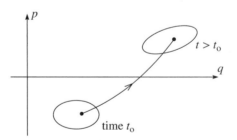

Figure 13.8   As time progresses, the closed surface at the lower left moves through phase space. Any point that is initially inside the surface remains inside for all time.

Changing Volumes

The two mathematical results that we need hold in spaces with any number of dimensions. We shall need to apply them in the $2n$-dimensional phase space. Nevertheless,

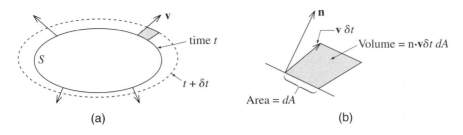

Figure 13.9   (a) The surface S moves with the fluid from its initial position (solid curve) at time t to a new position (dashed curve) at time $t + \delta t$. The change in V is just the volume between the two surfaces. (b) Enlarged view of the shaded volume of part (a). The vector $\mathbf{n}$ is a unit vector normal to the surface S, pointing outward.

let us consider them first in the familiar context of our everyday three-dimensional space. We imagine a three-dimensional space filled with a moving fluid. The fluid at each point $\mathbf{r}$ is moving with velocity $\mathbf{v}$. For each position $\mathbf{r}$ there is a unique velocity $\mathbf{v}$, but $\mathbf{v}$ can, of course, have different values at different points $\mathbf{r}$; thus, we can write $\mathbf{v} = \mathbf{v}(\mathbf{r})$. This is exactly analogous to the situation in phase space, where each phase point $\mathbf{z}$ is moving with a velocity that is uniquely determined by its position in phase space, $\dot{\mathbf{z}} = \dot{\mathbf{z}}(\mathbf{z})$.

Let us now consider a closed surface S in the fluid at a certain time t. We can imagine marking this surface with dye so that we can follow its motion as it moves with the fluid. The question we have to ask is this: If V denotes the volume contained inside S, how fast does V change as the fluid moves? Figure 13.9(a) shows two successive positions of the surface S at two successive moments a short time δt apart. The change in V during the interval δt is the volume between these two surfaces. To evaluate this, consider first the contribution of the small shaded volume in Figure 13.9(a), which is enlarged in Figure 13.9(b). This small volume is a cylinder whose base has an area dA. The side of the cylinder is given by the displacement $\mathbf{v}\,\delta t$, so the cylinder's height is the component of $\mathbf{v}\,\delta t$ normal to the surface. If we introduce a unit vector $\mathbf{n}$ in the direction of the outward normal to S, then the height of our cylinder is $\mathbf{n} \cdot \mathbf{v}\,\delta t$ and its volume is $\mathbf{n} \cdot \mathbf{v}\,\delta t\, dA$. The total change in the volume inside our surface S is found by adding up all these small contributions to give

$$\delta V = \int_S \mathbf{n} \cdot \mathbf{v}\,\delta t\, dA, \tag{13.53}$$

where the integral is a surface integral running over the whole of our closed surface S. Dividing both sides by δt and letting $\delta t \to 0$, we get the first of our two key results:

$$\frac{dV}{dt} = \int_S \mathbf{n} \cdot \mathbf{v}\, dA. \tag{13.54}$$

Figure 13.9(a) showed a fluid flow with the velocity $\mathbf{v}$ everywhere outward, as would be the case for the expanding air in a balloon whose temperature is rising. With $\mathbf{v}$ outward, the scalar product $\mathbf{n} \cdot \mathbf{v}$ is positive (since $\mathbf{n}$ was defined as the outward normal) and the integral (13.54) is positive, implying that V is increasing, as it should. If $\mathbf{v}$ were everywhere inward (air in a balloon whose temperature is falling), then $\mathbf{n} \cdot \mathbf{v}$ would be negative and V would be decreasing. In general, $\mathbf{n} \cdot \mathbf{v}$ can be positive on parts of the surface S and negative on others. In particular, if the fluid is incompressible, so that V can't change, then the contributions from positive and negative values of $\mathbf{n} \cdot \mathbf{v}$ must exactly cancel so that dV/dt can be zero.

We have derived the result (13.54) for a surface S and volume V in three dimensions, but it is equally valid in any number of dimensions. In an m-dimensional space both of the vectors $\mathbf{n}$ and $\mathbf{v}$ have m components and their scalar product is $\mathbf{n} \cdot \mathbf{v} = n_1 v_1 + \cdots + n_m v_m$. With this definition, the result (13.54) is valid whatever the value of m.

The Divergence Theorem

The second mathematical result that we need is called the **divergence theorem** or **Gauss's theorem**. This is one of the standard results of vector calculus (similar to Stokes's theorem that we used in Chapter 4) and you can find its proof in any text on vector calculus,[12] although its proof is quite straightforward and instructive in various simple cases. (See Problem 13.37.) The theorem involves the vector operator called the divergence. For any vector $\mathbf{v}$, the **divergence** of $\mathbf{v}$ is defined as

$$\nabla \cdot \mathbf{v} = \frac{\partial v_x}{\partial x} + \frac{\partial v_y}{\partial y} + \frac{\partial v_z}{\partial z} ; \tag{13.55}$$

here $\mathbf{v}$ can be any vector (a force or an electric field, for instance), but for us it will always be a velocity. If you have not met the divergence operator before, you might like to practice with some of Problems 13.31, 13.32, or 13.34.

The divergence theorem asserts that the surface integral in (13.54) can be expressed in terms of $\nabla \cdot \mathbf{v}$:

$$\int_S \mathbf{n} \cdot \mathbf{v} \, dA = \int_V \nabla \cdot \mathbf{v} \, dV. \tag{13.56}$$

Here the integral on the right is a volume integral over the volume V interior to the surface S. This theorem is an amazingly powerful tool. It plays a crucial role in applications of Gauss's law of electrostatics. It often allows us to perform integrals that would otherwise be hard to evaluate. In particular, there are many interesting fluid flows with the property that $\nabla \cdot \mathbf{v} = 0$; for such flows, the integral on the left may be very awkward to evaluate directly, but thanks to the divergence theorem we can see immediately that the integral is in fact zero. Since this is how we shall be using the divergence theorem, let us look at an example of such a flow right away.

---

[12] See, for example, Mary Boas, *Mathematical Methods in the Physical Sciences* (Wiley, 1983), p. 271.

EXAMPLE 13.7 A Shearing Flow

The velocity of flow of a certain fluid is

$$\mathbf{v} = ky\hat{\mathbf{x}} \tag{13.57}$$

where k is a constant; that is, $v_x = ky$ and $v_y = v_z = 0$. Describe this flow and sketch the motion of a closed surface that starts out as a sphere. Evaluate the divergence $\nabla \cdot \mathbf{v}$ and show that the volume enclosed by any closed surface S moving with the fluid cannot change; that is, the fluid flows incompressibly.

The velocity $\mathbf{v}$ is everywhere in the x direction and depends only on y. Thus all the points in any plane $y = $ constant move like a sliding rigid plate — a pattern called laminar flow.[13] The speed increases with y, so each plane moves a little faster than the planes below it, in a *shearing motion*, as indicated by the three thin arrows in Figure 13.10. If we consider a closed surface moving with the fluid and initially spherical, then its top will be dragged along a little faster than its bottom and it will be stretched into an ellipsoid of ever-increasing eccentricity as shown.

We can easily evaluate $\nabla \cdot \mathbf{v}$ using the definition (13.55)

$$\nabla \cdot \mathbf{v} = \frac{\partial v_x}{\partial x} + \frac{\partial v_y}{\partial y} + \frac{\partial v_z}{\partial z} = \frac{\partial ky}{\partial x} + 0 + 0 = 0. \tag{13.58}$$

Now, combining our two results (13.54) and (13.56), we see that the rate of change of the volume inside any closed surface is

$$\frac{dV}{dt} = \int_V \nabla \cdot \mathbf{v}\, dV. \tag{13.59}$$

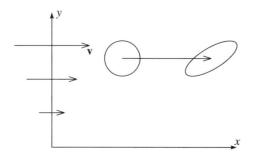

Figure 13.10   The fluid flow described by (13.57) is a laminar, shearing flow. The planes parallel to the plane $y = 0$ all move rigidly in the x direction with speed proportional to y. This shearing motion stretches the sphere into an ellipsoid.

---

[13] From the latin *lamina* meaning a thin layer or plate, from which we also get the verb "to laminate."

> Since we have shown that $\mathbf{\nabla \cdot v} = 0$ everywhere, it follows that $dV/dt = 0$ and
> the volume inside any closed surface moving with the fluid is constant for the
> flow of (13.57).

If you have never met the divergence theorem before, you might want to try one or both of the Problems 13.33 and 13.37. Here let me say just one more thing about the significance of $\mathbf{\nabla \cdot v}$. If we apply (13.59) to a sufficiently small volume V, then $\mathbf{\nabla \cdot v}$ will be approximately constant throughout the region of integration, and the right side of (13.59) becomes just $(\mathbf{\nabla \cdot v})V$ and (13.59) itself implies that

$$\mathbf{\nabla \cdot v} = \frac{1}{V}\frac{dV}{dt}$$

for any small volume V. Since dV/dt can be called the outward flow of $\mathbf{v}$, we can say that $\mathbf{\nabla \cdot v}$ is the *outward flow per volume*. If $\mathbf{\nabla \cdot v}$ is positive at a point $\mathbf{r}$, then there is an outward flow around $\mathbf{r}$ and any small volume around $\mathbf{r}$ is expanding (like the gas in a balloon that is heating up); if $\mathbf{\nabla \cdot v}$ is negative, then there is an inward flow and any small volume around $\mathbf{r}$ is contracting (like the gas in a balloon that is cooling down). In our example, $\mathbf{\nabla \cdot v}$ was zero, and any volume moving with the fluid was constant.

The divergence generalizes easily to any number of dimensions. In an m-dimensional space with coordinates $(x_1, \cdots, x_m)$ the divergence of a vector $\mathbf{v} = (v_1, \cdots, v_m)$ is defined as

$$\mathbf{\nabla \cdot v} = \frac{\partial v_1}{\partial x_1} + \cdots + \frac{\partial v_m}{\partial x_m}$$

and, with these definitions, the crucial result (13.59) takes exactly the same form, except that the integral is an integral over an m-dimensional region with volume element $dV = dx_1\,dx_2\cdots dx_m$.

Liouville's Theorem

We are finally ready to prove the main goal of this section — Liouville's theorem. This is a theorem about motion in phase space, the $2n$-dimensional space with coordinates $\mathbf{z} = (\mathbf{q}, \mathbf{p}) = (q_1, \cdots, q_n, p_1, \cdots, p_n)$. To simplify the notation, I shall consider just a system with one degree of freedom, so that $n = 1$ and the phase space is just two-dimensional with phase points $\mathbf{z} = (q, p)$. The general case goes through in almost exactly the same way, as you can check for yourself by doing Problem 13.36.

Each phase point $\mathbf{z} = (q, p)$ moves through the 2-dimensional phase space in accordance with Hamilton's equations, with velocity

$$\mathbf{v} = \dot{\mathbf{z}} = (\dot{q}, \dot{p}) = \left(\frac{\partial \mathcal{H}}{\partial p}, -\frac{\partial \mathcal{H}}{\partial q}\right). \tag{13.60}$$

We consider an arbitrary closed surface S moving through phase space with the phase points, as was illustrated in Figure 13.8. The rate at which the volume inside S changes is given by (13.59), where now the volume is a two-dimensional volume (really an area) and

$$\mathbf{\nabla} \cdot \mathbf{v} = \frac{\partial \dot{q}}{\partial q} + \frac{\partial \dot{p}}{\partial p} = \frac{\partial}{\partial q}\left(\frac{\partial \mathcal{H}}{\partial p}\right) + \frac{\partial}{\partial p}\left(-\frac{\partial \mathcal{H}}{\partial q}\right) = 0, \qquad (13.61)$$

which is zero because the order of the two differentiations in the double derivatives is immaterial. Since $\mathbf{\nabla} \cdot \mathbf{v} = 0$, it follows that $dV/dt = 0$, and we have proved that the volume V enclosed by any closed surface is a constant as the surface moves around in phase space. This is **Liouville's theorem**.

There is another way to state Liouville's theorem: We saw before that the number N of dots, representing identical systems, inside any given volume V cannot change. (No dot can cross the boundary S from the inside to the outside or vice versa.) We have now seen that the volume V cannot change. Therefore, the density of dots, $\rho = N/V$ cannot change either. This statement is sometimes paraphrased to say that the cloud of dots moves through phase space like an incompressible fluid. However, it is important to be aware what this statement means. The density ρ can, of course, be different at different phase points $\mathbf{z} = (q, p)$; all we are claiming is that as we follow a phase point along its orbit, the density at this point does not change.

Unfortunately we cannot pursue here the consequences of Liouville's theorem, but I can just mention one example: We saw in Chapter 12 that when motion is chaotic, two identical systems that start out with nearly identical initial conditions move rapidly apart in phase space. Thus if we consider a small initial volume in phase space, such as is shown in Figure 13.8, and if the motion is chaotic, then at least some pairs of points inside the volume must move rapidly apart. But we have now seen that the total volume V cannot change. Therefore, as the volume grows in one direction, it must contract in another direction, becoming something like a cigar. Now it frequently happens that the region in which the phase points can move is bounded. (For example, conservation of energy has this effect for the harmonic oscillator of Figure 13.5.) In this case, as the volume V gets longer and thinner, it has to become intricately folded in on itself, adding another twist to the already fascinating story of chaotic motion.

Finally, a couple of points on the validity of Liouville's theorem: First, in all of the examples of this chapter, we have assumed that the Hamiltonian is not explicitly dependent on time, $\partial \mathcal{H}/\partial t = 0$, and that the forces are conservative and the coordinates "natural," so that $\mathcal{H} = T + U$. However, none of these assumptions is necessary for the truth of Liouville's theorem. The proof given in this section depends only on the validity of Hamilton's equations, and any system which obeys Hamilton's equations also obeys Liouville's theorem. For example, a charged particle in an electromagnetic field obeys Hamilton's equations (Problem 13.18) and hence also Liouville's theorem, even though $\partial \mathcal{H}/\partial t$ may be nonzero and $\mathcal{H}$ is certainly not equal to $T + U$. Second, Liouville's theorem applies to the Hamiltonian phase space with coordinates $(\mathbf{q}, \mathbf{p})$, and there is no corresponding theorem for the Lagrangian state space with coordinates $(\mathbf{q}, \dot{\mathbf{q}})$. This is one of the most important advantages of the Hamiltonian over the Lagrangian approach.

Principal Definitions and Equations of Chapter 13

The Hamiltonian

If a system has generalized coordinates $\mathbf{q} = (q_1, \cdots, q_n)$, Lagrangian $\mathcal{L}$, and generalized momenta $p_i = \partial \mathcal{L} / \partial \dot{q}_i$, its **Hamiltonian** is defined as

$$\mathcal{H} = \sum_{i=1}^{n} p_i \dot{q}_i - \mathcal{L}, \qquad \text{[Eq. (13.22)]}$$

always considered as a function of the variables $\mathbf{q}$ and $\mathbf{p}$ (and possibly t).

Hamilton's Equations

The time evolution of a system is given by Hamilton's equations

$$\dot{q}_i = \frac{\partial \mathcal{H}}{\partial p_i} \qquad \text{and} \qquad \dot{p}_i = -\frac{\partial \mathcal{H}}{\partial q_i} \qquad [i = 1, \cdots, n]. \quad \text{[Eq. (13.25)]}$$

Phase Space and Phase-Space Orbits

The **phase space** of a system is the $2n$-dimensional space with points $(\mathbf{q}, \mathbf{p})$ defined by the n generalized coordinates q_i and the n corresponding momenta p_i. A **phase-space orbit** is the path traced in phase space by a system as time evolves.

[Sections 13.5 & 13.6]

Liouville's Theorem

If we imagine a large number of identical systems launched at the same time with slightly different initial conditions, the phase-space points that represent the systems can be seen as forming a fluid. Liouville's theorem states that the density of this fluid is constant in time (or, equivalently, that the volume occupied by any group of points is constant).

[Section 13.7]

Problems for Chapter 13

Stars indicate the approximate level of difficulty, from easiest (★) to most difficult (★★★).

SECTION 13.2 Hamilton's Equations for One-Dimensional Systems

13.1 ★ Find the Lagrangian, the generalized momentum, and the Hamiltonian for a free particle (no forces at all) confined to move along the x axis. (Use x as your generalized coordinate.) Find and solve Hamilton's equations.

13.2 ★ Consider a mass m constrained to move in a vertical line under the influence of gravity. Using the coordinate x measured vertically down from a convenient origin O, write down the Lagrangian $\mathcal{L}$

and find the generalized momentum $p = \partial \mathcal{L}/\partial \dot{x}$. Find the Hamiltonian $\mathcal{H}$ as a function of x and p, and write down Hamilton's equations of motion. (It is too much to hope with a system this simple that you would learn anything new by using the Hamiltonian approach, but do check that the equations of motion make sense.)

13.3 ⋆ Consider the Atwood machine of Figure 13.2, but suppose that the pulley is a uniform disc of mass M and radius R. Using x as your generalized coordinate, write down the Lagrangian, the generalized momentum p, and the Hamiltonian $\mathcal{H} = p\dot{x} - \mathcal{L}$. Find Hamilton's equations and use them to find the acceleration $\ddot{x}$.

13.4 ⋆ The Hamiltonian $\mathcal{H}$ is always given by $\mathcal{H} = p\dot{q} - \mathcal{L}$ (in one dimension), and this is the form you should use if in doubt. However, if your generalized coordinate q is "natural" (relation between q and the underlying Cartesian coordinates is independent of time) then $\mathcal{H} = T + U$, and this form is almost always easier to write down. Therefore, in solving any problem you should quickly check to see if the generalized coordinate is "natural," and if it is you can use the simpler form $\mathcal{H} = T + U$. For the Atwood machine of Example 13.2 (page 527), check that the generalized coordinate was "natural." [*Hint:* There are one generalized coordinate x and two underlying Cartesian coordinates x and y. You have only to write equations for the two Cartesians in terms of the one generalized coordinate and check that they don't involve the time, so it's safe to use $\mathcal{H} = T + U$. This is ridiculously easy!]

13.5 ⋆⋆ A bead of mass m is threaded on a frictionless wire that is bent into a helix with cylindrical polar coordinates (ρ, ϕ, z) satisfying $z = c\phi$ and $\rho = R$, with c and R constants. The z axis points vertically up and gravity vertically down. Using ϕ as your generalized coordinate, write down the kinetic and potential energies, and hence the Hamiltonian $\mathcal{H}$ as a function of ϕ and its conjugate momentum p. Write down Hamilton's equations and solve for $\ddot{\phi}$ and hence $\ddot{z}$. Explain your result in terms of Newtonian mechanics and discuss the special case that $R = 0$.

13.6 ⋆⋆ In discussing the oscillation of a cart on the end of a spring, we almost always ignore the mass of the spring. Set up the Hamiltonian $\mathcal{H}$ for a cart of mass m on a spring (force constant k) whose mass M is *not* negligible, using the extension x of the spring as the generalized coordinate. Solve Hamilton's equations and show that the mass oscillates with angular frequency $\omega = \sqrt{k/(m + M/3)}$. That is, the effect of the spring's mass is to add $M/3$ to m. (Assume that the spring's mass is distributed uniformly and that it stretches uniformly.)

13.7 ⋆⋆⋆ A roller coaster of mass m moves along a frictionless track that lies in the xy plane (x horizontal and y vertically up). The height of the track above the ground is given by $y = h(x)$. **(a)** Using x as your generalized coordinate, write down the Lagrangian, the generalized momentum p, and the Hamiltonian $\mathcal{H} = p\dot{x} - \mathcal{L}$ (as a function of x and p). **(b)** Find Hamilton's equations and show that they agree with what you would get from the Newtonian approach. [*Hint:* You know from Section 4.7 that Newton's second law takes the form $F_{\text{tang}} = m\ddot{s}$, where s is the distance measured along the track. Rewrite this as an equation for $\ddot{x}$ and show that you get the same result from Hamilton's equations.]

SECTION 13.3 Hamilton's Equations in Several Dimensions

13.8 ⋆ Find the Lagrangian, the generalized momenta, and the Hamiltonian for a free particle (no forces at all) moving in three dimensions. (Use x, y, z as your generalized coordinates.) Find and solve Hamilton's equations.

13.9 ⋆ Set up the Hamiltonian and Hamilton's equations for a projectile of mass m, moving in a vertical plane and subject to gravity but no air resistance. Use as your coordinates x measured horizontally and y measured vertically up. Comment on each of the four equations of motion.

13.10 ★ Consider a particle of mass m moving in two dimensions, subject to a force $\mathbf{F} = -kx\hat{\mathbf{x}} + K\hat{\mathbf{y}}$, where k and K are positive constants. Write down the Hamiltonian and Hamilton's equations, using x and y as generalized coordinates. Solve the latter and describe the motion.

13.11 ★ The simple form $\mathcal{H} = T + U$ is true only if your generalized coordinates are "natural" (relation betweeen generalized and underlying Cartesian coordinates is independent of time). If the generalized coordinates are not "natural," you must use the definition $\mathcal{H} = \sum p_i \dot{q}_i - \mathcal{L}$. To illustrate this point, consider the following: Two children are playing catch inside a railroad car that is moving with varying speed V along a straight horizontal track. For generalized coordinates you can use the position (x, y, z) of the ball relative to a point fixed in the car, but in setting up the Hamiltonian you must use coordinates in an inertial frame — a frame fixed to the ground. Find the Hamiltonian for the ball and show that it is not equal to $T + U$ (neither as measured in the car, nor as measured in the ground-based frame).

13.12 ★ Same as Problem 13.11, but use the following system: A bead of mass m is threaded on a frictionless, straight rod, which lies in a horizontal plane and is forced to spin with constant angular velocity ω about a fixed vertical axis through the midpoint of the rod. Find the Hamiltonian for the bead and show that it is not equal to $T + U$.

13.13 ★★ Consider a particle of mass m constrained to move on a frictionless cylinder of radius R, given by the equation $\rho = R$ in cylindrical polar coordinates (ρ, ϕ, z). The mass is subject to just one external force, $\mathbf{F} = -kr\hat{\mathbf{r}}$, where k is a positive constant, r is its distance from the origin, and $\hat{\mathbf{r}}$ is the unit vector pointing away from the origin, as usual. Using z and ϕ as generalized coordinates, find the Hamiltonian $\mathcal{H}$. Write down and solve Hamilton's equations and describe the motion.

13.14 ★★ Consider the mass confined to the surface of a cone described in Example 13.4 (page 533). We saw there that there have to be maximum and minimum heights $z_{\max}$ and $z_{\min}$, beyond which the mass cannot stray. When z is a maximum or minimum, it must be that $\dot{z} = 0$. Show that this can happen if and only if the conjugate momentum $p_z = 0$, and use the equation $\mathcal{H} = E$, where $\mathcal{H}$ is the Hamiltonian function (13.33), to show that, for a given energy E, this occurs at exactly two values of z. [*Hint:* Write down the function $\mathcal{H}$ for the case that $p_z = 0$ and sketch its behavior as a function of z for $0 < z < \infty$. How many times can this function equal any given E?] Use your sketch to describe the motion of the mass.

13.15 ★★ Fill in the details of the derivation of Hamilton's $2n$ equations (13.25) for a system with n degrees of freedom, starting from Equation (13.24). You can parallel the argument that led from (13.14) to (13.15) and (13.16), but you have $2n$ different derivatives to consider and lots of summations from $i = 1$ to n to contend with.

13.16 ★★ Starting from the expression (13.24) for the Hamiltonian, prove that $\partial \mathcal{H}/\partial t = -\partial \mathcal{L}/\partial t$. [*Hint:* Consider first a system with one degree of freedom, for which (13.24) simplifies to $\mathcal{H}(q, p, t) = p\dot{q}(q, p, t) - \mathcal{L}(q, \dot{q}(q, p, t), t)$.]

13.17 ★★★ Consider the mass confined to the surface of a cone described in Example 13.4 (page 533). We saw that there are solutions for which the mass remains at the fixed height $z = z_0$, with fixed angular velocity $\dot{\phi}_0$ say. **(a)** For any chosen value of p_ϕ, use (13.34) to get an equation that gives the corresponding value of the height z_0. **(b)** Use the equations of motion to show that this motion is stable. That is, show that if the orbit has $z = z_0 + \epsilon$ with ϵ small, then ϵ will oscillate about zero. **(c)** Show that the angular frequency of these oscillations is $\omega = \sqrt{3}\dot{\phi}_0 \sin\alpha$, where α is the half angle of the cone ($\tan\alpha = c$ where c is the constant in $\rho = cz$). **(d)** Find the angle α for which the frequency of oscillation ω is equal to the orbital angular velocity $\dot{\phi}_0$, and describe the motion for this case.

13.18 ★★★ All of the examples in this chapter and all of the problems (except this one) treat forces that come from a potential energy $U(\mathbf{r})$ [or occasionally $U(\mathbf{r}, t)$]. However, the proof of Hamilton's equations given in Section 13.3 applies to any system for which Lagrange's equations hold, and this can include forces not derivable from a potential energy. An important example of such a force is the magnetic force on a charged particle. **(a)** Use the Lagrangian (7.103) to show that the Hamiltonian for a charge q in an electromagnetic field is

$$\mathcal{H} = (\mathbf{p} - q\mathbf{A})^2/(2m) + qV.$$

(This Hamiltonian plays an important role in the quantum mechanics of charged particles.) **(b)** Show that Hamilton's equations are equivalent to the familiar Lorentz force equation $m\ddot{\mathbf{r}} = q(\mathbf{E} + \mathbf{v} \times \mathbf{B})$.

SECTION 13.4   Ignorable Coordinates

13.19 ★ In Example 13.3 (page 531) we saw that if we write the Hamiltonian for a two-dimensional central force problem in terms of polar coordinates r and ϕ, then the coordinate ϕ is ignorable. Write down the Hamiltonian for the same problem, but using rectangular coordinates x, y. Show that, with this choice, neither coordinate is ignorable. [The moral of this is that the choice of generalized coordinates calls for some care. In particular, you must look for any symmetries in a system and try to choose generalized coordinates to take advantage of them.]

13.20 ★ Consider a mass m moving in two dimensions, subject to a single force $\mathbf{F}$ that is independent of $\mathbf{r}$ and t. **(a)** Find the potential energy $U(\mathbf{r})$ and the Hamiltonian $\mathcal{H}$. **(b)** Show that if you use rectangular coordinates x, y with the x axis in the direction of $\mathbf{F}$, then y is ignorable. **(c)** Show that if you use rectangular coordinates x, y with neither axis in the direction of $\mathbf{F}$, then neither coordinate is ignorable. (Moral: Choose generalized coordinates carefully!)

13.21 ★★ Two masses m_1 and m_2 are joined by a massless spring (force constant k and natural length l_0) and are confined to move in a frictionless horizontal plane, with CM and relative positions $\mathbf{R}$ and $\mathbf{r}$ as defined in Section 8.2. **(a)** Write down the Hamiltonian $\mathcal{H}$ using as generalized coordinates X, Y, r, ϕ, where (X, Y) are the rectangular components of $\mathbf{R}$, and (r, ϕ) are the polar coordinates of $\mathbf{r}$. Which coordinates are ignorable and which are not? Explain. **(b)** Write down the 8 Hamilton equations of motion. **(c)** Solve the r equations for the special case that $p_\phi = 0$ and describe the motion. **(d)** Describe the motion for the case that $p_\phi \neq 0$ and explain physically why the r equation is harder to solve in this case.

13.22 ★★ In the Lagrangian formalism, a coordinate q_i is ignorable if $\partial\mathcal{L}/\partial q_i = 0$; that is, if $\mathcal{L}$ is independent of q_i. This guarantees that the momentum p_i is constant. In the Hamiltonian approach, we say that q_i is ignorable if $\mathcal{H}$ is independent of q_i, and this too guarantees p_i is constant. These two conditions must be the same, since the result "$p_i = $ const" is the same either way. Prove directly that this is so, as follows: **(a)** For a system with one degree of freedom, prove that $\partial\mathcal{H}/\partial q = -\partial\mathcal{L}/\partial q$ starting from the expression (13.14) for the Hamiltonian. This establishes that $\partial\mathcal{H}/\partial q = 0$ if and only if $\partial\mathcal{L}/\partial q = 0$. **(b)** For a system with n degrees of freedom, prove that $\partial\mathcal{H}/\partial q_i = -\partial\mathcal{L}/\partial q_i$ starting from the expression (13.24).

13.23 ★★★ Consider the modified Atwood machine shown in Figure 13.11. The two weights on the left have equal masses m and are connected by a massless spring of force constant k. The weight on the right has mass $M = 2m$, and the pulley is massless and frictionless. The coordinate x is the extension of the spring from its equilibrium length; that is, the length of the spring is $l_e + x$ where l_e is the equilibrium length (with all the weights in position and M held stationary). **(a)** Show that the total potential energy

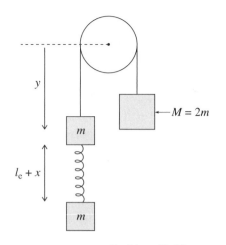

Figure 13.11   Problem 13.23

(spring plus gravitational) is just $U = \frac{1}{2}kx^2$ (plus a constant that we can take to be zero). **(b)** Find the two momenta conjugate to x and y. Solve for $\dot{x}$ and $\dot{y}$, and write down the Hamiltonian. Show that the coordinate y is ignorable. **(c)** Write down the four Hamilton equations and solve them for the following initial conditions: You hold the mass M fixed with the whole system in equilibrium and $y = y_o$. Still holding M fixed, you pull the lower mass m down a distance x_o, and at $t = 0$ you let go of both masses. [*Hint:* Write down the initial values of x, y and their momenta. You can solve the x equations by combining them into a second-order equation for x. Once you know $x(t)$, you can quickly write down the other three variables.] Describe the motion. In particular, find the frequency with which x oscillates.

SECTION 13.5   Lagrange's Equations vs. Hamilton's Equations

13.24 ⋆ Here is a simple example of a canonical transformation that illustrates how the Hamiltonian formalism lets one mix up the q's and the p's. Consider a system with one degree of freedom and Hamiltonian $\mathcal{H} = \mathcal{H}(q, p)$. The equations of motion are, of course, the usual Hamiltonian equations $\dot{q} = \partial\mathcal{H}/\partial p$ and $\dot{p} = -\partial\mathcal{H}/\partial q$. Now consider new coordinates in phase space defined as $Q = p$ and $P = -q$. Show that the equations of motion for the new coordinates Q and P are $\dot{Q} = \partial\mathcal{H}/\partial P$ and $\dot{P} = -\partial\mathcal{H}/\partial Q$; that is, the Hamiltonian formalism applies equally to the new choice of coordinates where we have exchanged the roles of position and momentum.

13.25 ⋆⋆⋆ Here is another example of a canonical transformation, which is still too simple to be of any real use, but does nevertheless illustrate the power of these changes of coordinates. **(a)** Consider a system with one degree of freedom and Hamiltonian $\mathcal{H} = \mathcal{H}(q, p)$ and a new pair of coordinates Q and P defined so that

$$q = \sqrt{2P} \sin Q \qquad \text{and} \qquad p = \sqrt{2P} \cos Q. \tag{13.62}$$

Prove that if $\partial\mathcal{H}/\partial q = -\dot{p}$ and $\partial\mathcal{H}/\partial p = \dot{q}$, it automatically follows that $\partial\mathcal{H}/\partial Q = -\dot{P}$ and $\partial\mathcal{H}/\partial P = \dot{Q}$. In other words, the Hamiltonian formalism applies just as well to the new coordinates as to the old. **(b)** Show that the Hamiltonian of a one-dimensional harmonic oscillator with mass $m = 1$ and force constant $k = 1$ is $\mathcal{H} = \frac{1}{2}(q^2 + p^2)$. **(c)** Show that if you rewrite this Hamiltonian in terms of the coordinates Q and P defined in (13.62), then Q is ignorable. [The change of coordinates (13.62) was cunningly chosen to produce this elegant result.] What is P? **(d)** Solve the Hamiltonian equation for $Q(t)$ and verify that, when rewritten for q, your solution gives the expected behavior.

SECTION 13.6 Phase-Space Orbits

13.26 ⋆ Find the Hamiltonian $\mathcal{H}$ for a mass m confined to the x axis and subject to a force $F_x = -kx^3$ where $k > 0$. Sketch and describe the phase-space orbits.

13.27 ⋆⋆ Figure 13.6 shows some phase-space orbits for a mass in free fall. The points A_o, B_o, C_o, D_o represent four different possible initial states at time 0, and A, B, C, D are the corresponding states at a later time. Write down the position $x(t)$ and momentum $p(t)$ as functions of t and use these to prove that $ABCD$ is a parallelogram with area equal to the rectangle $A_oB_oC_oD_o$. [This is an example of Liouville's theorem.]

13.28 ⋆⋆ Consider a mass m confined to the x axis and subject to a force $F_x = kx$ where $k > 0$. **(a)** Write down and sketch the potential energy $U(x)$ and describe the possible motions of the mass. (Distinguish between the cases that $E > 0$ and $E < 0$.) **(b)** Write down the Hamiltonian $\mathcal{H}(x, p)$, and describe the possible phase-space orbits for the two cases $E > 0$ and $E < 0$. (Remember that the function $\mathcal{H}(x, p)$ must equal the constant energy E.) Explain your answers to part (b) in terms of those to part (a).

SECTION 13.7 Liouville's Theorem*

13.29 ⋆ Figure 13.10 shows an initially spherical volume getting stretched into an ellipsoid by the shearing flow (13.57). Make a similar sketch for a volume that is initially spherical and centered on the origin.

13.30 ⋆ Figure 13.9 shows a fluid flow where the flow is everywhere outward (at least for all points on the surface S shown). This means that all contributions to the change δV in volume are positive, and V is definitely increasing. Sketch the corresponding picture for the case that the flow is outward on the upper part of S and inward on the lower part. Explain clearly why the contributions $\mathbf{n} \cdot \mathbf{v}\, \delta t\, dA$ to the change in V from the lower part of S are *negative*, and hence that δV can be of either sign, depending on whether the positive contributions outweigh the negative or vice versa.

13.31 ⋆ Evaluate the three-dimensional divergence $\nabla \cdot \mathbf{v}$ for each of the following vectors: **(a)** $\mathbf{v} = k\mathbf{r}$, **(b)** $\mathbf{v} = k(z, x, y)$, **(c)** $\mathbf{v} = k(z, y, x)$, **(d)** $\mathbf{v} = k(x, y, -2z)$, where $\mathbf{r} = (x, y, z)$ is the usual position vector and k is a constant.

13.32 ⋆⋆ Evaluate the three-dimensional divergence $\nabla \cdot \mathbf{v}$ for each of the following vectors: **(a)** $\mathbf{v} = k\hat{\mathbf{x}}$, **(b)** $\mathbf{v} = kx\hat{\mathbf{x}}$, **(c)** $\mathbf{v} = ky\hat{\mathbf{x}}$. We know that $\nabla \cdot \mathbf{v}$ represents the net outward flow associated with $\mathbf{v}$. In those cases where you found $\nabla \cdot \mathbf{v} = 0$, make a simple sketch to illustrate that the outward flow is zero; in those cases where you found $\nabla \cdot \mathbf{v} \neq 0$, make a sketch to show why and whether the outflow is positive or negative.

13.33 ⋆⋆ The divergence theorem is a remarkable result, relating the *surface* integral that gives the flow of $\mathbf{v}$ out of a closed surface S to the *volume* integral of $\nabla \cdot \mathbf{v}$. Occasionally it is easy to evaluate both of these integrals and one can check the validity of the theorem. More often, one of the integrals is much easier to evaluate than the other, and the divergence theorem then gives one a slick way to evaluate a hard integral. The following exercises illustrate both of these situations. **(a)** Let $\mathbf{v} = k\mathbf{r}$, where k is a constant and let S be a sphere of radius R centered on the origin. Evaluate the left side of the divergence theorem (13.56) (the surface integral). Next calculate $\nabla \cdot \mathbf{v}$ and use this to evaluate the right side of (13.56) (the volume integral). Show that the two agree. **(b)** Now use the same velocity $\mathbf{v}$, but let S be a sphere *not* centered on the origin. Explain why the surface integral is now hard to evaluate directly, but don't actually do it. Instead, find its value by doing the volume integral. (This second route should be no harder than before.)

13.34 ★★ (a) Evaluate $\nabla \cdot \mathbf{v}$ for $\mathbf{v} = k\hat{\mathbf{r}}/r^2$ using rectangular coordinates. (Note that $\hat{\mathbf{r}}/r^2 = \mathbf{r}/r^3$.) **(b)** Inside the back cover, you will find expressions for the various vector operators (divergence, gradient, etc.) in polar coordinates. Use the expression for the divergence in spherical polar coordinates to confirm your answer to part (a). (Take $r \neq 0$.)

13.35 ★★ A beam of particles is moving along an accelerator pipe in the z direction. The particles are uniformly distributed in a cylindrical volume of length L_0 (in the z direction) and radius R_0. The particles have momenta uniformly distributed with p_z in an interval $p_0 \pm \Delta p_z$ and the transverse momentum $p_\perp$ inside a circle of radius $\Delta p_\perp$. To increase the particles' spatial density, the beam is focused by electric and magnetic fields, so that the radius shrinks to a smaller value R. What does Liouville's theorem tell you about the spread in the transverse momentum $p_\perp$ and the subsequent behavior of the radius R? (Assume that the focusing does not affect either L_0 or Δp_z.)

13.36 ★★ Prove Liouville's theorem in the $2n$-dimensional phase space of a system with n degrees of freedom. You can follow closely the argument around Equations (13.60) and (13.61). The only difference is that now the phase velocity $\mathbf{v} = \dot{\mathbf{z}}$ is a $2n$-dimensional vector and $\nabla \cdot \mathbf{v}$ is a $2n$-dimensional divergence.

13.37 ★★★ The general proof of the divergence theorem

$$\int_S \mathbf{n} \cdot \mathbf{v}\, dA = \int_V \nabla \cdot \mathbf{v}\, dV \tag{13.63}$$

is fairly complicated and not especially illuminating. However, there are a few special cases where it is reasonably simple and quite instructive. Here is one: Consider a rectangular region bounded by the six planes $x = X$ and $X + A$, $y = Y$ and $Y + B$, and $z = Z$ and $Z + C$, with total volume $V = ABC$. The surface S of this region is made up of six rectangles that we can call S_1 (in the plane $x = X$), S_2 (in the plane $x = X + A$), and so on. The surface integral on the left of (13.63) is then the sum of six integrals, one over each of the rectangles S_1, S_2, and so forth. **(a)** Consider the first two of these integrals and show that

$$\int_{S_1} \mathbf{n} \cdot \mathbf{v}\, dA + \int_{S_2} \mathbf{n} \cdot \mathbf{v}\, dA = \int_Y^{Y+B} dy \int_Z^{Z+C} dz\ [v_x(X + A, y, z) - v_x(X, y, z)\,].$$

(b) Show that the integrand on the right can be rewritten as an integral of $\partial v_x/\partial x$ over x running from $x = X$ to $x = X + A$. **(c)** Substitute the result of part (b) into part (a), and write down the corresponding results for the two remaining pairs of faces. Add these results to prove the divergence theorem (13.63).

Collision Theory

The *collision experiment*, or *scattering experiment*, is the single most powerful tool for investigating the structure of atomic and subatomic objects. In this type of experiment one fires a stream of projectiles, such as electrons or protons, at a target object — an atom or atomic nucleus, for example — and, by observing the distribution of "scattered" projectiles as they emerge from the collision, one can gain information about the target and its interactions with the projectile. Perhaps the most famous collision experiment was the discovery by Ernest Rutherford (1871–1937) of the structure of the atom: Rutherford and his assistants fired streams of α particles (the positively charged nuclei of helium atoms) at a thin layer of gold atoms in a sheet of gold foil; by measuring the distribution of the scattered α particles, they were able to deduce that most of the mass of an atom is concentrated in a tiny, positively charged "nucleus" at the center of the atom. Since that time, most discoveries in atomic and subatomic physics (the discoveries of the neutron, of nuclear fission and fusion, of quarks, and many more) were made with the help of collision experiments, in which a stream of projectiles were directed at a suitable target and the outgoing particles carefully monitored.

You could imagine doing a scattering experiment with larger objects — scattering one billiard ball off another, or even a comet off the sun — but in these cases there are usually easier ways to find out about the target. Thus the main application of collision theory is at the atomic level and below. Since the correct mechanics for atomic and subatomic systems is quantum mechanics, this means that the most widely used form of collision theory is *quantum collision theory*. Nevertheless, many of the central ideas of quantum collision theory — total and differential scattering cross sections, lab and CM reference frames — already appear in the classical theory, which gives an excellent introduction to these ideas without the complications of quantum theory. This, then, is the main purpose of this chapter, to give an introduction to the main ideas of collision theory in the context of classical mechanics.

The main reason why collision theory is a rather complicated structure is that on the atomic and subatomic scale one cannot possibly follow the detailed orbit of the projectile as it interacts with the target. As we shall see, this means that we can learn very little by observing a single projectile. On the other hand, if we send in many projectiles, we can observe the number of them that get scattered in different directions, and we can learn lots from this. It is to handle the statistical distribution of the many scattered projectiles, in the many possible directions, that we have to introduce the idea of the *collision cross section*, which is the central concept of collision theory, both classical and quantum,[1] and is the main subject of this chapter.

14.1   The Scattering Angle and Impact Parameter

Before we introduce the central concept of the scattering cross section, it is helpful to introduce two other important parameters, the *scattering angle* and the *impact parameter*; and before we do this, it is good to have a couple of simple collision experiments in mind. All collision experiments start with a projectile approaching a target from so far away that it moves essentially freely and its energy is purely kinetic. In Figure 14.1, a fixed target exerts a force on the projectile, and, as the latter gets close to the target, the orbit curves, so that the projectile is "scattered" and moves away in a different direction. An example of this sort of collision is the famous Rutherford experiment, in which the projectile and target were both positively charged particles and the force that caused the scattering was the Coulomb repulsion between them. In Figure 14.2, the target is a hard sphere, which exerts a force on the projectile only when they come into contact (a *contact force*); thus, the projectile travels in a straight line until it hits the target (if it does) and then bounces and moves off in a different direction. A familiar example of this kind of event is the collision of two billiard balls (though in this case the projectile and target would be the same size).

With these two examples in mind, we can now define the first two key parameters of collision theory. The **scattering angle** θ is defined as the angle between the incoming and outgoing velocities of the projectile, as indicated in both Figures 14.1 and 14.2. In the absence of any target, the scattering angle would, of course, be zero. Thus $\theta = 0$ corresponds to *no scattering;* for example, in Figure 14.2 the projectile could miss the target entirely and then θ would be zero. The maximum possible value is $\theta = \pi$; in Figure 14.2, a head-on collision, in which the projectile comes in along the target's axis and bounces straight back, would give $\theta = \pi$.

The **impact parameter** b is defined as the perpendicular distance from the projectile's incoming straight-line path to a parallel axis through the target's center, as shown on the left in both figures. A second way to think about the impact parameter is illustrated toward the right in Figure 14.1: You can think of b as the distance that *would be* the distance of closest approach if there were no forces on the projectile, so that the orbit was just a straight line. In other words, the impact parameter tells

---

[1] Quantum mechanics is an intrinsically statistical theory, dealing with *probabilities* rather than definite predictable outcomes. Thus in quantum mechanics the need to discuss the distribution of different scattered directions is there from the outset — in contrast with the situation in classical mechanics, where the same need arises from the practical impossibility of observing the detailed orbit of a projectile. Nevertheless, the machinery for handling the problem — most notably the idea of the collision cross section — is very similar in the two theories.

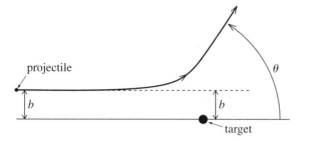

Figure 14.1   In this collision experiment the projectile approaches from the left moving like a free particle. When it begins to feel the force field of the target, its orbit curves and it moves away in a different direction. The scattering angle θ is the angle between the initial and final velocities. The impact parameter b is the perpendicular distance from the incoming straight-line orbit to a parallel axis through the center of the target.

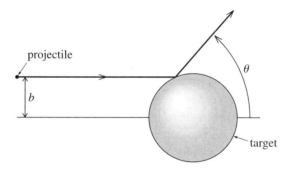

Figure 14.2   A scattering experiment in which the projectile and target interact only through a contact force, so that the projectile is deflected only when it actually hits the target and bounces off it. A collison occurs only if the impact parameter b is less than the radius of the target.

how closely the projectile was *aimed* at the target. If the impact parameter is very large, then the projectile will hardly feel the target, and θ will be small. In fact, in Figure 14.2, θ would be precisely zero for any value of b greater than the radius of the target. At the other extreme, the value $b = 0$ implies a head-on collision and often[2] corresponds to $\theta = \pi$. It is reasonably clear from Figures 14.1 and 14.2 that for a given value of b there will be a unique corresponding value of θ, and we shall find, in fact, that the main theoretical task in classical collision theory is to find the functional relation $\theta = \theta(b)$ between these two variables.

In atomic and subatomic physics (where collision theory has its greatest application), the *experimental* status of the scattering angle is totally different from that of

---

[2] In both our examples (the Rutherford experiment and hard-sphere scattering) $b = 0$ certainly implies $\theta = \pi$. On the other hand, if the force between the projectile and target is attractive, then a projectile with $b = 0$ will crash into the target, and — depending on the nature of the target — may never re-emerge, or may plow straight through and emerge with $\theta = 0$.

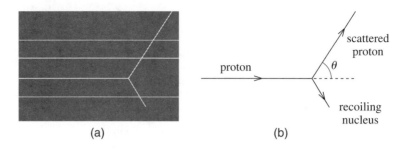

(a)   (b)

Figure 14.3   **(a)** In a cloud chamber, any moving charged particle leaves a visible track which can be photographed for later examination. Here four protons have entered the chamber from the left, three of them passing straight through undeflected. The fourth was deflected when it hit a nitrogen nucleus and left the chamber near the top right. The recoiling nucleus can be seen moving down and to the right. **(b)** Tracing of the tracks involved in the collision.

the impact parameter: The scattering angle θ is rather easily measured, whereas the impact parameter can never be measured directly. Let us address these two points in turn. There are many ways to measure the scattering angle, one of the most transparent of which is illustrated in Figure 14.3. This could represent a photograph made in a cloud chamber of a collision of a proton with a nitrogen nucleus. These particles are of course far too small to be photographed directly, but the cloud chamber is one of several devices that let one record the *track* of a moving charged particle — even when the particle is far too small to be seen itself. In the cloud chamber, a charged particle moving through a cloud of supersaturated water vapor ionizes some of the atoms that it passes, and these ionized atoms cause some of the water vapor to condense, creating a visible trail, somewhat like the vapor trail left in the sky by some aircraft. In Figure 14.3 you can clearly see the trail of the incoming proton. The nucleus which the proton eventually strikes is initially invisible, since it is not moving. However, when the proton strikes the nucleus, the proton's track makes an abrupt change of direction, and the nucleus recoils leaving its own track. From pictures like this, the scattering angle is easily measured.

On the other hand, the impact parameter can never be directly measured. The problem is that impact parameters of interest are of atomic or subatomic size — around 0.1 nanometers or less. A cloud-chamber track like that of Figure 14.3 has a width of order 1 millimeter, some 10 million times greater than the largest impact parameters of interest. Obviously direct and meaningful measurements of the impact parameter are out of the question. As we shall see in the next section, it is the impossibility of measuring the impact parameter that leads us to the notion of the collision cross section.

14.2   The Collision Cross Section

Let us imagine first that we were to observe a single collision like that shown in Figure 14.3. If we knew the impact parameter, then we could obviously deduce something about the size of the target or the strength and range of the force that it exerts on the projectile. But given that we do *not* know the impact parameter, there is remarkably little we can learn from a single event like this; in fact, about the only thing we can conclude from this one event is that there *is* some kind of obstacle in the projectile's path. On the other hand, if we can observe many different collisions of similar projectiles and targets, then we can begin to investigate the nature of the projectile and target and their interactions. To explore this important idea, I shall consider several simple examples.

Consider first an experiment like that shown in Figure 14.2, where a projectile of negligible size impinges on a hard sphere of radius R, with which it interacts only on contact. If the projectile hits the target, it will bounce off it and emerge in a different direction; if the projectile misses the target, it will pass straight through undeflected. When observed experimentally, the collision illustrated in Figure 14.2 will *look* something like Figure 14.3, and the observation of one such event tells us only that the target is there.

Suppose however we could repeat the same experiment many times. In practice this is accomplished in two ways: Instead of having just one target, we have many targets in a single target assembly — for example, many gold nuclei in a gold foil or many helium atoms in a tank of helium gas. And, instead of firing in a single projectile, we fire in a whole beam of projectiles. To begin, let us consider a single projectile passing through an assembly of hard-sphere targets. As "seen" by the incoming projectile, the target assembly looks something like Figure 14.4. Since we don't know the projectile's precise line of approach, we can't say whether it will hit one of the targets or not. However, we can calculate the *probability* that it will make a hit, as follows: If the targets are randomly positioned and sufficiently numerous,[3] we can speak of the **target density** n_{tar} as the number of targets per area, as viewed from the incident direction. If A is the total area of the target assembly, then the total number of targets is $n_{\text{tar}}A$. Next, we denote by $\sigma = \pi R^2$ the cross-sectional area, or just **cross section** of each target (as seen from the front), so that the total area of all the targets is $n_{\text{tar}}A\sigma$. Therefore,

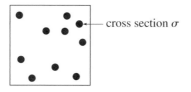

Figure 14.4   A target assembly with several hard-sphere targets, as seen head-on by the incoming projectile. The cross section σ is the area of any one target, perpendicular to the incident direction. The target density n_{tar} is the density (number/area) of targets as seen from the incident direction, as here.

---

[3] It is important to have lots of targets for statistical considerations to apply. On the other hand, the targets must not be *too* numerous, or some targets may get hidden in the "shadow" of others, and the projectile may make several collisions. It was to avoid multiple collisions that Rutherford used a *thin* foil in his famous experiment.

the probability that any one projectile makes a hit as it passes through the assembly on a random path is the ratio

$$\text{(probability of a hit)} = \frac{\text{area occupied by targets}}{\text{total area}} = \frac{n_{\text{tar}} A \sigma}{A} = n_{\text{tar}} \sigma. \quad (14.1)$$

If we now send in a beam containing a large number (call it N_{inc}) of incident projectiles, then the actual number of projectiles that get scattered (N_{sc}) should be the product of the probability (14.1) and N_{inc}:

$$N_{\text{sc}} = N_{\text{inc}} \, n_{\text{tar}} \, \sigma. \quad (14.2)$$

This is the basic relation of collision theory. Since we can measure the numbers N_{sc} and N_{inc} and the target density n_{tar}, Equation (14.2) lets us find the size (or cross section) σ of the target. In what follows, we shall see that the notion of the cross section gets generalized considerably and can become quite complicated, but the essential idea is always the same: By counting the number of scatterings (or reactions, or absoptions, or other processes) that result from a large number of similar collisions, one can use the analog of (14.2) to find the parameter σ, which is always the *effective area of the target for interacting with the projectile*.

EXAMPLE 14.1 Shooting Crows in an Oak Tree

A hunter observes 50 crows settling randomly in an oak tree, where he can no longer see them. Each crow has a cross-sectional area $\sigma \approx \frac{1}{2}$ ft^2, and the oak has a total area (as seen from the hunter's position) of 150 square feet. If the hunter fires 60 bullets at random into the tree, about how many crows would he expect to hit?

This situation closely parallels our simple scattering experiment. The target density is $n_{\text{tar}} = $ (number of crows)/(area of tree) $= 50/150 = 1/3$ ft^{-2}. The number of incident projectiles is $N_{\text{inc}} = 60$, so, by the analog of (14.2), the expected number of hits is

$$N_{\text{hit}} = N_{\text{inc}} \, n_{\text{tar}} \, \sigma = 60 \times \left(\tfrac{1}{3} \text{ ft}^{-2} \right) \times \left(\tfrac{1}{2} \text{ ft}^2 \right) = 10.$$

In practice, one often uses a steady stream of projectiles, and it may be more convenient to divide the incident *number* N_{inc} by the time Δt, to give the incident *rate* $R_{\text{inc}} = N_{\text{inc}}/\Delta t$. Similarly, the scattered rate is $R_{\text{sc}} = N_{\text{sc}}/\Delta t$. Dividing both sides of (14.2) by Δt, we get the completely equivalent relation

$$R_{\text{sc}} = R_{\text{inc}} \, n_{\text{tar}} \, \sigma$$

for these rates.

Since the cross section σ is an area, the SI unit of cross section is, of course, the square meter. Atomic and nuclear cross sections are inconveniently small when measured in square meters. In particular, typical nuclear dimensions are around 10^{-14} m, so nuclear cross sections are conveniently measured in units of 10^{-28} m². This area has come to be called a **barn** (as in "it's as big as a barn"),

$$1 \text{ barn} = 10^{-28} \text{ m}^2.$$

EXAMPLE 14.2 Scattering of Neutrons in an Aluminum Foil

If 10,000 neutrons are fired through an aluminum foil 0.1 mm thick and the cross section of the aluminum nucleus is about 1.5 barns,[4] how many neutrons will be scattered? (Specific gravity of aluminum = 2.7.)

The number of scatterings is given by (14.2), and we already know that $N_{\text{inc}} = 10^4$ and $\sigma = 1.5 \times 10^{-28}$ m². Thus all we need to find is the target density n_{tar}, the number of aluminum nuclei per area of the foil. (Of course, the foil contains lots of atomic electrons as well, but these do not contribute appreciably to the scattering of neutrons.) The density of aluminum (mass/volume) is $\varrho = 2.7 \times 10^3$ kg/m³. If we multiply this by the thickness of the foil ($t = 10^{-4}$ m), this will give the mass per area of the foil, and dividing this by the mass of an aluminum nucleus ($m = 27$ atomic mass units), we will have n_{tar}:

$$n_{\text{tar}} = \frac{\varrho t}{m} = \frac{(2.7 \times 10^3 \, \text{kg/m}^3) \times (10^{-4} \, \text{m})}{27 \times 1.66 \times 10^{-27} \, \text{kg}} = 6.0 \times 10^{24} \, \text{m}^{-2}. \quad (14.3)$$

Substituting into (14.2) we find for the number of scatterings

$$N_{\text{sc}} = N_{\text{inc}} \, n_{\text{tar}} \, \sigma = (10^4) \times (6.0 \times 10^{24} \, \text{m}^{-2}) \times (1.5 \times 10^{-28} \, \text{m}^2) = 9.$$

Here, we used the given cross section σ to predict the number N_{sc} of scatterings we should observe. Alternatively, we could have used the observed value of N_{sc} to *find* the cross section σ.

14.3 Generalizations of the Cross Section

The relation (14.2), with its many generalizations, is the fundamental relation of collision theory. Theorists calculate the cross section σ using assumed models of the target, and experimenters then use (14.2) to measure σ and compare with the predicted value. However, the projectile and target, and their interactions, are generally much more complicated than the simple point projectile and hard-sphere target that we used in the last section. In this section, we'll look at a few slightly more interesting cases.

---

[4] As we shall see shortly, the cross section of a target can be different for different projectiles. Thus, strictly speaking, I should say that the cross section of aluminum *for scattering neutrons* is about 1.5 barns. Moreover, the cross section can depend on the energy of the projectiles; the number given here is valid for energies from about 0.1 eV to about 1000 eV.

Scattering of Two Hard Spheres

Let us imagine a target that is a hard sphere of radius R_2 and a projectile that is another hard sphere of radius R_1, as shown in Figure 14.5. The two spheres make contact if and only if the impact parameter b is less than or equal to the *sum* of the two radii, $b \leq R_1 + R_2$; that is, the center of the projectile must lie inside a circle centered on the target with radius $R_1 + R_2$ and area $\sigma = \pi (R_1 + R_2)^2$. We can now go through exactly the same argument as in the last section, finding the probablility of any one projectile getting scattered, and thence the total number of projectiles scattered. The only difference is that the area of the target $\sigma = \pi R^2$ must now be replaced by $\sigma = \pi (R_1 + R_2)^2$. Thus we arrive at the same conclusion

$$N_{\mathrm{sc}} = N_{\mathrm{inc}} \, n_{\mathrm{tar}} \, \sigma \tag{14.4}$$

except that now

$$\sigma = \pi (R_1 + R_2)^2. \tag{14.5}$$

The main moral of this example is that we can continue to use the usual relation (14.4), but you should no longer see σ as the cross section of the target. Rather, σ is a property of the target *and* projectile and should be thought of as the *effective area of the former for scattering the latter*. In particular, the cross section of a particular target for scattering one kind of projectile may be very different from that for the same target with a different projectile.

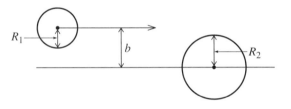

Figure 14.5   A hard-sphere projectile of radius R_1 approaches a hard-sphere target of radius R_2, with impact parameter b. A collision occurs only if $b \leq R_1 + R_2$.

EXAMPLE 14.3 Mean Free Path of an Air Molecule

The N_2 and O_2 molecules in the air around us behave very much like hard spheres of radius $R \approx 0.15$ nm. Use this to estimate the mean free path of an air molecule at STP.

The mean free path λ of a molecule in a gas is an important parameter that determines several properties of the gas — the conductivity, viscosity, and diffusion rates, for example. It is defined as the average distance that a molecule travels between collisions with other molecules. To estimate this, let us follow one chosen molecule as it moves through the gas, starting immediately after a collision. To simplify our discussion, I shall assume that all the other molecules are stationary. (This approximation changes the answer a little, but we shall still get a reasonable estimate.) We can think of our chosen molecule as a projectile moving through a target assembly of all the other stationary molecules, and our problem is to find the average distance x it will travel before making a collision. The cross section for collisions is given by (14.5) as $\sigma = \pi(2R)^2 = 4\pi R^2$. If we imagine first a thin slice of the gas with thickness dx (in the direction of our projectile's velocity), then the "target" density of this slice is

$$n_{\text{tar}} = \frac{N}{V}dx$$

where N is the total number of molecules and V their total volume, so that N/V is the number density (number/volume). Thus the probablity that our molecule will make a collision in any thin slice of thickness dx is given by (14.1) as

$$\text{prob(collision in } dx) = \frac{N\sigma}{V}dx. \tag{14.6}$$

Let us denote by prob(x) the probability that the projectile travels a distance x without making any collisions. The probability that it travels a distance x without colliding and then *does* collide in the next dx is the product of prob(x) and (14.6):

$$\text{prob(first collision between } x \text{ and } x + dx) = \text{prob}(x) \cdot \frac{N\sigma}{V}dx. \tag{14.7}$$

On the other hand, this same probabilty is just

$$\text{prob(first collision between } x \text{ and } x + dx) = \text{prob}(x) - \text{prob}(x + dx)$$

$$= -\frac{d}{dx}\text{prob}(x)\,dx. \tag{14.8}$$

Comparing (14.7) with (14.8), we get a differential equation for prob(x):

$$\frac{d}{dx}\text{prob}(x) = -\frac{N\sigma}{V}\,\text{prob}(x),$$

from which we see that prob(x) decreases exponentially with x,

$$\text{prob}(x) = e^{-(N\sigma/V)x}. \tag{14.9}$$

[Here I have used the initial condition that prob(0) is obviously 1.] The mean free path is the average value of x (that is, $\lambda = \langle x \rangle$), and, to find this average, we must multiply x by the probability (14.8) and integrate over all possible values of x:

$$\lambda = \langle x \rangle = \int_0^\infty x \left[\frac{N\sigma}{V} e^{-(N\sigma/V)x} \right] dx = \frac{V}{N\sigma}. \tag{14.10}$$

At STP we know that 22.4 liters of air contain Avogadro's number of molecules (one mole). Therefore,

$$\lambda = \frac{V}{N_A (4\pi R^2)} = \frac{22.4 \times 10^{-3}\,\text{m}^3}{(6.02 \times 10^{23}) \times 4\pi (0.15 \times 10^{-9}\,\text{m})^2}$$

$$= 1.3 \times 10^{-7}\,\text{m} = 130\,\text{nm}.$$

We see that the mean free path of an air molecule is considerably more than the inter-molecular spacing (around 3 nm) and vastly bigger than the molecular size (around 0.3 nm).

Different Processes and Targets

So far we have considered collisions in which the most that can happen is that the projectile is deflected by the target and emerges in a different direction. There are several other possibilities: Consider a collision between a point projectile and a target consisting of a ball of putty. If the putty is sufficiently absorbent, then any projectile that hits it will burrow in and never re-emerge. That is, the target will *capture*, or *absorb*, the projectile. Exactly our previous argument would then give the number of projectiles captured as $N_{\text{cap}} = N_{\text{inc}}\, n_{\text{tar}}\, \sigma$. We can easily make matters more complicated. For example, part of the target's surface could be absorbent and part hard. Projectiles that hit the absorbent surface would be captured and those that hit the hard part would be scattered (that is, re-emerge traveling in a different direction). In this case, we would have two separate relations analogous to (14.4), one to give the number of captures and the second for the number of scatterings:

$$N_{\text{cap}} = N_{\text{inc}}\, n_{\text{tar}}\, \sigma_{\text{cap}} \qquad \text{[capture]} \qquad (14.11)$$

and

$$N_{\text{sc}} = N_{\text{inc}}\, n_{\text{tar}}\, \sigma_{\text{sc}} \qquad \text{[scattering]}. \qquad (14.12)$$

Here σ_{cap} is the area of that part of the target that absorbs the projectile and σ_{sc} the area of the part that scatters it. The total cross section of the target is, of course,

$$\sigma_{\text{tot}} = \sigma_{\text{cap}} + \sigma_{\text{sc}}.$$

An example of a real collision in which both capture and scattering are possible is the collision of an electron and an atom, such as chlorine, that can capture an extra electron. In this case, we can no longer identify a particular area of the target that will capture the projectile, but it will still be true that the number of projectiles captured is proportional to N_{inc}, the number of incident projectiles, and to n_{tar}, the target density. Thus we can use the exact same equation (14.11) to define the **capture cross section** σ_{cap} as the relevant constant of proportionality. With this definition you can (and should) view σ_{cap} as the *effective area of the target for capturing the projectile*. In the same way, we can use (14.12) to define the **scattering cross section** σ_{sc} and then view σ_{sc} as the effective area of the target for scattering the projectile.

In collisions of electrons and atoms, there are other possibilities besides scattering and capture. For example, if the incident electron has enough energy, it may be able to *ionize* the atom, knocking one or more of the atomic electrons free. The number of ionizations can be written as

$$N_{\text{ion}} = N_{\text{inc}} \, n_{\text{tar}} \, \sigma_{\text{ion}} \qquad \text{[ionization]} \qquad (14.13)$$

and this defines the **ionization cross section** σ_{ion} as the effective area of the target atom for ionization by the incoming electron. Again, when a neutron collides with a nucleus of ^{235}U, it can cause the nucleus to *fission*, splitting into two much smaller nuclei and releasing some 200 MeV of kinetic energy, and we can define the **fission cross section** σ_{fis} as the effective area of a ^{235}U nucleus for fission by neutron bombardment, satisfying the equation $N_{\text{fis}} = N_{\text{inc}} \, n_{\text{tar}} \, \sigma_{\text{fis}}$.

There is one other classification that deserves mention. The word "scattering" is generally reserved for a process in which the projectile is deflected and moves off, leaving behind the *same* target — the same atom, the same nucleus, or whatever. This usage excludes processes like capture, in which the projectile does not emerge from the collision at all. It also excludes processes like ionization, in which an electron is knocked off the target atom, or like fission, in which the target nucleus is broken into pieces. If the internal motions of the target are left unchanged, the scattering is said to be **elastic**; if the internal motions of the target are changed by the collision, then the scattering is called **inelastic**. Consider, for example, the scattering of an electron by a stationary atom. To simplify our discussion, let us suppose that the atom is fixed (an excellent approximation, since an atom is so much heavier than an electron) and that the atomic electrons are initially in their **ground state** — the lowest possible energy level[5] and the atom's most stable state. When the incoming electron scatters off the atom, there are two possibilities: The electron may scatter elastically, emerging with its kinetic energy unchanged and leaving the target in its original state of internal motion. Or it can scatter inelastically, giving some of its kinetic energy to the atom and raising the target's internal motion to a higher energy level.[6] This latter process of **atomic excitation** was first observed by the German physicists James Franck (1882–1964) and Gustav Hertz (1887–1975)and gave compelling evidence for the existence of atomic energy levels (and won them the 1925 Nobel Prize).

When a projectile scatters off a target, we can, if we wish, distinguish between the two types of process, elastic and inelastic. The total number of scatterings N_{sc} in a given experiment is the sum of the elastic and inelastic scatterings, $N_{\text{sc}} = N_{\text{el}} + N_{\text{inel}}$, and we can define corresponding cross sections satisfying $\sigma_{\text{sc}} = \sigma_{\text{el}} + \sigma_{\text{inel}}$.

For a given target and given projectile, we can enumerate all the possible outcomes of a collision — scattering, capture, ionization, fission, and so on — and for each

---

[5] Recall that atoms can exist only in certain discrete "energy levels." Since an isolated atom eventually finds its way to the lowest energy level (ground state), the target atom in a collision is most often in its ground state.

[6] In general, both projectile and target can move, and they may both have internal structure and hence energy of internal motion — as in the collision of two molecules in a gas. In this case, an elastic scattering is defined as one in which the internal motions of both projectile and target are unchanged. This means the total kinetic energies $T_{\text{proj}} + T_{\text{tar}}$ are the same before and after the encounter.

outcome, we can define a corresponding cross section. The sum of all these partial cross sections is called the **total cross section** σ_{tot}. For example, for electrons colliding with an atom, it may be that there are just three possible outcomes, scattering, capture, or ionization; in this case, the total cross section would be

$$\sigma_{\text{tot}} = \sigma_{\text{sc}} + \sigma_{\text{cap}} + \sigma_{\text{ion}}.$$

By adding up the three equations (14.12), (14.11), and (14.13), we can see that the total cross section gives the total number of projectiles removed from the incident beam by all three possible processes:

$$N_{\text{tot}} = N_{\text{inc}} \, n_{\text{tar}} \, \sigma_{\text{tot}} \qquad \text{[total]}.$$

That is, σ_{tot} is the effective area of the target for interacting with the projectile in any of the possible ways.

In most cases, the various cross sections we have defined are found to vary with the energy of the incident projectile. For an example that is easily understood, consider the ionization cross section for an electron colliding with an atom. To ionize the atom, the incoming electron must have the minimum energy needed to knock one of the atomic electrons free. If the incident electron's energy is less than this **ionization energy**, then ionization is impossible; therefore, $N_{\text{ion}} = 0$, and the ionization cross section σ_{ion} defined by (14.13) is exactly zero. Above the ionization energy, ionization is possible and both N_{ion} and σ_{ion} are normally nonzero. Obviously σ_{ion} varies with energy. Although it is not always as obvious, we find in practice that almost all cross sections are likewise energy dependent.

14.4 The Differential Scattering Cross Section

In defining the cross section σ_{sc} in (14.12) we counted the total number of projectiles that were scattered, regardless of the particular direction in which they emerged. Obviously we could obtain more information if we chose to monitor these directions as well, and this leads us to the notion of the *differential cross section*, as we now discuss.

To simplify matters, let us consider a collision where the only possible interaction is elastic scattering. For example, you could consider the scattering of a point projectile off a hard sphere (Figure 14.2) or the Rutherford scattering of a positively charged alpha particle off a heavy positive nucleus (Figure 14.1). In either case, there are just two possibilities: Either the projectile misses the target entirely and emerges unscattered, or it scatters elastically.[7]

---

[7] If the alpha particle of the Rutherford experiment had enough energy, it could also raise the nucleus to a higher energy level or knock it apart. However, at the low energies available to Rutherford, this was not a possibility, and we can confine our attention to these low energies for now. There is another subtlety in Rutherford scattering that is possibly worth mentioning: The Coulomb force $F = kqQ/r^2$ of a truly isolated nucleus would extend all the way to infinity, and all alpha

If we are to monitor how many particles emerge in a given direction, we must agree on how to measure directions. It is standard to take the direction of the incident beam as our z axis, and then to specify the direction of any one scattered projectile by giving its polar angles θ and ϕ. Since these angles form an infinite continuum, we cannot speak of the number of particles scattered in the *exact* direction (θ, ϕ). Rather we must count the number of particles emerging in some narrow cone around (θ, ϕ). To characterize the size of this narrow cone, we use the notion of *solid angle*, which is defined as follows.

Solid Angle

To understand the definition of the solid angle of a cone it helps to recall the definition of the ordinary angle between two lines in a plane. This is illustrated in Figure 14.6(a): If the two lines meet at O, we draw a circle of any convenient radius r centered at O. The two lines define an arc of length s on the circle, and we define the angle $\Delta\theta$ (in radians) as $\Delta\theta = s/r$. (Since s is proportional to r, this definition is independent of our choice of r.) In a similar way, if a three-dimensional cone has its apex at O, we draw a sphere of radius r centered at O as in Figure 14.6(b). The cone intersects the sphere in a spherical surface of area A (proportional to r^2), and we define the **solid angle** of the cone as

$$\Delta\Omega = A/r^2. \tag{14.14}$$

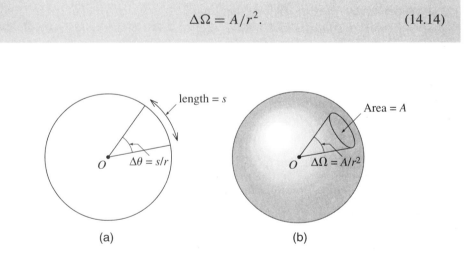

(a)   (b)

Figure 14.6   **(a)** The ordinary two-dimensional angle $\Delta\theta$ subtended by an arc length s of a circle is defined as $\Delta\theta = s/r$ where r is the radius of the circle and $\Delta\theta$ is in radians. **(b)** The solid angle $\Delta\Omega$ of a cone subtended by an area A on a sphere is defined as $\Delta\Omega = A/r^2$. Here r is the radius of the sphere and $\Delta\Omega$ is in steradians.

particles, however large their impact parameter, would be deflected a tiny bit; in other words, *all* of the incident alphas would be scattered. In practice, however, the Coulomb force of the nucleus is always screened at large distances by the atomic electrons, and those alphas that pass outside the whole atom are not scattered.

The unit of solid angle defined in this way is called the **steradian**, abbreviated **sr**. If the cone includes all possible directions, then since the area of the whole sphere is $4\pi r^2$, the solid angle is 4π. That is, the solid angle corresponding to all possible directions in three dimensions is 4π steradians, just as the ordinary angle corresponding to all directions in two dimensions is 2π radians. The cone shown in Figure 14.6(b) was a circular cone, but the definition $\Delta\Omega = A/r^2$ works equally for any shape of cone. For instance, we shall need to consider the narrow cone with polar angles in the ranges θ to $\theta + d\theta$ and ϕ to $\phi + d\phi$; this cone intersects the sphere in a "rectangular" surface of area $r^2 \sin\theta\, d\theta\, d\phi$, and so has solid angle

$$d\Omega = \sin\theta\, d\theta\, d\phi. \tag{14.15}$$

The Differential Cross Section

Armed with the notion of solid angle, we are ready to define the differential scattering cross section. We imagine our usual experiment, in which a large number N_{inc} of projectiles are directed at a target assembly with density n_{tar}. For any chosen cone of solid angle $d\Omega$ in any chosen direction (θ, ϕ), we now monitor the number of projectiles scattered into this $d\Omega$, as sketched in Figure 14.7. We denote this number by $N_{\text{sc}}(\text{into } d\Omega)$, and, by the familiar argument it must be proportional to N_{inc} and to n_{tar}, so can be written as

$$N_{\text{sc}}(\text{into } d\Omega) = N_{\text{inc}}\, n_{\text{tar}}\, d\sigma(\text{into } d\Omega) \tag{14.16}$$

where $d\sigma(\text{into } d\Omega)$ is the effective cross-sectional area of the target for scattering into the solid angle $d\Omega$. Since this is proportional to $d\Omega$, it is traditional to rewrite it as

$$d\sigma(\text{into } d\Omega) = \frac{d\sigma}{d\Omega} d\Omega$$

where the factor $d\sigma/d\Omega$ is called the **differential scattering cross section**. In terms of it, we can rewrite (14.16) as

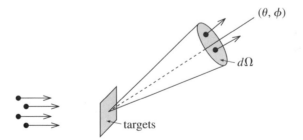

Figure 14.7   Projectiles are incident from the left on a rectangular target assembly, and we monitor the number $N_{\text{sc}}(\text{into } d\Omega)$ emerging into a cone of solid angle $d\Omega$ in the direction (θ, ϕ).

$$N_{sc}(\text{into } d\Omega) = N_{inc}\, n_{tar}\, \frac{d\sigma}{d\Omega}(\theta, \phi)\, d\Omega \qquad (14.17)$$

where I have added the argument (θ, ϕ) to emphasize that the differential cross section will (in general) depend on the direction of observation. Equation (14.17) can be taken as the definition of the differential cross section $d\sigma/d\Omega$. It is the experimentalist's job to measure $d\sigma/d\Omega$ using (14.17), and it is the theorist's job to predict $d\sigma/d\Omega$ based on some assumed model of the interactions between the projectile and the target.

If we add up the numbers $N_{sc}(\text{into } d\Omega)$ for all possible solid angles $d\Omega$ we will recover the total number of scatterings, N_{sc}. That is, integrating (14.17) over all solid angles will give N_{sc}. Since $N_{sc} = N_{inc} n_{tar} \sigma$, where σ is the total scattering cross section, we conclude that

$$\sigma = \int \frac{d\sigma}{d\Omega}(\theta, \phi)\, d\Omega = \int_0^\pi \sin\theta\, d\theta \int_0^{2\pi} d\phi\, \frac{d\sigma}{d\Omega}(\theta, \phi) \qquad (14.18)$$

where the second form follows from (14.15). That is, the total cross section[8] is the integral over all solid angles of the differential cross section.

EXAMPLE 14.4   Angular Distribution of Scattered Neutrons

At an incident energy of several MeV (million electron volts), the differential cross section for scattering of neutrons off a heavy nucleus might have the form

$$\frac{d\sigma}{d\Omega}(\theta, \phi) = \sigma_0 (1 + 3\cos\theta + 3\cos^2\theta) \qquad (14.19)$$

where σ_0 is a constant that could be about 30 millibarns per steradian (mb/sr). Describe the angular distribution of the scattered neutrons and find the total scattering cross section.

The most prominent feature of (14.19) is that it is independent of ϕ. (As we shall see in the next section, this is a fairly common occurrence.) This means that the distribution of scattered neutrons is *axially symmetric*, which makes visualization of the angular distribution much simpler, since we have only to worry about its dependence on θ. Figure 14.8 shows $d\sigma/d\Omega$ as a function of θ from $\theta = 0$ to π. We see that $d\sigma/d\Omega$ is largest at $\theta = 0$. That is, in this example at least, a particle that is scattered is most likely to be scattered near to the "forward direction" $\theta = 0$. (In quantum scattering, especially at high energies, this is often the case.) In this example, there is also a much smaller maximum in the backward direction at $\theta = \pi$, and the probability for scattering directly back at $\theta = \pi$ is greater than for any other direction in the backward hemisphere $\pi/2 < \theta < \pi$.

---

[8] In general, one should say the total *scattering* cross section, σ_{sc}; here we are assuming that scattering is the only possible interaction, so they are the same.

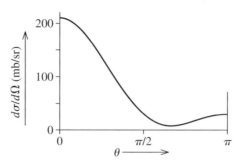

Figure 14.8   The differential cross section (14.19) for scattering of neutrons off a nucleus, plotted as a function of the scattering angle θ. The vertical axis shows $d\sigma/d\Omega$ in units of millibarns per steradian.

The total scattering cross section is found by integrating the differential cross section, as in (14.18). Since the integrand is independent of ϕ, the integration over ϕ is trivial and gives a factor of 2π, so

$$\sigma = \int \frac{d\sigma}{d\Omega}(\theta, \phi)\, d\Omega = 2\pi\sigma_{\rm o} \int_0^{\pi} \sin\theta\, d\theta (1 + 3\cos\theta + 3\cos^2\theta)$$

$$= 8\pi\sigma_{\rm o} = 754 \text{ mb.} \tag{14.20}$$

14.5 Calculating the Differential Cross Section

To simplify the calculation of the differential cross section, I shall assume that the scattering is axially symmetric. This is certainly the case if the target is spherically symmetric (like the hard sphere of Figure 14.2 or any target which exerts a spherically symmetric force field), since spherical symmetry implies axial symmetry. It means that the differential cross section is independent of ϕ and allows us to include all different values of ϕ in our discussion at the same time. We imagine a projectile incident on the target with impact parameter b. By calculating the projectile's trajectory, we can, in principle at least, find the corresponding scattering angle $\theta = \theta(b)$ as a function of b. Alternatively, by solving for b, we can express b as a function of θ, that is, $b = b(\theta)$.

Let us next consider all those projectiles that approach the target with impact parameters between b and $b + db$. These are incident on the annulus (the shaded ring shape) shown on the left of Figure 14.9. This annulus has cross sectional area

$$d\sigma = 2\pi b\, db. \tag{14.21}$$

These same particles emerge between angles θ and $\theta + d\theta$ in a solid angle

$$d\Omega = 2\pi \sin\theta\, d\theta, \tag{14.22}$$

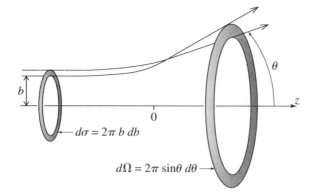

Figure 14.9   All projectiles incident between b and $b + db$ are scattered between angles θ and $\theta + d\theta$. The area on which these particles impinge is $d\sigma = 2\pi b\, db$, and the solid angle into which they scatter is $d\Omega = 2\pi \sin\theta\, d\theta$.

as indicated in Figure 14.9. The differential cross section $d\sigma/d\Omega$ is now found by simply dividing (14.21) by (14.22) to give

$$\frac{d\sigma}{d\Omega} = \frac{b}{\sin\theta}\left|\frac{db}{d\theta}\right| \tag{14.23}$$

where I have inserted the absolute value signs to ensure that $d\sigma/d\Omega$ is positive. (Because θ often decreases as b increases, $db/d\theta$ may be negative.)

In summary: To calculate the differential cross section for a projectile scattering off a given target, we must first calculate the projectile's trajectory, to find the scattering angle θ as a function of the impact parameter b (or vice versa). Then $d\sigma/d\Omega$ is found by simply differentiating b with respect to θ, as in (14.23).

EXAMPLE 14.5   Hard Sphere Scattering

As a first example of the use of (14.23), find the differential cross section for scattering of a point projectile off a fixed rigid sphere of radius R. Integrate your result over all solid angles to find the total cross section.

Our first task is to find the trajectory of a scattered projectile, as shown in Figure 14.10. The crucial observation is that when the projectile bounces off the hard sphere, its angles of incidence and reflection (both shown as α in the picture) are equal. (This "law of reflection" follows from conservation of energy and angular momentum — see Problem 14.13.) Inspection of the picture shows that the impact parameter is $b = R \sin\alpha$, and the scattering angle is $\theta = \pi - 2\alpha$. Combining these two equations we find that

$$b = R \sin\frac{\pi - \theta}{2} = R\cos(\theta/2), \tag{14.24}$$

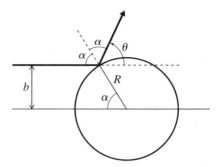

Figure 14.10   A point projectile bouncing off a fixed rigid sphere obeys the law of reflection, that the two adjacent angles labelled α are equal. The impact parameter is $b = R \sin \alpha$, and the scattering angle is $\theta = \pi - 2\alpha$.

and from (14.23), we find the differential cross section

$$\frac{d\sigma}{d\Omega} = \frac{b}{\sin\theta}\left|\frac{db}{d\theta}\right| = \frac{R\cos(\theta/2)}{\sin\theta}\frac{R\sin(\theta/2)}{2} = \frac{R^2}{4}. \tag{14.25}$$

The most striking thing about this result is that the differential cross section is isotropic; that is, the number of particles scattered into a solid angle $d\Omega$ is the same in all directions. To find the total cross section, we have only to integrate this result over all solid angles:

$$\sigma = \int \frac{d\sigma}{d\Omega}d\Omega = \int \frac{R^2}{4}d\Omega = \pi R^2,$$

which is, of course, the cross-sectional area of the target sphere.

14.6   Rutherford Scattering

Perhaps the most famous collision experiment of all time was Rutherford's experiment, in which he and his assistants observed the scattering of alpha particles off the gold nuclei in a thin gold foil and used the observed distribution to argue for the nuclear model of the atom. According to this model, the force of a nucleus (charge Q) on an alpha (charge q) is

$$F = \frac{kqQ}{r^2} = \frac{\gamma}{r^2}. \tag{14.26}$$

The alphas are scattered appreciably only if they approach close to the nucleus, well inside the orbiting atomic electrons. Therefore we can ignore the force of the latter, and (14.26) is the only force on the alphas. Therefore, as we saw in Chapter 8, the orbit of an alpha is a hyperbola, with the nucleus (which we'll treat as fixed for the moment) at its focus, as shown in Figure 14.11. If $\mathbf{u}$ denotes the unit vector pointing from the target to the alpha's point of closest approach, the orbit is symmetric about

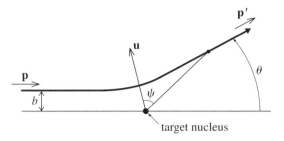

Figure 14.11   The Rutherford scattering of an alpha particle off a fixed atomic nucleus. The orbit is a hyperbola, which is symmetric about the line labelled by the fixed unit vector **u**. The position of the particle can be labelled by its angle ψ measured from **u**. As the particle moves away $(t \to \infty)$, $\psi \to \psi_o$, and as $t \to -\infty$, $\psi \to -\psi_o$. Therefore the scattering angle is $\theta = \pi - 2\psi_o$.

the direction of **u**, and it is convenient to label the alpha's position by the polar angle ψ, measured from **u** (see Figure 14.11). Let us denote by ψ_o the limit of ψ as the scattered alpha moves far away, so that the total angle subtended by the alpha's orbit is $2\psi_o$ and the scattering angle is

$$\theta = \pi - 2\psi_o. \qquad (14.27)$$

Our job now is to relate the scattering angle θ to the impact parameter b. We can do this by evaluating in two ways the change in the momentum of the projectile,

$$\Delta \mathbf{p} = \mathbf{p}' - \mathbf{p}, \qquad (14.28)$$

where **p** and **p**' are the momentum long before and long after the encounter. First, by conservation of energy, **p** and **p**' have equal magnitudes, so that the triangle shown in Figure 14.12 is isosceles, and

$$|\Delta \mathbf{p}| = 2p \sin(\theta/2). \qquad (14.29)$$

On the other hand, from Newton's second law, $\Delta \mathbf{p} = \int \mathbf{F} dt$. Comparing Figures 14.12 and 14.11, you can see that $\Delta \mathbf{p}$ is in the same direction as the unit vector **u**. Thus

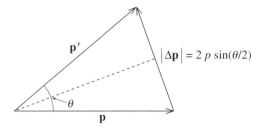

Figure 14.12   The change in momentum of the projectile is $\Delta \mathbf{p} = \mathbf{p}' - \mathbf{p}$. Since, $|\mathbf{p}| = |\mathbf{p}'|$, it is easily seen that $|\Delta \mathbf{p}| = 2p \sin(\theta/2)$.

the magnitude of $\Delta \mathbf{p}$ is given by the same integral, with $\mathbf{F}$ replaced by its component F_u in the direction of $\mathbf{u}$,

$$|\Delta \mathbf{p}| = \int_{-\infty}^{\infty} F_u \, dt.$$

From Figure 14.11 you can see that $F_u = (\gamma/r^2) \cos \psi$. Using the now-familiar trick, we can write $dt = d\psi/\dot{\psi}$, where, since $mr^2\dot{\psi} = \ell = bp$ (see Figure 14.11 again), we can replace $\dot{\psi}$ by bp/mr^2. Putting all of this together, we find

$$|\Delta \mathbf{p}| = \int_{-\psi_o}^{\psi_o} \frac{\gamma \cos \psi}{r^2} \frac{d\psi}{bp/mr^2} = \frac{\gamma m}{bp} 2 \sin \psi_o = \frac{2\gamma m}{bp} \cos(\theta/2). \quad (14.30)$$

[To understand the limits in the integral, recall that as $t \to \pm\infty$, so $\psi \to \pm\psi_o$. In the last step I used (14.27) to replace ψ_o by $(\pi - \theta)/2$ and hence $\sin \psi_o$ by $\cos(\theta/2)$.] Equating the two expressions (14.29) and (14.30) for $|\Delta \mathbf{p}|$, we can solve for b to give

$$b = \frac{\gamma m}{p^2} \frac{\cos(\theta/2)}{\sin(\theta/2)} = \frac{\gamma}{mv^2} \cot(\theta/2) \quad (14.31)$$

where in the last equality I replaced p by mv, and v is the projectile's incident speed.

Having found the impact parameter b as a function of the scattering angle θ, we can now use the result (14.23) to give the differential cross section

$$\frac{d\sigma}{d\Omega} = \frac{1}{\sin \theta} \cdot b \cdot \left|\frac{db}{d\theta}\right| = \frac{1}{2 \sin(\theta/2) \cos(\theta/2)} \cdot \frac{\gamma}{mv^2} \cot(\theta/2) \cdot \frac{\gamma}{mv^2} \frac{1}{2 \sin^2(\theta/2)}$$

or, replacing γ by kqQ,

$$\frac{d\sigma}{d\Omega} = \left(\frac{kqQ}{4E \sin^2(\theta/2)}\right)^2 \quad (14.32)$$

where E is the energy of the incident projectiles, $E = \frac{1}{2}mv^2$. This is the celebrated **Rutherford scattering formula**. It gives the differential cross section for scattering of a charge q, with energy E, off a fixed target of charge Q. While it is still today, nearly a century after its derivation by Rutherford, a much-used result, its great historical importance is that it was used to prove the existence of the atomic nucleus, as we now discuss briefly.[9]

_____

[9] Since the atom is a microscopic system, for which quantum, not classical, mechanics should be used, you may be surprised that the classical Rutherford formula worked so well for Rutherford and his assistants. It is one of the most amazing accidents in the history of physics that the quantum formula for scattering of two charged particles agrees exactly with Rutherford's classical formula. (This is certainly not true for other force laws.)

The Experiment of Geiger and Marsden

The best known and most important Rutherford-scattering experiment was performed by Rutherford's assistants Hans Geiger (inventor of the Geiger counter, 1882–1945) and Ernest Marsden (1889–1970) and published in 1913. Their goal was to test Rutherford's "planetary" model of the atom, according to which most of the atomic mass is concentrated in a tiny, positively charged nucleus.[10] As we have seen, this model leads to the cross section (14.32) for scattering of alpha particles, with several very specific predictions: The scattering probablility should be inversely proportional to $\sin^4 \theta/2$, inversely proportional to the energy squared E^2, and proportional to the nuclear charge squared (Q^2). Geiger and Marsden were able to verify all of these predictions with amazing precision, and hence to contribute to the rapid acceptance of Rutherford's nuclear atom. They used alpha particles coming from radon gas ("radium emanation" as it was called then), with energy around 6.5 MeV. (1 MeV = 10^6 electron volts, and 1 eV = 1.6×10^{-19} joules.) They directed a narrow "pencil" of these at a thin metal foil and counted the scattered particles using a small zinc sulphide screen. Any alpha particle striking this screen caused a tiny flash of light or "scintillation," which could be observed through a microscope. In this way, it was possible to count up to about 90 alpha particles per minute (a job needing great patience and concentration!). To observe the angular dependence of the scattering, they could swing the screen and microcope around to angles in the range $5° \leq \theta \leq 150°$. To test the dependence on incident energy, they passed the incident particles through thin sheets of mica, to slow them down and hence vary their energy. And to test the dependence on nuclear charge, they used various different target foils (gold, platinum, tin, silver, copper, and aluminum).

EXAMPLE 14.6 Angular Dependence

To isolate its angular dependence write the Rutherford cross section (14.32) as

$$\frac{d\sigma}{d\Omega}(\theta) = \frac{\sigma_0(E)}{\sin^4 \theta/2} \tag{14.33}$$

and find $\sigma_0(E)$ for scattering of 6.5 MeV alphas off gold. Find the differential cross section at 150° and 5° (Geiger and Marsden's largest and smallest angles). Find the number of alphas they would have had to count in a minute assuming the following values: The number of incident alphas in one minute, $N_{\text{inc}} = 6 \times 10^8$; the thickness of the gold foil, $t = 1 \, \mu\text{m}$; area of zinc sulphide screen = 1 mm^2; and distance of screen from target = 1 cm. Make a useable plot of the differential cross section as a function of scattering angle θ.

---

[10] Initially, the sign of the nuclear charge (positive or negative) was not clear, but it was soon found to be positive, with an equal negative charge carried by the orbiting electrons.

The charge of the alpha particle is $q = 2e$ and that of the gold nucleus is $Q = 79e$, so

$$\sigma_o(E) = \left(\frac{2 \times 79 \times ke^2}{4E} \right)^2.$$

This is easily evaluated in SI units, though a slicker way is to use the useful combination $ke^2 = 1.44$ MeV·fm (where fm stands for femtometer or 10^{-15} m). Either way, we find that

$$\sigma_o = 76.6 \times 10^{-30} \text{ m}^2/\text{sr} = 0.766 \text{ barns/sr}.$$

Substituting into (14.33) we get

$$\frac{d\sigma}{d\Omega}(150°) = 0.88 \text{ barns/sr} \quad \text{and} \quad \frac{d\sigma}{d\Omega}(5°) = 2.1 \times 10^5 \text{ barns/sr}. \quad (14.34)$$

The huge difference between these — more than 5 orders of magnitude — presents considerable practical difficulties, as we shall see. Before we can substitute into (14.17) to give the actual numbers counted we need to calculate n_{tar} and $d\Omega$. As usual, we can find n_{tar} in terms of the density of gold (specific gravity 19.3) and its atomic mass (197):

$$n_{\text{tar}} = \frac{\varrho t}{m} = \frac{(19.3 \times 10^3 \text{ kg/m}^3) \times (10^{-6} \text{ m})}{197 \times 1.66 \times 10^{-27} \text{ kg}} = 5.90 \times 10^{22} \text{ m}^{-2}.$$

Geiger and Marsden's screen had area $A = 1 \text{ mm}^2$ and was at a distance $r = 10$ mm from the target. Therefore, it subtended a solid angle

$$d\Omega = \frac{A}{r^2} = 0.01 \text{ sr}.$$

Putting all of this together, we find for the number of alphas hitting their screen at 150° in a minute

$$N_{\text{sc}}(\text{at } 150°) = N_{\text{inc}} \, n_{\text{tar}} \frac{d\sigma}{d\Omega}(150°) \, d\Omega$$

$$= (6 \times 10^8) \times (5.90 \times 10^{22} \text{ m}^{-2}) \times (0.88 \times 10^{-28} \text{ m}^2/\text{sr}) \times (0.01 \text{ sr})$$

$$= 31,$$

a number that they could count easily and accurately. On the other hand, the same calculation gives

$$N_{\text{sc}}(\text{at } 5°) = 7.5 \times 10^6,$$

a number that they could not possibly count or even estimate. Obviously measuring the cross section at small angles required them to use a much, much weaker source than at large angles.

Because of the huge variation of the cross section as the scattering angle varies, a straightforward linear plot of $d\sigma/d\Omega$ is not especially useful. If we choose a scale to show the small angles, the cross section for large angles will

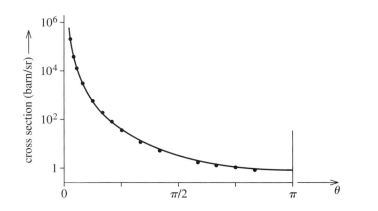

Figure 14.13   Semilog plot of the Rutherford differential cross section as a function of angle θ. The dots are the measurements of Geiger and Marsden.

appear to be zero; if we arrange to show the large angles, the cross section for small angles will disappear off scale. The solution to this is to make a semilog plot, that is, to graph the log of the cross section against θ. This gives the curve shown in Figure 14.13, where you can see clearly the variation by more than 5 orders of magnitude between 15° and 180°. The dots in this figure are the original data of Geiger and Marsden and show clearly why Rutherford's model of the atom gained such quick acceptance.

14.7   Cross Sections in Various Frames*

*As usual, sections marked with an asterisk can be omitted on a first reading.

For the most part, we have so far discussed collisions in which the target particle is *fixed*. While this is an excellent approximation if the target is very heavy compared to the projectile (as in scattering of electrons off an atom, for example), we must nevertheless recognize that there is no such thing as a truly fixed particle, and we must learn to treat collisions of two particles both of which can move. Fortunately, we already know how to do this: If we observe the motion in the CM frame (the reference frame where the center of mass is at rest), then the motion of the relative coordinate $\mathbf{r} = \mathbf{r}_1 - \mathbf{r}_2$ is precisely the same as that of a single particle with mass equal to the reduced mass $\mu = m_1 m_2 / M$. Thus, if we view the collision in the CM frame, then our problem is reduced back to the motion of a single "equivalent particle" in a fixed force field. The only remaining difficulty is this: The CM frame is usually not the frame in which we do experiments. Thus we must learn how to relate cross sections calculated in the CM frame to their corresponding values in the **lab frame**, the frame of the laboratory in which the experiment is to be performed. In particular, we are going to want to find the relation between the differential cross sections $(d\sigma/d\Omega)_{\text{cm}}$ of the CM frame and the corresponding $(d\sigma/d\Omega)_{\text{lab}}$ measured in the lab.

The CM Variables

Before we take up the problem of translating between frames, I need to mention two more elegant features of the CM frame. First, recall that the Lagrangian for two particles, when written in terms of the CM and relative coordinates, has the form (8.13)

$$\mathcal{L} = \tfrac{1}{2}M\dot{\mathbf{R}}^2 + \tfrac{1}{2}\mu\dot{\mathbf{r}}^2 - U(r).$$

(14.35)

Differentiating $\mathcal{L}$ with respect to the three components of $\dot{\mathbf{r}}$, we find that the generalized momentum corresponding to $\mathbf{r}$ is

$$\mathbf{p} = \mu\dot{\mathbf{r}}.$$

(14.36)

That is (as you would probably have guessed) the momentum for the relative motion is just that of a single particle of mass μ and velocity $\dot{\mathbf{r}}$.

The second property concerns the momenta of the two particles, as measured in the CM frame. Recall from (8.9) that

$$\mathbf{r}_1 = \mathbf{R} + \frac{m_2}{M}\mathbf{r} \qquad \text{and} \qquad \mathbf{r}_2 = \mathbf{R} - \frac{m_1}{M}\mathbf{r}.$$

(14.37)

We can differentiate these to get the two velocities. In particular, in the CM frame, $\dot{\mathbf{R}} = 0$, so we find that $\dot{\mathbf{r}}_1 = (m_2/M)\dot{\mathbf{r}}$. Multiplying by m_1, we find for the momentum of the projectile $\mathbf{p}_1 = \mu\dot{\mathbf{r}} = \mathbf{p}$. That is, in the CM frame, the projectile's momentum $\mathbf{p}_1$ is the same as the momentum $\mathbf{p}$ of the relative motion. In the same way we can prove that $\mathbf{p}_2 = -\mathbf{p}$, so that

$$\mathbf{p}_1 = -\mathbf{p}_2 = \mathbf{p} = \mu\dot{\mathbf{r}} \qquad \text{(in the CM frame)}.$$

(14.38)

The first equality here confirms that the total momentum in the CM frame is zero. The second is useful in the evaluation of cross sections: In measuring the differential cross section, we count the number of times the projectile emerges with its momentum $\mathbf{p}_1$ inside some solid angle $d\Omega$. Since $\mathbf{p}_1 = \mathbf{p}$ (in the CM frame), this is the same thing as the number of times the relative momentum $\mathbf{p}$ emerges in $d\Omega$. Therefore we can find the differential cross section in the CM frame just as if a single particle of mass μ were scattering off a fixed target. Thus, for example, the Rutherford formula (14.32), for scattering of one particle off a fixed target, also gives the differential cross section for scattering of two charged particles in their CM frame, provided we replace m by μ.

General Relation between Cross Sections in Different Frames

In the CM frame the projectile and target approach one another with equal and opposite momenta. In the lab frame of the traditional collision experiment (such as the Rutherford experiment) the target is initially at rest. In many modern colliding-beam experiments the projectile and target are both moving, in opposite directions.

In all of these cases, the initial momenta are collinear, and to simplify matters, I shall restrict our discussion to this case.[11]

To see how the cross section transforms between two different frames, we have only to look at its definition. Let us start with the total cross section, which was defined in (14.2) so that

$$N_{sc} = N_{inc} \, n_{tar} \, \sigma.$$

(We'll continue to assume that the only possible outcome of a collision is elastic scattering, so we don't have to worry about other processes like absorption or ionization.) We can use this same definition in either frame. Thus we define the CM total cross section σ_{cm} by

$$N_{sc}^{cm} = N_{inc}^{cm} \, n_{tar}^{cm} \, \sigma_{cm} \tag{14.39}$$

where all four quantities are measured in the CM frame. In exactly the same way we define σ_{lab} by

$$N_{sc}^{lab} = N_{inc}^{lab} \, n_{tar}^{lab} \, \sigma_{lab} \tag{14.40}$$

where all four quantities are measured in the lab frame. While any particular scattering event may look very different as viewed from the two different frames, the total number of events must be the same in either frame. Therefore,

$$N_{sc}^{cm} = N_{sc}^{lab}.$$

In the same way, the number of incident particles is the same, as seen in either frame, so that $N_{inc}^{cm} = N_{inc}^{lab}$. The target density n_{tar} is the density of target particles (number/area) seen from the incident direction, as illustrated in Figure 14.4. Since this is unaffected by any forward (or backward) motion of the target, $n_{tar}^{cm} = n_{tar}^{lab}$. Comparing now (14.39) and (14.40), we see that each of the first three terms of (14.39) is equal to the corresponding term in (14.40). Therefore the final terms must also be equal, and we have the elegantly simple result that the total scattering cross sections in the CM and lab frames are equal,

$$\sigma_{cm} = \sigma_{lab} \qquad \text{[total scattering cross section]}. \tag{14.41}$$

If other outcomes were possible, such as absorption or ionization, then exactly the same argument would lead to exactly corresponding results; for example, the total absorption cross section is the same in the CM and lab frames.

The differential cross section is a little more complicated. The scattering angle, measured as θ_{cm} in the CM frame, will generally have a different value θ_{lab} in the lab frame, and a given solid angle measured as $d\Omega_{cm}$ in the one will be measured as $d\Omega_{lab}$

---

[11] Some colliding beams are not perfectly collinear. Some experiments in atomic physics use beams that intersect at a large angle, and in the collisions of molecules in a gas, the two particles can approach one another at any angle. Although such oblique collisions are not especially difficult to handle, I shall consider only collinear collisions here.

in the other. Apart from these complications, we can use the same argument as before. The definition of the differential cross section (14.17),

$$N_{\text{sc}}(\text{into } d\Omega) = N_{\text{inc}}\, n_{\text{tar}}\, \frac{d\sigma}{d\Omega}\, d\Omega, \tag{14.42}$$

can be used in either frame. Just as before, the numbers N_{inc} and n_{tar} have the same values in either frame. Further, the number scattered in any chosen solid angle, called $d\Omega_{\text{cm}}$, in the CM frame is the same as the number going into the corresponding solid angle, called $d\Omega_{\text{lab}}$, in the lab frame. Thus, as before, the first three terms in (14.42) have the same values in both frames, and the same must therefore be true of the final product; that is

$$\left(\frac{d\sigma}{d\Omega}\right)_{\text{cm}} d\Omega_{\text{cm}} = \left(\frac{d\sigma}{d\Omega}\right)_{\text{lab}} d\Omega_{\text{lab}} \tag{14.43}$$

or

$$\left(\frac{d\sigma}{d\Omega}\right)_{\text{lab}} = \left(\frac{d\sigma}{d\Omega}\right)_{\text{cm}} \frac{d\Omega_{\text{cm}}}{d\Omega_{\text{lab}}}. \tag{14.44}$$

We see that the differential cross sections are not the same in the two frames, but only because a given solid angle has different values ($d\Omega_{\text{cm}}$ and $d\Omega_{\text{lab}}$) according to the frame used.

Since $d\Omega = \sin\theta\, d\theta\, d\phi = -d(\cos\theta)\, d\phi$, and since the azimuthal angle ϕ of the outgoing momentum is the same in both frames, we can rewrite (14.44) in the more useful, if perhaps less transparent, form

$$\left(\frac{d\sigma}{d\Omega}\right)_{\text{lab}} = \left(\frac{d\sigma}{d\Omega}\right)_{\text{cm}} \left|\frac{d(\cos\theta_{\text{cm}})}{d(\cos\theta_{\text{lab}})}\right|. \tag{14.45}$$

(The absolute value signs are needed since both cross sections are by definition positive, whereas the derivative on the right can sometimes be negative.) The problem of transforming the differential cross section from the CM to the lab frame is now reduced to the kinematic problem of finding θ_{cm} in terms of θ_{lab}, or vice versa, and then taking the indicated derivative.

14.8 Relation of the CM and Lab Scattering Angles*

*As usual, sections marked with an asterisk can be omitted on a first reading.

To relate the CM and lab scattering angles, we need to look at the momenta of the particles in both frames. Let us add subscripts "cm" and "lab" for these, and use a

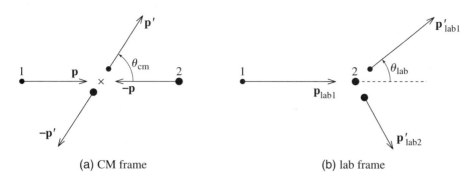

(a) CM frame   (b) lab frame

Figure 14.14   **(a)** An elastic collision as seen in the CM frame. The two particles approach the CM (shown as a cross) with equal and opposite momenta. **(b)** The same collision as seen in the lab frame, where particle 2 is initially at rest.

prime to indicate the outgoing values (and "unprime" for the incoming). The CM momenta are given by (14.38) as

$$\mathbf{p}_{cm1} = -\mathbf{p}_{cm2} = \mathbf{p} \qquad \text{(initial)} \tag{14.46}$$

and

$$\mathbf{p}'_{cm1} = -\mathbf{p}'_{cm2} = \mathbf{p}' \qquad \text{(final)} \tag{14.47}$$

where, as usual, $\mathbf{p}$ and $\mathbf{p}'$ denote the relative momentum ($\mathbf{p} = \mu\dot{\mathbf{r}}$). These values are illustrated in Figure 14.14(a). Notice that, by conservation of energy, all four momenta have equal magnitudes in the CM frame.

To be definite, I shall confine our discussion to the case that the "lab frame" is the traditional lab frame, in which particle 2 (the target) is initially at rest. The various momenta as seen in this frame are illustrated in Figure 14.14(b). To find these momenta, we can return to the two equations (14.37):

$$\mathbf{r}_1 = \mathbf{R} + \frac{m_2}{M}\mathbf{r} \qquad \text{and} \qquad \mathbf{r}_2 = \mathbf{R} - \frac{m_1}{M}\mathbf{r}. \tag{14.48}$$

Since particle 2 is initially at rest, the second of these implies that

$$\dot{\mathbf{R}} = \frac{m_1}{M}\dot{\mathbf{r}} = \frac{\mu}{m_2}\dot{\mathbf{r}} = \frac{\mathbf{p}}{m_2}. \tag{14.49}$$

This is the velocity of the center of mass, as seen in the lab frame, and allows us to relate any of the lab momenta to their corresponding CM values. In particular, differentiating the first of Equations (14.48), we find that

$$\mathbf{p}_{lab1} = m_1\dot{\mathbf{r}}_1 = m_1\dot{\mathbf{R}} + \mu\dot{\mathbf{r}} = \frac{m_1}{m_2}\mathbf{p} + \mathbf{p}$$

or

$$\mathbf{p}_{lab1} = \lambda\mathbf{p} + \mathbf{p} \tag{14.50}$$

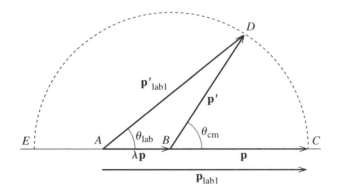

Figure 14.15   The relation of initial and final momenta in the CM and lab frames. The number λ is the mass ratio m_1/m_2 and was chosen to be 0.5 for this picture. The momenta shown as $\mathbf{p}$ and $\mathbf{p}'$ are the initial and final relative momenta, which are the same as $\mathbf{p}_{cm1}$ and $\mathbf{p}'_{cm1}$.

where I have introduced the important **mass ratio**

$$\lambda = \frac{m_1}{m_2}. \tag{14.51}$$

In exactly the same way, the final momentum $\mathbf{p}'_{lab1}$ is

$$\mathbf{p}'_{lab1} = \lambda \mathbf{p} + \mathbf{p}'. \tag{14.52}$$

The two results (14.50) and (14.52) are illustrated in Figure 14.15, where the lines BC and BD represent the initial and final momenta of particle 1 in the CM frame, while AC and AD are the corresponding lab values. By dropping a perpendicular from the point D to the line AC, you can check (Problem 14.25) that

$$\tan \theta_{lab} = \frac{\sin \theta_{cm}}{\lambda + \cos \theta_{cm}} \tag{14.53}$$

which tells us θ_{lab} in terms of θ_{cm}. Before we use this result to give us the lab cross section in terms of the corresponding CM value, let us use Figure 14.15 to establish a few other results.

Because $|\mathbf{p}| = |\mathbf{p}'|$, the point D lies on a circle with center B and radius p, as indicated. It is easy to see that, unless $\theta_{cm} = 0$ or π, θ_{lab} is always less than θ_{cm}, as you might expect. The details of Figure 14.15 depend on the relative sizes of the two masses: Suppose first that $\lambda < 1$ (that is, the projectile is lighter than the target, $m_1 < m_2$). In this case, the point A lies inside the circle as shown in Figure 14.15. (That figure was drawn using a mass ratio $\lambda = 0.5$.) If $\theta_{cm} = 0$, then the point D coincides with C and θ_{lab} is zero as well. If we imagine θ_{cm} increasing continuously from 0 to π, the point D moves continuously around the semicircle from C to E, with θ_{lab} always less than θ_{cm}, until $\theta_{cm} = \pi$, at which point θ_{lab} is also equal to π. (That is, if the projectile bounces straight back in the CM frame, the same is true in the lab — at least if $\lambda < 1$.) This behavior is illustrated in Figure 14.16, which is a graph of θ_{lab} against θ_{cm} for a mass ratio $\lambda = 0.5$.

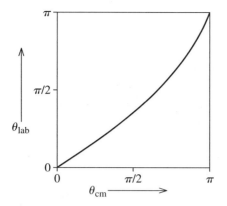

Figure 14.16   The lab scattering angle θ_{lab} as a function of the CM angle θ_{cm} (14.53) for a mass ratio of 0.5 (that is, $m_1 = 0.5 m_2$). The two angles are equal at 0 and π, but $\theta_{\text{lab}} < \theta_{\text{cm}}$ everywhere else.

If $\lambda = 1$ (that is, the projectile and target have equal masses), the behavior of θ_{lab} as a function of θ_{cm} is surprisingly different. In this case, (14.53) reduces to

$$\theta_{\text{lab}} = \tfrac{1}{2} \theta_{\text{cm}} \tag{14.54}$$

(see Problem 14.27). Thus, as θ_{cm} varies from 0 to π, θ_{lab} runs from 0 to $\pi/2$; in particular, in the lab frame of an equal-mass collision, the scattering angle can never exceed 90°. If $\lambda > 1$, the situation is different again, as you can explore in Problem 14.31.

To find the lab differential cross section, we need to find the derivative $d(\cos \theta_{\text{cm}})/d(\cos \theta_{\text{lab}})$ in (14.45). Using (14.53) it is a reasonably straightforward exercise (Problem 14.26) to show that

$$\frac{d(\cos \theta_{\text{lab}})}{d(\cos \theta_{\text{cm}})} = \frac{1 + \lambda \cos \theta_{\text{cm}}}{(1 + 2\lambda \cos \theta_{\text{cm}} + \lambda^2)^{3/2}}. \tag{14.55}$$

Substituting into (14.45), we find that

$$\left(\frac{d\sigma}{d\Omega} \right)_{\text{lab}} = \left(\frac{d\sigma}{d\Omega} \right)_{\text{cm}} \frac{(1 + 2\lambda \cos \theta_{\text{cm}} + \lambda^2)^{3/2}}{|1 + \lambda \cos \theta_{\text{cm}}|}. \tag{14.56}$$

EXAMPLE 14.7   Hard Sphere Scattering Again

Find the CM and lab differential cross sections for scattering of a point projectile of mass m_1 off a hard sphere of radius R and mass $m_2 = 2m_1$, and plot each as a function of the appropriate scattering angle.

The CM cross section is the same as that for a particle with mass equal to the reduced mass μ, scattering off a fixed target. In Example 14.5 (page 573) we found the latter to be just $R^2/4$. Therefore,

$$\left(\frac{d\sigma}{d\Omega} \right)_{\text{cm}} = \frac{R^2}{4}$$

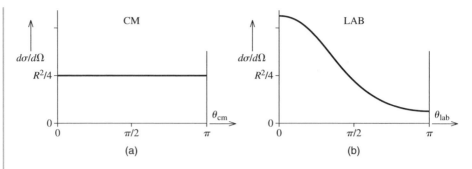

Figure 14.17   **(a)** The differential cross section for scattering off a hard sphere is isotropic in the CM frame. **(b)** In the lab frame, it is peaked in the forward direction.

and the lab cross section can be written down immediately from (14.56) (with $\lambda = 0.5$). The two cross sections are plotted as functions of their respective angles in Figure 14.17.[12] As we already knew, the CM cross section is isotropic. The lab cross section is markedly skewed in the forward direction.

Principal Definitions and Equations of Chapter 14

The Scattering Angle and Impact Parameter

The **scattering angle** is the angle θ by which a projectile is deflected in its encounter with a target. The **impact parameter** is the distance b by which the projectile would have missed the center of the target if it had been undeflected.   [Section 14.1]

The Collision Cross Section

The **cross section** σ_{oc} for a particular outcome "oc" (elastic scattering, absorption, reaction, fission) is defined by

$$N_{oc} = N_{inc} n_{tar} \sigma_{oc} \qquad \text{[Sections 14.2 \& 14.3]}$$

where N_{oc} is the number of outcomes of the type considered, N_{inc} is the number of incident projectiles, and n_{tar} is the density (number/area) of targets.

The Differential Cross Section

The **differential cross section** $\dfrac{d\sigma}{d\Omega}(\theta, \phi)$ for scattering in a direction (θ, ϕ) is defined by

---

[12] Equation (14.56) gives the lab cross section as a function of the CM angle θ_{cm}. To express it as an explicit function of θ_{lab}, one would have to solve (14.53) for θ_{cm} in terms of θ_{lab}. To make the plot of Figure 14.17(b), a much simpler procedure is to treat both θ_{lab} and $(d\sigma/d\Omega)_{lab}$ as functions of the parameter θ_{cm} and make a parametric plot with θ_{cm} running from 0 to π.

$$N_{sc}(\text{into } d\Omega) = N_{inc}\, n_{tar}\, \frac{d\sigma}{d\Omega}(\theta, \phi)\, d\Omega. \qquad \text{[Eq.(14.17)]}$$

Calculating the Differential Cross Section

If you can find the scattering angle θ as a function of the impact parameter b (or vice versa), then

$$\frac{d\sigma}{d\Omega} = \frac{b}{\sin\theta}\left|\frac{db}{d\theta}\right|. \qquad \text{[Eq. (14.23)]}$$

The Rutherford Formula

The differential cross section for scattering a charge q off a fixed charge Q is given by the **Rutherford formula**

$$\frac{d\sigma}{d\Omega} = \left(\frac{kqQ}{4E\sin^2(\theta/2)}\right)^2. \qquad \text{[Eq. (14.32)]}$$

The CM and Lab Cross Sections

The **lab frame** is generally understood to be the frame in which the target is at rest; the **CM frame** is that in which the CM is at rest. The differential cross sections in the two frames satisfy

$$\left(\frac{d\sigma}{d\Omega}\right)_{lab} = \left(\frac{d\sigma}{d\Omega}\right)_{cm}\left|\frac{d(\cos\theta_{cm})}{d(\cos\theta_{lab})}\right|. \qquad \text{[Eq. (14.45)]}$$

Problems for Chapter 14

Stars indicate the approximate level of difficulty, from easiest (⋆) to most difficult (⋆⋆⋆).

SECTION 14.2   The Collision Cross Section

14.1 ⋆ A blueberry pancake has diameter 15 cm and contains 6 large blueberries, each of diameter 1 cm. Find the cross section σ of a blueberry and the "target" density n_{tar} (number/area) of berries in the pancake, as seen from above. What is the probability that a skewer, jabbed at random into the pancake, will hit a berry (in terms of σ and n_{tar}, and then numerically)?

14.2 ⋆ **(a)** A certain nucleus has radius 5 fm. (1 fm = 10^{-15} m.) Find its cross section σ in barns. (1 barn = 10^{-28} m^2.) **(b)** Do the same for an atom of radius 0.1 nm. (1 nm = 10^{-9} m.)

14.3 ⋆ A beam of particles is directed through a tank of liquid hydrogen. If the tank's length is 50 cm and the liquid density is 0.07 gram/cm^3, what is the target density (number/area) of hydrogen atoms seen by the incident particles?

14.4 ⋆⋆ The cross section for scattering a certain nuclear particle by a copper nucleus is 2.0 barns. If 10^9 of these particles are fired through a copper foil of thickness 10 μm, how many particles are scattered? (Copper's density is 8.9 gram/cm^3 and its atomic mass is 63.5. The scattering by any atomic electrons is completely negligible.)

14.5 ★★ The cross section for scattering a certain nuclear particle by a nitrogen nucleus is 0.5 barns. If 10^{11} of these particles are fired through a cloud chamber of length 10 cm, containing nitrogen at STP, how many particles are scattered? (Use the ideal gas law and remember that each nitrogen molecule has two atoms. The scattering by any atomic electrons is completely negligible.)

14.6 ★★ Our definition of the scattering cross section, $N_{sc} = N_{inc} n_{tar} \sigma$, applies to an experiment using a narrow beam of projectiles all of which pass through a wide target assembly. Experimenters sometimes use a wide incident beam, which completely engulfs a small target assembly (the beam of photons from a car's headlamp, directed at a small piece of plastic, for example). Show that in this case $N_{sc} = n_{inc} N_{tar} \sigma$, where n_{inc} is the density (number/area) of the incident beam, viewed head-on, and N_{tar} is the total number of targets in the target assembly.

SECTION 14.4   The Differential Scattering Cross Section

14.7 ★ Calculate the solid angles subtended by the moon and by the sun, both as seen from the earth. Comment on your answers. (The radii of the moon and sun are $R_m = 1.74 \times 10^6$ m and $R_s = 6.96 \times 10^8$ m. Their distances from earth are $d_m = 3.84 \times 10^8$ m and $d_s = 1.50 \times 10^{11}$ m.)

14.8 ★ In their famous experiment, Rutherford's assistants, Geiger and Marsden, detected the scattered alpha particles using a zinc sulphide screen, which produced a tiny flash of light when struck by an alpha particle. If their screen had area 1 mm^2 and was 1 cm from the target, what solid angle did it subtend?

14.9 ★ By integrating the element of solid angle (14.15), $d\Omega = \sin\theta \, d\theta \, d\phi$, over all directions, verify that the solid angle corresponding to all directions is 4π steradians.

14.10 ★ By evaluating the necessary integral, verify the result (14.20) for the total cross section of Example 14.4 (page 571). (This is very easy if you change variables to $u = \cos\theta$.)

14.11 ★★ The differential cross section for scattering 6.5-MeV alpha particles at 120° off a silver nucleus is about 0.5 barns/sr. If a total of 10^{10} alphas impinge on a silver foil of thickness 1 μm and if we detect the scattered particles using a counter of area 0.1 mm^2 at 120° and 1 cm from the target, about how many scattered alphas should we expect to count? (Silver has a specific gravity of 10.5, and atomic mass of 108.)

14.12 ★★★ [Computer] In quantum scattering theory, the differential cross section is equal to the absolute value squared of a complex number $f(\theta)$, called the scattering amplitude:

$$\frac{d\sigma}{d\Omega} = |f(\theta)|^2. \tag{14.57}$$

The scattering amplitude can in turn be written as an infinite series

$$f(\theta) = \frac{\hbar}{p} \sum_{\ell=0}^{\infty} (2\ell + 1) \, e^{i\delta_\ell} \sin\delta_\ell \, P_\ell(\cos\theta). \tag{14.58}$$

Here $\hbar = 1.05 \times 10^{-34}$ J · s is called "h bar" and is Planck's constant divided by 2π, and p is the momentum of the incident projectile. The real numbers δ_ℓ are called the **phase shifts** and depend on the nature of the projectile and target and on the incident energy. And $P_\ell(\cos\theta)$ is the so-called Legendre polynomial. ($P_0 = 1$, $P_1 = \cos\theta$, etc.)

The **partial-wave series** (14.58) is especially useful at low energies where only a few of the phase shifts are different from zero. **(a)** Write down this series for the case of 10-MeV neutrons (mass $m = 1.675 \times 10^{-27}$ kg) scattering off a certain heavy nucleus for which $\delta_0 = -30°$, $\delta_1 = 150°$, and all other phase shifts are negligible. **(b)** Find an expression for the differential cross section (in terms of $\hbar$, m, the incident energy E, and the two nonzero phase shifts) and plot it for $0 \le \theta \le 180°$. **(c)** Find the total scattering cross section.

SECTION 14.5 Calculating the Differential Cross Section

14.13 ★★ In deriving the cross section for scattering by a hard sphere, we used the "law of reflection," that the angles of incidence and reflection of a particle bouncing off a hard sphere are equal, as in Figure 14.10. Use conservation of energy and angular momentum to prove this law. (The definition of "hard-sphere scattering" is that a projectile bounces with its kinetic energy unchanged. That the force is spherically symmetric implies, as usual, that angular momentum about the sphere's center is conserved.)

14.14 ★★ One can set up a two-dimensional scattering theory, which could be applied to puck projectiles sliding on an ice rink and colliding with various target obstacles. The cross section σ would be the effective width of a target, and the differential cross section $d\sigma/d\theta$ would give the number of projectiles scattered in the angle $d\theta$. **(a)** Show that the two-dimensional analog of (14.23) is $d\sigma/d\theta = |db/d\theta|$. (Note that in two-dimensional scattering it is convenient to let θ range from $-\pi$ to π.) **(b)** Now consider the scattering of a small projectile off a hard "sphere" (actually a hard disk) of radius R pinned down to the ice. Find the differential cross section. (Note that in two dimensions, hard "sphere" scattering is not isotropic.) **(c)** By integrating your answer to part (b), show that the total cross section is $2R$ as expected.

14.15 ★★★ [Computer] Consider a point projectile moving in a fixed, spherical force whose potential energy is

$$U(r) = \begin{cases} -U_o & (0 \le r \le R) \\ 0 & (R < r) \end{cases} \tag{14.59}$$

where U_o is a positive constant. This so-called spherical well represents a projectile which moves freely in either of the regions $r < R$ and $R < r$, but, when it crosses the boundary $r = R$, receives a radially inward impulse that changes its kinetic energy by $\pm U_o$ ($+U_o$ going inward, $-U_o$ going outward). **(a)** Sketch the orbit of a projectile that approaches the well with momentum p_o and impact parameter $b < R$. **(b)** Use conservation of energy to find the momentum p of the projectile inside the well ($r < R$). Let ζ denote the momentum ratio $\zeta = p_o/p$ and let d denote the projectile's distance of closest approach to the origin. Use conservation of angular momentum to show that $d = \zeta b$. **(c)** Use your sketch to prove that the scattering angle θ is

$$\theta = 2\left(\arcsin\frac{b}{R} - \arcsin\frac{\zeta b}{R}\right). \tag{14.60}$$

This gives θ as a function of b, which is what you need to get the cross section. The relation depends on the momentum ratio ζ, which in turn depends on the incoming momentum p_o and the well depth U_o. Plot θ as a function of b for the case that $\zeta = 0.5$. **(d)** By differentiating θ with respect to b, find an expression for the differential cross section as a function of b, and make a plot of $d\sigma/d\Omega$ against θ for the case that $\zeta = 0.5$. Comment. [*Hint:* To plot as a function of θ you don't need to solve for b in terms of θ; instead, you can make a parametric plot of the point $(\theta, d\sigma/d\Omega)$ as a function of the parameter b running from 0 to R.] **(e)** By integrating $d\sigma/d\Omega$ over all directions, find the total cross section.

SECTION 14.6   Rutherford Scattering

14.16 ★ One of the specific predictions of Rutherford's model of the atom was that the cross section should be inversely proportional to E^2 or, equivalently, to v^4. To test this, Geiger and Marsden varied the speed v by passing the incident alphas through thin sheets of mica to slow them down. According to Rutherford's prediction, the product of N_{sc} and v^4 should be the same whatever the incident speed (provided all other variables were held constant). Add a row showing $N_{sc}v^4$ to this table of their data and see how well Rutherford's prediction was confirmed.

Number of mica sheets	0	1	2	3	4	5	6
Counts, N_{sc} (per min)	24.7	29	33.4	44	81	101	255
Speed, v (arbitrary units)	1	0.95	0.90	0.85	0.77	0.69	0.57

14.17 ★ Another specific prediction of Rutherford's model of the atom was that the cross section should be proportional to the nuclear charge squared, that is, to Z^2, where Z is the atomic number, the number of protons in the nucleus. To test this, Geiger and Marsden counted the number of scatterings off various different targets (holding all other variables fixed), with the following results:

Target	Gold	Platinum	Tin	Silver	Copper	Aluminum
N_{sc}	1319	1217	467	420	152	26
Z	79	78	50	47	29	13

Add a row to this table to show the ratio N_{sc}/Z^2 and see how well Rutherford's prediction was confirmed. (At the time the atomic number was not known with certainty, nor was it well understood. Rutherford had guessed, correctly, that the nuclear charge was roughly equal to half the atomic mass, and this is what they used in place of Z.) The relatively poor agreement for the case of aluminum is probably due to our neglect of the target recoil, which is more important for the lighter targets.)

14.18 ★★ One the first observations that suggested his nuclear model of the atom to Rutherford was that several alpha particles got scattered by metal foils into the backward hemisphere, $\pi/2 \leq \theta \leq \pi$ — an observation that was impossible to explain on the basis of rival atomic models, but emerged naturally from the nuclear model. In an early experiment, Geiger and Marsden measured the fraction of incident alphas scattered into the backward hemisphere off a platinum foil. By integrating the Rutherford cross section (14.33) over the backward hemisphere, show that the cross section for scattering with $\theta \geq 90°$ should be $4\pi\sigma_0(E)$. Using the following numbers, predict the ratio $N_{sc}(\theta \geq 90°)/N_{inc}$: thickness of platinum foil $\approx 3\,\mu$m, density = 21.4 gram/cm^3, atomic weight = 195, atomic number = 78, energy of incident alphas = 7.8 MeV. Compare your answer with their estimate that "of the incident α particles about 1 in 8000 was reflected" (that is, scattered into the backward hemisphere). Small as this fraction is, it was still far larger than any rival model of the atom could explain.

14.19 ★★ An important simplification in our derivation of the Rutherford cross section was that the projectile's orbit is symmetric about the direction **u** of closest approach. (See Figure 14.11.) Prove that this is true of almost any conservative central force, as follows: **(a)** Assume that the effective potential (actual plus centrifugal) behaves as in Figure 8.4; that is, it approaches zero as $r \to \infty$ and approaches

$+\infty$ as $r \to 0$.[13] Use this to prove that any projectile that comes in from infinity must reach a minimum value r_{min} and then move out to infinity again. **(b)** This implies that a projectile must visit any value of r between r_{min} and infinity exactly twice, once on the way in and again on the way out. Prove that the values of $\dot{r}$ at these two points are equal and opposite, and that the values of $\dot{\psi}$ are equal (where ψ is the polar angle defined in Figure 14.11). **(c)** Use these results to prove that the orbit is symmetric under reflection about the direction **u**.

14.20 ★★ The derivation of the Rutherford cross section was made simpler by the fortuitous cancellation of the factors of r in the integral (14.30). Here is a method of finding the cross section which works, in principle, for any central force field: The general appearance of the scattering orbit is as shown in Figure 14.11. It is symmetric about the direction **u** of closest approach. (See Problem 14.19.) If ψ is the projectile's polar angle, measured from the direction **u**, then $\psi \to \pm\psi_0$ as $t \to \pm\infty$ and the scattering angle is $\theta = \pi - 2\psi_0$. (See Figure 14.11 again.) The angle ψ_0 is equal to $\int \dot{\psi}\, dt$ taken from the time of closest approach to ∞. Using the now-familiar trick you can rewrite this as $\int (\dot{\psi}/\dot{r})\, dr$. Next rewrite $\dot{\psi}$ in terms of the angular momentum ℓ and r, and rewrite $\dot{r}$ in terms of the energy E and the effective potential U_{eff} defined in Equation (8.35). Having done all this you should be able to prove that

$$\theta = \pi - 2 \int_{r_{min}}^{\infty} \frac{(b/r^2)\, dr}{\sqrt{1 - (b/r)^2 - U(r)/E}}. \tag{14.61}$$

Provided this integral can be evaluated, it gives θ in terms of b, and hence the cross section (14.23). For examples of its use, see Problems 14.21, 14.22, and 14.23.

14.21 ★★ Use the general relation (14.61) from Problem 14.20 to rederive the relation (14.24) for scattering by a hard sphere.

14.22 ★★★ Use the general relation (14.61) from Problem 14.20 to rederive the relation (14.31) for Rutherford scattering.

14.23 ★★★ Consider the scattering of a particle with energy E by a fixed, repulsive $1/r^3$ force field, with potential energy $U = \gamma/r^2$. Use the relation (14.61) from Problem 14.20 to find θ in terms of b and hence show that the differential cross section is

$$\frac{d\sigma}{d\Omega} = \frac{\gamma}{E} \frac{\pi^2(\pi - \theta)}{\theta^2(2\pi - \theta)^2 \sin\theta}. \tag{14.62}$$

To refresh your memory as to how to find r_{min} you might look at Figure 8.5 (the case $E > 0$). You should be able to solve your equation for θ in terms of b to get b in terms of θ and thence the cross section.

SECTION 14.8   Relation of the CM and Lab Scattering Angles *

14.24 ★★ Consider the scattering of two particles of equal mass (for example, scattering of protons off protons). In this case, $\theta_{lab} = \frac{1}{2}\theta_{cm}$. (See Problem 14.27.) **(a)** Use this result in (14.45) to prove that

---

[13] Although this behavior is definitely the norm, there are a few force fields for which it is not true. If the actual potential energy is strongly attractive near $r = 0$ (for example, $U(r) = -1/r^3$), then it dominates the centrifugal potential near $r = 0$, and the effective potential does not approach $+\infty$ as $r \to 0$. Our argument also breaks down for the special case that $b = 0$, in which case the projectile may smash directly into the target.

when the projectile and target have equal masses

$$\left(\frac{d\sigma}{d\Omega}\right)_{\text{lab}} = 4\cos\theta_{\text{lab}}\left(\frac{d\sigma}{d\Omega}\right)_{\text{cm}}. \tag{14.63}$$

(b) Write down the lab cross section for the scattering of two equal-mass hard spheres. (We know that in the CM frame, the differential cross section is $R^2/4$ where $R = R_1 + R_2$.) By integrating over all directions, verify that the total cross section in the lab frame is πR^2, as it has to be.

14.25 ★★ Using Figure 14.15, prove Equation (14.53), that

$$\tan\theta_{\text{lab}} = \frac{\sin\theta_{\text{cm}}}{\lambda + \cos\theta_{\text{cm}}}. \tag{14.64}$$

14.26 ★★ Using suitable trig identities, rewrite (14.64) to give $\cos\theta_{\text{lab}}$ as a function of $\cos\theta_{\text{cm}}$, and verify the derivative (14.55) that was essential for finding the relation between the CM and lab cross sections.

14.27 ★★ **(a)** By setting the mass ratio $\lambda = 1$ in Equation (14.64) of Problem 14.25, prove that in equal-mass scattering, $\theta_{\text{lab}} = \frac{1}{2}\theta_{\text{cm}}$. **(b)** Redraw Figure 14.15 for the case that $\lambda = 1$ ($m_1 = m_2$) and explain why the maximum value of θ_{lab} is $\pi/2$.

14.28 ★★ It is often interesting to know about the momentum of the recoiling target particle in the lab frame. Let us denote by ξ_{lab} the recoil angle, defined as the angle between the recoil momentum p'_{lab2} and the incident direction (the angle below the dashed line in Figure 14.14). **(a)** Show that in Figure 14.15 the recoil momentum is represented by the vector DC. Deduce that $\xi_{\text{lab}} = (\pi - \theta_{\text{cm}})/2$. **(b)** Show that, in the special case of equal masses ($m_1 = m_2$), $\xi_{\text{lab}} + \theta_{\text{lab}} = \pi/2$; that is, the angle between the two outgoing particles in an equal-mass, elastic collision is 90°. **(c)** Prove this last result directly using just conservation of momentum and energy (in an elastic collision).

14.29 ★★ An elastic collision is defined as one in which the total kinetic energy of the two particles is the same before and after the collision. **(a)** Show that in the CM frame, the individual kinetic energies of the two particles are separately conserved in an elastic collision. **(b)** Explain clearly why the same result is obviously *not* true in the lab frame. (Think about the energy of the target particle.) **(c)** Let ΔE denote the energy gained by the target particle in the collision (and hence the energy lost by the projectile). Using Figure 14.15, show that the fractional energy lost by the projectile (in the lab frame) is

$$\frac{\Delta E}{E} = \frac{4\lambda}{(1+\lambda)^2}\sin^2(\theta_{\text{cm}}/2)$$

where, as usual, λ is the mass ratio m_1/m_2. (Note that in Figure 14.15 the line DC represents the recoil momentum of the target.) **(d)** For a given mass ratio λ, what sort of collision gives the largest fractional energy loss? What value of λ maximizes this energy loss? (Your answer is important in situations where one wants a particle to lose energy as quickly as possible — as in a nuclear reactor, for example.)

14.30 ★★★ If you have not already done so, do Problem 14.24. **(a)** Now consider the scattering of two equal-mass hard spheres, A and B, with B initially stationary. Write down the standard expression (14.42) for the number of projectiles A scattered into a solid angle $d\Omega$ at a chosen angle Θ. Call this number $N(A \text{ into } d\Omega \text{ at } \Theta)$. Now suppose that we monitor for the number of target particles B recoiling into the same solid angle $d\Omega$ at the *same* angle Θ. Find $N(B \text{ into } d\Omega \text{ at } \Theta)$, the number of B's that will be observed. How does this compare with the number of A's?

14.31 ★★★ [Computer] Consider the elastic scattering of a projectile that is heavier than the target, that is, $m_1 > m_2$ or $\lambda > 1$. **(a)** Draw the analog of Figure 14.15 for this case. Show clearly that there are two different values of the CM angle θ_{cm} corresponding to each value of θ_{lab}. **(b)** What are the two CM angles that correspond to $\theta_{lab} = 0$? In terms of this example, explain why there is this two-fold ambiguity in θ_{cm} when $m_1 > m_2$. **(c)** Plot θ_{lab} as a function of θ_{cm} for the case that $\lambda = 2$. **(d)** Use your picture from part (a) to find an expression for the maximum possible value of θ_{lab} for a given value of λ. Check that your answer is correct for the case that $\lambda = 1$.

Special Relativity

From its publication in 1687 until 1905, Newtonian mechanics reigned supreme. It was applied to more and more systems, almost always with complete success. In those rare instances where Newtonian ideas appeared to fail, it was found that some complication had been overlooked, and, when this complication was included, Newton could again account for all the observations.[1] Newton's formulation was supplemented with new ideas (such as the notion of energy) and recast in different guises (by Lagrange and Hamilton), but the foundations seemed unshakeable. Then, toward the end of the nineteenth century, a few observations were made that seemed inconsistent with the classical, Newtonian, ideas. Heroic efforts were made to bring these observations into line with classical physics, but in 1905, Albert Einstein (1879–1955) published his first paper on the theory that we now call relativity, in which he showed that particles with speeds approaching the speed of light require a completely new form of mechanics, as I describe in this chapter. Even at slower speeds, Newtonian mechanics is only an approximation to the new "relativistic mechanics," but the difference is usually so small as to be undetectable. In particular, at the speeds usually encountered on earth, Newtonian mechanics is completely satisfactory, which explains why it is still a crucial and interesting part of physics (and justifies the other 15 chapters of this book).[2]

---

[1] Perhaps the greatest such triumph for Newton was the prediction and discovery of the planet Neptune: Calculations of the orbit of Uranus (taking account of the other known planets, and based, of course, on Newtonian mechanics) disagreed with the observed position by some 1.5 minutes of arc. In 1846, it was shown independently by the English astronomer John Couch Adams (1819–1892) and the Frenchman Urbain Le Verrier (1811–1877) that this discrepancy could be explained by the presence of a hitherto unnoticed planet outside the orbit of Uranus. Within a few months, the new planet, now called Neptune, was discovered by the German Johann Galle (1812–1910) at exactly its predicted position.

[2] In writing this chapter on relativity (particularly in the opening sections and the problems), it was sometimes difficult to resist borrowing ideas from the relativity chapters of *Modern Physics*, by Chris Zafiratos, Michael Dubson, and myself (second edition, Prentice Hall, 2003). I am grateful to Prentice Hall for giving me permission to do so.

15.1   Relativity

Let us first consider the significance of the name "relativity." A moment's thought should convince you that most physical measurements are made *relative* to a chosen reference system. That the position of a particle is $\mathbf{r} = (x, y, z)$ means that its position vector has components (x, y, z) *relative* to some chosen origin and a chosen set of axes. That an event occurs at time $t = 5$ s means that t is 5 seconds *relative* to a chosen origin of time, $t = 0$. If we measure the kinetic energy T of a car, it makes a big difference whether T is measured relative to a reference frame fixed on the road or to one fixed in the car. Almost all measurements require the specification of a reference frame, relative to which the measurement is to be made, and we can refer to this fact as the *relativity of measurements*.

The theory of relativity is the study of the consequences of the relativity of measurements. At first thought, this would seem unlikely to be a very interesting topic, but Einstein showed that a careful study of how measurements depend on the choice of coordinate system revolutionizes our whole conception of space and time, and requires a complete rethinking of Newtonian mechanics.

Einstein's relativity is really two theories. The first, called special relativity, is "special" in that it focuses primarily on unaccelerated frames of reference. The second, called general relativity, is "general" in that it includes accelerated reference frames. Einstein found that the study of accelerated frames leads naturally to a theory of gravitation, and general relativity turns out to be the relativistic theory of gravity. In practice, general relativity is required only in situations where its predictions differ appreciably from those of Newtonian gravity. These include the study of the intense gravity of black holes, of the large-scale universe, and of the effect of the earth's gravity on the extremely accurate time measurements needed for the global positioning system. In nuclear and particle physics, where we consider particles that move near the speed of light, but where gravity is usually completely negligible, special relativity is normally all that is needed. In this chapter, I shall treat only the special theory of relativity.[3]

15.2   Galilean Relativity

Many of the ideas of relativity are present in classical physics, and we have in fact met several in earlier chapters. Let us review these ideas and recast some of them in a form more suitable for our discussion of Einstein's relativity.

As we discussed in Chapter 1, Newton's laws hold in many different reference frames, namely, the so-called inertial frames, any one of which moves at constant velocity relative to any other. We can rephrase this to say that, in classical physics,

---

[3] To cover general relativity would require another book. Some good references are: R. Geroch, *General Relativity from A to B*, University of Chicago Press, 1978; I. R. Kenyon, *General Relativity*, Oxford University Press, 1990; B.F.A.Schutz, *A First Course in General Relativity*, Cambridge University Press, 1985; and James B. Hartle, *Gravity: An Introduction to Einstein's General Relativity*, Addison-Wesley, 2003.

Newton's laws are **invariant** (that is, unchanged) as we transfer our attention from one inertial frame to another. The classical transformation from one frame to a second, moving at constant velocity relative to the first, is called the **Galilean transformation**, so a compact way to say the same result is that Newton's laws are invariant under the Galilean transformation. Let us first review this claim.

The Galilean Transformation

For simplicity, consider first two frames S and S' that are oriented the same way; that is, the x' axis is parallel to the x axis, y' parallel to y, and z' parallel to z. Suppose further that the velocity $\mathbf{V}$ of S' relative to S is along the x axis. It was a fundamental assumption of Newtonian mechanics that there is a single universal time t. Thus if the observers in S and S' agree to synchronize their clocks (and to use the same unit of time), then $t' = t$. Finally, we can choose our origins O and O' so that they coincide at the time $t = t' = 0$. This configuration is illustrated in Figure 15.1, where S is a frame fixed to the ground. (We'll assume that a frame fixed to the earth is inertial — that is, we'll ignore the slow rotation of the earth.) The frame S' is fixed in a train that is traveling with velocity $\mathbf{V}$ along the x axis.

Consider now some event, such as the explosion of a small firecracker. As measured by observers in S this occurs at position $\mathbf{r} = (x, y, z)$ and time t; as measured in S' it occurs at $\mathbf{r}' = (x', y', z')$ and time t'. Our first (and very simple) task is to establish the mathematical relation between the coordinates (x, y, z, t) and (x', y', z', t'). A moment's inspection of Figure 15.1 should convince you that $x' = x - Vt$, and that $y' = y$ and $z' = z$. By the classical assumption concerning time, $t' = t$, so the required relations are

$$\left.\begin{array}{l} x' = x - Vt \\ y' = y \\ z' = z \\ t' = t. \end{array}\right\} \tag{15.1}$$

These four equations are called the **Galilean transformation**. They give the coordinates (x', y', z', t') of any event as measured in S' in terms of the corresponding coordinates (x, y, z, t) of the same event as measured in S. They are the mathematical expression of the classical ideas about space and time.

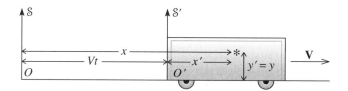

Figure 15.1   The frame S is fixed to the ground, while S' is fixed in a railroad car traveling with constant velocity $\mathbf{V}$ in the x direction. The two origins coincide, $O = O'$, at time $t = t' = 0$. The star indicates an event, such as a small explosion.

The Galilean transformation (15.1) relates the coordinates measured in two frames arranged with corresponding axes parallel and with relative velocity along the x axis, as shown in Figure 15.1 — an arrangement we can call the **standard configuration**. This is not, of course, the most general configuration. For example, if the relative velocity $\mathbf{V}$ is in an arbitrary direction, it is easy to see that (15.1) can be rewritten compactly as

$$\mathbf{r}' = \mathbf{r} - \mathbf{V}t \qquad \text{and} \qquad t' = t. \tag{15.2}$$

This is still not the most general form of the Galilean transformation, since we could rotate the axes, so that corresponding axes were no longer parallel, and we could displace the origins O or O' and the origins of time. However, (15.2) is general enough for our present purposes.

Using the Galilean transformation (15.2) we can immediately relate the velocities of an object, as measured in the two frames. If $\mathbf{v}(t) = \dot{\mathbf{r}}(t)$ is the velocity of the object as measured in $\mathcal{S}$ and $\mathbf{v}'(t)$ is likewise for $\mathcal{S}'$ then by differentiating (15.2) we find immediately that (remember that $\mathbf{V}$ is constant)

$$\mathbf{v}' = \mathbf{v} - \mathbf{V}. \tag{15.3}$$

This is the **classical velocity-addition formula**, which asserts that, according to the ideas of classical physics, relative velocities add (or subtract) according to the normal rules of vector arithmetic.

Galilean Invariance of Newton's Laws

To prove the invariance of Newton's laws under the Galilean transformation, suppose that the second law holds in frame $\mathcal{S}$; that is, that $\mathbf{F} = m\mathbf{a}$, with all three variables measured in $\mathcal{S}$. Now it is an experimental fact (at least in the domain of classical mechanics) that measurements of the mass of any object give the same results in all inertial frames. Thus the mass m' measured in $\mathcal{S}'$ is the same as that measured in $\mathcal{S}$, and $m' = m$. The proof that the same is true for the net force depends, to some extent, on one's definition of force. If we take the view that forces are defined by the readings on spring balances, then it is clear that the force $\mathbf{F}'$ measured in $\mathcal{S}'$ is the same as that measured in $\mathcal{S}$, and $\mathbf{F}' = \mathbf{F}$. Finally, differentiating (15.3) with respect to time (and remembering that $\mathbf{V}$ is constant, by assumption) we see that $\mathbf{a}' = \mathbf{a}$. We have now proved that each of the variables $\mathbf{F}'$, m', and $\mathbf{a}'$ of frame $\mathcal{S}'$ is equal to the corresponding variable $\mathbf{F}$, m, and $\mathbf{a}$ of frame $\mathcal{S}$. Therefore, if it is true that $\mathbf{F} = m\mathbf{a}$, it is also true that $\mathbf{F}' = m'\mathbf{a}'$. That is, Newton's second law is invariant under the Galilean transformation. I leave it as an exercise (Problem 15.1) to prove that the same is true of the first and third laws. The invariance of the laws of mechanics under the Galilean transformation was known to Galileo, who used it to argue that no experiment could tell whether the earth was "really" moving or "really" at rest, and hence that Copernicus's sun-centered view of the solar system was just as reasonable as the traditional earth-centered view.

Galilean Relativity and the Speed of Light

While Newton's laws are invariant under the Galilean transformation, the same is not true of the laws of electromagnetism. Whether we write them in their compact form as Maxwell's four equations, or in their original form (as Coulomb's law, Faraday's law, and so on), they can be true in one inertial frame, but if they are, and *if the Galilean transformation were the correct relation between different inertial frames*, then they could not be true in any other inertial frame. By far the quickest way to verify this claim is to recall that Maxwell's equations imply that light (and, more generally, any electromagnetic wave) propagates through the vacuum in any direction with speed

$$c = \frac{1}{\sqrt{\epsilon_o \mu_o}} = 3.00 \times 10^8 \text{ m/s}, \tag{15.4}$$

where ϵ_o and μ_o are the permittivity and permeability of the vacuum. Thus if Maxwell's equations hold in frame $\mathcal{S}$, then light must travel at the same speed c in any direction, as measured in $\mathcal{S}$. But now consider a second frame $\mathcal{S}'$, traveling at speed V along the x axis of $\mathcal{S}$, and imagine a beam of light traveling in the same direction. In $\mathcal{S}$ the light's speed is $v = c$. Therefore, in $\mathcal{S}'$ its speed is given by the classical velocity-addition formula (15.3) as

$$v' = c - V,$$

as shown on the left of Figure 15.2. Similarly, a beam of light traveling to the left will have speed $v = c$ in $\mathcal{S}$, but $v' = c + V$ in $\mathcal{S}'$. Depending on its direction, any beam of light will have speed v' (as measured in $\mathcal{S}'$) that varies anywhere between $c - V$ and $c + V$. Therefore, Maxwell's equations cannot hold in the inertial frame $\mathcal{S}'$.

Figure 15.2   Two frames $\mathcal{S}$ and $\mathcal{S}'$ in the standard configuration with relative velocity **V**. Two beams of light approach the car from opposite directions. If, as measured in $\mathcal{S}$, the light has speed c in either direction, then the classical velocity-addition formula implies that, as measured in $\mathcal{S}'$, it has speed $c - V$ traveling to the right, and $c + V$ traveling to the left.

If the Galilean transformation were the correct transformation between inertial frames, then although Newton's laws would hold in all inertial frames, there could only be one frame in which Maxwell's equations hold. This supposed unique frame, in which light would travel at the same speed in all directions, is sometimes called the ether frame.[4]

---

[4] The origin of the name is this: It was assumed that light must propagate through a medium, in much the same way that sound travels through the air. Since no one had ever detected this medium

The Michelson–Morley Experiment

The state of affairs just described, with the laws of mechanics valid in all inertial frames, but the laws of electromagnetism valid in a unique frame, was well understood toward the end of the nineteenth century. It was regarded by some (most notably Einstein) as unpleasing, and it was eventually shown by Einstein to be wrong. Nevertheless, it was logically consistent, and most physicists took for granted that there could be only one frame in which the speed of light had the same value c in all directions. Since the earth travels at a considerable speed in a continually changing direction around the sun, it seemed obvious that the earth must spend most of its time moving relative to the ether frame and hence that the speed of light as measured on earth should be different in different directions. The effect was expected to be very small. (The earth's orbital speed is $V \approx 3 \times 10^4$ m/s, large by terrestrial standards, but very small compared to $c = 3 \times 10^8$ m/s. Thus the fractional variation, between $c - V$ and $c + V$, was expected to be very small.) Nevertheless, in 1880, the American physicist Albert Michelson (1852–1931), later assisted by the chemist Edward Morley (1838–1923), devised an interferometer that should have easily detected the expected differences in the speed of light. To their surprise and dismay they found absolutely no variation.

Their experiments, and many different experiments with the same objective, have been repeated and have never found any reproducible evidence of variations in the speed of light relative to the earth. With hindsight, it is easy to draw the right conclusion: Contrary to all expectations, the speed of light is the same in all directions relative to an earth-based frame, even though the earth has different velocities at different times of year. In other words, it is not true that there is only one frame in which light has the same speed in all directions.

This conclusion is so surprising that is was not taken seriously for twenty years. Instead, several ingenious theories were advanced to explain the Michelson–Morley result while preserving the idea of a unique ether frame. For example, the so-called ether-drag theory held that the ether — the medium through which light was supposed to propagate — was dragged along with the earth, in much the same way the atmosphere *is* dragged along. This would imply that earth-bound observers are at rest relative to the ether and should measure the same speed of light in all directions. However, the ether-drag theory was incompatible with the phenomenon of stellar aberration.[5] None of these alternative theories was able to explain all of the observed facts (at least, not in a reasonable and economical way), and today almost all physicists accept that there is no unique ether frame and that the speed of light is a universal constant, with the same value in all directions in all inertial frames. The first person to

---

and since light could travel through seemingly empty space, the medium clearly had most unusual properties, and was named "ether" after the Greek for the stuff of the heavens. The "ether frame" was the frame in which the supposed ether was at rest.

   [5] The ether-drag theory would require that light entering the earth's envelope of ether would be bent. This would contradict stellar aberration, in which the apparent position of any one star moves around a small circle as the earth moves around its circular orbit — in a way that makes clear that the light from the star travels in a straight line as it approaches the earth.

accept this surprising idea whole-heartedly was Einstein, as we now discuss. In particular, we shall see that the universality of the speed of light requires us to reject the Galilean transformation and the classical picture of space and time on which it was based. This, in turn, will require us to modify much of our Newtonian mechanics.

15.3   The Postulates of Special Relativity

The special theory of relativity is based on the acceptance of the universality of the speed of light, as suggested by the Michelson–Morley experiment.[6] Einstein proposed two postulates, or axioms, expressing his conviction that *all* the laws of physics should hold in all inertial frames, and from these postulates, he developed his special theory of relativity.

Before we discuss the postulates of relativity, it would be good to agree on what we mean by an inertial frame:

Definition of an Inertial Frame

An inertial frame is any reference frame (that is, a system of coordinates x, y, z and time t) in which all the laws of physics hold in their usual form.

Notice that I have not yet specified what "all the laws of physics" are. Following Einstein, we shall use the postulates of relativity to help us decide what the laws of physics could be. (As always, the ultimate test will be whether they agree with experiment.) It will turn out that one of the classical laws that carries over into relativity is the law of inertia, Newton's first law. Thus our newly defined inertial frames are in fact the familiar "unaccelerated" frames, where an object subject to no forces travels with constant velocity. As before, a frame fixed to the earth is (to a good approximation) inertial; a frame fixed to an accelerating rocket or a spinning turntable is not. The big difference between the inertial frames of relativity and those of classical mechanics is the mathematical relation between different frames. In relativity, we shall find that the classical Galilean transformation must be replaced by the so-called Lorentz transformation.

Notice also that I have specified that an inertial frame is one where the physical laws hold "in their usual form." As we saw in Chapter 9, one can sometimes modify physical laws so that they hold in noninertial frames as well. (For example, by introducing the centrifugal and Coriolis forces, we could use Newton's second law in a rotating frame.) It is to exclude such modifications that I added the qualifier "in their usual form."

---

[6] Whether Einstein actually knew about the Michelson–Morley result when he was formulating his theory is not clear. There is some evidence that he did, but it seems clear that his main motivation was the conviction that Maxwell's equations should hold in all inertial frames. Whether he knew or not affects neither Einstein's amazing accomplishment nor the importance of the Michelson–Morley result as beautifully clear evidence in favor of Einstein's assumptions.

The first postulate of relativity asserts the existence of many different inertial frames, traveling at constant velocity relative to one another:

First Postulate of Relativity

If S is an inertial frame and if a second frame S' moves with constant velocity relative to S, then S' is also an inertial frame.

Another way to say this is that the laws of physics are invariant as we transfer our attention from one frame to a second one moving at constant velocity relative to the first. This is what we proved for the laws of mechanics, but we are now claiming it for *all* the laws of physics.

Another popular statement of the first postulate is that "there is no such thing as absolute motion." To understand this, consider two frames, S attached to the earth and S' attached to a rocket coasting at constant velocity relative to the earth. A natural question is whether there is any meaningful sense in which we could say that S is really at rest and S' is really moving (or vice versa). If the answer were "yes," then we could say that S is absolutely at rest and that anything moving relative to S is in absolute motion. However, this would contradict the first postulate of relativity: All of the laws observable by scientists in S are equally observable by scientists in S'; any experiment that can be performed in S can equally be performed in S'. Therefore, no experiment can show which frame is *really* moving. Relative to the earth, the rocket is moving; relative to the rocket, the earth is moving; and this is all we can say.

Yet another statement of the first postulate is that among all the inertial frames, there is no *preferred frame*. The laws of physics single out no one frame as being in any way more special than any other.

The second postulate specifies one of the laws that holds in all inertial frames:

Second Postulate of Relativity

The speed of light (in vacuum) has the same value c in every direction in all inertial frames.

This is, of course, the Michelson–Morley result.

Although the second postulate flies in the face of our everyday experience, it is by now a firmly established experimental fact. As we explore the consequences of Einstein's postulates we are going to encounter several surprising predictions, all of which seem to contradict our experience (for example, the phenomenon called time dilation, described in the next section). If you have difficulty accepting these predictions, there are two points to bear in mind: First, they are all logical consequences of the second postulate. Thus, once you have accepted the latter (surprising, but indisputably true), you *have* to accept all of its logical consequences, however counterintuitive they may seem. Second, all of these surprising phenomena (including the second postulate itself) have the subtle property that they become important only when objects travel with speeds comparable to the speed of light. In everyday life, with all speeds much less than c, these phenomena simply do not show up. In this sense, none of the surprising consequences of Einstein's postulates really conflict with our everyday experience.

15.4   The Relativity of Time; Time Dilation

Measurement of Time in a Single Frame

We are going to find that the second postulate forces us to abandon the classical notion of a single universal time. Instead, we shall find that the time of any one event, as measured in two different inertial frames, is in general different. This being the case, we need first to be quite clear what we mean by time, as measured in a single frame.

I shall take for granted that we have at our disposal lots of reliable tape measures and clocks. The clocks need not be identical, but they must have the property that, when brought together at the same point, at rest in the same inertial frame, they agree with one another. Let us now consider a single inertial frame $\mathcal{S}$, with origin O. We can station a chief observer at O with one of our clocks, and she can easily time any nearby event, such as a small explosion, since she will see it essentially instantaneously. To time an event farther away from the origin is harder, since light from the event has to travel to O before she can sense it. If she knew how far away the event occurred, then she could calculate how long the signal took to reach her (she knows that light travels at speed c) and subtract this from the time of arrival to give the time of the event. A simpler way to proceed (in principle anyway) is to employ a large number of helpers stationed at regular intervals throughout the region of interest and each with his own clock. The helpers can measure their distances from O, and we can check that their clocks are synchronized with the clock at O by having the chief observer send out a light signal at an agreed time (on her clock). Each helper can calculate the time taken by the signal to reach him and (allowing for this transit time) check that his clock agrees with the clock at O.

With enough helpers, stationed closely enough together, there will be a helper close enough to any event to time it essentially instantaneously. Once he has timed it, he can, at his leisure, inform everyone else of the result by any convenient means (such as a telephone). In this way, any event can be assigned a unique and well-defined time t as measured in the frame $\mathcal{S}$. In what follows, I shall assume that any inertial frame $\mathcal{S}$ comes with a set of rectangular axes $Oxyz$ and a team of helpers stationed at rest throughout $\mathcal{S}$ and equipped with synchronized clocks. This allows us to assign a position (x, y, z) and a time t to any event, as observed in the frame $\mathcal{S}$.

Time Dilation

Let us now compare measurements of times made by observers in two different inertial frames. Consider our familiar two frames, $\mathcal{S}$ anchored to the ground and $\mathcal{S}'$ traveling with a train in the x direction at speed V relative to $\mathcal{S}$. We now examine a **thought experiment** (or **gedanken experiment**, from the German) in which an observer on the train sets off a flashbulb on the floor of the train. The light travels to the roof, where

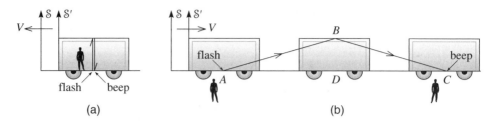

Figure 15.3   (a) The thought experiment as seen in frame S'. The light travels straight up and down again, and the flash and beep occur at the same place. (b) As seen in S, the flash and beep are separated by a distance $V \Delta t$. Notice that in S two observers are needed to time the two events, since they occur in different places.

it is reflected back and returns to its starting point, where it strikes a photocell and causes an audible "beep." We wish to compare the times, Δt and $\Delta t'$, as measured in the two frames, between the flash as the light leaves the floor and the beep as it returns.

As seen in the frame S', our experiment is shown in Figure 15.3(a). If the height of the train is h, then, as seen in S', the light travels a total distance $2h$ at speed c (second postulate) and so takes a time

$$\Delta t' = \frac{2h}{c}. \tag{15.5}$$

This is the time between the flash and the beep, as measured by an observer in S' (provided, of course, his clock is reliable).

As seen in S, our experiment is shown in Figure 15.3(b). In particular, the same beam of light is seen to travel along the two sides AB and BC of a triangle. If Δt is the time between the flash and the beep (as measured in S), the side AC has length $V \Delta t$. Thus the triangle ABD has sides[7] h, $V \Delta t/2$, and $c\Delta t/2$. (Notice that this is where we use the second postulate, that the speed of light is c in either frame.) Therefore,

$$(c \, \Delta t/2)^2 = h^2 + (V \Delta t/2)^2,$$

which we can solve to give

$$\Delta t = \frac{2h}{\sqrt{c^2 - V^2}} = \frac{2h}{c} \frac{1}{\sqrt{1 - \beta^2}} \tag{15.6}$$

where I have introduced the useful abbreviation

$$\beta = \frac{V}{c}, \tag{15.7}$$

which is just the speed V measured in units of c.

---

[7] I take for granted that the height of the train is the same in either frame. We'll prove this shortly.

The striking thing about the two results (15.5) and (15.6) is that they are not equal. The time between the same two events (the flash and the beep) has different values as measured in the two different inertial frames. Specifically,

$$\Delta t = \frac{\Delta t'}{\sqrt{1 - \beta^2}}. \qquad (15.8)$$

We derived this result for a thought experiment with a flash of light reflected back to its source by a mirror on the ceiling of the railroad car, but the conclusion applies to *any* two events that occur at the same place in the train. Suppose, for instance, an observer at rest in S' were to shout "Good" and a moment later "Grief." In principle, we could ignite a flashbulb at the "good," and arrange a mirror which would reflect the light back to arrive at the moment of "grief." Therefore, the relation (15.8) must apply to these two events, the "good" and the "grief." Since the timing of the two events cannot depend on whether we actually did the experiment with the light and the beeper, we conclude that the relation (15.8) must apply to *any* two events that occur at the same place in the frame S'.

You should avoid thinking that the clocks in one of our frames are somehow running incorrectly — on the contrary, it was essential to our argument that all clocks, in both frames, were running correctly. Further, it makes no difference what particular kinds of clock we used, so the conclusion (15.8) applies to all (accurate) clocks. That is, *time itself*, as measured in the two frames, is different in accordance with (15.8). As we shall discuss shortly, this surprising conclusion has been verified repeatedly.

If the frame S' is actually at rest (relative to S), then $V = 0$, so $\beta = 0$, and (15.8) reduces to $\Delta t' = \Delta t$. That is, there is no difference in the times unless S' is actually moving relative to S. Moreover, at normal terrestrial speeds, $V \ll c$, so $\beta \ll 1$ and the denominator in (15.8) is very close to one. That is, at the speeds of our everyday experience, the two times are very nearly equal — so close that it would be almost impossible to detect any difference, as the following example shows.

EXAMPLE 15.1   Time Differences for a Jet Plane

Suppose that the pilot of a jet traveling at a steady $V = 300$ m/s arranges to set off a flashbulb at intervals of exactly one hour (as measured in his reference frame). If we arrange two observers on the ground to check this, what would they measure for the time Δt between two successive flashes? (Take the ground to be an inertial frame; that is, ignore effects of the earth's rotation.)

The required interval is given by (15.8) with $\Delta t' = 1$ hour and $\beta = V/c = 10^{-6}$. So

$$\Delta t = \frac{\Delta t'}{\sqrt{1 - \beta^2}} = \frac{1 \text{ h}}{\sqrt{1 - 10^{-12}}}$$

$$\approx 1 \text{ h} \times (1 + \tfrac{1}{2} \times 10^{-12}) = 1 \text{ h} + 1.8 \times 10^{-9} \text{ s}$$

where in going to the second line, I have used the binomial approximation.[8] In this experiment, the time difference is less than 2 nanoseconds ($1 \text{ ns} = 10^{-9} \text{ s}$). It is not hard to see why classical physicists had failed to detect such differences!

As we increase V, the difference between the times in (15.8) gets bigger, and if we let V approach c, we can make the difference as large as we please. For example, if $V = 0.99c$, then $\beta = 0.99$ and (15.8) gives $\Delta t \approx 7\Delta t'$. Speeds this high are routinely achieved by the accelerators at particle-physics labs, and the predicted time difference is precisely confirmed.

If we put $V = c$ (that is, $\beta = 1$) in (15.8), we would get the absurd result $\Delta t = \Delta t'/0$, and if we put $V > c$ (that is, $\beta > 1$), we would get an imaginary value for Δt. These results suggest that V must always be less than c,

$$V < c,$$

a suggestion that proves correct and is one of the most profound results of relativity: The relative speed of two inertial frames can never equal or exceed c. That is, the speed of light, in addition to being the same in all inertial frames, is also the universal speed limit for the relative motion of any two inertial frames.

The factor $1/\sqrt{1 - \beta^2}$ occurs so often in relativity, it is usually given its own name, γ,

$$\gamma = \frac{1}{\sqrt{1 - \beta^2}}. \tag{15.9}$$

It is useful to remember that this new factor always satisfies $\gamma \geq 1$, and as $\beta \to 1$ (that is, $V \to c$) $\gamma \to \infty$.

In terms of the parameter γ the result (15.8) can be written a little more compactly as

$$\Delta t = \gamma \, \Delta t' \geq \Delta t'. \tag{15.10}$$

The asymmetry of this result (that $\Delta t'$ is never more than Δt) seems at first glance to violate the postulates of relativity, since it suggests a special role for the frame $\mathcal{S}'$ — namely, that $\mathcal{S}'$ is the special frame in which the time interval is minimum. However, this is just as it should be, since in our thought experiment $\mathcal{S}'$ *is* special, because it is the frame where the two events in question (the flash and the beep) occur at the same place. (This asymmetry was implicit in Figure 15.3, which showed one observer measuring $\Delta t'$, but two measuring Δt.) To emphasize this asymmetry, the time $\Delta t'$ is often renamed Δt_o and (15.10) rewritten as

$$\Delta t = \gamma \, \Delta t_o \geq \Delta t_o. \tag{15.11}$$

---

[8] This is a nice example of a calculation where one almost *has to* use the binomial approximation, since most calculators cannot tell the difference between 1 and $1 - 10^{-12}$.

The subscript on Δt_o is to emphasize that Δt_o is the time elapsed on a clock at rest in the special frame where the two events in question occurred at the same place. This time is often called the **proper time** between the two events. In (15.11), Δt is the corresponding time measured in *any* frame and is always greater than or equal to the proper time Δt_o. For this reason, the effect implied by (15.11) is called **time dilation** and can be loosely stated by saying that *a moving clock is observed to run slow*. As measured by observers on the ground, a clock in the moving train is found to run slow.

Finally, I should emphasize the fundamental symmetry between any two inertial frames. We chose to do our thought experiment in a way that gave the frame S' a special role. (It was the frame in which the flash and beep occurred at the same place.) But we could have done the experiment the other way round, with the flashbulb, mirror and beeper at rest on the ground, and in this case, we would have found the opposite effect, that $\Delta t' = \gamma \, \Delta t$. The advantage of writing the time-dilation formula in the form (15.11) is that it avoids the problem of remembering which is frame S, and which S'; the subscript on Δt_o always flags the proper time — the time measured in the frame in which the two events were at the same place.

Evidence for Time Dilation

Time dilation was predicted in 1905 but was not experimentally verified until 1941, by B. Rossi and D. B. Hall.[9] The problem was, of course, to get a clock traveling sufficiently fast to show a measurable dilation. Rossi and Hall exploited the natural clocks that come with unstable subatomic particles, which decay (on average) after a definite time, characteristic of the particle. The lifetime of an unstable particle can be specified by its **half-life**, $t_{1/2}$, the time in which half of a large number of the particles will decay. The muon is an unstable particle that is created in the earth's upper atmosphere when cosmic ray particles (mostly protons and alpha particles) from outer space collide with atmospheric atoms. Many of these muons have speeds quite close to the speed of light, and they live long enough to find their way down to the earth's surface. The muon had been discovered in 1935 by Carl Anderson in his studies of cosmic rays. By 1941 its half-life was known to be about $t_{1/2} = 1.5 \, \mu$s, meaning that half of a sample of muons *at rest* would decay in this time. If time dilation is correct, the half-life for a moving muon (as measured by earth-bound observers) should be larger by the factor γ as in (15.11). For example, if the muon had speed $0.8c$, then $\gamma = 1.67$, and the muon's half-life should be

$$t_{1/2}(\text{at speed } 0.8c) = 1.67 \times t_{1/2}(\text{at rest}) = 2.5 \, \mu\text{s}.$$

Rossi and Hall were able to separate out cosmic-ray muons according to their speed and they could find their half-lives by measuring how many of them survived the journey through the atmosphere. Although their measurements had quite large experimental errors, they were nonetheless good enough to verify Einstein's prediction (15.11) and to exclude the classical assumption of a single universal time.

---

[9] B. Rossi and D. B. Hall, *Physical Review,* vol. 59, p. 223 (1941).

A test of time dilation using man-made clocks had to await the development of superaccurate atomic clocks. In 1971 four portable atomic clocks were synchronized with a reference clock at the U. S. Naval Observatory in Washington DC and then flown around the world in a jet plane and returned to the Naval Observatory. The observed discrepancy between the reference clock and the portable clocks was (273 ± 7) ns (averaged over the four clocks) in excellent agreement with the predicted value (275 ± 21) ns.[10]

Tests of time dilation — using both the natural clocks of unstable particles and man-made atomic clocks — have been repeated with ever-increasing precision, and there is now no doubt that the relativity of time, as embodied in (15.11), is true. Another important test that is carried out thousands of times every day is the Global Positioning System (GPS). This system, which is used by airplanes, ships, cars, and hikers to find their positions within a few meters, times the arrival of signals from 24 GPS overhead satellites at the observer's receiver and calculates the receiver's position from the known positions of the satellites. To find the position within a few meters requires an accuracy of a few nanoseconds, which requires that allowances be made for the relativistic differences between the times of the satellite and earth-bound reference frames. The success of the GPS is a daily tribute to the correctness of relativity.[11]

15.5   Length Contraction

The postulates of relativity have forced us to the conclusion that time is relative — the time between two given events is different when measured in different inertial frames — and, even more important, this conclusion is born out by experiment. This, in turn, implies that the length of an object is likewise dependent on the frame in which it is measured. To see this, we'll conduct a second thought experiment with the train of Figure 15.3, this time measuring its length. For an observer (let's call him Q) on the ground (frame S) the simplest procedure is probably to measure the time Δt for the train to pass him and calculate the length as[12]

$$l = V \Delta t. \tag{15.12}$$

------

[10] See J. C. Hafele and R. E. Keating, *Science*, vol. 177, p. 166 (1972). Two trips were made, one going west and the other going east, both with satisfactory results. The numbers quoted here are for the more decisive westward trip. This experiment was actually a test of general, as well as special, relativity, since the predicted discrepancy has an appreciable contribution from gravitational effects.

[11] For a readable account of the large role of relativity in the GPS, see N. Ashby, *Physics Today*, May 2002, p. 41. As described there, there are important contributions from general, as well as special, relativity. Thus, the success of the GPS is a test of both theories.

[12] With so many of the familiar classical ideas being questioned, you are entitled to ask if it is legitimate to use the classical formula (15.12). However, this is just the definition of velocity (velocity = distance/time), and is certainly valid in any one reference frame (as long as we measure all quantities in this same frame).

To find the length l' of the train as measured in the train's rest frame, an observer on the train could simply use a long tape measure. However, for comparison with (15.12), it is convenient to use a different method. We can station two observers on the train, one at the front and another at the back, and have them record the times at which they pass the observer Q on the ground. The difference $\Delta t'$ between these two times is the time (as measured in frame S') for the train to pass observer Q, so the length of the train (again as measured in S') is just

$$l' = V \Delta t'. \tag{15.13}$$

Notice that we are making an important assumption here, that the speed of frame S relative to S' is the same as the speed V of S' relative to S. (The relative velocities are in opposite directions, but their magnitudes are the same.) This is true in classical mechanics, and it is also true in relativity, where it follows from the two postulates. The details of the argument require some care, but the gist is this: Consider the transformation from frame S to S'. We'll denote it by $(S \rightarrow S')$ temporarily. Suppose that, before making this transformation, we were to rotate our axes through $180°$ about the y (or z) axis, then make the transformation, and then rotate back again. The effect of the rotations is to reverse the direction of the x axis (and finally rotate it back again). The net effect of all three operations is precisely the transformation $(S' \rightarrow S)$. Since the rotations certainly don't change any speeds, we've proved that the speed of S' relative to S is the same as that of S relative to S'.

Comparing (15.12) with (15.13), we see that, since the times Δt and $\Delta t'$ are unequal, the same has to be true of the lengths l and l'. To quantify the difference, we must be careful to get the relation between Δt and $\Delta t'$ the right way around. These two times are the times (as measured in S and S') between two events: "front of train opposite observer Q" and "back of train opposite observer Q." These two events occur at the same place in frame S, so Δt is the proper time, and $\Delta t' = \gamma \, \Delta t$. Inserting this into (15.13) and comparing with (15.12), we see that $l' = \gamma l$ or

$$l = \frac{l'}{\gamma} \leq l'. \tag{15.14}$$

The length of the train as measured in S is less than that measured in S' (unless $V = 0$).

Like time dilation, the effect (15.14) is asymmetric, reflecting the asymmetry of the experiment. The frame S' is special, since it is the unique frame where the object being measured (the train) is at rest. [We could, of course, have done the experiment the other way round. If we had measured the length of a building that is at rest on the ground, then the roles of l and l' would have been reversed.] To avoid confusion as to which frame is which, it is common to rewrite (15.14) as

$$l = \frac{l_0}{\gamma} \leq l_0 \tag{15.15}$$

where l_o denotes the length of an object measured in the object's **rest frame** (the frame in which the object is at rest), while l is the length in *any* frame. The length l_o is called the object's **proper length**. Since $l < l_o$ (if $V \neq 0$), this difference in lengths is called the **length contraction** (or the Lorentz contraction, or Lorentz–Fitzgerald contraction, after the two physicists — the Dutch Hendrik Lorentz, 1853–1928, and the Irish George Fitzgerald, 1851–1901 — who first suggested there must be some such effect.) The result can be loosely paraphrased by saying that *a moving body is observed to be contracted.*

Like time dilation, length contraction is a real effect, well established by experiment. Since the two effects are so intimately connected, any evidence for one can be taken as evidence for the other. In particular, the decay of a high-speed unstable particle, when viewed in the particle's rest frame, can be interpreted as clear evidence for length contraction. (See Problem 15.12.)

Lengths Perpendicular to the Relative Velocity

The length contraction just derived applies to lengths in the direction of the relative velocity, such as the length of a train in the direction of motion. It is easy to see that there can be no analogous contraction or expansion of lengths perpendicular to the motion, such as the height of the train. Suppose for example there were a contraction and imagine two observers Q standing at rest in $\mathcal{S}$ and Q' in $\mathcal{S}'$. Suppose further that Q and Q' are equally tall (when at rest) and that Q' is holding a knife exactly level with the top of his head. If there is a contraction, then as measured by Q, observer Q' will be shortened as he rushes past, and Q will be scalped, or worse, as the knife goes by. But, unlike our previous thought experiments, this experiment is completely symmetric between the two frames: There is just one observer in each frame, and the only difference is the direction of the relative velocities. Therefore, it must also be that, as seen by Q', it is Q who is contracted; so the knife misses Q, and Q is not scalped. The assumption of a contraction has led us to a contradiction and there can be no contraction. A similar argument excludes the possibility of expansion, and, in fact, the knife just scrapes past Q as seen in either frame. We conclude that lengths perpendicular to the relative motion are unchanged. The length-contraction formula (15.15) applies only to lengths parallel to the relative velocity.

15.6   The Lorentz Transformation

According to the classical notions of space and time, we saw that the mathematical relation between coordinates in two inertial frames $\mathcal{S}$ and $\mathcal{S}'$ is the Galilean transformation (15.1). In relativity, this cannot be the correct relation. (For example, time dilation contradicts the equation $t = t'$.) However, we can deduce the correct relation using an argument similar to the one that we used in connection with Figure 15.1 to derive the Galilean result. We imagine two frames, $\mathcal{S}$ attached to the ground and $\mathcal{S}'$ attached to a train moving with speed V relative to $\mathcal{S}$. We imagine, further, the explosion of a firecracker, which leaves a burn mark on the wall of the railroad car at a point

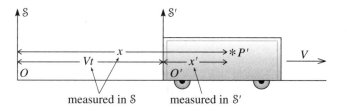

Figure 15.4 The coordinate x' is the horizontal distance, measured in S', between the origin O' and the burn mark at P'. The distances x and Vt are both measured in S at the time t (measured in S) of the explosion.

P'. The coordinates of this explosion are (x, y, z, t) as measured by observers in S and (x', y', z', t') in S'. Our object is to find formulas for x', y', z', and t' in terms of x, y, z, and t. The thought experiment is illustrated in Figure 15.4, which is just like Figure 15.1 except that we now know we must be very careful to identify the frames (S or S') relative to which the various distances are measured.

Since lengths perpendicular to the relative velocity are the same in both frames, we can immediately write

$$y' = y \qquad \text{and} \qquad z' = z \qquad (15.16)$$

exactly as with the Galilean transformation. The coordinate x' is the horizontal distance between the origin O' and the burn mark at P', as measured in S'. The same distance as measured in S is $x - Vt$, since x and Vt are the distances from O to P' and from O to O' at the instant t of the explosion (measured in S). Therefore, by the length-contraction formula (15.15) (x' is the proper length here)

$$x - Vt = x'/\gamma$$

or

$$x' = \gamma(x - Vt). \qquad (15.17)$$

This is the third of the four equations that we need. Notice that if $V \ll c$ then $\gamma \approx 1$ and (15.17) reduces to the Galilean relation $x' = x - Vt$.

Finally, to get an equation for t' we can use a simple trick. We could repeat the previous argument with the roles of S and S' exchanged. That is, we could let the explosion burn a mark at a point P on a wall fixed in S. Arguing as before, we would get the result

$$x = \gamma(x' + Vt'). \qquad (15.18)$$

(Notice that we could get this result directly from (15.17) by exchanging the primed and unprimed variables and replacing V by $-V$.) Substituting (15.17) into (15.18), we can eliminate x' and solve for t', to give (as you should check)

$$t' = \gamma(t - Vx/c^2). \qquad (15.19)$$

This is the required equation for t'. When $V \ll c$, we can neglect the second term and $\gamma \approx 1$, so (15.19) reduces to the Galilean relation $t' = t$.

Collecting together the results (15.16), (15.17), and (15.19), we get the required four equations:

The Lorentz Transformation

$$
\left.
\begin{aligned}
x' &= \gamma(x - Vt)\\
y' &= y\\
z' &= z\\
t' &= \gamma(t - Vx/c^2).
\end{aligned}
\right\}
\tag{15.20}
$$

These four equations are called the **Lorentz transformation** or the **Lorentz–Einstein transformation**, in honor of Lorentz, who first proposed them, and Einstein, who first interpreted them correctly. The Lorentz transformation gives the coordinates (x', y', z', t') of an event, as measured in S', in terms of its coordinates (x, y, z, t) as measured in S. It is the correct relativistic version of the classical Galilean transformation (15.1).

If we wanted to know the coordinates (x, y, z, t) in terms of (x', y', z', t'), we could solve the four equations (15.20), but a simpler way is just to exchange primed and unprimed variables and replace V by $-V$. Either way, the result is the **inverse Lorentz transformation**

$$
\left.
\begin{aligned}
x &= \gamma(x' + Vt')\\
y &= y'\\
z &= z'\\
t &= \gamma(t' + Vx'/c^2).
\end{aligned}
\right\}
\tag{15.21}
$$

The Lorentz transformation expresses all of the properties of space and time that follow from the postulates of relativity. Using it, one can calculate all of the kinematic relations between measurements made in different inertial frames. There are several examples of its use in the problems at the end of this chapter and here are a couple more.

EXAMPLE 15.2   Rederiving Length Contraction

Use the Lorentz transformation to rederive the length contraction formula (15.15). (Note that this will not give an alternative derivation of length contraction, since length contraction was used in deriving the Lorentz transformation. Rather we shall just get a consistency check.)

Consider our usual two frames, S fixed to the ground and S' fixed to a train traveling along the x axis with speed V relative to S. We wish to compare the lengths of the train as measured in S and S'. The measurement in S' is easy, since the train is at rest in this frame. An observer can, at his leisure, measure the x'

coordinates x_1' and x_2' of the back and front of the train, and its length is just the difference $l' = x_2' - x_1'$. This length is the proper length of the train, so

$$l_o = l' = x_2' - x_1'. \tag{15.22}$$

The measurement in S is harder since the train is moving. We could, with enough care, station two observers Q_1 and Q_2 beside the track so that the back of the train passes Q_1 at the exact same instant ($t_1 = t_2$) that the front passes Q_2. The length as measured in S is then just

$$l = x_2 - x_1.$$

Now, applying the Lorentz transformation (15.20) to the event "front of train passes Q_2" we get

$$x_2' = \gamma (x_2 - V t_2)$$

and, for the event "back of train passes Q_1,"

$$x_1' = \gamma (x_1 - V t_1).$$

Subtracting and remembering that $t_2 = t_1$, we find

$$l_o = x_2' - x_1' = \gamma (x_2 - x_1) = \gamma l$$

or $l = l_o/\gamma$, which is the length contraction (15.15).

Our next example is one of the many seeming paradoxes of relativity.

EXAMPLE 15.3   A Relativistic Snake

A relativistic snake, of proper length 100 cm, is traveling across a table at $V = 0.6c$. To tease the snake, a physics student holds two cleavers 100 cm apart and plans to bounce them simultaneously on the table so that the left one lands just behind the snake's tail. The student reasons as follows: "The snake is moving with $\beta = 0.6$, so its length is contracted by the factor $\gamma = 5/4$ (check this) and its length measured in my frame is 80 cm. Therefore, the cleaver in my right hand bounces well ahead of the snake, which is unhurt." This scenario is shown in Figure 15.5. Meanwhile the snake reasons thus: "The cleavers are approaching me at $\beta = 0.6$, so the distance between them is contracted to 80 cm, and I shall certainly be cut to pieces when they fall." Use the Lorentz transformation to resolve this paradox.

Let us choose frames S and S' in the usual way. The student is at rest in S, with the cleavers at $x_L = 0$ and $x_R = 100$ cm. The snake is at rest in S', with its tail at $x' = 0$ and its head at $x' = 100$. To resolve the dispute, we must find where and when the two cleavers fall, as observed in S and in S'.

In S the cleavers fall simultaneously at $t = 0$. At this time the snake's tail is at $x = 0$. Since his length is 80 cm, his head has to be at $x = 80$ cm. [You can

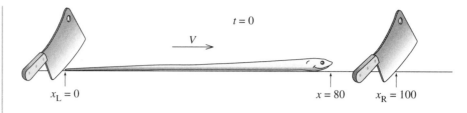

Figure 15.5   The snake paradox, as seen in the student's frame S. The cleavers fall simultaneously at time $t = 0$.

check this, if you want, using the transformation equation $x' = \gamma(x - Vt)$; with $x = 80$ cm and $t = 0$, this gives the correct value $x' = 100$ cm.] As observed in S, the experiment is as shown in Figure 15.5. The right cleaver falls comfortably ahead of the snake, the student is right, and the snake is unharmed.

What is wrong with the snake's reasoning? To answer this, we must examine the coordinates and times at which the two cleavers bounce, as observed in S'. The left cleaver falls at $t_L = 0$ and $x_L = 0$. According to the Lorentz transformation (15.20), the coordinates of this event, as seen in S' are

$$t'_L = \gamma(t_L - Vx_L/c^2) = 0$$

and

$$x'_L = \gamma(x_L - Vt_L) = 0.$$

As expected, the left cleaver falls just behind the snake's tail, at time $t'_L = 0$, as shown in Figure 15.6(a).

So far there are no surprises. However, the right cleaver falls at $t_R = 0$ and $x_R = 100$ cm. Therefore, as seen in S', it falls at a time given by the Lorentz transformation as

$$t'_R = \gamma(t_R - Vx_R/c^2) = -2.5 \text{ ns}.$$

(Check the numbers yourself.) The crucial point is that, as seen in S', *the two cleavers do not fall at the same time.* Since the right cleaver falls *before* the left one, it does not necessarily hit the snake, even though they are only 80 cm apart

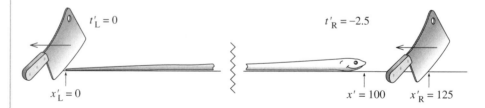

Figure 15.6   The snake paradox, as measured in the snake's frame S'. The cleavers move to the left with speed V, and the right one falls 2.5 ns *before* the left one. Even though the cleavers are only 80 cm apart, this lets them land 125 cm apart.

(in this frame). In fact, the position at which the right cleaver falls is given by the Lorentz transformation as

$$x'_R = \gamma(x_R - Vt_R) = 125 \text{ cm}.$$

The right cleaver does indeed miss the snake!

The resolution of this paradox, and many similar paradoxes, is that two events that are simultaneous in one frame are not necessarily simultaneous in a different frame — an effect sometimes called the **relativity of simultaneity**. As soon as we recognize that the two cleavers fall at different times in the snake's frame, there is no longer any problem understanding how they can both contrive to miss the snake.

15.7   The Relativistic Velocity-Addition Formula

As our next, and very important, application of the Lorentz transformation, let us use it to derive the relativistic velocity-addition formula. This formula is the answer to the following question: If an object — an electron, a baseball, a planet — is moving with velocity $\mathbf{v}$ relative to an inertial frame $\mathcal{S}$, how can we calculate its velocity $\mathbf{v}'$ relative to some other frame $\mathcal{S}'$? In classical physics, the answer to this question is the classical velocity-addition formula: If $\mathbf{V}$ denotes the velocity of $\mathcal{S}'$ relative to $\mathcal{S}$, then $\mathbf{v}' = \mathbf{v} - \mathbf{V}$. (Presumably, whoever named this formula wrote it as $\mathbf{v} = \mathbf{v}' + \mathbf{V}$.) For the special case that the axes of $\mathcal{S}$ and $\mathcal{S}'$ are parallel and $\mathbf{V}$ is in the x direction (our "standard" configuration), this becomes

$$v'_x = v_x - V, \qquad v'_y = v_y, \qquad \text{and} \qquad v'_z = v_z. \tag{15.23}$$

Our task now is to find the corresponding relativistic result.

Consider a particle moving with position $\mathbf{r}(t)$ or $\mathbf{r}'(t')$, as seen in $\mathcal{S}$ or $\mathcal{S}'$. The definition of the velocity $\mathbf{v}$ is the derivative

$$\mathbf{v} = \frac{d\mathbf{r}}{dt} \tag{15.24}$$

where $d\mathbf{r} = \mathbf{r}_2 - \mathbf{r}_1$ is the infinitesimal displacement between the positions at times t_1 and $t_2 = t_1 + dt$. Now, we can write down the Lorentz transformation for (x_2, y_2, z_2, t_2) and (x_1, y_1, z_1, t_1), and taking differences, we find

$$dx' = \gamma(dx - V dt), \quad dy' = dy, \quad dz' = dz, \quad dt' = \gamma(dt - V dx/c^2). \tag{15.25}$$

(Notice that $d\mathbf{r}$ and dt satisfy exactly the same transformation equations as $\mathbf{r}$ and t. This is because the Lorentz transformation turned out to be linear.) Using the definition (15.24), we can write down the components of $\mathbf{v}'$, and substituting (15.25) we find for v'_x

$$v'_x = \frac{dx'}{dt'} = \frac{\gamma(dx - V dt)}{\gamma(dt - V dx/c^2)}$$

or, canceling the factors of γ and dividing top and bottom by dt,

$$v'_x = \frac{v_x - V}{1 - v_x V/c^2}. \tag{15.26}$$

Similarly,

$$v'_y = \frac{dy'}{dt'} = \frac{dy}{\gamma(dt - V\,dx/c^2)}.$$

Dividing top and bottom by dt, we find for v'_y (and similarly v'_z)

$$v'_y = \frac{v_y}{\gamma(1 - v_x V/c^2)} \quad \text{and} \quad v'_z = \frac{v_z}{\gamma(1 - v_x V/c^2)}. \tag{15.27}$$

Notice that $v'_y \neq v_y$ even though $dy' = dy$. This is because $dt' \neq dt$. Notice also that γ is the factor pertaining to the speed V of $\mathcal{S}'$ relative to $\mathcal{S}$; that is $\gamma = 1/\sqrt{1 - V^2/c^2}$.

The three equations in (15.26) and (15.27) are the **relativistic velocity-addition formulas** or the relativistic velocity transformation. If all velocities are much less than c, then $\gamma \approx 1$ and we can ignore the second term in the denominators, and we recover the classical results (15.23). However, when the velocities concerned approach c, the relativistic velocity transformation can have some surprising results, as the following examples illustrate.

EXAMPLE 15.4   Adding Two Velocities Close to c

A rocket travelling at speed $0.8c$ relative to the earth shoots forward bullets with speed $0.6c$ (relative to the rocket). What is the bullets' speed relative to the earth?

If we choose frames in the usual way, with $\mathcal{S}$ fixed to the earth and $\mathcal{S}'$ fixed to the rocket, then $V = 0.8c$ and $v' = 0.6c$. Our task is to find v. The classical answer is of course $v = v' + V = 1.4c$. The relativistic answer is given by the inverse of (15.26), which we can find by the usual trick of exchanging primed and unprimed variables and reversing the sign of V. The result (from which I omit the subscripts x since all velocities are in the x direction) is

$$v = \frac{v' + V}{1 + v'V/c^2} \tag{15.28}$$

$$= \frac{0.6c + 0.8c}{1 + 0.8 \times 0.6} = \frac{1.4}{1.48}c \approx 0.95c.$$

The striking thing about this is that when we "add" $0.8c$ to $0.6c$ we get an answer that is less than c. In fact, it is fairly easy to prove that for any velocity with $v' < c$, the corresponding v is also, automatically, less than c. (See Problem 15.43.) That is, anything that travels with speed less than c in one frame has speed less than c in all frames.

EXAMPLE 15.5   Adding Two Velocities One of Which Equals c

The rocket of Example 15.4 shoots forward a signal (a pulse of light, for instance) with speed c relative to the rocket. What is the signal's speed relative to the earth?

Here $v' = c$, so (15.28) becomes

$$v = \frac{v' + V}{1 + v'V/c^2} = \frac{c + V}{1 + V/c} = c. \tag{15.29}$$

That is, a signal traveling in the x direction with speed c relative to $\mathcal{S}'$ also has speed c relative to $\mathcal{S}$. This result is actually true whatever the direction of travel, as you can prove in Problem 15.43. It asserts that the speed of light is invariant under the Lorentz transformation, in obedience to the second postulate of relativity (which led us to the Lorentz transformation in the first place).

15.8   Four-Dimensional Space—Time; Four-Vectors

The Lorentz transformation (15.20) mixes up space and time, in the sense that each of the equations for x' and t' involves both x and t. The Russian–German mathematician Hermann Minkowski (1864–1909) suggested that this mixing of space and time implies that time should be combined with the three spatial coordinates to form a four-dimensional *space–time*, on which the Lorentz transformations act as a kind of rotation. Before we examine this suggestion, let us review a couple of facts about the ordinary rotations of our everyday three-dimensional space.

Rotations of Ordinary Three-Dimensional Space

In discussing the vectors of ordinary three-dimensional space, it will be convenient to change our notation a bit: For any choice of orthogonal axes, I shall use the notation mentioned in Chapter 1, with the three unit vectors labeled $\mathbf{e}_1, \mathbf{e}_2, \mathbf{e}_3$. The components of a general vector $\mathbf{q}$ we'll call q_1, q_2, q_3, so that

$$\mathbf{q} = q_1\mathbf{e}_1 + q_2\mathbf{e}_2 + q_3\mathbf{e}_3 = \sum_{i=1}^{3} q_i\mathbf{e}_i \tag{15.30}$$

where, as usual,

$$q_i = \mathbf{e}_i \cdot \mathbf{q}. \tag{15.31}$$

To conform with this notation, I shall, from now on, rename the position vector $\mathbf{r} = (x, y, z)$ as $\mathbf{x} = (x_1, x_2, x_3)$.

Now consider a rotation which carries the axes defined by $\mathbf{e}_1, \mathbf{e}_2, \mathbf{e}_3$ into a second set with unit vectors $\mathbf{e}'_1, \mathbf{e}'_2, \mathbf{e}'_3$. The components q'_i of the same vector with respect to the new axes are easily found:

$$q'_i = \mathbf{e}'_i \cdot \mathbf{q} = \mathbf{e}'_i \cdot \sum_{j=1}^{3} q_j \mathbf{e}_j = \sum_{j=1}^{3} (\mathbf{e}'_i \cdot \mathbf{e}_j) q_j . \tag{15.32}$$

This equation expresses each of the coordinates q'_i in the new coordinate system as a sum over the coordinates q_j in the old system. (That is, it does for rotations what the Lorentz transformation does when we pass between frames in relative motion.) The coefficients in this sum are the scalar products $\mathbf{e}'_i \cdot \mathbf{e}_j$ of the unit vectors of the new and old systems.[13]

We can express (15.32) more compactly if we adopt the matrix notation of Chapter 10: Let $\mathbf{R}$ denote the 3×3 square matrix with elements

$$R_{ij} = \mathbf{e}'_i \cdot \mathbf{e}_j \tag{15.33}$$

and let $\mathbf{q}$ and $\mathbf{q}'$ denote the 3×1 columns made up of the coordinates

$$\mathbf{q} = \begin{bmatrix} q_1 \\ q_2 \\ q_3 \end{bmatrix} \quad \text{and} \quad \mathbf{q}' = \begin{bmatrix} q'_1 \\ q'_2 \\ q'_3 \end{bmatrix} . \tag{15.34}$$

[As in Chapter 10, it is good to stay a little relaxed about this notation. When there is any danger of confusion, we'll agree that $\mathbf{q}$ is a *column matrix* as in (15.34), but, when there is no such danger, we'll continue to call $\mathbf{q}$ a "vector" and even write it as the row (q_1, q_2, q_3).] With these notations, the rotation (15.32) takes the compact form

$$\mathbf{q}' = \mathbf{R}\mathbf{q}. \tag{15.35}$$

The effect of rotating our axes is to multiply the column $\mathbf{q}$ of coordinates q_i by a certain 3×3 **rotation matrix R**.

EXAMPLE 15.6   A Simple Rotation about One Axis

Consider a set of rectangular axes with the $x_1 x_2$ plane chosen horizontal and the x_3 axis vertically upward, and suppose we rotate these axes about the x_2 axis through an angle θ, to give a new set of axes as shown in Figure 15.7. Find the rotation matrix $\mathbf{R}$ for this rotation.

The required matrix $\mathbf{R}$ is easily written down using (15.33). Inspection of Figure 15.7 let's one evaluate the necessary scalar products to give

---

[13] The numbers $\mathbf{e}'_i \cdot \mathbf{e}_j$ are often called the **direction cosines** of the new axes with respect to the old, since $\mathbf{e}'_i \cdot \mathbf{e}_j = \cos\theta_{ij}$ where θ_{ij} is the angle between $\mathbf{e}'_i$ and $\mathbf{e}_j$.

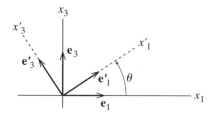

Figure 15.7 The primed axes are obtained from the unprimed by a counter-clockwise rotation about the x_2 axis (into the page) through an angle θ. The x_2 direction is unaffected by this rotation and the unit vectors $\mathbf{e}_2$ and $\mathbf{e}'_2$ both point into the page.

$$\mathbf{R} = \begin{bmatrix} \cos\theta & 0 & \sin\theta \\ 0 & 1 & 0 \\ -\sin\theta & 0 & \cos\theta \end{bmatrix}. \tag{15.36}$$

The effect of the rotation $\mathbf{R}$ on the coordinates of any point is $\mathbf{x}' = \mathbf{Rx}$ or

$$\left.\begin{aligned} x'_1 &= (\cos\theta)x_1 + (\sin\theta)x_3 \\ x'_2 &= x_2 \\ x'_3 &= (-\sin\theta)x_1 + (\cos\theta)x_3. \end{aligned}\right\} \tag{15.37}$$

One of the best arguments for regarding the three numbers x_1, x_2, x_3 as coordinates in a single three-dimensional space is that rotations can mix them up as in (15.37). One could imagine people taking the view that vertical distances (x_3) were somehow fundamentally different from horizontal ones (x_1 or x_2).[14] But surely such people would be dissuaded from this view when they noticed that (15.37) mixes x_1 and x_3 together (and, for $\theta = \pi/2$, simply exchanges their roles). We shall now argue similarly that Lorentz transformations are a kind of rotation that mixes the space and time coordinates in a four-dimensional space–time.

Lorentz Transformations as "Rotations" of Space–Time

A glance at the Lorentz transformation (15.20) should convince you that it mixes x and t in somewhat the same way that the rotation (15.37) mixes x_1 and x_3. We can make this parallel surprisingly close by polishing our notation. First, we'll rename our space coordinates x_1, x_2, x_3 as above, and we'll introduce a fourth coordinate

$$x_4 = ct \tag{15.38}$$

---

[14] Bizarre as such a view may appear, it does seem to be endorsed by some standard practices. For example, people in the business of storing water behind dams measure the volume of stored water in acre·feet, with horizontal areas measured in acres, but vertical depths in feet.

where the factor c guarantees that x_4 has the same dimensions as x_1, x_2, and x_3. Recalling the definition $\beta = V/c$, we can rewrite the Lorentz transformation (15.20) as

$$\left.\begin{aligned} x_1' &= \gamma x_1 - \gamma \beta x_4 \\ x_2' &= x_2 \\ x_3' &= x_3 \\ x_4' &= -\gamma \beta x_1 + \gamma x_4. \end{aligned}\right\} \tag{15.39}$$

We can further improve the parallel with the rotation (15.37) if we note that since $\gamma \geq 1$ we can define an "angle" ϕ such that $\gamma = \cosh \phi$. Some simple algebra (Problem 15.30) should convince you that this makes $\gamma \beta = \sinh \phi$ and (15.39) becomes

$$\left.\begin{aligned} x_1' &= (\cosh \phi)x_1 - (\sinh \phi)x_4 \\ x_2' &= x_2 \\ x_3' &= x_3 \\ x_4' &= (-\sinh \phi)x_1 + (\cosh \phi)x_4 \end{aligned}\right\} \tag{15.40}$$

and our parallel is as close as it can be. It is important to understand that no one would claim that the Lorentz transformation (15.40) mixes x_1 and x_4 in *exactly* the way that the rotation (15.37) mixes x_1 and x_3 — the trig functions of (15.37) have become hyperbolic functions in (15.40) (and a sign has changed). Nevertheless, the parallel is close and is a powerful argument for regarding $x_4 = ct$ as the fourth coordinate in a **four-dimensional space–time** or just **four-space**.

Four-Vectors

The four numbers x_1, x_2, x_3 and $x_4 = ct$ constitute a vector in four-dimensional space–time. Such vectors are called **four-vectors** to distinguish them from the three-dimensional, vectors, such as the position three-vector $\mathbf{x} = (x_1, x_2, x_3)$. Unfortunately, several different notations are used for four-vectors. I shall use ordinary italic letters for four-vectors; for example,

$$x = (x_1, x_2, x_3, x_4) = (\mathbf{x}, ct)$$

for the position–time vector just discussed. We shall be meeting several other four-vectors (for example, the four-momentum p, to be defined shortly). My notation for an arbitrary four-vector will be

$$q = (q_1, q_2, q_3, q_4) = (\mathbf{q}, q_4),$$

where the bold face $\mathbf{q}$, comprising the first three components of q, is called the *spatial component* of q and the fourth component q_4 is called the *time component*.[15]

---

[15] This notation has two drawbacks you should be alert to: (1) Since we'll still be using italic symbols for various scalars (for example, m for mass), you will need to tell from the context whether an italic symbol is a four-vector or a scalar. (2) We can no longer use the convention that q is the

As with three-vectors, it is often convenient to understand four-vectors to be 4×1 *columns*,

$$x = \begin{bmatrix} x_1 \\ x_2 \\ x_3 \\ x_4 \end{bmatrix} \quad \text{and} \quad q = \begin{bmatrix} q_1 \\ q_2 \\ q_3 \\ q_4 \end{bmatrix}. \tag{15.41}$$

With this notation, the Lorentz transformation (15.39) can be written in matrix form as

$$x' = \Lambda x \tag{15.42}$$

where Λ is the 4×4 matrix

$$\Lambda = \begin{bmatrix} \gamma & 0 & 0 & -\gamma\beta \\ 0 & 1 & 0 & 0 \\ 0 & 0 & 1 & 0 \\ -\gamma\beta & 0 & 0 & \gamma \end{bmatrix} \qquad \text{[standard boost]}. \tag{15.43}$$

This is not the most general Lorentz transformation. It is the transformation between two frames in what we have called standard configuration, with corresponding axes parallel and with the velocity of S' relative to S along the x axis. For many purposes, this **standard transformation** is the only one we need to consider, but we should take a moment to discuss more general transformations.

Any Lorentz transformation which leaves corresponding axes parallel is called a **pure boost** or just **boost**, since all it does is "boost" us from one frame to another traveling at constant velocity relative to the first, without any rotation. The general transformation involves some rotation as well. If the transformation is a *pure* rotation (no relative motion, just a change of orientation) then of course $t' = t$ and only the three spatial coordinates get changed. Thus we can write a pure rotation in the form (15.42), where the 4×4 matrix Λ has the block form

$$\Lambda = \Lambda_{\mathrm{R}} = \left[\begin{array}{ccc|c} & & & 0 \\ & \mathbf{R} & & 0 \\ & & & 0 \\ \hline 0 & 0 & 0 & 1 \end{array} \right] \qquad \text{[pure rotation]} \tag{15.44}$$

where $\mathbf{R}$ is the 3×3 matrix of the given rotation. (If you've never seen this kind of block matrix before, write out the equation $x' = \Lambda_{\mathrm{R}} x$ in all its detail, and notice how it rotates the three spatial coordinates but leaves the fourth component unchanged, so $t' = t$.)

If we want to write down a pure boost Λ_{B} in an arbitrary direction $\mathbf{u}$, we can construct it from a couple of rotations plus a standard boost: First we rotate so that our new x_1 axis points along the required direction $\mathbf{u}$; next we make the standard boost (15.43);

---

magnitude of the three-vector $\mathbf{q}$. Instead, we'll just use $|\mathbf{q}|$ (though I'll continue to use r for the magnitude of the position vector, $r = |\mathbf{x}|$).

and then we rotate back to our original orientation. Finally, any Lorentz transformation Λ can be expressed as the product of a boost followed by a suitable rotation, $\Lambda = \Lambda_R \Lambda_B$.[16] For some practice at handling different Lorentz transformations, see Problems 15.32 to 15.34.

A four-vector is defined as anything that transforms in the same way as the space–time vector $x = (\mathbf{x}, ct)$ under all Lorentz transformations (15.42). The formal definition is this:

Definition of a Four-Vector

In each inertial frame S, a four-vector is specified by a set of four numbers $q = (q_1, q_2, q_3, q_4)$ such that the values in two frames S and S' are related by the equation $q' = \Lambda q$, where Λ is the Lorentz transformation connecting S and S'.

Obviously the space–time vector $x = (\mathbf{x}, ct)$ fits this definition, and we shall find several more examples in the next few sections (including the four-momentum p mentioned above).

The great merit of the notion of four-vectors is that it often allows one to check with almost no effort whether a proposed physical law is relativistically invariant. Suppose for example we believe that there should be a law of the form

$$q = p \tag{15.45}$$

where we know that q and p are four-vectors. (The law of conservation of momentum has this form, $p_{\text{fin}} = p_{\text{in}}$, as we shall see.) Suppose further that the law is true in one frame S. Since the corresponding values in any other frame S' are $q' = \Lambda q$ and $p' = \Lambda p$, we have only to multiply both sides of (15.45) by Λ and we see that $q' = p'$. That is, the truth of our proposed four-vector law (15.45) in one frame S assures its truth in any other frame S'. (Of course, this doesn't guarantee that the law *is* true — only experiment can test that — but it does guarantee that the law would be consistent with the postulates of relativity.)

Any single quantity that is invariant under rotations is called a **rotational scalar** or a **three-scalar**; for example, the mass m of an object is a three-scalar, and so is the time t. In the same way, any single quantity that is invariant under Lorentz transformations is called a **Lorentz scalar** or **four-scalar**. For example, we shall see that the mass m (if suitably defined) is a four-scalar; that is, the mass of any object has the same value in all inertial frames. On the other hand, the time t is not a four-scalar; rather, as we have already seen, it is the fourth component of a four-vector.

---

[16] Actually we still don't have the *most* general transformation, since the spatial origins of our two frames still coincide at $t = t' = 0$. This can be taken care of by shifting one of the origins, but this additional possibility need not concern us here.

15.9   The Invariant Scalar Product

A set of transformations — such as the set of all rotations in three-space, or all Lorentz transformations in four-space — can often be characterized by the quantities that they leave invariant. Our main interest here is, of course, in the Lorentz transformations, but, for a little guidance as to what we should expect, let us look first at rotations.

The Invariant Scalar Product in Three-Space

One of the most obvious properties of rotations in three-space is that they do not change the length of any vector. If we define

$$s = \mathbf{x} \cdot \mathbf{x} = x_1^2 + x_2^2 + x_3^2 \tag{15.46}$$

as the length squared of any vector $\mathbf{x}$, then the value of $s = \mathbf{x} \cdot \mathbf{x}$ in one coordinate system is the same as its value $s' = \mathbf{x}' \cdot \mathbf{x}'$ in any other system obtained from the first by rotation. (In the terminology just introduced, s is a rotational scalar.) Since this is true for any $\mathbf{x}$, we can replace $\mathbf{x}$ by $\mathbf{x} = \mathbf{a} + \mathbf{b}$ (where $\mathbf{a}$ and $\mathbf{b}$ are any two other vectors) and the invariance of $(\mathbf{a} + \mathbf{b})^2$ implies that

$$\mathbf{a} \cdot \mathbf{a} + 2\mathbf{a} \cdot \mathbf{b} + \mathbf{b} \cdot \mathbf{b} = \mathbf{a}' \cdot \mathbf{a}' + 2\mathbf{a}' \cdot \mathbf{b}' + \mathbf{b}' \cdot \mathbf{b}'. \tag{15.47}$$

Canceling the terms that we already know to be equal, we find that

$$\mathbf{a} \cdot \mathbf{b} = \mathbf{a}' \cdot \mathbf{b}'. \tag{15.48}$$

In other words, invariance of the length of any three-vector under rotation implies that the scalar product of any two three-vectors is invariant. We shall employ a similar argument for the scalar product of four-space in a moment.

The Invariant Scalar Product in Four-Space

We can construct a scalar product in four-space with several of the properties of its three-dimensional analog. For any four-vector $x = (x_1, x_2, x_3, x_4) = (\mathbf{x}, ct)$, let us define

$$s = x_1^2 + x_2^2 + x_3^2 - x_4^2 = r^2 - c^2 t^2. \tag{15.49}$$

This s is obviously a generalization of the three-dimensional length squared, but note well the minus sign on the fourth term. (It is because of this minus sign that Lorentz transformations of space–time are not exactly analogous to rotations of ordinary space.) The quantity s is invariant under Lorentz transformations, as we can easily prove: Consider first the standard boost (15.39). Under this transformation, the quantity s becomes

$$s' = x_1'^2 + x_2'^2 + x_3'^2 - x_4'^2$$
$$= \gamma^2(x_1 - \beta x_4)^2 + x_2^2 + x_3^2 - \gamma^2(-\beta x_1 + x_4)^2$$
$$= \gamma^2(1 - \beta^2)x_1^2 + x_2^2 + x_3^2 - \gamma^2(1 - \beta^2)x_4^2$$
$$= s$$

where the last equality follows because $\gamma^2(1 - \beta^2) = 1$. Therefore, the quantity s is invariant under the standard boost. But we have seen that any Lorentz transformation can be built up from a standard boost and rotations, and s is certainly unchanged by a rotation (since r^2 and t are separately invariant under rotation). Therefore s is invariant under *any* Lorentz transformation.

Before we discuss the significance of the new invariant quantity s, we can use it to define an **invariant scalar product** in four-space. For any two four-vectors $x = (x_1, x_2, x_3, x_4)$ and $y = (y_1, y_2, y_3, y_4)$, we define[17]

$$x \cdot y = x_1 y_1 + x_2 y_2 + x_3 y_3 - x_4 y_4. \tag{15.50}$$

(Again, note well the minus sign on the fourth component — this "scalar product" is a little different from the usual scalar product in ordinary space.) Obviously the invariant s of (15.49) is just $s = x \cdot x$. And, as with rotations, the argument leading from (15.47) to (15.48) implies that because $x \cdot x$ is invariant for any one four-vector x, the scalar product $x \cdot y$ is invariant for any two four-vectors x and y. We shall see that the scalar product $x \cdot y$ plays as big a role in relativity as the ordinary product $\mathbf{a} \cdot \mathbf{b}$ does in classical physics. The scalar product of any four-vector x with itself is often written as $x \cdot x = x^2$, and can be called the "invariant length squared" of x, but you must not be misled by this terminology into thinking that x^2 is positive. On the contrary, x^2 can obviously be positive, negative, or zero.

For future reference, we can rewrite the invariance of $x \cdot y$ as a property of the Lorentz matrices Λ: We know that $x \cdot y = x' \cdot y'$ whatever the values of x and y, provided $x' = \Lambda x$ and $y' = \Lambda y$ where Λ is any Lorentz transformation. Therefore, we can say that, for any Lorentz transformation Λ,

$$x \cdot y = (\Lambda x) \cdot (\Lambda y) \tag{15.51}$$

for any two columns of four numbers x and y.

To understand where the scalar product came from, consider an experiment in which a flash bulb at the origin $\mathbf{x} = 0$ is fired at time $t = 0$ in a frame $\mathcal{S}$. The light

---

[17] Be warned! Physicists are fairly evenly divided between those who use the definition (15.50) and those who put a minus sign in front of the whole expression. Both conventions have their advantages and disadvantages.

from the flash will spread out at speed c, so that at any later time t it occupies the sphere $r^2 = c^2t^2$. Using our new notation, we can say that the spreading wavefront is located by the condition $x \cdot x = r^2 - c^2t^2 = 0$. Now, the invariance of $x \cdot x$ implies that $x \cdot x = 0$ if and only if $x' \cdot x' = 0$ in any other frame S'. Therefore a spherical wave spreading with speed c as seen in S will be a spherical wave spreading with speed c in S', and vice versa. We see that the invariance of the scalar product $x \cdot x$ is a reflection of the second postulate, that the speed of light is the same in all inertial frames.

15.10   The Light Cone

The scalar product $x \cdot x$ lets us divide space–time into five physically distinct regions. To help visualize this, it is convenient to ignore one of the spatial dimensions (x_3 say), so we can plot the remaining two spatial dimensions horizontally and $x_4 = ct$ vertically up, as in Figure 15.8. (Mathematically, this amounts to confining our attention to the "plane" $x_3 = 0$.) Consider again the light from a flashbulb fired at the space–time origin O (that is, fired at $\mathbf{x} = 0$ when $t = 0$). As time passes, the light moves outward in the x_1x_2 plane on an expanding circle with $r^2 = c^2t^2$, and this sweeps out the upper half-cone shown in Figure 15.8. This cone, called the **forward light cone**, is therefore the set of all points in space–time that would be visited by light released from the origin O. Mathematically, it is the set of all space–time points $x = (\mathbf{x}, ct)$ satisfying $x \cdot x = r^2 - c^2t^2 = 0$ and $t > 0$.

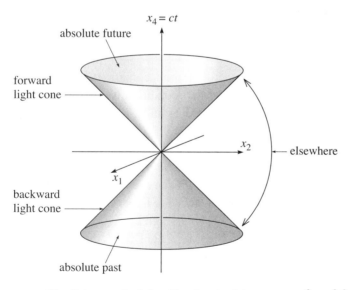

Figure 15.8   The light cone is defined by the condition $x \cdot x = r^2 - c^2t^2 = 0$ and divides space–time into five distinct parts: the forward and backward light cones, with $t > 0$ and $t < 0$ respectively; the interiors of the forward and backward light cones, called the absolute future and the absolute past; and the outside of the cone, labeled "elsewhere."

The lower half-cone shown in Figure 15.8 is called the **backward light cone** and is the set of all space–time points $x = (\mathbf{x}, ct)$ with the property that light released from x could subsequently pass through the origin, O. The whole light cone (forward and backward) is made up of the straight lines representing the path of any light ray that passes through the origin. Since light travels at speed c, and x_4 was cunningly chosen to be $x_4 = ct$, these lines have slope 1, so the surface of the light cone (if drawn to scale) makes an angle of 45°with the time axis. Since the light cone is defined by the condition that $x \cdot x = 0$ and since $x \cdot x$ is invariant (has the same value in all frames), it follows that the light cone is itself an invariant concept. That is, observers in any two frames will always agree as to which points lie on the light cone.

Interior of the Light Cone; Future and Past

Consider next a space–time point P, with coordinates $x = (\mathbf{x}, ct)$, that lies *inside* the forward light cone. This obviously has $t > 0$ and $r^2 < c^2 t^2$ or

$$\left. \begin{array}{l} x_4 > 0, \text{ and} \\ x_1^2 + x_2^2 + x_3^2 < x_4^2 \qquad (\text{or } x \cdot x < 0). \end{array} \right\} \tag{15.52}$$

These two conditions have a remarkable consequence: Notice first that since $t > 0$, we can assert that any event that occurs at P is later than any event at O, at least as observed in the frame $\mathcal{S}$ in which our coordinates x are measured. But what about some other frame $\mathcal{S}'$? To answer this, note that the second condition in (15.52) is just that $x \cdot x < 0$, and we know that $x \cdot x$ is invariant under Lorentz transformations. Therefore $x' \cdot x'$ is also negative, and the second condition is satisfied in $\mathcal{S}'$ as well. To see that the first condition is also satisfied in $\mathcal{S}'$, suppose first that $\mathcal{S}'$ is related to $\mathcal{S}$ by the standard Lorentz boost (15.39), under which

$$x_4' = \gamma(x_4 - \beta x_1). \tag{15.53}$$

Now, we know that $|\beta| < 1$ and the second condition in (15.52) guarantees that $|x_1| < x_4$. Therefore $x_4' > 0$, and the first condition is also satisfied in $\mathcal{S}'$. Since any Lorentz transformation can be made up of a standard boost and rotations (and rotations don't change x_4 at all), we conclude that both conditions (15.52) hold in all frames if they hold in one frame. In other words, the statement that P lies inside the forward light cone is a Lorentz-invariant statement. In particular, if P lies inside the forward light cone, then all observers will agree that an event that occurs at P is later than one at O. For this reason, the inside of the forward light cone is often called the **absolute future** — "absolute" because all observers agree that P is in the future of O. In a similar way, if P lies inside the backward light cone, then P is *earlier* than O as measured by all inertial observers, and this region is called the **absolute past**. (See Problem 15.39.)

So far we have considered the light cone with its vertex at the space–time origin O. This is defined by the light rays which happen to pass through O. If we considered instead light which passes through some other space–time point, Q, this would define a light cone with its vertex at Q — the light cone of Q. This cone would look just like Figure 15.8, except that the vertex would be at the arbitrary point Q rather

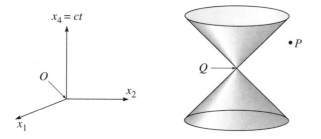

Figure 15.9   The light cone of an arbitrary space–time point Q with coordinates $x_Q = (\mathbf{x}_Q, ct_Q)$ is made up of all light rays that pass through Q. The point shown as P lies outside the light cone of Q.

than the origin O, as shown in Figure 15.9. Any point P on this cone must satisfy $(\mathbf{x}_P - \mathbf{x}_Q)^2 = c^2(t_P - t_Q)^2$, so that

$$(x_P - x_Q)^2 = 0 \qquad (P \text{ on light cone of } Q). \tag{15.54}$$

The inside of the forward light cone of Q is the absolute future of Q, all of whose points are later than Q as seen by all inertial observers. That is, for any point P inside the forward light cone of Q, all observers agree that $t_P > t_Q$.

Exterior of the Light Cone; Space-Like Vectors

The situation is entirely different for a point P that lies *outside* the light cone, as in Figure 15.9. First, the condition for P to be outside is that

$$(\mathbf{x}_P - \mathbf{x}_Q)^2 > c^2(t_P - t_Q)^2, \tag{15.55}$$

or, equivalently

$$(x_P - x_Q)^2 > 0 \qquad (P \text{ outside light cone of } Q). \tag{15.56}$$

This condition is symmetric between P and Q. Thus if P is outside the light cone of Q, then Q is outside the light cone of P, and vice versa. It is clear from Figure 15.9 that there are points P outside the light cone of Q, with P later than Q (that is, $t_P > t_Q$), and others for which P and Q are simultaneous ($t_P = t_Q$), and still others with P earlier that Q ($t_P < t_Q$). There is nothing remarkable about this claim; it is a straightforward consequence of the geometry of Figure 15.9. What *is* remarkable is the following proposition (which I shall prove directly):

Proposition
Let P be any given space–time point outside the light cone of a second given point Q. Then
    (1) there exist frames $\mathcal{S}$ in which $t_P > t_Q$
but
    (2) there also exist frames $\mathcal{S}'$ in which $t'_P = t'_Q$
and
    (3) there also exist frames $\mathcal{S}''$ in which $t''_P < t''_Q$.

This startling proposition implies that the time ordering of any two given events, each outside the other's light cone, can be different in different frames: Where one observer says that event A occurred before event B, a second observer can find them the other way around (and a third can find them to be simultaneous). This has profound implications related to the notion of **causality**: If one event A (an explosion, for instance) is the cause of another event B (the collapse of a distant building), then A must obviously occur first in time, since causes always precede their effects. According to our proposition, if the space–time point P is outside the light cone of Q, then neither Q nor P is unambiguously first in time. (In some frames it's one, and in some frames it's the other.) Therefore, nothing that happens at Q can be the cause of anything that happens at P, nor the other way around. Now, any kind of signal traveling from Q to P would have to travel with speed greater than c. [This follows from (15.55).] Conversely, if a signal emanating from Q had speed greater than c, then it could travel to some point P outside Q's light cone. It follows that *no causal influence can travel faster than the speed of light.*[18] Because the region outside the light cone of Q is completely immune to anything that happens at Q, this region is sometimes called the "**elsewhere**" of Q.

To simplify the proof of our proposition, let us put the point Q at the origin O, and abbreviate the coordinates of P to $x = (\mathbf{x}, ct)$. (The general case is no harder; it is just a bit messier notationally.) By making a rotation if necessary, we can put $\mathbf{x}$ on the positive x_1 axis, so that

$$x = (x_1, 0, 0, x_4). \tag{15.57}$$

Now let us assume that statement (1) above is true (so that $x_4 > 0$), and prove statements (2) and (3). [Obviously one of the statements must be true, and you can easily check that our arguments work equally well starting from either (2) or (3).] Let us now make the standard boost (15.39) to a new frame S' in which

$$x_4' = \gamma(x_4 - \beta x_1). \tag{15.58}$$

Since P is outside the light cone of O, $\mathbf{x}^2 > c^2 t^2$ which for the vector (15.57) means that $x_1 > x_4$. Therefore we can choose $\beta = x_4/x_1 < 1$, and, according to (15.58), $x_4' = 0$. That is, $t' = 0$, and we have proved statement (2) above for the case that $Q = O$. If P was outside the light cone of an arbitrary point Q, the corresponding boost would give a vector $x_P' - x_Q'$ with its fourth component equal to zero, and hence $t_P' = t_Q'$, again as required.

A four-vector whose fourth component is zero can be described as a pure-space vector, and one which can be brought into this form by a Lorentz transformation is called **space-like**. That is, a four-vector is space-like if there is a frame in which it is a pure-space vector, with zero fourth component. With this terminology, we can say

---

[18] If the causal signal could have speed greater than c, then it could travel from Q to some P outside Q's light cone, but we have just seen that this is impossible.

that the outside of the light cone is made up of all space-like vectors. Similarly, we can rephrase the result about causal relations to say that, if the separation $x_P - x_Q$ of two points P and Q is space-like, then nothing that happens at P can influence what happens at Q, nor vice-versa.

To prove statement (3) [from the assumed truth of statement (1)] we have only to look at the transformation (15.58) again. Since $x_1 > x_4$ we can choose β a little bigger than x_4/x_1 but still less than 1, and, with this choice, we get a frame (S'', say) in which $t'' < 0$ (or, in general, $t_P'' < t_Q''$) as required.

Time-Like Vectors

An argument similar to that just given for space-like vectors (Problem 15.44) shows that if a four-vector q lies inside the light cone (that is $q \cdot q < 0$), then there exists a frame S' in which it has the pure-time form $q' = (0, 0, 0, q_4')$. Naturally, therefore, we describe vectors inside the light-cone as being **time-like**. These can then be subdivided into forward time-like vectors (with $q_4 > 0$) and backward time-like (with $q_4 < 0$).

An important example of a forward time-like vector is the displacement four-vector dx of any material particle in a time dt. As we shall discuss shortly, a material particle can be defined as any particle with positive mass ($m > 0$). Equivalently (as we shall see), it is any particle for which, at any given time, there exists a rest frame; that is, a frame in which the particle is at rest, with $v = 0$. It is a matter of experience that all of the normal constituents of matter — electrons, protons, neutrons — have this property, and likewise all composites, such as atoms, molecules, baseballs, and stars.[19] Suppose now that, between the times t and $t + dt$, a material particle moves from $\mathbf{x}$ to $\mathbf{x} + d\mathbf{x}$, and consider the four-vector displacement

$$dx = (d\mathbf{x}, c\, dt) = (\mathbf{v}, c)\, dt.$$

In the particle's rest frame, $d\mathbf{x} = 0$, and dx has the pure-time form $dx = (0, 0, 0, cdt)$. Therefore, dx is time-like in all frames, and $dx^2 < 0$. Since

$$dx^2 = (\mathbf{v}^2 - c^2)\, dt^2 < 0,$$

we conclude that $\mathbf{v}^2 < c^2$ in all frames; that is, material particles cannot travel with speeds greater than or equal to the speed of light. Notice that this is the third sense in which we have proved the speed of light acts as a universal speed limit: (1) The relative speed of any two inertial frames is always less than c. (2) In any one inertial frame the speed of any causal signal is always less than or equal to c. And now (3) in any inertial frame, the speed of any material particle is less than c.

---

[19] As we shall discuss in Section 15.16, the only common particle which does not have this property is the photon, the particle of light. Since this travels at speed c in all frames, it certainly does not have a rest frame. Naturally, we do not regard a photon as a material particle.

15.11   The Quotient Rule and Doppler Effect

As a beautiful application of the properties of four-vectors, we next discuss the Doppler effect — the change in frequency of a wave due to the motion of the wave's source or the observer. Before we do this, we need to derive one more important property of four-vectors, the quotient rule.

The Quotient Rule

Suppose that we find a quantity k that is specified by four numbers $k = (k_1, k_2, k_3, k_4)$ in every inertial frame $\mathcal{S}$. It is naturally tempting to think that k is a four-vector, but this is not necessarily so. For example, in discussing the motion of an object of mass m, charge q, volume V, and temperature T, we could define

$$k = (m, q, V, T),$$

and this set of four numbers would be defined in every frame but is fairly obviously not a four-vector; that is, its value in one frame $\mathcal{S}$ is not related to the value in another frame $\mathcal{S}'$ by the Lorentz transformation $k' = \Lambda k$. Although this example may seem a bit artificial, it does show clearly that not every quantity k with four components is necessarily a four-vector. On the other hand, k *is* a four vector if it satisfies the conditions of the following theorem:

The Quotient Rule

Suppose that x is known to be a four-vector and that, in every inertial frame, $k = (k_1, k_2, k_3, k_4)$ is a set of four numbers, and suppose further that for every value of x the quantity $\phi = k \cdot x = k_1 x_1 + k_2 x_2 + k_3 x_3 - k_4 x_4$ is found to have the same value in all frames (that is, $\phi = k \cdot x$ is a four-scalar), then k is a four-vector.

The proof of this rule is surprisingly simple: First, that ϕ is a four-scalar implies that

$$k \cdot x = k' \cdot x'. \tag{15.59}$$

But from (15.51) we know that for any Lorentz transformation Λ

$$k \cdot x = (\Lambda k) \cdot (\Lambda x).$$

Now, by assumption, x is a four-vector, so we can replace Λx by x', to give

$$k \cdot x = \Lambda k \cdot x'. \tag{15.60}$$

Comparing (15.59) and (15.60), we see that

$$k' \cdot x' = (\Lambda k) \cdot x'.$$

This equation is true for any choice of x'. If we choose $x' = (1, 0, 0, 0)$, then it tells us that the first component of k' is equal to the first component of Λk, and continuing in this way we can show that all four components are equal, so that

$$k' = \Lambda k,$$

which shows that k (the "quotient" of the scalar ϕ and the vector x) is indeed a four-vector. Armed with this quotient rule, let us return to the Doppler effect.

Doppler Effect

When you think of the Doppler effect, you probably think of the Doppler shift of sound. The sound of a police siren rushing toward us has a higher pitch, and then — as it passes us and speeds away — a lower pitch, than when the siren is stationary; as a train speeds past a crossing bell, the passengers hear first a higher pitch as the train approaches the bell, then a lower pitch as the train moves away. There is a corresponding effect with light and all other forms of electromagnetic waves. The famous "red shift" of light from stars is used routinely by astronomers to find how fast a star is moving away from us (and, indirectly, how far away the star is); in the "Doppler cooling" of atoms, the Doppler shift of laser light "seen" by a moving atom is used to selectively slow down fast moving atoms and hence bring groups of atoms to very low temperatures; and on the highway, the Doppler shift of radar bounced off the front of your car is used by the police to measure your speed. To derive the formula for the Doppler shift of light we must work relativistically. (Light travels at the speed of light!) Armed with our knowledge of four-vectors, we shall find that the derivation is surprisingly easy.

Any sinusoidal plane wave has the form

$$\phi = A \cos(\mathbf{k} \cdot \mathbf{x} - \omega t - \delta). \tag{15.61}$$

Here the nature of the function ϕ depends on the wave under consideration; for a sound wave, it could be taken to be the pressure change produced by the sound; for light, it could be any component of the electromagnetic field. The vector $\mathbf{k}$ is called the **wave vector**; its direction is the direction of propagation of the wave and its magnitude is $|\mathbf{k}| = 2\pi/\lambda$, where λ is the wavelength; ω is the angular frequency, $\omega = 2\pi\nu$, where ν is the ordinary frequency; and δ is a (usually not very interesting) constant phase shift. The speed of the wave is $v = \omega/|\mathbf{k}|$; for light in a vacuum, this is of course c, so $\omega = c|\mathbf{k}|$.

Our main concern with the plane wave (15.61) is the phase $\mathbf{k} \cdot \mathbf{x} - \omega t$. It is impossible to resist writing this as a four-dimensional scalar product

$$\mathbf{k} \cdot \mathbf{x} - \omega t = k \cdot x, \tag{15.62}$$

where as always $x = (\mathbf{x}, ct)$ and k denotes the **wave four-vector**,

$$k = (\mathbf{k}, \omega/c). \tag{15.63}$$

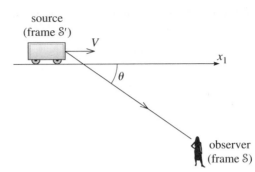

Figure 15.10   The Doppler experiment. A source of light is moving along the x_1 axis with speed V relative to the frame $\mathcal{S}$. The observer in frame $\mathcal{S}$ sees the light from the source traveling at an angle θ with the x_1 axis. The light's frequency is ω as measured in $\mathcal{S}$ and $\omega' = \omega_0$ as measured in the source's rest frame $\mathcal{S}'$.

To prove that k defined this way really is a four-vector, we note that the phase $k \cdot x$ at any point x determines the position on the wave relative to the troughs or crests of the wave. Since this has to be the same in any frame, it follows that $k \cdot x$ is a four-scalar, and since x is certainly a four-vector, the quotient rule guarantees that k is a four-vector as its name implies. Since the fourth component of k is ω/c and since we now know how k transforms, we are ready to find the frequency of a light signal as measured in a frame relative to which the source is moving.

The experiment we have in mind is shown in Figure 15.10. An observer at rest in the frame $\mathcal{S}$ observes a railroad car moving at speed V along the x_1 axis. The car is emiting light of angular frequency $\omega' = \omega_0$, as measured in the car's rest frame $\mathcal{S}'$. If the light reaching the observer travels at an angle θ with the x_1 axis, we want to know its frequency ω, as measured by the observer.

The wave four-vector k of the light reaching the observer has the form $k = (\mathbf{k}, k_4)$ where $k_4 = \omega/c = |\mathbf{k}|$. According to the standard Lorentz boost,

$$k_4' = \gamma(k_4 - \beta k_1).$$

Setting $k_4' = \omega'/c$, $k_4 = \omega/c$, and $k_1 = |\mathbf{k}| \cos\theta = (\omega/c) \cos\theta$, we find that

$$\omega' = \gamma\omega(1 - \beta\cos\theta).$$

Solving for ω and replacing ω' by ω_0, we get the **relativistic Doppler formula for light**

$$\omega = \frac{\omega_0}{\gamma(1 - \beta\cos\theta)}, \tag{15.64}$$

where ω_0 is the frequency of the light in the rest frame of the source, ω is the frequency observed in a frame where the source has velocity $\mathbf{V}$, and θ is the angle between $\mathbf{V}$ and the observed light.

15.12   Mass, Four-Velocity, and Four-Momentum

You may be becoming impatient that in eleven sections we have so far discussed only the kinematics of relativity. In fact, this reflects a truth about relativity — that many of its most interesting features, such as time dilation and the impossibility of causal signals traveling faster than the speed of light, are purely kinematic. Nevertheless, it is high time we took up relativistic dynamics, and this is what we now do.

In this section, I shall introduce the relativistic definitions of the mass and momentum of an object. It is important to recognize that there is really no such thing as the "correct" definition of a concept like mass or momentum when we move into the *terra incognita* of a new subject like relativity. Like Humpty Dumpty, we are, in principle, entitled to define words however we want.[20] Nevertheless, there are certain requirements of reasonableness that we can hope to impose: Any definition of momentum should coincide as closely as possible with the nonrelativistic definition in the domain where the latter has proved useful — namely when the speed of the object is much less than c. And we would like our new definitions to share with their nonrelativistic counterparts any properties that seem essential to the concept concerned. For example, we shall seek a definition of relativistic momentum with the property that the total momentum of an isolated system is conserved.

Mass in Relativity

There are, in fact, two different definitions of mass in relativity, both of which meet our requirements of reasonableness, and both with their supporters. The definition I shall use can be described as the *invariant mass* and is favored by the majority of practising physicists. The other, called the *variable mass*, is favored mostly by popularizers of relativity, since it makes some ideas seem easier at first. I shall describe the variable mass briefly later, but throughout this chapter I shall use the invariant mass, which is defined as follows: Given any object at rest (or moving with speed much, much less than c), we know that the nonrelativistic definition of mass produces a well-defined and useful quantity. To emphasize its definition, this mass is often called the **rest mass**. To define the mass of the same object traveling at high speed, we shall adopt the following, embarrassingly simple definition:

> ### Definition of Invariant Mass
> The mass, m, of an object, whatever its speed, is defined to be its rest mass.

If observers in an inertial frame $\mathcal{S}$ see an object sailing by at half the speed of light and want to know its mass, they must somehow bring the object to rest (or move themselves into a frame moving with the object) and then measure its mass using

---

[20] "When *I* use a word," Humpty Dumpty said in a rather scornful tone, "it means just what I choose it to mean — neither more nor less." Lewis Carroll, *Alice's Adventures in Wonderland*.

any convenient technique of nonrelativistic mechanics. The equivalence of all inertial frames guarantees that this procedure will produce the same answer in all frames, so the resulting mass can be called the invariant mass. However, since it is the only definition we shall be using, we shall generally call it just the mass. Since the mass defined this way has the same value in all frames, it is a Lorentz scalar.

The Proper Time of a Body

Before we take up the definition of relativistic momentum, it is convenient to introduce two more important kinematic quantities. The three-dimensional position $\mathbf{x}(t)$ of a body at time t defines a point $x = (\mathbf{x}(t), ct)$ in space–time, and, as time advances, this point traces a path, called the body's **world line**. We have seen that the separation dx between neighboring points x and $x + dx$ on the world line of a material body is a time-like vector. This means that there is a frame (namely the body's rest frame) where the separation is pure time-like, with the form $dx_o = (0, 0, 0, c\, dt_o)$. (The subscript "o" indicates the rest frame.) Since the two positions in three-space are equal ($\mathbf{x}_o = \mathbf{x}_o + d\mathbf{x}_o$), the time dt_o is the proper time between the two points on the body's world line. To find this proper time, we don't actually have to go to the rest frame. In any other frame, the separation has the form

$$dx = (\mathbf{v}\, dt, c\, dt)$$

and since $dx_o^2 = dx^2$, it follows that $-c^2 dt_o^2 = (v^2 - c^2)dt^2$, which we can solve for dt_o to give

$$dt_o = dt\sqrt{1 - v^2/c^2} = \frac{dt}{\gamma(v)} \tag{15.65}$$

where $\gamma(v)$ is the familiar γ factor $1/\sqrt{1 - v^2/c^2}$, calculated for the body's speed v. [You will recognize (15.65) as the time-dilation formula (15.11), so we didn't really need to go through this calculation.] We can apply (15.65) in any frame S and will, of course, get the same value for dt_o. That is, the proper time dt_o is a Lorentz scalar, which makes it a convenient quantity to work with, as we shall see.

The Four-Velocity

We have seen that the three-dimensional velocity $\mathbf{v}$ of a body transforms according to the rather complicated velocity-addition formulas (15.26) and (15.27). The reason for the complication is easy to see: The three-velocity $\mathbf{v} = d\mathbf{x}/dt$ is the quotient of a three-vector $d\mathbf{x}$ and the fourth component dt of a four-vector. So, little wonder that it transforms awkwardly. Having recognized the problem, we can easily construct a related vector which transforms more simply. If we considered $\mathbf{u} = d\mathbf{x}/dt_o$ instead of $d\mathbf{x}/dt$, then at least the denominator would be a scalar. In fact, while we're about it, we may as well consider the four-vector

$$u = \frac{dx}{dt_o} = \left(\frac{d\mathbf{x}}{dt_o}, c\frac{dt}{dt_o}\right). \tag{15.66}$$

Since this **four-velocity** is the quotient of a four-vector and a four-scalar, it clearly *is* a four-vector. If we use (15.65) to replace dt_0 by dt/γ, we find that

$$u = \gamma \left(\frac{d\mathbf{x}}{dt}, c\frac{dt}{dt} \right) = \gamma (\mathbf{v}, c). \tag{15.67}$$

The most prominent feature of this result is that the three-velocity $\mathbf{v}$ is *not* the spatial part of the four-velocity u (which is why I called the latter u rather than v). However, if our body is moving much slower than c, then $\gamma \approx 1$ and the spatial part of the four-velocity is indistinguishable from the ordinary three-velocity $\mathbf{v}$. As we shall see directly, the fact that u is a four-vector makes it useful in our efforts to construct a relativistic mechanics.

Relativistic Momentum

We are now ready to address the next definition in our relativistic mechanics — the definition of the momentum $\mathbf{p}$ of a body with mass m and velocity $\mathbf{v}$. We obviously want our definition to agree with the classical definition ($\mathbf{p} = m\mathbf{v}$), at least at nonrelativistic speeds, $|\mathbf{v}| \ll c$. What else we ask of our definition depends on what classical property of momentum we regard as so important that it should carry over to relativity. It would be hard to name the single most important property of momentum in classical mechanics, but the conservation of momentum is surely a strong candidate, and we shall look for a definition of $\mathbf{p}$ with the property that the total momentum $\mathbf{P} = \sum \mathbf{p}$ of an isolated system of bodies is conserved. To be consistent with the postulates of relativity, this law, if true at all, must be true in all inertial frames.

The simplest possibility would be that we could continue to use the classical definition $\mathbf{p} = m\mathbf{v}$, but we can dismiss this possibility fairly easily. With a little ingenuity one can construct a thought experiment in which the total classical momentum $\sum m\mathbf{v}$ is conserved in one frame $\mathcal{S}$, but not in a second frame $\mathcal{S}'$. One example, shown in Figure 15.11, is an elastic collision of two equal-mass particles a and b. As seen in frame $\mathcal{S}$, the two particles approach the origin with equal and opposite velocities in the $x_1 x_2$ plane, and emerge with the x_2 components of their velocities reversed. Obviously, the total classical momentum, as measured in $\mathcal{S}$, is zero ($\sum m\mathbf{v} = 0$) before and after the collision, and classical momentum is conserved. Figure 15.11(b) shows the same experiment, as seen in a frame $\mathcal{S}'$ which travels along the x_1 axis of $\mathcal{S}$ with speed V equal to the component v_1 of particle a, so that particle a travels straight up and then down the x_2' axis as seen in $\mathcal{S}'$. Using the relativistic velocity transformations (15.26) and (15.27), one can find the four velocities as measured in $\mathcal{S}'$. Since these calculations, although reasonably straightforward, are quite messy, I shall leave them as an exercise for the reader (Problem 15.54), but the important conclusion is easily stated: When we substitute the velocities of $\mathcal{S}'$ we find that the total classical momentum is *not* conserved in frame[21] $\mathcal{S}'$; that is, $\sum m\mathbf{v}_{\text{in}}' \neq \sum m\mathbf{v}_{\text{fin}}'$. Evidently, if we were

---

[21] The reason is actually fairly easy to see. The transformation (15.27) of the y component of velocity depends on the x component. Since particles a and b have different x components of velocity

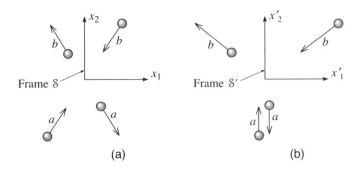

Figure 15.11   An elastic collision between two equal-mass particles, a and b. (a) In frame $\mathcal{S}$, the incoming particles approach with equal but opposite velocities and emerge with their x_2 components reversed. The total classical momentum $\sum m\mathbf{v}$ is zero before and after the collision. (b) The frame $\mathcal{S}'$ has velocity equal to the x_1 component of a's initial velocity in $\mathcal{S}$. Using the relativistic velocity-addition formula, it is easy to show that the total classical momentum $\sum m\mathbf{v}'$ is not conserved in this frame.

to adopt the classical definition of momentum, a law of conservation of momentum would be inconsistent with the postulates of relativity.

The problem with the classical definition of momentum, $\mathbf{p} = m\mathbf{v}$, derives from the awkward transformation of the three-velocity $\mathbf{v}$, and this suggests a more promising approach to our problem. Instead of using the three-velocity $\mathbf{v}$, suppose we used the four-velocity u to define the **four-momentum** of any object of mass m as

$$p = mu = (\gamma m \mathbf{v}, \gamma mc) \qquad \text{[definition of four-momentum].} \qquad (15.68)$$

[The last expression follows from (15.67).] Since m is a four-scalar and u is a four-vector, this defines p as a four-vector. If, in the usual way, we write

$$p = (\mathbf{p}, p_4) \qquad (15.69)$$

then this defines the **three-momentum**, $\mathbf{p}$, as the spatial part of the four-vector p or, comparing with (15.68),

$$\mathbf{p} = m\mathbf{u} = \gamma m \mathbf{v} \qquad \text{[definition of three-momentum].} \qquad (15.70)$$

If our object is traveling slowly ($|\mathbf{v}| \ll c$), then $\gamma \approx 1$ and our new definition of $\mathbf{p}$ agrees with the classical one, $\mathbf{p} = m\mathbf{v}$. In general, however, the two definitions differ by a factor of γ, and it is the new definition (15.70) that proves useful.

---

in $\mathcal{S}$, their y components wind up with different magnitudes in $\mathcal{S}'$. Thus $\sum m v'_y$ is nonzero and actually changes sign when the velocities reverse in the collision.

What about conservation of momentum? Since we are now saddled with a four-dimensional momentum vector, it seems clear that conservation of momentum, if it is to be true at all, should be a four-dimensional law, which we could write as

$$\sum p_{\text{fin}} = \sum p_{\text{in}}. \tag{15.71}$$

This is really four equations. The first three would be the law of conservation of the newly defined three-momentum, and the fourth would be the conservation of something else, namely the fourth component $\sum p_4$. Obviously, we need to find out quickly what this fourth component is, but I shall give this important question a section of its own. Briefly, though, we shall find that the fourth component of the new four-momentum is the energy (actually E/c), so that the law (15.71) is a wonderfully compact combination of the old laws of momentum *and* energy conservation.

Here, what I want to emphasize is this: Since the four-momentum p is a four-vector, the same is true of both sides of (15.71). Therefore, if (15.71) is true in one frame $\mathcal{S}$, it is automatically true in all frames; that is, our proposed law of conservation of four-momentum is compatible with the postulates of relativity. Whether the law is actually true must, of course, be decided by experiment. As you have no doubt guessed, the verdict is clear: Countless experiments have shown that the total four-momentum of an isolated system is constant.

Variable Mass

Some physicists like to rewrite the definition (15.70) of the relativistic three-momentum by introducing a **variable mass**

$$m_{\text{var}} = \gamma(v)m. \tag{15.72}$$

With this definition the three-momentum becomes

$$\mathbf{p} = m_{\text{var}}\mathbf{v}. \tag{15.73}$$

This has the advantage that it makes the relativistic momentum *look* like its nonrelativistic counterpart, $\mathbf{p} = m\mathbf{v}$. Nevertheless, it has important disadvantages, which have led the majority of practising physicists to avoid the use of the variable mass. First, it is not necessarily a good idea to make a new definition look like its older counterpart when there are, in fact, important differences. Second, the introduction of the variable mass fails to achieve a complete parallel with classical mechanics. For example, it is not true that the relativistic kinetic energy (which we shall define in the next section) is equal to $\frac{1}{2}m_{\text{var}}v^2$, nor is it true (in general) that $\mathbf{F} = m_{\text{var}}\mathbf{a}$. (See Problems 15.59 and 15.79.) Third, unlike the invariant mass, the variable is not a Lorentz scalar. For all of these reasons, I shall not use the variable mass here.

15.13   Energy, the Fourth Component of Momentum

The conservation of four-momentum means that, for any isolated system, there are four conserved quantities. The first three comprise the newly defined three-momentum **p**. But what about the fourth component? We are going to find that, for any freely moving object, the fourth component of the four-momentum defined by (15.68) is the energy divided by c. This is such an important result that I'll state it as a formal definition and then discuss its justification and consequences. Accordingly, we make the definition

---

Definition of Relativistic Energy

The energy E of a freely moving object with four-momentum $p = (\mathbf{p}, p_4)$ is

$$E = p_4 c = \gamma mc^2, \tag{15.74}$$

---

where the second expression follows from (15.68), which implies that $p_4 = \gamma mc$. With this definition, we can rewrite the four-momentum p as

$$p = (\mathbf{p}, E/c), \tag{15.75}$$

which explains why the four-momentum p is also called the **momentum–energy four-vector**.

In partial justification of the definition (15.74), notice first that (15.74) does at least have the dimensions of energy, namely [mass $\times$ speed^2]. Next, let us look at E for a nonrelativistic object to see if it looks like the nonrelativistic energy. With $v \ll c$, we can expand γ using the binomial series to give

$$\gamma = [1 - (v/c)^2]^{-1/2} = 1 + \tfrac{1}{2}(v/c)^2 + \cdots, \tag{15.76}$$

so that (15.74) becomes

$$E \approx mc^2 + \tfrac{1}{2}mv^2 \tag{15.77}$$

provided $v \ll c$. In nonrelativistic mechanics, mass was believed to be absolutely conserved, so the term mc^2 would have been considered to be constant. Since the zero of energy was arbitrary, a classical physicist would have interpreted (15.77) to say that the newly defined E is just the classical kinetic energy plus an irrelevant constant.

To illustrate the result (15.77), consider for a moment an elastic collision. Suppose, for example, that two atoms a and b, with masses m_a^{in} and m_b^{in} and nonrelativistic speeds v_a^{in} and v_b^{in} collide and re-emerge with final masses m_a^{fin} and m_b^{fin} and speeds v_a^{fin} and v_b^{fin}. (Of course, in classical mechanics, the initial and final masses would be the same, but we'll use different labels to avoid prejudging this question.) In any two-body collision, conservation of the newly defined relativistic energy (15.74) implies that

$$E_a^{\text{in}} + E_b^{\text{in}} = E_a^{\text{fin}} + E_b^{\text{fin}}$$

or, if the collision is nonrelativistic,

$$\left[m_a^{\text{in}} c^2 + \tfrac{1}{2} m_a^{\text{in}} (v_a^{\text{in}})^2 \right] + \left[m_b^{\text{in}} c^2 + \tfrac{1}{2} m_b^{\text{in}} (v_b^{\text{in}})^2 \right]$$

$$= \left[m_a^{\text{fin}} c^2 + \tfrac{1}{2} m_a^{\text{fin}} (v_a^{\text{fin}})^2 \right] + \left[m_b^{\text{fin}} c^2 + \tfrac{1}{2} m_b^{\text{fin}} (v_b^{\text{fin}})^2 \right].$$

Regrouping terms, we can write this as

$$M^{\text{in}} c^2 + T^{\text{in}} = M^{\text{fin}} c^2 + T^{\text{fin}} \qquad (15.78)$$

where M^{in} denotes the initial total mass, T^{in} the initial total kinetic energy, and so on.[22] Now, according to classical ideas, mass is conserved, so $M^{\text{in}} = M^{\text{fin}}$. Therefore, the two mass terms in (15.78) would cancel, leaving just

$$T^{\text{in}} = T^{\text{fin}}.$$

That is, the total kinetic energy would be conserved — which is precisely what we know to be true in an elastic collision. Thus, in the context of nonrelativistic elastic collisions, conservation of the newly defined relativistic energy (15.74) coincides with the familiar conservation of classical energy. This is perhaps the simplest and strongest single argument for regarding the definition (15.74) as an appropriate generalization of the classical notion of energy (together, of course, with the experimental fact that energy defined this way *is* conserved).

The argument of the last paragraph gives us a reassuringly familiar result in the case of elastic collisions. However, it is going to lead us to the first big surprise of our relativistic mechanics. We know that, even in the context of nonrelativistic mechanics, there are *inelastic* processes, in which total kinetic energy is *not* conserved. For example, in the case of our two atoms, the collision could disturb the internal motion of the atomic electrons, changing the internal energy of one (or both) of the atoms. In this case we know that the atoms would emerge with changed total kinetic energy, so that $T^{\text{fin}} \neq T^{\text{in}}$. (That such processes can occur is a well-established experimental fact; the Franck–Hertz experiment, in which electrons collided inelastically with mercury atoms, was a famous example.) Now, the argument leading to (15.78) depends only on the (true) assumption that relativistic energy is conserved, so (15.78) must apply to any possible nonrelativistic collision. In particular, in an inelastic collision with $T^{\text{fin}} \neq T^{\text{in}}$, (15.78) implies that the total mass of the two atoms has to change, $M^{\text{fin}} \neq M^{\text{in}}$. If we imagine an inelastic collision in which one of the atoms has its internal energy changed while the other is completely unchanged, then for the first atom we can say this: If the atom gains internal energy (is "excited"), then $T^{\text{fin}} < T^{\text{in}}$ and hence, from (15.78), $M^{\text{fin}} > M^{\text{in}}$; that is, when an atom gains internal energy it has to gain mass. Conversely, if the atom loses internal energy, it has to lose mass.

If relativistic energy is conserved (as it is), then it follows logically that mass cannot be conserved. The first question that must be addressed is why this nonconservation

---

[22] By "kinetic energy" I mean here the kinetic energy $\tfrac{1}{2} m v^2$ of the translational motion of either atom as a whole. Of course, an atom may also have kinetic energy of its electrons as they orbit the nucleus, but we shall include that as part of the atom's internal energy.

of mass had not been discovered much sooner. To answer this, notice that (15.78) can be written as

$$\Delta M c^2 = -\Delta T. \tag{15.79}$$

The brief answer to our questions is this: By everyday standards, c^2 is an exceedingly large quantity, so that, even if ΔT is fairly large, $\Delta M = \Delta T / c^2$ is still very small — in most cases, unobservably so — as the following example illustrates.

EXAMPLE 15.7   Mass Change in the Franck–Hertz Experiment

In their famous experiment in 1914, James Franck and Gustav Hertz fired electrons through a container of mercury vapor. The mercury atom has a state whose internal energy is 4.9 eV (1 eV = 1.6×10^{-19} J) higher than the atom's normal "ground state." In some of the collisions between the electrons and the mercury atoms, a mercury atom was excited into this state, with the result that the final kinetic energy of the emerging particles was 4.9 eV less than the initial; that is, $\Delta T = -4.9$ eV. By how much was the mass of the mercury atom increased by the collision?

According to (15.79), the increase of mass is

$$\Delta M = -\frac{\Delta T}{c^2} = \frac{4.9 \text{ eV}}{c^2} = 8.7 \times 10^{-36} \text{ kg.}$$

(Check the conversion and the arithmetic yourself.) The electron emerges with its mass unchanged (it is still just an electron), so all of this mass increase goes to the mercury atom (which is now an excited mercury atom). The increase is fantastically small by everyday standards, but the real question is this: How big is the mass increase compared to the original mass of the mercury atom, which is 200.6 atomic mass units or 3.3×10^{-25} kg? This fractional change in the mass of the atom is

$$\frac{\Delta m}{m} = \frac{8.7 \times 10^{-36}}{3.3 \times 10^{-25}} = 2.6 \times 10^{-11}.$$

This fractional change is far too small to be detected by any direct measurement of masses.

The energy released in typical chemical reactions is also of order a few eV per atom. For example, the burning of hydrogen in oxygen can be thought of as an inelastic collision

$$H_2 + H_2 + O_2 \rightarrow H_2O + H_2O \tag{15.80}$$

in which the total kinetic energy of the final two molecules is about 5 eV more than that of the original three.[23] Conservation of relativistic energy requires that the outgoing

---

[23] It is not a coincidence that the energy release (or loss) in chemical reactions is about the same as that in the Franck–Hertz experiment of Example 15.7. In both cases the change originates in

water molecules have less total mass than the initial hydrogens and oxygen, but, again, the difference is far too small to be detected by direct mass measurements.

In nuclear reactions, the kinetic energy released can be much greater. For example, in the neutron-induced fission

$$n + {}^{235}U \rightarrow {}^{90}Kr + {}^{143}Ba + n + n + n,$$

the kinetic energy increases by about 200 MeV, and the fractional loss of mass is about 1 part in 1000 — still not very large, but large enough to measure directly for many nuclear reactions. As we shall see later there are processes in which the mass change is even larger, but the evidence from nuclear physics is already sufficient to confirm the relativistic prediction (15.79) beyond reasonable doubt.

Mass Energy

We have seen that the term mc^2 in the result (15.77), $E \approx mc^2 + \frac{1}{2}mv^2$, is certainly not the "irrelevant constant" that a classical physicist would have taken it to be. In fact, in relativity, unlike classical mechanics, the energy does not contain an arbitrary constant. This is because we certainly want the four-momentum $p = (\mathbf{p}, E/c)$ to be a four-vector, and the addition of a constant to E would destroy this desirable property. (See Problem 15.66.) Bearing this in mind, let us look again at the relativistic definition of the energy of an object, $E = \gamma mc^2$. Even if the object is at rest, with $\gamma = 1$, the object still has some energy, given by $E = mc^2$ (perhaps the most famous equation in all of physics). This energy is naturally called the **rest energy** of the object or, since it is associated with the mass m, the **mass energy**.

The concept of mass energy lets us interpret the inelastic processes discussed above as processes in which some mass energy is converted into kinetic energy or vice versa. In the processes of atomic and nuclear physics, this conversion typically involves only a tiny fraction of the total mass energy, but there are processes in which there is 100% conversion. For example, in a collision between an electron (e^-) and its "antiparticle," the positron e^+, both particles can be annihilated,

$$e^- + e^+ \rightarrow radiation$$

with 100% of their mass energy becoming the energy of electromagnetic radiation.

When an object is moving, $\gamma > 1$ and its energy $E = \gamma mc^2$ is greater than its rest energy mc^2. This suggests that we define a quantity T by the equation

$$E = mc^2 + T. \tag{15.81}$$

This T is the additional energy that an object has by virtue of its motion and is naturally called the **kinetic energy**,

$$T = E - mc^2 = (\gamma - 1)mc^2. \tag{15.82}$$

---

differences in the energy levels of the electrons in the atoms or molecules, and these differences are almost always of order a few eV.

When the object is moving slowly, we have seen that $T \approx \frac{1}{2}mv^2$ [this follows from (15.77)], but in general the nonrelativistic result is incorrect and we must use the relativistic definition (15.82).

Three Useful Relations

There are three useful relations among the parameters m, $\mathbf{v}$, $\mathbf{p}$, and E that characterize an object's motion. First, since $p = \gamma m(\mathbf{v}, c)$ and also $p = (\mathbf{p}, E/c)$, we see at once that

$$\beta \equiv \frac{\mathbf{v}}{c} = \frac{\mathbf{p}c}{E}. \tag{15.83}$$

This relation lets you find the velocity of an object if you know its three-momentum $\mathbf{p}$ and energy E.

Consider next the invariant "length squared" $p^2 = p \cdot p$. In the object's rest frame p has the form $p = (0, 0, 0, mc)$, so that $p \cdot p = -(mc)^2$. Since both sides of this equation are invariant, it immediately follows that the same relation holds in any frame: For any object with four-momentum p and mass m,

$$p \cdot p = -(mc)^2 \tag{15.84}$$

in any inertial frame. This relation is well worth memorizing and can greatly simplify several calculations, as we shall see.

Finally, it is sometimes useful to rewrite the result (15.84) in terms of the three-momentum and the energy. Since $p = (\mathbf{p}, E/c)$, (15.84) becomes $\mathbf{p}^2 - (E/c)^2 = -(mc)^2$ or

$$E^2 = (mc^2)^2 + (\mathbf{p}c)^2. \tag{15.85}$$

This shows that the three quantities E, mc^2, and $|\mathbf{p}|c$ are related like the sides of a right triangle, with E as the hypotenuse, as indicated in Figure 15.12. At this stage there is no deep geometrical significance to this statement, but it does give an easy way to remember and visualize the relation (15.85). If the speed is much less than c, then $\gamma \approx 1$, so $E \approx mc^2$; in this case, the hypotenuse and base of the triangle are nearly equal, so $T \ll mc^2$ and the triangle is very low (height $\ll$ base). On the other hand, if v is very close to c, then $\gamma \gg 1$, so $E \gg mc^2$; in this case, $T \gg mc^2$, so the energy is mostly kinetic, and the triangle is very tall (height $\gg$ base) with $E \approx |\mathbf{p}|c$.

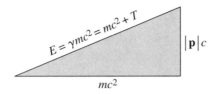

Figure 15.12   The three parameters E, mc^2, and $|\mathbf{p}|c$ are related like the sides of a right triangle with E as the hypotenuse.

EXAMPLE 15.8 Energy and Momentum of an Electron

The rest energy of an electron is about 0.5 MeV (actually 0.511, but for many purposes 0.5 is good enough). What is the electron's mass in the SI unit (the kilogram) and in MeV/c^2? If its kinetic energy is $T = 0.8$ MeV, what is its total energy E and what is the magnitude $|\mathbf{p}|$ of its three-momentum in MeV/c? What is its speed?

The given rest energy tells us that

$$mc^2 = 0.5 \text{ MeV}. \tag{15.86}$$

Solving for m, converting eV to joules, and putting in the value of c, we find (as you can check) that $m \approx 9 \times 10^{-31}$ kg (more precisely, 9.11×10^{-31}). Straightforward as this calculation is in kilograms, it is even easier, and often much more convenient, in MeV/c^2. We simply divide both sides of (15.86) by c^2 and we have the answer directly:

$$m = 0.5 \text{ MeV}/c^2.$$

Evidently, the mass in MeV/c^2 is numerically the same as mc^2 in MeV. This is so simple that it takes a little getting used to. If you've never used the unit MeV/c^2 before, just keep reminding yourself that the statement $m = 0.5$ MeV/c^2 is precisely equivalent to the statement that $mc^2 = 0.5$ MeV. The reason this is such a convenient way to specify masses is that our real concern is often *not* with the mass itself, but rather with the corresponding energy mc^2, and the most convenient unit for the latter is often the MeV.

If $T = 0.8$ MeV, then clearly $E = T + mc^2 = 1.3$ MeV, and by the "useful relation" (15.85)

$$|\mathbf{p}|c = \sqrt{E^2 - (mc^2)^2} = \sqrt{1.3^2 - 0.5^2} \text{ MeV} = 1.2 \text{ MeV}.$$

Once again, we could get an answer in SI units by making the necessary conversions, but a simpler, and often more convenient, course is just to divide both sides by c to give

$$|\mathbf{p}| = 1.2 \text{ MeV}/c.$$

Finally, according to the useful relation (15.83), the electron's dimensionless speed $\beta = v/c$ is

$$\beta = \frac{|\mathbf{p}|c}{E} = \frac{1.2 \text{ MeV}}{1.3 \text{ MeV}} = 0.92;$$

that is, $v = 0.92c$.

Notice how nicely the factors of c cancel if we measure masses in MeV/c^2 and momenta in MeV/c when using the relations (15.83) and (15.85). This is because m and $\mathbf{p}$ enter these relations only through the combinations mc^2 and $\mathbf{p}c$. For some practice at using these relations and the new units see Problems 15.61 to 15.63.

15.14   Collisions

The laws of conservation of energy and momentum play a key role in the analysis of collisions. In this section, I shall illustrate this claim with a couple of examples.

EXAMPLE 15.9   Collision of Two Lumps of Putty

A relativistic ball of putty, with mass m_a, energy E_a, and velocity $\mathbf{v}_a$, collides with a stationary ball of mass m_b, as shown in Figure 15.13. If the two balls fuse to form a single lump, what is the lump's mass m and with what velocity $\mathbf{v}$ does it move off?

To find the final mass, we have only to recall that the invariant "length squared" of the four-momentum of any object is $-m^2c^2$. If we denote the final four-momentum by p_{fin}, then

$$(p_{\text{fin}})^2 = -m^2c^2. \tag{15.87}$$

By conservation of momentum-energy, $p_{\text{fin}} = p_{\text{in}}$, where p_{in} is the total initial momentum; that is, $p_{\text{in}} = p_a + p_b$, from which

$$(p_{\text{in}})^2 = (p_a + p_b)^2 = p_a^2 + p_b^2 + 2p_a \cdot p_b$$

$$= -m_a^2c^2 - m_b^2c^2 - 2E_a m_b \tag{15.88}$$

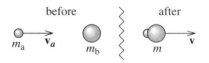

Figure 15.13   Two balls of putty collide and form a single lump.

where the last term comes about because $p_b = (0, 0, 0, m_b c)$. Comparing (15.87) and (15.88), we find that the mass of the final lump is

$$m = \sqrt{m_a^2 + m_b^2 + 2E_a m_b / c^2}.$$

Notice that if the original motion was nonrelativistic, then $E_a \approx m_a c^2$, and we recover the nonrelativistic result $m = m_a + m_b$; but, in general, $m > m_a + m_b$.

According to the useful relation (15.83), the final velocity is $\mathbf{v} = \mathbf{p}_{\text{fin}} c^2 / E_{\text{fin}}$. By conservation of four-momentum, we can replace the components of p_{fin} by those of p_{in} to give

$$\mathbf{v} = \frac{\mathbf{p}_a c^2}{E_a + m_b c^2} = \frac{\gamma_a m_a \mathbf{v}_a}{\gamma_a m_a + m_b} \tag{15.89}$$

where γ_a denotes the γ factor for the incoming ball a. Notice that if $v_a \ll c$, this reduces to the familiar nonrelativistic result $\mathbf{v} = m_a \mathbf{v}_a / (m_a + m_b)$.

The CM Frame

In nonrelativistic mechanics, we have seen that a very useful concept is that of the CM or center-of-mass frame — the frame in which the center of mass of a system is at rest. Alternatively, this frame can be characterized as the frame in which the total momentum is zero, $\mathbf{P} = \sum \mathbf{p} = 0$. (So you can think of "CM" as standing for "center of momentum" if you like.) This alternative definition carries over directly to relativistic mechanics.[24] We have seen that the four-momentum p of any material particle is forward time-like (lies inside the forward light cone).[25] Now, it is a simple exercise to prove (Problem 15.69) that the sum of any number of forward time-like vectors is itself forward time-like. Therefore, the total momentum $P = \sum p$ of any collection of particles is also time-like, and this guarantees that there exists a frame in which P has the form $P = (0, 0, 0, P_4)$. Naturally, we define this frame, in which the total three-momentum $\mathbf{P} = 0$, to be the CM frame of the system.

It often happens that a collision problem is very easy to solve in the CM frame. Thus, if we need to solve the same problem in some other frame $\mathcal{S}$, the simplest procedure is often to transform from $\mathcal{S}$ to the CM frame, solve the problem there, and then transform back to $\mathcal{S}$, as the following two examples illustrate.

---

[24] Oddly enough, the notion of center of mass does not carry over satisfactorily into relativity. Thus, it is better to think of the CM frame as the center-of-momentum frame.

[25] So far, we are considering only material particles, that is, particles with mass $m > 0$. We shall soon discuss the case of massless particles, for which p lies *on* the light cone. Fortunately (with one small exception) the same results apply even when some of the particles are massless. See Problems 15.88 and 15.89.

EXAMPLE 15.10   An Elastic Head-On Collision

Consider an elastic head-on collision between a projectile, with mass m_a and velocity $\mathbf{v}_a$, and a stationary target of mass m_b, as shown in Figure 15.14(a). (That the collision is head-on means that the two particles emerge from the collision both moving along the line of the incident velocity $\mathbf{v}_a$, as shown). What is the final velocity $\mathbf{v}_b$ of the target particle b?

Let us denote by $\mathcal{S}$ the lab frame, in which the experiment actually occurs (with b initially at rest), and take the direction of the incident velocity to be the x axis. To solve the problem directly in frame $\mathcal{S}$, we would write down the equations of conservation of energy and momentum and solve for the requested final velocity. Unfortunately, the equations are extremely messy, and a much simpler course is to transform to the CM frame $\mathcal{S}'$. In the CM frame the two incoming three-momenta are equal and opposite, and it is easy to see (Problem 15.68) that the collision simply reverses them both. Thus our procedure will be this: (1) Transform p_b to the CM frame $\mathcal{S}'$. (2) Reverse its spatial part $\mathbf{p}_b$. And (3) transform back to $\mathcal{S}$ and calculate the velocity. Before we do this, we need to find the velocity of the CM frame $\mathcal{S}'$ relative to $\mathcal{S}$. Since the total four-momentum is

$$P = \left(\mathbf{p}_a, \frac{E_a + m_b c^2}{c} \right),$$

the required (dimensionless) velocity $\boldsymbol{\beta}$ to transform to the CM frame is

$$\boldsymbol{\beta} = \frac{\mathbf{p}_a c}{E_a + m_b c^2}. \tag{15.90}$$

(In the nonrelativistic limit, with $\mathbf{p}_a \approx m_a \mathbf{v}_a$ and $E_a \approx m_a c^2$, this corresponds to a velocity $\mathbf{v} = m_a \mathbf{v}_a / (m_a + m_b)$, which is the velocity of the center of mass, as expected.)

We can now follow our three steps to solve the problem. In the lab frame, the initial four-momentum of the target b is

$$p_b^{\text{in}} = (0, 0, 0, m_b c) \qquad \text{[lab frame, initial]}. \tag{15.91}$$

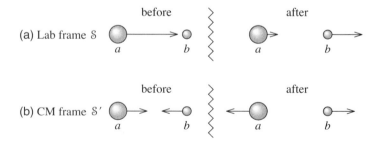

Figure 15.14   **(a)** An elastic head-on collision as seen in the lab frame $\mathcal{S}$, where the target b is initially at rest. **(b)** In the CM frame $\mathcal{S}'$, all of the three-momenta have the same magnitude. The only effect of the collision is to reverse the three-momentum of each particle. (The arrows represent momenta.)

Applying the standard Lorentz boost, with velocity (15.90), we find the corresponding CM momentum

$$p_b'^{\text{in}} = \gamma m_b c(-\beta, 0, 0, 1) \qquad \text{[CM frame, initial]}. \qquad (15.92)$$

In the CM frame the collision simply reverses the spatial components of this momentum. Thus the corresponding final momentum is

$$p_b'^{\text{fin}} = \gamma m_b c(\beta, 0, 0, 1) \qquad \text{[CM frame, final]}. \qquad (15.93)$$

Finally, transforming back to the lab frame, we find

$$p_b^{\text{fin}} = \gamma^2 m_b c \left(2\beta, 0, 0, (1+\beta^2)\right) \qquad \text{[lab frame, final]}. \qquad (15.94)$$

The corresponding dimensionless velocity is just the ratio $2\beta/(1+\beta^2)$, so the actual final velocity of the target is

$$\mathbf{v}_b = \frac{2\beta}{1+\beta^2} c \qquad (15.95)$$

with $\boldsymbol{\beta}$ given by (15.90).

The answer (15.95), although easily found, is not especially illuminating in general. In the special case that the two masses are equal, it is easy to show (Problem 15.73) that (15.95) reduces to $\mathbf{v}_b = \mathbf{v}_a$; that is, the target b emerges with precisely the velocity of the incoming projectile a, and the projectile therefore comes to a dead stop. This behavior is well known to students of nonrelativistic mechanics (and to billiards players) and shows that, in the case that $m_a = m_b$, the relativistic result (15.95) agrees exactly with the familiar nonrelativistic one.

Whether or not the two masses are equal, it is easily shown that in the limit $v_a \ll c$, the relativistic result (15.95) approaches the corresponding nonrelativistic one. (See Problem 15.73.)

Threshold Energies

Most of the elementary particles that have been discovered in the last seventy years or so were found when they were produced in collisions of other particles. For example, the negative pion, π^-, can be created in a collision of a proton and a neutron,

$$p + n \rightarrow p + p + \pi^-.$$

Similarly, the first antiproton to be observed was produced by a proton–proton collision in the reaction

$$p + p \rightarrow p + p + p + \bar{p}. \qquad (15.96)$$

(The negatively charged antiproton $\bar{p}$ is the "antiparticle" of the proton, with the same mass but opposite charge.) A quantity of great concern to any experimenter hoping to observe this kind of reaction is the **threshold energy**, defined as the minimum energy of the initial particles for which the reaction can occur.

Let us consider a reaction of the form

$$a + b \rightarrow d + \cdots + g.$$

In the traditional collision experiment, one of the original particles (let's say b) was usually at rest — defining what we have been calling the lab frame. Thus, the concern was to know the threshold energy for a reaction of this kind in the lab frame. At first glance this would seem a simple matter to calculate. The minimum possible energy of the final particles is just their total rest energy $\sum m_{\text{fin}} c^2$, the energy they have when all at rest. Surely, then, the threshold energy for the reaction is just $\sum m_{\text{fin}} c^2$. Unfortunately this plausible argument is wrong. The trouble is that in the lab frame the total initial three-momentum is nonzero. (Particle b is at rest, so a has to be moving to bring in the necessary energy.) Conservation of three-momentum requires that the final three-momentum be nonzero, and so the final particles cannot all be at rest. Therefore, the threshold energy is more than just $\sum m_{\text{fin}} c^2$. But what is it?

The easiest way to answer this question is to observe that in the CM frame, where the total three-momentum *is* zero, all of the final particles can be at rest. Thus

$$E_{\text{cm}} \geq \sum m_{\text{fin}} c^2 \tag{15.97}$$

and the equality here *is* possible, with all final particles at rest. We can now find the threshold energy in the lab, by comparing the total four-momenta in the two frames. In the CM frame the total four-momentum has the form $P_{\text{cm}} = (0, 0, 0, E_{\text{cm}}/c)$. In the lab frame, it is $P_{\text{lab}} = p_a + p_b$, where $p_a = (\mathbf{p}_a, E_a/c)$ and $p_b = (0, 0, 0, m_b c)$ are the momenta of the two orginal particles. Now by the invariance of the scalar product, $P_{\text{cm}}^2 = P_{\text{lab}}^2$, so

$$-E_{\text{cm}}^2/c^2 = (p_a + p_b)^2 = p_a^2 + p_b^2 + 2 p_a \cdot p_b = -m_a^2 c^2 - m_b^2 c^2 - 2 E_a m_b$$

or, solving for E_a,

$$E_a = \frac{E_{\text{cm}}^2 - m_a^2 c^4 - m_b^2 c^4}{2 m_b c^2}.$$

Inserting the minimum value of E_{cm} from (15.97), we find for the minimum energy of the projectile a in the lab frame

$$E_a^{\text{min}} = \frac{(\sum m_{\text{fin}})^2 - m_a^2 - m_b^2}{2 m_b} c^2. \tag{15.98}$$

A famous example of the use of this equation was in the design of the experiment to verify the existence of the antiproton using the reaction (15.96). In this reaction $\sum m_{\text{fin}} = 4 m_p$, while $m_a = m_b = m_p$, so the minimuum energy (15.98) is $7 m_p c^2$; that is, the minimum kinetic energy for protons to produce antiprotons by the reaction (15.96) was $6 m_p c^2 \approx 5600$ MeV. The reaction in question was first observed at Berkeley, using protons accelerated to this energy by a machine called the Bevatron, which had been specifically designed to accelerate protons to about 6000 MeV, just enough more than the threshold to be sure to do the job.

An important feature of (15.98) is that the leading term is proportional to $(\sum m_{\text{fin}})^2$. Thus if the particle one is hoping to produce is very heavy, $E_a^{\min}$ may be prohibitively large. For example, the particle called the ψ (or J/ψ) has a mass of about 3100 MeV/c^2 and was discovered in the reaction

$$ e^+ + e^- \rightarrow \psi $$

where the positron and electron have mass about 0.5 MeV/c^2 each. Putting these numbers into (15.98), we see that, if this reaction was to be produced by firing positrons at stationary electrons (or vice versa), the minimum incident energy would have had to be a fantastic $E^{\min} \approx 10^7$ MeV — well out of reach of any electron or positron accelerator in existence today. The way around this seemingly hopeless obstacle was to use colliding beams, with the electrons and positrons approaching one another with approximately equal and opposite momenta. That is, the experiment was done in the CM frame. From (15.97), you can see that the threshold in this case is only $E_{\text{cm}} \approx 3100$ MeV. In this experiment, the advantages of this much smaller threshold energy far outweighed the disadvantages of having to work with two beams of high-energy particles.

15.15   Force in Relativity

We have not yet introduced the concept of force into our relativistic mechanics. One of the reasons for this is that force plays a much smaller role in relativity than in nonrelativistic mechanics. Another is that the concept of force is much more complicated in relativity. The most obvious complication is that (like several other parameters, such as mass and velocity) force can be defined in several different ways. A second complication arises from the possibility that the rest mass of an object can change. As we have seen, an inelastic collision of an electron with an atom can give the atom additional internal energy and increase its rest mass. For a macroscopic example of the same effect, imagine holding a flame under a metal object; the heat absorbed increases the object's internal energy and its rest mass. Like most introductory texts, I shall avoid the complication of such "heat-like forces" by confining attention to forces that *do not change the rest masses of the objects on which they act.*[26] Fortunately, these include many of the important forces in special relativity, including the Lorentz force

$$ \mathbf{F} = q(\mathbf{E} + \mathbf{v} \times \mathbf{B}) \tag{15.99} $$

on a charge q in electric and magnetic fields $\mathbf{E}$ and $\mathbf{B}$.

---

[26] For a careful and clear discussion of "heat-like" forces, see the excellent book of Wolfgang Rindler, *Introduction to Special Relativity*, Oxford University Press, second edition, 1991, but be warned that Rindler tends to use m to denote the variable mass (what we would call γm).

Of the several conceivable definitions of force in relativity, the single most useful is probably the **three-force** defined as

$$\mathbf{F} = \frac{d\mathbf{p}}{dt} \tag{15.100}$$

where $\mathbf{p}$ denotes the relativistic three-momentum $\mathbf{p} = \gamma m \mathbf{v}$. This is not, of course, the same as the nonrelativistic force, since $\mathbf{p}$ is not the nonrelativistic momentum, but it does have the essential merit that it agrees with the nonrelativistic definition when $v \ll c$ (and $\gamma \approx 1$). A second property that recommends the definition (15.100) is that experiment shows that, with this definition, the force on a charge q in an electromagnetic field is given by the Lorentz equation (15.99). Thirdly, with the definition (15.100), we can prove the analog of the work-KE theorem, as follows: Recall the useful relation (15.85), that $E^2 = (\mathbf{p}c)^2 + (mc^2)^2$. Differentiating both sides with respect to time, we see that (remember we're assuming that the rest mass m doesn't change, so there is no term in dm/dt, but see Problem 15.85)

$$E \frac{dE}{dt} = \mathbf{p}c^2 \cdot \frac{d\mathbf{p}}{dt} = \mathbf{p}c^2 \cdot \mathbf{F}$$

or, dividing both sides by E and recalling that $\mathbf{p}c^2/E = \mathbf{v}$,

$$\frac{dE}{dt} = \mathbf{v} \cdot \mathbf{F}. \tag{15.101}$$

Multiplying both sides by dt we find that

$$dE = \mathbf{F} \cdot d\mathbf{x}, \tag{15.102}$$

where $d\mathbf{x}$ denotes the displacement $d\mathbf{x} = \mathbf{v} \, dt$. Finally, since $E = mc^2 + T$ and we are assuming that the mass m does not change, we find that

$$dT = \mathbf{F} \cdot d\mathbf{x} \tag{15.103}$$

which is precisely the work-KE theorem, generalized to include relativistic energies and forces.

EXAMPLE 15.11   Motion with a Constant Force

An object of fixed rest mass m is acted on by a uniform, constant force $\mathbf{F}$ (for example, the force on a charge in a uniform electrostatic field) and is released from rest at the origin at $t = 0$. Find the object's three-momentum $\mathbf{p}$, its three-velocity $\mathbf{v}$, and its position $\mathbf{x}$, all as functions of time.

Integrating (15.100) (with $\mathbf{F}$ constant), we find immediately

$$\mathbf{p} = \mathbf{F}t. \tag{15.104}$$

From (15.85) it is easy to see that $\gamma^2 = 1 + \mathbf{p}^2/(mc)^2$, so

$$\gamma = \sqrt{1 + (Ft/mc)^2}$$

and

$$\mathbf{v} = \frac{\mathbf{p}}{m\gamma} = \frac{\mathbf{F}t}{m\sqrt{1 + (Ft/mc)^2}}. \tag{15.105}$$

When t is small, we can neglect the second term inside the square root, and we recover the nonrelativistic answer $\mathbf{v} = \mathbf{F}t/m$, but when t gets large, the second term in the square root dominates and we find that v approaches c, without ever quite reaching it. This is consistent with our knowledge that no material particle can have speed greater than or equal to the speed of light.

To find the object's position $\mathbf{x}$, we have only to integrate (15.105) to give

$$\mathbf{x} = \frac{\mathbf{F}}{m}\left(\frac{mc}{F}\right)^2 \left(\sqrt{1 + \left(\frac{Ft}{mc}\right)^2} - 1\right). \tag{15.106}$$

As you can easily check, when t is small, this reduces to the familiar nonrelativistic result $\mathbf{x} = \frac{1}{2}\mathbf{F}t^2/m$ (that is, $\frac{1}{2}\mathbf{a}t^2$); when $t \to \infty$, it is asymptotic to $(ct + \text{const})$, in the direction of $\mathbf{F}$, as the speed approaches the speed of light (Problem 15.82).

Potential Energy

It can happen that, at least in one frame S, the force $\mathbf{F}$ on an object is the gradient of a function $U(\mathbf{x})$; that is, $\mathbf{F} = -\nabla U(\mathbf{x})$ and the force is conservative. This is the case, for example, for a charge q moving in an electrostatic field. When this happens, the work done on the object as it moves through a displacement $d\mathbf{x}$ is $\mathbf{F} \cdot d\mathbf{x} = -\nabla U \cdot d\mathbf{x} = -dU$. Combining this with the work–KE theorem (15.103), we find that $dT = -dU$ or $d(T + U) = 0$; that is, just as in nonrelativistic mechanics, if the force on an object is conservative, $T + U$ is conserved.

The Four-Force

The three-force $\mathbf{F} = d\mathbf{p}/dt$ is not the spatial part of a four-vector. (The trouble is that, although $d\mathbf{p}$ *is* the spatial part of a four-vector, dt is not a scalar.) In this respect, the three-force is like the three-velocity $\mathbf{v} = d\mathbf{x}/dt$, and the transformation of $\mathbf{F}$ from one frame to another is somewhat similar to that of $\mathbf{v}$ (Problem 15.83). Just as with the velocity, it is easy to see how to define a four-force that is closely related to the three-force. We can define the **four-force** on an object as the derivative of p with respect to the proper time t_o measured along the object's world line:

$$K = \frac{dp}{dt_o}. \tag{15.107}$$

(There is no widely accepted notation for the four-force, but K is one of the several notations used.) Since dp is a four-vector and dt_o is a four-scalar, K is automatically a four-vector. Since $dt_o = dt/\gamma$, we can rewrite K as

$$K = (\mathbf{K}, K_4) = \gamma \left(\frac{d\mathbf{p}}{dt}, \frac{1}{c}\frac{dE}{dt} \right) = \gamma (\mathbf{F}, \mathbf{v} \cdot \mathbf{F}/c) \tag{15.108}$$

where the last equality follows from (15.101). We see that the spatial part of the four-force is γ times the three-force $\mathbf{F}$, just as the spatial part of the four-velocity u is the γ times the usual three-velocity $\mathbf{v}$, as in (15.67).

The advantages of the four-force stem from its being a four-vector. This means that its transformation from one frame to another is just the familiar Lorentz transformation. It also means that the Lorentz invariance of any physical law formulated in terms of the four-force is easy to check. The main *disadvantage* of the four-force is that it gives the time derivative of momentum with respect to the proper time, where the three-force gives the derivative with respect to the time of any one inertial frame. Since our main interest is usually in the motion of an object in terms of the time in one particular frame (as opposed to the proper time of the moving object), the three-force is, in this respect, more useful.

15.16   Massless Particles; the Photon

A surprising consequence of relativity is the possibility of particles with zero mass, $m = 0$. In nonrelativistic mechanics, the notion of a massless particle makes no sense at all. The definitions $\mathbf{p} = m\mathbf{v}$ and $T = \frac{1}{2}mv^2$ show clearly that a particle with $m = 0$ would have no momentum and no kinetic energy and, hence, would presumably be nothing at all. At first glance, the same argument might seem to apply in relativity: If we let $m \to 0$ in the relativistic definitions

$$\mathbf{p} = \gamma m\mathbf{v} \quad \text{and} \quad E = \gamma mc^2 \tag{15.109}$$

we would seem to get the same conclusion — that a massless particle would have no momentum or energy and hence would not exist. Let us shelve this difficulty for a moment and look at the two relations (15.85) and (15.83)

$$E^2 = (mc^2)^2 + (\mathbf{p}c)^2 \quad \text{and} \quad \frac{\mathbf{v}}{c} = \frac{\mathbf{p}c}{E}. \tag{15.110}$$

If there were to be a particle with $m = 0$ (and if these relations still applied) then the first relation would become

$$E = |\mathbf{p}|c \qquad [\text{if } m = 0] \tag{15.111}$$

and this, combined with the second relation, would imply that the particle's speed had to be c,

$$v = c \qquad [\text{if } m = 0]. \tag{15.112}$$

In other words, if there were to be a massless particle, then the usual relations of relativistic mechanics would require that it always travel with speed c. If we return now to our original definitions (15.109) of $\mathbf{p}$ and E, we see that as $m \to 0$, so $v \to c$ and $\gamma \to \infty$. Thus for a massless particle, the two definitions would take the form $\infty \times 0$, which is undefined and hence does not actually rule out the existence of particles with $m = 0$.

Evidently, relativistic mechanics has room for massless particles, always traveling at speed c. Whether such particles exist is, of course, a question for experiment, and experiment tells us unambiguously that they do. The photon is the particle that carries the energy and momentum of electromagnetic waves; and experiment shows that, for a photon, E and $\mathbf{p}$ do satisfy (15.111) and that photons do always travel (no surprise!) at the speed of light.[27]

Notice that with $m = 0$, the four-momentum of a photon satisfies

$$p^2 = 0 \, ; \tag{15.113}$$

its invariant length squared is zero. We have seen that the four-momentum of a material particle (that is, a particle with $m > 0$) is always forward time-like. By contrast, that of any massless particle lies on the forward light cone and is **forward light-like**.

Since the definitions (15.109) are no longer meaningful, the question naturally arises how the energy and momentum of a massless particle are defined. In principle, at least, they can be defined using the conservation laws. Consider, for example, the emission of a photon by an atom X:

$$X^* \to X + \gamma \tag{15.114}$$

where X^* denotes an excited state of the atom, and X its ground state, and γ is the standard symbol for a photon. Since we already know how to define and measure the energy and momentum of either state of the atom, we can find the corresponding values for the photon from conservation of energy and momentum. This definition must, of course, be checked for consistency. Would a different process yield the same answers

---

[27] It used to be thought that the neutrino was another example of a massless particle, but current evidence shows that its mass, though small, is definitely nonzero, perhaps about 10^{-6} times the electron mass. (For comparison, the experimental limit on a possible photon mass is of order 10^{-20} times the electron mass.) On theoretical grounds it is generally assumed there must be a massless particle called the graviton, which does for gravity what the photon does for electromagnetism, but there is no direct evidence for the graviton.

for the same photon? For example, suppose we allowed the photon of (15.114) to collide with a second atom Y and eject an electron (the photoelectric effect):

$$\gamma + Y \rightarrow Y^+ + e \qquad (15.115)$$

where Y^+ denotes the positive ion of Y with one electron removed. Using this process, we could measure the energy and momentum of the photon, and these second measurements should yield the same answers as the first. Experiment has shown repeatedly that the energy and momentum of the photon are consistently defined in this way.

In fact there is a second way to find the energy and momentum of the photon. One of the first discoveries (due to Max Planck and Einstein) in the unfolding of quantum mechanics was that the energy of a photon is related to the frequency of its associated electromagnetic wave by the famous relation

$$E = \hbar\omega \qquad (15.116)$$

where $\hbar$ is Planck's constant (actually the original Planck constant h divided by 2π, $\hbar = h/2\pi = 1.05 \times 10^{-34}$ J·s) and omega is the angular frequency of the wave. Similarly, the momentum of a photon is given by

$$\mathbf{p} = \hbar\mathbf{k} \qquad (15.117)$$

where $\mathbf{k}$ is the wave vector of the wave. Thus, E and $\mathbf{p}$ can be found by measuring the frequency and wave vector of the corresponding wave. It is a pleasing fact that the two relations (15.116) and (15.117) can be combined into a single four-vector relation. Since $p = (\mathbf{p}, E/c)$ and the wave four-vector is $k = (\mathbf{k}, \omega/c)$, the two relations imply that

$$p = \hbar k. \qquad (15.118)$$

Since both sides of this equation are four-vectors, this relation is relativistically invariant; that is, the two quantum relations (15.116) and (15.117) are consistent with the principles of relativity.

The relation (15.117) is often rewritten in terms of the wavelength λ. Since $|\mathbf{k}| = 2\pi/\lambda$,

$$|\mathbf{p}| = \hbar|\mathbf{k}| = \frac{2\pi\hbar}{\lambda} = \frac{h}{\lambda}. \qquad (15.119)$$

In this form the relation is often called the **de Broglie relation** in honor of the French physicist Louis de Broglie (1892–1987), who first proposed that this relation should apply to the quantum wave associated with *any* particle — not just photons. For future reference, I'll rewrite a photon's four-momentum as follows:

$$p = \hbar k = \hbar\left(\mathbf{k}, \frac{\omega}{c}\right) = \frac{\hbar\omega}{c}(\hat{\mathbf{k}}, 1) \qquad (15.120)$$

where the last equality holds since $|\mathbf{k}| = \omega/c$, so $\mathbf{k} = (\omega/c)\hat{\mathbf{k}}$.

The Compton Effect

Historically, the most influential and persuasive evidence for the existence of massless photons obeying the relations of the last few paragraphs was the experiment of the American physicist Arthur Compton (1892–1962) in 1923. Compton fired X-ray photons at stationary electrons[28] and measured the increase in wavelength of the scattered photons. Classical theory would require that the wavelength of the scattered radiation should be exactly the same as that of the incident waves, but an increase in wavelength is easily explained if the radiation is carried by particle-like photons. When a photon collides with an electron, the electron recoils, taking some of the photon's original energy. Thus the emerging photons must have less energy and hence less momentum than those in the incident beam. According to (15.119) less momentum means longer wavelength, and the increase in wavelength is explained. Using the relations of the last few paragraphs, Compton was able to calculate the expected shift in wavelength and his experiment triumphantly confirmed his calculations.

The Compton experiment is shown schematically in Figure 15.15. The photon comes in from the left with four-momentum $p_{\gamma o}$ and emerges at angle θ with four-momentum p_γ, where according to (15.120)

$$p_{\gamma o} = \frac{\hbar\omega_o}{c}(\hat{\mathbf{k}}_o, 1) \qquad \text{and} \qquad p_\gamma = \frac{\hbar\omega}{c}(\hat{\mathbf{k}}, 1). \qquad (15.121)$$

The electron's initial four-momentum is

$$p_o = (0, 0, 0, mc) \qquad (15.122)$$

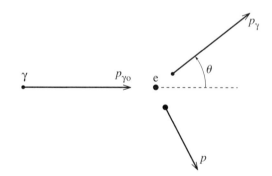

Figure 15.15   A photon, labeled γ, with four-momentum $p_{\gamma o}$ collides with a stationary electron. The photon emerges at an angle θ with four-momentum p_γ, and the electron recoils with four-momentum p.

---

[28] The target electrons were actually the valence electrons in the carbon atoms of a piece of graphite. Thus the electrons were certainly not perfectly stationary, but their kinetic energy (a few eV) was negligible compared to the energy of the X-ray photons (many thousands of eV). By the same token, the binding of the electrons (binding energy ≈ a few eV) was unimportant with such high photon energies.

and it recoils with four-momentum p. Now, by conservation of four-momentum, $p_0 + p_{\gamma 0} = p + p_\gamma$, or

$$p_0 + (p_{\gamma 0} - p_\gamma) = p.$$

Squaring both sides we find that

$$p_0^2 + 2p_0 \cdot (p_{\gamma 0} - p_\gamma) + \left(p_{\gamma 0}^2 - 2p_{\gamma 0} \cdot p_\gamma + p_\gamma^2 \right) = p^2.$$

Since $p_0^2 = p^2 = -m^2 c^2$, these two terms cancel, and since $p_{\gamma 0}^2 = p_\gamma^2 = 0$, these two terms drop out, and we are left with

$$p_0 \cdot (p_{\gamma 0} - p_\gamma) = p_{\gamma 0} \cdot p_\gamma. \tag{15.123}$$

Substitution of (15.122) and (15.121), followed by a little algebra, yields (as you should check)

$$\omega_0 - \omega = \frac{\hbar}{mc^2}(1 - \cos\theta)\omega_0 \omega$$

or

$$\frac{1}{\omega} - \frac{1}{\omega_0} = \frac{\hbar}{mc^2}(1 - \cos\theta).$$

Finally, replacing ω by $2\pi c / \lambda$, we find the desired shift in wavelength,

$$\Delta\lambda = \lambda - \lambda_0 = \frac{h}{mc}(1 - \cos\theta), \tag{15.124}$$

where I have replaced $2\pi\hbar$ by the original Planck constant h. This is the celebrated Compton formula for the shift in the wavelength of scattered radiation. That Compton's data agreed with this prediction at several angles gave strong support to the idea that the energy and momentum of radiation are carried by photons obeying the laws of relativistic mechanics, but with $m = 0$.

15.17   Tensors *

*As usual, sections marked with an asterisk can be omitted on a first reading.

In the next and final section of this chapter, I shall give a very brief account of the relativistic form of electromagnetic theory. Unfortunately, even this brief introduction requires some knowledge of the properties of tensors in four-dimensional space–time, and these four-tensors are the main subject of the present section. A complete account of four-tensors requires a rather elaborate machinery of "covariant" and "contravariant" vectors, but, for our present purposes we can manage without this formalism. Nevertheless, you should be aware that if you wish to pursue the study of relativity much further, you will need to master this more elaborate machinery.[29] Here I shall start my account by examining the transformation properties of three-dimensional vectors and tensors.

Vectors and Tensors in Three Dimensions

A three-vector $\mathbf{a}$ is characterized by its three components (in each frame) and by its transformation properties under rotations, as in (15.35),

$$\mathbf{a}' = \mathbf{Ra}, \tag{15.125}$$

where $\mathbf{a}$ and $\mathbf{a}'$ denote the columns made up of the three components in each of the two frames, and $\mathbf{R}$ is the (3×3) rotation matrix connecting them. In detail, this matrix equation reads

$$a'_i = \sum_j R_{ij} a_j \tag{15.126}$$

where the sum runs from $j = 1$ to 3.

For future reference we need to establish an important property of the rotation matrices $\mathbf{R}$. We know, of course, that any rotation leaves the scalar product $\mathbf{a} \cdot \mathbf{b}$ of any two vectors invariant; that is, $\mathbf{a} \cdot \mathbf{b} = \mathbf{a}' \cdot \mathbf{b}'$, with $\mathbf{a}'$ given by (15.125), and likewise $\mathbf{b}'$. Now, in matrix notation, the scalar product is

$$\mathbf{a} \cdot \mathbf{b} = \tilde{\mathbf{a}}\mathbf{b} \tag{15.127}$$

where we must now insist that $\mathbf{b}$ denotes the *column* of numbers b_1, b_2, b_3 and $\tilde{\mathbf{a}}$ the *row* of numbers a_1, a_2, a_3. Thus, in matrix notation, the equation $\mathbf{a} \cdot \mathbf{b} = \mathbf{a}' \cdot \mathbf{b}'$ reads

$$\tilde{\mathbf{a}}\mathbf{b} = (\mathbf{Ra})\tilde{\ }(\mathbf{Rb}) = \tilde{\mathbf{a}}(\tilde{\mathbf{R}}\mathbf{R})\mathbf{b}. \tag{15.128}$$

[Here I used the result that $(\mathbf{Ra})\tilde{\ } = \tilde{\mathbf{a}}\tilde{\mathbf{R}}$ — see Problem 15.94.] Since this must hold for any choices of $\mathbf{a}$ and $\mathbf{b}$, it is easy to show (Problem 15.95) that

$$\tilde{\mathbf{R}}\mathbf{R} = \mathbf{1}, \tag{15.129}$$

where, as before, $\mathbf{1}$ denotes the 3×3 unit matrix. Any matrix satisfying this condition is said to be an **orthogonal matrix**, so we have proved that rotations are given by 3×3 orthogonal matrices.

A three-dimensional tensor $\mathbf{T}$ comprises nine elements T_{ij} (in each three-dimensional Cartesian reference frame), where i and j both take values from 1 to 3. (Strictly speaking, a tensor with nine elements T_{ij} is a *second-rank* tensor. More generally, a tensor of rank n has 3^n elements, but we shall only be concerned with

---

[29] For a clear and reasonably simple account see David J. Griffiths, *Introduction to Electrodynamics*, third edition (Prentice Hall, 1999), pp. 501 and 535.

the case $n = 2$.) To find the transformation properties of a tensor, let us consider the simple example of a tensor with elements

$$T_{ij} = a_i b_j \tag{15.130}$$

where a_i and b_j are the components of any two vectors. (For example, $\mathbf{a}$ could be the position of a particle and $\mathbf{b}$ its velocity.) This obviously has the requisite nine elements and is, in a sense, the prototypical tensor.

The tranformation of the tensor (15.130) follows immediately from that of the two vectors from which it is constructed:

$$T'_{ij} = a'_i b'_j = \left(\sum_k R_{ik} a_k \right) \left(\sum_l R_{jl} b_l \right)$$
$$= \sum_{k,l} R_{ik} R_{jl} T_{kl}. \tag{15.131}$$

This transformation is the defining characteristic of a tensor. It closely parallels the transformation (15.126) for a vector, except that the tensor, with its two indices, gets two rotation matrices, one for each index.

We can write the transformation (15.131) in matrix form if we note that $(\mathbf{R})_{jl} = (\tilde{\mathbf{R}})_{lj}$, where $\tilde{\mathbf{R}}$ denotes the tranpose of $\mathbf{R}$, so that (15.131)can be written

$$\mathbf{T}' = \mathbf{R}\mathbf{T}\tilde{\mathbf{R}}. \tag{15.132}$$

Any three-dimensional (second-rank) tensor transforms according to this equation, and any set of nine elements which transforms in this way is a tensor.

One of the most important operations one can perform with tensors is to multiply them by vectors. For example, the angular momentum $\mathbf{L}$ of a rigid body is given by the product $\mathbf{L} = \mathbf{I}\boldsymbol{\omega}$ of the moment of inertia tensor $\mathbf{I}$ and the angular velocity vector $\boldsymbol{\omega}$. It is important that any product of this form is, as we would expect, a vector, and this is easy to prove: Let $\mathbf{T}$ be any tensor and $\mathbf{a}$ any vector, and in every reference frame let the column of three numbers $\mathbf{b}$ be defined as $\mathbf{b} = \mathbf{T}\mathbf{a}$. To show that, with this definition, $\mathbf{b}$ is a vector, we use the properties (15.132) and (15.125) of $\mathbf{T}$ and $\mathbf{a}$ as follows:

$$\mathbf{b}' = \mathbf{T}'\mathbf{a}' = (\mathbf{R}\mathbf{T}\tilde{\mathbf{R}})(\mathbf{R}\mathbf{a}) = \mathbf{R}\mathbf{T}(\tilde{\mathbf{R}}\mathbf{R})\mathbf{a} = \mathbf{R}\mathbf{T}\mathbf{a} = \mathbf{R}\mathbf{b}, \tag{15.133}$$

where the fourth equality follows because, according to (15.129), $\tilde{\mathbf{R}}\mathbf{R} = \mathbf{1}$. We conclude that $\mathbf{b}$, which is clearly a column of three numbers, transforms like a vector; that is, $\mathbf{b} = \mathbf{T}\mathbf{a}$ *is* a vector. We can turn this argument around and prove (Problem 15.99) that if $\mathbf{a}$ and $\mathbf{b}$ are known to be vectors and in every frame it is found that $\mathbf{b} = \mathbf{T}\mathbf{a}$, where $\mathbf{T}$ is a 3×3 array of numbers (in every frame), then $\mathbf{T}$ satisfies (15.132) and is therefore a tensor.

Vectors and Tensors in Four-Dimensional Space–Time

The discussion of vectors and tensors in four-dimensional space–time closely parallels that just given for three dimensions, with the only complications coming from the minus sign in the invariant scalar product. A four-vector is given by a column a of four numbers (in each frame), which transform under the Lorentz transformation $a' = \Lambda a$ as we move from one inertial frame S to another, S'. This transformation leaves invariant the scalar product, which we can write in matrix notation as

$$a \cdot b = a_1 b_1 + a_2 b_2 + a_3 b_3 - a_4 b_4 = \tilde{a} G b \tag{15.134}$$

where b denotes the column for the four-vector b, $\tilde{a}$ is the row for a, and G is the 4×4 matrix

$$G = \begin{bmatrix} +1 & 0 & 0 & 0 \\ 0 & +1 & 0 & 0 \\ 0 & 0 & +1 & 0 \\ 0 & 0 & 0 & -1 \end{bmatrix}. \tag{15.135}$$

This **metric matrix**, inserted between $\tilde{a}$ and b in (15.134), simply changes the sign of the fourth component of b and so inserts the needed minus sign in the scalar product.

The scalar product (15.134) is invariant when we replace a and b by Λa and Λb, and exactly the argument that led to (15.129) shows (Problem 15.98) that

$$\tilde{\Lambda} G \Lambda = G \tag{15.136}$$

— the relativistic analog of the condition $\tilde{\mathbf{R}}\mathbf{R} = \mathbf{1}$ for rotations. The set of all 4×4 matrices that satisfy the condition (15.136) is called the **Lorentz group**, since all Lorentz transformations must satisfy this condition.

A four-tensor (strictly speaking a four-tensor of rank 2) is defined as a set of sixteen numbers $T_{\mu\nu}$ (defined for every inertial frame S), where the indices μ and ν run from 1 to 4, which, when formed into a 4×4 matrix T, satisfy

$$T' = \Lambda T \tilde{\Lambda} \tag{15.137}$$

— a property that exactly parallels Equation (15.132) for three-tensors. Just as we form the scalar (or dot) product (15.134) of two four-vectors by inserting the matrix G between the two appropriate matrices, so we can form a dot product of a tensor and a vector in the same way:

$$T \cdot a = T G a. \tag{15.138}$$

It is a straightforward exercise to show that, if T is any tensor and a any vector, then $b = T \cdot a$ is a four-vector. [The proof parallels (15.133) for three dimensions — see Problem 15.96.] Conversely, you can prove (Problem 15.99) a "quotient rule" that if a and b are known to be four-vectors and if $b = T \cdot a$ in every frame S, then T (the "quotient" of b and a) is a four-tensor.

Armed with these definitions and properties of four-tensors, we are ready for our brief venture into relativistic electrodynamics.

15.18 Electrodynamics and Relativity

That light travels at speed c in all directions is a consequence of the laws of classical electromagnetism, and special relativity grew out of the realization that the speed of light c is the same in all inertial frames. These two observations suggest that classical electromagnetism might already be consistent with the principles of relativity. The simplest way to *prove* this suggestion is to show that the familiar laws of electrodynamics can be written in terms of four-scalars, four-vectors, and four-tensors, so that their invariance under Lorentz transformations is self-evident. Here I shall do this for just one law, the centrally important Lorentz-force law

$$\mathbf{F} = q(\mathbf{E} + \mathbf{v} \times \mathbf{B}). \tag{15.139}$$

In the process, we shall find the transformation rules for the electric and magnetic fields $\mathbf{E}$ and $\mathbf{B}$. Before we address the properties of the fields $\mathbf{E}$ and $\mathbf{B}$, you need to know that it is an experimental fact that the charge q of any particle has the same value in all inertial frames; that is, q is a Lorentz scalar.

I shall take the view that the Lorentz equation (15.139) is an observed fact, valid in all inertial frames (which it definitely is). In the form (15.139), it certainly does not *look* relativistically invariant, and our task is to rewrite it so that it does. Our first clue for how to do this is to notice that (15.139) defines $\mathbf{F}$ as a linear function of $\mathbf{v}$. The next, and very natural, step is to rewrite this linear relation in terms of the four-force K and the four-velocity u. Recall that

$$K = \left(\gamma \mathbf{F}, \frac{\gamma \mathbf{v} \cdot \mathbf{F}}{c} \right) \qquad \text{and} \qquad u = (\gamma \mathbf{v}, \gamma c). \tag{15.140}$$

Multiplying both sides of (15.139) by γ, you can see that K is a linear function of u. The simplest such relation would have the form $K = q\mathcal{F} \cdot u$, where $\mathcal{F}$ would be some as-yet unknown four-tensor (and I have inserted a separate factor of q since K is obviously proportional to q). In matrix form, this relation would read $K = q\mathcal{F}Gu$ [where K and u must now be seen as 4×1 columns, and G is the metric matrix (15.135)]. The 16 elements of the matrix $\mathcal{F}G$ can be found by writing out the components of K one at a time. For example, from (15.140) and (15.139), the first component of K is

$$K_1 = \gamma q(E_1 + v_2 B_3 - v_3 B_2) = q[B_3 u_2 - B_2 u_3 + (E_1/c)u_4]. \tag{15.141}$$

The coefficients of $u_1, \cdots, u_4$ are just the first row of the matrix $\mathcal{F}G$ in the proposed relation $K = q\mathcal{F}Gu$. Proceeding in this way, we find the whole matrix $\mathcal{F}G$ to be

$$\mathcal{F}G = \begin{bmatrix} 0 & B_3 & -B_2 & E_1/c \\ -B_3 & 0 & B_1 & E_2/c \\ B_2 & -B_1 & 0 & E_3/c \\ E_1/c & E_2/c & E_3/c & 0 \end{bmatrix}. \tag{15.142}$$

[Compare the first row of this matrix with the coefficients in (15.141); for more details, see Problem 15.104.] Finally, since $G^2 = 1$ (the 4×4 unit matrix), we can multiply by G on the right to give the **electromagnetic field tensor**

$$\mathcal{F} = \begin{bmatrix} 0 & B_3 & -B_2 & -E_1/c \\ -B_3 & 0 & B_1 & -E_2/c \\ B_2 & -B_1 & 0 & -E_3/c \\ E_1/c & E_2/c & E_3/c & 0 \end{bmatrix} \tag{15.143}$$

in terms of which the Lorentz force takes the beautifully simple form

$$K = q\mathcal{F} \cdot u. \tag{15.144}$$

We are taking the view that the Lorentz-force law is an experimental fact, valid in all inertial frames. In the equation (15.144), K and u are known to be four-vectors, and the charge q is a scalar. It follows from the quotient rule quoted below Equation (15.138) that $\mathcal{F}$ is a four-tensor, which will let us find the behavior of the electric and magnetic fields under any Lorentz transformation.[30] For future reference, notice that $\mathcal{F}$ is an *antisymmetric* tensor; that is, the matrix $\mathcal{F}$ is antisymmetric, satisfying $\tilde{\mathcal{F}} = -\mathcal{F}$.

Lorentz Transformation of Electric and Magnetic Fields

The field tensor $\mathcal{F}$ specifies the fields $\mathbf{E}$ and $\mathbf{B}$ in any given inertial frame $\mathcal{S}$. Since $\mathcal{F}$ is a four-tensor, its value in any other frame $\mathcal{S}'$ is given by (15.137) as

$$\mathcal{F}' = \Lambda\mathcal{F}\tilde{\Lambda}. \tag{15.145}$$

For any given Lorentz transformation Λ, it is a straightforward, though tedious, exercise to work out the right side of (15.145), and comparing the result to the definition (15.143) of $\mathcal{F}'$, we can write down the transformed fields. For example, for the standard boost with velocity v along the x_1 axis, one finds (Problem 15.105)

$$\begin{array}{llll} E_1' = E_1, & E_2' = \gamma(E_2 - \beta c B_3), & E_3' = \gamma(E_3 + \beta c B_2) \\ B_1' = B_1, & B_2' = \gamma(B_2 + \beta E_3/c), & B_3' = \gamma(B_3 - \beta E_2/c). \end{array} \tag{15.146}$$

The most striking feature of the transformations (15.146) is that they mix up the electric and magnetic fields. A configuration whose fields are purely electric in one frame $\mathcal{S}$ ($\mathbf{B} = 0$ everywhere, as for any static charge distribution), will inevitably have nonzero magnetic components in some other frames $\mathcal{S}'$ ($\mathbf{B}' \neq 0$). Thus we can say that

---

[30] There are many different ways to arrive at the field tensor $\mathcal{F}$, some of which *define* $\mathcal{F}$ to be a four-tensor. In such an approach, the Lorentz-force equation (15.144) automatically has the form "four-vector = four-vector," which guarantees the Lorentz invariance of the Lorentz-force equation.

in relativity the existence of electric fields *requires* the existence of magnetic fields and vice versa.

An important advantage of knowing the transformation properties of the electromagnetic fields is this: In seeking the fields due to a certain charge and current distribution in a frame S, it may be possible to find a frame S' in which the fields are more easily evaluated. If this happens, then our simplest course may be to write down the fields in S' and then transform them back to the original frame S, as in the following example.

EXAMPLE 15.12   Fields of a Long Straight Current

Find the E and B fields of an infinitely long, uniform line charge with density λ (measured in coulombs/meter), placed on the z axis of frame S and traveling with speed v in the $+z$ direction.

The moving line charge constitutes a current $I = \lambda v$ along the z axis, so our problem is to find the combined fields of a line charge and a line current. Let us recognize first that this can be done by elementary methods, without leaving the frame S: Using Gauss's law we can show that the E field of the line charge is $E = 2k\lambda/\rho$ radially outward from the z axis. (Here $k = 1/4\pi\epsilon_0$ is the Coulomb force constant and ρ is the perpendicular distance from the z axis, that is, the first of the coordinates ρ, ϕ, z of cylindrical polar coordinates.) Similarly, using Ampère's law, we can show the B field of the current is $B = (\mu_0/2\pi)I/\rho$ in the direction given by the right-hand rule, where μ_0 is the so-called permeability of space. We can express these two well-known results compactly using the unit vectors of cylindrical polar coordinates:

$$\mathbf{E} = \frac{2k\lambda}{\rho}\hat{\boldsymbol{\rho}} \qquad \text{and} \qquad \mathbf{B} = \frac{\mu_0}{2\pi}\frac{I}{\rho}\hat{\boldsymbol{\phi}}. \tag{15.147}$$

Both of these fields are sketched in Figure 15.16(a).

While the derivation using Gauss's and Ampère's laws is perfectly straightforward, it is instructive to rederive the same results by transforming to a frame S' traveling with the charges. In S', there is no current, so the only field is the radial electric field, $E' = 2k\lambda'/\rho'$, as shown in Figure 15.16(b). This field is in the direction of the unit vector $\hat{\boldsymbol{\rho}}' = (x'/\rho', y'/\rho', 0)$, so can be written as

$$\mathbf{E}' = \frac{2k\lambda'}{\rho'}\hat{\boldsymbol{\rho}}' = \frac{2k\lambda'}{\rho'^2}(x', y', 0). \tag{15.148}$$

Before we transform this back to the original frame S, we must recognize that the charge densities λ and λ' are not equal: The total charge contained in any given segment of the z axis must be the same in either frame (invariance of charge), so that $\lambda \, \Delta z = \lambda' \Delta z'$, but, because of length contraction, $\Delta z = \Delta z'/\gamma$. Therefore

$$\lambda = \gamma\lambda'. \tag{15.149}$$

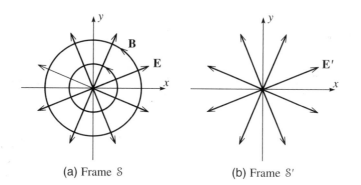

(a) Frame $\mathcal{S}$ (b) Frame $\mathcal{S}'$

Figure 15.16   The fields produced by a line charge on the z axis. **(a)** In frame $\mathcal{S}$, the line charge is moving up, out of the page. This constitutes a current, which produces a B field looping around the z axis — in addition to the E field, which is radially out from the z axis. **(b)** The frame $\mathcal{S}'$ is the rest frame of the charges, so there is no current and hence no B field — just the radial E field.

We must now transform the fields $\mathbf{E}'$, given by (15.148), and $\mathbf{B}' = 0$ back to frame $\mathcal{S}$. To do this we note first that $\mathcal{S}'$ is traveling along the z axis of $\mathcal{S}$ (not the x axis, as in the standard boost). Thus we must first rewrite (15.146) for a boost along the z axis. We must then find the inverse of this transformation (since we want the unprimed fields in terms of the primed). The result is, as you can easily check,

$$E_1 = \gamma(E_1' + \beta c B_2'), \qquad E_2 = \gamma(E_2' - \beta c B_1'), \qquad E_3 = E_3' \qquad (15.150)$$
$$B_1 = \gamma(B_1' - \beta E_2'/c), \qquad B_2 = \gamma(B_2' + \beta E_1'/c), \qquad B_3 = B_3'.$$

Substituting $\mathbf{B}' = 0$ and the components of $\mathbf{E}'$ from (15.148), we find that

$$\mathbf{E} = \gamma \frac{2k\lambda'}{\rho^2}(x, y, 0) = \frac{2k\lambda}{\rho}\hat{\boldsymbol{\rho}}. \qquad (15.151)$$

In writing the first equality I used the fact that x and y, and hence $\rho = \sqrt{x^2 + y^2}$, are invariant under a boost in the z direction; in the second, I replaced $\gamma\lambda'$ by λ. This agrees exactly with the E field found in (15.147).

Similarly, substituting $\mathbf{B}' = 0$ and (15.148) into the expressions for $\mathbf{B}$ in (15.150), we find the magnetic field

$$\mathbf{B} = \gamma\beta \frac{2k\lambda'}{c\rho^2}(-y, x, 0).$$

If we make the replacements $\gamma\lambda' = \lambda$, $\beta = v/c$, $k/c^2 = 1/(4\pi\epsilon_0 c^2) = \mu_0/4\pi$, and $(-y/\rho, x/\rho, 0) = \hat{\boldsymbol{\phi}}$, this becomes

$$\mathbf{B} = \frac{\mu_0}{2\pi}\frac{\lambda v}{\rho}\hat{\boldsymbol{\phi}} \qquad (15.152)$$

and, since $\lambda v = I$, the current, this is exactly the same as the B field in (15.147). The remarkable feature of this derivation of **B** is that it made no reference to Ampere's law. Gauss's law in the frame $\mathcal{S}'$, combined with the Lorentz transformation of the fields, has given us the result that we normally see as an expression of Ampere's law.

With this striking example of the behavior of the electromagnetic field under the Lorentz transformation, I must end our brief foray into relativistic electrodynamics. You can explore a few more aspects in the problems at the end of this chapter, and after that you could read the excellent books of Griffiths and Jackson.[31]

Principal Definitions and Equations of Chapter 15

Time Dilation

If two events, as observed in frame $\mathcal{S}_0$, occur at the same place and are separated by a time Δt_0, then the time between them as measured in any other frame $\mathcal{S}$ is

$$\Delta t = \gamma \, \Delta t_0 \qquad \text{[Eq. (15.11)]}$$

where $\gamma = 1/\sqrt{1 - \beta^2}$, $\beta = V/c$, and V is the speed of $\mathcal{S}$ relative to $\mathcal{S}_0$.

Length Contraction

If, as observed in frame $\mathcal{S}_0$, a body is at rest and has length l_0, then its length measured in a frame $\mathcal{S}$ traveling with velocity **V** in the direction of the length is

$$l = l_0/\gamma. \qquad \text{[Eq. (15.15)]}$$

Lengths perpendicular to **V** are unchanged.

The Lorentz Transformation

The coordinates of any one event as measured in two frames (in standard configuration) are related by the **Lorentz transformation**:

$$\left. \begin{aligned} x' &= \gamma(x - Vt) \\ y' &= y \\ z' &= z \\ t' &= \gamma(t - Vx/c^2). \end{aligned} \right\} \qquad \text{[Eq. (15.20)]}$$

---

[31] Chapter 12 of David J. Griffiths, *Introduction to Electrodynamics*, (third edition, Prentice Hall, 1999) is at approximately the level of this book but naturally emphasizes electrodynamics much more heavily. J. D. Jackson's *Classical Electrodynamics* (third edition, John Wiley, 1998) is a graduate text, which you could tackle after reading Griffiths' book.

The inverse Lorentz transformation is obtained by exchanging primed and unprimed variables and changing the sign of V.

The Velocity-Addition Formula

The velocities of a single object as measured in two frames (in standard configuration) are related by the **velocity-addition formula**

$$v'_x = \frac{v_x - V}{1 - v_x V/c^2}, \quad v'_y = \frac{v_y}{\gamma(1 - v_x V/c^2)}, \quad \text{and} \quad v'_z = \frac{v_z}{\gamma(1 - v_x V/c^2)}.$$

[Eqs. (15.26) & (15.27)]

Four-Vectors

If we rewrite the coordinates (x, y, z) as (x_1, x_2, x_3) and introduce $x_4 = ct$, then the four-vectors $x = (x_1, x_2, x_3, x_4)$ label points in a four-dimensional **space–time**. If we agree to arrange the components of x in a 4×1 column, then Lorentz transformations become "rotations" of the form $x' = \Lambda x$, where Λ is a 4×4 matrix. A **four-vector** is any set of four numbers, $q = (q_1, q_2, q_3, q_4)$ (one set for each inertial frame) which transform this way,

$$q' = \Lambda q.$$

[Section 15.8]

The Invariant Scalar Product

The **scalar product** of two four-vectors x and y is defined as

$$x \cdot y = x_1 y_1 + x_2 y_2 + x_3 y_3 - x_4 y_4$$

[Eq. (15.50)]

and is invariant under all Lorentz transformations. The scalar product of a vector with itself is often written as $x \cdot x = x^2$.

The Light Cone

The light cone of a point Q in space–time consists of all light rays through Q; equivalently, it contains all points P with $(x_P - x_Q)^2 = 0$.   [Section 15.10]

The Relativistic Doppler Effect

Light from a source traveling with velocity $\mathbf{V}$ relative to frame $\mathcal{S}$ is observed at an angle θ (θ = angle between $\mathbf{V}$ and the ray of light). If the frequency of the light, as measured in the source's rest frame is ω_0 the frequency observed in $\mathcal{S}$ is

$$\omega = \frac{\omega_0}{\gamma(1 - \beta \cos\theta)}.$$

[Eq. (15.64)]

Mass, Four-Velocity, Momentum, and Energy

The (invariant) **mass** of an object is defined to be its rest mass. The **four-velocity** is

$$u = \frac{dx}{dt_o} = \gamma(\mathbf{v}, c). \qquad \text{[Eqs. (15.66) \& (15.67)]}$$

The four-momentum is

$$p = mu = (\gamma m\mathbf{v}, \gamma mc) = (\mathbf{p}, E/c). \qquad \text{[Eqs. (15.68), (15.70), \& (15.75)]}$$

Three Useful Relations

$$\boldsymbol{\beta} = \mathbf{p}c/E, \quad p \cdot p = -(mc)^2, \quad \text{and} \quad E^2 = (mc^2)^2 + (\mathbf{p}c)^2. \quad \text{[Eqs. (15.83)–(15.85)]}$$

Three-Force and Four-Force

The **three-force F** and **four-force** K on a particle are

$$\mathbf{F} = \frac{d\mathbf{p}}{dt} \quad \text{and} \quad K = \frac{dp}{dt_o}. \qquad \text{[Eqs. (15.100) \& (15.107)]}$$

Massless Particles

With $m = 0$, a massless particle has

$$E = |\mathbf{p}|c, \quad v = c, \quad \text{and} \quad p^2 = 0. \qquad \text{[Eqs. (15.111)–(15.113)]}$$

Transformation of the Electromagnetic Fields

Under the standard boost, the electric and magnetic fields transform as follows:

$$\begin{aligned} E_1' &= E_1, & E_2' &= \gamma(E_2 - \beta c B_3), & E_3' &= \gamma(E_3 + \beta c B_2) \\ B_1' &= B_1, & B_2' &= \gamma(B_2 + \beta E_3/c), & B_3' &= \gamma(B_3 - \beta E_2/c). \end{aligned} \qquad \text{[Eq. (15.146)]}$$

Problems for Chapter 15

Stars indicate the approximate level of difficulty, from easiest (⋆) to most difficult (⋆⋆⋆).

SECTION 15.2   Galilean Relativity

15.1 ⋆ Using arguments similar to those of Section 15.2, prove that Newton's first and third laws are invariant under the Galilean transformation.

15.2 ★★ Consider a classical inelastic collision of the form $A + B \rightarrow C + D$. (For example, this could be a collision such as $Na + Cl \rightarrow Na^+ + Cl^-$ in which two neutral atoms exchange an electron and become oppositely charged ions.) Show that the law of conservation of classical momentum is invariant under the Galilean transformation if and only if total mass is conserved — as is certainly true in classical mechanics. (We shall find in relativity that the classical definition of momentum has to be modified and that total mass is *not* conserved.)

SECTION 15.4 The Relativity of Time; Time Dilation

15.3 ★ A low-flying earth satellite travels at about 8000 m/s. What is the factor γ for this speed? As observed from the ground, by how much would a clock traveling at this speed differ from a ground-based clock after one hour (as measured by the latter)? What is the percent difference?

15.4 ★ What is the factor γ for a speed of $0.99c$? As observed from the ground, by how much would a clock traveling at this speed differ from a ground-based clock after one hour (one hour as measured by the latter, that is)?

15.5 ★ A space explorer A sets off at a steady $0.95c$ to a distant star. After exploring the star for a short time, he returns at the same speed and gets home after a total absence of 80 years (as measured by earth-bound observers). How long do A's clocks say that he was gone, and by how much has he aged as compared to his twin B who stayed behind on earth?

[*Note:* This is the famous "twin paradox." It is fairly easy to get the right answer by judicious insertion of a factor of γ in the right place, but to understand it, you need to recognize that it involves *three* inertial frames: the earth-bound frame $\mathcal{S}$, the frame $\mathcal{S}'$ of the outbound rocket, and the frame $\mathcal{S}''$ of the returning rocket. Write down the time dilation formula for the two halves of the journey and then add. Notice that the experiment is *not* symmetrical between the two twins: B stays at rest in the single inertial frame $\mathcal{S}$, but A occupies at least two different frames. This is what allows the result to be unsymmetrical.]

15.6 ★ When he returns his Hertz rent-a-rocket after one week's cruising in the galaxy, Spock is shocked to be billed for three weeks' rental. Assuming that he traveled straight out and then straight back, always at the same speed, how fast was he traveling? (See note to Problem 15.5.)

15.7 ★★ The muons created by cosmic rays in the upper atmosphere rain down more-or-less uniformly on the earth's surface, although some of them decay on the way down, with a half-life of about 1.5 μs (measured in their rest frame). A muon detector is carried in a balloon to an altitude of 2000 m, and in the course of an hour detects 650 muons traveling at $0.99c$ toward the earth. If an identical detector remains at sea level, how many muons should it register in one hour? Calculate the answer taking account of the relativistic time dilation and also classically. (Remember that after n half-lives, 2^{-n} of the original particles survive.) Needless to say, the relativistic answer agrees with experiment.

15.8 ★★ The pion (π^+ or π^-) is an unstable particle that decays with a proper half-life of 1.8×10^{-8} s. (This is the half-life measured in the pion's rest frame.) **(a)** What is the pion's half-life measured in a frame $\mathcal{S}$ where it is traveling at $0.8c$? **(b)** If 32,000 pions are created at the same place, all traveling at this same speed, how many will remain after they have traveled down an evacuated pipe of length $d = 36$ m? Remember that after n half-lives, 2^{-n} of the original particles survive. **(c)** What would the answer have been if you had ignored time dilation? (Naturally it is the answer (b) that agrees with experiment.)

15.9 ★★ One way to set up the system of synchronized clocks in a frame $\mathcal{S}$, as described at the beginning of Section 15.4, would be for the chief observer to summon all her helpers to the origin O and synchronize their clocks there, and then have them travel to their assigned positions *very slowly*. Prove this claim as follows: Suppose a certain observer is assigned to a position P at a distance d from the origin. If he travels at constant speed V, when he reaches P how much will his clock differ from the chief's clock at O? Show that this difference approaches 0 as $V \to 0$.

15.10 ★★★ Time dilation implies that when a clock moves relative to a frame $\mathcal{S}$, careful measurements made by observers in $\mathcal{S}$ will find that the clock is running slow. This is not at all the same thing as saying that a single observer in $\mathcal{S}$ will *see* the clock running slow, and this latter statement is not always true. To understand this, remember that what we see is determined by the light as it arrives at our eyes. Consider an observer standing close beside the x axis as a clock approaches her with speed V along the axis. As the clock moves from position A to B, it will register a time Δt_0, but as measured by the observer's helpers, the time between the two events ("clock at A" and "clock at B") is $\Delta t = \gamma \, \Delta t_0$. However, since B is closer to the observer than A is, the light from the clock at B will reach the observer in a shorter time than will the light from A. Therefore, the time Δt_{see} between the observer's *seeing* the clock at A and *seeing* it at B is less than Δt. **(a)** Prove that

$$\Delta t_{see} = \Delta t \, (1 - \beta) = \Delta t_0 \sqrt{\frac{1 - \beta}{1 + \beta}}$$

(which is less than Δt_0). Prove both equalities. **(b)** What time will the observer see once the clock has passed her and is moving away?

The moral of this problem is that you must be careful how you state or think about time dilation. It's fine to say "Moving clocks are observed, or measured, to run slow," but it is definitely wrong to say "Moving clocks are seen to run slow."

SECTION 15.5   Length Contraction

15.11 ★ As a meter stick rushes past me (with velocity $\mathbf{v}$ parallel to the stick), I measure its length to be 80 cm. What is v?

15.12 ★★ Consider the experiment of Problem 15.8 from the point of view of the pions' rest frame. What is the half-life of the pions in this frame? In part (b), how long is the pipe as "seen" by the pions and how long does it take to pass the pions? How many pions remain at the end of this time? Compare with the answer to Problem 15.8 and describe how the two different arguments led to the same result.

15.13 ★★ **(a)** A meter stick is at rest in frame $\mathcal{S}_0$, which is traveling with speed $V = 0.8c$ in the standard configuration relative to frame $\mathcal{S}$. **(a)** The stick lies in the $x_0 y_0$ plane and makes an angle $\theta_0 = 60°$ with the x_0 axis (as measured in $\mathcal{S}_0$). What is its length l as measured in $\mathcal{S}$, and what is its angle θ with the x axis? [*Hint:* It may help to think of the stick as the hypotenuse of a 30–60–90 triangle of plywood.] **(b)** What is l if $\theta = 60°$? What is θ_0 in this case?

15.14 ★★★ Like time dilation, length contraction cannot be *seen* directly by a single observer. To explain this claim, imagine a rod of proper length l_0 moving along the x axis of frame $\mathcal{S}$ and an observer standing away from the x axis and to the right of the whole rod. Careful measurements of the rod's length at any one instant in frame $\mathcal{S}$ would, of course, give the result $l = l_0/\gamma$. **(a)** Explain clearly why the light which reaches the observer's eye at any one time must have left the two ends A and B of the rod at *different times*. **(b)** Show that the observer would see (and a camera would record) a length more than l. [It helps to imagine that the x axis is marked with a graduated scale.] **(c)** Show that if the observer is standing close beside the track, he will see a length that is actually more than l_o; that is, the length contraction is distorted into an expansion.

SECTION 15.6   The Lorentz Transformation

15.15 ★ Solve the Lorentz transformation equations (15.20) to give x, y, z, t in terms of x', y', z', t'. Verify that you get the inverse Lorentz tranformation (15.21). Observe that you could have found the same result by interchanging primed and unprimed variables and changing V to $-V$.

15.16 ★ Consider two events that occur at positions $\mathbf{r}_1$ and $\mathbf{r}_2$ and times t_1 and t_2. Let $\Delta\mathbf{r} = \mathbf{r}_2 - \mathbf{r}_1$ and $\Delta t = t_2 - t_1$. Write down the Lorentz transformation for $\mathbf{r}_1$ and t_1, and likewise for $\mathbf{r}_2$ and t_2, and deduce the transformation for $\Delta\mathbf{r}$ and Δt. Notice that differences $\Delta\mathbf{r}$ and Δt transform in exactly the same way as $\mathbf{r}$ and t. This important property follows from the linearity of the Lorentz transformation.

15.17 ★ Consider two events that occur simultaneously at $t = 0$ in frame $\mathcal{S}$, both on the x axis at $x = 0$ and $x = a$. **(a)** Find the times of the two events as measured in a frame $\mathcal{S}'$ traveling in the positive direction along the x axis with speed V. **(b)** Do the same for a second frame $\mathcal{S}''$ traveling at speed V but in the negative direction along the x axis. Comment on the time ordering of the two events as seen in the three different frames. This startling result is discussed further in Section 15.10.

15.18 ★★ Use the inverse Lorentz transformation (15.21) to rederive the time-dilation formula (15.8). [*Hint:* Consider again the thought experiment of Figure 15.3, with the flash and the beep that occur at the same positions as seen in frame $\mathcal{S}'$.]

15.19 ★★ A traveler in a rocket of proper length $2d$ sets up a coordinate system $\mathcal{S}'$ with its origin O' anchored at the exact middle of the rocket and the x' axis along the rocket's length. At $t' = 0$ she ignites a flashbulb at O'. **(a)** Write down the coordinates x'_F, t'_F and x'_B, t'_B for the arrival of the light at the front and back of the rocket. **(b)** Now consider the same experiment as observed from a frame $\mathcal{S}$ relative to which the rocket is traveling with speed V (with $\mathcal{S}$ and $\mathcal{S}'$ in the standard configuration). Use the inverse Lorentz transformation to find the coordinates x_F, t_F and x_B, t_B for the arrival of the two signals. Explain clearly in words why the two arrivals are simultaneous in $\mathcal{S}'$ but not in $\mathcal{S}$. This phenomenon is called the relativity of simultaneity.

SECTION 15.7   The Relativistic Velocity-Addition Formula

15.20 ★ Newton's first law can be stated: If an object is isolated (subject to no forces), then it moves with constant velocity. We know that this is invariant under the Galilean transformation. Prove that it is also invariant under the Lorentz transformation. [Assume that it is true in an inertial frame $\mathcal{S}$, and use the relativistic velocity-addition formula to show that it is also true in any other $\mathcal{S}'$.]

15.21 ★ A rocket traveling at speed $\frac{1}{2}c$ relative to frame $\mathcal{S}$ shoots forward bullets traveling at speed $\frac{3}{4}c$ relative to the rocket. What is the speed of the bullets relative to $\mathcal{S}$?

15.22 ★ A rocket is traveling at speed $0.9c$ along the x axis of frame $\mathcal{S}$. It shoots a bullet whose velocity $\mathbf{v}'$ (measured in the rocket's rest frame $\mathcal{S}'$) is $0.9c$ along the y' axis of $\mathcal{S}'$. What is the bullet's velocity (magnitude and direction) as measured in $\mathcal{S}$?

15.23 ★ As seen in frame $\mathcal{S}$, two rockets are approaching one another along the x axis traveling with equal and opposite velocities of $0.9c$. What is the velocity of the rocket on the right as measured by observers in the one on the left? [This and the previous two problems illustrate the general result that in relativity the "sum" of two velocities that are less than c is always less than c. See Problem 15.43.]

15.24 ★ A robber's getaway vehicle, which can travel at an impressive $0.8c$, is pursued by a cop, whose vehicle can travel at a mere $0.4c$. Realizing that he cannot catch up with the robber, the cop tries to shoot him with bullets that travel at $0.5c$ (relative to the cop). Can the cop's bullets hit the robber?

15.25 ★ A rocket is traveling at speed V along the x axis of frame S. It emits a signal (for example, a pulse of light) that travels with speed c along the y' axis of the rocket's rest frame S'. What is the speed of the signal as measured in S?

15.26 ★ Two objects A and B are approaching one another, traveling in opposite directions along the x axis of frame S with speeds v_A and v_B. At time $t = 0$, they are at positions $x = 0$ and $x = d$. Write down their positions for an arbitrary time t and show that they meet at time $t = d/(v_A + v_B)$. Notice that this implies that the relative velocity of the two objects is $v_A + v_B$ as measured in the frame S in which they are both moving. This may seem surprising at first thought, since we can clearly choose values of v_A and v_B for which this relative velocity is larger than c.[32]

15.27 ★★★ Frame S' travels at speed V_1 along the x axis of frame S (in the standard configuration). Frame S'' travels at speed V_2 along the x' axis of frame S' (also in the standard configuration). By applying the standard Lorentz transformation twice find the coordinates x'', y'', z'', t'' of any event in terms of x, y, z, t. Show that this transformation is in fact the standard Lorentz transformation with velocity V given by the relativistic "sum" of V_1 and V_2.

15.28 ★★★ The relativistic velocity-addition formula is the answer to the following question: If $\mathbf{u}$ is the velocity of an inertial observer B relative to an observer A, and $\mathbf{v}$ is the velocity of C relative to B, what is the velocity $\mathbf{w}$ of C relative to A? Let us denote the answer by $\mathbf{w} = "\mathbf{u} + \mathbf{v}."$ In classical physics, this is just the ordinary vector sum of $\mathbf{u}$ and $\mathbf{v}$; in relativity, it is given by the inverse of the velocity addition formulas (15.26) and (15.27) (at least for the case that $\mathbf{u}$ points along the x axis). Taking $\mathbf{u} = (u, 0, 0)$ and $\mathbf{v} = (0, v, 0)$, write down the components of "$\mathbf{u} + \mathbf{v}$" and also of "$\mathbf{v} + \mathbf{u}$." [Be careful to distinguish between the γ factors γ_u and γ_v pertaining to u and v.] Show that "$\mathbf{u} + \mathbf{v}$" $\neq$ "$\mathbf{v} + \mathbf{u}$," but that the two vectors have equal magnitudes and differ only by a rotation about the z axis. This rotation is sometimes called the Wigner rotation and is the cause of the so-called Thomas precession, which has an important effect on the fine structure of atomic energy levels.

SECTION 15.8   Four-Dimensional Space–Time; Four-Vectors

15.29 ★ (a) Find the 3×3 matrix $\mathbf{R}(\theta)$ that rotates three-dimensional space about the x_3 axis, so that $\mathbf{e}_1$ rotates through angle θ toward $\mathbf{e}_2$. **(b)** Show that $[\mathbf{R}(\theta)]^2 = \mathbf{R}(2\theta)$, and interpret this result.

15.30 ★ The "angle" ϕ introduced in connection with Equation (15.40) has several useful properties. For any speed $v < c$ (with corresponding factors β and γ) we can define ϕ so that $\gamma = \cosh \phi$. Defined in this way, ϕ is called the **rapidity** corresponding to v. Prove that $\sinh \phi = \beta \gamma$ and that $\tanh \phi = \beta$.

15.31 ★ Here is a handy property of the rapidity introduced in Problem 15.30: Suppose that observer B has rapidity ϕ_1 as measured by A and that C has rapidity ϕ_2 as measured by B (with both velocities along the x axis). That is, the speed of B relative to A has $\beta_1 = \tanh \phi_1$ and so on. Prove that the rapidity of C as measured by A is just $\phi = \phi_1 + \phi_2$.

---

[32] Nevertheless, this does not violate any principle of relativity. We shall see that no single object can have speed greater than c relative to any inertial frame, but there is nothing to prohibit *two* objects from having relative speed greater than c as measured in a frame where both objects are moving.

15.32 ★ In Section 15.8, I claimed that the 4×4 matrix Λ_R corresponding to a pure rotation has the block form (15.44). Verify this claim by writing out the separate components of the equation $x' = \Lambda_R x$ and showing that the spatial part (x_1, x_2, x_3) is rotated, while x_4 is unchanged.

15.33 ★★ **(a)** By exchanging x_1 and x_2, write down the Lorentz transformation for a boost of velocity V along the x_2 axis and the corresponding 4×4 matrix Λ_{B2}. **(b)** Write down the 4×4 matrices Λ_{R+} and Λ_{R-} that represent rotations of the $x_1 x_2$ plane through $\pm\pi/2$, with the angle of rotation measured counterclockwise. **(c)** Verify that $\Lambda_{B2} = \Lambda_{R-}\Lambda_{B1}\Lambda_{R+}$, where Λ_{B1} is the standard boost along the x_1 axis, and interpret this result.

15.34 ★★ Let $\Lambda_B(\theta)$ denote the 4×4 matrix that gives a pure boost in the direction that makes an angle θ with the x_1 axis in the $x_1 x_2$ plane. Explain why this can be found as $\Lambda_B(\theta) = \Lambda_R(-\theta)\Lambda_B(0)\Lambda_R(\theta)$, where $\Lambda_R(\theta)$ denotes the matrix that rotates the $x_1 x_2$ plane through angle θ and $\Lambda_B(0)$ is the standard boost along the x_1 axis. Use this result to find $\Lambda_B(\theta)$ and check your result by finding the motion of the spatial origin of the frame $\mathcal{S}$ as observed in $\mathcal{S}'$.

15.35 ★★ Prove the following useful result, called the *zero-component theorem*: Let q be a four-vector, and suppose that one component of q is found to be zero in all inertial frames. (For example, $q_4 = 0$ in all frames.) Then all four components of q are zero in all frames.

SECTION 15.9 The Invariant Scalar Product

15.36 ★ We have seen that the scalar product $x \cdot x$ of any four-vector x with itself is invariant under Lorentz transformations. Use the invariance of $x \cdot x$ to prove that the scalar product $x \cdot y$ of any two four-vectors x and y is likewise invariant.

15.37 ★ Verify directly that $x' \cdot y' = x \cdot y$ for any two four-vectors x and y, where x' and y' are related to x and y by the standard Lorentz boost along the x_1 axis.

15.38 ★★ As an observer moves through space with position $\mathbf{x}(t)$, the four-vector $(\mathbf{x}(t), ct)$ traces a path through space–time called the observer's world line. Consider two events that occur at points P and Q in space–time. Show that if, as measured by the observer, the two events occur at the same time t, then the line joining P and Q is orthogonal to the observer's world line at the time t; that is, $(x_P - x_Q) \cdot dx = 0$, where dx joins two neighboring points on the world line at times t and $t + dt$.

SECTION 15.10 The Light Cone

15.39 ★ Suppose that a point P in space–time with coordinates $x = (\mathbf{x}, x_4)$ lies inside the backward light cone as seen in frame $\mathcal{S}$. This means that $x \cdot x < 0$ and $x_4 < 0$ at least in frame $\mathcal{S}$. Prove that these two conditions are satisfied in all frames. Since this means that all observers agree that $t < 0$, this justifies calling the inside of the backward light cone the absolute past.

15.40 ★ Show that the statement that a point x in space–time lies on the forward light cone is Lorentz invariant.

15.41 ★ In the proposition on page 627, it is obvious that at least one of the three statements has to be true. In the proof given there, I showed that if statement (1) is true, then so are statements (2) and (3). To complete the proof, show that (2) implies (1) and (3). [Strictly speaking you should also check that (3) implies (1) or (2), but this is so similar to the argument already given that you needn't bother.]

15.42 ★ Prove that if x is time-like and $x \cdot y = 0$, then y is space-like.

15.43 ★ **(a)** Show that if a body has speed $v < c$ in one inertial frame, then $v < c$ in all frames. [*Hint:* Consider the displacement four-vector $dx = (d\mathbf{x}, c\,dt)$, where $d\mathbf{x}$ is the three-dimensional displacement in a short time dt.] **(b)** Show similarly that if a signal (such as a pulse of light) has speed c in one frame, its speed is c in all frames.

15.44 ★★ **(a)** Show that if q is time-like, there is a frame S' in which it has the form $q' = (0, 0, 0, q_4')$. **(b)** Show that if q is forward time-like in one frame S, then it is forward time-like in all inertial frames.

SECTION 15.11   The Quotient Rule and Doppler Effect

15.45 ★ The quotient rule derived at the start of Section 15.11 is only one of several similar quotient rules. Here is another. Suppose that k and x are both known to be four-vectors and that in every inertial frame k is a multiple of x. That is, $k = \lambda x$ in frame S, and $k' = \lambda' x'$ in frame S', and so on. Then the factor λ (the "quotient" of k and x) is in fact a four-scalar with the same value in all frames, $\lambda = \lambda'$. Prove this quotient rule.

15.46 ★ **(a)** Show that in the case that the source is approaching the observer head on, the Doppler formula (15.64) can be rewritten as $\omega = \omega_0\sqrt{(1+\beta)/(1-\beta)}$. **(b)** What is the corresponding result for the case that the source is moving directly away from the observer?

15.47 ★ Consider the tale of the physicist who is ticketed for running a red light and argues that because he was approaching the intersection, the red light was Doppler shifted and appeared green. How fast would he have to have been going? ($\lambda_{red} \approx 650$ nm and $\lambda_{green} \approx 530$ nm.)

15.48 ★★ The factor γ in the Doppler formula (15.64), which can be ascribed to time dilation, means that even when $\theta = 90°$ there is a Doppler shift. (In classical physics there is no Doppler shift when $\theta = 90°$ and the source has zero velocity in the direction of the observer.) This *transverse Doppler shift* is therefore a test of time dilation, and has yielded some very accurate tests of the theory. However, except when the source is moving very close to the speed of light, the transverse shift is quite small. **(a)** If $V = 0.2c$, what is the percentage shift when $\theta = 90°$? **(b)** Compare this with the shift when the source approaches the observer head-on.

SECTION 15.12   Mass, Four-Velocity, and Four-Momentum

15.49 ★ Show that the four-velocity of any object has invariant length squared $u \cdot u = -c^2$.

15.50 ★ For any two objects a and b, show that the scalar product of their four-velocities is $u_a \cdot u_b = -c^2\gamma(v_{rel})$, where $\gamma(v)$ denotes the usual γ factor, $\gamma(v) = 1/\sqrt{1 - v^2/c^2}$, and v_{rel} denotes the speed of a in the rest frame of b or vice versa.

15.51 ★★ **(a)** For the collision shown in Figure 15.11, verify that all four components of the total four-momentum $p_a + p_b$ [with the individual momenta defined relativistically as in (15.68)] are conserved in the frame S of part (a). **(b)** In two lines or less, prove that total four-momentum is conserved in the frame S' of part (b). [This problem does not, of course, prove that the law of conservation of four-momentum is generally true, but it does at least show that the law is consistent with the collision of Figure 15.11.]

15.52 ★★ **(a)** Suppose that the total three-momentum $\mathbf{P} = \sum \mathbf{p}$ of an isolated system is conserved in all inertial frames. Show that if this is true (which it is), then the fourth component P_4 of the total four-momentum $P = (\mathbf{P}, P_4)$ *has* to be conserved as well. **(b)** Using the zero-component theorem of Problem 15.35, you can prove the following stronger result very quickly: If any one component of the total four-momentum P is conserved in all frames, then *all four* components are conserved.

15.53 ⋆⋆ For any two objects a and b, show that

$$p_a \cdot p_b = m_a E_b = m_b E_a = m_a m_b c^2 \gamma(v_{\text{rel}})$$

where m_a is the mass of a, and E_b is the energy of b in a's rest frame, and vice versa, and v_{rel} is the speed of a in the rest frame of b (or vice versa).

15.54 ⋆⋆⋆ **(a)** Using the correct relativistic velocity-addition formulas make a table showing the four velocities as seen in frame $\mathcal{S}'$ of the collision of Figure 15.11(b) in terms of the initial velocity of a in $\mathcal{S}$. [Give the latter some simple name, such as $\mathbf{v}_a = (\xi, \eta, 0)$.] **(b)** Add a column showing the total classical momentum $m_a \mathbf{v}'_a + m_b \mathbf{v}'_b$ before and after the collision, and show that the y component of the classical momentum is *not* conserved in $\mathcal{S}'$.

15.55 ⋆⋆⋆ Since the four-velocity $u = \gamma(\mathbf{v}, c)$ is a four-vector its transformation properties are simple. Write down the standard Lorentz boost for all four components of u. Use these to deduce the relativistic velocity-addition formula for $\mathbf{v}$.

SECTION 15.13 Energy, the Fourth Component of Momentum

15.56 ⋆ When oxygen combines with hydrogen in the reaction (15.80) about 5 eV of energy is released (that is, the kinetic energy of the two final molecules is 5 eV more than that of the initial three molecules). **(a)** By how much does the total rest mass of the molecules change? **(b)** What is the fractional change in total mass? **(c)** If one were to form 10 grams of water by this reaction, what would be the total change in mass?

15.57 ⋆ When a radioactive nucleus of astatine 215 decays at rest, the whole atom is torn into two in the reaction

$$^{215}\text{At} \rightarrow {}^{211}\text{Bi} + {}^{4}\text{He}.$$

The masses of the three atoms are (in order) 214.9986, 210.9873, and 4.0026, all in atomic mass units. (1 atomic mass unit $= 1.66 \times 10^{-27}$ kg $= 931.5$ MeV/c^2.) What is the total kinetic energy of the two outcoming atoms, in joules and in MeV?

15.58 ⋆ **(a)** What is a particle's speed if its kinetic energy T is equal to its rest energy? **(b)** What if its energy E is equal to n times its rest energy?

15.59 ⋆ If one defines a variable mass $m_{\text{var}} = \gamma m$, then the relativistic momentum $\mathbf{p} = \gamma m \mathbf{v}$ becomes $m_{\text{var}} \mathbf{v}$ which looks more like the classical definition. Show, however, that the relativistic kinetic energy is *not* equal to $\frac{1}{2} m_{\text{var}} v^2$.

15.60 ⋆ A particle of mass m_a decays at rest into two identical particles each of mass m_b. Use conservation of momentum and energy to find the speed of the outgoing particles.

15.61 ⋆ A particle of mass 3 MeV/c^2 has momentum 4 MeV/c. What are its energy (in MeV) and speed (in units of c)?

15.62 ⋆ A particle of mass 12 MeV/c^2 has a kinetic energy of 1 MeV. What are its momentum (in MeV/c) and its speed (in units of c)?

15.63 ⋆ **(a)** What is a mass of 1 MeV/c^2 in kilograms? **(b)** What is a momentum of 1 MeV/c in kg·m/s?

15.64 ★ As measured in the inertial frame S, a proton has four-momentum p. Also as measured in S, an observer at rest in a frame S' has four-velocity u. Show that the proton's energy, as measured by this observer, is $-u \cdot p$.

15.65 ★★ The relativistic kinetic energy of a particle is $T = (\gamma - 1)mc^2$. Use the binomial series to express T as a series in powers of $\beta = v/c$. **(a)** Verify that the first term is just the nonrelativistic kinetic energy, and show that to lowest order in β the difference between the relativistic and nonrelativistic kinetic energies is $3\beta^4 mc^2/8$. **(b)** Use this result to find the maximum speed at which the nonrelativistic value is within 1% of the correct relativistic value.

15.66 ★★ In nonrelativistic mechanics, the energy contains an arbitrary additive constant — no physics is changed by the replacement $E \to E +$ constant. Show that this is not the case in relativistic mechanics. [*Hint:* Remember that the four-momentum p is supposed to tranform like a four-vector.]

SECTION 15.14 **Collisions**

15.67 ★ Two balls of equal masses (m each) approach one another head-on with equal but opposite velocities of magnitude $0.8c$. Their collision is perfectly inelastic, so they stick together and form a single body of mass M. What is the velocity of the final body and what is its mass M?

15.68 ★ Consider the elastic, head-on collision of Example 15.10, in which two particles (masses m_a and m_b) approach one another traveling along the x axis, collide, and emerge traveling along the same axis. In the CM frame (by its definition) $\mathbf{p}_a^{\text{in}} = -\mathbf{p}_b^{\text{in}}$. Use conservation of momentum and energy to prove that $\mathbf{p}_a^{\text{fin}} = -\mathbf{p}_a^{\text{in}}$; that is, the momentum of particle a (and likewise b) just reverses itself in the CM frame.

15.69 ★ **(a)** Show that the four-momentum of any material particle ($m > 0$) is forward time-like. **(b)** Show that the sum of any two forward time-like vectors is itself forward time-like, and hence that the sum of any number of forward time-like vectors is itself forward time-like.

15.70 ★ **(a)** Use the results of Problem 15.69 to prove that for any number of material particles there exists a CM frame, that is, a frame in which the total three-momentum is zero. **(b)** Relative to an arbitrary frame S, show that the velocity of the CM frame is given by $\boldsymbol{\beta} = \sum \mathbf{p}c / \sum E$.

15.71 ★ One way to create exotic heavy particles is to arrange a collision between two lighter particles

$$a + b \to d + e + \cdots + g$$

where d is the heavy particle of interest and $e, \cdots, g$ are other possible particles produced in the reaction. (A good example of such a process is the production of the ψ particle in the process $e^+ + e^- \to \psi$, in which there are no other particles $e, \cdots, g$.) **(a)** Assuming that m_d is much heavier that any of the other particles, show that the minimum (or threshold) energy to produce this reaction in the CM frame is $E_{\text{cm}} \approx m_d c^2$. **(b)** Show that the threshold energy to produce the same reaction in the lab frame, where the particle b is initially at rest, is $E_{\text{lab}} \approx m_d^2 c^2/2m_b$. **(c)** Calculate these two energies for the process $e^+ + e^- \to \psi$, with $m_e \approx 0.5$ MeV/c^2 and $m_\psi \approx 3100$ MeV/c^2. Your answers should explain why particle physicists go to the trouble and expense of building colliding-beam experiments.

15.72 ★ A mad physicist claims to have observed the decay of a particle of mass M into two identical particles of mass m, with $M < 2m$. In response to the objections that this violates conservation of energy, he replies that if M was traveling fast enough it could easily have energy greater than $2mc^2$ and hence could decay into the two particles of mass m. Show that he is wrong. [He has forgotten that both energy and momentum are conserved. You can analyse this problem in terms of these two conservation laws, but it is much simpler to go to the rest frame of M.]

15.73 ★★ Consider the head-on elastic collision of Example 15.10, in which the final velocity $\mathbf{v}_b$ of particle b is given by (15.95). **(a)** Show that, in the special case that the masses are equal ($m_a = m_b$), $v_b = v_a$, the initial velocity of particle a. Show that in this case the final velocity of a is zero. [This result for equal-mass collisions is well known in classical mechanics; you have now shown that it extends to relativity.] **(b)** Show that in the nonrelativistic limit (15.95) reduces to $v_b = 2v_a m_a / (m_a + m_b)$. By doing the necessary nonrelativistic calculations, show that this agrees with the nonrelativistic answer for elastic head-on collisions.

15.74 ★★ A particle a traveling along the positive x axis of frame S with speed $0.5c$ decays into two identical particles, $a \rightarrow b + b$, both of which continue to travel on the x axis. **(a)** Given that $m_a = 2.5 m_b$, find the speed of either b particle in the rest frame of particle a. **(b)** By making the necessary transformation on the result of part (a), find the velocities of the two b particles in the original frame S.

15.75 ★★ A particle of unknown mass M decays into two particles of known masses $m_a = 0.5\ \text{GeV}/c^2$ and $m_b = 1.0\ \text{GeV}/c^2$, whose momenta are measured to be $\mathbf{p}_a = 2.0\ \text{GeV}/c$ along the x_2 axis and $\mathbf{p}_b = 1.5\ \text{GeV}/c$ along the x_1 axis. ($1\ \text{GeV} = 10^9\ \text{eV}$.) Find the unknown mass M and its speed.

15.76 ★★ Particle a is pursuing particle b along the x_1 axis of a frame S. The two masses are m_a and m_b and the speeds are v_a and v_b (with $v_a > v_b$). When a catches up with b they collide and coalesce to form a single particle of mass m and speed v. Show that

$$m^2 = m_a^2 + m_b^2 + 2m_a m_b \gamma(v_a) \gamma(v_b)(1 - v_a v_b / c^2)$$

and find v.

15.77 ★★★ Consider the elastic head-on collision of Example 15.10, in which particle a collides with a stationary particle b. Assuming that $m_a \neq m_b$, show that the final kinetic energy of particle a satisfies $T_a^{\text{fin}} < (m_a - m_b)^2 c^2 / 2m_b$. [*Hint:* Look at the CM frame where you can show that the four-vector $p_a^{\text{fin}} - p_b^{\text{in}}$ is time-like, so that $(p_a^{\text{fin}} - p_b^{\text{in}})^2 < 0$.] **(b)** The result of part (a) implies that if T_a^{in} is large, almost all this incoming energy is lost to b. This is quite different from the nonrelativistic situation. Prove that in nonrelativistic mechanics the proportion of kinetic energy retained by a is fixed, independent of T_a^{in}. Specifically, $T_a^{\text{fin}} = T_a^{\text{in}} (m_a - m_b)^2 / (m_a + m_b)^2$.

15.78 ★★★ Consider the elastic collision shown in Figure 15.17. In the lab frame S, particle b is initially at rest; particle a enters with four-momentum p_a and scatters through an angle θ; particle b recoils at an angle ψ. In the CM frame S', the two particles approach and emerge with equal and opposite momenta, and particle a scatters through an angle θ'. **(a)** Show that the velocity of the CM frame relative to the lab frame is $\mathbf{V} = \mathbf{p}_a c^2 / (E_a + m_b c^2)$. **(b)** By transforming the final momentum of a back from the CM to the lab frame, show that

$$\tan \theta = \frac{\sin \theta'}{\gamma_V (\cos \theta' + V / v_a')} \tag{15.153}$$

where v_a' is the speed of a in the CM frame. **(c)** Show that in the limit that all speeds are much smaller than c, this result agrees with the nonrelativistic result (14.53) (where $\lambda = m_a / m_b$). **(d)** Specialize now to the case that $m_a = m_b$. Show that, in this case, $V / v_a' = 1$, and find a formula like (15.153) for $\tan \psi$.

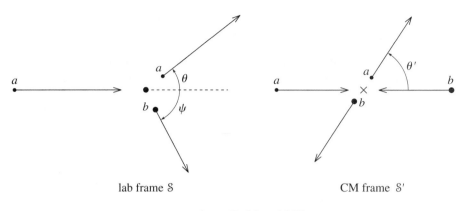

Figure 15.17   Problem 15.78.

(e) Show that the angle between the two outgoing momenta is given by $\tan(\theta + \psi) = 2/(\beta_V^2 \gamma_V \sin \theta')$. Show that in the limit that $V \ll c$, you recover the well-known nonrelativistic result that $\theta + \psi = 90°$.

SECTION 15.15   Force in Relativity

15.79 ★ Consider an object of mass m (which you may assume is constant), acted on by a force $\mathbf{F}$. From the definition (15.100) prove that

$$\mathbf{F} = \gamma m \mathbf{a} + (\mathbf{F} \cdot \mathbf{v})\mathbf{v}/c^2,$$

where $\mathbf{a} = d\mathbf{v}/dt$ is the object's acceleration. Notice that it is certainly not true in relativity that $\mathbf{F} = m\mathbf{a}$. Nor is it true that $\mathbf{F} = m_{\text{var}}\mathbf{a}$, where m_{var} is the variable mass $m_{\text{var}} = \gamma m$, except in the special case that $\mathbf{F}$ happens to be perpendicular to $\mathbf{v}$. In general, $\mathbf{F}$ and $\mathbf{a}$ are not even in the same direction.

15.80 ★ A particle of mass m and charge q moves in a uniform, constant magnetic field $\mathbf{B}$. Show that if $\mathbf{v}$ is perpendicular to $\mathbf{B}$, the particle moves in a circle of radius

$$r = |\mathbf{p}/qB|. \qquad (15.154)$$

[This result agrees with the nonrelativistic result (2.81), except that $\mathbf{p}$ is now the relativistic momentum $\mathbf{p} = \gamma m \mathbf{v}$.]

15.81 ★ An electron (mass 0.5 MeV/c^2) moves with speed $0.7c$ in a circular path in a magnetic field of 0.02 teslas. Using the relativistic result (15.154) of Problem 15.80, find the radius of the electron's orbit. What would your answer have been if you used the classical definition of momentum? [Needless to say, the relativistic result is confirmed by experiment, and this gave some of the first evidence of the correctness of relativistic mechanics.[33]]

15.82 ★ (a) Verify the result (15.106) for the position of a particle moving in a uniform electric field, by integrating the expression (15.105). (b) When t is small, the particle should be moving slowly and

---

[33] In a paper in *Göttingen Nachrichten*, p. 143 (1901), Walter Kaufmann showed that the electron's "apparent mass" (what we would call its variable mass) in a magnetic field seemed to increase with speed in rough accord with the relativistic formula $m_{\text{var}} = \gamma(v)m$. Notice that this predated Einstein's first paper on relativity by some four years.

(15.106) should agree with the nonrelativistic result $\mathbf{x} = \frac{1}{2}\mathbf{a}t^2$. Verify that it does. **(c)** Show that when t is large, $\mathbf{x} \approx \hat{\mathbf{F}}(ct + \text{const})$ and explain this result.

15.83 ★ Starting from the definition (15.100) of the force $\mathbf{F}$ on an object, prove that the transformation of the components of $\mathbf{F}$ as we pass from a frame $\mathcal{S}$ to a second frame $\mathcal{S}'$, traveling at speed V in the standard configuration relative to $\mathcal{S}$, is

$$F_1' = \frac{F_1 - \beta\mathbf{F}\cdot\mathbf{v}/c}{1 - \beta v_1/c}, \qquad F_2' = \frac{F_2}{\gamma(1 - \beta v_1/c)}, \qquad F_3' = \frac{F_3}{\gamma(1 - \beta v_1/c)} \qquad (15.155)$$

where $\beta = \beta(V)$ and $\gamma = \gamma(V)$ relate to the relative speed of the two frames and $\mathbf{v}$ is the velocity of the object as measured in $\mathcal{S}$.

15.84 ★★ A mass m is thrown from the origin at $t = 0$ with initial three-momentum p_o in the y direction. If it is subject to a constant force F_o in the x direction, find its velocity $\mathbf{v}$ as a function of t, and by integrating $\mathbf{v}$ find its trajectory. Check that in the nonrelativistic limit the trajectory is the expected parabola.

15.85 ★★ We have seen that there are processes in which the mass of an object varies with time. **(a)** Starting from (15.85), prove that $dm/dt_o = -u \cdot K/c^2$, where t_o is the object's proper time, u is its four-velocity, and K is the four-force on the object. **(b)** This means that the necessary and sufficient condition that a force doesn't change an object's mass is that $u \cdot K = 0$. It is an experimental fact that if a charged particle is at rest in an electromagnetic field (even instantaneously) then $dE/dt = 0$. Use this to argue that electromagnetic forces do not cause a particle's mass to change.

SECTION 15.16   Massless Particles; the Photon

15.86 ★ The neutral pion π^0 is an unstable particle (mass $m = 135$ MeV/c^2) that can decay into two photons, $\pi^0 \rightarrow \gamma + \gamma$. **(a)** If the pion is at rest, what is the energy of each photon? **(b)** Suppose instead that the pion is traveling along the x axis and that the photons are observed also traveling along the x axis, one forward and one backward. If the first photon has three times the energy of the second, what was the pion's original speed v?

15.87 ★ A neutral pion (Problem 15.86) is traveling with speed v when it decays into two photons, which are seen to emerge at equal angles θ on either side of the original velocity. Show that $v = c \cos\theta$.

15.88 ★ Two particles a and b with masses $m_a = 0$ and $m_b > 0$ approach one another. Prove that they have a CM frame (that is, a frame in which their total three-momentum is zero). [*Hint:* As you should explain, this is equivalent to showing that the sum of two four-vectors, one of which is forward light-like and one forward time-like, is itself forward time-like.]

15.89 ★ Show that any two zero-mass particles have a CM frame, provided their three-momenta are not parallel. [*Hint:* As you should explain, this is equivalent to showing that the sum of two forward light-like vectors is forward time-like, unless the spatial parts are parallel.]

15.90 ★★ The first positrons to be observed were created in electron–positron pairs by high-energy cosmic-ray photons in the upper atmosphere. **(a)** Show that an isolated photon cannot convert to an electron–positron pair in the process $\gamma \rightarrow e^+ + e^-$. [Show that this process inevitably violates conservation of four-momentum.] **(b)** What actually occurs is that a photon collides with a stationary nucleus with the result

$$\gamma + \text{nucleus} \rightarrow e^+ + e^- + \text{nucleus}.$$

Convince yourself that the formula (15.98) can be used to find the minimum energy for a photon to induce this reaction. [The derivation of (15.98) assumed that the incident particle had $m > 0$.] Show that, provided the mass of the nucleus is much greater than that of the electron, the minimum photon energy to induce this reaction is approximately $2m_ec^2$. [This is exactly the energy one would have calculated for the process $\gamma \to e^+ + e^-$ and shows that the role of the nucleus is just as a "catalyst" that can absorb some three-momentum.]

15.91 ★★ An excited state X^* of an atom at rest drops to its ground state X by emitting a photon. In atomic physics it is usual to assume that the energy E_γ of the photon is equal to the difference in energies of the two atomic states, $\Delta E = (M^* - M)c^2$, where M and M^* are the rest masses of the ground and excited states of the atom. This cannot be exactly true, since the recoiling atom X must carry away some of the energy ΔE. Show that in fact $E_\gamma = \Delta E[1 - \Delta E/(2M^*c^2)]$. Given that ΔE is of order a few eV, while the lightest atom has M of order 1 GeV/c^2, discuss the validity of the assumption that $E_\gamma = \Delta E$.

15.92 ★★ A positive pion decays at rest into a muon and neutrino, $\pi^+ \to \mu^+ + \nu$. The masses involved are $m_\pi = 140$ MeV/c^2, $m_\mu = 106$ MeV/c^2, and $m_\nu = 0$. (There is now convincing evidence that m_ν is not exactly zero, but is small enough that you can take it to be zero for this problem.) Show that the speed of the outgoing muon has $\beta = (m_\pi^2 - m_\mu^2)/(m_\pi^2 + m_\mu^2)$. Evaluate this numerically. Do the same for the much rarer decay mode $\pi^+ \to e^+ + \nu$, ($m_e = 0.5$ MeV/c^2).

15.93 ★★★ Consider a head-on elastic collision between a high-energy electron (energy E_0 and speed $\beta_0 c$) and a photon of energy $E_{\gamma 0}$. Show that the final energy E_γ of the photon is

$$E_\gamma = E_0 \frac{1 + \beta_0}{2 + (1 - \beta_0)E_0/E_{\gamma 0}}.$$

[*Hint:* Use (15.123).] Show that $E_\gamma < E_0$, but that if $\beta_0 \to 1$, then $E_\gamma/E_0 \to 1$; that is, a very high-energy electron loses almost all its energy to the photon in a head-on collision. What fraction of its original energy would the electron retain if $E_0 \approx 10$ TeV and the photon was in the visible range, $E_{\gamma 0} \approx 3$ eV? (Remember that the mass of the electron is about 0.5 MeV/c^2; 1 TeV = 10^{12} eV.)

SECTION 15.17   Tensors

15.94 ★ Prove that for any two matrices A and B, where A has as many columns as B has rows, the transpose of AB satisfies $\widetilde{(AB)} = \tilde{B}\tilde{A}$.

15.95 ★ By making suitable choices for the n-dimensional vectors $\mathbf{a}$ and $\mathbf{b}$, show that if $\tilde{\mathbf{a}}C\mathbf{b} = \tilde{\mathbf{a}}D\mathbf{b}$ for any choices of $\mathbf{a}$ and $\mathbf{b}$ (where C and D are $n \times n$ matrices), then $\mathbf{C} = \mathbf{D}$.

15.96 ★ Prove that if T and a are respectively a four-tensor and a four-vector, then $b = T \cdot a = TGa$ is a four-vector; that is, it transforms according to the rule $b' = \Lambda b$.

15.97 ★ (a) A tensor T is said to be symmetric if $T_{\mu\nu} = T_{\nu\mu}$. Prove that if T is symmetric in one inertial frame, then it is symmetric in all inertial frames. (b) T is antisymmetric if $T_{\mu\nu} = -T_{\nu\mu}$. Prove that if T is antisymmetric in one inertial frame, then it is antisymmetric in all inertial frames. (An example of the latter property is the electromagnetic field tensor, which is antisymmetric in all frames.)

15.98 ★★ (a) Use the invariance of the scalar product $a \cdot b = \tilde{a}Gb$ to prove that the 4×4 Lorentz transformation matrices Λ must satisfy the condition (15.136), $\tilde{\Lambda}G\Lambda = G$. (b) Verify that the standard Lorentz boost (15.43) does satisfy this condition.

15.99 ★★ A useful form of the quotient rule for three-dimensional vectors is this: Suppose that **a** and **b** are known to be three-vectors and suppose that for every orthogonal set of axes there is a 3×3 matrix **T** with the property that $\mathbf{b} = \mathbf{Ta}$ for every choice of **a**, then **T** is a tensor. **(a)** Prove this. **(b)** State and prove the corresponding rule for four-vectors and four-tensors.

15.100 ★★★ **(a)** The statement that ∇ is a vector operator means that if $\phi(\mathbf{x})$ is any scalar, then the three components of $\nabla\phi = (\partial\phi/\partial x_1, \partial\phi/\partial x_2, \partial\phi/\partial x_3)$ transform according to the three-vector transformation law (15.126). Prove this last statement. [*Hint:* Remember the chain rule, that $\partial\phi/\partial x_i = \sum_j (\partial x'_j/\partial x_i)\partial\phi/\partial x'_j$.] **(b)** Prove that in four-dimensional space–time, if ϕ is any four-scalar, the quantity $\Box\phi$ defined with the components

$$\Box\phi = \left(\frac{\partial\phi}{\partial x_1}, \frac{\partial\phi}{\partial x_2}, \frac{\partial\phi}{\partial x_3}, -\frac{\partial\phi}{\partial x_4}\right) \tag{15.156}$$

(note well the minus sign on the fourth component) is a four-vector. This result is crucial in writing down Maxwell's equations for the electromagnetic field.

SECTION 15.18   Electrodynamics and Relativity

15.101 ★ **(a)** Prove that $\mathbf{E} \cdot \mathbf{B}$ and $E^2 - c^2B^2$ are both invariant under any Lorentz tranformation. [Use the transformation equations (15.146) to prove the required results for the standard boost and then explain why if either quantity is invariant under the standard boost then it is invariant under *any* Lorentz transformation.] Use these results to prove the following two propositions: **(b)** If **E** and **B** are perpendicular in frame $\mathcal{S}$, then they are perpendicular in any other frame $\mathcal{S}'$, and **(c)** if $E > cB$ in a frame $\mathcal{S}$, then there cannot exist a frame in which $E = 0$.

15.102 ★ **(a)** Starting from the transformation equations (15.146) for the standard boost along the x_1 axis, find the corresponding boost along the x_3 axis. **(b)** Write down the inverse of this transformation and then verify the results (15.151) and (15.152) for the fields of a moving line charge.

15.103 ★ **(a)** Using the transformation equations (15.146), show that if $\mathbf{E} = 0$ in frame $\mathcal{S}$, then $\mathbf{E}' = \mathbf{v} \times \mathbf{B}'$ in $\mathcal{S}'$. **(b)** Similarly, show that if $\mathbf{B} = 0$ in frame $\mathcal{S}$, then $\mathbf{B}' = -\mathbf{v} \times \mathbf{E}'/c^2$.

15.104 ★★ We defined the electromagnetic field tensor by the equation $K = q\mathcal{F} \cdot u \equiv q\mathcal{F}Gu$, where K is the four-force on a charge q and u is its four-velocity. **(a)** Starting from the Lorentz force (15.139) write down the four components of K [as in (15.141)]. **(b)** Use these to find the matrix $\mathcal{F}G$ and show that the tensor $\mathcal{F}$ has the form claimed in (15.143).

15.105 ★★ Since $\mathcal{F}$ is a four-tensor it has to transform according to the rule (15.145), $\mathcal{F}' = \Lambda\mathcal{F}\tilde\Lambda$. Using the form (15.143) for $\mathcal{F}$ and the standard Lorentz boost for Λ, find the matrix $\mathcal{F}'$ and verify the transformation equations (15.146) for the electromagnetic fields.

15.106 ★★ Derive the Lorentz-force law from Coulomb's law as follows: **(a)** If a charge q is at rest in frame $\mathcal{S}'$, then Coulomb's law tells us that the force on q is $\mathbf{F}' = q\mathbf{E}'$. Use the inverse of the force transformation (15.155) in Problem 15.83 to write down the force **F** as seen in $\mathcal{S}$. (Answer in terms of $\mathbf{E}'$ for now.) **(b)** Now use the field transformation (15.146) to rewrite your answer in terms of **E** and **B** and show that $\mathbf{F} = q(\mathbf{E} + \mathbf{v} \times \mathbf{B})$.

15.107 ★★ It is a result well known in classical electromagnetism that one can introduce a three-scalar potential ϕ and a three-vector potential **A** such that the fields **E** and **B** can be written as

$$\mathbf{E} = -\nabla\phi - \frac{\partial\mathbf{A}}{\partial t} \quad \text{and} \quad \mathbf{B} = \nabla \times \mathbf{A}. \tag{15.157}$$

In relativity these potentials are combined to form a single four-potential $A = (\mathbf{A}, \phi/c)$. Prove that

$$\mathcal{F}_{\mu\nu} = \Box_\mu A_\nu - \Box_\nu A_\mu$$

where $\Box$ is the four-dimensional gradient operator defined in (15.156) of Problem 15.100. (If we accept that A really is a four-vector, this gives an alternative proof that $\mathcal{F}$ is a four tensor.)

15.108 ⋆⋆ Consider an electric charge distribution, with charge density ϱ, moving with velocity $\mathbf{v}$ relative to a frame $\mathcal{S}$. **(a)** Show that $\varrho = \gamma \varrho_0$, where ϱ_0 is the charge density in the rest frame. (Notice that $\mathbf{v}$ can vary with position, so different parts of the distribution will have different rest frames, but that's all right.) **(b)** The three-current density is defined as $\mathbf{J} = \varrho \mathbf{v}$. Show that the four-current density, defined as $J = (\mathbf{J}, c\varrho)$, is a four-vector. **(c)** It is a well-known result in electromagnetism that conservation of charge implies the so-called equation of continuity, $\nabla \cdot \mathbf{J} + \partial \varrho / \partial t = 0$ (where $\nabla \cdot \mathbf{J} = \sum \partial J_i / \partial x_i$ is the so-called divergence of $\mathbf{J}$). Show that this condition is equivalent to the manifestly invariant condition $\Box \cdot J = 0$, where $\Box$ is the four-dimensional gradient defined in (15.156) of Problem 15.100.

15.109 ⋆⋆⋆ Two equal charges q are moving side-by-side in the positive x direction in frame $\mathcal{S}$. The distance between them is r and their speed is v. Find the force on either charge due to the other in two ways: **(a)** Find the force in their rest frame $\mathcal{S}'$ and transform back to $\mathcal{S}$, using the force transformation (15.155) of Problem 15.83. Note that the force in $\mathcal{S}$ is less than in the rest frame. **(b)** Find the electric and magnetic fields in $\mathcal{S}'$ and thence in $\mathcal{S}$, using the field transformation (15.146). Use these fields (in $\mathcal{S}$) to write down the Lorentz force on either charge in $\mathcal{S}$. Note that in $\mathcal{S}$ there is an attractive electric force and a repulsive magnetic force. As $\beta \to 1$ they become nearly equal and their resultant approaches zero.

15.110 ⋆⋆⋆ A charge q is moving with constant speed v along the x axis of frame $\mathcal{S}$ with position $vt\hat{\mathbf{x}}$. **(a)** Write down the electric and magnetic fields in the charge's rest frame $\mathcal{S}'$. **(b)** Use the inverse of the field transformation (15.146) to write down the electric field in the original frame $\mathcal{S}$. [In the first instance, you will find $\mathbf{E}$ in terms of the primed variables x', y', z', t', but you can use the standard Lorentz transformation to eliminate them in favor of x, y, z, t.] Show that the field at position $\mathbf{r}$ and time t is

$$\mathbf{E} = \frac{kq(1-\beta^2)}{(1-\beta^2 \sin^2\theta)^{3/2}} \frac{\hat{\mathbf{R}}}{R^2} \tag{15.158}$$

where $\mathbf{R} = \mathbf{r} - vt\hat{\mathbf{x}}$ is the vector pointing from the charge's position to the point of observation $\mathbf{r}$, and θ is the angle between $\mathbf{R}$ and the x axis. **(c)** Sketch the behavior of the field strength as a function of θ for fixed R, and make a sketch of the electric field lines at one fixed time t.

15.111 ⋆⋆⋆ Two of Maxwell's four equations read

$$\nabla \times \mathbf{B} - \frac{1}{c^2}\frac{\partial \mathbf{E}}{\partial t} = \mu_0 \mathbf{J} \quad \text{and} \quad \nabla \cdot \mathbf{E} = \frac{1}{\epsilon_0}\varrho \tag{15.159}$$

where $\mathbf{J}$ and ϱ are the current and charge densities that gave rise to the fields. Show that these two equations can be written as the single four-vector equation $\Box \cdot \mathcal{F} = -\mu_0 \tilde{J}$, where $\Box$ is the four-dimensional gradient operator introduced in Problem 15.100, J is the four-current $(\mathbf{J}, c\varrho)$, and the scalar product $\Box \cdot \mathcal{F} = \tilde{\Box} G \mathcal{F}$.

CHAPTER 16

Continuum Mechanics

We can divide classical mechanics into three main areas, in order of increasing complexity. **(1)** The **mechanics of point masses**. Occasionally, these point masses are elementary particles, such as the electron, whose mass is (as far as we know) concentrated at a point; but usually they are extended objects whose mass is certainly not localized at a point but which, for the purposes at hand, can be approximated as if it were. Thus in treating the flight of baseballs, or finding the orbits of the planets, it is an often an excellent approximation to treat them as point masses. In this case, the configuration of any system is given by a finite set of coordinates, three for each point mass. **(2)** The **mechanics of rigid bodies**. Here we acknowledge that the mass of interest is spread out over some nonzero volume, but we assume that the relative positions of the various parts of any one body are fixed; that is, all bodies are rigid. As we saw, we have to allow for the possible rotational motion of such bodies, but the configuration of any system is still specified by a discrete and finite set of coordinates; for example, for a single rigid body we need just six coordinates, three for the CM position and three more for the body's orientation. The notion of a rigid body is an idealization — all real bodies can be deformed — but for many systems it is a reasonable and extremely useful approximation. **(3)** Finally, there is **continuum mechanics**, in which we acknowledge that the mass in a system can be distributed over some region *and* that the relative positions of the various parts can change in a continuous, but otherwise arbitrary, manner. Clearly the motion of any fluid — the flow of air past an airplane wing or of water down a pipe — is a problem in continuum mechanics. But so also is the motion of a solid when the independent motions of its parts are important, as for example in the flexing of a heavily loaded steel beam or the vibration of the earth's crust in response to an earthquake. In continuum mechanics a system comprises a continuously infinite number of parts, and the specification of its configuration requires an infinite number of coordinates.

So far in this book we have treated only the first two topics, the mechanics of point masses and rigid bodies — what we could call discrete mechanics — with a discrete, finite number of coordinates. This final chapter is intended as the briefest of introductions to continuum mechanics. A thorough introduction would require another book, but I hope that I can give at least a sense of some of the central ideas. Specifically, we'll see how the passage from discrete to continuum mechanics changes the ordinary differential equations of the former to *partial differential equations*. We'll see how these partial differential equations often lead to the *wave equation*, which governs the behavior of sound waves in liquids and gases, of seismic waves in the earth's crust, and of many other waves, most notably electromagnetic waves such as light and microwaves. The first three sections deal with one-dimensional continuous systems, and then, in Section 16.4 we move out into three dimensions. Perhaps the biggest complication in three dimensions is that the forces and displacements involve two tensors, the *stress* and *strain tensors*. One of the main objectives of this chapter is to introduce these two important concepts. Section 16.7 introduces the stress tensor for fluids and solids, and Section 16.8 the strain tensor for solids. Section 16.9 gives the relation between the two tensors, the generalized Hooke's law. Sections 16.10 and 16.11 derive the equation of motion for an elastic solid and use it to analyze the longitudinal and transverse waves in a solid. The last two sections are a very brief introduction to the mechanics of inviscid fluids. In Section 16.12, we'll derive the equation of motion and the so-called continuity equation, and in Section 16.13 we'll use them to analyze the possible waves in a fluid.

Before we get started, I should emphasize that the notion of a continuous distribution of matter is itself an idealization. The properties of the air flowing in a wind tunnel certainly appear to vary continuously and smoothly from place to place. For example, we are used to assuming that air has a density $\varrho(\mathbf{r})$ which gives the mass $\varrho(\mathbf{r})dV$ in a small volume dV, and, while $\varrho(\mathbf{r})$ may certainly vary with $\mathbf{r}$, we assume that it does so smoothly. However, we know perfectly well that, when viewed under a super-microscope that could resolve fractions of nanometers, the air would be seen to consist of individual molecules, and the density $\varrho(\mathbf{r})$ would vary wildly between large values near to each molecule and zero in the huge spaces in between. Fortunately, the scale of these wild variations is tiny compared to the scales of normal interest. For example, even if we were interested in regions as small as 1 mm^3, this volume contains some 10^{16} molecules. Thus the density $\varrho(\mathbf{r})$ that we actually work with is the density *averaged over this huge number of molecules* and does indeed vary smoothly with $\mathbf{r}$. The idea that, at a scale of millimeters or more, matter can be treated as continuous, with parameters such as the density being averages over many molecules, is called the **continuum hypothesis**. The success of continuum mechanics is ample justification for this hypothesis, which we shall adopt throughout this chapter.

16.1   Transverse Motion of a Taut String

As our first example of a continuous system, let us consider a taut string lying along the x axis. In equilibrium, we assume that the string lies exactly on the x axis, but now suppose that it is undergoing a small motion (perhaps an oscillatory motion) in the y direction. One simple way to specify the string's configuration at any one time is to give its displacement $u(x)$ from the x axis, as shown in Figure 16.1(a). Specifically, at any one time, a small element of the string whose equilibrium position was at x on

Figure 16.1   **(a)** The position of a continuous string, at any one time, is specified by the function $u(x)$ which gives the string's displacement from its equilibrium position on the x axis. **(b)** A set of n point masses joined by massless strings has its configuration specified by the discrete set of displacements u_i with $i = 1, \cdots, n$.

the axis is now located a distance $y = u(x)$ above the axis. This scheme needs little explanation, but is worth contrasting with a related discrete system, namely a set of n point masses $m_1, \cdots, m_n$ joined by a taut massless string, which lies in equilibrium on the x axis. If these masses are allowed to move in the y direction as in Figure 16.1(b), their configuration can be specified by their displacements $u_1, \cdots, u_n$ from the axis. Where the discrete system is specified by these n variables u_i, with $i = 1, \cdots, n$, our continuous system is specified by the *continuous function* $u(x)$. The role of the discrete index i attached to u_i is now played by the continuous variable x in $u(x)$. Where the index i specifies which of the n masses is at position $y = u_i$, the variable x specifies which of the infinitely many pieces of the string is at $y = u(x)$.

If the systems of Figure 16.1 are moving, the displacements u depend on the time t. In the discrete case they become $u_i(t)$ and in the continuous case we must write $u(x, t)$, a function of two variables. In the discrete case, Newton's second law becomes a set of ordinary differential equations for the $u_i(t)$ (for example the coupled differential equations of Chapter 11). In the continuous case, Newton's law becomes a *partial differential equation* for $u(x, t)$, involving partial derivatives with respect to both x and t, as we now show.

To explore the motion of the string, we shall apply Newton's second law to a small segment AB of the string, between x and $x + dx$ as shown in Figure 16.2. To simplify our discussion we shall ignore gravity, and we shall assume that the displacement $u(x, t)$ remains so small for all x and all t, that the string remains nearly parallel to the x axis. This guarantees that the string's length is essentially unchanged and hence that the tension T remains the same for all x and all t. The net force on the segment AB is then just $\mathbf{F}^{\text{net}} = \mathbf{F}_1 + \mathbf{F}_2$, where $\mathbf{F}_1$ and $\mathbf{F}_2$ are the tension forces due to the adjacent sections of string, as shown in Figure 16.2. If ϕ denotes the angle between the string and the x axis, the x component of this net force is

$$F_x^{\text{net}} = T \cos(\phi + d\phi) - T \cos \phi$$

but, since ϕ and $\phi + d\phi$ are both small, both cosines are very close to 1 and F_x^{net} is negligible, consistent with our assumption that the motion is in the y direction only. On the other hand, the y component is certainly not negligible:

$$F_y^{\text{net}} = T \sin(\phi + d\phi) - T \sin \phi = T \cos \phi \, d\phi. \tag{16.1}$$

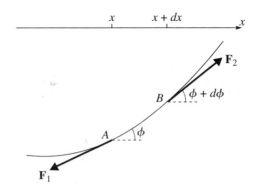

Figure 16.2   The two forces on the small element AB of string are the tension forces $\mathbf{F}_1$ and $\mathbf{F}_2$ exerted by the adjacent sections of string.

Since ϕ is small we can replace $\cos\phi$ by 1, and we can write $d\phi = (\partial\phi/\partial x)dx$. [The derivative is a partial derivative since $\phi = \phi(x, t)$ depends on x and t.] Finally, again since ϕ is small, $\phi = \partial u/\partial x$, the slope of the string. Therefore,

$$F_y^{\text{net}} = T\frac{\partial\phi}{\partial x}dx = T\frac{\partial^2 u}{\partial x^2}dx. \tag{16.2}$$

By Newton's second law, $F_y^{\text{net}} = ma_y$, where a_y is the acceleration $a_y = \partial^2 u/\partial t^2$ and m is the mass of the segment AB, equal to $\mu\, dx$ if we use μ to denote the linear mass density of the string. Therefore,

$$F_y^{\text{net}} = \mu\frac{\partial^2 u}{\partial t^2}dx. \tag{16.3}$$

Equating the two expressions (16.2) and (16.3), we arrive at the equation of motion of our taut string:

$$\frac{\partial^2 u}{\partial t^2} = c^2\frac{\partial^2 u}{\partial x^2}. \tag{16.4}$$

Here I have introduced the important constant

$$c = \sqrt{\frac{T}{\mu}}, \tag{16.5}$$

where T is the tension in our string and μ is its linear mass density (mass/length).

The equation of motion (16.4) is called the **one-dimensional wave equation** since its solutions are waves traveling along the string, as we shall see. As anticipated, it is a partial differential equation, involving derivatives with respect to x and t. The constant

c has the dimensions of speed (as you should check) and is the speed with which the waves travel. The wave equation (16.4) governs the motion of many different waves — waves on a string, waves of sound or light, seismic waves, and many more. Therefore, I shall give it a new section of its own.

16.2   The Wave Equation

We are going to show that there are just three kinds of solution of the wave equation (16.4): **(1)** a disturbance $u(x, t)$ that travels rigidly along the string from left to right; **(2)** a disturbance $u(x, t)$ that travels rigidly along the string from right to left; and **(3)** any combination of these two. The proof of this claim is startlingly simple, although it depends on a trick that you probably wouldn't think of right away. We change variables from x and t to

$$\xi = x - ct \qquad \text{and} \qquad \eta = x + ct. \qquad (16.6)$$

It is a straightforward exercise (Problem 16.4) to show that

$$\frac{\partial^2 u}{\partial t^2} - c^2 \frac{\partial^2 u}{\partial x^2} = -4c^2 \frac{\partial}{\partial \xi} \frac{\partial u}{\partial \eta}, \qquad (16.7)$$

so, written in terms of the new variables, the wave equation (16.4) becomes simply

$$\frac{\partial}{\partial \xi} \frac{\partial u}{\partial \eta} = 0. \qquad (16.8)$$

To solve this equation, let us temporarily write $\partial u / \partial \eta = h$, so that (16.8) becomes $\partial h / \partial \xi = 0$. This states simply that h does not depend on ξ, although it can, of course, depend on η. Therefore, we can write $h = h(\eta)$, and we now have

$$\frac{\partial u}{\partial \eta} = h(\eta).$$

For any given value of ξ we can integrate this equation to give $u = \int h(\eta)\, d\eta +$ "constant," where the "constant" may be different for different values of ξ. If we call this "constant" $f(\xi)$ and set the integral $\int h(\eta)\, d\eta = g(\eta)$, then we have proved that every solution of (16.8) must have the form

$$u = f(\xi) + g(\eta). \qquad (16.9)$$

By substituting into the left side of (16.8), you can see that a function of this form is a solution of (16.8) for any choice of the two functions $f(\xi)$ and $g(\eta)$. Thus, (16.9) is the general solution of (16.8).

Reverting to the original variables x and t, we have shown that the general solution of the wave equation (16.4) has the form

$$u(x, t) = f(x - ct) + g(x + ct) \qquad (16.10)$$

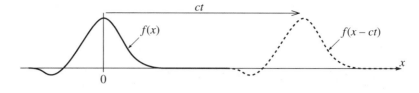

Figure 16.3 Motion of the wave (16.11). At time 0, the disturbance is given by $u = f(x)$. At a later time t, it is given by $u = f(x - ct)$, which has the same shape but has moved rigidly to the right by a distance ct.

where f and g are any two functions. To see what these solutions represent, let us consider first the case that the function $g = 0$, so that our solution is just

$$u(x, t) = f(x - ct). \tag{16.11}$$

What does this solution look like? Notice first that at time $t = 0$, the solution is $u(x, 0) = f(x)$; that is, the function $f(x)$ is just the disturbance at time $t = 0$. Figure 16.3 shows a possible such function. The solid curve, with a large maximum at $x = 0$ and a small dip on the left, shows the function $f(x)$, the shape of the disturbance at $t = 0$. At a later time t, the disturbance is given by $f(x - ct)$. Since $f(x)$ has its maximum at $x = 0$, it follows that $f(x - ct)$ has its maximum when $x - ct = 0$. Therefore, the maximum that was at $x = 0$ is now at $x = ct$. Since a similar argument applies to any point of the curve (for example, the minimum on the left), we conclude that the whole disturbance has moved bodily to the right by a distance ct. That is, the disturbance is a wave traveling rigidly to the right with speed c.

A similar argument shows that a solution of the form $u(x, t) = g(x + ct)$ represents a wave traveling rigidly to the *left* with speed c, and the general solution (16.10) is a superposition of two waves, one traveling to the right and the other to the left. The functions f and g that appear in the general solution (16.10) are determined by the initial conditions of any particular problem. As you might guess, to determine a particular solution we need to specify the position u and the initial velocity $\dot{u} = \partial u / \partial t$ at one initial time, as in the following example.

EXAMPLE 16.1   Evolution of a Triangular Wave

A short segment of a long taut string is pulled aside and released from rest at $t = 0$, so that its initial displacement is

$$u(x, 0) = u_0(x) \tag{16.12}$$

where $u_0(x)$ is the triangular function shown in Figure 16.4(a). Find the disturbance $u(x, t)$ for any later time t.

The solution must have the form (16.10), where the two functions f and g are to be determined by the initial conditions. The given initial displacement (16.12) implies that

$$f(x) + g(x) = u_0(x). \tag{16.13}$$

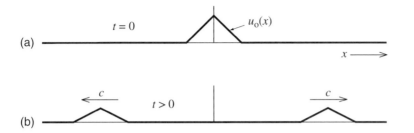

Figure 16.4 (a) The initial displacement of a string at $t = 0$ is given by the triangular function $u_0(x)$. (b) At any later time, the wave consists of two triangles, each half as high as the original, one traveling to the right and one to the left.

This does not, by itself, determine f and g separately, and we must also look at the initial velocity. Differentiating (16.10) with respect to t, we see that the initial velocity of the string (in the y direction, of course) is

$$\left[\frac{\partial u}{\partial t}\right]_{t=0} = -cf'(x) + cg'(x)$$

where the prime denotes differentiation of a function with respect to its argument. In our case, the string is released from rest, so $f'(x) - g'(x) = 0$. Integrating with respect to x, we conclude that[1]

$$f(x) - g(x) = 0. \tag{16.14}$$

Solving (16.13) and (16.14), we conclude that

$$f(x) = g(x) = \tfrac{1}{2}u_0(x)$$

and the actual disturbance (16.10), at any time t, is

$$u(x, t) = f(x - ct) + g(x + ct) = \tfrac{1}{2}u_0(x - ct) + \tfrac{1}{2}u_0(x + ct). \tag{16.15}$$

Our original triangle has separated into two triangles, half as high, traveling outward in opposite directions with speed c, as shown in Figure 16.4(b).

It is interesting to let the solution (16.15) evolve backward to times $t < 0$. At these times it represents two triangles approaching the origin from opposite sides. When t is close to 0, the two triangles meet and start to interfere. At $t = 0$ they overlap exactly and interfere to produce a triangular wave twice their individual heights, and then, as t increases past 0, they separate again and move apart as in Figure 16.4(b).

---

[1] Strictly speaking there should be a constant of integration in (16.14), but as you can easily check, it cancels out of $u = f + g$, so we can just as well choose it to be zero.

An important special case of the solution (16.10) is the case that the functions f and g are sinusoidal. If $g = 0$, then the disturbance takes the form

$$u(x, t) = A \sin[k(x - ct)] = A \sin(kx - \omega t) \qquad (16.16)$$

where A and k are arbitrary constants, and $\omega = kc$. This is a sinusoidal wave traveling to the right with amplitude A, wave number k (or wavelength $\lambda = 2\pi/k$) and angular frequency ω (or period $\tau = 2\pi/\omega$). If we replace $x - ct$ with $x + ct$, we obtain a similar sinusoidal wave

$$u(x, t) = A \sin[k(x + ct)] = A \sin(kx + \omega t) \qquad (16.17)$$

traveling to the left. The sum of these two solutions is itself a solution,

$$u(x, t) = A \sin(kx - \omega t) + A \sin(kx + \omega t) = 2A \sin(kx) \cos(\omega t). \quad (16.18)$$

(Use the relevant trig identities to check this.) This solution has the remarkable property that it is not traveling at all (neither to right nor left). Instead it simply oscillates up and down like $\cos(\omega t)$, with amplitude (at any one point x) equal to $2A \sin(kx)$. In particular, at those points (the **nodes**) where kx is an integer multiple of π ($kx = n\pi$), the string does not move at all, as shown in Figure 16.5. We see that by superposing two carefully chosen traveling waves we have formed a **standing wave**. As we shall see in the next section, these standing waves play an important role in the oscillations of a finite length of string and are, in fact, the continuum analog of the normal modes of a system of coupled oscillators.

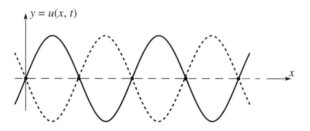

Figure 16.5   The standing wave (16.18) at three successive times, $t = 0$ (solid curve), $t = \tau/4$ (long dashes), and $t = \tau/2$ (short dashes), where τ is the period. The small dots on the x axis are the nodes, where $kx = n\pi$ and the string does not move at all. Half way between any two successive nodes is an antinode, where the string oscillates up and down with maximum amplitude $2A$.

16.3   Boundary Conditions; Waves on a Finite String*

* As usual, sections marked with an asterisk can be omitted on a first reading.

So far we have been assuming, implicitly, that our string is infinitely long, or at least that it is so long that we can ignore any effects of its ends. Real strings are, of course, finite in length and have ends. The motion of the string itself is governed by the same wave equation (16.4) as before, but the existence of the ends imposes additional **boundary conditions** on its solutions. These boundary conditions vary with the nature of the string's ends; for instance, the string may be tied down at an end, or it may simply flap freely, and the boundary conditions appropriate to these two situations are quite different. Here we shall consider just one type of boundary condition and one method of solving for it. Specifically, we'll consider a string that is tied down at both ends (at $x = 0$ and $x = L$), and we shall solve this problem by a method analogous to our discussion of normal modes in Chapter 11.

Normal Modes

The problem that we have to solve is this: For $0 < x < L$, the displacement $u(x, t)$ of our string must satisfy the wave equation (16.4)

$$\frac{\partial^2 u}{\partial t^2} = c^2 \frac{\partial^2 u}{\partial x^2}, \tag{16.19}$$

with initial conditions that fix the position u and velocity $\dot{u}$ at $t = 0$; in addition, it must satisfy the boundary conditions at $x = 0$ and $x = L$ that

$$u(0, t) = u(L, t) = 0 \tag{16.20}$$

for all times t. Of the several ways to solve this problem, we shall follow the approach of Chapter 11; that is, we shall start by looking for solutions that vary sinusoidally in time, with the form

$$u(x, t) = X(x) \cos(\omega t - \delta), \tag{16.21}$$

where the function $X(x)$ and the constants ω and δ are to be determined. As usual, there is nothing to stop us seeking solutions with this form, and, as before, we shall find that such solutions do exist and that *any* solution of our problem can be written in terms of them.

Substitution of the assumed form (16.21) into the wave equation (16.19) reduces the latter to the form

$$-\omega^2 X(x) \cos(\omega t - \delta) = c^2 \frac{d^2 X(x)}{dx^2} \cos(\omega t - \delta)$$

or

$$\frac{d^2 X(x)}{dx^2} = -k^2 X(x) \tag{16.22}$$

where

$$k = \frac{\omega}{c}. \tag{16.23}$$

We see that the assumption of the sinusoidal time dependence (16.21) has reduced the partial differential equation (16.19) to the ordinary differential equation[2] (16.22) — an equation, moreover, whose solutions we can easily write down.

The general solution of (16.22) is just

$$X(x) = a\cos(kx) + b\sin(kx) \tag{16.24}$$

and this yields a solution of the wave equation (16.19) for any choice of the constants a and b. However, we have still to satisfy the boundary conditions (16.20), which require that

$$X(0) = X(L) = 0. \tag{16.25}$$

The condition that $X(0) = 0$ requires simply that the coefficient a in (16.24) be zero, so that $X(x) = b\sin(kx)$. Thus, the condition that $X(L) = 0$ requires that either $b = 0$ or that $\sin(kL) = 0$. In the former case, our solution is identically zero, and the string doesn't move — a solution, but a trivial one. If $\sin(kL) = 0$, then kL must be an integer multiple of π, and we get a nontrivial solution,

$$u(x, t) = \sin(kx)A\cos(\omega t - \delta), \tag{16.26}$$

where the boundary conditions have forced k to have one of the values

$$k = k_n = n\frac{\pi}{L} \qquad [n = 1, 2, 3, \cdots]. \tag{16.27}$$

By (16.23), $\omega = ck$, so the corresponding frequency ω must have the form

$$\omega = \omega_n = n\frac{\pi c}{L} \qquad [n = 1, 2, 3, \cdots]. \tag{16.28}$$

We conclude that there are indeed solutions in which the string oscillates sinusoidally with a single frequency ω, provided ω has one of the values (16.28). This result is reminiscent of Chapter 11, where we found that a system of n coupled oscillators could oscillate in any of various sinusoidal *normal modes* with frequencies $\omega_1, \cdots, \omega_n$. The main difference is that the systems of Chapter 11 had a finite number of degrees of freedom and an equal number of normal frequencies. Here, our string has an infinite number of degrees of freedom and an infinite number of normal frequencies as in (16.28). Figure 16.6 shows the three normal modes of lowest frequency for our string — the **fundamental** and the first two **overtones**. If you compare these pictures with Figure 16.5, you will see that each of the normal modes of our finite string is just a section of a standing wave on an infinite string. That our finite string is fixed at its two ends means that the points $x = 0$ and $x = L$ must be nodes, which requires that the length L must equal an integer number of half wavelengths, $L = n\lambda/2$. Since $\lambda = 2\pi/k$, this explains the condition (16.27) that $k = n\pi/L$.

The allowed frequencies (16.28) of our string are all integer multiples of the lowest frequency, $\omega_n = n\omega_1$. They are, among many other things, the frequencies at which any

---

[2] Our method of solution here is closely related to the method of *separation of variables*. See Problem 16.9.

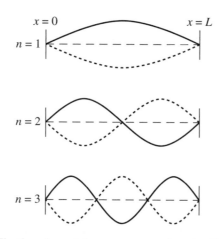

Figure 16.6   The three lowest-frequency normal modes (16.26) of a string of length L fixed at both ends. In each picture the solid curve, the long dashes, and the short dashes show the string at three successive times, separated by a quarter cycle. The $n = 1$ mode is called the fundamental.

stringed musical instrument, such as a piano or guitar, can vibrate. The corresponding modes, including the fundamental, are called the **harmonics** of the string, since they "harmonize" well (that is, make a pleasing sound to the ear) when played together.

The General Solution

The normal modes (16.26) determine *all* possible motions of our finite string, in the sense that any possible motion can be expanded in terms of the normal-mode solutions. To see this we shall need to use some of the properties of Fourier series described in Section 5.7. First, let us note that any motion of the string is given by a function $u(x, t)$ that satisfies the wave equation (16.19) and the boundary conditions (16.20), and is determined by its initial position $u(x, 0) = u_0(x)$ and velocity $\dot{u}(x, 0) = \dot{u}_0(x)$. To see that any such solution can be expanded in terms of the normal modes, we first rewrite the normal-mode solution (16.26) in the "sine plus cosine" form as

$$u(x, t) = \sin k_n x (B_n \cos \omega_n t + C_n \sin \omega_n t). \tag{16.29}$$

Our claim is that any possible motion can be written as a linear combination of these normal-mode solutions:

$$u(x, t) = \sum_{n=1}^{\infty} \sin k_n x (B_n \cos \omega_n t + C_n \sin \omega_n t). \tag{16.30}$$

To prove this, we note first that this linear combination certainly satisfies the wave equation and the boundary conditions that u vanish at $x = 0$ and $x = L$. At time $t = 0$, the claimed solution is just

$$u(x, 0) = \sum_{n=1}^{\infty} B_n \sin k_n x. \tag{16.31}$$

This is a Fourier sine series, and the coefficients B_n can be chosen so the $u(x, 0)$ matches any given initial value $u_o(x)$.[3] Similarly, the velocity of the proposed solution (16.30) is

$$\dot{u}(x, 0) = \sum_{n=1}^{\infty} \omega_n C_n \sin k_n x \qquad (16.32)$$

and we can choose the coefficients C_n so that this is equal to any given initial velocity $\dot{u}_o(x)$. We conclude that the proposed solution satisfies the equation of motion and the boundary conditions, and that by choice of the coefficients B_n and C_n can match any given initial conditions. Therefore, any possible motion of the string can be expanded in terms of the normal modes as in (16.30).

Since all of the frequencies in (16.30) are integer multiples of the lowest frequency ($\omega_n = n\omega_1$), every term in (16.30) is periodic with period $\tau = 2\pi/\omega_1$. Therefore, every possible motion of our finite string is periodic with this period. [Of course, if certain coefficients in (16.30) are zero, the motion may be periodic with a shorter period as well, but *every* solution has the period of the fundamental mode.]

EXAMPLE 16.2   A Triangular Wave on a Finite String

A string of length $L = 8$ is fixed at both ends. It is given a small triangular displacement, as in Figure 16.7 and released from rest at $t = 0$. Find the Fourier coefficients B_n and C_n in the expansion (16.30) and using some reasonable finite number of terms to approximate the infinite series, plot the position of the string at five equally spaced times from $t = 0$ to $t = \tau/2$, where τ is the period of the motion.

Since the string is initially at rest, all of the coefficients C_n are zero. The coefficients B_n are given by the integral

$$B_n = \frac{2}{L} \int_0^L u_o(x) \sin \frac{n\pi x}{L} \, dx. \qquad (16.33)$$

[This is nearly, but not quite, the standard formula (5.84); for details, see Problem 16.13.] It is easy to see that this is zero when n is even. When n is odd, we can write $n = 2m + 1$ and you can check (Problem 16.10) that

$$B_{2m+1} = (-1)^m \frac{32}{(2m + 1)^2 \pi^2} \left(1 - \cos \frac{(2m + 1)\pi}{8}\right). \qquad (16.34)$$

Putting these coefficients into the expansion (16.30) and choosing some reasonable finite number of term, we can get a good approximation for the displacement $u(x, t)$ for all times t. Using just the first five or so terms, we get a moderate approximation, but since it is very little trouble (for a computer) to use more terms, I chose to use the sum of the first twenty. The results are shown in

---

[3] There is a small complication that I am ignoring here. The series (16.31) is not the usual Fourier series since it is missing any cosine terms. However, it contains twice as many sine terms as the usual Fourier series since $k_n = n\pi/L$ (as opposed to the usual $2n\pi/L$), and one can prove that this series can be used to expand any (reasonable) function on the interval $0 \leq x \leq L$. See Problem 16.13.

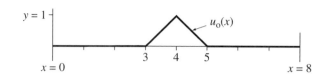

Figure 16.7   A string is released from rest at $t = 0$ in the triangular position shown.

Figure 16.8. Each of these five plots deserves careful attention. The first shows the initial displacement of Figure 16.7, as approximated by the first 20 terms of its Fourier series. The approximation is very good, although it fails to reproduce the sharp turn at the apex. (Obviously, no finite sums of sines or cosines can actually have an instantaneous change of slope.) In the second picture, the initial triangle has split into two separate triangles, traveling in opposite directions. This is exactly the behavior we saw in Example 16.1 (Figure 16.4); because neither of the waves has reached the boundaries at $x = 0$ and L, the motion is so far unaffected by the presence of the boundaries. Skipping for a moment over the third picture, you can see in the fourth that each triangle has been reflected by the walls, and is now traveling back toward the center, although they are now inverted. In the third picture, the original and the reflected waves are both present and are interfering destructively to produce zero net displacement. Finally, in the last picture, the two reflected waves have coalesced momentarily into a single inverted triangle. If we were to follow the motion further, we would see the two reflected waves continue on until they hit the opposite walls and reflect again. (See Problem 16.11.)

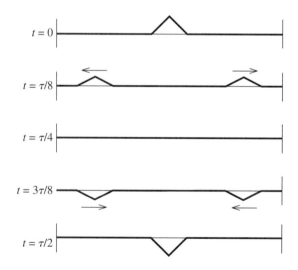

Figure 16.8   Five successive snapshots of a string that was released from the initial position of Figure 16.7, calculated using the first twenty nonzero terms of the Fourier series (16.30). The first picture shows the initial position (approximated by 20 terms of its Fourier series). The four succeeding pictures show the position at intervals of $\tau/8$, where τ is the period of the fundamental.

16.4 The Three-Dimensional Wave Equation

We live in a three-dimensional world, and the wave equation (16.4)

$$\frac{\partial^2 u}{\partial t^2} = c^2 \frac{\partial^2 u}{\partial x^2} \tag{16.35}$$

needs to be generalized to three dimensions. It is not hard to guess what the proper generalization should be. If $p = p(x, y, z, t) = p(\mathbf{r}, t)$ denotes some sort of disturbance in a three-dimensional system (for example, the pressure in a sound wave traveling through the air), then we would surely guess that the appropriate generalization of (16.35) should be

$$\frac{\partial^2 p}{\partial t^2} = c^2 \left(\frac{\partial^2 p}{\partial x^2} + \frac{\partial^2 p}{\partial y^2} + \frac{\partial^2 p}{\partial z^2} \right). \tag{16.36}$$

We shall meet a couple of examples of disturbances that do indeed satisfy this *three-dimensional wave equation* later in this chapter. In particular, I shall prove in Section 16.13 that the pressure[4] in any inviscid fluid (for example, air) is an example, for which the wave speed c is given by

$$c = \sqrt{\frac{\text{BM}}{\varrho_0}}, \tag{16.37}$$

where BM denotes the *bulk modulus* of the fluid, and ϱ_0 is the equilibrium density. (I'll define the bulk modulus shortly. For the moment, just see it as a parameter that characterizes the resistance of the fluid to compression.)

It is usual to streamline the notation in the wave equation (16.36): If, as usual, we take the view that ∇ is the "vector" with components

$$\nabla = \left(\frac{\partial}{\partial x}, \frac{\partial}{\partial y}, \frac{\partial}{\partial z} \right)$$

then the scalar product of ∇ with itself is obviously

$$\nabla^2 = \nabla \cdot \nabla = \left(\frac{\partial}{\partial x} \right)^2 + \left(\frac{\partial}{\partial y} \right)^2 + \left(\frac{\partial}{\partial z} \right)^2. \tag{16.38}$$

You may well have met this differential operator before, perhaps in your study of electromagnetism. It plays a huge role in many subjects — electromagnetism, quantum mechanics, fluid mechanics, elasticity, thermodynamics, and more — and

---

[4] Strictly speaking the pressure p discussed throughout this section is the *incremental* pressure, the difference between the total pressure and the equilibrium atmospheric pressure.

is called the **Laplacian** for its role in Laplace's equation of electrostatics. With this notation, we can rewrite the **three-dimensional wave equation** (16.36) as

$$\frac{\partial^2 p}{\partial t^2} = c^2 \, \nabla^2 p. \tag{16.39}$$

Plane Waves

The equation (16.39) has many, many solutions, of which the simplest are the so-called plane waves. A simple example of a plane wave is a solution of (16.39) [or (16.36)] which is independent of y and z,

$$p(\mathbf{r}, t) = p(x, t).$$

Obviously a disturbance with this form has the same value at all points in any plane $x = $ constant. If we substitute this form into (16.36), the derivatives with respect to y and z drop out, and we are left with the one-dimensional wave equation

$$\frac{\partial^2 p}{\partial t^2} = c^2 \frac{\partial^2 p}{\partial x^2}$$

whose most general solution we already know to be $p = f(x - ct) + g(x + ct)$. In particular, a solution $p = f(x - ct)$ is a plane disturbance (p constant in any plane perpendicular to the x axis) that is traveling in the x direction with speed c (hence the name "plane wave").

Similarly, a solution of the form $p = f(y - ct)$ or $f(z - ct)$ is a plane wave traveling in the y or z direction. More generally, if $\mathbf{n}$ denotes an arbitrary unit vector, then a disturbance of the form

$$p = f(\mathbf{n} \cdot \mathbf{r} - ct) \tag{16.40}$$

satisfies the wave equation (16.39), is constant in any plane perpendicular to $\mathbf{n}$, and travels in the direction $\mathbf{n}$ with speed c. (See Problem 16.15.) If the function f is a sinusoidal function, $f(\xi) = \cos k\xi$, say, the wave (16.40) is the sinusoidal plane wave

$$p = \cos[k(\mathbf{n} \cdot \mathbf{r} - ct)], \tag{16.41}$$

whose crests lie in planes perpendicular to $\mathbf{n}$ and travel with speed c in the direction of $\mathbf{n}$, as shown in Figure 16.9. This kind of wave is easier to visualize in two dimensions. You could think, for example, of "plane" waves on the surface of a pond.

A plane wave is a mathematical idealization that never occurs in practice, since no real disturbance can be constant over an infinite plane. Nevertheless, it is frequently a very useful approximation. The light shining on us from the sun is well approximated as a plane wave, as is the sound from a distant explosion.

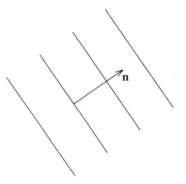

Figure 16.9   The sinusoidal plane wave (16.41). The wave crests (or wavefronts) are planes perpendicular to the unit vector **n** and they travel with speed c in the direction of **n**.

Spherical Waves

Another important solution of the three-dimensional wave equation is a spherical wave, for example, a sound wave traveling radially outward from a small, omnidirectional loudspeaker. If we assume that such a wave is spherically symmetric, then it must have the form $p = p(r, t)$. (That is, in spherical polar coordinates, p is independent of θ and ϕ.) It is not hard to show that for a function of this form

$$\nabla^2 p = \frac{1}{r}\frac{\partial^2}{\partial r^2}(rp). \qquad (16.42)$$

[The obvious way to prove this is to evaluate the left side using the definition (16.38) of ∇^2; the simplest is to look up the expression for ∇^2 in polar coordinates inside the back cover. See Problem 16.16.] Therefore, the wave equation becomes

$$\frac{\partial^2 p}{\partial t^2} = c^2 \frac{1}{r}\frac{\partial^2}{\partial r^2}(rp)$$

or, multiplying both sides by r,

$$\frac{\partial^2}{\partial t^2}(rp) = c^2 \frac{\partial^2}{\partial r^2}(rp). \qquad (16.43)$$

We see that for a spherical wave the function $rp(r, t)$ satisfies the one-dimensional wave equation with respect to r and t. Therefore the general solution has the form

$$rp(r, t) = f(r - ct) + g(r + ct).$$

In particular, if the function g is zero, the disturbance has the form

$$p(r, t) = \frac{1}{r}f(r - ct). \qquad (16.44)$$

The factor $f(r - ct)$ represents a disturbance traveling rigidly outward. Since this is what one might have guessed for a radially spreading wave, the question is, "Why the factor of $1/r$?" To answer this we need a result that you may recall from an introductory physics class. The **intensity** of any wave in three dimensions is defined as the power delivered by the wave to unit area perpendicular to the direction of propagation, and the intensity of a sound wave is proportional to p^2. (For guidance in proving this, see Problem 16.37.) Thus the factor $1/r$ in (16.44) implies that the intensity is proportional to $1/r^2$, which is exactly what is required for conservation of energy: At a distance r from the source, the energy of the wave is spread out over an area $4\pi r^2$. Therefore, as the wave moves radially outward, the intensity *has* to fall off like $1/r^2$ to keep the total power constant.

If the function f in (16.44) is sinusoidal, $f(\xi) = \cos k\xi$, say, then (16.44) represents a sinusoidal wave whose crests are traveling radially outward from the origin with speed c, as illustrated in Figure 16.10.

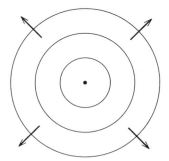

Figure 16.10   A spherical wave. The wave crests are spheres moving outward from the origin. To help visualize what this shows, it may help to think of it as a two-dimensional wave, such as the ripples created on a pond by a stick which moves in and out of the water at the origin.

16.5   Volume and Surface Forces

Our next objective is to see how to find the equation of motion of a three-dimensional continuous system. In general, this is a very complicated problem, and the details depend strongly on the precise nature of the system. For example, the equations of motion of a fluid are very different from those of an elastic solid. Nevertheless, there are some reasonably simple general principles that apply to many different continuous systems, and these are what I shall describe next.

As you would probably guess, we find the equation of motion of a continuous body by applying Newton's second law to an arbitrary small element dV of the body. (I use the word "body" here to mean any chunk of matter, solid, liquid, or gas.) This is exactly analogous to what we did in Section 16.1, where we applied Newton's second law to a length dx of the one-dimensional string, but the three-dimensional case is naturally more complicated. We need first to discuss the geometry of the volume element dV and then the specification of the forces on, and resulting displacement of, dV.

Elements of Volume

The shape of the volume element on which we focus is arbitrary. It could be spherical, or rectangular (in the shape of a brick), or anything else our whim dictates. For simplicity, you could have in mind a simple rectangular volume, as in Figure 16.11. The volume on which we focus will usually be an infinitesimal volume, as the label dV is intended to suggest. The surface that bounds dV could be the actual boundary of the continuous body (for example, the walls of the cylinder containing a gas), but it will usually be an "imaginary" surface, that is, an arbitrary surface interior to the whole body. The whole bounding surface is naturally a *closed surface* that divides all of space into exactly two parts, the "inside" (namely, dV) and the "outside" (everything else). This means that we can specify the orientation of any part of the surface, such as the face S in Figure 16.11, by the unit vector **n**, normal to the surface and pointing outward from dV.[5]

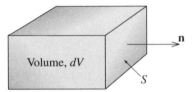

Figure 16.11   A small element of volume dV of a continuous body can have any shape, but a convenient choice is the rectangular shape shown here. The orientation of any part S of the surface (the right-hand end here) is specified by a unit vector **n** that points outward from dV.

Forces on the Volume Element

The two most important kinds of force on a volume element dV of a continuous system are called *volume forces* and *surface forces*. An example of a volume force is the force of gravity, $\mathbf{F} = \varrho \mathbf{g} dV$, where ϱ is the mass density of the material and **g** is the acceleration of gravity. A second example is the electrostatic force $\mathbf{F} = \varrho \mathbf{E} dV$ of an electric field **E** on a material with charge density ϱ. The definition of a **volume force** is simply that it is a force proportional to the volume dV. Volume forces are generally the result of an external field (such as gravity), and to be definite I shall usually assume that the only volume force is that of gravity. In any event, the body forces are almost always known and well understood. Therefore, our main concern is with the surface forces.

A familiar example of a surface force is the force $p\,dA$ of the pressure p in a fluid on a small surface element of area dA. The definition of a **surface force** is a force proportional to the area dA of the surface on which it acts. Surface forces are generally

---

[5] When one is concerned with a nonclosed surface, the orientation of a small part S can still be specified by a unit vector **n** that is normal to S, but this leaves an ambiguous sign (since **n** and $-\mathbf{n}$ both fit the definition). Fortunately, in this chapter S will always be a part of a closed surface, so we can insist, simply and unambiguously, that **n** point *outward*.

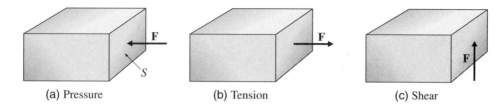

(a) Pressure   (b) Tension   (c) Shear

Figure 16.12   Three different surface forces on the face S of a rectangular volume. **(a)** Hydrostatic pressure acts normally to the surface and inward, so that $\mathbf{F} = -p\mathbf{n}\,dA$, with the minus sign since $\mathbf{F}$ is inward whereas $\mathbf{n}$ is outward. **(b)** A simple tension in the direction normal to S. **(c)** By definition, a shearing force acts tangentially to S.

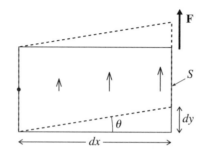

Figure 16.13   A shearing force $\mathbf{F}$ applied to the face S of the rectangular solid of Figure 16.12 (seen here from directly in front). If the opposite face is held fixed, the shear produces the motion shown, in which the planes parallel to S all move in the direction of $\mathbf{F}$, changing the originally rectangular cross section into a parallelogram. The distances dx and dy are used to define the shearing strain in Equation (16.54) below.

the result of intermolecular forces of the molecules just outside the surface acting on those just inside. Figure 16.12 shows three important special cases of surface forces, a pressure force, a simple tension, and a shearing force. Notice that both the pressure force and the tension act normally to the surface S, whereas the shearing force, by definition, acts tangentially to S. The tendency of a shearing force is to produce the shearing motion shown in Figure 16.13.

When Is Pressure Isotropic?

To conclude this section, I shall prove a result that you probably learned in an introductory physics class, that the pressure in any static fluid acts equally in all directions, or, briefly, that the pressure is isotropic. Actually, the result is a bit more general than this, and I shall prove it in its greater generality. A characteristic property of any fluid is that it can support no shearing forces in equilibrium, and the absence of shearing forces is in fact the essential feature that leads to the isotropy of pressure. I shall prove in a moment that in any substance where there are no shearing forces the pressure is isotropic. Clearly the result applies to any static fluid, but it also applies

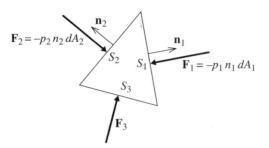

Figure 16.14   The surfaces S_1 and S_2 (seen edge on) are normal to the two arbitrary unit vectors $\mathbf{n_1}$ and $\mathbf{n_2}$. They are identical rectangles and, together with S_3, form an isosceles prism (seen end on). The three forces $\mathbf{F}_1$, $\mathbf{F}_2$, and $\mathbf{F}_3$ are the surface forces on the three faces and are normal to the surfaces since, by assumption, there are no shearing forces.

to a moving fluid, provided there are no shearing forces. Since the cause of shearing forces in fluids is viscosity, we can say that pressure is isotropic even in moving fluids, *provided the viscosity is zero*. Of course, there are very few fluids whose viscosity is exactly zero, but there are plenty of situations where the viscosity is small enough to be negligible, and in such situations the pressure is effectively isotropic. Our result is obviously very useful, but, more important, the method of proof has many other applications, as we shall see.

Let us consider, then, any medium in which there are no shearing forces, and let $\mathbf{n_1}$ and $\mathbf{n_2}$ be any two directions. At any particular point in the medium, let us construct two small, equal rectangular surfaces, S_1 normal to $\mathbf{n_1}$ and S_2 normal to $\mathbf{n_2}$, so as to form a small triangular prism, as in Figure 16.14. The surface forces on the three faces shown are normal to the faces and can be written as $\mathbf{F}_1 = -p_1 \mathbf{n}_1 \, dA_1$ and so on, where we write the pressures on the three faces as p_1, p_2, and p_3 to allow for the possibility that they might be different. (Our aim is to prove that, in fact, $p_1 = p_2 = p_3$.) These pressures can, of course, vary from point to point in the medium, but by considering a small enough volume we can ensure that they vary by a negligible amount within our volume. We are now ready to apply Newton's second law to our small prism. The mass of the prism is $m = \varrho \, dV$. The net force on the prism is $\mathbf{F}_1 + \mathbf{F}_2 + \mathbf{F}_3 + \mathbf{F}_{\text{vol}}$ where $\mathbf{F}_{\text{vol}}$ denotes the total volume force (for example, the weight $\mathbf{F}_{\text{vol}} = \varrho \mathbf{g} \, dV$). Thus the equation $\mathbf{F} = m\mathbf{a}$ becomes[6]

$$\mathbf{F}_1 + \mathbf{F}_2 + \mathbf{F}_3 + \mathbf{F}_{\text{vol}} = m\mathbf{a}$$

which we can rewrite as

$$\mathbf{F}_1 + \mathbf{F}_2 + \mathbf{F}_3 = m\mathbf{a} - \mathbf{F}_{\text{vol}}. \tag{16.45}$$

---

[6] Strictly speaking we should include the two pressure forces on the two ends of our prism, but we shall be concerned only with the components of this equation in the plane of the picture in Figure 16.14, so we can ignore these.

Now comes the supreme act of cunning. Equation (16.45) applies to the small prism of Figure 16.14, but it would certainly also apply to a smaller prism. Let us therefore shrink our prism by a factor of λ in all three directions. The three surface terms on the left of (16.45) are proportional to area, so they will decrease by a factor of λ^2. Both the mass and volume force on the right are proportional to dV and must decrease by a factor of λ^3. Thus the counterpart of (16.45) for our smaller prism is

$$\lambda^2(\mathbf{F}_1 + \mathbf{F}_2 + \mathbf{F}_3) = \lambda^3(m\mathbf{a} - \mathbf{F}_{\text{vol}})$$

or, dividing both sides by λ^2,

$$(\mathbf{F}_1 + \mathbf{F}_2 + \mathbf{F}_3) = \lambda(m\mathbf{a} - \mathbf{F}_{\text{vol}}). \tag{16.46}$$

This equation holds for any value of λ (smaller than 1). In particular, we can let λ approach zero, and we reach the surprising conclusion that the three surface terms must sum to zero by themselves,

$$\mathbf{F}_1 + \mathbf{F}_2 + \mathbf{F}_3 = 0. \tag{16.47}$$

It is easy to check that, because the triangle in Figure 16.14 is isosceles, this requires that $\mathbf{F}_1$ and $\mathbf{F}_2$ must have equal magnitudes, $F_1 = F_2$. (Just take components perpendicular to $\mathbf{F}_3$ to check this.) Since $F_1 = p_1 dA_1$ and $F_2 = p_2 dA_2$, and $dA_1 = dA_2$, we conclude that

$$p_1 = p_2. \tag{16.48}$$

Since the directions $\mathbf{n}_1$ and $\mathbf{n}_2$ were arbitrary, we have proved that pressure is independent of direction in any medium where there are no shearing forces. In particular, pressure is isotropic in any static fluid and also in any moving fluid that has negligible viscosity.

16.6   Stress and Strain: The Elastic Moduli

As we shall see in the next section, the surface forces inside a continuous body (solid, liquid, or gas) can be expressed in terms of a three-dimensional tensor called the stress tensor. In Section 16.8, we'll see that the resulting displacements of the body can be expressed in terms of a second tensor called the strain tensor. Finally, before we can write down the equation of motion we need to establish the relationship between the two tensors. [This last statement is the continuum analog of the familiar requirement that to write down the equation of motion of a mass on a spring, we need to know Hooke's law ($F = kx$), which relates the tension in the spring (F) to the extension (x).] The general theory of the stress and strain tensors is quite complicated, so in this section I shall mention a few simple special cases before we plunge into the general case.

Stress

Since any surface force F is proportional to the area A of the surface on which it acts, it is natural to consider the ratio F/A, and this ratio is called the **stress**. As we shall see in the next section, in general we need to discuss a *stress tensor*, but a simple example that lacks this complication is the pressure force in a static fluid, for which the stress is just the pressure:

$$\text{stress} = \frac{F}{A} = \text{pressure, } p \qquad \text{[in a static fluid].} \qquad (16.49)$$

Similarly, you may recall that the stress in a wire or rod subjected to a simple tension is defined as

$$\text{stress} = \frac{\text{tension}}{\text{area}} \qquad \text{[for a wire in tension]} \qquad (16.50)$$

where the area is the cross-sectional area of the wire. For a simple shearing force, like that in Figure 16.13, the shearing stress is defined as

$$\text{stress} = \frac{\text{shearing force}}{\text{area}} \qquad \text{[for a shear].} \qquad (16.51)$$

However complicated the situation, we shall find that the stress (or any component of the stress tensor) can be defined as the ratio of a surface force (or one of its components) to the area of the surface on which it acts. In particular, stress always has the dimensions of [force/area].

Strain

The result of a stress is almost always a deformation, or change in the dimensions, of the body on which the stress acted — a change in the volume of a liquid, or the length of a wire, for instance. When this change is expressed as a *fractional change* it is called the **strain**. For example, in a static fluid, the strain would be the fractional change in volume,

$$\text{strain} = \frac{dV}{V} \qquad \text{[in a static fluid].} \qquad (16.52)$$

For a wire under tension, the strain would be the fractional change in length,

$$\text{strain} = \frac{dl}{l} \qquad \text{[for a wire in tension].} \qquad (16.53)$$

For the simple shearing force of Figure 16.13, the strain is defined as

$$\text{strain} = \frac{dy}{dx} \qquad \text{[for a shear]} \qquad (16.54)$$

where dy is the displacement in the direction of shear and dx is the perpendicular distance across which the shear occurs (see Figure 16.13).

Relation of Stress to Strain: The Elastic Moduli

When the stresses in a medium are not too large, we would expect that the resulting strain would be linear in the stress. In the case of a stretched wire, this relation is written as

$$\text{stress} = (\text{Young's modulus}) \times \text{strain} \qquad \text{or} \qquad \frac{dF}{A} = \text{YM} \cdot \frac{dl}{l} \quad (16.55)$$

where YM is **Young's modulus** for the material of the wire.[7] [If you rewrite this equation as $F = (A\,\text{YM}/l)dl$, you will recognize it as Hooke's law, with the force constant $k = (A\,\text{YM}/l)$. The advantage of writing it in terms of Young's modulus is that YM, unlike k, is characteristic of the material and independent of the dimensions A and l.] In Equation (16.55), dl is the extension caused by an increment dF in the tension.

For any material subject to hydrostatic pressure only, a small increase dp in pressure will cause a change in volume given by

$$\text{stress} = (\text{bulk modulus}) \times \text{strain} \qquad \text{or} \qquad dp = -\text{BM} \cdot \frac{dV}{V} \quad (16.56)$$

where BM is the **bulk modulus** for the material, and the minus sign is because an increase in pressure causes a decrease in volume. For a shearing stress,

$$\text{stress} = (\text{shear modulus}) \times \text{strain} \qquad \text{or} \qquad \frac{F}{A} = \text{SM} \cdot \frac{dy}{dx} \quad (16.57)$$

where SM is the **shear modulus** for the material.

To summarize this section, *stress* characterizes the surface forces in a continuous medium,

$$\text{stress} = \frac{\text{force}}{\text{area}} \quad (16.58)$$

while *strain* characterizes the resulting deformation,

$$\text{strain} = \text{fractional deformation.} \quad (16.59)$$

---

[7] There are almost as many notations for the various elastic moduli as there are books on the subject. The notations used here are unconventional, but you will, I hope, be able to remember which is which.

While stress always has the units of pressure (force/area), strain is always dimensionless.[8] The various elastic moduli (Young's, bulk, and shear) are the ratios of stress to the corresponding strain,

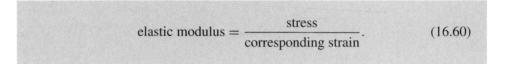

$$\text{elastic modulus} = \frac{\text{stress}}{\text{corresponding strain}}. \qquad (16.60)$$

16.7   The Stress Tensor

In this section I shall derive the general expression for the surface force on a small area dA of a closed surface S in a continuous medium. As usual, we shall use $\mathbf{n}$ to denote the unit outward normal to S at the location of dA. To streamline the notation, I shall define a vector $d\mathbf{A}$ in the direction of $\mathbf{n}$, with magnitude dA. That is,

$$d\mathbf{A} = \mathbf{n}\, dA. \qquad (16.61)$$

This vector tells us the orientation and size of the small piece of surface under consideration. Our first task is to show that the surface force $\mathbf{F}(d\mathbf{A})$ on the surface element specified by $d\mathbf{A}$ is in fact a linear function of $d\mathbf{A}$, that is, that

$$\mathbf{F}(\lambda_1\, d\mathbf{A}_1 + \lambda_2\, d\mathbf{A}_2) = \lambda_1\mathbf{F}(d\mathbf{A}_1) + \lambda_2\mathbf{F}(d\mathbf{A}_2) \qquad (16.62)$$

where λ_1 and λ_2 are any two real numbers and $d\mathbf{A}_1$ and $d\mathbf{A}_2$ are any two vectors.

The force $\mathbf{F}(d\mathbf{A})$ is independent of the precise shape of surface element. On the other hand, it is proportional to the area dA, so

$$\mathbf{F}(\lambda\, d\mathbf{A}) = \lambda\mathbf{F}(d\mathbf{A}) \qquad (16.63)$$

for any positive number λ (not too large). If we were to replace $d\mathbf{A}$ by $-d\mathbf{A}$, this would interchange the inside and outside of our surface, and by Newton's third law this would change the sign of the surface force; that is, $\mathbf{F}(-d\mathbf{A}) = -\mathbf{F}(d\mathbf{A})$. Therefore, Equation (16.63) actually holds for negative, as well as positive, values of λ.

Let us next consider any two small vectors $d\mathbf{A}_1$ and $d\mathbf{A}_2$. At any point in our continuous medium, consider two small rectangular surfaces touching along one common edge, with orientations and areas given by $d\mathbf{A}_1$ and $d\mathbf{A}_2$, as shown in Figure 16.15. Consider now the triangular prism defined by these two rectangles and the third rectangle labeled as $d\mathbf{A}_3$ in the figure. This prism has two remarkable properties. First,

---

[8] There is no very obvious way to remember which is stress and which strain. One possibility is this: In everyday language we say "stress causes strain" and likewise "force causes deformation." Thus "stress" goes with "force" and "strain" with "deformation." Alternatively, note that alphabetically, "stress" comes after "strain" just as "force" comes after "deformation."

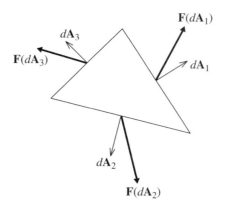

Figure 16.15   The arbitrary small vectors $d\mathbf{A}_1$ and $d\mathbf{A}_2$ define two rectangular surfaces (seen edge on) which meet at one common edge (bottom right). These two rectangles define a triangular prism (seen end on) whose third rectangular face is labeled by the vector $d\mathbf{A}_3$. The surface forces on the three rectangular faces are shown as $\mathbf{F}(d\mathbf{A}_1)$ and so on.

because the three edges shown form a closed triangle, the same is true of the three vectors $d\mathbf{A}_1$, $d\mathbf{A}_2$, and $d\mathbf{A}_3$. Therefore

$$d\mathbf{A}_1 + d\mathbf{A}_2 + d\mathbf{A}_3 = 0.$$

Second, by the same argument as we used to prove the isotropy of pressure in an inviscid fluid [Equation (16.47)], we can prove that[9]

$$\mathbf{F}(d\mathbf{A}_1) + \mathbf{F}(d\mathbf{A}_2) + \mathbf{F}(d\mathbf{A}_3) = 0.$$

Exploiting these last two equations in turn, we find that

$$\mathbf{F}(d\mathbf{A}_1 + d\mathbf{A}_2) = \mathbf{F}(-d\mathbf{A}_3) = -\mathbf{F}(d\mathbf{A}_3)$$
$$= \mathbf{F}(d\mathbf{A}_1) + \mathbf{F}(d\mathbf{A}_2). \tag{16.64}$$

Finally, combining (16.63) and (16.64), we can immediately verify (16.62), and we have proved that the force $\mathbf{F}(d\mathbf{A})$ is linear in $d\mathbf{A}$.

It is a fundamental result of linear algebra, that if one vector ($\mathbf{F}$ in this case) is a linear function of a second vector ($d\mathbf{A}$), then the components of the first are related to those of the second by a linear relation of the form[10]

$$F_i(d\mathbf{A}) = \sum_{j=1}^{3} \sigma_{ij} dA_j \tag{16.65}$$

---

[9] I am again ignoring the forces on the two triangular ends. If we denote the two ends by $d\mathbf{A}_4$ and $d\mathbf{A}_5$, then $d\mathbf{A}_4 = -d\mathbf{A}_5$, so, by (16.63), $\mathbf{F}(d\mathbf{A}_4) = -\mathbf{F}(d\mathbf{A}_5)$ and these two forces cancel each other.

[10] The proof is quite easy: Suppose that $\mathbf{u}$ is a linear function of $\mathbf{v}$. Since $u_i = \mathbf{e}_i \cdot \mathbf{u}$ and $\mathbf{v} = \sum_j \mathbf{e}_j v_j$, it follows that $u_i(\mathbf{v}) = \mathbf{e}_i \cdot \mathbf{u}(\sum_j \mathbf{e}_j v_j) = \sum_j [\mathbf{e}_i \cdot \mathbf{u}(\mathbf{e}_j)] v_j$, which has the advertized form (16.65) with $\sigma_{ij} = \mathbf{e}_i \cdot \mathbf{u}(\mathbf{e}_j)$.

or, in matrix form,

$$\mathbf{F}(d\mathbf{A}) = \Sigma\, d\mathbf{A}. \qquad (16.66)$$

In this second relation (16.66), Σ denotes a (3×3) matrix,[11] made up of the nine numbers σ_{ij} of (16.65). The matrix Σ defines a second-rank, three-dimensional tensor called the **stress tensor**. The stress tensor can, of course, vary from point to point in the medium, but its significance at each point $\mathbf{r}$ is this: For every point $\mathbf{r}$ in the medium (and at each given time t), there is a unique (3×3) matrix Σ which gives the force on any surface element $d\mathbf{A}$ at $\mathbf{r}$ via the relation (16.66).

The Elements of the Stress Tensor

The mathematical significance of the stress tensor Σ and its elements σ_{ij} cannot be more succinctly expressed than in the relations (16.66) and (16.65), but to get a feel for their physical significance it helps to look at some special cases. For example, let us consider a small area dA normal to the x axis, for which $d\mathbf{A} = \mathbf{e}_1 dA$. Since only one of the components of $d\mathbf{A}$ is nonzero (namely, the first component), the sum in (16.65) reduces to a single term. For example, the x component of (16.65) reads

$$F_1(\text{on area } dA \text{ normal to } \mathbf{e}_1) = \sigma_{11} dA.$$

Turning this around, we can say that σ_{11} is the first component of the force per area on a surface perpendicular to the first (x) axis. In the same way, we conclude that σ_{ii} is the ith component of the force per area on a surface perpendicular to the ith axis. To put this another way, a diagonal element σ_{ii} of the stress tensor, Σ, gives the normal component of the force per area on a surface perpendicular to the ith axis.

The off-diagonal elements σ_{ij} $(i \neq j)$ can be similarly interpreted. Consider again the case of a small area dA normal to the x axis, for which the second component of (16.65) reads

$$F_2(\text{on area } dA \text{ normal to } \mathbf{e}_1) = \sigma_{21} dA.$$

Evidently σ_{21} is the second (y) component of the force per area on a surface perpendicular to the first (x) axis. A similar argument applies to σ_{31}, and we can say that σ_{21} and σ_{31} are the two components of the tangential or shearing force per area on a surface perpendicular to the first axis. More generally the six off-diagonal elements σ_{ij} $(i \neq j)$ tell us the six shearing forces on the three coordinate planes through the point under consideration.

---

[11] Don't confuse the boldface capital Greek "sigma," Σ, in (16.66) with the summation sign in (16.65).

The Stress Tensor Is Symmetric

The six off-diagonal elements σ_{ij} are not actually independent, since they are equal in pairs. Specifically, the stress tensor Σ is symmetric, so that $\sigma_{ij} = \sigma_{ji}$, as we can now prove using, yet again, the argument introduced at the end of Section 16.5 to prove the isotropy of pressure in an inviscid fluid. This time, we'll consider a small prism, whose axis is in the z direction and whose cross section is a square, parallel to the xy plane, as shown in Figure 16.16. The prism's angular momentum about its axis satisfies

$$\frac{dL_3}{dt} = \Gamma_3, \tag{16.67}$$

where Γ_3 is the z component of the net torque on the prism. The four forces (actually components of forces) that contribute to Γ_3 are the shearing forces shown in the figure as F_a, F_b, F_c, and F_d. From (16.65), we see that $F_a = \sigma_{12}\,dA$, while $F_b = \sigma_{21}\,dA$. The forces F_c and F_d are equal in magnitude to F_a and F_b respectively, but in the opposite directions. Thus the total torque Γ_3 is

$$\Gamma_3 = F_b l - F_a l = (\sigma_{21} - \sigma_{12})l\,dA. \tag{16.68}$$

Using the now-familiar trick, we next reduce our prism by a factor of λ in all three directions. In (16.67) this reduction multiplies Γ_3 by a factor of λ^3, but L_3 by a factor of λ^4. If we divide by λ^3 and let $\lambda \to 0$, we see that Γ_3 must in fact be zero. According to (16.68), this implies that $\sigma_{21} = \sigma_{12}$. Similar arguments take care of the other off-diagonal elements, and we conclude that

$$\sigma_{ij} = \sigma_{ji} \tag{16.69}$$

for all i and j. That is, the stress tensor Σ is symmetric.

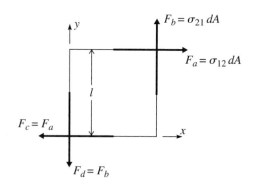

Figure 16.16   End view of a square prism with its axis parallel to the z axis. The four forces that contribute to the prism's rotation about its axis are shown as F_a, F_b, F_c, and F_d. The square ends have side l and the four faces that are seen end-on here have area dA.

EXAMPLE 16.3   The Stress Tensor in a Static Fluid

Write down the stress tensor $\mathbf{\Sigma}$ in a static fluid at a point where the pressure is p.

We know that in a static fluid there are no shearing forces and that, at any given point, the pressure is the same in all directions. Thus the surface force on a small surface element labeled by $d\mathbf{A}$ is just

$$\mathbf{F}(d\mathbf{A}) = -p\,d\mathbf{A}$$

where p is a constant (at any one point of the fluid), independent of $d\mathbf{A}$. (The minus sign is because $d\mathbf{A}$ points outward, whereas the pressure force is inward.) Comparing this with the definition (16.66) of the stress tensor, we see that

$$\mathbf{\Sigma} = -p\mathbf{1} \tag{16.70}$$

where $\mathbf{1}$ is the (3×3) unit matrix. This beautifully simple result expresses succinctly that in a static fluid (and also a moving fluid provided the viscosity is negligible) the only surface force is the pressure force, which is normal to the surface and independent of the surface's orientation.[12]

EXAMPLE 16.4   A Numerical Example of Stress

At a certain point P in a continuous medium the stress tensor has the value

$$\mathbf{\Sigma} = \begin{bmatrix} -1 & 2 & 0 \\ 2 & -2 & 0 \\ 0 & 0 & 1 \end{bmatrix}. \tag{16.71}$$

A small surface element at P has area dA and is parallel to the plane $x + y + z = 0$. What is the force on this surface element and what is the angle between this force and the normal to the surface element? To be definite, take P to be in the positive octant (x, y, z all positive), and assume that the outside of the surface is the side away from the origin.

Notice first that I have not specified the units of the components of $\mathbf{\Sigma}$, but they would of course be the units of pressure (force/area). If our medium was the water at the bottom of a river they might be kilopascals; for a rock in the earth's crust, they might be megapascals.

The force on our surface element is given by (16.66) with $d\mathbf{A} = \mathbf{n}\,dA$, where $\mathbf{n}$ is the unit normal to the surface. We are told that the surface is parallel to the plane $x + y + z = 0$, so[13]

---

[12] This example suggests a neat alternative proof of the isotropy of the pressure force. The absence of any shearing forces means that $\mathbf{\Sigma}$ must be diagonal (all off-diagonal elements equal to zero) *with respect to any choice of axes*. It is easy to show that the only tensor with this property is a multiple of the unit matrix.

[13] There are several ways to see this. One simple one is to note that the plane is given in the form $f(x, y, z) = $ constant, and it is a standard result of vector calculus that the vector ∇f is normal to a

$$\mathbf{n} = \frac{1}{\sqrt{3}} \begin{bmatrix} 1 \\ 1 \\ 1 \end{bmatrix}. \tag{16.72}$$

Therefore the force on the surface element is

$$\mathbf{F}(d\mathbf{A}) = dA \, \mathbf{\Sigma} \, \mathbf{n} = \frac{dA}{\sqrt{3}} \begin{bmatrix} 1 \\ 0 \\ 1 \end{bmatrix}. \tag{16.73}$$

The angle θ between the force and the normal is given by

$$\cos \theta = \frac{\mathbf{F} \cdot \mathbf{n}}{|\mathbf{F}| \cdot |\mathbf{n}|} = \frac{2/3}{\sqrt{2/3} \times 1} = \sqrt{\frac{2}{3}}.$$

So $\theta = \arccos(\sqrt{2/3}) = 35.3°$. This last illustrates the obvious fact that, in the presence of shearing forces, the force on a surface element is not necessarily normal to the surface.

16.8 The Strain Tensor for a Solid

In the last section we saw how the stress tensor expresses the surface forces inside a continuous medium, solid, liquid, or gas. This section presents a corresponding discussion of the strain tensor as a description of the displacements within the medium. Unfortunately, in this case, the analysis of solids is quite different from that of fluids, and for simplicity I shall confine myself to the discussion of solids.[14]

To specify the configuration of a continuous solid we must give the position of each of its continuously many constituent pieces. A convenient way to do this is to specify that the particular small volume dV that was "originally" at position $\mathbf{r}$ is now at position $\mathbf{r} + \mathbf{u}(\mathbf{r})$. The "original" position could be its equilibrium position or just the position at some convenient initial time t_o. Either way, the vector $\mathbf{u}(\mathbf{r})$ is the displacement needed to move the piece from its reference position $\mathbf{r}$ to its current position.

At first glance, you might think that $\mathbf{u}(\mathbf{r})$ would be a good measure of the strain of the body, but it isn't hard to see that this is not so. Consider for example, the possibility that $\mathbf{u}(\mathbf{r}) = \mathbf{u}_o$ is just a constant (independent of $\mathbf{r}$). Such a displacement simply moves our whole body rigidly through the vector $\mathbf{u}_o$ and requires no internal stresses at all. Stresses arise not so much from displacement of our solid as from *distortion*, and distortions require that different parts of the body are displaced by different amounts,

---

surface of this form (see Problem 4.18). Therefore, $\mathbf{n} = \pm \nabla f / |\nabla f|$. Since $\mathbf{n}$ must point away from the origin, the plus sign applies, and it is easy to check that this gives the result (16.72).

[14] It is easy to see that there has to be a big difference between solids and fluids. For example, a change in the shape of a solid certainly constitutes a strain and usually entails appreciable stresses; but a change in shape of a fluid (transferring milk from a square carton to a round bowl, for example) does not normally constitute a strain since it requires no stresses.

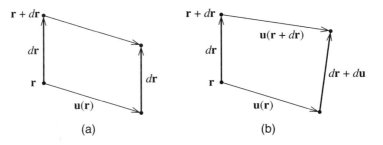

Figure 16.17   (a) In a rigid translation, all points in the body are displaced by the same amount. That is, $\mathbf{u}(\mathbf{r})$ is the same for all $\mathbf{r}$, and the separation $d\mathbf{r}$ of any two neighboring points is unchanged. (b) Any distortion of the body requires that $\mathbf{u}(\mathbf{r})$ vary from point to point. Here the points $\mathbf{r}$ and $\mathbf{r} + d\mathbf{r}$ move by different amounts, and their separation changes from $d\mathbf{r}$ to $d\mathbf{r} + d\mathbf{u}$.

as illustrated in Figure 16.17(b). Figure 16.17(a) shows a rigid translation, with $\mathbf{u}(\mathbf{r})$ the same for all $\mathbf{r}$, so that the separation, $d\mathbf{r}$, of any two neighboring points remains the same. In part (b), $\mathbf{u}(\mathbf{r})$ and $\mathbf{u}(\mathbf{r} + d\mathbf{r})$ are different, and the separation of the two neighboring points changes from $d\mathbf{r}$ to $d\mathbf{r} + d\mathbf{u}$. The change $d\mathbf{u}$ in their separation can be expressed in terms of the derivatives of $\mathbf{u}$ with respect to $\mathbf{r}$:

$$du_i = \sum_j \frac{\partial u_i}{\partial r_j} dr_j \tag{16.74}$$

or, in matrix notation,

$$d\mathbf{u} = \mathbf{D}\, d\mathbf{r} \tag{16.75}$$

where $\mathbf{D}$ is the **derivatives matrix** (or derivatives tensor) made up of the partial derivatives $\partial u_i / \partial r_j$:

$$\mathbf{D} = \begin{bmatrix} \partial u_1/\partial r_1 & \partial u_1/\partial r_2 & \partial u_1/\partial r_3 \\ \partial u_2/\partial r_1 & \partial u_2/\partial r_2 & \partial u_2/\partial r_3 \\ \partial u_3/\partial r_1 & \partial u_3/\partial r_2 & \partial u_3/\partial r_3 \end{bmatrix}. \tag{16.76}$$

The elements of the derivatives matrix $\mathbf{D}$ tell us how rapidly the displacement $\mathbf{u}(\mathbf{r})$ varies as we move around inside the solid, and you might reasonably guess that $\mathbf{D}$ would be a good measure of strain. Unfortunately there is one more complication to discuss. We have already noted that a rigid translation of our solid should not count as a strain, and the same is true of any rigid rotation. Thus, we must examine what form $\mathbf{D}$ would take for a rigid rotation and then, for an arbitrary displacement given by $\mathbf{D}$, somehow subtract out of $\mathbf{D}$ that part which corresponds to a rigid rotation, to leave something that truly represents what we want to mean by strain.

In what follows we shall be concerned only with small strains (meaning that all of the derivatives in $\mathbf{D}$ are much smaller than 1). So let us consider a small rigid rotation which we can label by a vector $\boldsymbol{\theta} = \theta\mathbf{u}$, where the unit vector $\mathbf{u}$ identifies the axis of rotation and θ is the (small) angle of rotation. It is not hard to calculate the resulting displacement of any point $\mathbf{r}$ from scratch, but we can save a little trouble by recalling Equation (9.22), $\mathbf{v} = \boldsymbol{\omega} \times \mathbf{r}$, for the velocity of a point $\mathbf{r}$ in a rigid body rotating with angular velocity $\boldsymbol{\omega}$. Multiplying both sides by a small time dt, we find that the displacement $\mathbf{u}(\mathbf{r})$ is[15]

$$\mathbf{u}(\mathbf{r}) = \mathbf{v}\,dt = \boldsymbol{\omega}\,dt \times \mathbf{r} = \boldsymbol{\theta} \times \mathbf{r} \tag{16.77}$$

since $\boldsymbol{\theta} = \boldsymbol{\omega}\,dt$. If you write out the components of this equation and differentiate, you can easily check that $d\mathbf{u} = \mathbf{D}\,d\mathbf{r}$ as in (16.75) where

$$\mathbf{D} = \begin{bmatrix} 0 & \theta_3 & -\theta_2 \\ -\theta_3 & 0 & \theta_1 \\ \theta_2 & -\theta_1 & 0 \end{bmatrix} \qquad \text{[any small rotation].} \tag{16.78}$$

That is, for any small rotation given by the vector $\boldsymbol{\theta} = (\theta_1, \theta_2, \theta_3)$ the derivatives matrix is given by the antisymmetric matrix (16.78). (A matrix $\mathbf{M}$ is antisymmetric if $\tilde{\mathbf{M}} = -\mathbf{M}$.) Conversely, any antisymmetric matrix has the form (16.78), so any antisymmetric matrix (with all its elements small) is the derivatives matrix for a small rotation. Therefore, if the derivatives matrix (16.76) is found to be antisymmetric, it corresponds to a rotation and should not be considered a strain.

To exploit the result of the last paragraph, we need to use an elementary theorem from matrix theory, that any square matrix $\mathbf{M}$ can be written as the sum of two matrices, one of which is antisymmetric and one symmetric, as we can easily verify from the following obvious identity:

$$\mathbf{M} = \tfrac{1}{2}(\mathbf{M} - \tilde{\mathbf{M}}) + \tfrac{1}{2}(\mathbf{M} + \tilde{\mathbf{M}}). \tag{16.79}$$

Since the first of these is clearly antisymmetric and the second symmetric, this proves the claimed theorem. The derivatives matrix, $\mathbf{D}$, of any displacement can be decomposed in this way as

$$\mathbf{D} = \mathbf{A} + \mathbf{E}. \tag{16.80}$$

Here $\mathbf{A}$ is the antisymmetric part of $\mathbf{D}$; it represents a rigid rotation and does not contribute to the strain. The second term $\mathbf{E}$ is called the **strain tensor**;[16] it is the symmetric part of $\mathbf{D}$,

$$\mathbf{E} = \tfrac{1}{2}(\mathbf{D} + \tilde{\mathbf{D}}) \tag{16.81}$$

---

[15] It is important that dt, and hence θ, be small. Otherwise, $\mathbf{v}$ will change appreciably during the rotation.

[16] The terminology here is hopelessly nonuniform. The "strain tensor" has many slightly different definitions and is denoted by many different symbols. About the best one can say of the usage here is that at least *some* other authors use it.

where **D** is the derivatives matrix, as defined in (16.76). The strain tensor defined in this way is a good measure of strain as the following two examples illustrate.

EXAMPLE 16.5   Dilatation

The strain tensor **E** at a certain point P in a solid is a multiple of the unit matrix,

$$\mathbf{E} = e\mathbf{1} \tag{16.82}$$

where e is a small number, positive or negative. Describe the displacement of points in the neighborhood of P (which, for convenience, we can take to be the origin).

Since we are not interested in overall rigid displacements or rotations, we may as well assume that the point P does not move and that the immediate neighborhood of P is unrotated. In this case, the antisymmetric part **A** of **D** in (16.80) is zero and $\mathbf{D} = \mathbf{E}$. Thus a point at position $d\mathbf{r}$ (relative to P) is displaced to $d\mathbf{r} + d\mathbf{u}$, as illustrated in Figure 16.18, where $d\mathbf{u} = \mathbf{E}\, d\mathbf{r} = e\, d\mathbf{r}$. That is, the point $d\mathbf{r}$ is moved to $(1 + e)d\mathbf{r}$. Since this statement is independent of the direction of $d\mathbf{r}$, we conclude that the whole sphere of any small radius dr is enlarged, or dilated, in all directions by the factor $1 + e$, and we refer to the strain (16.82) as a **spherical strain** or a **dilatation**. If e is positive, the sphere is actually enlarged; if e is negative, it is reduced.

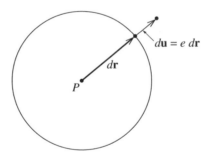

Figure 16.18   The strain (16.82) moves the point at $d\mathbf{r}$ radially out to $(1 + e)d\mathbf{r}$. Thus any small sphere centered on P is just *dilated* by a factor of $1 + e$ in all directions.

For future reference, notice that since any volume is stretched by a factor of $(1 + e)$ in all three directions, volumes are increased by a factor of $(1 + e)^3 \approx 1 + 3e$. (Remember that $e \ll 1$.) In other words, the dilatation $\mathbf{E} = e\mathbf{1}$ results in a fractional increase of $3e$ in any small volume; that is,

$$\frac{dV}{V} = 3e. \tag{16.83}$$

EXAMPLE 16.6   A Shearing Strain

The strain tensor $\mathbf{E}$ at a certain point P in a solid has the form

$$\mathbf{E} = \begin{bmatrix} 0 & \gamma & 0 \\ \gamma & 0 & 0 \\ 0 & 0 & 0 \end{bmatrix} \tag{16.84}$$

(with $\gamma \ll 1$) or, if we denote the elements of $\mathbf{E}$ by ϵ_{ij}, then $\epsilon_{12} = \epsilon_{21} = \gamma$ while all other ϵ_{ij} are zero. Describe the displacement of points in the neighborhood of P (which we can take to be the origin again).

As before we'll assume that there is no overall translation or rotation, so that $\mathbf{E}$ is the same as the derivatives matrix $\mathbf{D}$, whose only nonzero elements are

$$\frac{\partial u_1}{\partial r_2} = \frac{\partial u_2}{\partial r_1} = \gamma.$$

This implies that if we move out along the r_2 axis, the only component of $\mathbf{u}$ that changes is u_1. Therefore, any point on the r_2 axis is displaced sideways, in the r_1 direction, as shown in Figure 16.19. Similarly, a point on the r_1 axis is displaced upward in the r_2 direction. The net effect is a shear in which the two axes are tilted toward one another as shown in the figure. (This is if γ is positive; if γ is negative, they tilt the other way.) The angle through which both axes tilt is equal to the parameter γ, as long as γ is small.

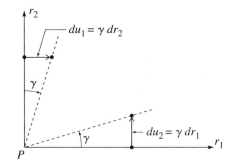

Figure 16.19   Under the strain (16.84), points on the r_2 axis move in the r_1 direction and vice versa. The result is a shear in which the two axes tilt as shown.

We see from this last example that the off-diagonal elements ϵ_{ij} of the strain tensor are associated with shearing strains. In the same way, the diagonal elements are associated with stretching along the axes. For instance if ϵ_{11} is nonzero, then points on the r_1 axis are displaced along the axis (in addition to any sideways displacement), and the whole axis is stretched by a factor $1 + \epsilon_{11}$. For this reason, the three diagonal elements ϵ_{11}, ϵ_{22}, and ϵ_{33} can be called the **stretching elements** of $\mathbf{E}$.

Decomposition of the General Strain Tensor

The last two examples lead us to our final maneuver with the strain tensor. We've seen that if $\mathbf{E}$ is diagonal and its diagonal elements are equal, $\epsilon_{11} = \epsilon_{22} = \epsilon_{33} = e$, so that $\mathbf{E} = e\mathbf{1}$, then the corresponding deformation is the simple dilatation of Example 16.5. Even if a given strain tensor $\mathbf{E}$ does not meet these conditions, you might guess that we could define e as its average stretch, that is, the average of the three diagonal elements,

$$e = \tfrac{1}{3}(\epsilon_{11} + \epsilon_{22} + \epsilon_{33}) \tag{16.85}$$

and that the simple dilatation $e\mathbf{1}$ would bear some useful relation to the original tensor $\mathbf{E}$. Before I show that this is the case, I should mention that, in the theory of matrices, the sum of the diagonal elements is an important concept called the **trace** of the matrix. That is, for any $(n \times n)$ matrix $\mathbf{M}$ with elements m_{ij}, we define its trace, $\operatorname{tr}\mathbf{M}$, as the sum

$$\operatorname{tr}\mathbf{M} = \sum_{i=1}^{n} m_{ii} = m_{11} + \cdots + m_{nn}. \tag{16.86}$$

Thus another way to state the definition (16.85) of the average stretch e of any strain tensor $\mathbf{E}$ is that it is 1/3 of the trace:

$$e = \tfrac{1}{3}\operatorname{tr}\mathbf{E}. \tag{16.87}$$

The matrix $e\mathbf{1}$ is a pure dilatation, which naturally changes the volume of any small region around the point of interest, and it can be shown (Problem 16.24) that it changes the volume by the same amount as the original strain $\mathbf{E}$. Therefore, if we write

$$\mathbf{E} = e\mathbf{1} + \mathbf{E}' \tag{16.88}$$

we have expressed $\mathbf{E}$ as the sum of two separate strains, the first of which is a pure dilatation that changes volumes by the same amount as $\mathbf{E}$ and the second of which gives the same shearing strains as $\mathbf{E}$ but causes no change of volumes. We can call the first term, $e\mathbf{1}$, the **spherical part** of $\mathbf{E}$. The second term, $\mathbf{E}' = \mathbf{E} - e\mathbf{1}$, is sometimes called the **strain deviator** or **deviatoric part** of $\mathbf{E}$, presumably because it is the amount by which $\mathbf{E}$ deviates from the corresponding pure dilatation. Mathematically we can characterize the decomposition (16.88) by saying that the first term is a multiple of the unit matrix with the same trace as $\mathbf{E}$ and the second has zero trace. We shall see in the next section that this decomposition of $\mathbf{E}$ plays an essential role in the relation between strain and stress.

EXAMPLE 16.7   A Numerical Example of Strain

The strain tensor at a certain point in a solid is found to be

$$\mathbf{E} = \begin{bmatrix} -0.01 & 0.02 & 0.05 \\ 0.02 & 0.03 & 0.04 \\ 0.05 & 0.04 & 0.04 \end{bmatrix}. \tag{16.89}$$

Decompose this strain as in (16.88) into its spherical and deviatoric parts.

The average stretch is easily seen to be $e = \frac{1}{3}\mathrm{tr}\,\mathbf{E} = 0.02$ and $\mathbf{E}'$ is then found by subtraction as $\mathbf{E}' = \mathbf{E} - e\mathbf{1}$. So

$$e\mathbf{1} = \begin{bmatrix} 0.02 & 0 & 0 \\ 0 & 0.02 & 0 \\ 0 & 0 & 0.02 \end{bmatrix} \quad \text{and} \quad \mathbf{E}' = \begin{bmatrix} -0.03 & 0.02 & 0.05 \\ 0.02 & 0.01 & 0.04 \\ 0.05 & 0.04 & 0.02 \end{bmatrix}. \quad (16.90)$$

It is easy to check that the original strain tensor $\mathbf{E}$ is indeed equal to $e\mathbf{1} + \mathbf{E}'$. Notice that the trace of $e\mathbf{1}$ is the same as that of $\mathbf{E}$, as it should be, while the trace of $\mathbf{E}'$ is zero.

16.9   Relation between Stress and Strain: Hooke's Law

The final step in writing down the equation of motion for a continuous solid is to find the relation between the stress and strain tensors, $\boldsymbol{\Sigma}$ and $\mathbf{E}$. This relation, sometimes called the **constitutive equation**, corresponds to Hooke's law for a mass on a spring, expressing the force (or stress) in terms of the extension (or strain). It is reasonable to assume that, at least for small disturbances, the relation between stress and strain should be linear, and this is what I shall assume here. There certainly are plenty of examples of materials that fit this assumption reasonably well — a chunk of metal, or even the rock of the earth's crust. When the required relation is linear, it is called the *generalized Hooke's law*. To simplify the discussion still further (and this is a huge simplification), I shall assume that the solid is isotropic, which implies that the relation between $\boldsymbol{\Sigma}$ and $\mathbf{E}$ is independent of our choice of axes, or rotationally invariant.

We wish to express the stress tensor $\boldsymbol{\Sigma}$ as a function $\boldsymbol{\Sigma} = f(\mathbf{E})$ of the strain tensor $\mathbf{E}$. The function f must be linear, and it must be rotationally invariant. Linearity is a familiar property. To be rotationally invariant means this: If R denotes any rotation of our coordinate axes and $\mathbf{M}_R$ the result of rotating a matrix $\mathbf{M}$ by the rotation R, then it must be true that

$$f(\mathbf{E}_R) = [f(\mathbf{E})]_R \qquad (16.91)$$

for any strain $\mathbf{E}$ and any rotation R. That is, the stress corresponding to the rotated strain $\mathbf{E}_R$ (left side of the equation) must be the same as the result of rotating the stress corresponding to $\mathbf{E}$ (right side of the equation). Now, we have seen that any strain tensor can be decomposed as the sum of its spherical and deviatoric parts:

$$\mathbf{E} = e\mathbf{1} + \mathbf{E}'. \qquad (16.92)$$

This decomposition has two important properties. First, it is rotationally invariant; that is, when we rotate axes, each part rotates separately into the corresponding part of the rotated tensor. (The spherical part of $\mathbf{E}$ rotates into the spherical part of $\mathbf{E}_R$, and likewise the deviatoric part.) Second, it is impossible to decompose **E** any further and retain this property.[17]

The crucial result, the proof of which is unfortunately beyond the scope of the mathematics I am assuming here, is this: If $\Sigma = f(\mathbf{E})$ where the function f is linear and rotationally invariant, and if **E** is decomposed as in (16.92), then the most general possible form of the function f is this:

$$\Sigma = \alpha e \mathbf{1} + \beta \mathbf{E}' \tag{16.93}$$

where α and β are two constants (which depend on the material of which our solid is made) and $e = \frac{1}{3} \operatorname{tr} \mathbf{E}$ as usual.[18] The relation (16.93) is called the **generalized Hooke's law**, or just **Hooke's law**, and any solid that obeys it is called an **elastic solid**. It is often convenient to rewrite Hooke's law (16.93) in terms of **E** (rather than $\mathbf{E}' = \mathbf{E} - e\mathbf{1}$), to give[19]

$$\Sigma = (\alpha - \beta)e\mathbf{1} + \beta \mathbf{E}. \tag{16.94}$$

We can solve this for **E** in terms of Σ in two steps. Taking the trace of (16.93) we find that $\operatorname{tr} \Sigma = 3\alpha e$, so $e = \operatorname{tr} \Sigma / 3\alpha$. Substituting this into (16.94) we find that

$$\mathbf{E} = \frac{1}{3\alpha\beta}[3\alpha\Sigma - (\alpha - \beta)(\operatorname{tr} \Sigma)\mathbf{1}]. \tag{16.95}$$

As you might guess, the constants α and β are related to the elastic moduli introduced at the end of Section 16.6 [Equations (16.55) to (16.57)]. Let's start with the bulk modulus.

---

[17] In the language of group theory, the two parts in (16.92) are irreducible. For a discussion of the necessary group theory see Chapter 10 of *Mathematics for Scientists and Engineers* by Harold Cohen, Prentice Hall (1992) or Chapter 16 of *Mathematical Methods of Physics* by Jon Mathews and R. L. Walker, W. A. Benjamin (1970).

[18] Within the framework of group representations, the proof is amazingly simple: The linear function f must commute with all rotations. The decomposition (16.92) splits the space of all symmetric matrices into two *irreducible* subspaces (of dimensions 1 and 5 respectively). By Schur's lemma, the restriction of f to either of these irreducible subspaces can be at most multiplication by a scalar (which we can call α or β) and we have proved (16.93).

[19] This equation is often written as $\Sigma = 3\lambda e\mathbf{1} + 2\mu\mathbf{E}$, in which case λ and μ are called the **Lamé constants**.

Bulk Modulus

Imagine a solid subject only to an isotropic pressure, p (no shear stresses). In this case we know that the stress tensor has the simple form $\boldsymbol{\Sigma} = -p\mathbf{1}$. Substituting this into Hooke's law (16.95) we find that

$$\mathbf{E} = \frac{1}{\alpha\beta}[-\alpha + (\alpha - \beta)]p\mathbf{1} = -\frac{p}{\alpha}\mathbf{1}. \qquad (16.96)$$

That is, the strain tensor $\mathbf{E}$ is also a multiple of the unit matrix, which we could write as $\mathbf{E} = e\mathbf{1}$, with $e = -p/\alpha$. But we know from (16.83) that $e = \frac{1}{3}dV/V$. Comparing these two expressions for e, we conclude that $p = -\frac{1}{3}\alpha\, dV/V$, and comparing this with the definition (16.56) of the bulk modulus, we see that

$$\alpha = 3\,\mathrm{BM}. \qquad (16.97)$$

Shear Modulus

Let us consider next the simple shearing strain $\mathbf{E}$ given in (16.84) of Example 16.6 and illustrated in Figure 16.19. Since this has zero trace, $e = 0$, and Hooke's law (16.94) reduces to the simple form

$$\boldsymbol{\Sigma} = \beta\mathbf{E}.$$

In particular, the stress responsible for the shear is

$$\frac{F}{A} = \sigma_{12} = \beta\epsilon_{12} = \beta\gamma. \qquad (16.98)$$

We need to compare this with the definition (16.57) of the shear modulus. Unfortunately, (16.57) refers to the strain of Figure 16.13, which is not quite the same as that of Figure 16.19. Specifically, in Figure 16.19 both axes tilt inward by an angle γ, whereas in Figure 16.13 the x axis turns through angle θ while the y axis is unchanged. A moment's thought should convince you that the displacement of Figure 16.13 is a combination of a simple shear, as in Figure 16.19, followed by a rotation to bring the y axis back to its original direction. This means that the angle θ of Figure 16.13 equals twice the angle γ of Figure 16.19, $\theta = 2\gamma$. Putting this into the definition (16.57) of the shear modulus, we find that

$$\frac{F}{A} = \mathrm{SM}\,\frac{dy}{dx} = \mathrm{SM}\,\theta = 2\,\mathrm{SM}\,\gamma,$$

and comparing with (16.98), we conclude that the constant β is just twice the shear modulus,

$$\beta = 2\,\mathrm{SM}. \qquad (16.99)$$

Young's Modulus

We can similarly identify Young's modulus in terms of the constants α and β, but I shall leave this as an exercise (Problem 16.27). The result is that

$$\text{YM} = \frac{3\alpha\beta}{2\alpha + \beta} = \frac{9\,\text{BM} \cdot \text{SM}}{3\,\text{BM} + \text{SM}}, \tag{16.100}$$

where the last expression results from substituting (16.97) for α and (16.99) for β. An interesting feature of this result is that it shows that only two of the three elastic moduli are independent. For instance, if we know BM and SM we can calculate YM. (See Problem 16.26.)

16.10   The Equation of Motion for an Elastic Solid

Let us consider an infinitesimal volume dV of an elastic solid. Its mass is $\varrho\,dV$, where ϱ is its density, and its position is $\mathbf{r} + \mathbf{u}(\mathbf{r}, t)$, where $\mathbf{r}$ is its equilibrium position and $\mathbf{u}(\mathbf{r}, t)$ is its displacement from equilibrium. [The displacement $\mathbf{u}(\mathbf{r}, t)$ depends on $\mathbf{r}$ and t, but the equilibrium position $\mathbf{r}$ of any given piece of the solid is fixed and independent of t.] Newton's second law applied to this volume reads

$$\varrho\,dV\,\frac{\partial^2 \mathbf{u}}{\partial t^2} = \mathbf{F}_{\text{vol}} + \mathbf{F}_{\text{sur}} \tag{16.101}$$

where $\mathbf{F}_{\text{vol}}$ denotes the volume force and $\mathbf{F}_{\text{sur}}$ the net surface force on dV. The volume force is proportional to dV, and to be definite, I shall assume it is just the force of gravity, so that

$$\mathbf{F}_{\text{vol}} = \varrho\,g\,dV. \tag{16.102}$$

We know that the surface force on any small element of the surface of our volume is $\boldsymbol{\Sigma}\,d\mathbf{A}$, where as usual $\boldsymbol{\Sigma}$ is the 3×3 stress tensor and $d\mathbf{A}$ is the vector labeling the surface element (considered to be a 3×1 column vector in the matrix product). Thus the net surface force is

$$\mathbf{F}_{\text{sur}} = \int \boldsymbol{\Sigma}\,d\mathbf{A} \tag{16.103}$$

where the integral runs over the closed surface of the volume under consideration. To cast this integral in a more convenient form we need to use the divergence theorem. This theorem was introduced in Chapter 13 on Hamiltonian mechanics, but in case you didn't read that chapter I'll review the main ideas here. (For more details, see Section 13.7 and Problems 13.31 to 13.34.) The divergence theorem states that any surface integral of the form $\int \mathbf{v} \cdot d\mathbf{A}$ over a closed surface S is equal to a certain volume integral, namely

$$\int_S \mathbf{v} \cdot d\mathbf{A} = \int_V \nabla \cdot \mathbf{v}\,dV. \tag{16.104}$$

Here **v** is any vector, V is the volume enclosed by the surface S, and $\nabla \cdot \mathbf{v}$ is the **divergence** of **v**,

$$\nabla \cdot \mathbf{v} = \frac{\partial v_x}{\partial x} + \frac{\partial v_y}{\partial y} + \frac{\partial v_z}{\partial z} = \sum_{j=1}^{3} \partial_j v_j \qquad (16.105)$$

where I have introduced the convenient shorthand

$$\partial_j = \frac{\partial}{\partial r_j}.$$

Rewritten in terms of components, the divergence theorem (16.104) reads

$$\int \sum_j v_j \, dA_j = \int \sum_j \partial_j v_j \, dV, \qquad (16.106)$$

and the ith component of the surface force (16.103) is

$$(\mathbf{F}_{\text{sur}})_i = \int \sum_j \sigma_{ij} \, dA_j. \qquad (16.107)$$

For each fixed value of i (1, 2, or 3), this has the form of the left side of (16.106), so can be replaced by a term with the form of the right side,

$$(\mathbf{F}_{\text{sur}})_i = \int \sum_j \partial_j \sigma_{ij} \, dV. \qquad (16.108)$$

We can write this in more compact vector form if we introduce a vector $\nabla \cdot \mathbf{\Sigma}$ defined so that its ith component is[20]

$$(\nabla \cdot \mathbf{\Sigma})_i = \sum_{j=1}^{3} \partial_j \sigma_{ji}. \qquad (16.109)$$

Note well that since ∇ is a vector operator while $\mathbf{\Sigma}$ is a tensor, $\nabla \cdot \mathbf{\Sigma}$ is a vector. Armed with this notation we can write (16.108) as

$$\mathbf{F}_{\text{sur}} = \int \nabla \cdot \mathbf{\Sigma} \, dV. \qquad (16.110)$$

This result is valid for the surface force on any volume, small or large. However, our interest is in an infinitesimal volume dV. For a small enough volume, the integrand is constant and the result becomes just

$$\mathbf{F}_{\text{sur}} = \nabla \cdot \mathbf{\Sigma} \, dV \qquad (16.111)$$

for any small volume dV.

---

[20] For an arbitrary tensor **M**, the elements m_{ij} and m_{ji} aren't necessarily the same, and we have to be careful of the order of the indices i and j. Luckily $\mathbf{\Sigma}$ is symmetric, so it doesn't matter which order we write them in.

Let us now return to the equation of motion (16.101) and substitute (16.102) for $\mathbf{F}_{\text{vol}}$ and (16.111) for $\mathbf{F}_{\text{sur}}$. When we do this, every term acquires a factor of dV, which we can cancel to give

$$\varrho \frac{\partial^2 \mathbf{u}}{\partial t^2} = \varrho \mathbf{g} + \nabla \cdot \mathbf{\Sigma}. \tag{16.112}$$

This important equation is easy to understand. The left side represents the $m\mathbf{a}$ of $m\mathbf{a} = \mathbf{F}$. The first term on the right represents the force of gravity (or more generally the body force) and the second the surface force.

Before we can use this equation of motion, we must use Hooke's law to replace the stress tensor $\mathbf{\Sigma}$ by the strain tensor $\mathbf{E}$. In the form (16.94), Hooke's law reads

$$\mathbf{\Sigma} = (\alpha - \beta)e\mathbf{1} + \beta\mathbf{E} \tag{16.113}$$

or, in terms of its components,

$$\sigma_{ji} = (\alpha - \beta)e\delta_{ji} + \beta\epsilon_{ji} \tag{16.114}$$

where δ_{ji} denotes the **Kronecker delta symbol**

$$\delta_{ji} = \begin{cases} 1 & \text{if } j = i \\ 0 & \text{if } j \neq i \end{cases} \tag{16.115}$$

Since $\epsilon_{ji} = \frac{1}{2}(\partial_j u_i + \partial_i u_j)$, the average stretch e is

$$e = \frac{1}{3}\sum_i \epsilon_{ii} = \frac{1}{3}\sum_i \partial_i u_i = \frac{1}{3}\nabla \cdot \mathbf{u}.$$

Putting these results into (16.114), we find

$$\sigma_{ji} = \frac{1}{3}(\alpha - \beta)\delta_{ji}\nabla \cdot \mathbf{u} + \frac{1}{2}\beta(\partial_i u_j + \partial_j u_i).$$

This lets us evaluate $\nabla \cdot \mathbf{\Sigma}$ for use in (16.112):

$$(\nabla \cdot \mathbf{\Sigma})_i = \sum_j \partial_j \sigma_{ji} = \frac{1}{3}(\alpha - \beta)\sum_j \delta_{ji}\partial_j(\nabla \cdot \mathbf{u}) + \frac{1}{2}\beta\sum_j \partial_j \partial_i u_j + \frac{1}{2}\beta\sum_j \partial_j \partial_j u_i.$$

Each of the terms in this ugly result simplifies. In the first term, notice that $\sum_j \delta_{ji}\partial_j = \partial_i$. In the second term, $\sum_j \partial_j \partial_i u_j = \partial_i \sum_j \partial_j u_j = \partial_i \nabla \cdot \mathbf{u}$, and in the third, $\sum_j \partial_j \partial_j = \nabla^2$. Therefore

$$\nabla \cdot \mathbf{\Sigma} = \frac{1}{3}(\alpha - \beta)\nabla(\nabla \cdot \mathbf{u}) + \frac{1}{2}\beta\nabla(\nabla \cdot \mathbf{u}) + \frac{1}{2}\beta\nabla^2\mathbf{u}$$

$$= (\tfrac{1}{3}\alpha + \tfrac{1}{6}\beta)\nabla(\nabla \cdot \mathbf{u}) + \tfrac{1}{2}\beta\nabla^2\mathbf{u}$$

$$= (\text{BM} + \tfrac{1}{3}\text{SM})\nabla(\nabla \cdot \mathbf{u}) + \text{SM}\nabla^2\mathbf{u} \tag{16.116}$$

where in the last line I have used (16.97) and (16.99) to rewrite α and β in terms of the bulk and shear moduli, BM and SM.

Finally we are ready to write the equation of motion of an elastic solid in a usable form. Sustituting (16.116) into (16.112), we find

$$\varrho \frac{\partial^2 \mathbf{u}}{\partial t^2} = \varrho \mathbf{g} + \left(\text{BM} + \tfrac{1}{3}\text{SM}\right) \nabla(\nabla \cdot \mathbf{u}) + \text{SM}\nabla^2 \mathbf{u}. \qquad (16.117)$$

In the next section, we'll use this equation, often called the **Navier equation** (after the French engineer Claude Navier, 1785–1836), to derive the two main kinds of wave in an elastic solid.

16.11   Longitudinal and Transverse Waves in a Solid

It is well known that there are two main kinds of wave in an elastic solid–longitudinal and transverse. To show this, we'll examine the Navier equation (16.117). We'll assume that gravity is unimportant and set $\mathbf{g} = 0$. (This is usually an excellent approximation, one exception being very slow — $\tau \gtrsim 200$ s — free oscillations of the earth, for which gravity is important.) Without loss of generality, we'll look for a plane wave propagating in the x (that is, r_1) direction, so that $\mathbf{u}$ depends only on x and t.

Let us first examine the possibility of a longitudinal disturbance, for which the displacement $\mathbf{u}$ would be in the direction of propagation,

$$\mathbf{u} = [u_x(x, t), 0, 0].$$

In this case, $\nabla \cdot \mathbf{u} = \partial u_x/\partial x$ and the only nonzero component of $\nabla(\nabla \cdot \mathbf{u})$ is its x component, which is $\partial^2 u_x/\partial x^2$. The only nonzero component of $\nabla^2 \mathbf{u}$ is likewise the x component, which is also equal to $\partial^2 u_x/\partial x^2$. Putting all of this together in the equation of motion (16.117), we obtain

$$\varrho \frac{\partial^2 u_x}{\partial t^2} = \left(\text{BM} + \tfrac{4}{3}\text{SM}\right) \frac{\partial^2 u_x}{\partial x^2}. \qquad (16.118)$$

This is the wave equation, with wave speed

$$c_{\text{long}} = \sqrt{\frac{\text{BM} + \tfrac{4}{3}\text{SM}}{\varrho}}. \qquad (16.119)$$

We conclude that longitudinal waves are indeed possible, with speed c_{long} given by (16.119).

If instead we look for a transverse (or shear) wave, traveling in the x direction but with the displacement in the y direction, then $\mathbf{u}$ would have the form

$$\mathbf{u} = [0, u_y(x, t), 0].$$

In this case $\nabla \cdot \mathbf{u} = 0$, while $\nabla^2 \mathbf{u}$ has only a y component equal to $\partial^2 u_y / \partial x^2$, so (16.117) becomes

$$\varrho \frac{\partial^2 u_y}{\partial t^2} = \text{SM} \frac{\partial^2 u_y}{\partial x^2}.$$

This is the wave equation, with wave speed

$$c_{\text{tran}} = \sqrt{\frac{\text{SM}}{\varrho}} \qquad (16.120)$$

and we conclude that transverse waves are possible, with speed c_{tran} given by (16.120).

Notice that $c_{\text{long}} > c_{\text{tran}}$. Therefore, if longitudinal and transverse signals set out simultaneously from some source, the longitudinal one will arrive first at a distant detector. For instance, it is well known in the earth sciences that the longitudinal waves from a distant earthquake arrive before the transverse ones, and this gives seismologists a way to measure how far away an earthquake or explosion occurred. For this reason, longitudinal waves are also called **primary** or **P waves** and transverse **secondary** or **S waves**.

EXAMPLE 16.8 Waves in Rock

The elastic moduli of the material of the earth's crust vary, but representative values would be BM $\approx$ 40 GPa and SM $\approx$ 25 GPa. (These are the approximate values for granite, whose density is about 2.7×10^3 kg/m^3, and 1 GPa $= 10^9$ Pa $= 10^9$ N/m^2.) What would be the speed of longitudinal and transverse seismic waves in rock with these values?

According to (16.119), the longitudinal speed is

$$c_{\text{long}} = \sqrt{\frac{(40 + \frac{4}{3} \times 25) \times 10^9 \text{ N/m}^2}{2.7 \times 10^3 \text{ kg/m}^3}} \approx 5.25 \text{ km/s}.$$

Similarly, from (16.120) we find the transverse speed to be

$$c_{\text{tran}} = \sqrt{\frac{25 \times 10^9 \text{ N/m}^2}{2.7 \times 10^3 \text{ kg/m}^3}} \approx 3.0 \text{ km/s}.$$

A striking feature of the formula (16.120) for the transverse wave speed is that if the shearing modulus SM is zero — as it is in fluids — then c_{tran} is zero. This suggests (correctly) that fluids cannot support transverse waves.[21] A beautiful application of this result is that tranverse seismic waves (unlike longitudinal) are found not to propagate through the center of the earth, showing that some region near the earth's center (namely, the outer core) is liquid.

---

[21] Of course this argument is not quite watertight since our derivation of (16.120) assumed a solid medium. Nevertheless the suggested conclusion is correct, and correct for essentially the right reason: Transverse waves require a transverse (shearing) restoring force, which a fluid cannot supply.

16.12   Fluids: Description of the Motion *

*As usual, sections marked with an asterisk can be omitted on a first reading.

In the last four sections we have focussed primarily on the motion of *solid* continuous media. In the final two sections of this chapter, I would like to give a brief introduction to the motion of fluids. Unfortunately, the analysis of a general fluid — particularly a viscous fluid — is complicated, and, to keep this chapter from growing unconscionably long, I shall restrict myself mostly to **inviscid** or **ideal fluids**, that is, fluids whose viscosity is negligible. One might argue that to neglect viscosity is to throw the baby out with the bathwater, but the fact is that there are many problems in fluid motion where it *is* reasonable to neglect viscosity, and, more important, all of the tools needed to discuss inviscid fluids are needed for the analysis of viscous fluids, so the discussion here is really an essential preliminary to any subsequent study of viscous fluids. Furthermore, several ideas discussed here — the convective derivative and the equation of continuity, in particular — are equally applicable to either case.

Material versus Spatial Descriptions

So far we have analyzed what is happening in a continuous medium by specifying that the piece of material that was originally at position $\mathbf{r}$ is now at position $\mathbf{r} + \mathbf{u}(\mathbf{r}, t)$. This approach is sometimes called the **material description**, since it focuses on a particular piece of the material. It turns out that in discussing fluids it is often more convenient *not* to follow the individual pieces of fluid, but rather to specify what is happening at each fixed point in space. Thus we might give the velocity $\mathbf{v}(\mathbf{r}, t)$, the density $\varrho(\mathbf{r}, t)$, and so forth, of the fluid at each fixed point $\mathbf{r}$ (and time t). This approach is often called the **spatial description**.[22]

Some advantages of the spatial approach when discussing a fluid are reasonably obvious: With a solid, each material piece of the solid usually has a well-defined equilibrium position $\mathbf{r}$, and this is what we use to label each piece [giving the piece's displacement $\mathbf{u}(\mathbf{r}, t)$ from $\mathbf{r}$]. In a fluid, the pieces of the fluid do not generally have a unique equilibrium position. We could, of course, use the piece's initial position as a reference, but in a problem of fluid flow, the pieces usually stray inconveniently far from their initial positions. Thus in discussing a fluid, it is usually more convenient simply to focus on what is happening at each fixed point in space — the spatial description.

The Material Derivative

The main disadvantage of the spatial description emerges most clearly when we try to use Newton's second law. The acceleration $\mathbf{a}$ in $\mathbf{F} = m\mathbf{a}$ is, of course, the acceleration of any small piece of the fluid. Unfortunately, if $\mathbf{v}(\mathbf{r}, t)$ is the fluid's velocity at a

---

[22] The material and spatial descriptions are often called the *Lagrangian* and *Eulerian* descriptions, respectively — names that are neither historically accurate (both methods being due to Euler) nor especially easy to remember.

point $\mathbf{r}$, then $\mathbf{a}$ is not just $\partial \mathbf{v}/\partial t$. The partial derivative $\partial \mathbf{v}/\partial t$ is the rate of change of the fluid velocity *at the fixed point* $\mathbf{r}$, whereas the acceleration we want is the rate of change of $\mathbf{v}$ *of a given piece of fluid as it moves*. This same distinction applies to any other parameter and is, perhaps, easier to visualize in the case of a scalar, such as the density or temperature. Consider, for example, the density of a small element of the fluid. If at time t the element is at position $\mathbf{r}$, then the required density is $\varrho(\mathbf{r}, t)$. But a short time dt later, the element will have moved to a new position $\mathbf{r} + d\mathbf{r}$, where $d\mathbf{r} = \mathbf{v}\,dt$, so the density is now $\varrho(\mathbf{r} + d\mathbf{r}, t + dt)$. (Note well how, if we wish to follow the material element, we must evaluate the new density at the new time $t + dt$ *and* the new position $\mathbf{r} + d\mathbf{r}$.) Thus the change in the density of our volume element is

$$d\varrho = \varrho(\mathbf{r} + d\mathbf{r}, t + dt) - \varrho(\mathbf{r}, t) = \frac{\partial \varrho}{\partial t}dt + d\mathbf{r} \cdot \nabla \varrho$$

$$= \frac{\partial \varrho}{\partial t}dt + (\mathbf{v}dt) \cdot \nabla \varrho.$$

Dividing both sides by dt we find for the time derivative of ϱ

$$\frac{d\varrho}{dt} = \frac{\partial \varrho}{\partial t} + \mathbf{v} \cdot \nabla \varrho. \tag{16.121}$$

This derivative is called the **material derivative** since it gives the rate of change of ϱ as we follow the motion of a material piece of the fluid.[23]

We can apply a similar argument to other parameters of the fluid. For example, dp/dt (defined in exactly the same way) would be the rate of change of the pressure as we follow the motion of a material element of fluid. In particular, we can examine each component of the fluid's velocity and putting the three components together find

$$\frac{d\mathbf{v}}{dt} = \frac{\partial \mathbf{v}}{\partial t} + \mathbf{v} \cdot \nabla \mathbf{v}. \tag{16.122}$$

This derivative is, of course, the acceleration of the volume element of fluid. Armed with this result, we are ready to write down the equation of motion for an inviscid fluid.

Equation of Motion for an Inviscid Fluid

Consider now a small volume element dV of fluid with mass $\varrho\,dV$ and acceleration given by (16.122). Newton's second law implies that

---

[23] Other names for the material derivative are *total* or *convective derivative*. Some authors use the symbol D/Dt (instead of d/dt) to emphasize the special character of this derivative.

$$\varrho \, dV \frac{d\mathbf{v}}{dt} = \mathbf{F} = \mathbf{F}_{\text{vol}} + \mathbf{F}_{\text{sur}}. \qquad (16.123)$$

I shall assume that the only volume force is gravity, so that $\mathbf{F}_{\text{vol}} = \varrho \, dV\mathbf{g}$, where $\mathbf{g}$ is the acceleration of gravity. We know from (16.111) that the surface force is $\mathbf{F}_{\text{sur}} = \nabla \cdot \Sigma dV$ and from (16.70) that $\Sigma = -p\mathbf{1}$. (This last was derived for a static fluid, but the argument needed only that the viscosity be negligible.) To evaluate $\nabla \cdot \Sigma$, we'll take components:

$$(\nabla \cdot \Sigma)_i = \sum_j \partial_j \sigma_{ji} = -\sum_j \partial_j (p\delta_{ji}) = -\partial_i p.$$

Therefore, $\nabla \cdot \Sigma = -\nabla p$. Putting all these results together in (16.123), we see that every term contains a factor of dV, and, canceling this factor, we arrive at the equation of motion for an inviscid fluid,

$$\varrho \frac{d\mathbf{v}}{dt} = \varrho\mathbf{g} - \nabla p. \qquad (16.124)$$

Bernoulli's Theorem

As a first simple application of the equation of motion (16.124), let us derive a familiar result from most introductory physics courses, Bernoulli's theorem. This theorem is usually stated for the case of steady flow of an incompressible, inviscid fluid, and this is the case I'll consider here. That the flow is steady means that the parameters p, $\mathbf{v}$, and ϱ at any fixed point $\mathbf{r}$ are constant; that is, the partial derivatives $\partial p/\partial t$ and so on are all zero. That the fluid is incompressible means that the density doesn't change, $d\varrho/dt = 0$. (Notice that in the case of the density, both $\partial\varrho/\partial t$ and $d\varrho/dt$ are zero.) I shall consider the case that $\mathbf{g}$ is uniform and in the negative z direction, so that we can write $\mathbf{g} = -\nabla(gz)$.[24] Let us make this replacement in (16.124), and then dot the whole equation with $\mathbf{v}$. This gives

$$\varrho\mathbf{v} \cdot \frac{d\mathbf{v}}{dt} + \varrho\mathbf{v} \cdot \nabla(gz) + \mathbf{v} \cdot \nabla p = 0. \qquad (16.125)$$

We can simplify all three terms in this equation. For the first term, note that $\mathbf{v} \cdot d\mathbf{v}/dt = \frac{1}{2}d(v^2)/dt$. For the second and third terms, note that $\mathbf{v} \cdot \nabla f = df/dt - \partial f/\partial t$, for any function f. For the functions of interest here, $\partial f/\partial t = 0$, so $\mathbf{v} \cdot \nabla f = df/dt$. With these replacements, (16.125) becomes

$$\frac{1}{2}\varrho\frac{dv^2}{dt} + \varrho\frac{d(gz)}{dt} + \frac{dp}{dt} = 0. \qquad (16.126)$$

---

[24] Even if $\mathbf{g}$ is not uniform, it is certainly conservative, so we can always introduce a function Φ (called the gravitational potential) such that $\mathbf{g} = -\nabla\Phi$. To avoid introducing unfamiliar symbols, I decided to treat the common case that $\Phi = gz$.

Finally, since $d\varrho/dt = 0$, we can bring the factors of ϱ inside the derivatives and we find that

$$\frac{d}{dt}\left(\tfrac{1}{2}\varrho v^2 + \varrho gz + p\right) = 0. \tag{16.127}$$

This asserts that the quantity $\Psi = \tfrac{1}{2}\varrho v^2 + \varrho gz + p$ is constant as we move along with any material element of the fluid. In other words, Ψ is constant along any streamline, which is precisely the content of Bernoulli's theorem for steady, incompressible, inviscid flow. Because the term $\tfrac{1}{2}\varrho v^2$ appears with the pressure p in the Bernoulli Ψ and is associated with the motion, $\tfrac{1}{2}\varrho v^2$ is sometimes called the *dynamic pressure*.

The Equation of Continuity

The conservation of mass implies an important relation, called the equation of continuity, between the density and velocity of any fluid, viscous or inviscid. If you have taken a course in electromagnetism, you may have met a corresponding relation that reflects the conservation of charge. To prove the relation, consider a small volume dV of fluid. As this volume moves, its mass $\varrho\,dV$ cannot change. Therefore,

$$\frac{d}{dt}(\varrho\,dV) = dV\frac{d\varrho}{dt} + \varrho\frac{d}{dt}(dV) = 0. \tag{16.128}$$

The rate of change of a moving volume was evaluated in Equation (13.59) of Section 13.7. (If you did not read Chapter 13, you could read just the proof of this result, starting at the subsection "Changing Volumes" on page 544.) For any volume V (small or large) the result is

$$\frac{dV}{dt} = \int_V \boldsymbol{\nabla} \cdot \mathbf{v}\,dV.$$

For an infinitesimal volume dV, this reduces to

$$\frac{d}{dt}(dV) = \boldsymbol{\nabla} \cdot \mathbf{v}\,dV.$$

Inserting this in (16.128) and canceling the common factor of dV, we arrive at the **equation of continuity**,

$$\frac{d\varrho}{dt} + \varrho\boldsymbol{\nabla} \cdot \mathbf{v} = 0 \tag{16.129}$$

or equivalently (see Problem 16.34)

$$\frac{\partial\varrho}{\partial t} + \boldsymbol{\nabla} \cdot (\varrho\mathbf{v}) = 0. \tag{16.130}$$

This relation plays a crucial role in fluid dynamics, as does the corresponding relation (with ϱ replaced by the charge density) in electrodynamics. We shall use it in the next section in deriving the speed of sound in a fluid.

16.13   Waves in a Fluid *

As usual, sections marked with an asterisk can be omitted on a first reading.

Armed with the equations of motion and continuity, we can now discuss the possibility of waves in an inviscid fluid. We imagine a fluid which undergoes a small disturbance from equilibrium, so that it acquires a small (presumably oscillatory) velocity $\mathbf{v}$ and its pressure and density become

$$p = p_0 + p' \tag{16.131}$$

and

$$\varrho = \varrho_0 + \varrho'. \tag{16.132}$$

Here p_0 and ϱ_0 are the equilibrium values, and p' and ϱ' are small increments. Notice first that in equilibrium, the equation of motion (16.124) implies simply that

$$\varrho_o \mathbf{g} - \nabla p_0 = 0. \tag{16.133}$$

For the actual situation, we insert (16.131) and (16.132) into the equation of motion to give

$$(\varrho_0 + \varrho') \left(\frac{\partial \mathbf{v}}{\partial t} + \mathbf{v} \cdot \nabla \mathbf{v} \right) = (\varrho_0 + \varrho')\mathbf{g} - \nabla(p_0 + p').$$

This ugly equation simplifies. First, by the equilibrium condition (16.133), the first and third terms on the right cancel exactly. Second, by the assumption that the disturbance is small, we can drop any terms that are second order or higher in the small quantities v, ϱ', or p', or their derivatives. Thus we can ignore the term $\mathbf{v} \cdot \nabla \mathbf{v}$, and likewise the terms involving ϱ' on the left. This leaves us with

$$\varrho_0 \frac{\partial \mathbf{v}}{\partial t} = \varrho'\mathbf{g} - \nabla p'. \tag{16.134}$$

Finally, it is not hard to show (Problem 16.38) by putting in realistic numbers that the term $\varrho'\mathbf{g}$ on the right is negligible compared to $\nabla p'$. Thus the equation of motion for our small disturbance is just

$$\varrho_0 \frac{\partial \mathbf{v}}{\partial t} = -\nabla p'. \tag{16.135}$$

We can treat the equation of continuity (16.130) in a similar way. Inserting (16.131) and (16.132), we get (as you should check)

$$\frac{\partial \varrho'}{\partial t} = -\varrho_0 \nabla \cdot \mathbf{v} - \mathbf{v} \cdot \nabla \varrho_0. \tag{16.136}$$

For essentially the same reasons that the first term on the right of (16.134) was negligible, the last term here can be neglected (see Problem 16.38 again), so the equation of continuity becomes

$$\frac{\partial \varrho'}{\partial t} = -\varrho_0 \nabla \cdot \mathbf{v}. \tag{16.137}$$

The equations of motion (16.135) and continuity (16.137) give two equations for the three variables $\mathbf{v}$, p', and ϱ'. We get a third equation by looking at the definition (16.56) of the bulk modulus, $p = \mathrm{BM}\,(-dV/V)$. As stated, this relates the pressure to the corresponding change of volume, but it applies equally to a *change* in pressure, in which case it would read $dp = \mathrm{BM}\,(-dV/V)$. (For a moment we'll call the volume of interest V, so dV can be its increment.) Now, the mass of a given element of fluid cannot change, so the quantity ϱV must be a constant. Therefore $\varrho\,dV + V\,d\varrho = 0$, or $dV/V = -d\varrho/\varrho$. Combining these two results we see that

$$dp = \mathrm{BM}\,\frac{d\varrho}{\varrho}. \tag{16.138}$$

Now, in our case, the change of pressure dp is what we've been calling p'. Likewise, $d\varrho$ is what we've been calling ϱ', and the original density is ϱ_0. Therefore, in our current notation

$$p' = \mathrm{BM}\,\frac{\varrho'}{\varrho_0}. \tag{16.139}$$

We can use this last result to eliminate ϱ' from the equation of continuity (16.137) to give

$$\frac{\partial p'}{\partial t} = -\mathrm{BM}\,\nabla \cdot \mathbf{v}.$$

If we differentiate this with respect to t and invoke the equation of motion (16.135), we obtain

$$\frac{\partial^2 p'}{\partial t^2} = -\mathrm{BM}\,\nabla \cdot \frac{\partial \mathbf{v}}{\partial t} = \frac{\mathrm{BM}}{\varrho_0}\,\nabla^2 p'$$

where in the second expression I interchanged the order of the spatial and time derivatives, and in the third expression I used the equation of motion (16.135). This result is the three-dimensional wave equation with wave speed

$$c = \sqrt{\frac{BM}{\varrho_0}}. \tag{16.140}$$

I'll show in just a moment that the waves in a fluid are necessarily longitudinal. Since longitudinal waves in a continuous medium are what we generally call sound waves, we have proved that the speed of sound in a fluid is $c = \sqrt{BM/\varrho_0}$.

EXAMPLE 16.9   Speed of Sound in Water

Given that the bulk modulus of water is 2.2 GPa, what is the speed of sound in water?

The density of water is, of course, 1000 kg/m^3, so the speed of sound as given by (16.140) is

$$c = \sqrt{\frac{BM}{\varrho_0}} = \sqrt{\frac{2.2 \times 10^9 \text{ N/m}^2}{10^3 \text{ kg/m}^3}} = 1.5 \text{ km/s}.$$

To show that waves in a fluid are longitudinal, let us imagine a wave traveling in the direction of a unit vector $\mathbf{n}$, so that

$$p' = f(\mathbf{n} \cdot \mathbf{r} - ct). \tag{16.141}$$

According to the equation of motion (16.135), this implies that

$$\frac{\partial \mathbf{v}}{\partial t} = -\frac{1}{\varrho_0} \nabla p' = -\frac{1}{\varrho_0} \nabla f(\mathbf{n} \cdot \mathbf{r} - ct) = -\frac{\mathbf{n}}{\varrho_0} f'(\mathbf{n} \cdot \mathbf{r} - ct)$$

where f' denotes the derivative of f with respect to its argument. This relation can be immediately integrated to give the velocity of the fluid:[25]

$$\mathbf{v} = -\frac{\mathbf{n}}{\varrho_0} \int f'(\mathbf{n} \cdot \mathbf{r} - ct) \, dt = \frac{\mathbf{n}}{c\varrho_0} f(\mathbf{n} \cdot \mathbf{r} - ct) = \frac{p'\mathbf{n}}{c\varrho_0}. \tag{16.142}$$

(If you don't see the integration here, try the change of variables $\xi = \mathbf{n} \cdot \mathbf{r} - ct$.) This result has two important features. First, we see that the fluid velocity is proportional to the pressure p'. In particular, in a sinusoidal wave, the velocity will oscillate in phase with p'. Second, the fluid velocity is in the direction of propagation; that is, the wave is longitudinal. As we anticipated in Section 16.11, a fluid cannot support transverse waves.

---

[25] In the third expression, I have omitted a constant of integration. This "constant" could depend on $\mathbf{r}$ (though certainly not on t) and could, in fact, be nonzero. For example, we could imagine a disturbance that included a constant uniform velocity superposed on the oscillatory wave motion. However, such a time-independent velocity is not part of what we would describe as the wave, so we can take it to be zero in our discussion of the wave motion.

Principal Definitions and Equations of Chapter 16

The One-Dimensional Wave Equation

The **one-dimensional wave equation** is

$$\frac{\partial^2 u}{\partial t^2} = c^2 \frac{\partial^2 u}{\partial x^2}. \qquad \text{[Eq. (16.4)]}$$

The general solution is

$$u(x, t) = f(x - ct) + g(x + ct) \qquad \text{[Eq. (16.10)]}$$

the first term of which represents a disturbance traveling to the right and the second one traveling to the left.

The Three-Dimensional Wave Equation

$$\frac{\partial^2 p}{\partial t^2} = c^2 \nabla^2 p. \qquad \text{[Eq. (16.39)]}$$

Stress, Strain, and the Elastic Moduli

$$\text{stress} = \frac{\text{force}}{\text{area}}, \quad \text{strain} = \text{fractional deformation.} \qquad \text{[Eqs. (16.58) \& (16.59)]}$$

$$\text{elastic modulus (YM, BM, SM)} = \frac{\text{stress}}{\text{corresponding strain}}. \qquad \text{[Eq. (16.60)]}$$

The Stress Tensor

The **stress tensor** is a 3×3 symmetric matrix Σ defined so that the surface force on a small element of area $d\mathbf{A}$ is

$$\mathbf{F}(d\mathbf{A}) = \Sigma \, d\mathbf{A}. \qquad \text{[Eq. (16.66)]}$$

The Strain Tensor for a Solid

If $\mathbf{u}(\mathbf{r})$ denotes the displacement of an element of solid from its original position $\mathbf{r}$, the **strain tensor** is the 3×3 symmetric matrix $\mathbf{E}$ with elements

$$\epsilon_{ij} = \frac{1}{2}\left(\frac{\partial u_i}{\partial r_j} + \frac{\partial u_j}{\partial r_i}\right) \qquad \text{[Eq. (16.81)]}$$

which can be decomposed as the sum of two terms:

$$\mathbf{E} = e\mathbf{1} + \mathbf{E}' \qquad \text{[Eq. (16.88)]}$$

where $e = \frac{1}{3}\text{tr}\mathbf{E}$.

Generalized Hooke's Law

For an isotropic solid, the most general relation between the stress and strain tensors is

$$\mathbf{\Sigma} = \alpha e \mathbf{1} + \beta \mathbf{E}' \qquad \text{[Eq. (16.93)]}$$

where the constants α and β can be related to the three elastic moduli.

$$\text{[Eqs. (16.97), (16.99) \& (16.100)]}$$

Equation of Motion for an Elastic Solid

The equation of motion for an isotropic elastic solid is the **Navier equation**:

$$\varrho \frac{\partial^2 \mathbf{u}}{\partial t^2} = \varrho \mathbf{g} + \left(\mathrm{BM} + \tfrac{1}{3}\mathrm{SM}\right) \mathbf{\nabla}(\mathbf{\nabla} \cdot \mathbf{u}) + \mathrm{SM}\nabla^2 \mathbf{u}. \qquad \text{[Eq. (16.117)]}$$

Waves in a Solid

The speeds of **longitudinal** (or **primary**, or **P**) waves and of **transverse** (or **shear**, or **secondary** or **S**) waves are

$$c_{\text{long}} = \sqrt{\frac{\mathrm{BM} + \tfrac{4}{3}\mathrm{SM}}{\varrho}} \qquad \text{and} \qquad c_{\text{tran}} = \sqrt{\frac{\mathrm{SM}}{\varrho}}. \qquad \text{[Eqs. (16.119) \& (16.120)]}$$

The Material Derivative in a Fluid

The time rate of change of a parameter ξ (the density, temperature, or velocity) as we follow a material element of a moving fluid is the **material derivative**

$$\frac{d\xi}{dt} = \frac{\partial \xi}{\partial t} + \mathbf{v} \cdot \mathbf{\nabla}\xi. \qquad \text{[Eq. (16.121)]}$$

Equation of Motion for an Inviscid Fluid

$$\varrho \frac{d\mathbf{v}}{dt} = \varrho \mathbf{g} - \mathbf{\nabla}p. \qquad \text{[Eq. (16.124)]}$$

Equation of Continuity

Conservation of mass implies the **equation of continuity**:

$$\frac{\partial \varrho}{\partial t} + \mathbf{\nabla} \cdot (\varrho \mathbf{v}) = 0. \qquad \text{[Eq. (16.130)]}$$

Waves in a Fluid

The speed of longitudinal waves in a fluid is

$$c = \sqrt{\frac{BM}{\varrho_o}}$$

[Eq. (16.140)]

but a fluid cannot support transverse waves.

Problems for Chapter 16

Stars indicate the approximate level of difficulty, from easiest (★) to most difficult (★★★).

SECTION 16.1   Motion of a Taut String

16.1 ★ Verify that the quantity $c = \sqrt{T/\mu}$ that appears in the wave equation for a string does indeed have the units of a speed.

16.2 ★★ The wave equation (16.4) is the equation of motion for a continuous string, as illustrated in Figure 16.1(a). You can obtain this equation as the limit as $n \to \infty$ of the equations for the n discrete masses of Figure 16.1(b). You need to be careful with the limiting process. As $n \to \infty$, the spacing b between the masses (see Figure 16.20) and the individual masses m must both go to zero in such a way that the linear mass density m/b approaches μ, the density of the continuous string. You can guarantee this by taking $m = \mu b$. Write down Newton's second law for the position u_i of the ith mass and show that it goes over to the wave equation as $b \to 0$.

SECTION 16.2   The Wave Equation

16.3 ★ Let $f(\xi)$ be an arbitrary (twice differentiable) function. Show by direct substitution that $f(x - ct)$ is a solution of the wave equation (16.4).

16.4 ★ Show that if we make the change of variables $\xi = x - ct$ and $\eta = x + ct$, then, as in (16.7),

$$\frac{\partial^2 u}{\partial t^2} - c^2 \frac{\partial^2 u}{\partial x^2} = -4c^2 \frac{\partial}{\partial \xi} \frac{\partial u}{\partial \eta}.$$

16.5 ★ **(a)** Show that $u = g(x + ct)$ is a solution of the wave equation (16.4) for any twice differentiable function $g(\xi)$. **(b)** Argue clearly that this solution represents a disturbance that travels undistorted to the left.

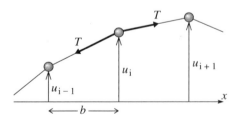

Figure 16.20   Problem 16.2

16.6 ★ There is a small flaw in Example 16.1 (page 686). In Equation (16.14) I omitted a constant of integration, so the equation should really have read $f(x) - g(x) = k$. Show that this still gives the same final answer (16.15).

16.7 ★★ [Computer] Make plots of the two triangular waves of Example 16.1 (page 686) at several closely spaced times and then animate them. Describe the motion. For the purposes of the plot you may as well take the speed c, the height of the triangle at time 0, and the half width of the base all equal to 1. Make your plots for lots of times ranging from $t = -4$ to 4.

16.8 ★★ [Computer] Make plots similar to Figure 16.5 of the standing wave (16.18) for several equally spaced times from $t = 0$ to τ, the period. Take $2A = 1$ and $k = \omega = 2\pi$. Animate your pictures and describe the motion.

SECTION 16.3   Boundary Conditions; Waves on a Finite String

16.9 ★★ The motion of a finite string, fixed at both ends, was determined by the wave equation (16.19) and the boundary conditions (16.20). We solved these by looking for a solution that was sinusoidal in time. A different, and rather more general, approach to problems of this kind is called **separation of variables**. In this approach, we seek solutions of (16.19) with the *separated* form $u(x, t) = X(x)T(t)$, that is, solutions that are a simple product of one function of x and a second of t. [As usual, there's nothing to stop us trying to find a solution of this form. In fact, there is a large class of problems (including this one) where this approach is known to produce solutions, and enough solutions to allow expansion of *any* solution.] **(a)** Substitute this form into (16.19) and show that you can rewrite the equation in the form $T''(t)/T(t) = c^2 X''(x)/X(x)$. **(b)** Argue that this last equation requires that both sides of this equation are separately equal to the same constant (call it K). It can be shown that K has to be negative.[26] Use this to show that the function $T(t)$ has to be sinusoidal — which establishes (16.21) and we're back to the solution of Section 16.3. The method of separation of variables plays an important role in several areas, notably quantum mechanics and electromagnetism.

16.10 ★★ Using the integral (16.33), show that the Fourier coefficients of the triangular wave of Figure 16.7 are zero for n even and given by (16.34) for n odd.

16.11 ★★ [Computer] Make plots similar to Figure 16.8 of the wave of Example 16.2 but from $t = 0$ to τ, the period, and for more closely spaced times. Animate your pictures and describe the motion.

16.12 ★★ Consider a semi-infinite string, fixed at the origin $x = 0$ and extending far out to the right. Let $f(\xi)$ be a function that is localized around the origin, such as the function of Figure 16.4(a). **(a)** Describe the wave given by the function $f(x + ct)$ for a large negative time t_0. **(b)** One way to solve for the subsequent motion of this wave on the semi-infinite string is called the **method of images** and is as follows: Consider the function $u = f(x + ct) - f(-x + ct)$. (The second term here is called the "image." Can you explain why?) Obviously this satisfies the wave equation for all x and t. Show that it coincides with the given wave of part (a) at the initial time t_0 and everywhere on the semi-infinite string. Show also that it obeys the boundary condition that $u = 0$ at $x = 0$. **(c)** It is a fact that there is a unique wave that obeys the wave equation and any given initial and boundary conditions. Therefore the wave of part (b) is *the* solution for all times (on our semi-infinite string). Describe the motion on the semi-infinite string for all times.

---

[26] Actually this isn't hard to show. Look at the equation $X''(x)/X(x) = K/c^2$. You can show that if $K > 0$ there are no solutions satisfying the boundary conditions that $X(0) = X(L) = 0$.

16.13 ⋆⋆ In connection with Equation (16.31), I claimed that *any* function on the interval $0 \leq x \leq L$ can be expanded in a Fourier series containing just sine functions. This is at first sight very surprising since one is used to the claim that the general Fourier series requires sines *and* cosines. In this problem, you'll prove this surprising claim. Let $f(x)$ be any function defined for $0 \leq x \leq L$. We can define a function $f(x)$ for *all* x by setting it equal to the given function in the original interval and requiring that

$$f(-x) = -f(x) \qquad \text{and} \qquad f(x + 2L) = f(x). \tag{16.143}$$

for all x. Prove that this defines a function which is (1) periodic with period $2L$, (2) odd, and (3) the same as the original $f(x)$ on the original interval. Write down the ordinary Fourier expansion for this new $f(x)$ and show that the coefficients of the cosine terms are all zero. This establishes the possibility of expanding the original function in terms of sines alone.[27] Bearing in mind that the period of the new function is $2L$, write down the standard formula (5.84) for the expansion coefficients and show that your answer agrees with (16.33). The Fourier sine series is especially convenient for discussing functions that are zero at the end points $x = 0$ and L.

16.14 ⋆⋆⋆ [Computer] A taut string of length $L = 1$ is released from rest at $t = 0$, with initial position

$$u(x, 0) = \begin{cases} 2x & [0 \leq x \leq \frac{1}{2}] \\ 2(1 - x) & [\frac{1}{2} \leq x \leq 1]. \end{cases} \tag{16.144}$$

Take the wave speed on the string to be $c = 1$. **(a)** Sketch this initial shape and find the coefficients B_n in its Fourier sine series (16.31). **(b)** Make plots of the sum of the first several terms for several closely spaced times between $t = 0$ and τ, the period. Animate your plots and describe the motion.

SECTION 16.4   The Three-Dimensional Wave Equation

16.15 ⋆⋆ Let $f(\xi)$ be any function with first two derivatives $f'(\xi)$ and $f''(\xi)$, and let $\mathbf{n}$ be an arbitrary fixed unit vector. **(a)** Show that $\nabla f(\mathbf{n} \cdot \mathbf{r} - ct) = \mathbf{n} f'(\mathbf{n} \cdot \mathbf{r} - ct)$. **(b)** Hence show that $f(\mathbf{n} \cdot \mathbf{r} - ct)$ satisfies the three-dimensional wave equation (16.38). **(c)** Argue that $f(\mathbf{n} \cdot \mathbf{r} - ct)$ represents a signal that is constant in any plane perpendicular to $\mathbf{n}$ (at any fixed time t) and propagates rigidly with speed c in the direction of $\mathbf{n}$.

16.16 ⋆⋆ Let $f(\mathbf{r})$ be any spherically symmetric function; that is, when expressed in spherical polar coordinates, (r, θ, ϕ), it has the form $f(\mathbf{r}) = f(r)$, independent of θ and ϕ. **(a)** Starting from the definition (16.38) of ∇^2, prove that

$$\nabla^2 f = \frac{1}{r} \frac{\partial^2}{\partial r^2} (rf).$$

(b) Prove the same result using the formula inside the back cover for ∇^2 in spherical polar coordinates. (Obviously, this second proof is much simpler, but the hard work is hidden in the derivation of the formula for ∇^2.)

---

[27] But note that it has the form $\sum B_n \sin(n\pi x/L)$. The usual Fourier series has sines and cosines, but their argument is $2n\pi x/L$. Thus the new Fourier sine series has, in a sense, twice as many terms to make up for having only sines.

SECTION 16.6   Stress and Strain: the Elastic Moduli

16.17 ★★ In Section 16.1 we derived the wave equation for transverse waves in a taut string. Here you will examine the possibility of longitudinal waves in the same string. Suppose that an element of string whose equilibrium position is x is displaced a short distance in the x direction to $x + u(x, t)$. **(a)** Consider a short piece of string of length l and use the definition (16.55) of Young's modulus YM to show that the tension is $F = A\,\text{YM}\,\partial u/\partial x$, where A is the cross sectional area of the string. [If the string is already in tension in its equilibrium position, this F is the *additional* tension, that is, $F = $ (actual $-$ equilibrium).] **(b)** Now consider the forces on a short section of string dx and show that u obeys the wave equation with wave speed $c = \sqrt{\text{YM}/\varrho}$ where ϱ is the density (mass/volume) of the string.

SECTION 16.7   The Stress Tensor

16.18 ★ Figure 16.15 is an end view of a triangular prism, whose three faces are labeled by the vectors $d\mathbf{A}_1$, etc. (The magnitude of $d\mathbf{A}_1$ is the area of the corresponding surface and the direction is normal to it. There are two more faces parallel to the plane of the paper, but these do not concern us.) The ends of the three faces form a closed triangle. Expain clearly why this implies that $d\mathbf{A}_1 + d\mathbf{A}_2 + d\mathbf{A}_3 = 0$.

16.19 ★ Let $\mathbf{n}_1$ and $\mathbf{n}_2$ be any two unit vectors and P a point in a continuous medium. $\mathbf{F}(\mathbf{n}_1\,dA)$ is the surface force on a small area dA at P with unit outward normal $\mathbf{n}_1$, so $\mathbf{n}_2 \cdot \mathbf{F}(\mathbf{n}_1\,dA)$ is the component of that force in the direction of $\mathbf{n}_2$. Prove *Cauchy's reciprocal theorem* that $\mathbf{n}_2 \cdot \mathbf{F}(\mathbf{n}_1\,dA) = \mathbf{n}_1 \cdot \mathbf{F}(\mathbf{n}_2\,dA)$.

16.20 ★★ It is found that the stress tensor at any point (x, y, z) in a certain continuous medium has the form (with an unspecified, convenient choice of units)

$$\mathbf{\Sigma} = \begin{bmatrix} xz & z^2 & 0 \\ z^2 & 0 & -y \\ 0 & -y & 0 \end{bmatrix}. \tag{16.145}$$

Find the surface force on a small area dA of the surface $x^2 + y^2 + 2z^2 = 4$ at the point $(1, 1, 1)$.

16.21 ★★ At any given point P of a continuous medium, the surface forces are given by the stress tensor, which is a real symmetric matrix $\mathbf{\Sigma}$. It is a well-known theorem of linear algebra (see the appendix) that any such matrix can be brought into diagonal form by a suitable rotation of the Cartesian coordinate axes. Use this to prove that at any point P there are three orthogonal directions (the **principal stress axes** at P) with the property that the surface force on any surface normal to one of these directions is exactly normal to the surface.

16.22 ★★★ Show that if the stress tensor $\mathbf{\Sigma}$ is diagonal (all off-diagonal elements zero) with respect to *any* choice of orthogonal axes, then it is in fact a multiple of the unit matrix. This gives an alternative and elegant proof that if there are no shearing stresses (in any coordinate system) then the pressure forces are independent of direction. To do this problem, you need to know how the elements of a tensor transform as we rotate our coordinate axes, as described in Section 15.17. Assume that with respect to one set of axes $\mathbf{\Sigma}$ is diagonal but that not all three diagonal elements are equal. (For example, $\sigma_{11} \neq \sigma_{33}$.) It is not hard to come up with a rotation — that of Equation (15.36) will do — such that in the rotated system $\sigma'_{13} \neq 0$.

SECTION 16.8   The Strain Tensor for a Solid

16.23 ★ An important tool in the development of the strain tensor was the decomposition (16.79) of a matrix $\mathbf{M}$ into its antisymmetric and symmetric parts. Prove that this decomposition is unique. [*Hint:* Show that if $\mathbf{M} = \mathbf{M}_A + \mathbf{M}_S$ where $\mathbf{M}_A$ and $\mathbf{M}_S$ are respectively antisymmetric and symmetric, then $\mathbf{M}_A = \frac{1}{2}(\mathbf{M} - \tilde{\mathbf{M}})$ and $\mathbf{M}_S = \frac{1}{2}(\mathbf{M} + \tilde{\mathbf{M}})$.]

16.24 ★ Write out the components of the displacement (16.77), $\mathbf{u}(\mathbf{r}) = \boldsymbol{\theta} \times \mathbf{r}$, for a small rotation $\boldsymbol{\theta}$ and verify that the derivatives matrix is given by Equation (16.78).

16.25 ★★★ At a certain point P (which you can choose to be your origin) in a continuous solid, the strain tensor is $\mathbf{E}$. Assume for simplicity that whatever displacements have occurred left P fixed and the neighborhood of P unrotated. **(a)** Show that the x axis near P is stretched by a factor of $(1 + \epsilon_{11})$. **(b)** Hence show that any small volume around P has changed by $dV/V = \mathrm{tr}\,\mathbf{E}$. This shows that any two strains that have the same trace dilate volumes by the same amount. In the decomposition $\mathbf{E} = e\mathbf{1} + \mathbf{E}'$ (16.88), the spherical part $e\mathbf{1}$ changes volumes by the same amount as $\mathbf{E}$ itself, while the deviatoric part $\mathbf{E}'$ doesn't change volumes at all.

SECTION 16.9   Relation between Stress and Strain: Hooke's Law

16.26 ★ The table below gives the three elastic moduli for several materials. According to (16.100) Young's modulus for any given material can be calculated if we know the bulk and shear moduli. Using the data for BM and SM, calculate YM for each of the materials and compare with the given values in the third column. (The densities will be needed for Problem 16.32.)

Elastic Moduli (in GPa) and Densities (in g/cm^3)

Material	BM	SM	YM	ϱ
Iron	90	40	100	7.8
Steel	140	80	200	7.8
Sandstone	17	6	16	1.9
Perovskite	270	150	390	4.1
Water	2.2	0	0	1.0

16.27 ★★★ Consider a taut wire or rod lying along the x axis. To define Young's modulus YM we apply a pure tension along the axis; that is, a stress with $\sigma_{11} > 0$ and all other $\sigma_{ij} = 0$. **(a)** Use Equation (16.95) to write down the corresponding strain tensor $\mathbf{E}$. **(b)** Argue from the definition (16.55) of Young's modulus to show that $\mathrm{YM} = \sigma_{11}/\epsilon_{11}$. **(c)** Combine these two results to verify the expression (16.100) for YM, showing in particular that $\mathrm{YM} = 9\,\mathrm{BM} \cdot \mathrm{SM}/(3\,\mathrm{BM} + \mathrm{SM})$.

16.28 ★★★ Consider again the wire or rod of Problem 16.27. In general, when one stretches the wire longitudinally it will contract in the transverse directions. The ratio of the transverse fractional contraction to the longitudinal fractional stretch is called **Poisson's ratio** (after the French mathematician and student of Laplace and Lagrange, 1781–1840) and is denoted by the Greek letter "nu," ν. **(a)** Show that $\nu = -\epsilon_{22}/\epsilon_{11}$. **(b)** Use the method of Problem 16.27 to show that $\nu = (3\,\mathrm{BM} - 2\,\mathrm{SM})/(6\,\mathrm{BM} + 2\,\mathrm{SM})$. **(c)** Calculate Poisson's ratio for the five materials listed in Problem 16.26. Comment on its value for materials with $\mathrm{BM} \gg \mathrm{SM}$.

16.29 ★★★ When we change our coordinate axes, the strain tensor changes in accordance with Equation (15.132), which we can rewrite as $\mathbf{E}_R = \mathbf{R}\mathbf{E}\tilde{\mathbf{R}}$, where $\mathbf{R}$ is the (3×3) orthogonal rotation matrix. Use the property (15.129) of orthogonal matrices to show that tr $\mathbf{E}_R$ = tr $\mathbf{E}$; that is, the trace of any tensor is rotationally invariant. Use this result to show that the decomposition $\mathbf{E} = e\mathbf{1} + \mathbf{E}'$ is rotationally invariant, in the sense described below Equation (16.92).

SECTION 16.10   The Equation of Motion for an Elastic Solid

16.30 ★ If δ_{ji} denotes the Kronecker delta symbol (16.115) and **a** is a vector with components a_j $(j = 1, 2, 3)$, prove that $\sum_j \delta_{ji} a_j = a_i$. In the same way, show that $\sum_j \delta_{ji} \partial_j = \partial_i$, a result we used in proving the important identity (16.116).

SECTION 16.11   Longitudinal and Transverse Waves in a Solid

16.31 ★ A seismograph records the signals arriving from a distant earthquake. If the S waves arrive 12 minutes after the P waves, how far away was the earthquake? Use the speeds found in Example 16.8 (page 722).

16.32 ★ [Computer] Using appropriate software, calculate the speeds of longitudinal and transverse waves in the five materials listed in Problem 16.26. Arrange for the software to give you a nice readable table of values.

SECTION 16.12   Fluids: Description of the Motion

16.33 ★ Write down the equation of motion (16.124) as applied to a *static* fluid. Assuming that **g** is uniform and ϱ is constant (independent of **r**), prove the well-known result from introductory physics that the pressure difference between two points $\mathbf{r}_1$ and $\mathbf{r}_2$ is just $\Delta p = \varrho g h$, where h is the vertical difference in elevations of $\mathbf{r}_1$ and $\mathbf{r}_2$.

16.34 ★★ Equations (16.129) and (16.130) are two different forms of the equation of continuity. Prove that they are equivalent.

SECTION 16.13   Waves in a Fluid

16.35 ★ A crucial step in showing that the waves in a fluid are necessarily longitudinal was the integral in (16.142). For an arbitrary function $f(\xi)$, with derivative $f'(\xi)$, prove that $\int f'(\mathbf{n} \cdot \mathbf{r} - ct) dt = -f(\mathbf{n} \cdot \mathbf{r} - ct)/c$.

16.36 ★★ To find the speed of sound in air using the result (16.140) requires a little care. (Even the great Newton got this one wrong!) The trouble is to decide on the correct value of the bulk modulus of air. Because the vibrations are so rapid, there is no time for heat transfer and the air expands and contracts *adiabatically*, so that pV^γ = constant, where γ is the so-called "ratio of specific heats," $\gamma = 1.4$ for air. **(a)** Show that the bulk modulus is BM = γp. **(b)** Use the ideal gas law, $pV = nRT$ to show that the density is $\varrho_0 = pM/RT$, where M is the average molecular mass of air ($M \approx 29$ grams/mole). **(c)** Put these results together to show that the speed of sound is $c = \sqrt{\gamma RT/M}$. Find the speed of sound at 0°C, and compare with the accepted value of 331 m/s.

16.37 ★★ Show that the intensity I of a sound wave is proportional to the square of the pressure increment p'. To do this consider a small sliver of fluid normal to the direction of propagation with area dA. Write down the rate at which this sliver does work on the fluid just in front of it, then divide by dA to show that the time-average intensity is $\langle I \rangle = \langle p'^2 \rangle / c\varrho_0$.

16.38 ★★★ A crucial step in deriving the wave equation for waves in a fluid was the neglect of the first term on the right of Equation (16.134). **(a)** Justify this by using (16.139) to rewrite the right side of (16.125) as $\varrho' \mathbf{g} - \mathrm{BM} \nabla \varrho' / \varrho_0$. Argue that the ratio of the first to the second term is of order $g \varrho_0 \lambda / \mathrm{BM}$, where λ is a typical distance over which ϱ' varies. (A good choice for λ would be the wavelength of the proposed wave — of order a centimeter or at most a few meters.) Using the values for water (BM = 2 GPa, etc.) show that the first term is negligible. (You would also reach the same conclusion for air.) **(b)** Show with a similar argument that the second term on the right of (16.136) is negligible.

Diagonalizing Real Symmetric Matrices

A.1 Diagonalizing a Single Matrix

In Chapter 10 we met the moment of inertia tensor $\mathbf{I}$ for a rigid body. With respect to an arbitrary set of orthogonal axes, $\mathbf{I}$ is a real symmetric 3×3 matrix which gives the body's angular momentum $\mathbf{L}$ in terms of its angular velocity $\boldsymbol{\omega}$ as $\mathbf{L} = \mathbf{I}\boldsymbol{\omega}$. We defined a *principal axis* as an axis with the property that if $\boldsymbol{\omega}$ points along the axis, then $\mathbf{L}$ is parallel to $\boldsymbol{\omega}$, that is

$$\mathbf{L} = \mathbf{I}\boldsymbol{\omega} = \lambda\boldsymbol{\omega}, \tag{A.1}$$

for some number λ. We saw that if a body has three orthogonal principal axes, then with respect to these axes $\mathbf{I}$ has the *diagonal form*

$$\mathbf{I} = \begin{bmatrix} \lambda_1 & 0 & 0 \\ 0 & \lambda_2 & 0 \\ 0 & 0 & \lambda_3 \end{bmatrix} \tag{A.2}$$

where λ_1, λ_2, and λ_3 are the moments of inertia about the three principal axes. Conversely, if $\mathbf{I}$ has this diagonal form, then the axes with respect to which $\mathbf{I}$ was calculated were principal axes. For this reason, the process of finding the principal axes is often referred to as **diagonalization of the inertia tensor**. I claimed in Section 10.4 that any rigid body, spinning about any origin O, does have three orthogonal principal axes. It is the main purpose of this appendix to prove that claim.

The process of diagonalizing a matrix comes up over and over again in many different branches of physics. For example, if you read Chapter 16, you know that both the stress and strain tensors are given by real, symmetric matrices, and it is frequently convenient to find axes with respect to which one of these is diagonal. [For example, the axes with respect to which the stress tensor (at a given point P) is diagonal are called the *principal axes of stress* and have the tidy property that the stress along each of these axes is a pure stretch.] In quantum mechanics, probably the most important thing to do with the operator that represents any given dynamical variable is to diagonalize it.[1] To emphasize the generality of the process, I shall denote the matrix that we wish to diagonalize by $\mathbf{A}$, and since we frequently have occasion to diagonalize an $n \times n$ matrix, where n is any integer, not necessarily 3, we'll suppose for now that $\mathbf{A}$ is an arbitrary $n \times n$ symmetric, real matrix. Nevertheless, the example that you may want to keep in mind is that $\mathbf{A} = \mathbf{I}$, the 3×3 moment of inertia tensor of a rigid body. In the context of classical mechanics, the matrices that we wish to diagonalize are almost always tensors (moment of inertia tensor, stress tensor, strain tensor, etc.), and in this section I shall assume that $\mathbf{A}$ is the matrix representing an n-dimensional tensor.

Before we prove our main result, let us pause to consider the effect of changing the axes with respect to which the tensor of interest is evaluated. In general, if the $n \times n$ matrix $\mathbf{A}$ represents an arbitrary n-dimensional tensor (with respect to a given set of axes), then the matrix $\mathbf{A}'$ that represents the same tensor with respect to a different set of axes is given by the orthogonal transformation $\mathbf{A}' = \mathbf{R}\mathbf{A}\tilde{\mathbf{R}}$, where $\mathbf{R}$ is the orthogonal rotation matrix that relates the two sets of axes, as discussed in connection with Equation (15.132). Fortunately, if you haven't yet studied the orthogonal transformations of tensors, you can still follow our main proof, if you are content to consider just the case that the matrix $\mathbf{A}$ is the moment of inertia tensor $\mathbf{A} = \mathbf{I}$. This matrix was defined by the sums (10.37) and (10.38) (or the corresponding integrals, for a continuous body). If we change our axes, the set of coordinates x, y, z is replaced by a different set x', y', z', and using these new coordinates naturally leads to a different 3×3 matrix $\mathbf{I}'$, and this is all you need to know about the relation of the two matrices $\mathbf{I}$ and $\mathbf{I}'$.

We are now ready to prove the following important theorem:

Diagonalization of a Real Symmetric Tensor

If $\mathbf{A}$ is a real symmetric $n \times n$ matrix representing an n-dimensional tensor, then there exist n orthogonal unit vectors $\mathbf{e}_1, \cdots, \mathbf{e}_n$ with the properties that (1) each $\mathbf{e}_i$ is an eigenvector of $\mathbf{A}$, that is

$$\mathbf{A}\mathbf{e}_i = \lambda_i \mathbf{e}_i \tag{A.3}$$

for some real eigenvalue λ_i and (2) with respect to the axes defined by these n unit vectors, the tensor is represented by the diagonal matrix

$$\mathbf{A}' = \begin{bmatrix} \lambda_1 & 0 & \cdots & 0 \\ 0 & \lambda_2 & \cdots & 0 \\ \vdots & \vdots & \ddots & \vdots \\ 0 & 0 & \cdots & \lambda_n \end{bmatrix} \tag{A.4}$$

---

[1] In quantum mechanics, the dynamical variables are represented by complex, Hermitian matrices (rather than real, symmetric matrices). However, the problem of diagonalizing them is very similar in the two cases. Here I shall discuss just the case of real symmetric matrices.

Before we prove this, notice that since the n unit vectors $\mathbf{e}_1, \cdots, \mathbf{e}_n$ are mutually orthogonal, they are certainly linearly independent. Therefore, *any* vector in the n-dimensional space can be expanded in terms of them; that is, $\mathbf{e}_1, \cdots, \mathbf{e}_n$ form an **orthonormal basis** of the space in which they lie.

The proof of this result proceeds in several steps:

Step 1. A has at least one eigenvalue and corresponding eigenvector.   We
have seen repeatedly that the eigenvalue equation (A.3) requires that
$\det(\mathbf{A} - \lambda\mathbf{1}) = 0$. This determinant is an nth degree polynomial in λ so it
certainly has at least one zero, $\lambda = \lambda_1$, say.[2] Now, it is a well-known result of
linear algebra[3] that, if $\det(\mathbf{B}) = 0$, then there exists at least one nonzero vector
$\mathbf{a}$ such that $\mathbf{B}\mathbf{a} = 0$. Therefore, since $\det(\mathbf{A} - \lambda_1\mathbf{1}) = 0$, there exists at least
one eigenvector $\mathbf{a}$ such that

$$\mathbf{A}\mathbf{a} = \lambda_1\mathbf{a}. \tag{A.5}$$

Step 2. The eigenvalue λ_1 is real.   Nothing we have said so far guarantees that
the eigenvalue λ_1 and eigenvector $\mathbf{a}$ are real. To show that λ_1 is, consider the
following: If we multiply (A.5) on the left by the row $\tilde{\mathbf{a}}^*$ (that is, the complex
conjugate of the transpose of the column $\mathbf{a}$), we find that

$$\lambda_1 = \frac{\tilde{\mathbf{a}}^*\mathbf{A}\mathbf{a}}{\tilde{\mathbf{a}}^*\mathbf{a}}. \tag{A.6}$$

Now, it is easy to see that both terms in this fraction are real: First,

$$\tilde{\mathbf{a}}^*\mathbf{a} = \sum_i a_i^* a_i = \sum_i |a_i|^2 > 0.$$

Meanwhile the numerator is

$$\tilde{\mathbf{a}}^*\mathbf{A}\mathbf{a} = (\tilde{\mathbf{a}}^*\mathbf{A}\mathbf{a})\tilde{} = \tilde{\mathbf{a}}\mathbf{A}\mathbf{a}^* = (\tilde{\mathbf{a}}^*\mathbf{A}\mathbf{a})^*$$

which shows that $\tilde{\mathbf{a}}^*\mathbf{A}\mathbf{a}$ is real. [For the first equality, I used the fact that the
left side is a 1×1 matrix and hence equal to its transpose; for the second I
used the well-known result that $(\mathbf{m}\mathbf{n}\mathbf{p})\tilde{} = \tilde{\mathbf{p}}\tilde{\mathbf{n}}\tilde{\mathbf{m}}$ and that our given matrix $\mathbf{A}$ is
symmetric; in the last, I used the fact that $\mathbf{A}$ is real.] Since both numerator and
denominator in (A.6) are real (and the denominator nonzero), it follows that
the eigenvalue λ_1 is real. (Notice that this argument applies to any eigenvalue
of $\mathbf{A}$; thus, *any* eigenvalue of a real symmetric matrix is real.)

Step 3. The eigenvector can be taken to be real.   One might be tempted to expect
that the eigenvectors of a real matrix are necessarily real, but this is actually

---

[2] In general, an nth degree polynomial has n zeroes, but some (or even all) of these can be equal. Nevertheless, we're certainly safe in claiming that there is at least one zero.

[3] See, for example, *Mathematical Methods for Scientists and Engineers* by Donald A. McQuarrie (University Science Books, 2003), page 434, or *Mathematical Methods in Physical Sciences* by Mary Boas (Wiley, 1983), page 133.

false, for, if a real vector $\mathbf{a}$ satisfies (A.5), then so too does $i\mathbf{a}$, which is certainly not real. Thus, an eigenvector of $\mathbf{A}$ can in general be complex. However, taking the complex conjugate of (A.5) and remembering that $\mathbf{A}$ and λ_1 are real, we see that, if $\mathbf{a}$ is an eigenvector, so is $\mathbf{a}^*$. This in turn means that both of the vectors $\mathbf{a} + \mathbf{a}^*$ and $i(\mathbf{a} - \mathbf{a}^*)$ are likewise. Since both of these are real and at least one is nonzero, we have shown that for every eigenvalue there is at least one *real* eigenvector. Therefore, we can, without loss of generality, assume that the eigenvector $\mathbf{a}$ is real.

Step 4. Choose a new basis including the eigenvector.   Our next step is to normalize the real eigenvector $\mathbf{a}$ and to choose a new orthonormal basis with this normalized eigenvector as its first unit vector. That is, we define the unit vector

$$\mathbf{e}_1 = \frac{\mathbf{a}}{|\mathbf{a}|} \tag{A.7}$$

which also satisfies the eigenvalue equation (A.5),

$$\mathbf{A}\mathbf{e}_1 = \lambda_1 \mathbf{e}_1, \tag{A.8}$$

and then choose $n-1$ more unit vectors orthogonal to $\mathbf{e}_1$ and to each other to define a new set of orthogonal axes.[4] With respect to this new basis, the vector $\mathbf{e}_1$ is represented by a column whose first entry is a 1 and all of whose other entries are zero. The eigenvalue equation (A.8) implies that (with respect to our new basis) the first column of the matrix representing $\mathbf{A}$ has λ_1 for its first entry and zeroes for all the rest. Since the matrix is symmetric, it follows that it has the form

$$(\text{new matrix } \mathbf{A} \text{ with respect to new basis}) = \begin{bmatrix} \lambda_1 & 0 & \cdots & 0 \\ \hline 0 & & & \\ \vdots & & \mathbf{A}_1 & \\ 0 & & & \end{bmatrix} \tag{A.9}$$

where $\mathbf{A}_1$ is an $(n-1) \times (n-1)$ real symmetric matrix.

Step 5. Repeat steps 1 through 4 on the matrix $\mathbf{A}_1$.   The matrix $\mathbf{A}_1$ can be viewed as acting on the $(n-1)$-dimensional subspace orthogonal to our first new basis vector $\mathbf{e}_1$. It has at least one real eigenvalue λ_2 and a corresponding eigenvector, which we can take to be real and normalize to give our second unit vector $\mathbf{e}_2$. We now choose an orthonormal basis comprising $\mathbf{e}_1$, $\mathbf{e}_2$, and $n-2$ other unit

---

[4] In three dimensions, it is easy to see that this is always possible. Given $\mathbf{e}_1$ we just choose any two unit vectors in the plane perpendicular to $\mathbf{e}_1$. In n dimensions the argument is essentially the same; to make it watertight, one can use the *Gram-Schmidt orthogonalization procedure*. See *Mathematical Methods for Scientists and Engineers* by Donald A. McQuarrie (University Science Books, 2003), page 448.

vectors, and with respect to this second new basis, the matrix representing our tensor has the form

(new matrix $\mathbf{A}$ with respect to second new basis) $=$

$$\begin{bmatrix} \lambda_1 & 0 & 0 & \cdots & 0 \\ 0 & \lambda_2 & 0 & \cdots & 0 \\ \hline 0 & 0 & & & \\ \vdots & \vdots & & \mathbf{A}_2 & \\ 0 & 0 & & & \end{bmatrix} \qquad (A.10)$$

where $\mathbf{A}_2$ is an $(n-2) \times (n-2)$ real symmetric matrix.

Step 6. Repeat steps 1 through 4 on $\mathbf{A}_2$, then $\mathbf{A}_3$, etc.   After $n-3$ further repetitions, the matrix representing our tensor will have the diagonal form claimed in (A.4), and this completes our proof.

A.2   Simultaneous Diagonalization of Two Matrices

In Chapter 11 we saw that a system with n degrees of freedom, oscillating about a position of stable equilibrium, obeys an equation of motion of the form

$$\mathbf{M\ddot{q}} = -\mathbf{Kq}, \qquad (A.11)$$

where $\mathbf{q}$ is a column of n generalized coordinates, and $\mathbf{M}$ and $\mathbf{K}$ are $n \times n$ real symmetric matrices, called the mass and spring-constant matrices respectively. In what follows it is important that both $\mathbf{M}$ and $\mathbf{K}$ are *positive definite* matrices. To see what this means, consider first the matrix $\mathbf{K}$: According to (11.53), the potential energy is $U = \frac{1}{2}\tilde{\mathbf{q}}\mathbf{Kq}$. This is zero at the equilibrium position $\mathbf{q} = 0$ and, since the equilibrium is stable, U must be greater than 0 for any $\mathbf{q} \neq 0$. Therefore, the matrix $\mathbf{K}$ must have the property that $\tilde{\mathbf{q}}\mathbf{Kq} > 0$ for any $\mathbf{q}$ not equal to zero — the defining property of a positive definite matrix. Similarly, according to (11.54), the kinetic energy is $T = \frac{1}{2}\tilde{\mathbf{q}}\mathbf{M\dot{q}}$, and this must be positive for any $\dot{\mathbf{q}} \neq 0$; that is, $\mathbf{M}$ also must be positive definite.

We defined a normal mode as any motion in which all n coordinates oscillate sinusoidally at the same frequency ω, so that $\mathbf{q}(t) = \text{Re}(\mathbf{a}e^{i\omega t})$, and we saw that a normal mode is possible if and only if ω and $\mathbf{a}$ satisfy

$$\mathbf{Ka} = \omega^2\mathbf{Ma}. \qquad (A.12)$$

I claimed (and in some specific examples we saw explicitly) that there are n independent solutions $\mathbf{a}$ of this generalized eigenvalue equation and hence that *any* possible motion can be expressed as a linear combination of the normal modes.

To prove this claim, notice that if we expand a solution of (A.11) in terms of the n independent solutions $\mathbf{a}$ of (A.12), then the new generalized coordinates $\mathbf{q}'$ satisfy[5]

$$\ddot{q}'_i = -\omega_i^2 q'_i. \tag{A.13}$$

Comparing (A.13) with (A.11), we see that we have to prove that there exists a basis of the n-dimensional configuration space with respect to which the matrices $\mathbf{M}$ and $\mathbf{K}$ are both diagonal, with the forms

$$\mathbf{M}' = \mathbf{1} = \begin{bmatrix} 1 & \cdots & 0 \\ \vdots & \ddots & \vdots \\ 0 & \cdots & 1 \end{bmatrix} \quad \text{and} \quad \mathbf{K}' = \begin{bmatrix} \omega_1^2 & \cdots & 0 \\ \vdots & \ddots & \vdots \\ 0 & \cdots & \omega_n^2 \end{bmatrix}. \tag{A.14}$$

In particular, in the new basis, the mass matrix is just the unit matrix. We shall prove this, although we shall see that in general the new basis is not orthonormal. The proof leans heavily on our previous proof and, like it, proceeds in several steps.

Step 1. Diagonalize M. Since $\mathbf{M}$ is real and symmetric, we can find an $n \times n$ orthogonal matrix $\mathbf{R}$ which diagonalizes $\mathbf{M}$. That is, $\mathbf{M}' = \mathbf{R}\mathbf{M}\tilde{\mathbf{R}}$ is diagonal, with the form

$$\mathbf{M}' = \mathbf{R}\mathbf{M}\tilde{\mathbf{R}} = \begin{bmatrix} \mu_1 & \cdots & 0 \\ \vdots & \ddots & \vdots \\ 0 & \cdots & \mu_n \end{bmatrix}. \tag{A.15}$$

If we define $\mathbf{q}' = \mathbf{R}\mathbf{q}$ and $\mathbf{K}' = \mathbf{R}\mathbf{K}\tilde{\mathbf{R}}$, then, with respect to the new coordinates, the eigenvalue equation (A.12) becomes $\mathbf{K}'\mathbf{a}' = \omega^2 \mathbf{M}'\mathbf{a}'$.

Step 2. Rescale coordinates so that $\mathbf{M}'' = \mathbf{1}$. In terms of the new coordinates $\mathbf{q}'$, the kinetic energy is

$$T = \tfrac{1}{2}\dot{\tilde{\mathbf{q}}}'\mathbf{M}'\dot{\mathbf{q}}' = \sum \mu_i \dot{q}_i'^2. \tag{A.16}$$

Since this must be positive for any $\dot{\mathbf{q}}' \neq 0$, all of the numbers μ_i in (A.15) must be positive. Therefore, we can scale each of the coordinates q_i' up by a factor of $\sqrt{\mu_i}$. Specifically, we'll define new coordinates $q_i'' = q_i'\sqrt{\mu_i}$. If, in addition, we define a diagonal (though not orthogonal) matrix

$$\mathbf{S} = \begin{bmatrix} 1/\sqrt{\mu_1} & \cdots & 0 \\ \vdots & \ddots & \vdots \\ 0 & \cdots & 1/\sqrt{\mu_n} \end{bmatrix} \tag{A.17}$$

and set

$$\mathbf{M}'' = \mathbf{S}\mathbf{M}'\tilde{\mathbf{S}} = \mathbf{1} \quad \text{and} \quad \mathbf{K}'' = \mathbf{S}\mathbf{K}'\tilde{\mathbf{S}}, \tag{A.18}$$

then, with respect to these new coordinates, the mass matrix is just the unit matrix, the kinetic energy has the simple form $T = \tfrac{1}{2}\sum \dot{q}''^2$, and, most important, since $\mathbf{M}'' = \mathbf{1}$,

---

[5] Notice that the coordinates that I am here calling q_i' are precisely the normal coordinates that I introduced in Section 11.7 (where I called them ξ_i).

the generalized eigenvalue equation $\mathbf{Ka} = \omega^2 \mathbf{Ma}$ has become the *ordinary* eigenvalue equation

$$\mathbf{K}''\mathbf{a}'' = \omega^2 \mathbf{a}''. \tag{A.19}$$

Step 3. Diagonalize $\mathbf{K}''$. According to Section A.1, there is an orthogonal matrix $\mathbf{T}$ which diagonalizes $\mathbf{K}''$. That is, if we define

$$\mathbf{K}''' = \mathbf{T}\mathbf{K}''\tilde{\mathbf{T}} \qquad \text{and} \qquad \mathbf{M}''' = \mathbf{T}\mathbf{M}''\tilde{\mathbf{T}} = \mathbf{1}, \tag{A.20}$$

then both $\mathbf{K}'''$ and $\mathbf{M}'''$ are diagonal matrices, with $\mathbf{M}'''$ still equal to $\mathbf{1}$. This proves the existence of the required n eigenvectors with the advertised properties.[6]

---

[6] One small point: Since the eigenvalues are supposed to be the squares of the normal frequencies, it is essential that they be positive. This is assured since $\mathbf{K}$ and hence $\mathbf{K}'''$ are positive definite.

Further Reading

Texts on classical mechanics at about the level of this book

- Ralph Baierlein, *Newtonian Mechanics* (McGraw-Hill, 1983)
- Vernon Barger and Martin Olsson, *Classical Mechanics: A Modern Perspective* (2nd edition, McGraw-Hill, 1995)
- Grant Fowles and George Cassiday, *Analytical Mechanics* (6th edition, Saunders, 1999)
- T. W. B. Kibble and F. H. Berkshire, *Classical Mechanics* (4th edition, Longman, 1996)
- Keith Symon, *Mechanics* (3rd edition, Addison-Wesley, 1971)
- Stephen Thornton and Jerry Marion, *Classical Dynamics of Particles and Systems* (5th edition, Thomson, 2004)

More advanced texts on classical mechanics

- Louis Hand and Janet Finch, *Analytical Mechanics* (Cambridge University Press, 1998) — an undergraduate text, but distinctly more advanced than any of the above.
- Herbert Goldstein, Charles Poole, and John Safko, *Classical Mechanics* (3rd edition, Addison-Wesley, 2002) — an astonishingly successful and enduring graduate text, first published in 1950.

Books on mathematical methods

- Mary Boas, *Mathematical Methods in the Physical Sciences* (2nd edition, Wiley, 1983) — an undergraduate text, beautifully written and comprehensive; still one of the best around.
- Donald McQuarrie, *Mathematical Methods for Scientists and Engineers* (University Science Books, 2003) — a new entry at the undergraduate level,

which has received glowing reviews — readable and very comprehensive, with 1161 pages.

- Jon Mathews and R. L. Walker, *Mathematical Methods of Physics* (2nd edition, W. A. Benjamin, 1970) — a graduate text, but very accessible.

Tables of integrals and other mathematical formulas

- M. Abramowitz and I. Stegun, *Handbook of Mathematical Functions* (Dover, 1965).
- H. B. Dwight, *Tables of Integrals and Other Mathematical Data* (4th edition, MacMillan, 1961)
- Alan Jeffrey, *Handbook of Mathematical Formulas and Integrals* (2nd edition, Academic Press, 2000)

Books on Chaos

- James Gleick, *Chaos, Making a New Science* (Viking-Penguin, 1987) — a highly readable, nontechnical history of chaos theory.
- Gregory Baker and Jerry Gollub, *Chaotic Dynamics: An Introduction* (2nd edition, Cambridge University Press, 1996) — a pioneering undergraduate text on chaos, which naturally covers much more ground than here.
- Steven H. Strogatz, *Nonlinear Dynamics and Chaos* (Addison-Wesley, 1994) — a beautiful, mathematical account of many aspects of chaos theory.

Books on relativity

- Albert Einstein, *Relativity* (15th edition, Crown, 1961) — a readable account by the great man himself.
- Wolfgang Rindler, *Introduction to Special Relativity* (Oxford University Press, 1991) — a very nice account that goes a bit further than here, by one of the experts.
- James Hartle, *Gravity: An Introduction to Einstein's General Relativity* (Addison-Wesley, 2003) — an excellent account of the general theory written for undergraduates.
- C. W. Misner, K. S. Thorne, and J. A. Wheeler, *Gravitation* (Freeman, 1970) — a classic, and still the most comprehensive text on general relativity.

Books on continuum mechanics

- Gerard Middleton and Peter Wilcock, *Mechanics in the Earth and Environmental Sciences* (Cambridge University Press, 1994)
- D. S. Chandrasekharaiah and Lokenath Debnath, *Continuum Mechanics* (Academic Press, 1994)
- Lawrence E. Malvern, *Introduction to the Mechanics of a Continuous Medium* (Prentice Hall, 1969)

Answers for Odd-Numbered Problems

Chapter 1

1.1 $\mathbf{b} + \mathbf{c} = 2\hat{\mathbf{x}} + \hat{\mathbf{y}} + \hat{\mathbf{z}}$, $\quad 5\mathbf{b} + 2\mathbf{c} = 7\hat{\mathbf{x}} + 5\hat{\mathbf{y}} + 2\hat{\mathbf{z}}$, $\quad \mathbf{b} \cdot \mathbf{c} = 1$, $\quad \mathbf{b} \times \mathbf{c} = \hat{\mathbf{x}} - \hat{\mathbf{y}} - \hat{\mathbf{z}}$.

1.5 $\theta = \arccos\sqrt{2/3} = 0.615$ rad or $35.3°$.

1.11 The particle moves counterclockwise around the ellipse $(x/b)^2 + (y/c)^2 = 1$ in the xy plane, making one complete orbit in a period $2\pi/\omega$.

1.23 $\mathbf{v} = (\lambda\mathbf{b} - \mathbf{b} \times \mathbf{c})/b^2$.

1.25 Any solution has the form $f(t) = Ae^{-3t}$, which contains one arbitrary constant.

1.27 As seen from the ground the puck travels straight across the turntable passing through the center O. As seen by an observer sitting on the turntable, the puck follows a curving path as shown.

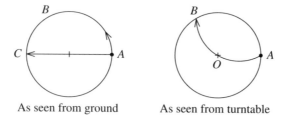

As seen from ground   As seen from turntable

1.35 The position is $\mathbf{r} = (v_0 t \cos\theta, 0, v_0 t \sin\theta - \frac{1}{2}gt^2)$. The time to return to ground is $t = (2v_0 \sin\theta)/g$, and the distance traveled is $(2v_0^2 \sin\theta \cos\theta)/g$.

1.37 (a) If we measure x straight up the slope, $x = v_0 t - \frac{1}{2}gt^2 \sin\theta$. **(b)** Time to return, $t = 2v_0/(g\sin\theta)$.

1.39 $x = v_0 t \cos\theta - \frac{1}{2}gt^2 \sin\phi$, $\quad y = v_0 t \sin\theta - \frac{1}{2}gt^2 \cos\phi$, $\quad z = 0$.

1.41 Tension $= m\omega^2 R$ (or mv^2/R).

1.47 (a) $\rho = \sqrt{x^2 + y^2}$, $\phi = \arctan(y/x)$ (chosen to lie in the correct quadrant), and z is the same as in Cartesians. The coordinate ρ is the perpendicular distance from P to the z axis. If we use r for the coordinate ρ, then r is not the same thing as $|\mathbf{r}|$ and $\hat{\mathbf{r}}$ is not the unit vector in the direction of $\mathbf{r}$ [see part (b)].

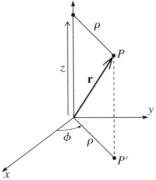

(b) The unit vector $\hat{\boldsymbol{\rho}}$ points in the direction of increasing ρ (with ϕ and z fixed), that is, directly away from the z axis; $\hat{\boldsymbol{\phi}}$ is tangent to a horizontal circle through P centered on the z axis (counterclockwise, seen from above); $\hat{\mathbf{z}}$ is parallel to the z axis. $\mathbf{r} = \rho\hat{\boldsymbol{\rho}} + z\hat{\mathbf{z}}$.
(c) $a_\rho = \ddot{\rho} - \rho\dot{\phi}^2$, $\quad a_\phi = \rho\ddot{\phi} + 2\dot{\rho}\dot{\phi}$, $\quad a_z = \ddot{z}$.

1.49 $\phi = \phi_\mathrm{o} + \omega t$ and $z = z_\mathrm{o} + v_\mathrm{oz}t - \frac{1}{2}gt^2$.

1.51 In the picture, the solid curve is a numerical solution of the differential equation (found with Mathematica's NDSolve). The dashed curve is the small-oscillation approximation (1.57) with the same initial condition ($\phi_\mathrm{o} = \pi/2$). Considering how large the initial angle is, the small-angle approximation does remarkably well. The only significant discrepancy is that the approximation oscillates somewhat too fast, as one would expect. (For large amplitudes, the true period is a little longer.)

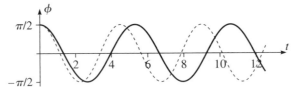

Chapter 2

2.1 The two forces are about equal when $v \approx 1$ cm/s, and if $v \gg 1$ cm/s the linear force is negligible. For a beachball, the corresponding speed is about 1 mm/s.

2.3 (b) $R \approx 0.01$, and it is very safe to neglect the quadratic drag.

2.5

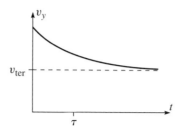

2.7 If $F = F_o$, a constant, $v = v_o + at$, where $a = F_o/m$.

2.11 (a) $v_y(t) = -v_{ter} + (v_o + v_{ter})e^{-t/\tau}$ and $y(t) = -v_{ter}t + (v_o + v_{ter})\tau(1 - e^{-t/\tau})$.
(b) $t_{top} = \tau \ln(1 + v_o/v_{ter})$ and $y_{max} = [v_o - v_{ter}\ln(1 + v_o/v_{ter})]\tau$.

2.13 $v = \pm\omega\sqrt{x_o^2 - x^2}$, where $\omega = \sqrt{k/m}$, and $x(t) = x_o\cos\omega t$.

2.15 Time of flight, $t = 2v_{yo}/g$.

2.19 (a) $y = \dfrac{v_{yo}}{v_{xo}}x - \dfrac{1}{2}g\left(\dfrac{x}{v_{xo}}\right)^2$.

2.23 (a) Terminal speed = 22 m/s (ballbearing), **(b)** 140 m/s (steel shot), **(c)** 107 m/s (parachutist in free fall).

2.27 Velocity, $v(t) = v_{ter}\tan\left(\arctan\dfrac{v_o}{v_{ter}} - \dfrac{cv_{ter}}{m}t\right)$; (time to top) $= \dfrac{m}{cv_{ter}}\arctan\dfrac{v_o}{v_{ter}}$, where $v_{ter} = \sqrt{mg\sin(\theta)/c}$.

2.29

time (sec)	0	1	5	10	20	30
actual speed (m/s)	0	9.7	37.7	48.1	50.0	50.0
speed in vacuum (m/s)	0	9.8	49.0	98.0	196.0	294.0

2.31 (a) Terminal speed, $v_{ter} = 20.2$ m/s. **(b)** Time to ground, $t = 2.78$ s (2.47 in vacuum), and speed at ground, $v = 17.7$ m/s (24.2 in vacuum).

2.33 (a)

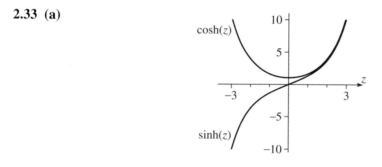

(b) $\sinh(z) = -i\sin(iz)$.
(c)

$$\frac{d\,\cosh(z)}{dz} = \sinh(z), \qquad \frac{d\,\sinh(z)}{dz} = \cosh(z),$$

$$\int \cosh(z)\,dz = \sinh(z), \qquad \int \sinh(z)\,dz = \cosh(z).$$

2.35 (b) At $t = 2\tau$ and 3τ, the speed v is 96% and 99.5% of its terminal value.

2.39 (a)

$$t = \frac{m}{\sqrt{f_{fr}c}} \left(\arctan \sqrt{\frac{c}{f_{fr}}} v_0 - \arctan \sqrt{\frac{c}{f_{fr}}} v \right).$$

(b)

v (m/s)	15	10	5	0
t (s)	6.3	18.4	48.3	142

The corresponding times if we neglect friction are (from Problem 2.26) 6.7, 20.0, 60.0, and ∞. To neglect friction, compared to the quadratic air resistance, is quite good at higher speeds, but terrible at very low speeds.

2.41 The velocity is $v(y) = \sqrt{(v_0^2 + v_{ter}^2)e^{-2gy/v_{ter}^2} - v_{ter}^2}$, where $v_{ter} = \sqrt{mg/c}$; $y_{max} = 17.7$ m, compared with 20.4 in the vacuum.

2.43 (a) The solid curve is the true trajectory, the dashed is that in a vacuum.

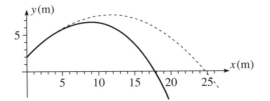

(b) The true range is 17.7 m and the range in vacuum 24.8 m.

2.45 (b) $z = 3 + 4i = 5e^{0.927i}$; **(c)** $z = 2e^{-i\pi/3} = 1 - i\sqrt{3}$.

2.47 (a) $z + w = 9 + 4i$, $z - w = 3 + 12i$, $zw = 50$, $z/w = -0.56 + 1.92i$; **(b)** $z + w = (4 + 2\sqrt{3}) + (4\sqrt{3} + 2)i$, $z - w = (4 - 2\sqrt{3}) + (4\sqrt{3} - 2)i$, $zw = 32i$, $z/w = \sqrt{3} + i$.

2.49 (b) $\cos 3\theta = \cos\theta(\cos^2\theta - 3\sin^2\theta)$ and $\sin 3\theta = \sin\theta(3\cos^2\theta - \sin^2\theta)$.

2.53 $m\dot{v}_x = qBv_y$, $m\dot{v}_y = -qBv_x$, $m\dot{v}_z = qE$.
The motion of x and y is the same as in Figure 2.15, clockwise motion around a circle at constant angular velocity $\omega = qB/m$. Meanwhile, $z = z_0 + v_{zo}t + \frac{1}{2}a_zt^2$, where $a_z = qE/m$. The particle moves in a helix or spiral of constant radius around the z axis, with an increasing pitch as the motion in the z direction accelerates.

2.55 (a) $\dot{v}_x = \omega v_y$, $\dot{v}_y = -\omega(v_x - E/B)$, and $\dot{v}_z = 0$.
(b) $v_{dr} = E/B$.
(c) $v_x = v_{dr} + (v_{xo} - v_{dr})\cos\omega t$, $v_y = -(v_{xo} - v_{dr})\sin\omega t$, and $v_z = 0$.
The transverse velocity (v_x, v_y) goes steadily around a circle of radius $(v_{xo} - v_{dr})$, with a constant drift v_{dr} in the x direction superposed.
(d) $x = v_{dr}t + R\sin\omega t$ and $y = R(\cos\omega t - 1)$
where $R = (v_{xo} - v_{dr})/\omega$. This trajectory is a cycloid, whose precise appearance depends on the initial velocity v_{xo}, as illustrated below for seven different values of v_{xo}. Notice, in particular, that if $v_{xo} = v_{dr}$ then $R = 0$ and the charge drifts straight through the fields, as we already knew. (The values of v_{xo} are shown as multiples of v_{dr} and distances as multiples of v_{dr}/ω.)

2.55 (cont)

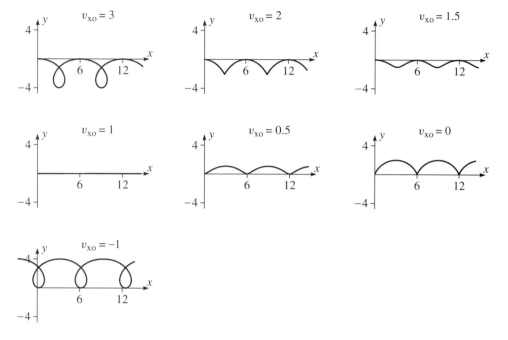

Chapter 3

3.3 The vectors $\mathbf{v}_2$ and $\mathbf{v}_3$ have equal magnitudes $v_2 = v_3 = \sqrt{2}\,v_0$, and are at 45° on either side of the initial direction.

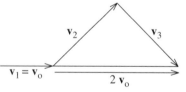

3.7 Final speed, $v \approx 2100$ m/s. Thrust $\approx 2.5 \times 10^7$ N, a little bigger than the initial weight $\approx 2.0 \times 10^7$ N.

3.9 Minimum exhaust speed ≈ 2400 m/s.

3.11 (b) and **(c)** $v = v_{\text{ex}} \ln(m_0/m) - gt \approx 900$ m/s, compared to 2100 m/s in zero gravity.

3.13 Height $\approx 4.0 \times 10^4$ m.

3.15 CM position, $\mathbf{R} = (1/6, 0, 0)$.

3.15 (cont)

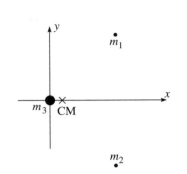

3.17 The CM is about 4.6×10^3 km from the earth's center.

3.19 (a) The CM would follow the same parabola as the unexploded shell. **(b)** The second piece hits the gun that fired it. **(c)** No.

3.21 $X = Z = 0$ and $Y = 4R/3\pi$.

3.23 Velocity of second piece is $\mathbf{v} - \Delta\mathbf{v}$. The CM (empty circles) is at the midpoint of the line joining the two fragments and clearly continues on the same parabola as the grenade followed before the explosion.

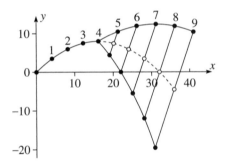

3.25 Final angular velocity, $\omega = \omega_0(r_0/r)^2$.

3.29 Final angular velocity, $\omega = \omega_0(R_0/R)^5 = \omega_0/32$.

3.31 Moment of inertia, $I = \frac{1}{2}MR^2$.

3.33 Moment of inertia, $I = \frac{2}{3}Mb^2$.

3.35 (a)

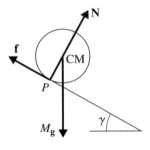

(b) and **(c)** Either way, $\dot{v} = \frac{2}{3}g \sin \gamma$.

3.37 (a)

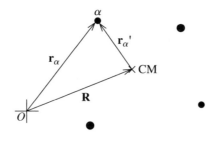

Chapter 4

4.3 (a) $W = 0$, **(b)** $W = 1$, **(c)** $W = \pi/2$.

4.7 (a) $W(\mathbf{r}_1 \rightarrow \mathbf{r}_2) = -(m\gamma/3)(y_2^3 - y_1^3); \quad U(\mathbf{r}) = (m\gamma/3)y^3$.
(b)

(c) $v_{\text{fin}} = \sqrt{2\gamma h^3/3}$.

4.9 (b) $x_o = mg/k$.

4.11

function	$\partial/\partial x$	$\partial/\partial y$	$\partial/\partial z$
$ay^2 + 2byz + cz^2$	0	$2ay + 2bz$	$2by + 2cz$
$\cos(axy^2z^3)$	$-ay^2z^3\sin(axy^2z^3)$	$-2axyz^3\sin(axy^2z^3)$	$-3axy^2z^2\sin(axy^2z^3)$
$ar = a\sqrt{x^2 + y^2 + z^2}$	ax/r	ay/r	az/r

4.13

function f	$\partial f/\partial x$	$\partial f/\partial y$	$\partial f/\partial z$	∇f
$\ln(r)$	x/r^2	y/r^2	z/r^2	$\hat{\mathbf{r}}/r$
r^n	nxr^{n-2}	nyr^{n-2}	nzr^{n-2}	$nr^{n-1}\hat{\mathbf{r}}$
$g(r)$	$g'(r)x/r$	$g'(r)y/r$	$g'(r)z/r$	$g'(r)\hat{\mathbf{r}}$

4.15 Using (4.35) we get $\Delta f \approx 0.44$, compared with the exact $\Delta f = 0.45$ (to two figures).

4.19 (a) The surface $x^2 + 4y^2 = K$ is an elliptical cylinder, centered on the z axis, with "radius" $\sqrt{K}$ in the x direction and half that in the y direction. **(b)** The unit normal to the surface is $\mathbf{n} = (1, 4, 0)/\sqrt{17}$ (or $-\mathbf{n}$). The direction of maximum increase is $\mathbf{n}$ (and maximum decrease is $-\mathbf{n}$).

4.21 The gravitational potential energy is $U(r) = -GMm/r$.

4.23 (a) F is conservative and $U = -\frac{1}{2}k(x^2 + 2y^2 + 3z^2)$. **(b) F** is conservative and $U = -kxy$. **(c) F** is not conservative. In (a) and (b), U was chosen to be zero at the origin.

4.29 (a)

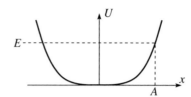

(b) The time to reach A is $t(0 \to A) = \sqrt{m/2k} \int_0^A dx/\sqrt{A^4 - x^4}$. The period is $4t(0 \to A)$.
(d) $\tau = 3.71$.

4.31 (a) Dropping all uninteresting constants, we have $E = \frac{1}{2}(m_1 + m_2)\dot{x}^2 - (m_1 - m_2)gx$.

4.33 (b)

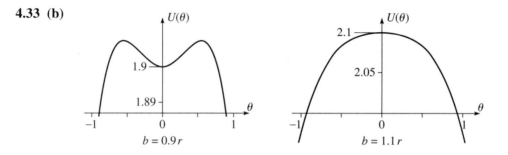

(c) For $b < r$ there can be two further equilibrium points (symmetrically placed on either side of $\theta = 0$), both of which are unstable.

4.35 (a) $E = \frac{1}{2}(m_1 + m_2 + I/R^2)\dot{x}^2 - (m_1 - m_2)gx$ (plus a constant that we may as well drop).
(b) Equation of motion is $(m_1 + m_2 + I/R^2)\ddot{x} = (m_1 - m_2)g$, either way.

4.37 (a) $U(\phi) = MgR(1 - \cos\phi) - mgR\phi$.
(b) There are equilibrium positions only if $m \le M$. If $m = M$, there is one equilibrium (unstable) at $\phi = 90°$. If $m < M$, there are two positions determined by the condition $m = M\sin\phi$, which has two solutions symmetrically placed above and below $\phi = 90°$. The lower position is stable, the upper unstable. [See the pictures of part (c)].
(c)

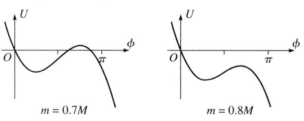

If $m = 0.7M$, the wheel swings up to a maximum $\phi < \pi$, then swings back to $\phi = 0$, and oscillates indefinitely. If $m = 0.8M$, the wheel swings past $\phi = \pi$ and continues to rotate counterclockwise until the string runs out. **(d)** The critical value of m/M is 0.725.

4.39 (c) If $\Phi = 45°$, this approximation gives $\tau = 1.037\tau_0$, which represents a 3.7% correction to the small-amplitude approximation (τ_0), and is itself within 0.3% of the exact answer ($1.040\tau_0$).

4.51 $U(\mathbf{r}_1, \mathbf{r}_2, \mathbf{r}_3, \mathbf{r}_4)$

$$= [U_{12}(\mathbf{r}_1 - \mathbf{r}_2) + U_{13}(\mathbf{r}_1 - \mathbf{r}_3) + U_{14}(\mathbf{r}_1 - \mathbf{r}_4) + U_{23}(\mathbf{r}_2 - \mathbf{r}_3)$$

$$+ U_{24}(\mathbf{r}_2 - \mathbf{r}_4) + U_{34}(\mathbf{r}_3 - \mathbf{r}_4)]$$

$$+ [U_1^{\text{ext}}(\mathbf{r}_1) + U_2^{\text{ext}}(\mathbf{r}_2) + U_3^{\text{ext}}(\mathbf{r}_3) + U_4^{\text{ext}}(\mathbf{r}_4)].$$

4.53 (b) $E = T_1 + T_2 + U_1 + U_2 + U_{12} = \frac{1}{2}mv_1^2 + \frac{1}{2}mv_2^2 - ke^2 \left(\dfrac{1}{r_1} + \dfrac{1}{r_2} - \dfrac{1}{r_{12}} \right).$

(c) Long before: $E = T_1 + T_2 + U_1 + 0 + 0 = T_2 - \dfrac{ke^2}{2r}.$

Long after: $E' = T_1' + T_2' + 0 + U_2' + 0 = T_1' - \dfrac{ke^2}{2r'}.$

By conservation of energy, $T_1' = T_2 + \frac{1}{2}ke^2 \left(\dfrac{1}{r'} - \dfrac{1}{r} \right).$

Chapter 5

5.3 $U(\phi) = mgl(1 - \cos\phi)$ and $k = mgl.$

5.5 (a) $B_1 = C_1 + C_2$ and $B_2 = i(C_1 - C_2).$
(b) $A = \sqrt{B_1^2 + B_2^2}$ and $\delta = \arctan(B_2/B_1)$, chosen in the right quadrant.
(c) $C = Ae^{-i\delta}$. **(d)** $C_1 = C/2$ and $C_2 = C^*/2.$

5.7 (a) $B_1 = x_0$ and $B_2 = v_0/\omega$. **(b)** $\omega = 10$ s^{-1}, $B_1 = 3$ m, $B_2 = 5$ m.

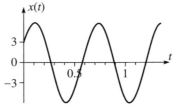

(c) The first time $x = 0$ is $t = 0.26$ s; the first time $\dot{x} = 0$ is $t = 0.10$ s.

5.9 Period, $\tau = 1.05$ s.

5.11
$$A = \sqrt{\dfrac{x_2^2 v_1^2 - x_1^2 v_2^2}{v_1^2 - v_2^2}} \qquad \text{and} \qquad \omega = \sqrt{\dfrac{v_1^2 - v_2^2}{x_2^2 - x_1^2}}.$$

5.13
$$r_0 = \lambda R \qquad \text{and} \qquad \omega = \sqrt{\dfrac{2U_0}{m\lambda R^2}}.$$

5.17 (a) If the fraction p/q is in its lowest terms, $\tau = 2\pi p/\omega_x.$

5.19 $k' = 2k(2a - l_0)/a.$

5.23 $dE/dt = \dot{x}(m\ddot{x} + kx).$

5.25 **(a)** and **(b)**

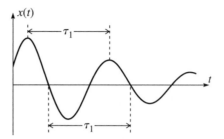

(c) With $\beta = \omega_\text{o}/2$, the amplitude shrinks by a factor of 0.027 in one period (much more than in the picture, for which β was chosen to be $\omega_\text{o}/10$ and the shrinkage factor is about 0.53).

5.29 $\tau_1 = 1.006$ sec, and $\beta = 0.110\omega_\text{o}$.

5.31 Each picture shows $x(t)$ as a function of t for the value of β indicated.

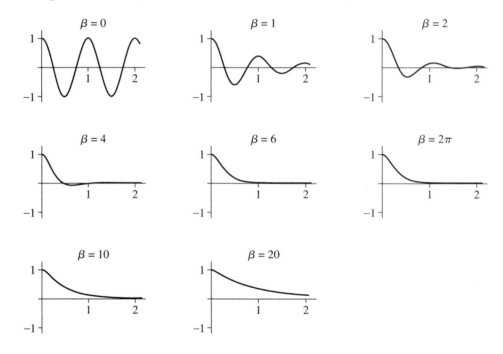

5.37 $A = 26.9$, $\delta = 3.04$ rad, $B_1 = 26.7$, and $B_2 = -6.18$.

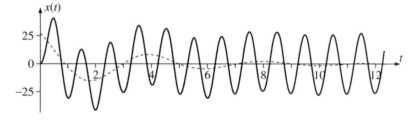

The solid curve is the actual motion; the dashed curve is the transient, homogeneous solution.

5.43 **(a)** $k \approx 4 \times 10^4$ N/m. **(b)** $f \approx 6$ Hz. **(c)** $v \approx 5$ m/s or roughly 10 mph.

5.49 $a_0 = f_{max}/2$ and $a_n = 4f_{max}/(n\pi)^2$, for n odd but 0 for n even $(n > 0)$; $b_n = 0$ for all n.

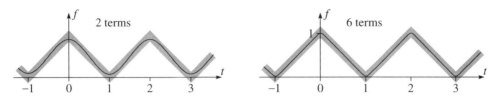

The left picture shows the sum of the first two terms (the constant term plus the first cosine) and the "sawtooth function" itself. The right picture shows the sum of the first six terms. This follows the sawtooth so closely that it is hard to tell them apart except at the corners.

5.53 $A_0 = 1/2\omega_o^2$, $A_n = 4/n^2\pi^2\sqrt{(\omega_o^2 - n^2\omega^2)^2 + (2\beta n\omega)^2}$ for n odd but zero for n even (> 0).
(a) With $\tau_o = 2$, $\omega_o = \pi$, and the first four coefficients A_n ($n = 0, 1, 2, 3$) are 0.0507, 0.6450, 0, and 0.0006.

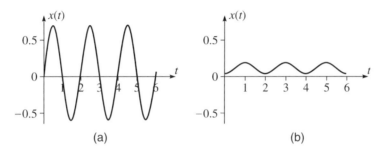

(b) With $\tau_o = 3$, $\omega_o = 2\pi/3$, and the first four coefficients A_n ($n = 0, 1, 2, 3$) are 0.1140, 0.0734, 0, and 0.0005.

5.57

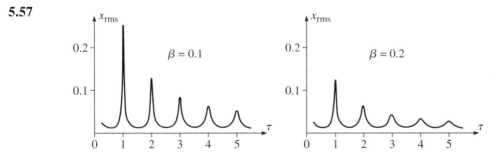

The left picture shows the data for this problem; the right shows the data of Figure 5.26 (though drawn here to a slightly different scale). Notice that the resonances with $\beta = 0.1$ are twice as high and half as wide as those for $\beta = 0.2$.

Chapter 6

6.3 Time to travel from P_1 via Q to P_2 is $\left(\sqrt{x^2 + y_1^2 + z^2} + \sqrt{(x - x_2)^2 + y_2^2 + z^2}\right)/c$.

6.5 Time to travel from A via P to B is $2\sqrt{2}(R/c)\cos(\theta/2)$.

6.7 $\phi = az + b$, where the constants a and b are chosen so the path passes through the given endpoints. In general there are many different paths of this form.

6.9 $y = \sinh(x)/\sinh(1)$.

6.11 The path is a parabola, $x = C + (y - D)^2/4C$, with C and D constant.

6.17 $\rho = \rho_0/\cos[(\phi - \phi_0)/\sqrt{1 + \lambda^2}]$. With $\lambda = 0$, the cone becomes a plane, and this equation becomes the equation of a straight line.

6.23 (a) $v = \sqrt{(v_o \cos\phi + Vy)^2 + (v_o \sin\phi)^2}$. **(c)** $y_{\max} = 366$ miles; (time saved) = 27 minutes.

Chapter 7

7.1 $\mathcal{L} = \frac{1}{2}m(\dot{x}^2 + \dot{y}^2 + \dot{z}^2) - mgz$. The three Lagrange equations are $0 = m\ddot{x}$, $0 = m\ddot{y}$, and $-mg = m\ddot{z}$.

7.3 $\mathcal{L} = \frac{1}{2}m(\dot{x}^2 + \dot{y}^2) - \frac{1}{2}k(x^2 + y^2)$. The two Lagrange equations are $m\ddot{x} = -kx$ and $m\ddot{y} = -ky$. This is the isotropic oscillator of Section 5.3.

7.5

$$(\nabla f)_r = \frac{\partial f}{\partial r} \quad \text{and} \quad (\nabla f)_\phi = \frac{1}{r}\frac{\partial f}{\partial \phi}.$$

7.7 (a) $m_\alpha \ddot{\mathbf{r}}_\alpha = -\nabla_\alpha U$, $[\alpha = 1, \cdots, N]$, **(b)** $\mathcal{L} = \sum_\alpha \frac{1}{2}m_\alpha \dot{\mathbf{r}}_\alpha^2 - U(\mathbf{r}_1, \cdots, \mathbf{r}_N)$.

7.9 $x = R\cos\phi$, $y = R\sin\phi$, and $\phi = \arctan(y/x)$, chosen to lie in the correct quadrant.

7.11 $x = A\cos\omega t + l\sin\phi$, $y = l\cos\phi$, and $\phi = \arctan[(x - A\cos\omega t)/y]$.

7.15 $\mathcal{L} = \frac{1}{2}(m_1 + m_2)\dot{x}^2 + m_2 gx$, $a = gm_2/(m_1 + m_2)$.

7.17 $\ddot{x} = g(m_1 - m_2)/(m_1 + m_2 + I/R^2)$.

7.21 $\mathcal{L} = \frac{1}{2}m(\dot{r}^2 + r^2\omega^2)$, and $r = Ae^{\omega t} + Be^{-\omega t}$.

7.23 $\mathcal{L} = \frac{1}{2}m(\dot{x} - A\omega\sin\omega t)^2 - \frac{1}{2}kx^2$.

7.27 (Acceleration of mass $4m$) = $g/7$ downward.

7.29 $\mathcal{L} = \frac{1}{2}m\left[R^2\omega^2 + l^2\dot{\phi}^2 + 2Rl\omega\dot{\phi}\sin(\phi - \omega t)\right] - mg(R\sin\omega t - l\cos\phi)$ and $l\ddot{\phi} = -g\sin\phi + \omega^2 R\cos(\phi - \omega t)$.

7.31 (a) $\mathcal{L} = \frac{1}{2}(m + M)\dot{x}^2 + \frac{1}{2}M\left(L^2\dot{\phi}^2 + 2\dot{x}L\dot{\phi}\cos\phi\right) - \frac{1}{2}kx^2 + MgL\cos\phi$. The x and ϕ equations are
$(m + M)\ddot{x} + ML(\ddot{\phi}\cos\phi - \dot{\phi}^2\sin\phi) = -kx$ and $M(L\ddot{\phi} + \ddot{x}\cos\phi) = -Mg\sin\phi$.
(b) With x and ϕ both small, these become
$(m + M)\ddot{x} + ML\ddot{\phi} = -kx$ and $M(L\ddot{\phi} + \ddot{x}) = -Mg\phi$.

7.33 $x(t) = x_o \cosh\omega t + (g/2\omega^2)(\sin\omega t - \sinh\omega t)$.

7.35 $\mathcal{L} = \frac{1}{2}mR^2\left[\omega^2 + (\dot{\phi} + \omega)^2 + 2\omega(\dot{\phi} + \omega)\cos\phi\right]$. For small oscillations about B, the angular frequency is ω.

7.37 (a) $\mathcal{L} = m\dot{r}^2 + \frac{1}{2}mr^2\dot{\phi}^2 - mgr$. **(b)** The r and ϕ equations are $mr\dot{\phi}^2 - mg = 2m\ddot{r}$ and $mr^2\dot{\phi} = $ const. **(c)** $r_o = [\ell^2/(m^2g)]^{1/3}$. **(d)** Angular frequency $= \sqrt{3/2}\,\ell/mr_o^2$.

7.39 (a) $\mathcal{L} = \frac{1}{2}m\left(\dot{r}^2 + r^2\dot{\theta}^2 + r^2\sin^2\theta\,\dot{\phi}^2\right) - U(r)$.
(b) The r, θ, and ϕ equations are

$$m\ddot{r} = mr\left(\dot{\theta}^2 + \sin^2\theta\,\dot{\phi}^2\right) - \partial U/\partial r$$

$$\frac{d}{dt}\left(mr^2\dot{\theta}\right) = mr^2\sin\theta\cos\theta\,\dot{\phi}^2$$

$$\frac{d}{dt}\left(mr^2\sin^2\theta\,\dot{\phi}\right) = 0.$$

(c) The motion remains in the equatorial plane $\theta = \pi/2$, consistent with our knowledge that the motion is confined to a plane.
(d) The motion remains in the longitudinal plane $\phi = \phi_o$.

7.41 $\mathcal{L} = \frac{1}{2}m\left(\dot{\rho}^2 + \rho^2\omega^2 + 4k^2\rho^2\dot{\rho}^2\right) - mgk\rho^2$, and the equation of motion is
$(1 + 4k^2\rho^2)\ddot{\rho} + 4k^2\rho\dot{\rho}^2 = (\omega^2 - 2gk)\rho$.
The bottom of the wire, $\rho = 0$, is an equilibrium, which is stable if $\omega^2 < 2gk$, but unstable if $\omega^2 > 2gk$. If $\omega^2 = 2gk$, the bead is in equilibrium at *any* ρ, but the equilibrium is unstable (except at $\rho = 0$).

7.43 (a) $\mathcal{L} = \frac{1}{2}(M + m)R^2\dot{\phi}^2 - MgR(1 - \cos\phi) + mgR\phi$, and the equation of motion is $(M + m)R\ddot{\phi} = -Mg\sin\phi + mg$.
(b)

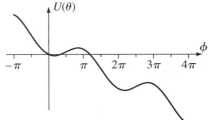

Notice that there are actually several equilibriums, separated by one or more complete revolutions.
(c)

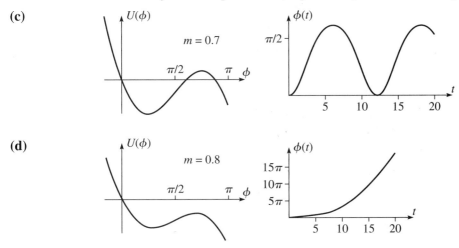

(d)

7.49 **(b)** $\mathcal{L} = \frac{1}{2}m\dot{\mathbf{r}}^2 + q\dot{\mathbf{r}}\cdot\mathbf{A} = \frac{1}{2}m(\dot{\rho}^2 + \rho^2\dot{\phi}^2 + \dot{z}^2) + \frac{1}{2}qB\rho^2\dot{\phi}$, and the three Lagrange equations are

$$m\ddot{\rho} = m\rho\dot{\phi}^2 + qB\rho\dot{\phi}, \qquad \frac{d}{dt}\left(m\rho^2\dot{\phi} + \frac{1}{2}qB\rho^2\right) = 0, \qquad \text{and} \qquad m\ddot{z} = 0.$$

7.51 $\mathcal{L}(x, y) = \frac{1}{2}m(\dot{x}^2 + \dot{y}^2) + mgy.$ **(a)** The two modified Lagrange equations are

$$\lambda\frac{x}{l} = m\ddot{x} \qquad \text{and} \qquad mg + \lambda\frac{y}{l} = m\ddot{y}.$$

Chapter 8

8.3 $y_1 = L + \frac{m_1}{M}v_0t - \frac{1}{2}gt^2 + \frac{m_2v_0}{M\omega}\sin\omega t$ and $y_2 = \frac{m_1}{M}v_0t - \frac{1}{2}gt^2 - \frac{m_1v_0}{M\omega}\sin\omega t.$

8.7 **(a)** Period $\tau = 2\pi r^{3/2}/\sqrt{Gm_2}$. **(b)** $\tau = 2\pi r^{3/2}/\sqrt{GM}$. These two answers are the same in the limit that $m_2 \to \infty$. **(c)** $\tau = 0.71$ years.

8.9 **(a)** $\mathcal{L} = \frac{1}{2}M(\dot{X}^2 + \dot{Y}^2) + \frac{1}{2}\mu(\dot{r}^2 + r^2\dot{\phi}^2) - \frac{1}{2}k(r - L)^2.$
(b) $M\ddot{X} = 0$ and $M\ddot{Y} = 0$, with solutions $\mathbf{R} = \mathbf{R}_o + \dot{\mathbf{R}}_o t$. **(c)** The r and ϕ equations are
$$\mu\ddot{r} = \mu r\dot{\phi}^2 - k(r - L) \qquad \text{and} \qquad \mu r^2\dot{\phi} = \text{const}.$$
If $r = $ const, then $\dot{\phi} = $ const, and $r = L + \mu r\dot{\phi}^2/k.$
If $\dot{\phi} = $ const, then $r = L + A\cos(\omega t - \delta)$, where $\omega = \sqrt{k/\mu} = \sqrt{2k/m_1}.$

8.13 **(a)**

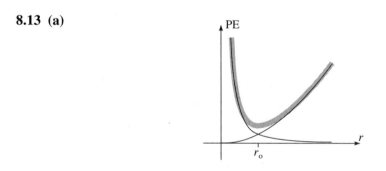

(b) $r_o = (\ell^2/k\mu)^{1/4}.$ **(c)** Angular frequency of oscillations, $\omega = \sqrt{4k/\mu}.$

8.15 Percent variation $\approx 0.1\%.$

8.19 Eccentricity, $\epsilon = 0.17$; (height when on y axis) $= 1424$ km.

8.21 **(a)** If $\ell \to 0$, then $a \to r_{\max}/2.$ **(b)** $\tau_{(\ell\to 0)} = (\pi/\sqrt{2GM})(r_{\max})^{3/2}.$
(c) $t = (\pi/2\sqrt{2GM})(r_{\max})^{3/2}.$ **(d)** and **(e)** $\tau_{(\ell=0)} = (2\pi/\sqrt{2GM})(r_{\max})^{3/2} = 2\tau_{(\ell\to 0)}.$

8.23 **(b)** $\beta = \sqrt{1 + m\lambda/\ell^2}$ and $c = \ell^2\beta^2/mk.$ **(c)** The orbit is closed if β is a rational number, $\beta = p/q$ (where p and q are integers). If $\lambda \to 0$, the orbit becomes a Kepler ellipse.

8.25 (a) $r_0 = 1$.

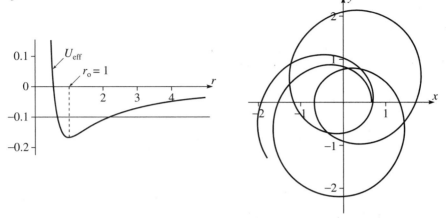

(b) You can see from the picture on the left that if $E = -0.1$, the inner turning point is at about $r_{min} = 0.7$. If we use this as a starting value for any equation-solving program (such as Mathematica's FindRoot), we find that this root of the equation $U_{eff}(r) = -0.1$ is actually $r_{min} = 0.6671$.
(c) Obviously the orbit shown on the right has not closed after 3.5 revolutions, and it clearly won't close for a long time. (In fact, it never does, but this is harder to prove.)

8.27 $c = 8.87 \times 10^7$ km, $\epsilon = 0.753$, $\delta = 1.72$ rad.

8.29 The new orbit would be a parabola, tangent to the old circular orbit at the point at which the great disappearance occurred. The earth would be just not bound.

8.31 The sketch shows the path of the relative position $\mathbf{r} = (x, y)$, that is, the orbit of particle 1 as seen from particle 2.

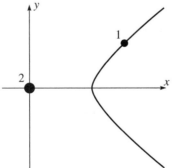

8.35 First thrust factor $= \sqrt{2/5}$;   second $= \sqrt{5/8}$.

Chapter 9

9.1 (Angle of tilt) $= \arctan(A/g)$ with vertical.

9.3 (a) $F_{tid}/mg \approx 1.1 \times 10^{-7}$. **(b)** Same magnitude, opposite direction.

9.9 $\mathbf{F}_{cor} = 2mv_0 \Omega \cos\theta$ due east; $F_{cor}/mg \approx 0.011$.

9.13 The maximum value of α is about 0.1°; the minimum is zero.

9.15 $g = g_0\sqrt{\cos^2\theta + \lambda^2\sin^2\theta}$.

9.19 **(a)** As seen from the ground, the puck moves in a straight line. As seen from the merry-go-round, its initial acceleration is radially outward; as it speeds up, it curves to its right and spirals outward from the center. **(b)** As seen from the ground, it remains stationary. As seen from the merry-go-round, it moves in a clockwise circle centered on the axis of the merry-go-round.

9.21

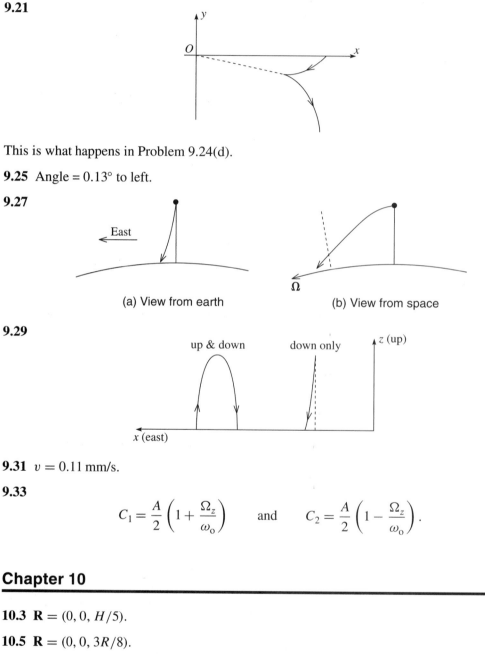

This is what happens in Problem 9.24(d).

9.25 Angle $= 0.13°$ to left.

9.27

(a) View from earth   (b) View from space

9.29

9.31 $v = 0.11$ mm/s.

9.33

$$C_1 = \frac{A}{2}\left(1 + \frac{\Omega_z}{\omega_0}\right) \qquad \text{and} \qquad C_2 = \frac{A}{2}\left(1 - \frac{\Omega_z}{\omega_0}\right).$$

Chapter 10

10.3 $\mathbf{R} = (0, 0, H/5)$.

10.5 $\mathbf{R} = (0, 0, 3R/8)$.

10.7 **(a)** $V = \frac{2}{3}\pi R^3(1 - \cos\theta_0)$; **(b)** $\mathbf{R} = (0, 0, Z)$ where $Z = \frac{3R}{16} \cdot \frac{1 - \cos 2\theta_0}{1 - \cos\theta_0}$.

10.9 $I = \frac{1}{2}MR^2$.

10.11 (a) $I(\text{solid}) = \dfrac{2}{5}MR^2$;   **(b)** $I(\text{hollow}) = \dfrac{2}{5}M\dfrac{b^5 - a^5}{b^3 - a^3}$.

10.13 (a) $\tau = 2\pi\sqrt{I/mga}$;   **(b)** $l = I/(ma)$.

10.15 $\omega = \sqrt{3g(\sqrt{2} - 1)/2a}$.

10.17 $I_{zz} = \frac{1}{5}M(a^2 + b^2)$.

10.23 All products of inertia involving z are automatically zero, $I_{xz} = I_{yz} = I_{zx} = I_{zy} = 0$.

10.25 (a) and **(b)** The inertia tensors $\mathbf{I}_{\text{cm}}$ about the CM and $\mathbf{I}_A$ about A are

$$\mathbf{I}_{\text{cm}} = \frac{1}{3}M\begin{bmatrix} b^2 + c^2 & 0 & 0 \\ 0 & c^2 + a^2 & 0 \\ 0 & 0 & a^2 + b^2 \end{bmatrix} \quad \text{and}$$

$$\mathbf{I}_A = \frac{1}{3}M\begin{bmatrix} 4(b^2 + c^2) & -3ab & -3ac \\ -3ba & 4(c^2 + a^2) & -3bc \\ -3ca & -3cb & 4(a^2 + b^2) \end{bmatrix}.$$

(c) $\mathbf{L} = \frac{1}{3}M\omega\left(4(b^2 + c^2), -3ab, -3ac\right)$.

10.27

$$\mathbf{I} = \frac{1}{4}M\begin{bmatrix} (R^2 + 2h^2) & 0 & 0 \\ 0 & (R^2 + 2h^2) & 0 \\ 0 & 0 & 2R^2 \end{bmatrix}.$$

10.35 (a)

$$\mathbf{I} = ma^2\begin{bmatrix} 10 & 0 & 0 \\ 0 & 6 & 1 \\ 0 & 1 & 6 \end{bmatrix}.$$

(b) The principal moments are $\lambda_1 = 10ma^2$, $\lambda_2 = 7ma^2$, and $\lambda_3 = 5ma^2$. The corresponding principal directions are $\mathbf{e}_1 = (1, 0, 0)$, $\mathbf{e}_2 = \frac{1}{\sqrt{2}}(0, 1, 1)$, and $\mathbf{e}_3 = \frac{1}{\sqrt{2}}(0, 1, -1)$.

10.37 (a)

$$\mathbf{I} = \begin{bmatrix} 2 & -1 & 0 \\ -1 & 2 & 0 \\ 0 & 0 & 4 \end{bmatrix}.$$

(b) $\lambda_1 = 1$, $\lambda_2 = 3$, and $\lambda_3 = 4$; $\mathbf{e}_1 = \frac{1}{\sqrt{2}}(1, 1, 0)$, $\mathbf{e}_2 = \frac{1}{\sqrt{2}}(1, -1, 0)$, and $\mathbf{e}_3 = (0, 0, 1)$.

10.39 $\Omega \approx 21$ rad/s or about 200 rpm.

10.47 About 1010 years.

10.53 $b^2 \gg 4ac$.

10.57 (a) $\mathcal{L} = \frac{1}{2}M(\dot{X}^2 + \dot{Y}^2 + R^2\dot{\theta}^2\sin^2\theta) + \frac{1}{2}\lambda_1^{\text{cm}}(\dot{\phi}^2\sin^2\theta + \dot{\theta}^2) + \frac{1}{2}\lambda_3^{\text{cm}}(\dot{\psi} + \dot{\phi}\cos\theta)^2 - MgR\cos\theta$ where λ_1^{cm} and λ_3^{cm} are the two principal moments about the CM. **(c)** The larger precession rate is bigger about the CM than about the tip. According to (10.111) the smaller rate is unchanged. [But note that (10.111) is an approximation; if we keep the next term in the approximation, we find that the smaller rate is slightly reduced.]

Chapter 11

11.1 $k_1(l_1 - L_1) = k_2(l_2 - L_2) = k_3(l_3 - L_3)$.

11.3 $\omega^2 = \frac{1}{2m_1m_2}\{m_1(k_2 + k_3) + m_2(k_1 + k_2)$

$$\pm\sqrt{m_1^2(k_2 + k_3)^2 + m_2^2(k_1 + k_2)^2 - 2m_1m_2(k_2k_3 + k_3k_1 + k_1k_2 - k_2^2)}\}.$$

11.5 **(a)** Let $m_1 = m_2 = m$ and $k_1 = k_2 = k$, and $\omega_o = \sqrt{k/m}$. Then the normal frequencies are

$$\omega_1 = \omega_o\sqrt{\frac{3 - \sqrt{5}}{2}} = 0.62\omega_o \qquad \text{and} \qquad \omega_2 = \omega_o\sqrt{\frac{3 + \sqrt{5}}{2}} = 1.62\omega_o.$$

(b) In mode 1, the two carts oscillate in phase, with the amplitude of m_2 equal to 1.62 times that of m_1. In mode 2, they oscillate exactly out of phase, with the amplitude of m_2 equal to 0.62 times that of m_1.

11.7 **(b)** $B_1 = A$ and $B_2 = C_1 = C_2 = 0$.

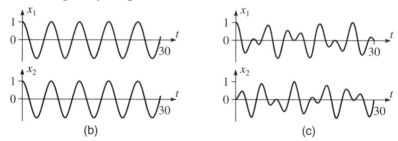

(b)   (c)

(c) $B_1 = B_2 = A/2$ and $C_1 = C_2 = 0$.

11.9 **(b)** $\xi_1 = A_1\cos(\omega_1 t - \delta_1)$ and $\xi_2 = A_2\cos(\omega_2 t - \delta_2)$, where $\omega_1 = \sqrt{k/m}$ and $\omega_2 = \sqrt{3k/m}$, and $A_1, \delta_1, A_2, \delta_2$ are all arbitrary constants. Hence

$$x_1 = A_1\cos(\omega_1 t - \delta_1) + A_2\cos(\omega_2 t - \delta_2)$$
$$x_2 = A_1\cos(\omega_1 t - \delta_1) - A_2\cos(\omega_2 t - \delta_2).$$

11.11 **(a)**
$$m\ddot{x}_1 = -2kx_1 + kx_2 - b\dot{x}_1 + F_o\cos\omega t$$
$$m\ddot{x}_2 = kx_1 - 2kx_2 - b\dot{x}_2$$

(c)
$$\xi_1(t) = A_1\cos(\omega t - \delta_1) + B_1e^{-\beta t}\cos(\omega_1 t - \delta_1^{tr})$$
$$\xi_2(t) = A_2\cos(\omega t - \delta_2) + B_2e^{-\beta t}\cos(\omega_2 t - \delta_2^{tr})$$

where the constants $A_1, A_2, \delta_1, \delta_2$ are given by Eqs. (5.64) and (5.65) (except that now $f_o = F_o/2m$ and, in the case of A_2 and δ_2, ω_o^2 is replaced by $3\omega_o^2$); the constants $B_1, B_2, \delta_1^{tr}, \delta_2^{tr}$ in the transient terms are arbitrary and determined by the initial conditions, and $\omega_1 = \sqrt{\omega_o^2 - \beta^2}$, while $\omega_2 = \sqrt{3\omega_o^2 - \beta^2}$.

11.13 **(b)** Let $\beta = b/2m$, $\omega_1 = \sqrt{k/m - \beta^2}$, and $\omega_2 = \sqrt{(k + 2k_2)/m - \beta^2}$. Then

$$\xi_1 = e^{-\beta t}(B_1\cos\omega_1 t + C_1\sin\omega_1 t) \qquad \text{and} \qquad \xi_2 = e^{-\beta t}(B_2\cos\omega_2 t + C_2\sin\omega_2 t)$$

where B_1, C_1, B_2, C_2 are arbitrary constants.

(c) With the given initial conditions (and $\beta \ll 1$)

$$x_1 = \tfrac{1}{2}Ae^{-\beta t}(\cos \omega_1 t + \cos \omega_2 t) \qquad \text{and} \qquad x_2 = \tfrac{1}{2}Ae^{-\beta t}(\cos \omega_1 t - \cos \omega_2 t).$$

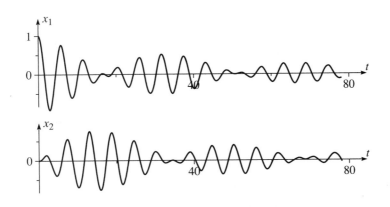

11.15 $\mathcal{L} = T - U$ where T and U are given in (11.38) and (11.37). The ϕ_1 equation is

$$(m_1 + m_2)L_1^2\ddot{\phi}_1 + m_2 L_1 L_2 \ddot{\phi}_2 \cos(\phi_1 - \phi_2) + m_2 L_1 L_2 \dot{\phi}_2^2 \sin(\phi_1 - \phi_2)$$
$$= -(m_1 + m_2)g L_1 \sin \phi_1$$

and the ϕ_2 equation is

$$m_2 L_1 L_2 \ddot{\phi}_1 \cos(\phi_1 - \phi_2) + m_2 L_2^2 \ddot{\phi}_2 - m_2 L_1 L_2 \dot{\phi}_1^2 \sin(\phi_1 - \phi_2)$$
$$= -m_2 g L_2 \sin \phi_2.$$

11.17 **(a)** $\omega_1^2 = \tfrac{3}{4}\omega_0^2$ and $\omega_2^2 = \tfrac{3}{2}\omega_0^2$, where $\omega_0 = \sqrt{g/L}$.

For the first mode, $\mathbf{a} = A\begin{bmatrix} 1 \\ 3 \end{bmatrix}$; for the second, $\mathbf{a} = A\begin{bmatrix} 1 \\ -3 \end{bmatrix}$.

(b) $\begin{bmatrix} \phi_1 \\ \phi_2 \end{bmatrix} = \dfrac{\alpha}{6}\left\{ \begin{bmatrix} 1 \\ 3 \end{bmatrix} \cos \omega_1 t - \begin{bmatrix} 1 \\ -3 \end{bmatrix} \cos \omega_2 t \right\}$, which is not periodic.

11.19 **(a)** $\mathcal{L} = \tfrac{1}{2}(m + M)\dot{x}^2 + ML\dot{x}\dot{\phi} + \tfrac{1}{2}ML^2\dot{\phi}^2 - \left(\tfrac{1}{2}kx^2 + \tfrac{1}{2}MgL\phi^2 \right)$. The x and ϕ equations are

$$(m + M)\ddot{x} + ML\ddot{\phi} = -kx \qquad \text{and} \qquad ML\ddot{x} + ML^2\ddot{\phi} = -MgL\phi.$$

(b) With the given numerical values, the normal frequencies are

$$\omega_1 = \sqrt{2 - \sqrt{2}} = 0.77 \qquad \text{and} \qquad \omega_2 = \sqrt{2 + \sqrt{2}} = 1.85.$$

In the first mode, the cart and bob oscillate in phase (both moving to the right and then both moving to the left), with the bob's amplitude (of motion relative to the cart) $\sqrt{2}$ times bigger than the cart's. In the second mode, the cart and bob oscillate exactly out of phase, again with the amplitude of the bob equal to $\sqrt{2}$ times that of the cart.

11.23 $\omega_2^2 = \dfrac{g}{L} + \dfrac{k}{m}$ and $\omega_3^2 = \dfrac{g}{L} + 3\dfrac{k}{m}$.

11.25 If $\omega_0 = \sqrt{k/m}$, then the normal frequencies are

$$\omega_1 = \omega_0 \sqrt{2 - \sqrt{2}}, \qquad \omega_2 = \omega_0 \sqrt{2}, \qquad \text{and} \qquad \omega_3 = \omega_0 \sqrt{2 + \sqrt{2}}.$$

In the first mode, all three carts oscillate in phase, with $a_1 = a_3 = a_2/\sqrt{2}$; in the second, the middle cart is stationary, while the first and third oscillate exactly out of phase, with $a_1 = -a_3$ and $a_2 = 0$; in the third, the left and right carts oscillate in phase, while the middle one is exactly out of phase, with $a_1 = a_3 = -a_2/\sqrt{2}$.

11.27 (a) $\mathcal{L} = \frac{1}{2}m(\dot{x}_1^2 + \dot{x}_2^2) - \frac{1}{2}k(x_1 - x_2)^2$. The normal frequencies are $\omega_1 = 0$ and $\omega_2 = \omega_0\sqrt{2}$ (with $\omega_0 = \sqrt{k/m}$). (b) In the second mode, the two carts oscillate with equal amplitudes, but exactly out of phase. (c) In the first mode, $x_1 = x_2 = x_0 + v_0 t$; that is, they move at constant velocity with the spring at its equilibrium length.

11.29 $\omega_1 = \sqrt{2k/m}$, $\omega_2 = \sqrt{6k/m}$, and $\omega_3 = \sqrt{g/r_0}$, where r_0 is the equilibrium value of r.

11.31 The three normal frequencies are 0, $\sqrt{2}\,\omega_0$, and $\sqrt{3}\,\omega_0$, where $\omega_0 = \sqrt{k/m}$.

11.35 (a) $\xi_1 = \phi_1 + \phi_2$ and $\xi_2 = \phi_1 - \phi_2$ (or any convenient multiple of these). (c) The first mode has $\phi_1 = \phi_2 = Ae^{-\beta t}\cos(\omega_1 t - \delta)$, and the second $\phi_1 = -\phi_2 = Ae^{-\beta t}\cos(\omega_2 t - \delta)$, where

$$\omega_1 = \sqrt{\frac{g}{L} - \beta^2}, \qquad \omega_2 = \sqrt{\frac{g}{L} + \frac{2k}{m} - \beta^2}, \qquad \text{and} \qquad \beta = \frac{b}{2m}.$$

Chapter 12

12.1 $x_1(t) = (t + k)^2 + 1$, for any constant k.

12.7

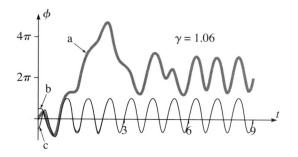

12.9

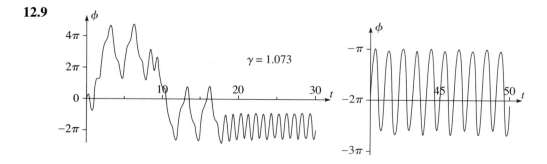

12.11 **(b)** The least-squares line has slope -1.54, giving $\delta = e^{1.54} = 4.66$, compared with the correct value 4.67.

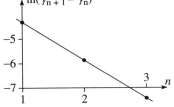

12.13 $\lambda \approx 0.64$.

12.15

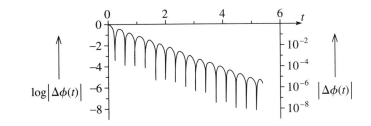

The picture confirms that $\Delta\phi(t)$ decreases exponentially.

12.17

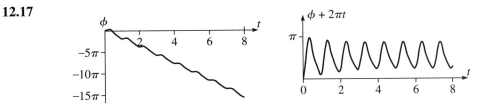

12.19 **(a)** $x = A\cos(\omega t - \delta)$, where $\omega = \sqrt{k/m}$, and A and δ are arbitrary constants.

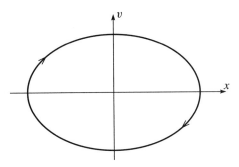

(b) $E = \frac{1}{2}kx^2 + \frac{1}{2}m\dot{x}^2 = \text{const.}$

12.21 **(a)**

x_0	x_1	x_2	x_3	x_4	x_5	$\cdots$	x_{28}	x_{29}	x_{30}
0	1.00	0.54	0.86	0.65	0.793	$\cdots$	0.7391	0.7391	0.7391
3	-0.99	0.55	0.85	0.66	0.791	$\cdots$	0.7391	0.7391	0.7391
100	0.86	0.65	0.80	0.70	0.765	$\cdots$	0.7391	0.7391	0.7391

(b) There is a single, stable attractor at $x^* = 0.739085$.

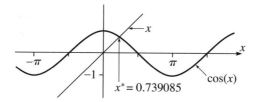

12.23 (a)

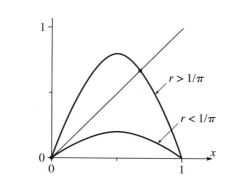

(b) $r_0 = 1/\pi = 0.318$. **(c)** $r_1 = 0.720$.

12.25

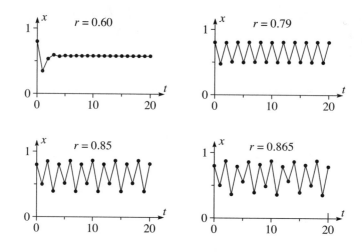

12.27 (c) $x_a = 0.5130$ and $x_b = 0.7995$.

12.29

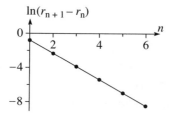

From the slope of the least-squares fit, we get $\delta = 4.69$.

12.31

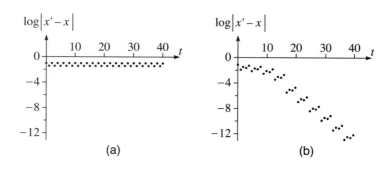

(a)   (b)

12.33 See Figure 12.41.

Chapter 13

13.1 $\mathcal{H} = p^2/2m$. The Hamilton equations are $\dot{x} = p/m$ and $\dot{p} = 0$, with solutions $p = p_0 = $ const, and $x = x_0 + v_0 t$, where $v_0 = p_0/m$.

13.3 The Hamiltonian is $\mathcal{H} = p^2/2(m_1 + m_2 + \frac{1}{2}M) - (m_1 - m_2)gx$. The Hamilton equations are $\dot{x} = p/(m_1 + m_2 + \frac{1}{2}M)$ and $\dot{p} = (m_1 - m_2)g$, and the acceleration is $\ddot{x} = g(m_1 - m_2)/(m_1 + m_2 + \frac{1}{2}M)$.

13.5 The Hamiltonian and the two Hamilton equations are
$$\mathcal{H} = \frac{p^2}{2m(c^2 + R^2)} + mgc\phi, \qquad \dot{\phi} = \frac{p}{m(c^2 + R^2)}, \qquad \text{and} \qquad \dot{p} = -mgc.$$
Combining the last two gives $\ddot{z} = c\ddot{\phi} = -g\, c^2/(c^2 + R^2)$.

13.7 (a) $\mathcal{H} = \dfrac{p^2}{2m[1 + h'(x)^2]} + mgh(x)$. (b) The Hamilton equations are
$$\dot{x} = \frac{p}{m[1 + h'(x)^2]} \qquad \text{and} \qquad \dot{p} = \frac{p^2 h'(x)h''(x)}{m[1 + h'(x)^2]^2} - mgh'(x).$$

13.9 The Hamiltonian is $\mathcal{H} = (p_x^2 + p_y^2)/2m + mgy$. The Hamilton equations are $\dot{x} = p_x/m$, $\dot{p}_x = 0$, $\dot{y} = p_y/m$, and $\dot{p}_y = -mg$.

13.11 The Hamiltonian is $\mathcal{H} = \dfrac{p_x^2 + p_y^2 + p_z^2}{2m} - p_x V + mgz$ (with x measured along the track and z vertically up).

13.13 $\mathcal{H} = \dfrac{1}{2m}\left(p_z^2 + \dfrac{p_\phi^2}{R^2}\right) + \frac{1}{2}k(R^2 + z^2)$, $z = A\cos(\omega t - \delta)$, with $\omega = \sqrt{k/m}$, and $\dot{\phi} = $ const.

13.17 (a) $z_0 = [p_\phi^2/(m^2 c^2 g)]^{1/3}$ (d) $\alpha = \arcsin(1/\sqrt{3}) = 35.3°$.

13.19 $\mathcal{H} = \dfrac{1}{2m}(p_x^2 + p_y^2) + U(\sqrt{x^2 + y^2})$.

13.21 (a) $\mathcal{H} = \dfrac{1}{2M}(P_x^2 + P_y^2) + \dfrac{1}{2\mu}\left(p_r^2 + \dfrac{p_\phi^2}{r^2}\right) + \frac{1}{2}k(r - l_0)^2$. X, Y, and ϕ are ignorable; r is not. (c) $r = l_0 + A\cos(\omega t - \delta)$, where $\omega = \sqrt{k/\mu}$ and A and δ are constants.

13.23 **(b)** The two conjugate momenta are $p_x = m(\dot{x} + \dot{y})$ and $p_y = m(\dot{x} + 4\dot{y})$. $\mathcal{H} = \dfrac{1}{2m}$ $\left[\frac{1}{3}(p_x - p_y)^2 + p_x^2\right] + \frac{1}{2}kx^2$. **(c)** $x = x_o \cos \omega t$ and $y = y_o + \frac{1}{4}x_o(1 - \cos \omega t)$, where $\omega = \sqrt{4k/3m}$.

13.25 **(c)** $P = \mathcal{H}$. **(d)** $Q = t + \text{const.}$

13.29

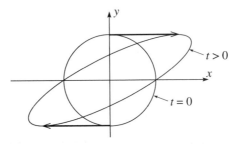

13.31 **(a)** $3k$,   **(b)** 0,   **(c)** k,   **(d)** 0.

13.33 **(a)** LHS $= 4\pi k R^3 =$ RHS. **(b)** Same.

Chapter 14

14.1 $\sigma = 0.79 \text{ cm}^2$, $n_{\text{tar}} = 0.034 \text{ cm}^{-2}$, probability $= 0.027$.

14.3 (Number density) $= 2.1 \times 10^{28} \text{ atoms/m}^2$.

14.5 $N_{\text{sc}} = 2.7 \times 10^7$.

14.7 $\Delta\Omega_{\text{moon}} = 6.45 \times 10^{-5}$ sr, and $\Delta\Omega_{\text{sun}} = 6.76 \times 10^{-5}$ sr.

14.11 $N_{\text{sc}} \approx 29$.

14.15 **(a)**

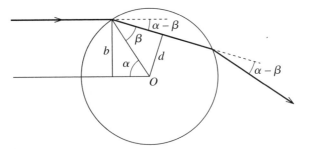

(b) $\sqrt{p_o^2 + 2mU_o}$. **(c)** See left picture.

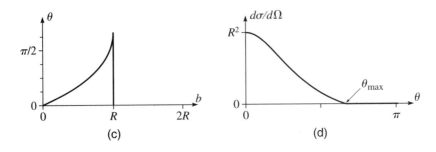

(d)

$$\frac{d\sigma}{d\Omega} = \frac{b}{(\sin\theta)d\theta/db} \quad \text{where} \quad \frac{d\theta}{db} = 2\left(\frac{1}{\sqrt{R^2 - b^2}} - \frac{\zeta}{\sqrt{R^2 - \zeta^2 b^2}}\right)$$

for $0 \leq \theta \leq \theta_{max} = \pi - \arcsin\zeta$; for $\theta > \theta_{max}$, $d\sigma/d\Omega = 0$. **(e)** $\sigma_{tot} = \pi R^2$.

14.17 $N_{sc}/Z^2 = 0.21, 0.20$, etc.

14.31 (b) If $\theta_{lab} = 0$, then θ_{cm} could be 0 or π.

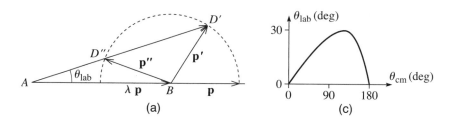

(a)   (c)

(d) $\theta_{lab}(max) = \arcsin(1/\lambda)$.

Chapter 15

15.3 $\gamma = 1 + 3.56 \times 10^{-10}$; difference $= (\Delta t_o - \Delta t) = -1.28\,\mu s$; percent difference $= -3.56 \times 10^{-8}\%$.

15.5 A's clocks read, and A has aged by, 25 years, compared with the 80 years by which B has aged.

15.7 Number expected, $N = 420$, taking account of time dilation; but $N = 29$, ignoring time dilation.

15.11 $v = \frac{3}{5}c$.

15.13 (a) $l = 91.7$ cm and $\theta = 70.9°$; **(b)** $l = 83.2$ cm and $\theta_o = 46.1°$.

15.17 (a) If we call the two events 1 and 2, then, as observed in $\mathcal{S}'$, $t_1' = 0$ but $t_2' = -\gamma\beta a/c$. **(b)** As observed in $\mathcal{S}''$, $t_1' = 0$ but $t_2' = +\gamma\beta a/c$.

15.19 (a) $x_F' = d, \quad t_F' = d/c, \quad x_B' = -d, \quad t_B' = d/c$.
(b) $x_F = \gamma(1 + \beta)d, \quad t_F = \gamma(1 + \beta)d/c, \quad x_B = -\gamma(1 - \beta)d, \quad t_B = \gamma(1 - \beta)d/c$.

15.21 $v = 0.91c$.

15.23 The velocity of the right rocket relative to the left one is $0.994c$ to the left.

15.25 (Speed as measured in $\mathcal{S}$) $= c$.

15.29 (a)

$$\mathbf{R}(\theta) = \begin{bmatrix} \cos\theta & \sin\theta & 0 \\ -\sin\theta & \cos\theta & 0 \\ 0 & 0 & 1 \end{bmatrix}.$$

15.33 (a)

$$\left. \begin{array}{l} x_1' = x_1 \\ x_2' = \gamma(x_2 - \beta x_4) \\ x_3' = x_3 \\ x_4' = \gamma(x_4 - \beta x_2) \end{array} \right\} \quad \text{whence} \quad \Lambda_{B2} = \begin{bmatrix} 1 & 0 & 0 & 0 \\ 0 & \gamma & 0 & -\gamma\beta \\ 0 & 0 & 1 & 0 \\ 0 & -\gamma\beta & 0 & \gamma \end{bmatrix}.$$

(b)

$$\Lambda_{R+} = \begin{bmatrix} 0 & 1 & 0 & 0 \\ -1 & 0 & 0 & 0 \\ 0 & 0 & 1 & 0 \\ 0 & 0 & 0 & 1 \end{bmatrix} \quad \text{and} \quad \Lambda_{R-} = \begin{bmatrix} 0 & -1 & 0 & 0 \\ 1 & 0 & 0 & 0 \\ 0 & 0 & 1 & 0 \\ 0 & 0 & 0 & 1 \end{bmatrix}.$$

15.47 $v = 0.20c$.

15.55 Since u is a four-vector

$$u_1' = \gamma(V)[u_1 - \beta(V)u_4], \quad u_2' = u_2, \quad u_3' = u_3, \quad u_4' = \gamma(V)[u_4 - \beta(V)u_1].$$

15.57 $T(\text{Bi}) + T(\text{He}) = 1.3 \times 10^{-12}\,\text{J} = 8.1\,\text{MeV}$.

15.61 $E = 5\,\text{MeV}; \quad v = 0.8c$.

15.63 $1\,\text{MeV}/c^2 = 1.78 \times 10^{-30}\,\text{kg}; 1\,\text{MeV}/c = 5.34 \times 10^{-22}\,\text{kg·m/s}$.

15.65 (b) $v_{\text{max}} = 0.12c$.

15.67 $v_{\text{f}} = 0; M = \frac{10}{3}m$.

15.71 (c) The minimum energies are $E_{\text{cm}} \approx 3100\,\text{MeV}$, but $E_{\text{lab}} \approx 9.6 \times 10^6\,\text{MeV}$.

15.75 $M = 2.95\,\text{GeV}/c^2; v = 0.65c$.

15.81 $r(\text{rel}) = 8.3\,\text{cm}; r(\text{nonrel}) = 5.9\,\text{cm}$.

15.93 If E denotes the final energy of the electron, $E/E_o \approx 0.002$.

15.109 (a) If we choose axes so that the two particles lie in the xy plane, then in their rest frame $\mathcal{S}'$, the force of one on the other is $\mathbf{F}' = (0, kq^2/r'^2, 0)$. According to (15.155) this transforms to $\mathbf{F} = (0, kq^2/\gamma r^2, 0)$ in $\mathcal{S}$. **(b)** In $\mathcal{S}'$ the fields of one charge at the position of the other are $\mathbf{E}' = (0, kq/r'^2, 0)$ and $\mathbf{B}' = (0, 0, 0)$. In $\mathcal{S}$, they are $\mathbf{E} = (0, \gamma kq/r^2, 0)$ and $\mathbf{B} = (0, 0, \gamma\beta kq^2/cr^2)$; these produce a force $\mathbf{F} = q\mathbf{v} \times \mathbf{B}$ which is easily seen to be the same as in part (a).

Chapter 16

16.7

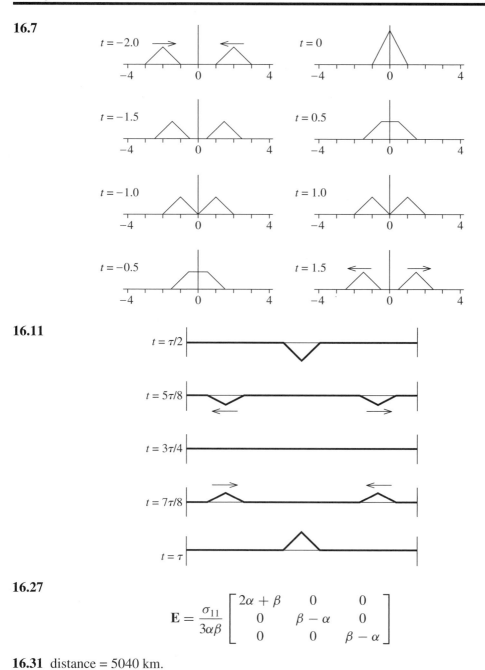

16.11

16.27

$$\mathbf{E} = \frac{\sigma_{11}}{3\alpha\beta} \begin{bmatrix} 2\alpha + \beta & 0 & 0 \\ 0 & \beta - \alpha & 0 \\ 0 & 0 & \beta - \alpha \end{bmatrix}$$

16.31 distance = 5040 km.

Index

All entries are identified by their page number. In addition, when a reference is to a whole section or chapter, I have indicated the section or chapter in parenthesis; for example, (Sec.1.1) or (Ch.1). Similarly, when a reference is primarily to a figure, example, problem, or footnote, I have added a parenthesis, such as (Fig.1.2), (Ex.1.3), (Pr.1.4), or just (Ft.).